Macrocycles

Macrocycles

Construction, Chemistry and Nanotechnology Applications

FRANK DAVIS

Cranfield Health, Cranfield University, Bedfordshire, UK

SÉAMUS HIGSON

Cranfield Health, Cranfield University, Bedfordshire, UK

A John Wiley and Sons, Ltd., Publication

Library of Congress Cataloging-in-Publication Data

Higson, Séamus.
  Macrocycles : construction, chemistry, and nanotechnology applications / Frank Davis, Séamus Higson.
      p. cm.
  Includes bibliographical references and index.
  ISBN 978-0-470-71462-1 (hardback) – ISBN 978-0-470-71463-8 (paper) – ISBN 978-1-119-98993-6 (ebook)
  1. Macrocyclic compounds. I. Davis, Frank, 1966- II. Title.
  QD400.H54 2011
  547′.5 – dc22

                                                                          2010045660

A catalogue record for this book is available from the British Library.

Print ISBN Cloth: 9780470714621
Print ISBN Paper: 9780470714638
ePDF ISBN: 9781119989936
oBook ISBN: 9780470980200
ePub ISBN: 9781119990291

Typeset in 10/12pt Times by Laserwords Private Limited, Chennai, India

Contents

Preface

The intention of this work is to serve as a detailed introduction to the field of macrocyclic chemistry. We will attempt to take the reader on a journey through this field, beginning with the simplest of systems and progressing to increasingly complex ones, showing their inherent beauty and aesthetic appeal. Macrocyclic compounds are becoming more and more useful with the passing of time and are employed in ever wider fields of application.

We will begin in Chapter 1 with the simplest of the cyclic systems, low-molecular-weight compounds such as cyclohexane and benzene, along with a brief discussion of bonding and aromaticity. Chapter 2 will then discuss larger systems such as annulenes and fused-aromatic-ring systems. Initially these will all be carbon rings, but as we progress the use of other elements will become more and more common, expanding the chemistry and binding interactions observed in these systems. This will begin in Chapter 3, which will not only discuss the syntheses and properties of these compounds, but will also introduce the concept of the template effect. The template effect is what makes the synthesis of many of the complex systems described within this work possible. What the template effect does is pre-organise the units that make up these macrocycles before covalent bond formation takes place; it can be based on a variety of interactions such as oxygen–metal interactions within crown ether formation. Other interactions that can be utilised to pre-organise the system include hydrogen bonding, metal–π interactions and interactions between electron-rich and electron-poor aromatic systems. These interactions are responsible for the good yields and clean reaction products obtained for many of these syntheses, many of which would be highly unlikely or impossible without these pre-organisation events.

As we progress through the book the macrocyclic systems become more complex. Chapter 4 discusses calixarenes, based on multiple phenol units arranged to form larger rings, while Chapter 5 looks at similar systems based on heteroaromatic units such as calixpyrroles. Chapter 6 discusses the naturally obtained cyclodextrins; these beautiful cyclic polysugars have been obtained in a range of sizes and substitution patterns and are now becoming widely used within various commercial applications. Further chapters discuss other synthetic systems such as the bowl-shaped cyclotriveratrylenes (Chapter 7) and the pumpkin-shaped cucurbituril macrocycles (Chapter 8). Besides the syntheses and structures of all these families of compounds, an account will be given of their binding of a wide variety of guests, utilising a range of interactions. Many of these compounds bind with a specificity and selectivity that can only be surpassed by biological interactions and has led to their use as sensors, extraction agents or selective encapsulation agents for applications such as drug delivery.

Finally the assembly of these systems into even larger and more complex arrangements will be discussed. Supramolecular interactions that lead to the formation of mechanically interlocked molecules such as the rotaxanes, where linear molecules are threaded through macrocycles, molecular knots, and the catenanes, where two or more macrocyclic rings are threaded through each other, are described in Chapter 9. Initial studies on the use of these systems as molecular machines are described in Chapter 10, such as molecular shuttles and switches, nano-valves and logic gates. The potential for using these types of systems in molecular computing could perhaps address the requirement for ever smaller transistors to increase computing power.

We hope that within this work we not only impart knowledge of these systems but also an appreciation of their inherent fascination for many scientists. The sheer elegance of some of the syntheses along with the simplicity of many others helps to impart some understanding of the intricate supramolecular interactions which occur within these systems. Besides their aesthetic appeal, the potential applications for many of these systems mean that the field of macrocyclic chemistry will continue to be a fascinating and important area of study for the foreseeable future.

Frank Davis and Séamus P.J. Higson
Cranfield, September 2010

1

Introduction

Ever since the dawn of man, humans have been chemists of one form or another. One of the first chemical reactions primitive man discovered was that certain materials could be burnt and the resultant heat released used to cook food and warm dwellings. As time progressed other chemistries were discovered, ranging from the smelting of metals to brewing to the use of plant extractions for dyeing textiles.

The ancient Greek philosopher Democritus proposed an atomistic theory of matter, which became popular again in the sixteenth and seventeenth centuries AD with the work of some of the great chemists of that time such as Boyle, Cavendish, Lavoisier and Priestley. Many elemental compounds were discovered by these and other workers, and later workers such as Kekulé and Frankland introduced concepts such as valence and molecular structure. One of the results of this work was the discovery of the tetravalence of carbon. Molecular structures for compounds such as alkanes, alcohols, acids and so on were also deduced around this time.

1.1 Simple Ring Compounds

Many early ring compounds were discovered and isolated and had their properties determined long before their actual physical structures were known. Once valency and the concept of the chemical bond were introduced, the structures of some alkanes were deduced to be cyclic, such as the simple hydrocarbon compound cyclohexane (Figure 1.1a). There are a huge number of simple ring carbons, ranging from the simple cyclopropane ring, the smallest of all hydrocarbon ring compounds, through to huge cyclic structures. These large structures are often termed 'macrocycles' to reflect their large size; it is with these compounds that this work is concerned.

There are many aliphatic hydrocarbon ring compounds, with cyclopropane being the smallest, followed by cyclobutane, cyclopentane, cyclohexane and so on. These simple cycloalkanes are similar to the corresponding linear alkanes in their general physical properties, but have higher melting/boiling points and densities. This is due to stronger intermolecular forces, since the ring shape allows for a larger area of contact between molecules. As chemical synthesis methods have improved, the number and ring size of these compounds have increased dramatically, as demonstrated by consideration of, for example, the crystal structures obtained for the cycloalkane $C_{288}H_{576}$.[1]

Macrocycles: Construction, Chemistry and Nanotechnology Applications, First Edition.   Frank Davis and Séamus Higson.
© 2011 John Wiley & Sons, Ltd. Published 2011 by John Wiley & Sons, Ltd.

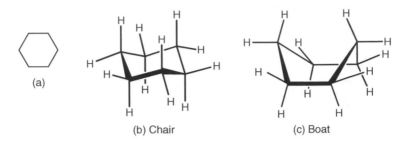

(a)

(b) Chair

(c) Boat

Figure 1.1   *Schematic, chair and boat structures of cyclohexane*

Of course, the formation of large rings is in no way limited to carbon. Many other elements can be incorporated into ring structures: sulfur for example usually exists as a cyclic S_8 compound, although other ring sizes from 6–20 and polymeric forms have been synthesised.[2] Se_8 is also known, although selenium tends to form polymeric chains. Cyclic siloxanes (with Si—O— chain repeat units) contain a variety of ring sizes and are common industrial chemicals.

Many other elements, although not capable of forming stable ring structures by themselves, can be incorporated into carbon-based rings. Typical examples include tetrahydrofuran, piperidine and ethylene sulphide. The incorporation of heteratoms into hydrocarbon rings has led to the development of several classes of macrocycles such as crown ethers and cryptands (see Chapter 2).

As drawn in Figure 1.1, the structure of cyclic alkanes such as cyclohexane appears to be a simple flat ring. However the real structures of these systems are far more complex. Since three points define a plane, cyclopropane is by definition flat. In cyclobutane however the carbon atoms adopt a puckered conformation, with three atoms in a plane and the fourth at an angle of about 25°. Cyclohexane, if it existed as a flat hexagon, would undergo considerable angle strain and as a consequence exists in a 'chair-like' conformation (Figure 1.1), with a carbon–carbon bond angle of 109.5°. A second, less energetically-favoured conformation is the 'boat' conformation, which cannot be isolated; there are also a number of other potential structures, such as the twist conformation. Substituted cyclohexanes can exist in the chair conformation with substituents which are either 'equatorial' or 'axial'; these two isomers tend to interconvert rapidly at room temperature. In the case of large substituents, these are mostly in the equatorial position, since this conformation is energetically more stable. Large ring structures have a multitude of possible conformers.

Multiple ring systems are also possible, either linked as in bicyclohexyl or fused as in decalin. Within the field of natural product chemistry, there are many examples of fused multiple aliphatic ring systems. A review of this is far beyond the scope of this work, but these include for example the steroid family of molecules such as cholesterol with fused cyclohexane and cyclopentane rings, as well as the multiple ring systems of adamantane. There are also a huge number of carbohydrates based on linked cyclic furanose and pyranose systems, such as sucrose. Within natural product chemistry, the five- and six-membered ring compounds tend to dominate, although there are many exceptions such as the terpenes, pinene with fused six- and four-membered rings, cembrene A with a 14-membered ring, and the penicillins, which contain a four-membered ring. An extreme example is the family of compounds known as ladderanes, which are formed by certain bacteria, an example of which is pentacycloammoxic acid,[3] with five fused cyclobutane rings. Figure 1.2 shows the structure of these compounds.

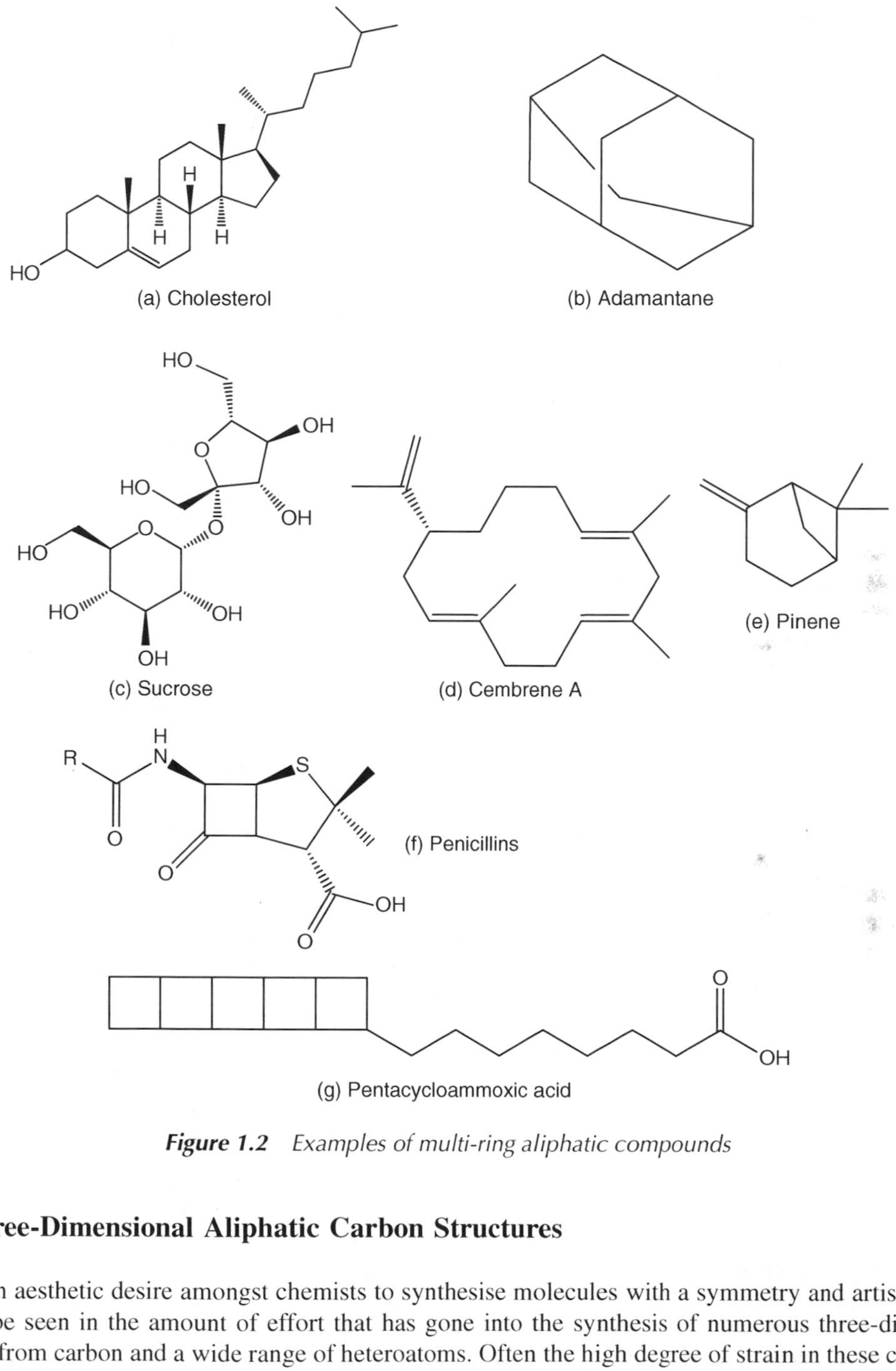

Figure 1.2   *Examples of multi-ring aliphatic compounds*

1.2   Three-Dimensional Aliphatic Carbon Structures

There is an aesthetic desire amongst chemists to synthesise molecules with a symmetry and artistic beauty. This can be seen in the amount of effort that has gone into the synthesis of numerous three-dimensional structures from carbon and a wide range of heteroatoms. Often the high degree of strain in these compounds

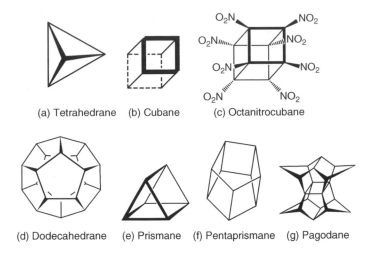

(a) Tetrahedrane   (b) Cubane   (c) Octanitrocubane

(d) Dodecahedrane   (e) Prismane   (f) Pentaprismane   (g) Pagodane

Figure 1.3   *Platonic and other strained hydrocarbons*

can lead to unusual forms of bonding and novel chemistries. Some of these small molecules will be detailed below.

Platonic solids are regular convex polyhedra in which all angles and side lengths are identical. The simplest of the five Platonic solids is the tetrahedron. Attempts have been made to synthesise tetrahedrane (Figure 1.3), C_4H_4, but with no success, and it seems unlikely that this molecule is stable enough to exist under normal laboratory conditions. However, derivatives of tetrahedrane where the hydrogen atoms are replaced by larger stabilising groups have been successfully synthesised. Derivatives of tetrahedrane substituted with either four tertiarybutyl[4] or four trimethylsilyl[5] groups have been successfully isolated as stable solids. In the case of the trimethylsilyl derivative, the C—C bonds were significantly shorter than typical C—C bonds and this compound could also be dimerised to form a ditetrahedrane with an extremely short (144 pm) bond connecting the two tetrahedra.[5]

Tetrahedral molecules also exist for other elements. White phosphorus is made up of P_4 tetrahedra, and As_4 tetrahedra are also known. A synthesis of a substituted silicon version of tetrahedrane with an Si_4 tetrahedron substituted with stabilising silyl groups has also been reported.[6]

The second of the Platonic solids is the cube, and its chemical equivalent, cubane C_8H_8, has been known[7] since 1964. Cubane (Figure 1.3) is a stable solid melting at 131 °C and has a very high density (1.29) for a hydrocarbon. The same group demonstrated the rich chemistry of cubane by synthesising nitrated versions with between four and all eight hydrogen atoms replaced by nitro groups;[8] this is a highly energetic compound with potential for use as a high explosive. Cubane-type structures do exist in nature, such as for example a number of iron-sulfur proteins containing cubane-type structures with Fe and S atoms at alternating corners.

A third Platonic solid is the dodecahedron. Dodecahedrane ($C_{20}H_{20}$) was first synthesised[9] in 1982. The structure was confirmed by NMR spectra, which showed that all carbon and hydrogen atoms were equivalent (Figure 1.3). When a sample of dodecahedrane was bombarded with helium ions, a small fraction of the dodecahedrane molecules were shown to form a so-called He@$C_{20}H_{20}$ compound, where the helium is not bound by a chemical bond (since it is a noble gas) but rather is encapsulated in the carbon cage and physically unable to escape.[10]

Two other Platonic solids exist, the octahedron and the icosahedron. However, octahedrane (which would have the formula C_6) is thought to be too highly strained to exist, especially as stabilising groups cannot be attached. A carbon icosahedron cannot exist since it would require each carbon to bind to five neighbouring carbons, which is ruled out by the tetravalency of carbon.

Apart from the platonic solids, a series of other symmetrical hydrocarbons also exists. Benzene has the formula C_6H_6, which would normally require a combination of multiple bonds and ring systems, but much of the chemistry of benzene does not fit with the presence of unsaturated groups. One proposed structure was that of Ladenburg, which had the carbon atoms forming a prism (Figure 1.3). Although later proved not to be the structure of benzene, the compound was eventually synthesised and named prismane.[11] Prismane is stable at room temperature but decomposes to benzene upon heating.[11] A wide range of other esoteric hydrocarbons have also been synthesised, including pentaprismane[12] and pagodane.[13]

1.3 Annulenes

Many compounds have been found to have properties which are not in keeping with their predicted structures. One such is benzene, for which a formula C_6H_6 was deduced from Faraday's work in 1825, which discovered the empirical formula and molecular weight of benzene. However, if the classical valencies of carbon and hydrogen were to be maintained, this would require the incorporation of multiple rings or bonds within the structure. A number of possible structures were proposed but the one that became most prevalent was that of Kekulé, who suggested that benzene was in fact cyclohexa-1,3,5-triene (Figure 1.4a), with a six-membered ring structure containing alternating double and single bonds.[14] This explained some of the properties of benzene, such as the fact that there was only one isomer for singly-substituted benzenes but three isomers of disubstituted rings.[15] Twenty-five years later at a meeting in his honour, Kekulé apparently spoke of how he had realised the structure of benzene during a dream in which he saw a snake biting its own tail.

However, it soon became obvious that benzene could not be the simple hexatriene originally postulated. First it was realised that there should be more isomers of disubstituted compounds than could be isolated. The three isomers of, for example, dimethyl benzene are the 1,2, 1,3 and 1,4 substituted derivatives. However, there should be two isomers of 1,2-dimethylbenzene, as shown in Figure 1.4b. These compounds ought to be separable, but only one isomer could be isolated. One possible explanation for this could be that benzene and substituted benzenes are mixtures of two rapidly equilibrating cyclohexatrienes.

Another fault with the postulated structure was that the chemistry of cyclohexatriene should be that of a highly reactive alkene; cyclohexatriene should for example decolourise bromine water, with the concurrent formation of highly brominated derivatives. However, for benzene this reaction does not occur under conditions under which many other alkenes readily brominate. This required the development of molecular orbital theory and the idea of resonance energy. Instead of alternating single and double bonds, a structure was proposed where the carbon–carbon bonds are intermediate between single and double bonds. Interactions between the p-orbitals lead to the formation of circular delocalised 'clouds' of electrons above

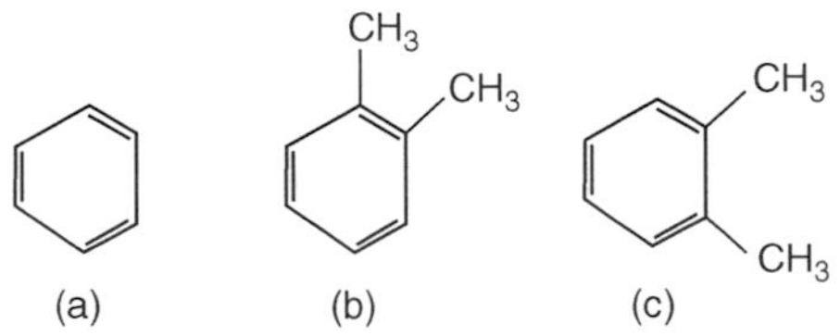

Figure 1.4 *Structures of 'cyclohexatriene' and isomers of 'dimethyl cyclohexatriene'*

and below the plane of the carbon atoms. The convention is to draw the six-membered benzene ring as a hexagon with a circle (symbolising the cyclic delocalised system) inside it (Figure 1.5a). The symmetry of such a structure was finally proved when X-ray crystallography could be utilised to determine the structure of the benzene ring. Work by Kathleen Lonsdale in 1929 on hexamethylbenzene[16] showed it to have a symmetrical flat hexagonal structure, with the C—C bonds of the ring having a length (142 pm) intermediate between those of carbon–carbon single and double bonds. The term 'aromatic' was coined for hydrocarbons of this type.

One consequence of molecular orbital theory was the Huckel $4n + 2$ rule. For a molecule to have aromatic properties it must follow three rules: it must have $4n + 2$ electrons in a circular conjugated bond system (for example, benzene has six, i.e. $n = 1$), it must be capable of assuming a planar (or almost planar) conformation, and finally each atom must be able to participate in the delocalised ring system by having either an unshared pair of electrons or a p-orbital.

Once this rule was formulated, interest was generated in synthesising analogues of benzene with alternating single/double bonds but of different ring sizes. These compounds have been grouped under the name '[n]annulene', where n is the number of atoms in the ring. Therefore benzene could be referred to as [6]annulene. A wide range of other annulenes have been synthesised.

The smallest annulene, cyclobutadiene or [4]annulene (Figure 1.5b), has four electrons available to participate in an aromatic system. This does not follow the $4n + 2$ rule and experimental measurements show that there is no aromaticity. Cyclobutadiene is rectangular rather than square and is highly reactive, forming a dimer with a reaction half-life measurable in seconds. However, metal–cyclobutadiene complexes[17] such as $(C_4H_4)Fe(CO)_3$ display much higher stabilities because the metal atom donates two electrons to the cyclobutadiene ring, giving it six electrons, which enables it to obey the $4n + 2$ rule. Similarly, the dilithium salt of a tetrasilylated cyclobutadiene dianion,[18] $C_4(SiMe_3)_4{}^{2-}$ $2Li^+$, has been shown to be relatively stable at room temperature and to contain a square, planar cyclobutadiene species.

Cyclooctatetraene or [8]annulene (Figure 1.5c) was found not to be aromatic and displayed the chemistry of a conjugated polyene. However, it is much more stable than cyclobutadiene and is available commercially. With eight electrons, cyclooctatetraene was not expected to have an aromatic structure, as confirmed by an X-ray study[19] which showed the molecule adopts a 'tub' shape with alternating single and double bonds. Reaction with potassium metal gives a dianion, however, which is highly stable, obeys the $4n + 2$ rule since it has 10 electrons – and has been shown to be planar and aromatic in nature by X-ray studies.[20]

A large range of higher annulenes have been synthesised and a comprehensive review is beyond the scope of this chapter. Early work has been reviewed by Sondiemer,[21] and much later work has also been reviewed.[22] Within this chapter we will provide a brief summary of work that has been carried out in this field.

[10]annulene has been synthesised. Since this obeys the Huckel $4n + 2$ rule, it would be expected to be aromatic. However, the NMR and reactivity of this compound are typical of a polyene-type structure rather than an aromatic one. This can be explained by the fact that aromatic systems need a high degree of planarity. A simple all-cis ring system such as that shown in Figure 1.5d would have C—C bond angles of 144° rather than the 120° found in benzene. This high degree of strain prevents a planar structure from forming; a possible structure with two trans double bonds would alleviate this but would display a high degree of steric repulsion between the hydrogen atoms shown in Figure 1.5e. This can be alleviated however by removing the two hydrogens and replacing them, for example with a methylene bridge[23] as shown in Figure 1.5f. This compound is still not completely planar but the NMR spectrum indicates considerable delocalisation. A triply bridged system (Figure 1.5g) with increased planarity has also been synthesised[24] and displays considerable aromatic chemistry.

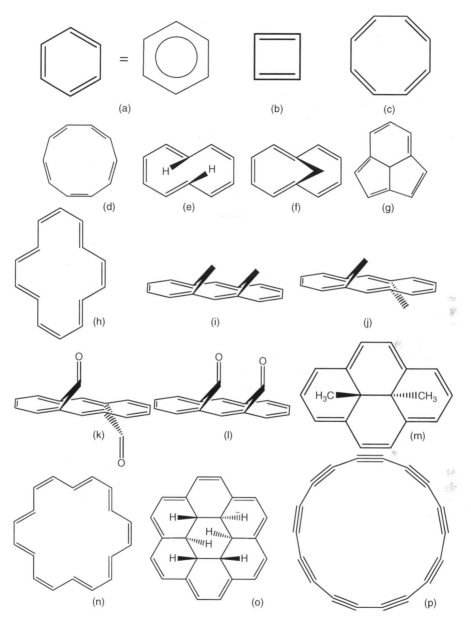

Figure 1.5 *Structures of the annulenes*

[12]annulene behaves similarly to cyclooctatetraene in that it is nonplanar and highly reactive. Reaction with lithium metal gives the dianion,[25] which does obey the $4n + 2$ rule and although nonplanar is much more stable than the parent compound, indicating some aromaticity. Very similar behaviour is observed for 16-annulene,[26] with the parent compound displaying polyene-type chemistry and the dianion being much more stable and almost planar.

[14]annulene (Figure 1.5h) obeys the $4n + 2$ rule and is therefore aromatic. However, there is some steric interference from hydrogen atoms located within the ring, which leads to deviations from planarity. X-ray crystallographic studies[27] demonstrate this, but there is no single/double bond alternation and other studies such as the NMR spectra also confirm the aromatic structure.[21,22] Attempts have been made to reduce the ring strain by replacing the internal hydrogens with bridging groups such as methylene. For example, compounds have been synthesised with two methylene bridges (Figure 1.5i). When the bridges are on the same side of the ring, a stable compound with an aromatic structure results,[28] whereas when the bridges are opposite to each other (Figure 1.5j) a puckered polyene structure is observed[29] and the compound reacts readily with oxygen. Similar behaviour occurs when carbonyl bridging groups are used,[30,31] with the *syn* (Figure 1.5k) isomer being highly stable and displaying a flat aromatic system, whereas the *anti* (Figure 1.5l) isomer is unstable and shows no evidence of aromaticity. More complex bridging units have also been utilised, such as in the dihydrodimethylpyrene molecule shown in Figure 1.5m, which has an outer 14-carbon ring with alternating double bonds and is strongly aromatic. X-ray studies show a structure in which all the peripheral bonds are essentially the same length and in the same plane.[32] A large number of compounds of this nature have been synthesised and their aromaticity has been investigated in detail.[22]

Apart from benzene itself, [18]annulene (Figure 1.5n) is the most stable of the annulenes[22] and it has the correct number of atoms to allow bond angles of $120°$, thereby eliminating ring strain. X-ray studies show that it has an approximately planar structure with C—C bond lengths varying from 0.138 to 0.142 nm throughout the structure,[33] although there are some minor deviations from planarity due to steric interactions and crystal packing. An interesting structure has been synthesised[34] in which bridging groups increase the rigidity of the ring, as shown in Figure 1.5o; this compound has been shown to have a higher ring current (88% of the predicted maximum), indicating more efficient conjugation than [18]annulene (56%).

The higher annulenes, [20]annulene,[35] [22]annulene[36] and [24]annulene,[37] have all been successfully synthesised. NMR spectra indicate as expected that [22]annulene is aromatic (unfortunately as yet no X-ray structures have been obtained to confirm this), whereas [20] and [24]annulene are not. Syntheses of [30]annulene have been reported[38] but yields were too low to adequately characterise the material, the product was quite unstable and no evidence for aromaticity could be obtained. Theoretical studies on annulenes containing up to 66 carbons have been carried out;[39] these indicate that for annulenes containing 30 or more carbon atoms, conformational flexibility will lead to a drop in electron delocalisation and nonaromatic structures with alternating single/double bonds will predominate.

A range of dehydroannulenes with one or more triple bonds within the ring system have been synthesised, often as intermediates in the process of making various annulenes.[21,22] These usually tend to show less aromaticity and be less stable than the annulenes themselves. However, they are systems of interest and have been the subject of several reviews[21,22,38,40]. One of the simplest dehydroannulenes is benzyne or didehydrobenzene, C_6H_4, which is an extremely reactive species that can be trapped by, for example, a Diels–Alder reaction with such species as cyclopentadiene or anthracene, and can be stabilised by complexation with transition-metal atoms. A hexadehydrobenzene species with alternating single and triple bonds would be highly unlikely to exist due to the extremely high ring strain within such a molecule, but the larger C_{18} ring has been predicted to be relatively stable, possibly as a polycumulene with all the bonds being C=C double bonds rather than with the alternating single/triple bond structure, as shown in Figure 1.5p.[41] The C_{18} ring system and larger C_{24} and C_{30} rings have not been synthesised and characterised as yet but evidence of the C_{18} ring has been detected in mass spectra[42] and by trapping it as a reaction product in a low-temperature glass.[43] It has also been postulated to be a component of interstellar clouds and to exist in the hearts of dying stars.[22]

1.4 Multi-Ring Aromatic Structures

Hexagons are one of the shapes that can pack perfectly without any intervening space, as shown for example by the structure of a honeycomb. This is exemplified in aromatic chemistry by the large number of fused-aromatic-ring-system compounds that have been discovered in natural substances or synthesised over the years. Although aromatic, not all of these compounds obey the Huckel $4n + 2$ rule, which appears not to be valid for many compounds containing more than three fused aromatic rings. Examples of some of these are shown in Figure 1.6.

[TI]Naphthalene consists of two benzene rings fused together and is commercially extracted from coal tar. Its major uses include as a fumigant, for example in mothballs, and as an intermediate in the synthesis

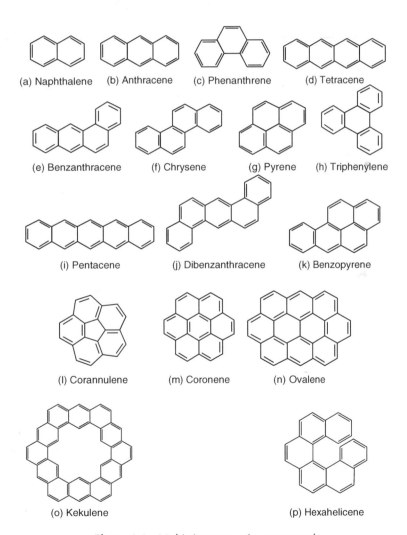

(a) Naphthalene   (b) Anthracene   (c) Phenanthrene   (d) Tetracene

(e) Benzanthracene   (f) Chrysene   (g) Pyrene   (h) Triphenylene

(i) Pentacene   (j) Dibenzanthracene   (k) Benzopyrene

(l) Corannulene   (m) Coronene   (n) Ovalene

(o) Kekulene   (p) Hexahelicene

Figure 1.6 *Multi-ring aromatic compounds*

of other industrial chemicals such as phthalic anhydride. The molecule is planar, with carbon–carbon bond lengths that are not all identical to each other but are close to those of benzene. Extended versions of naphthalene with three (anthracene), four (tetracene), five (pentacene) and more rings have been either isolated from products such as coal tar or synthesised in the laboratory.

Anthracene, with three fused benzene rings, is again commonly extracted from coal tar. Anthracene is planar and the central ring is much more reactive than the others. For example, the central positions are easily oxidised to give anthraquinone and the central rings participate readily in Diels–Alder type reactions with a variety of dienes. Irradiation with UV light causes anthracene to dimerise via a $4 + 4$ cycloaddition reaction of the central rings. Tetracene is a pale orange powder which can act as a molecular organic semiconductor. Again it is planar, prone to oxidation and readily participates in Diels–Alder reactions. Pentacene is a blue oxygen-sensitive compound and is being investigated for such purposes as use in organic thin film transistors[44] and photovoltaic devices.[45] Hexacene and heptacene cannot be isolated in bulk since they readily dimerise and are extremely oxygen-sensitive, although derivatives of these compounds have been isolated.

Other polycyclic aromatic hydrocarbons include the three-ring-system phenanthrene (Figure 1.6f), again with the central ring being the preferred site for a wide range of chemical reactions. Larger systems include pyrene, which is widely used as a fluorescent probe, and chrysene, which is similar in reactivity to phenanthrene. Many of these hydrocarbons have been found in tobacco smoke and some, such as benzopyrene, have been shown to be highly carcinogenic.[46]

Larger ring systems have also been studied, such as coronene, which occurs naturally in the mineral carpathite, and ovalene, which can be formed in deep-sea hydrothermal vents. One of the larger systems synthesised is kekulene, with its large inner cavity. X-ray experiments[47] have demonstrated that the structure of kekulene is that of a large flat ring, but not all of the bond lengths are equivalent and it appears it contains six discrete aromatic rings linked together, rather than being one large aromatic system. One of the largest systems synthesised contains 222 carbon atoms.[48] As these systems become larger, the compounds become less soluble and their properties approach those of graphite, which has a structure essentially of layers of infinite benzene rings, that is carbon atoms arranged in a hexagonal lattice with carbon–carbon distances of 0.142 nm, and planes separated by 0.335 nm. Single-graphite planes have been isolated; this material is known as graphene and is the subject of much current research due to its potential novel physical and electronic properties.[49]

The aromatic systems mentioned so far have been in the main planar or near-planar structures. Not all aromatic systems follow this rule. Hexahelicene (Figure 1.6p) consists of six aromatic rings and would be expected to have a planar structure. However, this would mean that the atoms at the extreme ends of the cyclic structure would have to occupy the same space. This is impossible, so the molecule is actually twisted into a spiral shape, meaning that it is chiral. Both of the isomers have been isolated[50] and display high optical rotation (3640°). Corannulene (Figure 1.6l) is not flat like the similar coronene structure, but is in fact bowl-shaped. The 'central' ring is five-rather than six-membered, which results in the loss of planarity. The ultimate example of the effect of five-membered rings on aromatic compounds is in buckminsterfullerene, where the presence of 12 five-membered rings along with 20 six-membered rings causes the C_{60} molecule to assume the shape of a sphere.

1.5 Porpyrins and Phthalocanines

Most of the ring systems described earlier in this chapter are simple hydrocarbons. However, there are a huge number of aromatic systems that include heteroatoms. These range from simple molecules such as pyridine, pyrrole, furan and thiophene through to much larger compounds. One very important class of compounds is the porphyrins.

The basic structure of the porphyrin unit is shown in Figure 1.7a; it consists of a large flat aromatic ring with four pyrrole units bound together by methane carbons. The parent macrocycle contains 22 electrons, thereby obeying the $4n + 2$ rule. Porphyrins are usually very highly coloured compounds due to the presence of this large aromatic system. There are many methods of synthesising porphyrins but the simplest involves cyclisation of pyrrole with substituted aldehydes, as shown in Figure 1.7b. Four aldehydes condense with four pyrrole units under acidic conditions to form a cyclic tetramer[51,52] (the initial tetramer formed is not actually aromatic but under the conditions of the reaction is readily oxidised to the porphyrin).

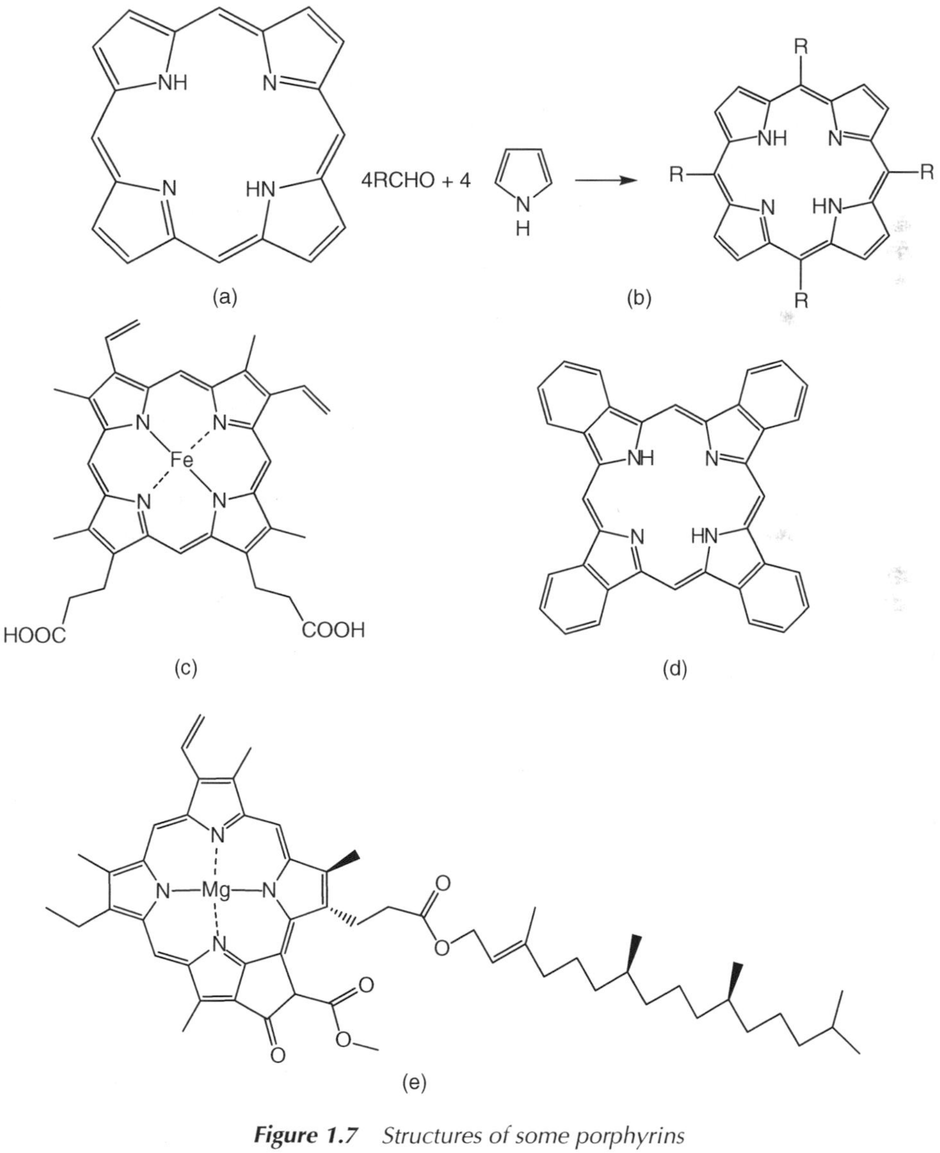

Figure 1.7 *Structures of some porphyrins*

The presence of the nitrogen atoms within the ring facilitates the binding of metal atoms to form metalloporphyrins. In the parent porphyrin structure (known as a free-base porphyrin), two of the nitrogen atoms have hydrogen atoms bound to them. Upon binding of metals these hydrogen atoms are lost and the metal is bound within the central N_4 cavity. One example of this metal binding can be found in the heme porphyrins such as heme B (Figure 1.7.c). These types of porphyrin reversibly form complexes with oxygen and are found within haemoglobin, the oxygen-carrying protein that makes up much of our red blood cells.

Many variations on the porphyrin theme are known. The aromatic system can be extended as in the tetrabenzoporphyrins (Figure 1.7d). Alternatively, more reduced forms or variations where one of the methane units is missing and replaced by a direct pyrrole–pyrrole connection are known. These systems generally do not obey the $4n + 2$ rule. These related porphyrin analogues include corrins (containing a direct pyrrole–pyrrole link and found in such natural products as vitamin B12), along with the reduced forms known as chlorins, bacteriochlorophylls and corphins. Bacteriochlorophylls and corphins are used as subunits in enzymes found in certain bacteria. Chlorins, which can be thought of as dihydroporphyrins, are widely found in nature. A magnesium chlorin (Figure 1.7e) is a typical example, known as chlorophyll A; this unit is vital to the process of photosynthesis and without this group of materials, green plants and therefore ultimately most other forms of life would not exist.

Phthalocyanines are synthetic analogues of tetrabenzoporphyrins, in which the methine bridges are replaced by nitrogen atoms. These are flat aromatic systems similar to porphyrins and can complex metal atoms in a similar manner, as shown for copper phthalocyanine in Figure 1.8. The phthalocyanines are highly coloured systems due to their large aromatic ring systems and tend towards the blues and greens. This, combined with their stability, has led to extensive use of substituted phthalocyanines within the dye industry (for instance, copper phthalocyanine is known as phthalocyanine blue BN). There are a wide variety of methods for the synthesis of phthalocyanines, mostly based on similar cyclisation reactions to those used for the porphyrins. An example is given in Figure 1.8b, which shows the condensation of four phthalonitrile units to form phthalocyanine.

The simplicity of the phthalocyanine synthesis and the wide variety of structural variations available have made these compounds the subject of widespread interest. Possible applications abound due to the novel physical, electronic and optical properties of these materials, along with their thermal and chemical stability. A review of this is outside the scope of this work, but we will mention that phthalocyanines and their derivatives have been investigated for use as optical switches, liquid crystals, sensors, organic photoelectric cells, nonlinear optical materials, electrochromic materials and optical information-recording media.[53]

The wide synthetic flexibility of these materials has led to a plethora of variations on the basic phthalocyanine unit. For example, the benzo units can be replaced with naphthalene or anthracene units to give extended phthalocyanines. Polymeric materials based on phthalocyanines have been made by linking the phthalocyanines edge to edge, or via a substituent group or via atoms complexed in the centre of the phthalocyanine cavity.[53,54] In addition to this, the presence of an N_8 unit (rather than N_4 for porphyrins) has allowed the binding of larger metal atoms such as the lanthanides. Due to the presence of the d-orbitals on these metals, they can bind to more substituent atoms. This has enabled the development of compounds such as the bis-phthalocyanines:[55] for example lutetium bisphthalocyanine, whose optical spectra change dramatically on exposure to various vapours, giving rise to potential sensor applications. Trisphthalocyanines are also available; in this context workers have for example sandwiched a lutetium atom between two phthalocyanine rings and then added a europium atom and a third phthalocyanine ring (Figure 1.8d) to make a triple-decker sandwich compound.[56] It has also been possible to make polymeric phthalocyanines via a central atom, with examples including polyphthalocyanines linked by a central Si—O chain (Figure 1.8e), which have been synthesised and deposited as ultrathin films[57] and shown to display novel liquid-crystalline properties.[58]

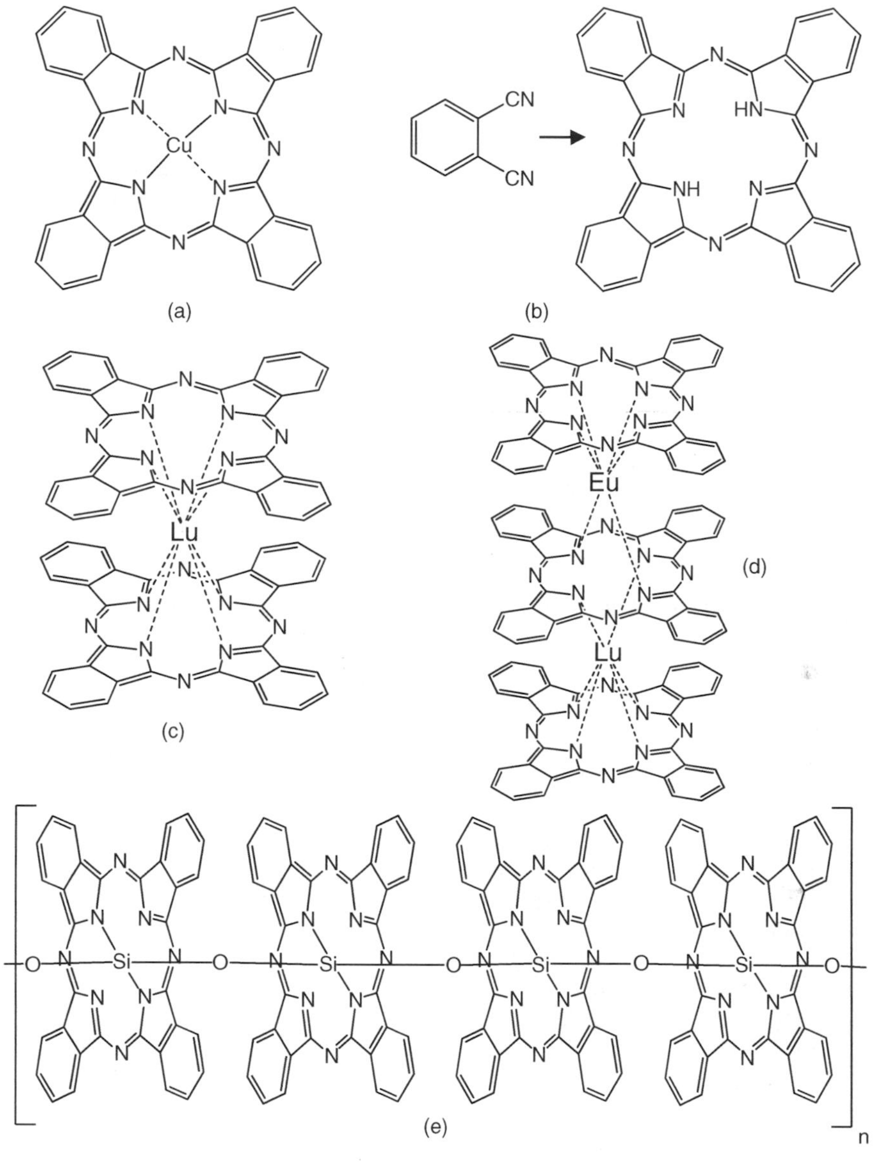

Figure 1.8   *Structures of mono, bis, tris and poly phthalocyanines*

1.6   Conclusions

This chapter has served to introduce some of the simpler ring systems, both aliphatic and aromatic in nature. The aromatic systems tend to have planar or distorted planar structures, which can limit their ability to form complexes (but not prevent it, as the examples of porphyrins and phthalocyanines prove). Further chapters will address the many ring systems that are nonplanar, which thus possess three-dimensional structures that allow for a richness and diversity of chemistry and complex formation.

References

1. Lieser G, Lee KS, Wegner G. Packing of long-chain cycloalkanes in various crystalline modifications: an electron diffraction investigation. *Coll Poly Sci.* 1988; **266**: 419–428.
2. Meyer B. Elemental sulfur. *Chem Rev.* 1976; **76**: 367–388.
3. Sinninghe Damsté JS, Strous M, Rijpstra WIC, Hopmans EC, Geenevasen JAJ, van Duin ACT, van Niftrik LA, Jetten MSM. Linearly concatenated cyclobutane lipids form a dense bacterial membrane. *Nature.* 2002; **419**: 708–712.
4. Maier G, Pfriem S, Schäfer U, Matusch R. Tetra-tert-butyltetrahedrane. *Angew Chem Int Ed.* 1978; **17**: 520–521.
5. Tanaka M, Sekiguchi A. Hexakis(trimethylsilyl)tetrahedranyl tetrahedrane. *Angew Chem Int Ed.* 2005; **44**: 5821–5823.
6. Ichinohe M, Toyoshima M, Kinjo R, Sekiguchi A. Tetrasilatetrahedranide: a silicon cage anion. *J Am Chem Soc.* 2003; **125**: 13328–13329.
7. Eaton PE, Cole TW. Cubane. *J Am Chem Soc.* 1964; **86**: 3157–3158.
8. Zhang M-X, Eaton PE, Gilardi R. Hepta- and octanitrocubanes. *Angew Chem Int Ed.* 2000; **39**: 401–404.
9. Paquette LA, Ternansky RJ, Balogh DW, Kentgen G. Total synthesis of dodecahedrane. *J Am Chem Soc.* 1983; **105**: 5446–5450.
10. Cross RJ, Saunders M, Prinzbach H. Putting helium inside dodecahedrane. *Org Lett.* 1999; **1**: 1479–1481.
11. Katz TJ, Acton N. Synthesis of prismane. *J Am Chem Soc.* 1973; **95**: 2738–2739.
12. Eaton PE, Or YS, Branca SJ. Pentaprismane. *J Am Chem Soc.* 1981; **103**: 2134–2136.
13. Fessner WD, Sedelmeier G, Spurr PR, Rihs G, Prinzbach H. Pagodane: the efficient synthesis of a novel, versatile molecular framework. *J Am Chem Soc.* 1987; **109**: 4626–4642.
14. Kekulé FA. Sur la constitution des substances aromatiques. *Bulletin de la Societe Chimique de Paris.* 1865; **3**: 98–110.
15. Kekulé FA. Benzolfest Rede. *Berichte def Deutschen Chemischen Gesellschaft.* 1890; **23**: 1302–1311.
16. Lonsdale K. The structure of the benzene ring in hexamethylbenzene. *Proc Roy Soc.* 1929; **123A**: 494–515.
17. Emerson GF, Watts L, Pettit R. Cyclobutadiene- and benzocyclobutadiene-iron tricarbonyl complexes. *J Am Chem Soc.* 1965; **87**: 131–133.
18. Sekiguchi A, Matsuo T, Watanabe H. Synthesis and characterization of a cyclobutadiene dianion dilithium salt: evidence for aromaticity. *J Am Chem Soc.* 2000; **122**: 5652–5653.
19. Kaufman HS, Fankuchen I, Mark H. Structure of cyclo-octatetraene. *Nature.* 1948; **161**: 165.
20. Katz TJ. The cyclooctatetraenyl dianion. *J Am Chem Soc.* 1960; **82**: 3784–3785.
21. Sondheimer F. The annulenes. *Proc Roy Soc A.* 1967; **297**: 173–204.
22. Kennedy RD, Lloyd D, McNab H. Annulenes 1980–2000. *J Chem Soc.* 2002. Perkin Trans I.
23. Vogel W, Klug W, Breuer A. 1-6-methano[10]annulene. *Org Synth Coll.* 1988; **6**: 731.
24. Gilchrist TL, Tuddenham D, McCague R, Moddy CJ, Ress CW. 7b-methyl-7bh-cyclopent[cd]indene, an unsubstituted tricyclic aromatic [10]annulene. *J Chem Soc Chem Commun.* 1982; **14**: 657–658.
25. Oth JFM, Schröder G. Annulenes. Part XII. The dianion of [12]annulene. *J Chem Soc B.* 1971; **5**: 904–907.
26. Stevenson CD, Kurth TL. Perturbations in aromatic and antiaromatic characters due to deuteration: the case of [16]annulene. *J Am Chem Soc.* 1999; **121**: 1623–1624.
27. Chiang CC, Paul IC. Crystal and molecular structure of [14]annulene. *J Am Chem Soc.* 1972; **94**: 4741–4743.
28. Vogel E, Sombroek J, Wagemann W. Syn-1,6:8,13-bismethano[14]annulene. *Angew Chem Int Ed Engl.* 1975; **14**: 564–565.
29. Vogel E, Haberland U, Günther H. Anti-1,6:8,13-bismethano[14]annulene. *Angew Chem Int Ed Engl.* 1970; **9**: 513–514.
30. Balci M, Schlalenback R, Vogel E. 15,16-dioxo-syn-1,6:8,13-bismethano[14]annulene. *Angew Chem Int Ed Engl.* 1981; **20**: 809–811.
31. Vogel E, Nitsche R, Krieg H-U. 15,16-dioxo-anti-1,6:8,13-bismethano[14]annulene. *Angew Chem Int Ed Engl.* 1981; **20**: 811–813.
32. Williams RV, Edwards WD, Vij A, Tolberts RW, Mitchell RH. Theoretical study and X-ray structure determination of dimethyldihydropyrene. *J Org Chem.* 1998; **63**: 3125–3127.

33. Gorter S, Keulemans-Rutten E, Krever M, Romers C, Cruickshank DW. [18]-annulene, $C_{18}H_{18}$, structure, disorder and Hueckel's 4n + 2 rule. *J Acta Crystallogr*. 1995; **B51**: 1036–1045.

34. Otsubo T, Gray R, Boekelheide V. Bridged [18]annulenes. 12b,12c,12d,12e,12f,12g-hexahydrocoronene and its mono- and dibenzo analogs. Ring-current contribution to chemical shifts as a measure of degree of aromaticity. *J Am Chem Soc*. 1978; **100**: 2449–2456.

35. Metcalf BW, Sondheimer F. Unsaturated macrocyclic compounds. LXXXVI. [20]annulene. *J Am Chem Soc*. 1971; **93**: 6675–6677.

36. McQuilkin RM, Metcalf BW, Sondheimer F. [22]annulene. *J Chem Soc D*. 1971; **7**: 338–339.

37. Calder IC, Sondheimer F. [24]annulene: dependence of nuclear magnetic resonance spectrum on temperature. *J Chem Soc Chem Commun*. 1966; 904–905.

38. Sondheimer F. Annulenes. *Acc Chem Res*. 1972; **5**: 81–91.

39. Choi CH, Kertesz M. Bond length alternation and aromaticity in large annulenes. *J Chem Phys*. 1998; **108**: 6681–6688.

40. Diederich F, Gobbi L. Cyclic and linear acetylenic molecular scaffolding. *Top Curr Chem*. 1999; **201**: 43–79.

41. Parasuk V, Almlof J, Feyereisen MW. The [18] all-carbon molecule: cumulene or polyacetylene? *J Am Chem Soc*. 1991; **113**: 1049–1050.

42. Diederich F, Rubin Y, Knobler CB, Whetten RL, Schriver KE, Houk KN, Li Y. All-carbon molecules: evidence for the generation of cyclo[18]carbon from a stable organic precursor. *Science*. 1989; **245**: 1088–1090.

43. Adamson GA, Rees CW. Towards the total synthesis of cyclo[n]carbons and the generation of cyclo[6]carbon. *J Chem Soc. Perkin Trans*. 1996; **13**: 1535–1543.

44. Koch N. Organic electronic devices and their functional interfaces. *Chem Phys Chem*. 2007; **8**: 1438–1455.

45. Nanditha DM, Dissanayake M, Adikaari AADT, Curry RJ, Hatton RA, Silva SRP. Nanoimprinted large area heterojunction pentacene-C60 photovoltaic device. *Appl Phys Lett*. 2007; **90**: 253–502.

46. Denissenko MF, Pao A, Tang M, Pfeifer GP. Preferential formation of benzo[a]pyrene adducts at lung cancer mutational hotspots in P53. *Science*. 1996; **274**(5286): 430–432.

47. Krieger CF, Diederich F, Schweitzer D, Staab HA. Molecular structure and spectroscopic properties of kekulene. *Angew Chem Int Ed Engl*. 1979; **18**: 699–701.

48. Simpson CD, Brand JD, Berresheim AJ, Przybilla L, Rader HJ, Mullen K. Synthesis of a giant 222 carbon graphite sheet. *Chem Eur J*. 2002; **8**: 1424–1429.

49. Freitag M. Graphene: nanoelectronics goes flat out. *Nature Nanotech*. 2008; **3**: 455–457.

50. Newman, MS, Lednicer D. The synthesis and resolution of hexahelicene. *J Am Chem Soc*. 1956; **76**: 4765–4770.

51. Rothemund P. Formation of porphyrins from pyrrole and aldehydes. *J Am Chem Soc*. 1935; **57**: 2010–2011.

52. Adler AD, Longo FR, Finarelli JD, Goldmacher J, Assour J, Korsakoff L. A simplified synthesis for meso-tetraphenylporphine. *J Org Chem*. 1967; **32**: 476.

53. Claessens CG, Hahn U, Torres T. Phthalocyanines: from outstanding electronic properties to emerging applications. *Chem Record*. 2008; **8**: 75–97.

54. McKeown NB. Phthalocyanine containing polymers. *J Mat Chem*. 2000; **9**: 1979–1995.

55. Snow AW, Barger WR. In: Leznoff CC, Lever ABP, editors. Phthalocyanines: Properties and Applications. Vol **1**. New York: VCH; 1989.

56. Pushkarev VE, Shulishov EV, Tomilov YV, Tomilova LG. The development of highly selective approaches to sandwich-type heteroleptic double- and triple-decker lutetium(III) and europium(III) phthalocyanine complexes. *Tet Lett*. 2007; **30**: 5269–5273.

57. Orthmann E, Wegner G. Preparation of ultrathin layers of molecularly controlled architecture from polymeric phthalocyanines by the Langmuir-Blodgett-technique. *Angew Chem Int Ed*. 1986; **25**: 1105–1107.

58. Adib ZA, Davidson K, Nooshin H, Tredgold RH. Magnetic orientation of phthalocyaninato-polysiloxanes. *Thin Solid Films*. 1991; **201**: 187–195.

2
Cyclophanes

2.1 Introduction

Chapter 1 described a number of multi-ring organic compounds, in which the rings are usually fused via adjacent carbon atoms. This can be seen in, for instance, tetralin (1,2,3,4-tetrahydronaphthalene; see Figure 2.1a), which can be thought of as a single benzene ring with a four-carbon aliphatic bridge linking the ortho positions. This arrangement is relatively unstrained and the presence of the aliphatic chain does not overtly affect the structure of the aromatic ring since the aliphatic chain and the ring do not interact. It is obvious however that should the chain bridge the 1,3 or 1,4 positions of the ring, there would be two effects: first strain would increase since the chain would have to bridge a longer distance and second the chain would have to pass 'above' the ring, thereby increasing the potential for interaction between the chain and the aromatic π system; this could potentially modify the properties of the aromatic ring.

Simple cyclophanes can be thought of as hydrocarbons containing a benzene ring combined with an aliphatic chain bridging two nonadjacent carbons. There are numerous variations on this theme, such as compounds with multiple aromatic units or aromatic systems larger than benzene rings. They are not limited to hydrocarbon structures, since heteroatoms can be included within the chains, or alternatively the benzene rings can be replaced with furans, thiophenes, pyridines and so on. There has been a wide study of cyclophanes since the strain inherent in many of these molecules can modify the conformations and reactivities of the aromatic rings. Much of the early work in these types of systems, especially the development of many of the classical cyclophane systems during the 1950s and 1960s, has been reviewed elsewhere, such as within the husband-and-wife team Donald and Jane Crams' elegantly named paper[1] on 'Bent and battered benzene rings', as well as within the immense number of papers published by this group.

2.2 Cyclophanes with One Aromatic System and Aliphatic Chain

As stated earlier, within materials such as tetralin, which could be thought of as a '1,2-cyclophane', there is no inherent strain; tetralin displays chemistry typical of a substituted aromatic system. However, the situation is very different when the chain bridges the 1,3 or 1,4 positions. The presence of a long bridging chain has only minimal effect on the aromatic ring structure but as the bridging chain gets shorter it begins

Macrocycles: Construction, Chemistry and Nanotechnology Applications, First Edition.   Frank Davis and Séamus Higson.
© 2011 John Wiley & Sons, Ltd. Published 2011 by John Wiley & Sons, Ltd.

to 'pull' the benzene ring into a nonplanar shape, since otherwise it would be unable to reach across the ring. This strain has a number of effects on the cyclophane's structure and behaviour, as detailed below.

2.2.1   Properties of the cyclophanes

Metacyclophanes have the general structure shown in Figure 2.1b, where an aliphatic chain bridges the 1,3 positions on the benzene rings. Usually they are named as [n]metacyclophanes, where n is the number of carbons in the aliphatic chain. As the chain becomes shorter, ring strain increases and the stability of the compound decreases. [4]metacyclophane is an unstable intermediate which, while it can be formed during processes such as flash-vacuum thermolysis of Dewar benzene derivatives,[2] cannot be isoglated but rather decomposes to products such as tetralin. UV irradiation of similar materials at low temperature led to the detection of [4]metacyclophane by UV spectroscopy.[3] [5]metacyclophane can be synthesised as a colourless oil,[4] but at room temperature it polymerises. When the chain is further lengthened, as in [6]metacyclophane and higher analogues, the compound becomes stable at room temperature.

Paracyclophanes are similar in behaviour except that the bridge is of course between opposite carbons on the benzene ring (Figure 2.1c). [4]paracyclophane was studied by the same groups who performed the work on metacyclophanes detailed above and these workers showed that [4]paracyclophane could be generated photochemically. On generation at $-20\,°C$, the cyclophane can either polymerise to form poly-p-xylylene or can be derivatised with trifluoroacetic acid.[5] Alternatively, photogeneration in a glass at 77 K allows measurement of the UV spectrum.[6] [5]paracyclophane can be synthesised in a similar manner[7] and decomposes at room temperature. [6]paracyclophane is stable, as are the larger cyclophanes. As the aliphatic chain length increases, its effects on stability and the benzene ring structure decrease due to decreasing strain. An extensive review of these compounds has recently been published.[8]

With the [n]metacyclophanes, the effect of the bridging chain on the aromatic ring is profound. Rather than adopting its preferred planar conformation, the benzene ring is forced to exist in a twisted-boat conformation, as shown in Figure 2.2a. As the chain length decreases, the angle of the bridgehead carbons from the plane of the ring and the distortion from the planar structure increases.[8] X-ray crystallographic data of crystalline derivatives of [6]metacyclophane show this angle to be of the order of 19.6° and 26.8° for the corresponding [5]metacyclophane.[8] No stable derivatives of [4]metacyclophanes exist, but theoretical calculations[8] give an angle of 40.6°. [6]paracyclophane has been shown to have a similar boat-structure (Figure 2.2b) angle (19–21°) by crystallographic studies, and this angle has been calculated to be 23.7° and 29.7° for the [5] and [4]paracyclophanes respectively.[8] What is interesting is that the carbon–carbon bonds within the ring are still approximately the same lengths, indicating that aromaticity is retained, rather than an alternating single/double-bond polyene structure.

The resultant nonplanarity of the benzene ring has a variety of effects on the chemical and physical properties of these compounds. Aromatic compounds adsorb strongly in the UV region of the spectrum and

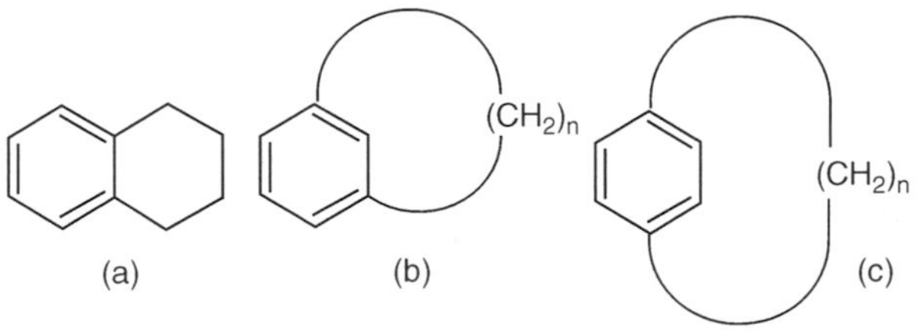

Figure 2.1   *Structures of (a) tetralin, (b) metacyclophanes and (c) paracyclophanes*

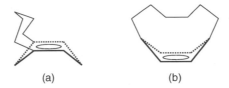

Figure 2.2   *Conformations of [6]meta- and [6]paracyclophanes*

both meta- and paracyclophanes show distinct red shifts compared to dialkyl benzenes.[9] As the bridging chains get longer this effect becomes less, and for $n > 8$ there is almost no effect at all. This effect appears to be consistent across a wide range of cyclophanes with varying aromatic groups and could be thought of as a measurement of the strain on the aromatic rings in these systems. NMR spectra are also affected somewhat by the distortions from planarity.[10,11] The ring protons, however, still give signals at high frequencies for the highly strained [5]metacyclophane (6.8–7.8 ppm), indicating a high ring current and an aromatic structure which is also correlated by low values for some of the chain protons, indicating shielding by an aromatic system is occurring.[12] [6]metacyclophane is similarly also shown to be aromatic. In the case of [6]metacyclophane, NMR studies show the aliphatic ring can flip from one side of the aromatic ring to the other; steric constraints do not allow this for [5]metacyclophane. The paracyclophanes give similar results (aromatic protons at 7.17 ppm, aliphatics at 2.49, 2.15 and 0.33, indicating some shielding occurs) at room temperature.[13] As the temperature is lowered, the spectrum becomes more complex due to 'freezing' of the structure and shielding effects increase (some methylene protons as low as −0.6 ppm). A substituted [4]paracyclophane with bulky side groups was stable enough for measurement of the NMR spectrum, which indicated considerable aromaticity[14] even though the system was still highly reactive. An extensive review has been published elsewhere on the NMR spectra of a wide range of cyclophane structures.[10]

2.2.2   Chemistry of the cyclophanes

Many reactions which progress slowly or not at all with simple alkylbenzenes are easily achievable with cyclophanes. A simple example of this is the instability of [4] and [5]cyclophanes compared to compounds such as tetralin, which can be distilled at 206–208 °C. The Diels–Alder reaction is the addition of alkenes to 1,3 dienes. Although the theoretical cyclohexatriene possesses alternating single/double bonds, benzene derivatives because of their aromatic nature do not participate in the Diels–Alder reaction unless a combination of vigorous reaction conditions and highly active dienophiles is used.[8] However, the strained cyclophanes react readily with dienophiles under much less forcing conditions. For metacyclophanes the reaction normally occurs across the 2 and 5 positions of the ring since this relieves most strain. Usually the more strained the system, the more reactive this will be. For example, the active dienophile dimethyl acetylene dicarbonate reacts readily at room temperature with [5]metacyclophane; it also reacts with [6]metacyclophane at room temperature, but more slowly.[8] Cyclophanes also react with dichlorocarbene, which does not occur with unstrained benzene derivatives. Protonation of cyclophanes with acids often catalyses the shift of the bridging ring to form a 1,2 type cyclophanes (such as a tetralin). The change in reactivity also occurs with substituent groups. Chlorobenzene in this context will not react with methoxide ion but a [5]metacyclophane with a chloro substituent at the 2 position will undergo nucleophilic substitution to give the 2-methoxy derivative. Similar behaviour is noted for the paracyclophanes, with for example [4]paracyclophane reacting with dienophiles, readily adding bromine[15] and rearranging to less strained isomers under acidic conditions. The [4] and [5] paracyclophanes display even more enhanced reactivities.

2.2.3   Synthesis of the cyclophanes

Metacyclophanes have been synthesised through a variety of routes, the most successful being the rearrangement of propellane-type derivatives, which is suitable for both [5] and [6]metacyclophanes. A variety of cyclophanes including the [6]metacyclophane were prepared[16] in 1975. [5]metacyclophane[17] was initially prepared using this method in 1977. Synthesis of the Dewar benzene isomers followed by photochemical catalysed rearrangement has been shown to be a suitable method for generating [4]metacyclophanes,[2,3] although it should be noted that these compounds need to be 'trapped' by further chemical reactions since they are not stable enough to be isolated.

There have been a wide variety of synthetic routes to the paracyclophanes, many of which are based on the synthesis of Dewar benzene-type derivatives, which can then be induced to undergo thermal or photochemical rearrangements. Photochemical methods have been prevalent in the synthesis of the unstable [4] and [5]paracyclophanes.[5-7] The stable [6]paracyclophane was initially synthesised[13] in 1974, as shown in Figure 2.3, from the cyclic ketone which was converted first into a hydrazone and then into its lithium salt. Thermal pyrolysis produced 5–10% of [6]paracyclophane via a carbine-type reaction but the necessity of isolation using gas chromatography greatly reduced the achieved yield. Other reactions such as rearrangement of Dewar benzenes by thermal or photochemical methods have also been used. A full review of the synthesis of cyclophanes is beyond the remit of this chapter, but much of the work up to the 1990s was extensively reviewed by Kane *et al.*[18]

The bridging ring is of course not limited to an alkyl chain. There are a wide variety of moieties capable of acting as a cyclophane bridge. Unsaturated chains are possible, an extreme example being [4]paracyclophane-1,3-diene (Figure 2.4), which can be synthesised by photolysis of the Dewar benzene isomer; although it is too unstable to be isolated, it can be trapped by a Diels–Alder reaction with cyclopentadiene.[19] A wide variety of bridging chains, a few of which will be further detailed within this

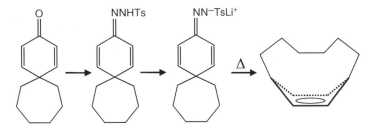

Figure 2.3   *Synthesis of [6]paracyclophane*

Figure 2.4   *Structure of [4]paracyclophane-1,3-diene*

work, have been utilised, containing such units as aromatic rings, double and triple bonds and heteroatoms such as N, O, S, Si and P.

2.3   Cyclophanes with More than One Aromatic Ring

Chemists have often attempted to design what the Crams called 'internally tortured molecules with inherent suicidal tendencies that skirt a fine line between stability and self-destruction'.[1] We have already seen some of these types of molecule earlier in this chapter; another variant on this theme are molecules where two or more aromatic systems are clamped together in close proximity, leading to a high degree of both bond strain and π−π interactions. Systems in which two benzene atoms are joined by long bridging groups tend to behave like simple open chain structures, however when the number of carbons in the bridges is four or less, as with the earlier cyclophanes, the structures and chemistry of these compounds are affected.

The most highly strained system would be [1,1']paracyclophane, the structure of which is shown in Figure 2.5a. This system can be synthesised by photochemical reaction of a bis-Dewar benzene isomer at 77 K,[8] although the extreme strain in this system once again renders it too unstable to be isolated. The compound was stable enough however at 77 K or in THF solution at −60 °C for spectroscopic investigations to be made.[21] As determined earlier, the strain causes red shifts in the UV/Vis spectrum of the cyclophane, with adsorption extending out as far as 450 nm. The NMR spectrum of the cyclophane gave peaks at 6.94 and 3.38 ppm, indicating that the benzene rings are still aromatic. Substituted versions have been synthesised which are stable enough to be crystallised.[22] X-ray studies show that the benzene rings adopt the boat conformation, with the angle between the bridgehead carbons and the plane of the other ring atoms being about 25°.

One of the most extensively studied cyclophanes has to be [2,2']paracyclophane[1] (Figure 2.5b). Although the benzene rings are connected by linkages just two carbons long (meaning this is a highly strained system), this compound and various substituted derivatives are stable and are now commercially used in the production of the various poly(xylylene) polymers.[20] Poly-p-xylylene (Figure 2.5c) can be synthesised by either pyrolysis of xylene or reaction of p-xylylene dichloride with sodium.[23] Extraction of this polymer led to the isolation of small amounts of [2,2']paracyclophane.[23] Other workers studied a range of chemical methods to make a variety of cyclophanes, usually starting with compounds with two phenyl groups linked by a two-or-more-carbon chain.[24] These phenyl groups had reactive groups *para* to the alkyl chain such as bromomethyl groups, which could then be ring-closed using the Wurtz reaction (Figure 2.6a). Other reactive groups included ester groups, which could be condensed to form a cyclic acyloin, which could then be reduced to give the hydrocarbon cyclophane[24] (Figure 2.6b). Another synthetic scheme (Figure 2.6c) involves the quaternisation of α-bromo-p-xylene with trimethylamine and conversion to the

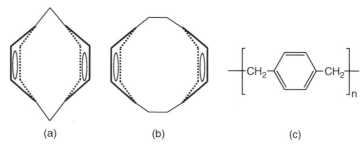

(a)   (b)   (c)

Figure 2.5   *Structures of (a) [1,1']paracyclophane, (b) [2,2']paracyclophane and (c) poly(p-xylylene)*

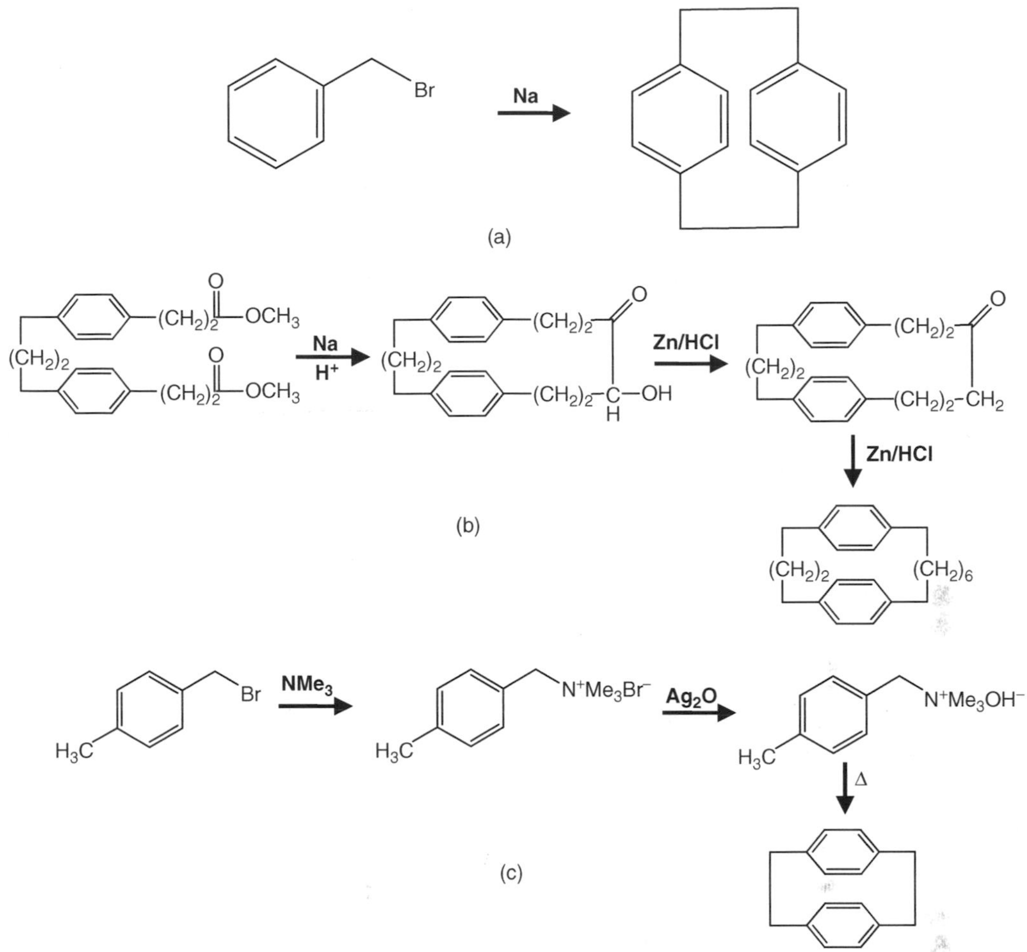

Figure 2.6   *Synthesis of [n,n']paracyclophanes*

hydroxide salt with silver oxide followed by a 1,6-Hoffman elimination by refluxing in toluene[25] to give the cyclophane with yields of up to 19% along with insoluble polymeric material.

As expected, the X-ray crystallographic structure[26] of [2,2']paracyclophane reveals that the benzene rings adopt the boat conformation, with the angle between the bridgehead carbons and the plane of the other ring atoms being about 12–13°. The distance between the bridgehead carbons of adjacent aromatic rings is only 0.278 nm, with 0.309 nm between the planes of the other aromatic carbons – much closer that the 0.335 nm separation of graphite planes or the 0.340 nm typical for stacked aromatic molecules. There is a slight (6°) staggering of the benzene rings;[1] also of interest is that the aromatic hydrogen atoms are bent *towards* the centre of the molecules. This is thought to be due to interactions between the two π systems forcing relatively more of the electron density towards the outside faces of the benzene rings.[1] The NMR spectrum shows the aromatic protons at 6.47 ppm and the benzylic protons at 3.01 ppm,[27] demonstrating that aromaticity is retained, and once again there is a noticeable red shift in the UV/Vis spectrum compared to open-chain analogues.[24] When longer bridging chains with greater flexibility are

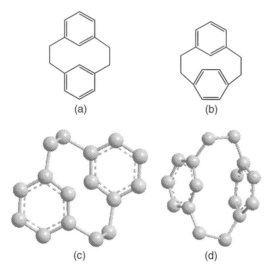

(a)   (b)

(c)   (d)

Figure 2.7   Structures of (a) [2,2′]metacyclophane and (b) [2,2′]metaparacyclophane, and (c, d) their relative crystal structures

used, these strain-related effects are decreased. [3,3′]paracyclophane[28] shows less of a red shift and has a structure in which the benzene rings are less bent (6.4°) and the rings are decentred with respect to each other. Longer-chain separated cyclophanes show less strain effect, while [6,6′]paracyclophane behaves as the open-chain analogue.[1]

Metacyclophanes and a number of mixed metaparacyclophanes are also known. [2,2′]metacyclophane (Figure 2.7a) was first synthesised in 1899, although its crystal structure (Figure 2.7c) was not determined until over 50 years later.[29] The benzene rings are distorted in a similar manner to [n]metacyclophanes and there is an extremely close approach (0.269 nm) of one ring to the other. The intermediate compound [2,2′]metaparacyclophane[30] (Figure 2.7b) can be synthesised by the acid-catalysed rearrangement of [2,2′]paracyclophane. This has also had its structure determined;[1] what is interesting is that the para ring suffers somewhat more distortion from planarity than in [2,2′]paracyclophane (Figure 2.7d).

Other variations on the basic [n,n′]paracyclophane structure have been investigated and a few examples will be given. Rather than alkane chains, a series of paracyclophanes with unsaturated bridging groups have been studied. The highly strained compound [2,2′]paracyclophanediene (Figure 2.8a) has been synthesised by bromination and dehydrobromination of the parent cyclophane.[31] This would be expected from a simple consideration of its structure to be highly conjugated. The bridging double bonds are however orthogonal to the benzene rings and this prevents their interaction with the aromatic system. This is clearly shown by various spectrographic methods, which demonstrate minimal interaction between alkenic and aromatic systems.[31] The crystal structure shows a typical paracyclophane, with bending of the aromatic rings[32] similar to that in the saturated version. Larger systems containing a 1,3-butadiene linkage (Figure 2.8b) or using the larger cyclophane [2,2,2]paracyclophane-triene (Figure 2.8c) have also been studied.[33]

Many substituted cyclophanes in which substituents are attached to one or both benzene rings are known. Once again a review of these is far beyond the scope of this chapter so only a few examples will be given. Some of the simplest systems involve just a single substituent on one of the benzene rings, such as [2,2′]paracyclophane (Figure 2.9a) with a single carboxylic acid substituent. What is interesting about this family of compounds is that there is an inherent asymmetry in the molecule, which means

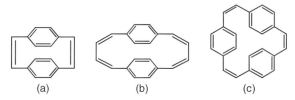

Figure 2.8 *Structures of (a) [2,2′]paracyclophanediene, (b) [4,4′]paracyclophanetetraene and (c) [2,2′,2″]-paracyclophanetriene*

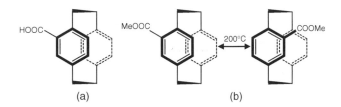

Figure 2.9 *Structures of [2′2]paracyclophane carboxylic acid and isomerisation of its methyl ester*

that the two isomers shown are actually enantiomers and display optical activity.[34] Similar behaviour is observed for the corresponding [3,3′] compound, but for the [4,4′] compound no resolution of enantiomers is possible, indicating that the longer four-carbon bridges allow facile rotation of the aromatic rings.[35] The [2,2′]metacyclophanes and [2,2′]metaparacyclophanes, when mono-substituted, also show this restricted rotation, and in the case of the metaparacyclophanes it can sometimes be found that one ring will rotate while the other does not.[1]

As found for the [n]cyclophanes, the ring strain has a considerable effect on the reactivity of the [n,n′]cyclophanes. Both the bridge and the aromatic system can be affected, as can be seen for example when the optically active ester derivative of [2,2′]paracyclophane (Figure 2.9b) is heated to 200 °C, causing it to racemise.[36] However, molecular models prove that it is impossible for the aromatic rings to rotate and instead the system undergoes cleavage and reformation of one of the bridging chains. Other rearrangements include the fact that when treated with HCl/AlCl₃ the [2,2′]paracyclophane rearranges to [2,2′]metaparacyclophane.[37] The ring strain also has effects on the chemistry of the aromatic rings; for instance, the acetylation or nitration of the aromatic rings of the [2,2′]paracyclophanes[35] proceeds much faster than similar reactions with cyclophanes with longer bridging chains. Also, the presence of a substituent on one aromatic ring can affect the reactivity of the opposing ring, with substitution usually taking place on the carbon just opposite the original functional group – the so-called 'pseudo gem effect'.[38]

Other novel reactions include the formation of inter-ring bonds, a brief example of which will be given. When a cyclophane containing one naphthalene and one brominated benzene ring (Figure 2.10) is treated with strong base, the benzene ring is dehydrobrominated to give a benzyne intermediate, which in turn undergoes an internal Diels–Alder reaction with the opposing naphthalene ring.[39] Chromium is well known to form organometallic compounds with benzene rings and many exohedral chromium–cyclophane complexes are known.[10] However, the presence of two benzene rings in such close proximity allows the formation of endohedral organometallic complexes with materials such as [2,2] and [3,3]paracyclophane[40,41] (Figure 2.11a,b), where the chromium atom is sandwiched between two rings of the same cyclophane. Another simple and symmetric structure is found in the complex[42] formed between silver triflate and [2,2′,2′]paracyclophane (Figure 2.11c).

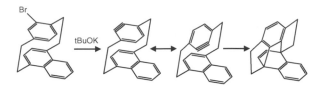

Figure 2.10   *Intermolecular reactions of paracyclophanes*

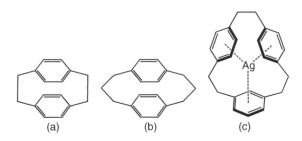

Figure 2.11   *Complexes of cyclophanes with metals*

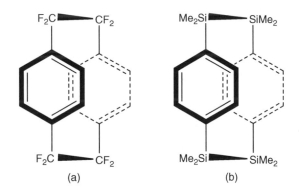

Figure 2.12   *Octafluoro[2,2′]paracyclophane and tetrasila[2,2′]paracyclophane*

Benzene rings linked by nonhydrocarbon bridges have been described. Examples include fluorinated bridges[43] such as octacafluoro[2,2′]paracyclophane (Figure 2.12a), which is used as a feedstock for the polymer parylene AF4 and the related trimer dodecafluoro[2,2′,2′]paracyclophane. Silicon bridges (Figure 2.12b) have also been utilised, with the much longer Si–Si bond bridge allowing for much less distortion of the benzene rings.[44]

Once the existence of such molecules as [2,2′]paracyclophane was shown, it was inevitable that chemists would attempt to construct larger, more complex systems. A 'triple-decker' cyclophane[45] (Figure 2.13a) was first synthesised in 1967. Synthesis of a wide range of multiple systems containing up to six benzene rings linked together by two-carbon bridges has been reported.[46] For example, the quadruple system (Figure 2.13b) was examined using X-ray crystallography, which showed the outer benzene rings to assume the boat conformation and the inner rings to adopt a twisted conformation due to the strain imposed by the four bridges.[46]

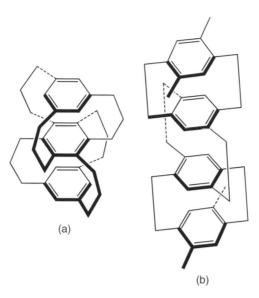

(a)

(b)

Figure 2.13   *Triple and quadruple cyclophanes*

Another field of interest was the incorporation of more connecting bridges between the two benzene rings. The logical conclusion of this course of action was finally realised in 1979 with the synthesis of the first 'superphane' with six two-carbon bridges linking together every aromatic carbon.[47] Superphane is a stable compound, melting point 325–327 °C, and X-ray crystallography shows it to have a distorted structure with the aromatic–aliphatic C–C bonds bent at an angle of about 20° to the ring, as well as the smallest ring–ring spacing reported for a cyclophane of 262 pm.[48]

2.4   Napthalenophanes and Other Aromatic Systems

Many other aromatic hydrocarbons have been included in cyclophanes, some examples of which will be given. Versions of [6]paracyclophane, for example, where the benzene ring is incorporated into a larger fused ring system, are known. Both naphthalene and anthracene were bridged across the 1,4-positions of the ring[49] to give the resultant [6](1,4)naphthalenophane (Figure 2.14a) as a colourless oil or [6](1,4)anthracenophane (Figure 2.14b) as a solid with a melting point of 161–163 °C. As found with the paracyclophanes, there were red shifts in the UV spectra compared to reference compounds and the NMR spectra indicated aromaticity was retained. X-ray studies[49] of the anthracenophane showed the bridged ring again existed in a boat-like conformation. Similar studies were carried out on the [6](9,10)anthracenophane (Figure 2.14c) and showed that the bridging and resultant distortion of the aromatic system led to a much more unstable compound than the 1,4-bridged analogue.[50]

Pyrene has also been bridged in a similar manner, with chains not much longer than those in the cyclophanes. When we consider that the smallest stable paracyclophane has a six-carbon bridge, even with a much larger pyrene moiety, both the (1,7)dioxa[7][51] and (1,8)dioxa[8](2,7)pyrenophane[52] have been isolated (Figure 2.14d,e). X-ray studies of [8]pyrenophane show that the ring, rather than being flat, has an overall bend of 87.8°, with the distortion being spread evenly across it. [7]pyrenophane is even more

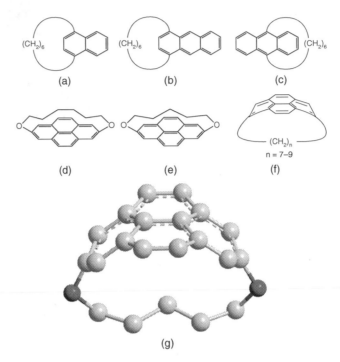

Figure 2.14   (a–f) [n]cyclophanes containing multiring systems, (g) crystal structure of 2.14e

strained,[51] with an overall bend of 109.1° being observed (Figure 2.14g). Similar hydrocarbon analogues[53] have also been synthesised (Figure 2.14f).

Naphthalene, with its two fused benzene rings, has been incorporated into a large number of cyclophanes. The earliest naphthalenophane appears to have been synthesised in 1951 using a Wurtz reaction on 2,7-dimromethylnaphthalene (Figure 2.15a), with the product being dehydrogenated to form coronene.[54] Analogues of [2,2′]cyclophane have been synthesised, for example by using the Hoffman elimination method. Cram and coworkers successfully synthesised [2,2′](1,4)naphthalenophane[55] (Figure 2.15b) and showed it existed in the *anti* rather than the *syn* conformer. A wide variety of napthalenophane isomers have been reported in the literature; the NMR behaviour of many are reviewed by Ernst.[10] Layered compounds are also possible, such as the triple napthalenophane shown in Figure 2.15c,[56] where the outer rings are distorted into a boat form and the inner ring is in a twist form, similar to the benzene-derived cyclophanes. Similarly, in 1961 other workers[57] synthesised [2,2′](9,10)anthracenophane (Figure 2.15d).

One point worth noting about the naphthalene and anthracene compounds above is their ability to undergo condensation reactions between the rings. For example, [2,2′](1,4)naphthalenophane when irradiated undergoes an intramolecular condensation to give the compound shown in Figure 2.16a.[58] Similar cyclisation reactions occurred for the anthracenophane,[57] except in the latter case cyclobutane rings were produced by the condensation reaction (Figure 2.16b). Phenanthrene has also been incorporated into cyclophanes, with for example three different isomers of [2,2′]phenanthrenophane (Figure 2.17) being successfully isolated[59] from the reaction products of vinyl-substituted paracyclophanes with benzyne (didehydrobenzene). [2,2′](2.7)pyrenophane has also been synthesised,[60] with the UV/Vis spectrum shown to be highly red-shifted compared to open-chain compounds. The [3,3′] and [4,4′]pyrenophanes display lesser red shifts, as expected due to the lower strain in these systems.[61]

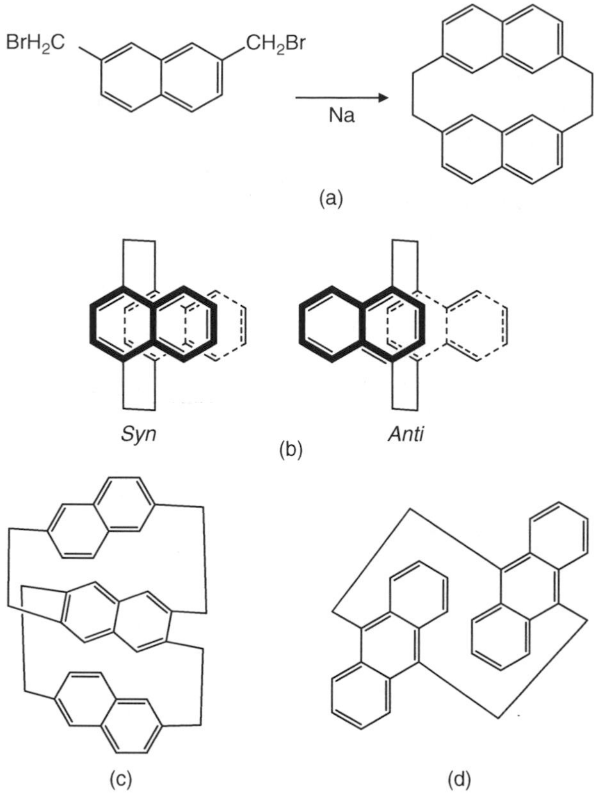

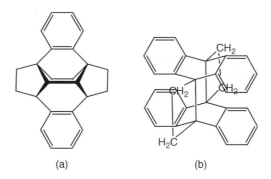

Figure 2.15 *[n, n']cyclophanes containing multi-ring systems*

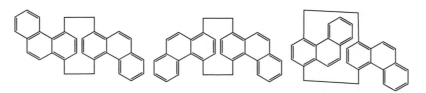

Figure 2.16 *Cyclisation products from [2,2']naphthalenophane and [2,2']anthracenophane*

Figure 2.17 *Three isomers of phenanthrenophane*

2.5 Cyclophanes Containing Heteroaromatic Systems

Most of the aromatic systems mentioned in this chapter have been hydrocarbons, but cyclophanes containing heteroaromatic rings have also been widely studied. [2,2′](2,5)furanophane and [2,2′](2,5)-thiophenophane (Figure 2.18a,b), for example, have been isolated and their structures determined by X-ray crystallography.[62] In both cases the aromatic ring is distorted into an envelope shape, with the heteroatom being out of plane with the carbon atoms; however, the thiophene rings are more distorted than the furan rings. What is also of interest is that the furanophane is conformationally flexible, whereas the thiophene system is rigid, indicating the nature of the heteroatom has a large effect on the system. Pyridinophanes[63] such as [2,2′](2,6)pyridinophane (Figure 2.18c) have also been studied and again the aromatic rings are shown to be distorted into a boat shape.

2.6 Ferrocenophanes

A large number of compounds have been synthesised which incorporate a variety of metal 'sandwich' complexes into cyclophanes, but a discussion of such a wide range of materials is outside the scope of this chapter and we will limit this section to ferrocene compounds. Ferrocene was the first known sandwich complex. It has an iron atom covalently bonded to two cyclopentadienyl ligands, as shown in Figure 2.19a. It is an aromatic compound and undergoes typical aromatic reactions such as Friedel–Crafts acylation, at rates approximately a million times more than would be observed for benzene. What makes this one of the most studied organometallic compounds is its ease of handing, with ferrocene being soluble in most common organic solvents and stable in air and at high temperatures. In ferrocene itself the two rings are seen to be parallel to one another. A ferrocenophane is a compound in which there is bridging between the rings of either a ferrocene molecule or two or more different ferrocene systems; in strained systems this often leads to the rings being tilted with respect to each other.

Mononuclear systems are those in which the two rings are bridged by either an atom, a molecular chain or another ring system. This can cause strain in the system, which often manifests itself as a tilting of the two rings relative to each other. The smallest systems are single-atom bridged ferrocenophanes, which have been reviewed.[64] No stable compound containing a single carbon-atom bridge has been isolated and it is thought that such [1]ferrocenophanes are too highly strained to exist. However, a range of compounds bridged by heteroatoms are known. One of the most highly strained ferrocenophanes is the

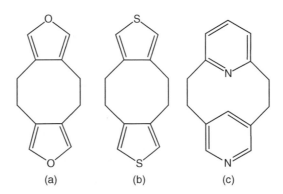

(a)   (b)   (c)

Figure 2.18 *[2,2′]heterocyclophanes*

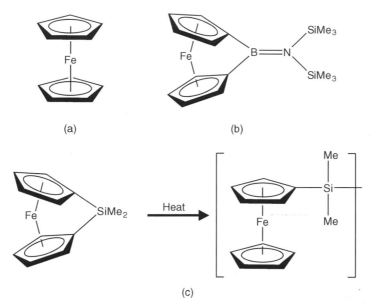

Figure 2.19 *(a) Ferrocene, (b) a boron-bridged [1]ferrocenophane and (c) polymerisation of [1,1']dimethyl-silaferrocenophane*

compound shown in Figure 2.19b,[65] in which the bridge is a single boron atom and the rings are tilted at an angle of 32° relative to each other. Other elements have been used to bridge the rings, such as S, P, Ge, Se, Sn and Zr. However, the majority of research on [1]ferrocenophanes has been on those with silicon bridges.[64] These were the first [1]ferrocenophanes synthesised in 1975,[66] with $Si(C_6H_5)_2$ bridging groups (Figure 2.19c). In these systems the strain leads to a structure in which the cyclopentadiene rings are tilted at about 20° to each other.

The high degree of strain in these compounds is reflected by their ability to form polymers via a ring-opening mechanism. Silicon-bridged ferrocenophanes were first polymerised thermally in 1992.[67] Other methods, such as anionic or transition-metal-catalysed polymerisations have been used.[64] The resultant polymers (Figure 2.19c) are stable thermoplastics which can be melt-processed and are soluble in a variety of organic solvents.

[2]ferrocenophanes can be synthesised with carbon bridges. The simple [2]ferrocenophanes shown in Figure 2.20a was the first carbon-bridged [2]ferrocenophane synthesised[68] and is a highly strained system with a tilt angle of 23°. A wide variety of other compounds of this nature have been made, such as for example the two compounds shown in Figure 2.20b,c, which were synthesised from a bisacetylene-substituted ferrocene,[69] and again exhibited high tilt angles. [2]ferrocenophanes have been shown to undergo ring-opening polymerisation.[70]

[3]ferrocenophanes and [4]ferrocenophanes have been widely studied since they are relatively unstrained and have been found to be relatively easy to synthesise.[64] This also allows the use of multiple bridges, culminating in the synthesis of a ferrocene derivative with five bridges[71] (Figure 2.20d). Multinuclear ferrocenes have also been examined, the simplest of which (Figure 2.20e) is the [0,0']ferrocenophane,[72] in which the rings are directly linked. This compound is highly insoluble. The [1,1']ferrocenes such as that shown in Figure 2.20e are much more amenable and much less strained than the systems in which bridging is from ring to ring.[73]

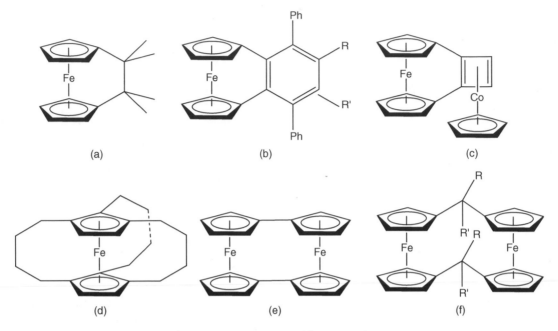

Figure 2.20   *Structures of ferrocenophanes*

2.7   Conclusions

In this chapter we have attempted to show how combinations of ring systems, both aliphatic and aromatic in nature, and the strain inherent in many of these systems can have effects on their structure, stability and reactivity. The distortion of aromatic systems away from planar structures effects their spectra and their physical and chemical properties. Understanding the effects that the three-dimensional structures of many of these systems have on their subsequent chemical and physical properties is crucial. The breadth of work on these cyclophane systems means that this chapter can only serve as an introduction to this subject and the interested reader is recommended to study the bibliography and references below.

Bibliography

Voegtle F. Cyclophanes I and II. Springer-Verlag; 1983.
Voegtle F. Cyclophanes Chemistry: Synthesis, Structures and Reactions. John Wiley and Sons; 1993.
Diederich F. Cyclophanes. Monographs in Supramolecular Chemistry. *Royal Society of Chemistry*; 1991.
Keehn PM, Rosenburg SM. Cyclophanes. Vol. 1 and 2. Academic Press Inc; 1984.
Gleiter R, Hoft H. Modern Cyclophane Chemistry. Wiley-VCH; 2004.

References

1. Cram DJ, Cram JM. Cyclophane chemistry: bent and battered benzene rings. *Acc Chem Res.* 1970; **46**: 204–213
2. Kostermans GBM, van Dansik P, de Wolf WH, Bickelhaupt F. The intermediate formation of [4]metacyclophane on flash vacuum thermolysis. *J Org Chem.* 1988; **53**: 4531–4534.

3. Tsuji T, Nishida S. Photochemical generation of [4]paracyclophanes from 1,4-tetramethylene Dewar benzenes: their electronic absorption spectra and reactions with alcohols. *J Am Chem Soc.* 1988; **110**: 2157–2164.

4. van Eis MJ, Wijsman GW, de Wolf WH, Bickelhaupt F, Rogers DW, Kooijman H, Spek AL. Hydrogenation of [5]- and [6]metacyclophane: reactivity and thermochemistry. *Chem Eur J.* 2000; **6**: 1537–1546.

5. Gerardus B, Kostermans M, Bobeldijk M, de Wolf WH, Bickelhaupt F. [4]paracyclophane intercepted. *J Am Chem Soc.* 1987; **109**: 2471–2475.

6. Tsuji T, Nishida S. [4]paracyclophane: electronic absorption spectrum and trapping by alcohols. *J Chem Soc, Chem Commun.* 1987; 1189–1190.

7. Jenneskens LW, de Kanter FJJ, Kraakman PA, Turkenburg LAM, Koolhaas WE, de Wolf WH, Bickelhaupt F, Tobe Y, Kakiuchi K, Odaira Y. [5]paracyclophane. *J Am Chem Soc.* 1985; **107**: 3716–3717.

8. Tsuji T. Highly strained cyclophanes. In: Gleiter R, Hopf R, editors. Modern cyclophane chemistry. Weinhein: Wiley-VCH; 2004.

9. Allinger NL, Sprague JT, Liljefors T. Conformational analysis. CIV. Structures, energies, and electronic absorption spectra of the [n]paracyclophanes. *J Am Chem Soc.* 1974; **96**: 5100–5104.

10. Ernst L. NMR studies of cyclophanes. *Prog Nucl Magn Reson Spectroc.* 2000; **37**: 45–190.

11. Ernst L, Ibrom K. NMR spectra of cyclophanes. In: Gleiter R, Hopf R, editors. Modern cyclophane chemistry. Weinhein: Wiley-VCH; 2004.

12. Turkenburg LAM, de Wolf WH, Bickelhaupt F, Cofino WP, Lammertsma K. [5]metacyclophanes: a spectroscopic and theoretical investigation of structure and conformation. *Tet Lett.* 1983; **24**: 1821–1824.

13. Kane VV, Wolf AD, Jones M. [6]paracyclophane. *J Am Chem Soc.* 1974; **96**: 2643–2644.

14. Tsuji T, Okuyama M, Ohkita M, Kawai H, Suzuki T. Functionalization and kinetic stabilization of the [4]paracyclophane system and aromaticity of its extremely bent benzene ring. *J Am Chem Soc.* 2003; **125**: 951–961.

15. Tobe Y, Jimbo M, Kobiro K, Kakiuchi K. Telomers of bent arenes. Acid-catalyzed dimerization and trimerization of the 1,4-hexamethylene-bridged arenes [6]paracyclophane, [6](1,4)naphthalenophane, and [6][(1,4)anthracenophane. *J Org Chem.* 1991; **56**: 5241–5243.

16. Hirano S, Hara H, Hiyama T, Fujita S, Nozaki H. Synthetic and structural studies of [6]-, [7]- and [10]metacyclophanes. *Tetrahedron.* 1975; **31**: 2219–2227.

17. van Straten JW, de Wolf WH, Bichelhaupt F. [5]metacyclophane. *Tet Lett.* 1977; **18**: 4447–4670.

18. Kane VV, de Wolf WH, Bichelhaupt F. Synthesis of small cyclophanes. *Tetrahedron.* 1994; **50**: 4575–4622.

19. Tsuji T, Nishida S, Okuyama M, Osawa E. Photochemical generation of bicyclo[4.2.2]decapentaene from [4.2.2]propellatetraene: experimental and theoretical study of the .pi.-bond-shift isomers of bicyclo[4.2.2]decapentaene, [4]paracyclophane-1,3-diene, and 1,6-ethenocycloocta-1,3,5,7-tetraene. *J Am Chem Soc.* 2003; **125**: 951–961.

20. Kramer P, Sharma AK, Hennecke EE, Yasuda H. Polymerization of para-xylylene derivatives (parylene polymerization). I. Deposition kinetics for parylene N and parylene C. *Journal of Polymer Science: Polymer Chemistry Edition.* 2003; **22**: 475–491.

21. Farthing AC. Lin-Poly-p-xylylene. Part I. Intermediates and polymers of low molecular weight. *J Chem Soc.* 1953; 3261–3264.

22. Tsuji T, Ohkita M, Konno T, Nishida S. [1.1]paracyclophane: photochemical generation from the corresponding bis(dewar benzene) derivative and theoretical study of its structure and strain energy. *J Am Chem Soc.* 1997; **119**: 8425–8431.

23. Kawai H, Suzuki T, Ohkita M, Tsuji T. A kinetically stabilized [1.1]paracyclophane: isolation and C-ray structural analysis. *Angew Chem Int Ed.* 1998; **37**: 817–819.

24. Cram DJ, Steinberg H. Macro rings. I. Preparation and spectra of the paracyclophanes. *J Am Chem Soc.* 1951; **73**: 5691–5704.

25. Winberg HE, Fawcett FS. [2.2]paracyclophane. *Organic Syntheses.* 1973; **5**: 883.

26. Brown CJ. Lin-poly-p-xylylene. Part II. The crystal structure of di-p-xylylene. *J Chem Soc.* 1953; 3265–3270.

27. Otsubo T, Mizogami S, Sakata Y, Misumi S. Layered compounds. XVI. NMR spectra of multilayered [2.2]paracyclophanes. *Bull Chem Soc J.* 1973; **46**: 3831–3835.

28. Ganzael PK, Trueblood KN. The crystal and molecular structure of [3,3′]paracyclophane. *Acta Cryst.* 1965; **18**: 958–968.

29. Brown CJ. The crystal structure of di-m-xylylene. *J Chem Soc.* 1953; 3278–3285.

30. Cram DJ, Helgeson RC, Lock D, Singer LA. [2.2]metaparacyclophane, a highly strained ring system. *J Am Chem Soc.* 1966; **88**: 1324–1325.

31. Dewhirst KC, Cram DJ. Macro rings. XVII. An extreme example of steric inhibition of resonance in a classically-conjugated hydrocarbon. *J Am Chem Soc.* 1958; **80**: 3115–3125.

32. Coulter CL, Trueblood KN. The crystal structure of the diolefin of [2.2]paracyclophane. *Acta Cryst.* 1963; **16**: 667–676.

33. Cram DJ, Dewhirst KC. Macro rings. XIX. Olefinic paracyclophanes. *J Am Chem Soc.* 1959; **81**: 5963–5971.

34. Cram DJ, Allinger NL. Macro rings. XII. Stereochemical consequences of steric compression in the smallest paracyclophane. *J Am Chem Soc.* 1955; **77**: 6289–6294.

35. Cram DJ, Wechter WJ, Kierstead RW. Macro rings. XVIII. Restricted rotation and transannular electronic effects in the paracyclophanes. *J Am Chem Soc.* 1958; **80**: 3126–3132.

36. Reich HJ, Cram DJ. Macro rings. XXXVI. Ring expansion, racemization, and isomer interconversions in the [2.2]paracyclophane system through a diradical intermediate. *J Am Chem Soc.* 1969; **91**: 3517–3532.

37. Cram DJ, Helgeson RC, Lock D, Singer LA. [2,2]metaparacyclophane: a highly strained ring system. *J Am Chem Soc.* 1966; **88**: 1324–1325.

38. Reich HJ, Cram DJ Macro rings. XXXV. Transannular directive influences in electrophilic substitution of mono-substituted[2.2]paracyclophanes. *J Am Chem Soc.* 1969; **91**: 3505–3516.

39. Mori N, Horiki M, Akimoto H. Evidence for aryne-ring rotation in isomeric 4,5-dehydro[2.2](1,4)naphthal-enoparacyclophanes. *J Am Chem Soc.* 1992; **114**: 7927–7928.

40. Elschenbrosch C, Mockel R, Zenneck U. (η^{12}-[2.2]paracyclophane)chromium(0). *Angew Chem Int Ed.* 1978; **17**: 531–532.

41. Koray AR, Ziegler ML, Blank NE, Harnel MW. (η^{12}-[3.3]paracyclopan)chrom(O) und (η^{12}-[3.3]paracyclophan) chrom(i)hexafluoro phosphat. *Tet Lett.* 1979; **20**: 2465–2466.

42. Pierre J-L, Baret P, Chautempts P, Armand M. [2.2.2]paracyclophane, a novel type of metal cation complexing agent (.pi.-prismand). *J Am Chem Soc.* 1981; **103**: 2986–2988.

43. Zhu S, Mao Y, Jin G, Quin C, Chu Q, Hu C. A convenient synthesis of octafluoro[2,2′]paracyclophane and dodecafluoro[2,2′,2′]paracyclophane. *Tet Lett.* 2002; **43**: 669–671.

44. Sakurai H, Hoshi S, Kamira A, Hosomi A, Kabuto C. Octamethyltetrasila[2,2′] paracyclophane. *Chem Lett.* 1986; 1781–1784.

45. Hubert AJ. Multimacrocyclic compounds. Part III. Attempts to prepare benzenoid cage compounds from novel polyacetylenes. *J Chem Soc.* 1967; **C**: 13–14.

46. Otsubo T, Mizogami S, Otsubo I, Tozuka Z, Sakagami A, Sakata Y, Misumi S. Layered compounds. XV. Synthesis and properties of multilayered cyclophanes. *Bull Chem Soc J.* 1973; **46**: 3519–3530.

47. Sekine Y, Boekelheide V. [2.2.2.2.2.2](1,2,3,4,5,6)cyclophane:superphane. *J Am Chem Soc.* 1979; **101**: 3126–3127.

48. Sekine Y, Brown M, Boekelheide V. A study of the synthesis and properties of [2.2.2.2.2.2](1,2,3,4,5,6)cyclo-phane:superphane. *J Am Chem Soc.* 1981; **103**: 1777–1785.

49. Tobe Y, Takahashi T, Ishikawa T, Yoshimura M, Suwa M, Kobiro K, Kakiuchi K, Gleiter R. Bent acenes: synthesis and molecular structure of [6](1,4)naphthalenophane and [6](1,4)anthracenophane. *J Am Chem Soc.* 1990; **112**: 8889–8894.

50. Tobe Y, Utsumi T'N, Saiki S, Naemura K. Synthesis and characterization of [6](9,10)anthracenophane. *J Org Chem.* 1994; **59**: 5516–5517.

51. Bodwell GJ, Bridson JN, Houghton TJ, Kennedy JWJ, Mannion R. 1,7-dioxa[8](2,7)pyrenophane: the pyrene moiety is more bent than that of C_{70}. *Chem Eur J.* 1999; **35**: 1823–1827.

52. Bodwell GJ, Bridson JN, Houghton TJ, Kennedy JWJ, Mannion R. 1,8-dioxa[8](2,7)pyrenophane, a severely distorted polycyclic aromatic hydrocarbon. *Angew Chem Int Ed.* 1996; **35**: 1320–1321.

53. Bodwell GJ, Fleming JJ, Mannion R, Miller DO. Nonplanar aromatic compounds. 3. A proposed new strategy for the synthesis of buckybowls: synthesis, structure and reactions of [7]-, [8]- and [9] (2,7)pyrenophanes. *J Org Chem.* 2000; **65**: 5360–5370.

54. Baker W, Glockling F, McOrmie JFW. Eight and higher membered ring compounds. Part V. Di-(naphthalene-2:7-methylene) and its conversion into coronene. *J Chem Soc.* 1951; 1118–1121.

55. Cram DJ, Dalton CK, Knox GR. Macro ring. XXIX. Stereochemistry of a 1,6-cycloaddition reaction. *J Am Chem Soc.* 1963; **85**: 1088–1093.

56. Otsubo T, Aso Y, Ogura F, Misumi S, Kawamoto A, Tanaka J. Synthesis, structure, and properties of triple-layered [2.2][2.2]naphthalenophane. *Bull Chem Soc J.* 1989; **62**: 164–170.

57. Golden JH. Bi(anthracene-9,10-dimethylene)(tetrabenzo-[2,2]-paracyclophane). *J Chem Soc.* 1961; 3741–3748.

58. Wasserman HH, Keehn PM. Dibenzoequinene: a novel heptacyclic hydrocarbon from the photolysis of [2.2]paracyclonaphthane. *J Am Chem Soc.* 1967; **89**: 2770–2772.

59. Aly AA, Hopf H, Ernst L. Cyclophanes. XLVII. Novel synthesis of phenanthrenoparacyclophanes and phenanthrenophanes and a study of their NMR properties. *Eur J Org Chem.* 2000; 3021–3029.

60. Staab HA, Kirrstetter RGH. [2.2](2,7)pyrenophan als Excimeren-Modell: Synthese und spektroskopische Eigenschaften. *Liebigs Ann.* 1979; 886–898.

61. Staab HA, Riegler N, Diederich F, Krieger C, Schweitzer D. [3.3]- and [4.4](2,7)pyrenophanes as excimer models: synthesis, molecular structure, and spectroscopic properties. *Chem Ber.* 1984; **117**: 246–259.

62. Pahor NB, Calligaris M, Randaccio L. Structural aspects of [2.2]heterophanes. Part II. Molecular structure of [2.2](2,5)thiophenophane and [2.2](2,5)furanophane. *J Chem Soc Perkin Trans.* 1978; **2**: 42–45.

63. Pahor NB, Calligaris M, Randaccio L. Structural aspects of [2.2]heterophanes. Part I. Molecular structure of [2.2](2,6)pyridinophane. *J Chem Soc Perkin Trans.* 1978; **2**: 38–42.

64. Manners I, Vogel U. Strained heteroatom-bridged metallocenophanes. In: Gleiter R, Hopf R, editors. Modern cyclophane chemistry. Weinhein: Wiley-VCH; 2004.

65. Berenbaum A, Braunschweig H, Dirk R, Englert U, Green JC, Jäkle F, Lough AJ, Manners I. Synthesis, electronic structure, and novel reactivity of strained, boron-bridged [1]ferrocenophanes. *J Am Chem Soc.* 2000; **122**: 5765–5774.

66. Osbourne AG, Whiteley RH. Silicon-bridged [1]ferrocenophanes. *J Organomet Chem.* 1975; **101**: C27–C28.

67. Foucher DA, Tang BZ, Manners I. Ring-opening polymerization of strained, ring-tilted ferrocenophanes: a route to high-molecular-weight poly(ferrocenylsilanes). *J Am Chem Soc.* 1992; **114**: 6246–6248.

68. Rinehart KL, Frerichs AK, Kittle PA, Westman LF, Gustafson DH, Pruett RL, McMahon JE. 1,1'-(tetramethylethylene)-ferrocene: obliquity and n.m.r. in bridged ferrocenes. *J Am Chem Soc.* 1960; **82**: 4111–4112.

69. Yasufuku K, Yamazaki H. Chemistry of mixed transition-metal complexes. VIII. Preparations of π-cyclopentadienyl (ferrocenylcyclobutadiene)cobalt complexes and 1,1'-(o-phenylene)ferrocenes. *J Organomet Chem.* 1977; **127**: 197–207.

70. Nelson JM, Nguyen PL, Petersen R, Rengel H, Macdonald PM, Lough AJ, Manners I, Raju NP, Greedan JE, Barlow S, O'Hare D. Thermal ring-opening polymerization of hydrocarbon-bridged [2]ferrocenophanes: synthesis and properties of poly(ferrocenylethylene)s and their charge-transfer polymer salts with tetracyanoethylene. *Chem Eur J.* 1997; **3**: 573–584.

71. Hisatome M., Watanabe J., Yamakawa K., Iitaka Y. [45](1,2,3,4,5)Ferrocenophane:superferrocenophane. *J Am Chem Soc.* 1986; **108**: 1333–1334.

72. LeVanda C, Bechgaard K, Cowan DO, Mueller-Westerhoff UT, Eilbracht P, Candela GA, Collins RL. Bis(fulvalene)diiron, its mono- and dications: intramolecular exchange interactions in a rigid system. *J Am Chem Soc.* 1976; **98**: 3181–3187.

73. Barr TH, Lentzner HL, Watts WE. Bridged ferrocenes. VI. Synthesis and stereochemistry of [1,1]ferrocenophanes. *Tetrahedron.* 1969; **25**: 6001–6013.

3

Crown Ethers, Cryptands and Other Compounds

3.1   Introduction

A huge number of hydrocarbon rings have been synthesised, some of which have been described in previous chapters. However, the chemistry of carbon, whilst an extremely rich and widely varied field of study, has many limitations. The chemistry, structural variations and other properties of macrocycles can be greatly enriched by the incorporation of other elements. The presence of these heteroatoms vastly extends the possible conformations, chemistries, self-organisations and bindings of macrocycles. These macrocycles will be detailed within this and later chapters. We will consider crown ethers, which are amongst the oldest synthetic heteromacrocycles, along with other materials such as cryptands, spherands and their derivatives. We will also attempt to introduce the concepts of supramolecular chemistry and molecular recognition.

Supramolecular chemistry can be thought of as an extension to classical chemistry. Over the past centuries chemists have developed a wide range of methods for making and breaking covalent bonds. This has led to the synthesis of such complex molecules as vitamin B12, polypeptides and artificial genes. However, covalent bonds are not solely responsible for linking together and shaping molecular entities; a whole range of noncovalent forces such as hydrogen bonding, CH$-\pi$ interactions, $\pi-\pi$ interactions, dipole interactions and Van der Waals forces are involved in this. It is these interactions which in nature allow the exquisite control of structure required for biological reactions to proceed and for life to exist. For example, many proteins can be written as a simple linear chain of amino acid residues. This does not give any impression of the complex three-dimensional nature of these moieties, which is vital to their function. The intricate structure of these molecules is based on a combination of covalent bonding, hydrogen bonding, dipolar and hydrophobic interactions. Similarly, a single DNA strand can be thought of as a linear polymer of deoxyribose phosphate with a mixture of four bases attached as a side chain. If we combine two chains with the correct sequences of bases, however, something spectacular occurs. Nobody can fail to be impressed by the complexity and sheer beauty of the DNA double helix, held together by multiple hydrogen bonding and other interactions. Compared to what nature has achieved, we are still stumbling in the dark.

Macrocycles: Construction, Chemistry and Nanotechnology Applications, First Edition.   Frank Davis and Séamus Higson.
© 2011 John Wiley & Sons, Ltd. Published 2011 by John Wiley & Sons, Ltd.

Many chemical reactions require very specific conditions to proceed, such as high temperatures, pressures or specific solvents; they are often nonselective, can give a mixture of products and can be vulnerable to the presence of common environmental materials such as water and oxygen. By contrast, biological reactions usually progress at high rates within narrow temperature and pressure ranges, often working on a single substrate and giving a single product – and all in aqueous solution. In fact, philosophers hypothesised that organic compounds possessed a 'vital force', that much of the chemistry of biological systems could not be duplicated by laboratory chemistry and that organic compounds would prove impossible to synthesise. Eventually this was proved to be incorrect, as chemists finally gained the knowledge to construct molecules such as amino acids and sugars.

The selectivity and efficiency of biological molecules are due to their extremely exact structures, which allow them to recognise certain substrates. These are then bound in a specific manner, which, in the case of enzymes for example, makes the substrate extremely susceptible to catalysis. Molecular interactions form the basis of this recognition and allow for such complex processes as the binding and oxidation of glucose by glucose oxidase, the highly selective interactions of antibodies with their antigens, and the storage and transcription of the genetic code. These processes depend on the existence of many molecular recognition events. Sometimes there can be a preorganised host into which a suitable guest will simply bind, the so-called lock and key model. Other systems utilise a more cooperative system, where the presence of the guest causes the host to organise around it. And of course in the case of two DNA strands, they can be thought of as cooperating equivalently in the organisation and binding process. Again, nature has shown the way with systems such as DNA, antibodies and enzymes.

Supramolecular chemistry is a relatively new field of science which attempts to construct artificial compounds capable of the highly specific binding required for such selective molecular interactions. The compounds described in the following chapters are among the many materials that have been developed as 'host' compounds, which in many cases bind certain 'guests' with high specificity. One of the simplest systems is that of the crown ethers.

3.2 Crown Ethers

Simple ethers have been known for many years; Paracelsus in the 16th century described the analgesic properties of diethyl ether, for example. Polyethers such as polyethylene oxide are commercially widespread and have a large number of applications. The interactions of ethers with potassium and its alloys with sodium were reported and it was noted that ethers with more than one oxygen atom tended to give the most intensely coloured blue solutions.[1] One of the more effective ethers used was the cyclic tetramer of propylene oxide (Figure 3.1a). We now know that the coloration observed was due to the formation of electrides, where the potassium atom dissociates into a cation and an electron.[2] Electrides tend to be unstable, but in the case of ethers, the ether chain can solvate the cation and prevent it from combining with the free electrons.

In 1960 Charles Pedersen was attempting to react dihydropyran with one of the hydroxyls of 1,2-dihydroxybenzene and then further react the product with *bis*-(chloroethyl ether).[3] After the first step he realised that about 10% of the starting material remained in the mixture but decided to continue anyway. The second reaction gave what he described as 'an unattractive goo', from which mixture (0.4%) a small amount of a white crystalline product could be obtained. At that moment he could have simply thrown it away, but instead started upon the path that eventually led to a share in the 1987 Nobel Prize for Chemistry, along with two others who will be mentioned a great deal within this and other chapters, Jean-Marie Lehn and Donald Cram.

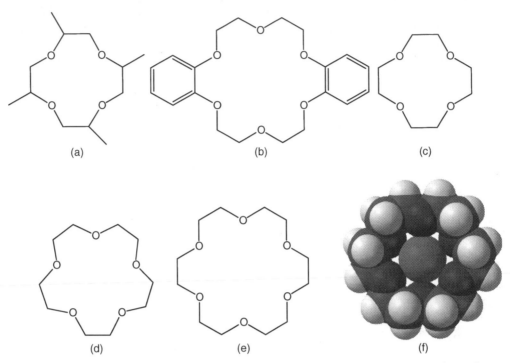

Figure 3.1   *Structures of (a) propylene glycol tetramer, (b) dibenzo-18-crown-6, (c) 12-crown-4, (d) 15-crown-5 and (e) 18-crown-6. (f) Space-filling model of 18-crown-6/K⁺ complex*

Amongst the things Pedersen observed about his compound was that it contained no free hydroxyl groups, and from its molecular weight a cyclic structure (Figure 3.1b) resulting from a 2 + 2 addition was deduced. This was highly unusual since most ring-forming reactions tend to give five-, six- or seven-membered rings, whereas this had a central 18-membered macrocycle. Interestingly, although it had only minimal solubility in methanol, addition of sodium salts allowed the compound to be dissolved. From this and molecular models of the compound which showed it to be doughnut-shaped, Pedersen realised that the positive sodium ion could sit inside the central ring, stabilised by the negative dipoles of the oxygen atoms.

Further work on this family of compounds led to publication of his ground-breaking paper on what he named crown ethers,[4] since their actual chemical names were long and cumbersome. His original compound, for example, was named dibenzo-18-crown-6. The system names first the substituent, then the ring size and finally the number of oxygen atoms present in the ring. Other workers have used the term 'corand' for these macrocyclic structures. Figure 3.1 shows the respective structures of (c) 12-crown-4, (d) 15-crown-5 and (e) 18-crown-6. This naming system is still preferred today since the full chemical names of this family of compounds are usually very long and cumbersome. Within his paper Pedersen describes the synthesis of a large range of crown ethers with different central ring sizes. He also varied the nature of the adjoining rings, utilising either one or two benzo, cyclohexyl, naphtho or decalin rings. Usually these were synthesised by combining the required amounts of 1,2-dihydroxybenzene or 2,3-dihydroxynapthalene with a linear polyether capped with chlorine atoms under basic conditions to give the aromatic crowns, which could be catalytically hydrogenated to give their aliphatic analogues. Later work tends to utilise the ditosylates rather than the dichlorides since this usually leads to higher yields.

3.3   Simple Complexes with Crown Ethers

These compounds were all screened for their ability to form complexes with various species. Stable complexes could be obtained with alkali metal salts, ammonium salts and a variety of transition metal and lanthanide salts. Complexation behaviour is a function of both ring size and substitution. For example, dicyclohexyl-12-crown-4 shows some affinity for lithium and none for larger alkali metals;[3,4] the corresponding 15-crown-5 solvates $Na > K > Cs > Li$; the 18-crown-6 prefers $K > Cs > Na > Li$; and the 21-crown-7 $K \approx Cs > Na > Li$. Also, 18-crown-6 is much more selective for potassium over other alkali metals than the corresponding dibenzo compound. Figure 3.1f displays a 3D model of the structure of the 18-crown-6 complex with potassium ion, showing how the potassium ion fits into the centre of the ring, with the six oxygen atoms all acting as donors due to their negative dipoles. The potential for these compounds as solvating agents for various ions in organic solvents was immediately realised. As an example, potassium permanganate is often used as an oxidant but is unsuitable for use in many organic solvents because of low solubility. The use of dibenzo-18-crown-6 allows potassium permanganate to be dissolved in solvents such as benzene and this approach can be used to oxidise compounds which will not dissolve in water.

The theory was that the presence of, effectively, a 'hole' in the centre of the polyether, varying from 0.12–0.15 nm for 12-crown-4 compounds up to 0.34–0.43 nm for 21-crown-7 macrocycles,[3,4] allows in effect a solvation of the guest within this central cavity. An interesting effect of this is that it appears that the guest is not just a passive part of this process but actually has its own part to play. This was shown from the yields of various reactions. The yields of dibenzo-18-crown-6, for example, are much higher (45%) when sodium or potassium hydroxide is used as a base during the ring-closing reaction than if lithium or ammonium hydroxide is used.[4,5] This is thought to be due to the intermediates in the reaction process interacting with the alkali metals and so being brought into an orientation that disposes them to undergo the final ring-closure reaction. This is an early example of templating chemical reactions, where the units required to form macrocyclic or other three-dimensional structures are brought into the correct orientation for a ring closure or other reaction to occur, often averting the need for high-dilution reaction conditions. The yields of dibenzo-18-crown-6 can be increased to 80% by a two-stage reaction in which two equivalents of the dihydroxy compound react with one dichloro-substituted chain to give the dibenzo-substituted chain, followed by addition of a second equivalent of dichloride and more base to cause cyclisation, thereby eliminating the formation of a 1 : 1 cyclisation byproduct.

The so-called 'template' effect is highly important in the synthesis of both of these compounds as well as many other of the macrocycles that will be mentioned within this work. Without the template effect, the synthesis of many of these compounds would either simply not occur or would have to be carried out under high-dilution conditions – which would increase the cost and complexity of such syntheses. Although ring-closing reactions are quite common, they usually involve the formation of five-, six- or seven-membered rings rather than an 18-atom macrocycle. For example, in Pedersen's synthesis[4,5] of the dibenzo-18-crown-6 compounds, the condensation reaction between a diphenol and dichloro compound would be expected to lead to a polymeric product with a mixture of molecular weights. The fact that the macrocycle can be obtained in such a good yield indicates that this cannot just be a simple condensation reaction and that some other process must also be occurring.

The explanation for the high yields of the crowns is based on the presence of a template metal ion, such as potassium, which can contributed by the base (KOH). The ethyleneoxy unit ($-CH_2CH_2O-$) is especially suitable for the formation of complexes since within a chain or ring composed of these compounds, each unit is flexible and can easily adopt the trans conformation. In the case of a crown ring, this means that the lone pairs of each oxygen atom can be directed towards the centre of the ring. In the case of the aromatic unit, the two oxygen atoms are already in the correct configuration. If we imagine two equivalents of the potassium salt of 1,2-dihydroxybenzene reacting with the *bis*-chloroethylether to give the dibenzo-substituted compound, it is likely that the resultant product will interact with potassium ions via dipolar interactions. Should a second equivalent of *bis*-chloroethylether be added and react at one end of this unit, this product will have six oxygen atoms in its chain and potentially coil around the potassium. This brings the remaining hydroxyl and chloro groups into close proximity, leading to a high possibility of reaction and cyclisation. This is shown in Figure 3.2a. Similar template effects occur in the synthesis of nonaromatic crowns, with for instance condensation of triethylene glycol with its ditosylate (Figure 3.2b) under basic conditions[6] demonstrating that yields of 18-crown-6 are dramatically increased in the presence of metal ions, especially potassium.

X-ray crystallographic techniques were applied to these systems to confirm their cyclic structures and ion complexation. Much of the early work was carried out by Mary Truter and collaborators in London. Their studies proved Pedersen's conclusion that the crown ethers enclosed the cations within the macrocyclic

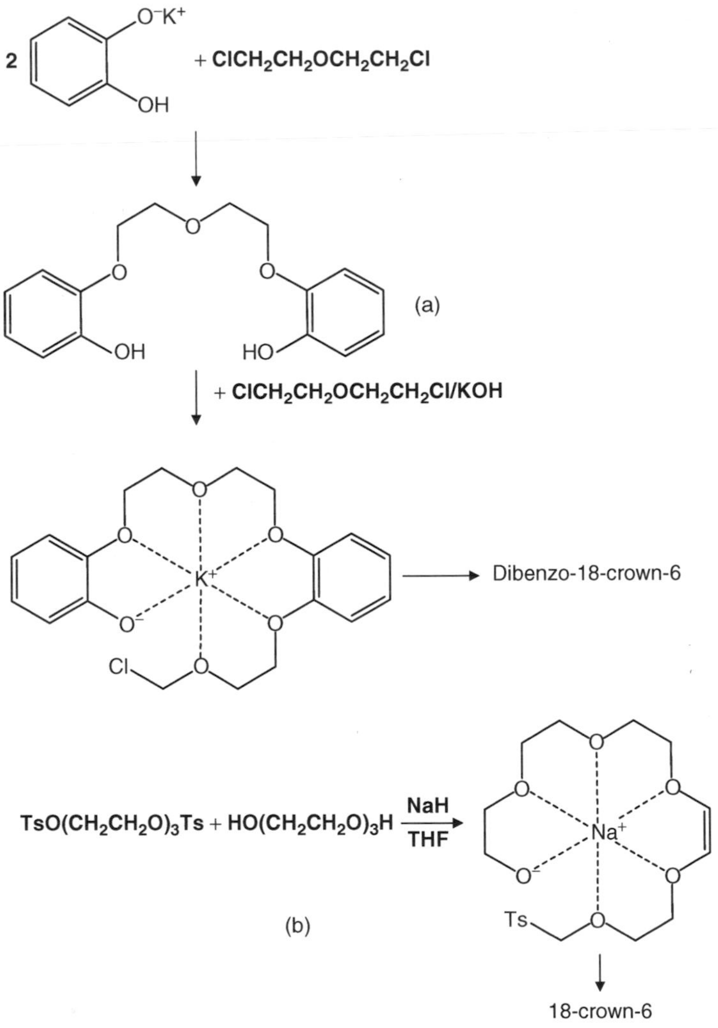

Figure 3.2   *Synthesis of (a) dibenzo-18-crown-6 and (b) 18-crown-6*

ring. The structure of the sodium complex of dibenzo-18-crown-6 was shown to have a central sodium atom in a hexagonal bipyramid environment, with the six oxygen atoms of the crown in the equatorial plane and two water molecules occupying the apices.[7] A similar complex, Na:benzo-15-crown-5, had a pentagonal bipyramid arrangement, again with two water molecules included.[7] Larger crowns give more complex structures; in the potassium complex with dibenzo-30-crown-10 (Figure 3.3a), for example, the macrocycle is described as being 'wrapped round the cation like the seam of a tennis ball',[7,8] obviously to maximise favourable $O-K^+$ interactions. This effectively shields the potassium ion from the environment and in fact this crown has good selectivity for potassium over other alkali metals. The larger crowns could also give 1:2 complexes such as that formed between dibenzo-24-crown-8 and two equivalents of potassium thiocyanate,[9] where the potassium ions complex with the ether oxygens and the thiocyanate ions serve as bridging ligands. A similar structure was found when potassium isothiocyanate was used (Figure 3.3b) and in both cases there was evidence of benzene rings from other ligands being in the correct

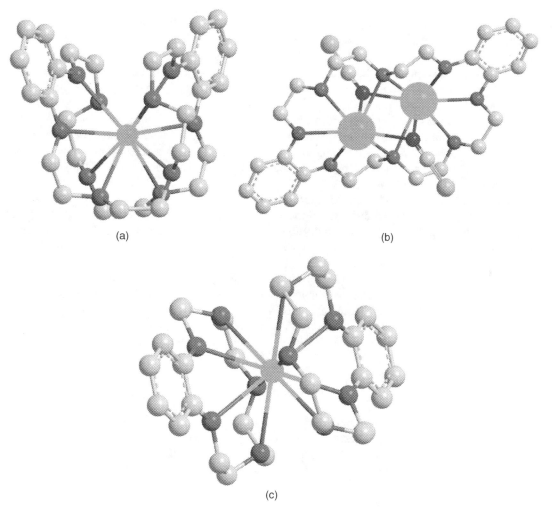

(a)

(b)

(c)

Figure 3.3   *(a) 1:1 complex of dibenzo-30-crown-10 with KI, (b) 1:2 complex of dibenzo-24-crown-8 with KNCS and (c) 2:1 complex of benzo-15-crown-5 with KI*

position to allow electrostatic attraction between the potassium ion and the π-electrons.[10] Alternatively, crystallisation of benzo-15-crown-5 with potassium iodide gave a 2 : 1 complex,[11] where the potassium ion is sandwiched between the two crowns with all the K–O distances being approximately equal (Figure 3.3c).

What is interesting is that the uncomplexed crown ethers often do not have the same structure as the complexed forms. For example, in the complex with potassium, 18-crown-6 has its oxygen atoms all in the same plane orientated inwards towards the guest. However, when no central ion is present, the compound adopts a structure that has two of the oxygens pointing outwards, showing that the oxygen–alkali metal ion interactions cause the macrocycle to rearrange and shape the complex.

Besides the alkali metals, numerous other species have been shown to form complexes with crown ethers. Within this chapter we will restrict ourselves to a general overview. For more data on the many cation–crown complexes studied, a detailed and extensive review was published in 1985.[12] Some general trends will be discussed here; for example, we have already mentioned the effects of the size of the crown ether on its interactions with alkali metals. Alkaline earths and other divalent metals can be bound even more strongly, with for example dicyclohexyl-18-crown-6 having almost no affinity for Ca^{2+} but binding Hg^{2+}, Sr^{2+} and Ba^{2+} much more strongly than K^+ and having a very high affinity for Pb^{2+} – almost a thousand times higher than for K^+. The corresponding dibenzo compound has also been shown to bind barium and strontium, as well as silver.[4] Lanthanides have also been studied; 18-crown-6 for example forms strong complexes with La^{3+} and Ce^{3+}, with the binding constants then dropping by up to 100-fold as we move along the lanthanide series as far as Gd^{3+}. Then there is an abrupt cessation of binding for any further members of the lanthanides.[13]

Organic cations can also be bound by crown ethers. For example, 18-crown-6 can bind ammonium ion (NH_4^+), whereas the larger 27-crown-9 molecule forms a 1 : 1 complex[14] with guanidinium ion $C(NH_2)_3^+$. Within the same paper, formations of complexes with 18-crown-6 and arenediazonium salts were also described. Neutral molecules can also be complexed, such as hydroquinone,[15] which can be crystallised as a 1 : 1 complex with 18-crown-6.

The result of Pedersen's groundbreaking paper was a surge of interest in crown ethers and other variations on this theme. One topic of research showed that a cyclic structure was not always necessary as expected, since oligo and polyethers displayed the ability to complex cations to some degree.[1] Linear analogues of crown ethers were synthesised and the name 'podands' was coined for these systems, with details of early studies on podands being reported in a 1979 review.[16] For example, a linear analogue of 18-crown-6, namely $CH_3O(CH_2CH_2O)_5CH_3$ (Figure 3.4a), was synthesised and was shown to complex alkali metal ions; however, this more open structure was much less effective, having a binding constant approximately 10 000 times lower than the corresponding cyclic compound. Binding constants can however be improved by variation of the end-groups, allowing more facile synthesis of derivatised podands than the corresponding crowns. Podands have been synthesised with a variety of interesting binding properties but since they are acyclic we will not discuss them within this work; the reader is referred to the Bibliography and the extensive and detailed review by Vögtle and Weber.[16]

One of the easier approaches to varying the crown ether structure is to utilise Pedersen's original method for the dibenzocrowns but to vary the aromatic diol. For example, 2,3-dihydroxy naphthalene can be used, along with many other dihydroxy compounds. A substituted [2,2′]paracyclophane could be utilised as the source material to synthesise the two compounds shown in Figure 3.4b,c, with the two crown ethers bridging either across the individual benzene rings or between them.[17] Heteroaromatic units were incorporated, such as the compound shown in Figure 3.4d, which is one of a wide range of pyridyl crown ethers synthesised by this group[18] and shown to have a higher binding constant for t-butyl ammonium ions than for 18-crown-6 itself. Replacing one of the oxygens of 18-crown-6 with a methylene group[19] reduces the binding constant by over a thousand. The same group also incorporated a xylyl group within the ring, with a reduction of binding constant by a factor of 700. Study of these compounds with different

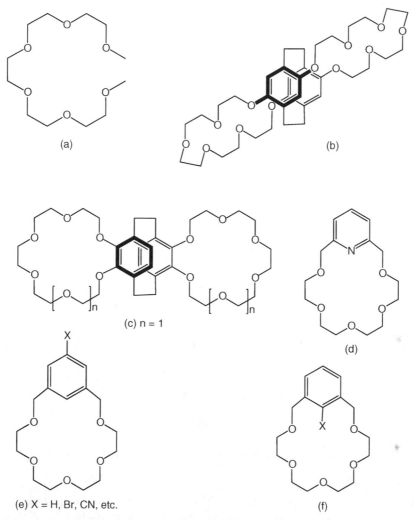

Figure 3.4   *Structures of (a) a linear analogue of 18-crown-6, (b,c) bridged [2,2′]paracyclophane crowns, (d) a pyridino crown ether and (e,f) substituted benzocrown ethers*

5-substituents on the aromatic ring, as shown in Figure 3.4e, showed that substitution of the aromatic ring had large (by a factor of 100) effects on the binding constant.[19] Similar work was carried out using aromatics substituted at the 2 position (Figure 3.4f) and showed that in some cases[20] the incorporation of H-bonding groups increased the binding constant by over an order of magnitude.

One of the most elegant pieces of work included the binaphthyl unit within a crown system. Binaphthol exists in two enantiomeric forms, as shown in Figure 3.5a, since it cannot undergo rotation around the C—C bond connecting the two naphthyl rings due to steric hindrance. The presence of the two hydroxyl groups thereby enables this moiety to be included in a Pedersen-type synthesis to give binaphthyl crown ethers, as shown in Figure 3.5b,c. If a single enantiomer of binaphthol is used then the resultant crown is also chiral. This field was studied by Cram's group, who synthesised a wide number of chiral crown ethers and other macrocycles[21] for study as hosts for chiral compounds. Compound 3d was shown to have a high

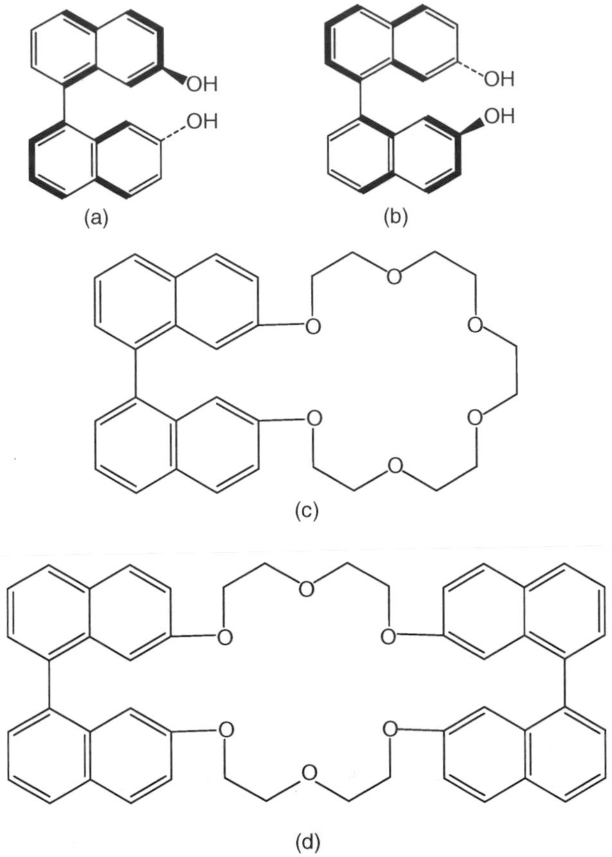

Figure 3.5   *Structures of (a,b) enantiomers of binaphthol, (c) mono-binaphthyl and (d) bis-binaphthyl crowns*

enantiomeric selectivity for amino acid salts and their esters. These crowns were shown to be suitable for use within chromatographic separations.[22] Both liquid–liquid phase (where the crown is dissolved in chloroform or dichloromethane) and solid–liquid phase (where the binaphthyl unit is substituted with a siloxy group, enabling coupling to silica gel) experiments were carried out and separation could be observed for the two enantiomers of various amino acids. NMR and X-ray crystallographic studies demonstrated the interactions occurring between the host and guest, and showed that when the binaphthyl units are in the R,R configuration the resultant complex with a D-amino acid is less sterically hindered than with the L enantiomer.[23]

3.4   Azacrowns, Cyclens and Cyclams

All the crowns described so far have been based on carbon–oxygen rings. The introduction of nitrogen into these systems adds a new degree of flexibility to the possible structures. Oxygen can only be divalent under most conditions and although this enables it to be included in a chain or macrocycle, very little further chemistry can be performed. Nitrogen however is either trivalent or tetravalent (in its charged ammonium form). This means it can be used to introduce new chemistry and substitution reactions to the macrocycle. Also, simple replacement of O with N—H or N—R within the macroring will have an effect on its binding properties since nitrogen and oxygen have different affinities for various ions, along with the ability of N—H to act as a hydrogen bond donor. The effects of pH on binding will also be affected by the much higher basicity of nitrogen.

Some of the earliest descriptions of crowns containing nitrogen atoms came from the group of Joyce Lockhart,[24] who utilised 2-aminophenol and 1,2-diaminobenzene units to incorporate nitrogen atoms in crowns such as those shown in Figure 3.6a–c, by reacting polyether dichlorides with the amine. Other methods involve protecting the nitrogen group of diethanolamine with benzyl groups and then reacting the resultant compound with a ditosylated oligoether to form the aza macrocycle, followed by removal of the benzyl group. A range of materials can be synthesised using a similar method, where an amine is reacted directly;[25] benzylamine, for example, can be reacted with triethylene glycol diiodide to give the 2 + 2 addition product, an N-benzyl-substituted diaza-18-crown-6. This can then have the benzyl group removed to give the diaza-18-crown-6 compound, and it is worth noting that monoaza crowns can also be synthesised by this method.

An alternative method was developed by Jean-Marie Lehn and coworkers,[26] where oligoethers containing either two amino or two acid chloride end groups are reacted together to give a cyclic polyether with two

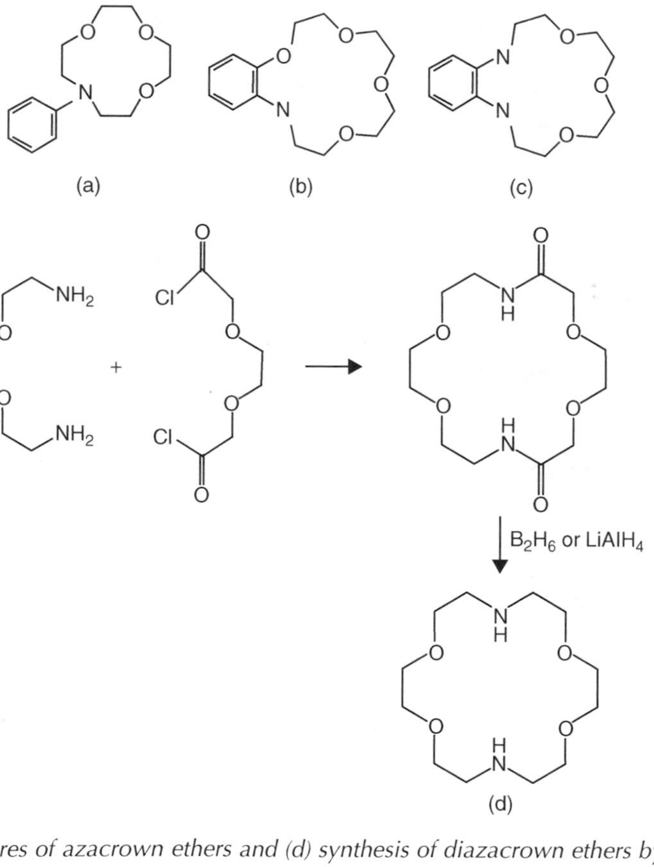

Figure 3.6 *(a–c) Structures of azacrown ethers and (d) synthesis of diazacrown ethers by amine/acid chloride method*

amide groups. These amides can then be reduced to give a diaza macrocycle such as that shown in Figure 3.6d. This is then used in the synthesis of cryptands, a series of compounds which will be described in Section 3.7. A recent review[27] gives further details of all of these and other types of synthesis.

The presence of nitrogen atoms affects the binding properties of these crowns. An example of this can be seen in the relative binding of potassium and silver. The log of the binding constant of 18-crown-6 with potassium is 6.1 and with silver is 1.6, indicating a large preference for potassium. For the mono-aza crown, the binding constants are 3.9 and 3.3 respectively, indicating that the potassium is bound much less and the silver more than in the case of the parent crown. In the case of the diazacrown (Figure 3.6d), the values are 2.0 and 7.8, indicating that now silver is bound with high selectivity.[28] The diaza-18-crown-6 has been reported to form complexes with transition metals such as copper,[26] as well as other metals including cobalt, nickel, zinc, cadmium, mercury and lead. A wide range of complexation studies with oxygen- and nitrogen-based crowns have been made: far too many to list within this work. However, a general rule can be drawn that oxygen atoms within a crown favour the binding of alkali and alkaline earth metals, whereas the presence of nitrogen atoms leads to preferential binding of transition metals.[27]

It is possible to synthesise all nitrogen analogues of crown ethers. One of the most common is the aza analogue of 12-crown-4, which is often given the trivial name 'cyclen'. This family of compounds has been very highly studied and can be synthesised in several different ways. One is via a condensation method similar to that of crown ethers, where first diethylene triamine ($H_2NCH_2CH_2NHCH_2CH_2NH_2$) is tosylated at all three positions and then the end amino groups are deprotonated with sodium ethoxide to give the dianion shown in Figure 3.7a. These can then be reacted with the tosylated derivative of diethanolamine to give the tosylated derivative of tetraaza-12-crown-4, which can be deprotected with sulfuric acid to give the parent amine[29] (Figure 3.7b). Similarly, the deprotonated salts can be reacted with ethylene carbonate to substitute the amines with —CH_2CH_2OH groups (Figure 3.7c). These hydroxy groups can of course then be tosylated again, reacted with more of the dianion (Figure 3.7a), and finally deprotected to give the hexaaza version[30] of 18-crown-6 (Figure 3.7d). A variety of macrocyclic amines up to the octaaza-24-crown-8 compound can be made using these methods.[29] A more recent synthesis[31] involves the condensation of triethylenetetraamine and dithiooxamide to form the *bis*-imidazoline compound (Figure 3.8a), which can then be reduced with an organoaluminium hydride to undergo a ring expansion to give the tetraaza-12-crown-4 compound.

Another method used with great versatility is the synthesis of Schiff bases. Dialdehydes and diamines can be condensed together to form macrocyclic Schiff bases; a variety of possible reaction products can occur – Figure 3.8b shows a schematic of a $2 + 2$ addition reaction. These can then easily be reduced to give the cyclic amines. This method is highly versatile and allows the incorporation of a wide variety of aliphatic and aromatic groups and heteroatoms within the macrocycle. The synthesis of cyclen-type compounds has recently been reviewed.[32]

There are a wide number of these cyclen-type compounds so we will restrict ourselves to two of the most highly studied. These are the aza analogues of 12-crown-4 (cyclen), which is the most studied of this family, and of 18-crown-6 (often known as hexacyclen). Hexacyclen has been shown to be a highly basic molecule, usually existing as the tetraprotonated compound under neutral pH conditions. The presence of the charged groups indicates that the azamacrocycles must have accompanying anions, which potentially could be selectively bound inside the macrocycle. Although hexacyclen does form complexes with anions such as nitrate, the anions do not occupy the centre of the ring in the same way 18-crown-6 complexes potassium, probably because in the case of the cyclen, the cavity is partially filled by the N-H protons. X-ray structures of, for example, the tetra(trifluoromethyl sulfonate) salt of hexacyclen[33] clearly show that although the hexacyclen adopts a puckered ring quite similar to the crown analogue, the anions all lie outside the ring system (Figure 3.9a). However, larger rings such as the cyclen analogue of 30-crown-10 (Figure 3.9b) can include complex anions such as $[PdCl_4]^{2-}$ within the ring.[34]

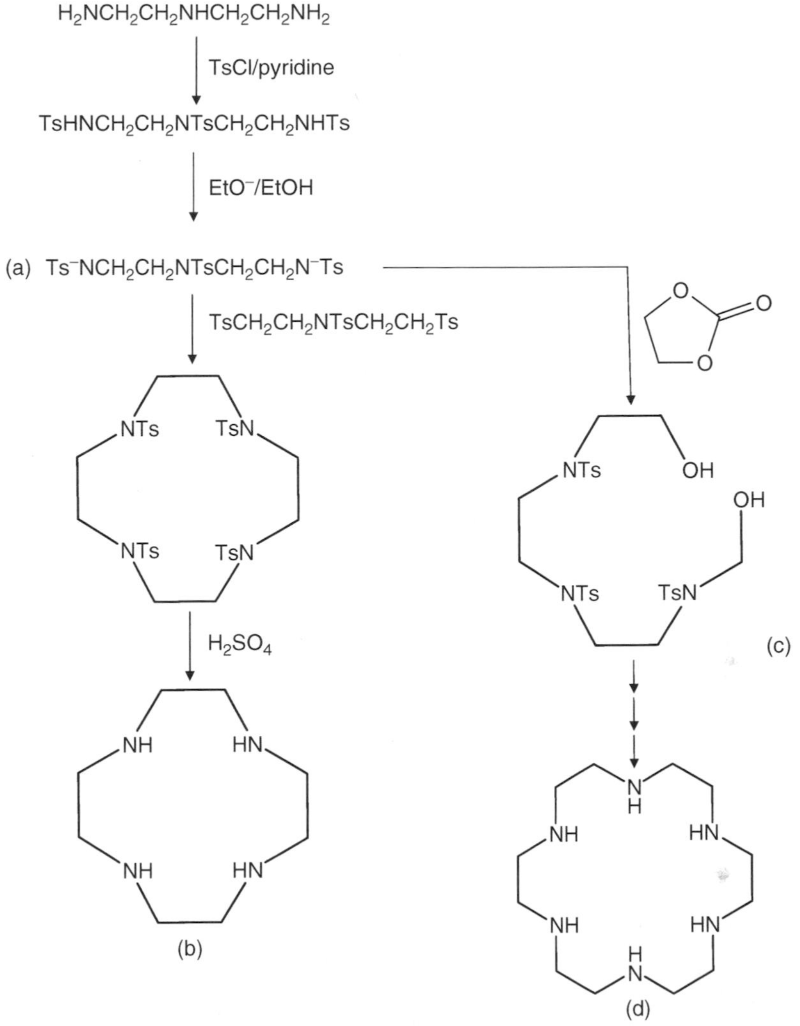

Figure 3.7 *Synthesis of azacrown ethers (cyclens)*

Metals are also complexed by cyclens. Cyclen itself forms complexes with many transition metals, but the metal ion is generally too small to fit within the ring system. Instead the ring usually adopts a nonplanar configuration, which enables it to occupy four of the coordination sites around a metal ion. For example, when cyclen and $Co(NO_3)_2$ are crystallised together, a complex is formed in which the cyclen adopts a puckered conformation as shown in Figure 3.10a, with three of its nitrogens in the equatorial positions around the cobalt ion and the fourth in an axial position, with the remaining axial and equatorial positions occupied by the nitrates.[35] Different structures can be obtained, for example when zinc cyclen perchlorate is crystallised from ethanol;[36] in this case the zinc atom is in a distorted square-pyramid conformation, with the four nitrogen atoms providing the base and an ethanol molecule at the apex. A similar structure has also been observed in the copper and nickel complexes of the *bis*-cyclen compound[37] shown in Figure 3.10b, with the cyclens providing the base of the pyramid and a water molecule its apex.

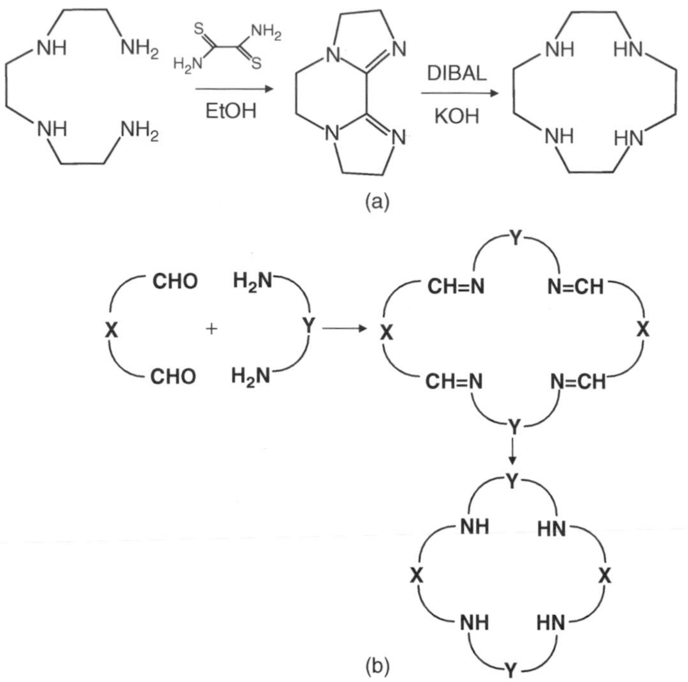

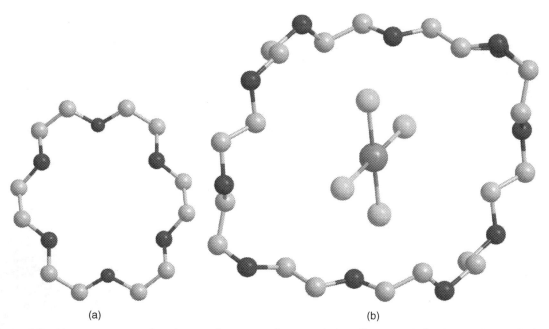

Figure 3.8   Synthesis of azacrown ethers

Figure 3.9   X-ray structures of (a) hexacyclen/tetra(trifluoromethyl) sulfonate and (b) decacyclen including a PdCl₄ unit

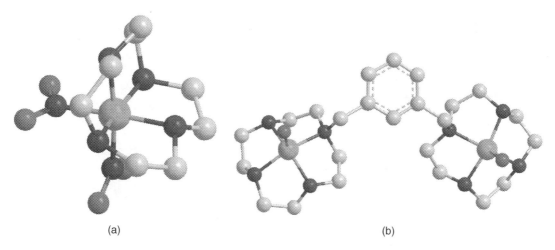

Figure 3.10 *X-ray structures of (a) cyclen/Co(NO₃)₂ and (b) a copper bis-cyclen*

Hexacyclen forms a wide series of compounds, especially with transition and precious metals. Liquid–liquid extraction techniques showed hexacyclen to be an efficient complexing agent for Ag^+, Cu^+, Cu^{2+}, Hg^{2+}, Pd^{2+} and Pt^{2+}, but not for Fe^{3+}.[38] Many of these metal ions prefer a hexacoordinate system such as an octahedral configuration. Hexacyclen contains six nitrogen atoms and is flexible enough to assume the required conformation. An example of this can be seen in Figure 3.11a, which shows the X-ray structure of one of the isomers of a Co^{3+} complex.[39] Other workers have shown that a Cr^{3+} complex has a distorted octahedral structure[40] and an Hg^{2+} complex has a distorted trigonal prismatic structure.[41] Lanthanides and actinides can also be complexed; Nd^{3+} for example forms a complex[42] in which the Nd ion is 10-coordinate with both the macrocycle and the nitrate groups involved in the complex, whereas UO_2^{2+} forms a hexagonal bipyramid form[43] in which the macrocycle nitrogens are in the equatorial positions (Figure 3.11b). Complexes in which the macrocycle binds two metal ions are also known, such as for example the di-copper complex[44] shown in Figure 3.11c.

Another very widely studied variation on this theme is the 'cyclam' series of compounds. Cyclam is similar to cyclen except that two of the ethylene bridges are replaced by propylene bridges (Figure 3.12a). Many of these compounds are synthesised from Schiff bases. These can be made by several different methods; amongst the earliest described was the nickel-templated reaction of acetone with the *tris*-ethylene diamine complex of nickel[45] (Figure 3.12b). Substituted Schiff bases can also be made by reacting materials such as 2,6-dicarbonyl pyridine compounds with diamino compounds of various chain lengths (Figure 3.12c), with iron salts being used as a template.[46] Ring size can often be controlled by selection of the metal ions and whether $1 + 1$ or $2 + 2$ addition products are formed. Polyethers with terminal amine groups can be utilised to give combined Schiff base/crown ether-type ligands.

Schiff bases do complex metal ions, but the reactions used to form these macrocycles are reversible, with the resultant products often being easily hydrolysed. Reduction of the Schiff base to the amine eliminates this problem. It has been noted that complexation with metals often improves the stability of the Schiff base. A recent review was published on the chemistry of these systems[47] and there is also an extensive review on their complexation behaviour with many metals.[48]

Synthesis of the Schiff bases and their reduction to the corresponding saturated macrocycles[49] has given rise to a wide range of compounds. The resultant cyclic amines are extremely good chelators of metals, as demonstrated by the binding constant of cyclam to Cu^{2+} being 10 000 times stronger than the

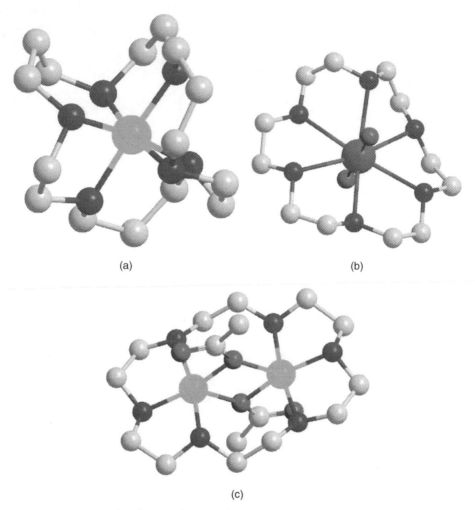

(a)

(b)

(c)

Figure 3.11 *X-ray structures of (a) hexacyclen/Co(ClO$_4$)$_3$, (b) hexacyclen/UO$_2$·2CF$_3$SO$_3^-$ and (c) hexacyclen/Cu$_2$·2CH$_3$CO$_2^-$ 2PF$_6^-$ (some counter-ions removed for clarity)*

corresponding linear analogue.[50] Similar macrocyclic effects have already been described for the crown ethers. The structures of many metal–cyclam systems have been obtained and show that the cyclam molecule is quite flexible and can adopt a variety of conformations depending on factors such as the chemical substitution of the macrocycle and the metal ion. For smaller metals such as Cu^{2+}, a survey of the Cambridge Structure database[51] showed that in the majority of cases the copper was in an octahedral or square planar configuration, although other configurations such as square pyramid or trigonal bipyramid are known. A range of metal perchlorates have been complexed with cyclam, with Zn^{2+} [52], Co[53] and Ni^{3+} [54] all having octahedral-type structures in which the macrocyclic nitrogens fill the equatorial positions. Pd^{2+} cyclam has a square planar structure,[55] as does the complex[56] with Ag$^+$.

Large metal ions cannot be encapsulated within the ring. The Pb(NO$_3$)$_2$/cyclam complex, for example, adopts a distorted *cis*-octahedral conformation[57] where instead of the two nitrate groups being on opposite sides of the metal ion they are *cis* to each other, with the amine nitrogens occupying the remaining four

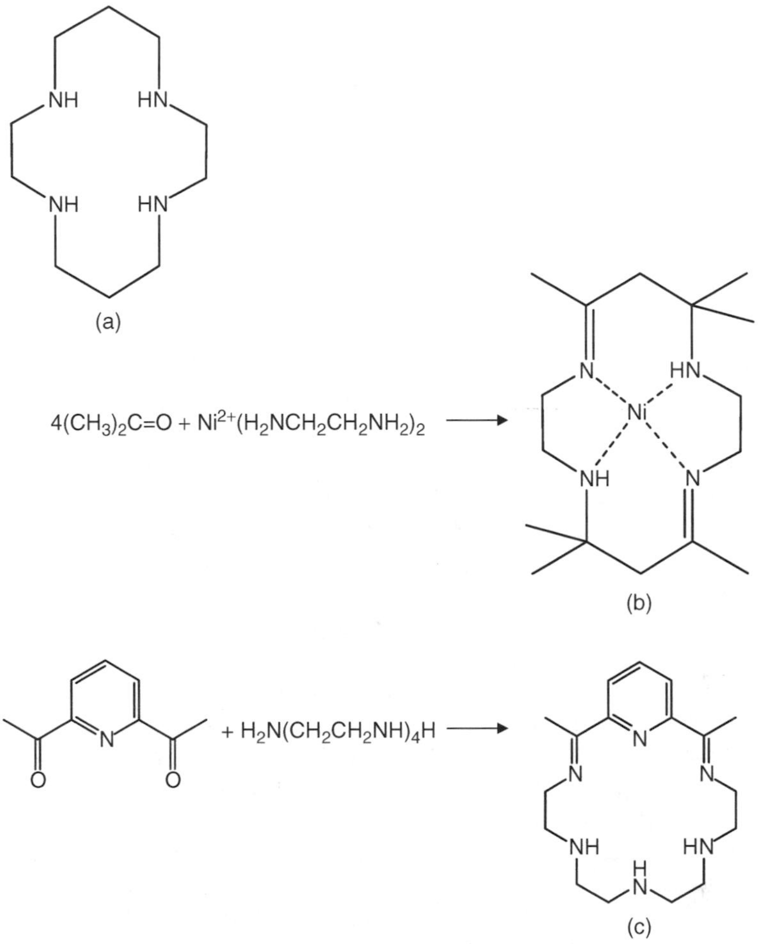

$$4(CH_3)_2C=O + Ni^{2+}(H_2NCH_2CH_2NH_2)_2 \longrightarrow$$

$$+ H_2N(CH_2CH_2NH)_4H \longrightarrow$$

Figure 3.12   *(a) Structure of 'cyclam' and (b,c) synthesis of Schiff bases*

positions. When complexed with cyclam,[58] Hg^{2+} adopts square-pyramid conformation in the solid state, with the macrocycle nitrogens occupying the base positions and a chloride ion the apex. Cadmium adopts a mixture of square-pyramid and trigonal-bipyramid conformations in solution and when Hg^{2+} or Cd^{2+} are crystallised with a cyclam in which the amine protons are replaced by methyl groups,[58] X-ray studies show the complexes to adopt a trigonal-bipyramid conformation.

The presence of nitrogen atoms within azacrowns and other macrocyclic systems allows the introduction of more substitution. This has been achieved in many ways, some of which will be discussed later. However, substitution of the N atoms can often reduce complexing ability, such as seen in the formation of the complex of Cu^{2+} with the tetra-N-methyl or tetra-N-acetyl cyclam being several orders of magnitude slower than that with the parent compound.[59] A similar effect is observed for Hg^{2+}, Pb^{2+}, Co^{2+} and Zn^{2+}. The tetraacetate also binds Zn^{2+} and Co^{2+} more strongly than the methyl compound. Since the amines are prone to protonation, pH also has an effect;[60] for example, Cd^{2+} complexes of cyclam were stable at pH > 8.2 but dissociated completely when the Ph was lowered to 5.3.

3.5   Crowns Containing Other Heteroatoms

Sulfur has also been incorporated into macrocyclic systems and has noticeable effects on their behaviour since it is larger than oxygen and is considered a relatively 'soft' atom while oxygen is considered a 'hard' atom. Some of Pedersen's early crown ethers were synthesised containing a mixture of oxygen and sulfur atoms, usually by replacing 1,2-dihydroxy benzene with 2-mercaptophenol or 1,2-dimercaptobenzene in his classical synthetic scheme.[61] Comparison of the extractions of potassium and of silver nitrate showed that although dicyclohexyl-18-crown-6 extracts both metals effectively, incorporation of two or four 'softer' sulfur atoms (Figure 3.13a–c) improved silver extraction but greatly reduced the affinity for potassium.

Cyclic thioethers are synthesised by a variety of methods, many of which involve the reaction of thioethers with alkyl halides under basic catalysis.[62] There can be problems in that some of the compounds that might be potential building blocks can be quite dangerous, such as $ClCH_2CH_2SCH_2CH_2Cl$, also known as 'mustard gas' and classified as a chemical weapon. All sulfur analogues of 12-crown-4, 15-crown-5 and 18-crown-6 (Figure 3.13d) have been synthesised and their X-ray crystal structures obtained.[63] In the case of hexathia-18-crown-6, studies showed some of the sulfur atoms pointing towards the centre of the ring and some away. This type of compound appears to bind metal in a very different manner than the conventional crown ethers. X-ray structures of many 18-crown-6 complexes show the binding to be electrostatic in nature, with the crown adopting an essentially planar form, with the metal atom in the centre, since the electrostatic bonding is essentially nondirectional. However sulfur atoms tend to form dative bonds to metals and this leads to much greater directionality. The earliest complex for which an X-ray structure was obtained was that of hexathia-18-crown-6 with nickel,[64] which shows that the macrocycle wraps around the nickel atom so that the sulfur atoms are in an octahedral arrangement. Further papers by Hintsa *et al*. show that Cu^{2+} and Co^{2+} also adopt octahedral conformations. However, there are exceptions to this rule: $CrCl_3$ forms a 1 : 1 complex with the chromium ion in an octahedral environment, although only three of the S atoms are complexed to the metal, with the remaining three positions being occupied by chloride ions.[65] Thia versions of other ligands such as the tetrathia analogue of cyclam, where sulfur atoms replace the nitrogen atoms, are known. Again, these form complexes with a wide variety of metals, such as for example Hg^{2+}, which can be bound in either a square pyramidal structure[66] when it has perchlorate counterions or a tetrahedral structure when it has chloride counterions.

Other heteroatoms have been utilised in crown formation, and for example selenium compounds similar to cyclam and crown ethers (such as shown in Figure 3.13e) have been synthesised.[67] The tetraseleno compound (Figure 3.13e) has been shown to form complexes[68] with both Cu^+ and Cu^{2+}. Mixed selena–thia macrocycles were also reported. A tetraseleno compound (Figure 3.13f) and a hexaseleno compound (Figure 3.13g) have also been shown to form 1 : 1 and 1 : 2 complexes respectively[69] with Pd^{2+}. In both cases the palladium ions exist in a square planar configuration, as shown by X-ray crystallography. Crown ethers containing tellurium[70] and its platinum complex have also been reported. A range of phosphorus-containing crowns,[71] which also contain oxy, aza or thia groups (Figure 3.13h), have also been described and shown to form complexes with transition metals.

3.6   Lariat and Bibracchial Crown Ethers

From some of the crystal structures shown previously, it can be seen that the crown ethers form an equatorial band around a metal ion, with counter-ions being complexed at the axial sites. Additional complexation between the macrocycle and the ion can be introduced by the synthesis of compounds such as lariat ethers

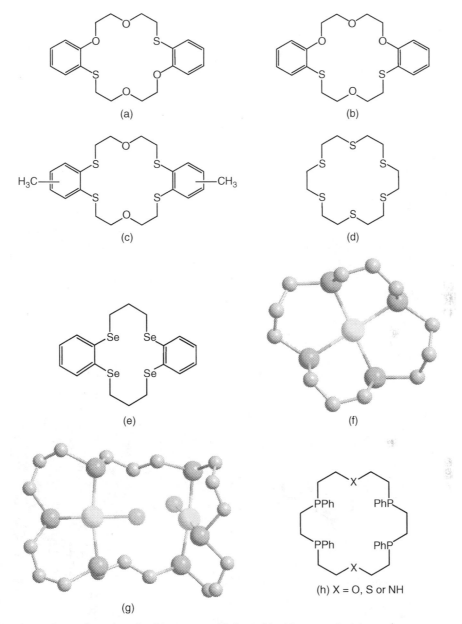

Figure 3.13   *(a–c) Examples of early thiacrowns, (d) hexathia-18-crown-6, (e) a seleno crown, (f,g) crystal structures of seleno crown/Pd complexes and (h) a phosphorus-containing crown*

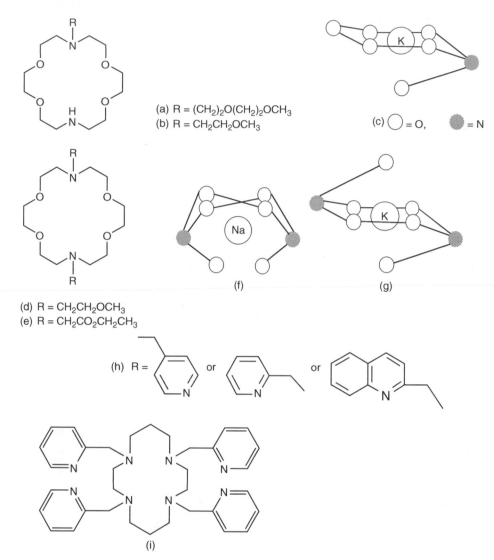

(a) R = (CH$_2$)$_2$O(CH$_2$)$_2$OCH$_3$
(b) R = CH$_2$CH$_2$OCH$_3$

(c) ◯ = O,   ● = N

(d) R = CH$_2$CH$_2$OCH$_3$
(e) R = CH$_2$CO$_2$CH$_2$CH$_3$

(f)   (g)

(h) R =   or   or

(i)

Figure 3.14   *(a,b) Lariat ethers, (c) schematic of the coordination around K$^+$ for 3.14b, (d,e) bibracchial lariat ethers, (f) schematic of the coordination of 3.14d around Na$^+$, (g) schematic of the coordination of 3.14d around K$^+$, (h) azacrown ethers and (i) cyclam substituted with pyridine units*

(taking their name from the lariats or lassos used by cattle herders). These consist of a crown ether with a side chain that can also form complexes with the central metal ion. A typical side chain might be a short polyether chain and can be attached to the nitrogen atom in an azacrown ether. An example is shown in Figure 3.14a, where a short chain containing oxygen atoms is attached to the aza group. This enables further interaction and enhances the binding between metal and macrocycle. Attachment of the side chain to one of the carbons of the crown ether is also possible.

Many of the early studies on lariat ethers have been reviewed by Gokel.[72] A range of ring sizes and side groups were utilised and their complexation with various ions was studied. One of the important results of this work showed that it is not the ring size per se that determines the binding strength but rather the number of binding groups available; for example, a 15-membered crown with a —$(CH_2CH_2O)_2CH_3$ side chain had essentially the same binding constant as the corresponding 18-membered crown (Figure 3.14b) with a —$CH_2CH_2OCH_3$ side chain.[72] Addition of a —$(CH_2CH_2O)_3CH_3$ chain to the corresponding 12-membered crown also greatly improved its binding to sodium and potassium. The lariats bind much more strongly than the corresponding azacrowns without the side chains and a list of the binding constants for several systems has been published.[73] This shows that for the 12-crown-4 derivative, binding constants for sodium and potassium increase dramatically with addition of a —$CH_2CH_2OCH_3$ side chain: additional —CH_2CH_2O— units cause further but smaller increases up to chains containing four oxygens. For the aza-15-crown-5 system the compound with a single —$CH_2CH_2OCH_3$ side chain again shows dramatic enhancement of binding constant (>100-fold for sodium and potassium) over the unsubstituted crown and severalfold further enhancement upon doubling the side-chain length, although this decreases slightly if more ether units are added. For the aza-18-crown-6 compounds, a single —$CH_2CH_2OCH_3$ side chain has the most dramatic enhancements (about 100-fold for sodium and potassium), while addition of a second ether unit causes a small increase for potassium but not sodium. Other ions such as ammonium and calcium were also studied, but in these cases the effects of side chains on their binding were much less pronounced.

Since the addition of a single sidearm was shown to enhance binding of anions, crown ethers with two or three side chains were studied. These were known as bibracchial lariat ethers (BIBLES) and tribracchial lariat ethers (TRIBLES). Again, enhancements were seen for these compounds: addition of two —$CH_2CH_2OCH_3$ side chains to diaza-18-crown-6 (Figure 3.14d) caused >1000-fold increases in the binding constants for sodium and potassium ions, and addition of an ester side chain (Figure 3.14e) led to 10 000-fold enhancements compared to the unsubstituted diazacrown. Calcium was also 100 times more strongly bound by the ester side chain than the ether.[73] Triaza systems have also been synthesised based on a triaza 18-crown-6 ring system, but binding constants were found to be lower than the corresponding diaza compounds and it is obvious from molecular models that steric crowding prohibits any further binding interactions than for two chains. The subject of these multiple armed systems has been reviewed elsewhere.[72,74]

Solid-state studies were also made of these compounds and their complexes. One of the first successfully analysed complexes[75] was that between the compound shown in Figure 3.14b and potassium. The five-ring oxygen atoms adopt an almost identical equatorial binding conformation to that seen for 18-crown-6, with the nitrogen atom being somewhat bent out of the plane, with the side chain bending back over the centre of the ring, allowing the oxygen to interact with the metal ion (Figure 3.14c). When a bibracchial system is used, however, the structure is highly dependent on the metal ion. When the compound shown in Figure 3.14d binds to sodium, the ring and pendent groups are shown by X-ray crystallography[75] to wrap around the sodium ion, with both pendent groups on the same side of the macrocycle (Figure 3.14e). However, when the complex with potassium is crystallised, it has a structure with the potassium ion within the macrocycle and the pendent groups *trans* to each other,[76] as shown in Figure 3.14f.

Other macrocyclic systems and sidearm groups have also been used as the basis for lariat-type compounds, and for example both diaza-18-crown-6 and cyclam have had their nitrogen groups substituted with pyridine moieties[77] to give the structures shown in Figure 3.14h,i. The crowns showed little enhancement for sodium binding but significant increases in their affinity for calcium and copper ions. Similar furan-substituted crowns showed good affinity for potassium and barium, but the pyridine-substituted cyclam was considerably poorer. Also, whereas cyclam itself is a poor transporter of ammonium ion, a cyclam substituted with four furan groups showed a greatly enhanced affinity for ammonium.[78]

3.7   Cryptands

Crown ethers and their derivatives are usually flexible molecules capable of accommodating a wide variety of guests, which is one of the reasons they can often display poor selectivity. It should also be noted that many of the crowns are restricted in the manner in which they can arrange themselves around a guest, possibly reducing potential binding interactions. Jean-Marie Lehn, who shared the Nobel Prize with Pedersen and Cram, deduced that a flexible molecule which could present donor atoms in a three-dimensional array could potentially be capable of actually encapsulating a guest.

   We have already mentioned Lehn's method (Figure 3.6d) for the synthesis of diamide-containing crowns from diamines and diacid chlorides, followed by reduction to the diazacrown.[26] This material can then serve as a feedstock for a further amidation reaction (Figure 3.15a), which after reduction leads to the addition of another diether chain.[79] These materials were first reported in 1969 and the term 'cryptand' was coined for them. 'Cryptate' is the name for a complex formed of a guest and a cryptand. One of the most widely used cryptands (Figure 3.15a,b) is 1,10-diaza-4,7,13,16,21,24-hexaoxabicyclo[8.8.8]hexacosane,

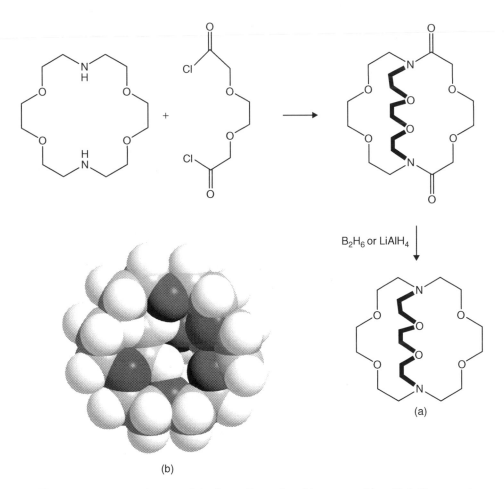

(a)

(b)

Figure 3.15   *(a) Synthesis and (b) three-dimensional structure of free [2,2,2]cryptand*

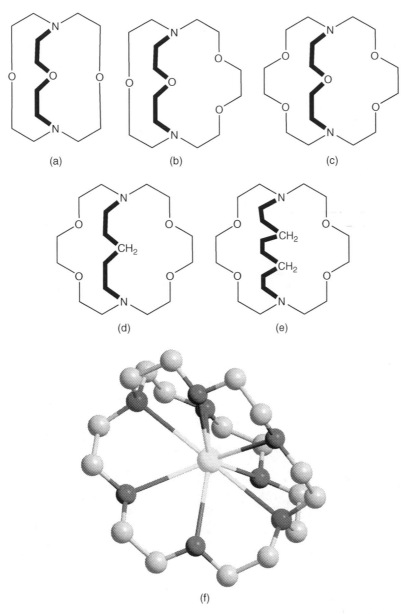

Figure 3.16 *Structures of (a) [1,1,1]cryptand, (b) [2,1,1]cryptand, (c) [2,2,1]cryptand, (d,e) hydrocarbon-bridged cryptands; (f) crystal structure of [2,2,2]cryptand complexed with rubidium*

also termed [2,2,2]cryptand. The figures describe the number of ether oxygens in the chains and so the compound shown in Figure 3.16a is [1,1,1]cryptand.

The more rigid three-dimensional structures of cryptands (Figure 3.15b) confer higher selectivity and specificity than similar-sized crown ethers. This is probably due to a combination of cryptands having a more defined 'hole' size than the more flexible crown ethers, meaning that they cannot constrict or expand to accommodate ions of the wrong size. Also, the three-dimensional cryptands are already more preorganised into a binding conformation and therefore there is less rearrangement of the macrocycle as it goes from an uncomplexed to a complexed state. This means there are less unfavourable entropic and enthalpic obstacles to binding.

As with the crown ethers, the cryptands are shown to complex alkali metals.[80] However, the binding coefficients are much higher. The binding constant for [2,2,2]cryptand for potassium is some 10^4 times higher than for 18-crown-6. Much higher selectivities are also noted, and for example in water [2,1,1]cryptand (Figure 3.16b) has a binding constant for lithium over 100 times higher than that for sodium. Similar selectivities are observed for [2,2,1]cryptand (Figure 3.16c), which has a high preference for sodium (binding constant approximately 30 times that for potassium), and [2,2,2]cryptand, which prefers potassium (binding constant approximately 30 times that for sodium). As the cryptands get larger than these, the binding coefficients in water tend to drop, but good binding is still observed in methanol.[80]

One possibility is that the increase in binding strength for cryptands over crowns is simply due to the presence of more donor atoms (eight for [2,2,2]cryptand against six for 18-crown-6). However, a cryptand containing just six donor atoms and a bridging five-carbon chain (Figure 3.16d) was shown to bind potassium over a thousand times more strongly than the corresponding diaza-18-crown-6 compound[73] and had a binding constant only 40 times less than the corresponding [2,2,1]cryptand. This appears to prove that the incorporation of a bridging chain increases the binding ability, possibly due to the preorganisation of the cryptate and better definition of the cavity. We have already discussed the macrocyclic effects of crown ethers, cryptands appear to display an even stronger 'macrobicyclic effect'.

Similar results were obtained for complexes with alkaline earth metals with even higher binding constants. An enhancement of binding coefficient of about 10^5 was observed for barium-[2,2,2]cryptand in comparison to a similar crown ether.[80] Again, selectivity was observed, with [2,1,1]cryptand showing high Ca/Mg selectivity and [2,2,2]cryptand showing high selectivity for Sr/Ca and Ba/Ca (ratios of binding constants are 4000 and 10^5 respectively). An interesting piece of work involved the study of the complexation of barium and potassium with various cryptands.[81] What was notable was that [2,2,2]cryptand has a binding constant for Ba^{2+} 10^4 times greater than that for K^+, but for a similar hydrocarbon-bridged compound (Figure 3.16e) the selectivities are reversed, with the binding constant for potassium being over 200 times greater than that for barium.[81]

The structure of the cryptates was examined by X-ray crystallography, with for example the rubidium complex of [2,2,2]cryptand being shown to have a structure (Figure 3.16f) in which the rubidium ion is entirely encapsulated within the cavity.[82] Further work showed that the sodium, potassium and caesium complexes all had similar structures and that the cryptand adopts a conformation intermediate between a bicapped trigonal prism and a bicapped trigonal antiprism.[83] Encapsulation was also observed for sodium and potassium complexes with the smaller [2,2,1]cryptand.[84]

As for the crown ethers, various alternative heteroatoms to oxygen have been incorporated into cryptands. The compound shown in Figure 3.17a exists as a highly protonated species in aqueous solution and forms a complex with azide ion as shown.[85] Sulfur atoms[86] have also been successfully incorporated into cryptands, as shown in Figure 3.17b–d. The compound shown in Figure 3.17b has been shown to form complexes with potassium, barium, silver, lead and thallium. Disulfide groups have also been included into cryptands to give ellipsoidal-shaped receptors.[87]

3.8   Spherands

Within this chapter we have seen how improving the organisation of the host and constructing molecules with higher levels of structural definition can increase the binding strength and selectivity of the host.

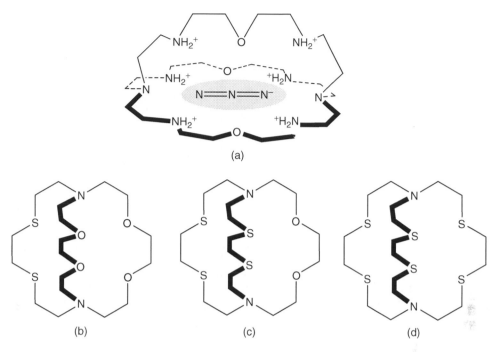

Figure 3.17 *(a) Macrocyclic amine ligand complex with azide and (b–d) cryptands containing sulfur atoms*

The cyclic crown ethers display much higher binding constants than the linear podands and are in turn exceeded by the three-dimensional cryptands. For this reason, workers addressed the possibility of synthesising molecules with preordered structures that would match as closely as possible potential hosts. These structures have been named 'spherands' due to their approximately spherical shape and it was thought that being preorganised into the optimal shape for binding would both remove any need for the host to reorganise and minimise any unfavourble entropic or enthalphic penalties.

We will consider examples of two types of spherand. The first is a development of the cryptand synthesis in which azacrown ethers are bridged with two chains, thereby creating a tricyclic structure. The structure of one of these compounds can be seen in Figure 3.18a, while a space-filling model in Figure 3.18b clearly shows its approximately spherical structure. This compound has been shown to form complexes with K^+, Cs^+, Ba^{2+} and NH_4^+ cations in chloroform.[88] It was noted that exchange of bound cations was relatively slow compared to crown ethers and cryptands. In acidic solution the same compound protonates and forms a strong complex with chloride ion.[89] Modelling shows that chloride is the best fit for the internal cavity, bromide and fluoride fit less well and have lower binding constants, while iodide, nitrate and perchlorate do not form complexes at all. X-ray crystal structures[90] have been obtained for the ammonium and the chloride complexes, and it has been shown in both cases that the guest resides inside the central cavity of the spherand.

The second type of spherand was first reported by Donald Cram, the third of the 1987 Nobel Prize winners for work on supramolecular chemistry. Whereas the systems described so far have been relatively flexible, this group utilised a completely different type of compound. Starting from *p*-cresol,[91] the much more rigid system shown in Figure 3.19a was synthesised. This contains a permanent cavity of the correct size and geometry to include guests such as alkali metal ions. The idea for this was that the presence

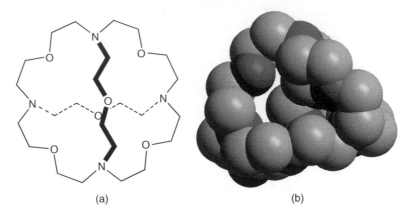

Figure 3.18   (a) Macrotricyclic compound and (b) its three-dimensional structure

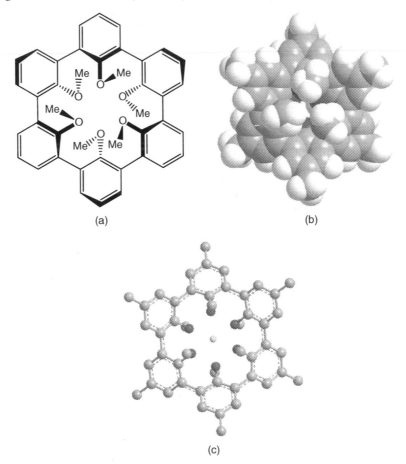

Figure 3.19   (a) An aromatic-based spherand, (b) its three-dimensional structure and (c) the X-ray structure of the complex with Li$^+$

of the permanent binding site would facilitate ion binding since there would be no requirement for the host to rearrange its structure to accommodate a guest. The compound shown is capable of complexing sodium or lithium (but not larger metals), with sodium being preferred. However, the 'monomer' of this system, *p*-methylanisole, is only a poor ligand for metal ions. Also, the linear version of the cyclic compound[92] displays much lower binding to metal ions (binding constants differ by a factor of 10^{12}) due to the lack of a cyclic structure and of a preorganised cavity. X-ray studies[93] show that the spherand has a practically identical structure, both in its unoccupied form and as its lithium (Figure 3.19c) or sodium complex.

Larger versions of these compounds containing eight rings were synthesised and shown to host alkali metals, with caesium having the highest binding constant.[94] In the same work a number of spherands bearing cyano groups instead of methoxy groups were also developed and were shown to be hosts for alkali metals and ammonium salts.[95] Chiral systems have also been developed in which three, four or five binapthol units, each of which is chiral, are linked to form a macrocycle.[96]

Intermediate systems containing a rigid unit combined with a flexible system such as a crown ether were also developed and are known as 'hemispherands'. Similar cryptand/spherand conjugates were named 'cryptaspherands'. A variety of these were synthesised in an attempt to utilise the high selectivity of the spherands but extend it to different ions. It appears that the selectivities of these compounds fall into a series, with spherands > cryptaspherands ~ cryptands > hemispherands > crown ethers > podands. The hemispherands containing three benzene rings bridged with a variety of groups were the first to be synthesised[97] and showed high binding constants for many ions, which were dependent on the molecular structure. A typical hemispherand is shown in Figure 3.20a.

Cryptaspherands have also been synthesised[98] by bridging a diazacrown ether with an aromatic unit to give structures such as that shown in Figure 3.20b and their crystal structures have been obtained. Complexation studies[99] demonstrated that they were stronger complexing agents than the cryptands. Selectivity was dependent on structure, with the cryptaspherand in Figure 3.20b showing Na/Li selectivities of >4000 for example and other compounds showing Rb/Cs selectivities of >100. Cram's group was to take this idea for using rigid preorganised compounds as hosts for a wide variety of inorganic and organic species much further, utilising systems such as calixarenes and resorcinarenes to make cavitands and hemicavitands – molecules which will be discussed in Chapter 4.

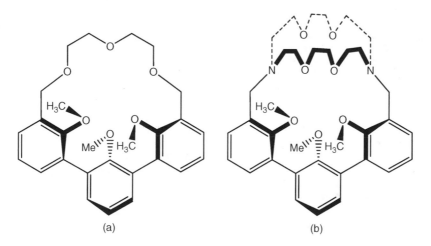

(a)   (b)

Figure 3.20   *(a) A hemispherand and (b) a cryptaspherand*

3.9   Combined and Multiple Systems

The hemispherands and cryptashperands are examples of systems in which different binding moieties are combined in a single molecule. There are many other such systems, a few of which will be discussed here. One of the earliest studies linked together 15-crown-5 units to form dimers and polymers as shown in Figure 3.21a–c, and similar studies were carried out on the 18-crown-6 ethers.[100] The *bis* and poly crowns were found to be more efficient extractants than the monomer (Figure 3.21a) and this was thought to be due to the formation of sandwich complexes between two crown units and a metal ion.

What makes *bis* crown ethers of such interest is the potential to join them together with an active group. For example, when two crown ether units are joined by an azobenzene group, light irradiation can switch the azo unit from its *trans* form (Figure 3.21d) to the *cis* form.[101] In the *cis* form the two crown units are much closer to each other, which makes it a more effective complexant for potassium, which is too large to comfortably fit in a 15-crown-5 unit but can form a sandwich compound with two units. Irradiation of this crown noticeably increases its transport rate of potassium picrate. Evidence for the formation of sandwich complexes came in further work,[102] which showed that larger cations, especially rubidium, were transported more effectively and also that the steady-state ratio under irradiation of *cis/trans* was affected by the metal ion, being 52:48 with no metal and 98:2 when rubidium was present.

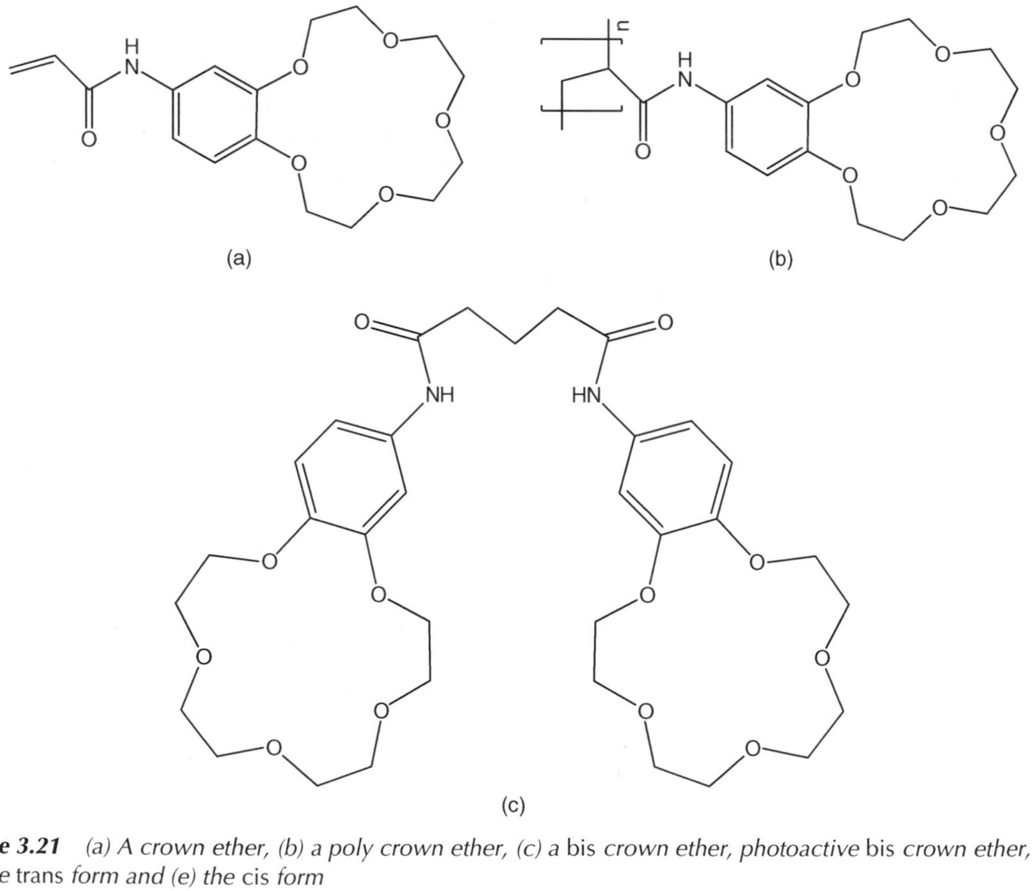

(a)   (b)

(c)

Figure 3.21   *(a) A crown ether, (b) a poly crown ether, (c) a bis crown ether, photoactive bis crown ether, with (d) the trans form and (e) the cis form*

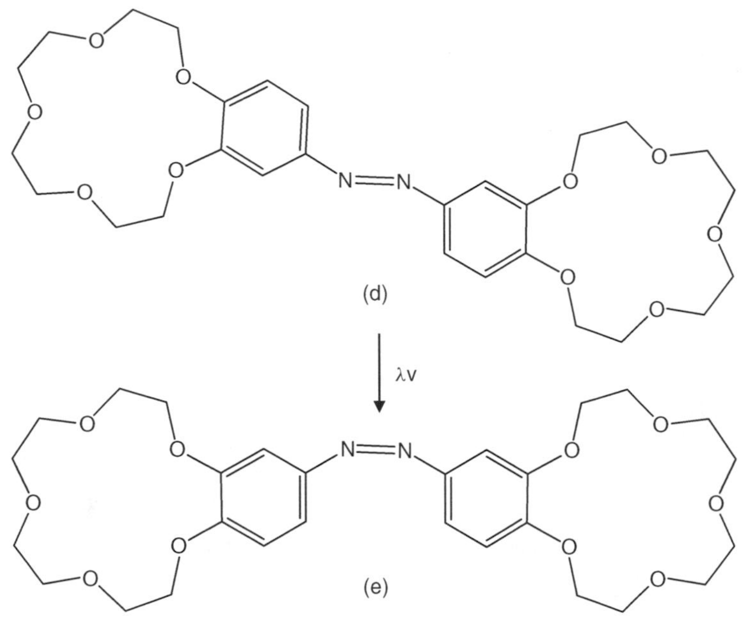

(d)

λv

(e)

Figure 3.21   *(continued)*

Chiral spacers have been used[103] as bridges between two crown units and the resultant *bis* crown has been shown to selectively transport chiral diammonium compounds. Metallocenes have also been utilised as bridging groups, as shown in Figure 3.22a, where ruthenocene or ferrocene[104] is used to link two azacrown ether units. The ferrocene derivative is shown to selectively form complexes with K^+ over Na^+ and Cs^+; the crystal structure of the 1:1 complex with potassium is shown in Figure 3.22b and a space-filling representation is shown in Figure 3.22c; together these clearly demonstrate the formation of a sandwich complex. Since ferrocene can be electrochemically switched between the neutral form and ferricinium ion, compounds of this type open up the possibility of electrochemical control of binding. Another system which has been synthesised is one in which diazacrown units can be linked by ferrocene bridges,[105] which has been shown to give both a ferrocene-bridged 1 + 1 addition cryptand and the 2 + 2 addition product (Figure 3.23a). This compound[105] has been shown by X-ray studies to have a large central cavity, as has the similar thiacryptand[106] compound (Figure 3.23b).

Another series of versatile ligands (Figure 3.24a–c) was synthesised by joining together two diazacrowns with a variety of aromatic groups.[107] When mixed with methylammonium ions, 2:1 ion:ligand complexes were formed, whereas larger substituted ammonium cations only formed a 1:1 complex. Much stronger binding constants were observed for α,ω-alkyl diammonium cations, for which NMR studies showed that 1:1 complexes were formed with the guest contained within the host. Triple-bridged dimers could be obtained from triaza-18-crown-6 and were shown to display similar behaviour.[108] Variation of the aromatic linkers meant that different guests could be bound selectively, with for example the naphthyl derivative (Figure 3.24c) binding the pentyl and hexyl diammonium compounds but not the butyl or octyl. The crystal structure[109] of the pentyl complex is shown in Figure 3.24d. Use of the chiral binaphthyl unit as a linker[110] allowed the synthesis of chiral *bis* crown ethers, which could differentiate between enantiomers of chiral ammonium compounds. Recently, other workers[111] have synthesised a *bis* crown ether with a linking unit

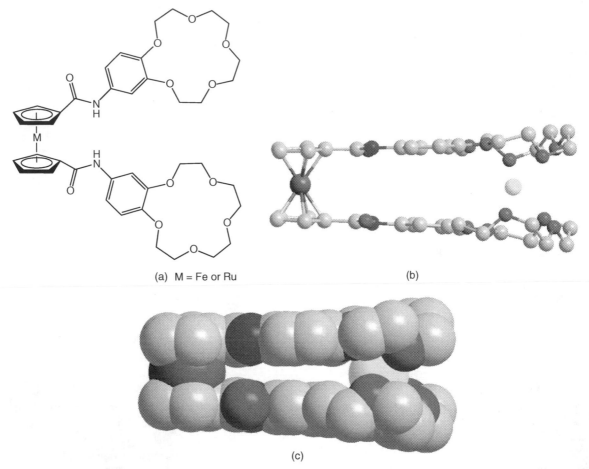

Figure 3.22 *(a) Metallocene* bis *crown ethers, (b) their X-ray structure and (c) a space-filling model of the ferrocene compound*

containing photosensitive stilbazole units (Figure 3.25a); this could be incorporated into ultra-thin films and shown by fluorescence studies to selectively bind 1,3-diaminopropane salts. Flexible oligopropylene oxide spacers have also been used to construct a number of *bis* crown ethers which again show simple binding of small alkali metals and sandwich-complex formation with larger ions.[112] Many other systems have been synthesised containing other units. These include for example the *bis*-cyclen compound,[37] as mentioned earlier (Figure 3.10b), and a wide variety of mixed systems with crowns, cryptands and other compounds, such as calixarenes – some of which will be mentioned in later chapters.

Larger assemblies containing crown units have been made. One, two and three benzocrown ether units have been attached to a benzene ring,[113] as shown in Figure 3.26a, which shows the triply-substituted system with benzo-15-crown-5 units attached. The benzo-18-crown-6 analogue was also synthesised. The 15-crown-5 compounds showed increasing K/Na selectivity as the number of attached crown units increased. Unlike some of the previous *bis* crowns in this section, sandwich complexes were not formed with large ions; instead alkali metal-bridged dimers were formed. Other workers used hexahydrotriazine[114] as a central unit to assemble *tris* crown ethers which could capture alkai metals and form sandwich

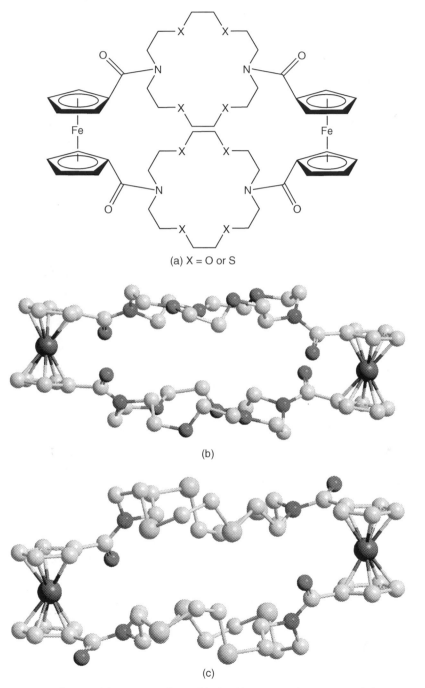

(a) X = O or S

(b)

(c)

Figure 3.23   (a) Bis-metallocene bis crown ethers, (b) the X-ray crystal structure of the oxa and (c) the thia bis-crown ethers

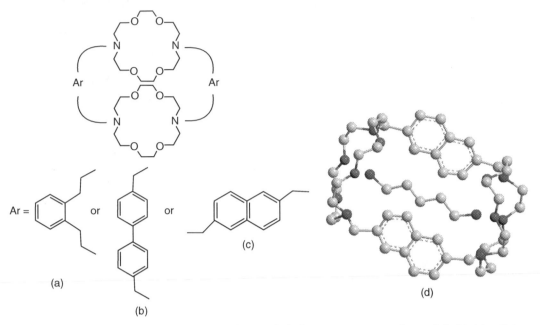

Ar =   or   or

(a)   (b)   (c)   (d)

Figure 3.24   *(a–c) Aromatic linked* bis *crown ethers and (d) the X-ray structure of a naphthyl-bridged* bis *crown ether containing a 1,5-diaminopentane guest*

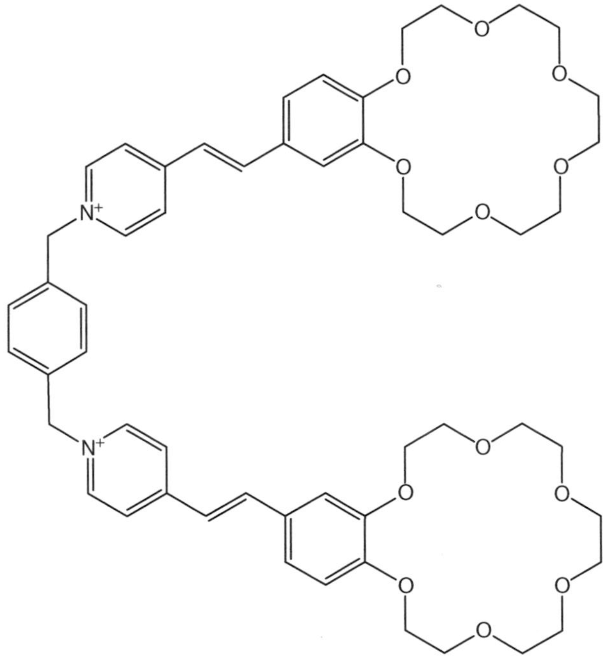

Figure 3.25   *A stilbene-linked* bis *crown ether*

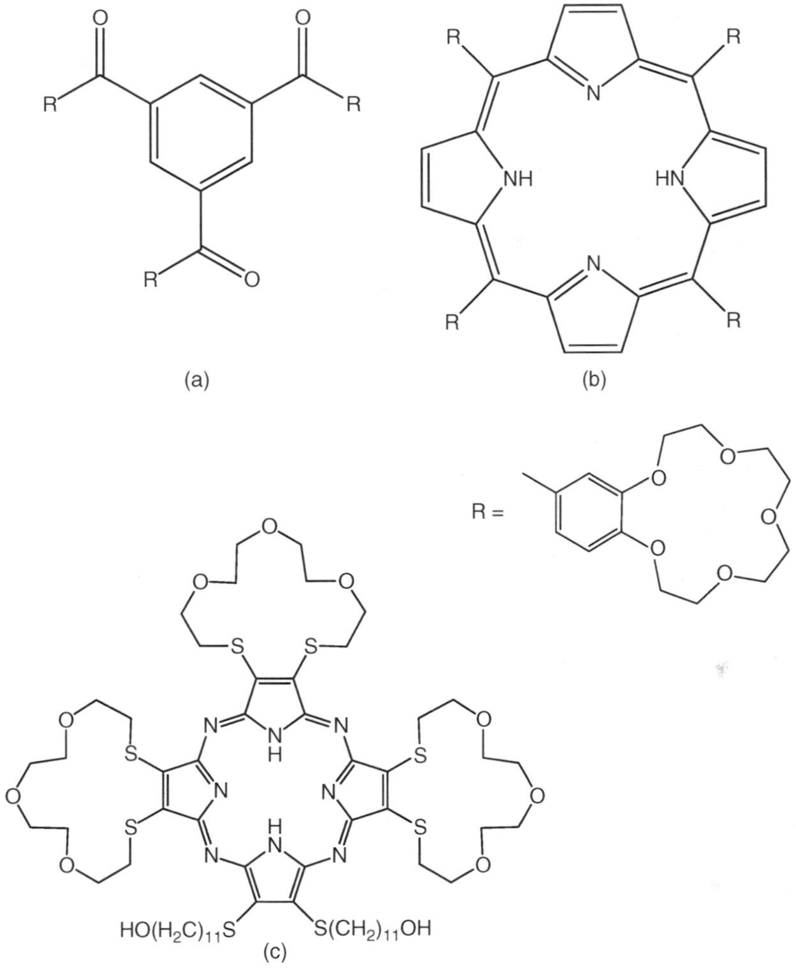

R =

Figure 3.26   *(a) Benzene-, (b) porphyrin- and (c) porphrazine-linked crown ethers*

complexes. We have already discussed porphyrins in Chapter 1; these have been used as units to link four crown ethers together[115] (Figure 3.26b). The resulting compounds complex a variety of ions. The porphyrin shown has a high selectivity for potassium and tends to form sandwich complexes which promote the dimerisation of the porphyrin and quenching of its fluorescence. The dimerisation of the porphyrin and its metallo derivatives is also shown to affect the electrochemistry of these compounds.[116] Other work has utilised porphyrins bridged with a single crown ether unit in a face-to-face arrangement and the resultant fluorescent molecule is quenched by the binding of Cu^{2+} or Zn^{2+} ions.[117] Phthalocyanines[118] have also been used as scaffolds to assemble up to four crown ether units. Porphrazines substituted with dithia-15-crown-5 units (Figure 3.26c) have been shown to give specific optical responses to silver and mercury ions.[119]

Polymer crown ethers have also been utilised, as mentioned earlier in this section.[100] A wide variety of polymers have been synthesised, one of the simplest (Figure 3.27a) being poly(vinylbenzo-18-crown-6), which has been shown to have a high affinity for both alkali metal cations and organic compounds

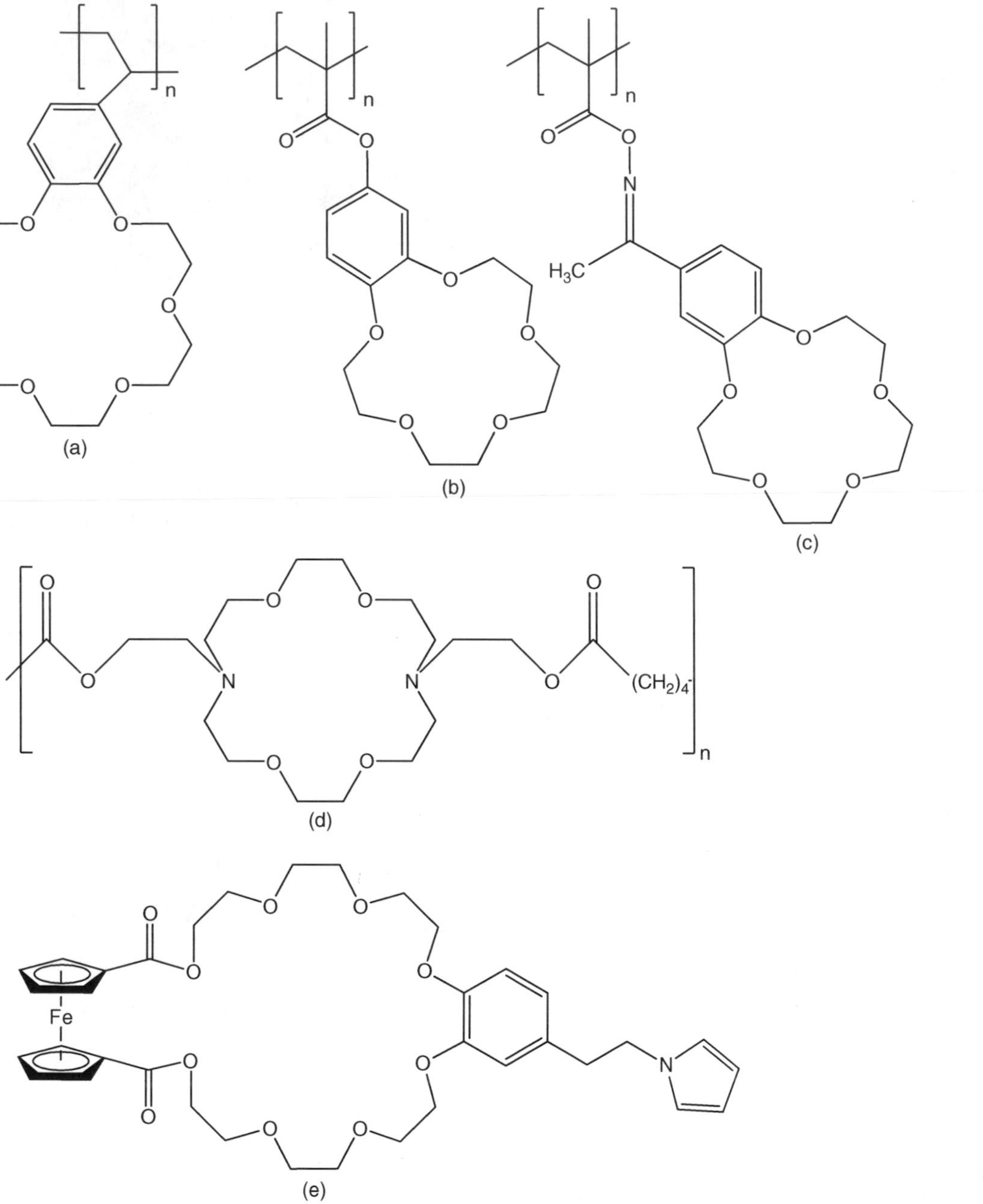

Figure 3.27   *(a) Poly(vinylbenzo-18-crown-6), (b,c) polymethyl methacrylate substituted with crown ethers, (d) a diaza-18-crown-6 condensation polymer and (e) a electropolymerisable crown ether*

(such as dyes) in water.[120] This is thought to be due to its polysoap-like structure, with a hydrophobic backbone and the relatively hydrophilic crown ether units. Polyacrylamides and *bis*-acrylamides with pendant benzo crown ethers were blended with polyvinyl chloride (PVC) and utilised as potassium-selective membranes.[121] The materials could also be grafted onto silica[122] and used as solid phases in ion-chromatography separations, being capable of separating the alkali and alkaline earth metals depending on affinity. Methyl methacrylate polymers with crown ether side chains (Figure 3.27b) have also been developed and their complexes with lithium perchlorate have been shown to display ionic conductivity,[123] indicating potential uses as a solid-state electrolyte. Photoresponsive stibazole groups have been used to link crown ethers to polymethyl methacylate backbones (Figure 3.27c) and the resulting materials have been shown to complex alkali metal picrates and then be capable of releasing them into solution upon irradiation.[124]

Condensation polymers have also been constructed. Diaza-18-crown-6, with its reactive amine groups, for example, can be condensed with a diacid chloride to form a 'Nylon'-type polymer (Figure 3.27d). These polymers[125] can be mixed with alkali metal perchlorates and shown to have ionic conductivity in the solid state, the complex with $KClO_4$ having the lowest conductivity, indicating the strongest crown–ion interaction. The same group also utilised diazacrown ethers containing side chains, thereby synthesising poly(lariat ethers).[126] Another method has been to synthesise electropolymerisable monomers such as that shown in Figure 3.27e, which can be electropolymerised through the pyrrole group to give a film of a polypyrrole–ferrocene crown ether.[127] This material, when interrogated by impedance spectroscopy, has been shown to selectively recognise calcium and barium ions. A chiral binaphthyl crown ether can also be deposited electrochemically[128] and shown to recognise certain neurotransmitters. Dendrimers have also been utilised as substrates for the attachment of crown ethers. A carbosilane dendrimer for instance was recently grafted with 15-crown-5 moieties and utilised in the production of an ammonium-selective electrode.[129]

Most of the work on these polymeric systems appears to have used crown ethers, but other moeities have been incorporated into polymeric systems, with both crown and cryptand units, for example, having been successfully grafted onto chloromethylated crosslinked polystyrene beads[130] and used as polymeric extraction agents for lanthanides or phase-transfer catalysts. Similar cryptand polymers have also been used for several separation technologies, including nitrogen-isotope separation.[131]

There have also been numerous reports on the incorporation of crown ethers and similar materials into a number of supramolecular assemblies via noncovalent interactions. Crowns have been utilised in the construction of such assemblies as rotaxanes and catenanes; these will be described in Chapter 9.

3.10   Applications of Crown Ethers and Related Compounds

Although they have only been around for about 50 years, the use of crown ethers and related compounds has become widespread within both the academic and the industrial world. We will attempt to give a brief overview of how these compounds are being used today.

One of the earliest applications was the solvation of ionic salts in organic media. Many ionic materials simply will not dissolve in solvents, such as hexane or toluene. Conversely, many organic compounds are insoluble in water. This becomes a problem when we wish to react such a compound with a salt. Some solvents such as dimethylformamide or dimethyl sulfoxide will dissolve both organic and inorganic compounds, but these are high-boiling solvents which can often be difficult to remove afterwards. Crown ethers and cryptands offer a solution to this problem since they can complex metal ions, encasing them in an organic sheath and rendering them soluble in organic solvents. An example of this is potassium permanganate, a powerful oxidising agent which is completely insoluble in benzene; however, upon addition of 18-crown-6 and a trace of water we obtain the formation of an intensely purple solution which can be used to oxidise many water-insoluble organic compounds. Another group of materials which is very insoluble in organic solvents is the superoxides; again this insolubility can be overcome by using crown ethers or cryptands.

Another example is the synthesis of alkyl cyanides, which can be performed using a simple nucleophilic substitution reaction of an alkyl halide with a cyanide salt. The most economic way to perform this reaction would be to dissolve the salt in a liquid alkyl halide without any solvent, but a material such as sodium cyanide will not dissolve in a long-chain alkyl halide. One solution is phase-transfer catalysis, where the salt is dissolved in water and a small amount of a crown ether or cryptand is added as a catalyst. Addition of the alkyl halide results in a two-phase system. However, the catalyst binds the metal ion, allowing it to be solvated in the alkyl halide so that some of the salt is transported into the organic phase. There it reacts to give the alkyl cyanide and a halide ion. This transport phenomenon is repeated and eventually all the alkyl halide is converted. Another factor in using these types of method is that quite often the anion can become activated as the encapsulation of the cation hinders any association in solution; also, the anion is relatively poorly solvated by the organic solvent. This poor solvation can often greatly increase the reactivity of the anion and the rates of any reactions in which it participates. A wide range of crown ethers, cryptands and so on are commercially available for exactly this purpose. To aid separation, many of these macrocycles have been attached to polymeric supports such as crosslinked polystyrene, and these materials are also commercially available.

We have already mentioned how the preferential binding of some materials enables the use of many materials as selective media. Crown ethers and cryptands have been utilised in a variety of ion-selective electrodes and specific crowns are available commercially, for example several of the commercial Selectophore range of compounds are based on crown ethers or other macrocyclic compounds and are incorporated into various sensor or chromatographic applications. We have reported within this chapter on how macrocycles can be utilised in chromatographic separations, such as for example in ion-chromatography or Donald Cram's work on enantiomeric separation using chiral crown ethers.

One field in which this range of compounds has excelled is the sensing of a wide variety of species. For any sensor, there has to be a recognition event (e.g. binding of a metal) followed by a measurable transduction event. Although the macrocyclic systems can be highly selective binding agents, one problem is that the binding interactions are essentially noncovalent and there is often no easily measurable side product such as a detectable molecule or the production or consumption of electrons. This problem can be addressed by combining, usually in the same molecule, the binding macrocycle with a reporter group that can be interrogated. Binding of a substrate affects the binding macrocycle and its reporter group, leading to a change in the overall nature of the sensing molecule. Various reporter groups have already been mentioned within this chapter, such as ferrocene units, which will have their electrochemical behaviour modified by a binding event. Other reporter groups include photochemically active groups such as the porphyrins or stilbazole units, which have their fluorescence behaviour modified by binding of substrates. Incorporation of the macrocycles into conductive polymer films has also been studied.

Responsive materials in which chemical events can be modulated by electrical or optical stimulation have also been addressed. Within this chapter we have mentioned polymeric crowns which bind and release ions upon optical stimulation. Optically controlled transfer of substrates using photoresponsive crown ethers has also been discussed. For other materials such as cryptands, cyclens or azacrown ethers, the presence of the amine groups means that the binding of a variety of substrates can be controlled by the pH. Other workers have used lariat-type ether compounds in which the side chain can be photochemically switched, which can affect the electron density within the overall molecule and modulate binding. Similar work, for example using ferrocene-bridged crown ethers and cryptands, has produced electrochemically switchable systems in which the redox state of the bridge controls binding. The potential is for these types of materials to be used as controlled binding and release agents.

The ion-binding nature of these compounds has also led to their investigation as solid-state electrolytes, for use in such applications as lithium-ion batteries and fuel cells. They have also been investigated as potential artificial ion channels, transporting ions across hydrophobic membranes. Liquid crystalline macrocycles whose mesomorphic properties change upon binding of guests have also been studied.

One extreme example of the stabilising ability of these macrocyclic materials can be seen in the synthesis of alkalides and electrides.[2] The characteristic chemistry of alkali metals is of a loss of an electron to give the cation. If there is no corresponding species to accept this electron, for example to form sodium chloride, the atom can still potentially disassociate into a cation and an electron, but will almost immediately recombine to give the neutral atom. However, when the cation can be encapsulated in a macrocyclic compound, the cation–electron pair can be stabilised and actually exist as a compound. For example, caesium electride, in which the caesium cation exists as a sandwich complex between two 15-crown-5 ether molecules, has been synthesised as an air-stable solid.[2] A similar range of compounds can be synthesised in which a second alkali metal atom can actually exist as an anion, for example hexacyclen can be used to stabilise caesium sodide [Cs$^+$(hexacyclen)Na$^-$].

There has also been much attention paid to possible medical applications for a whole range of macrocyclic compounds. For example, the natural antibiotic nonactin (Figure 3.28a) is an oxygen-rich macrocycle which acts by transporting metal ions across bacterial cell membranes until the resulting osmotic pressure build-up ruptures the cell wall. This has led to much research into the development of artificial ions transporting materials, and macrocyclic compounds have been widely studied due to their ion-complexing

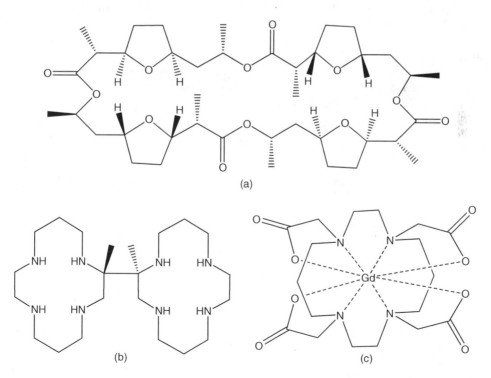

(a)

(b)   (c)

Figure 3.28   *(a) Nonactin, (b)* bis-*cyclam with anti-HIV activity and (c) Prohance*

abilities. Another family is the cyclams, which have many potential medical applications, as reviewed by Liang and Sadler.[132] A detailed discussion of the potential medical uses of these materials is beyond the scope of this chapter, but we will mention that cyclams, and especially the *bis*-cyclam depicted in Figure 3.28b, have displayed strong anti-HIV activity. They have also been used in the radiopharmaceutical field as encapsulants for radionuclides such as ^{67}Cu, ^{90}Y, ^{99}Tc, ^{111}In and ^{186}Re, which are used in both the diagnosis and therapy of a wide variety of conditions, including cancer, liver disorders and numerous neurological and cardiological conditions. They are also widely used as contrast agents for magnetic resonance imaging (MRI); the commercial MRI contrast agent Prohance (Figure 3.28c), for example, is based on a cyclam-encapsulated gandolinum compound.

3.11   Conclusions

We have looked at several series of compounds: the crown ethers, the cryptands, the cyclens and so on. This chapter has served to introduce the concept of the template effect, which simplifies the synthesis of many of these compounds, and the macrocyclic effect, which explains their enhanced binding strengths and specificities compared to simpler linear systems. We have also seen how the structure of the compounds – factors such as the macrocyclic ring size, the presence of side groups or bridging rings, the effects of heteroatoms such as N or S and the ionisation state – affects binding capability. We have also introduced the concepts of macrocyclic rearrangement to accommodate a guest, and of utilising more rigid systems to effectively construct cavities of precisely tailored shapes and sizes. It is clear that the physics and chemistry of these macrocyclic systems are greatly affected by all these structural factors. This opens up the possibility of designer molecules with reactivities, selectivities and specificities that approach those synthesised by that greatest of chemists, nature.

We would like to finish this chapter with a quote from Pedersen's Nobel Prize lecture[3]: 'But whether it be in biology or some other field, it is my fervent wish that before too long it matters not by whom the crown ethers were discovered but rather that something of great benefit to mankind will be developed about which it will be said that were it not for the crown compounds it could not be.' It appears that his hopes are beginning to or perhaps already have come true.

Bibliography

Gokel GW. Crown Ethers and Cryptands. Monographs in Supramolecular Chemistry. Cambridge: Royal Society of Chemistry; 1991.
Lehn J-M. Supramolecular Chemistry. New York: Wiley-VCH; 1995.
Cragg P. A Practical Guide to Supramolecular Chemistry. Wiley-Blackwell; 2005.
Steed JW, Atwood JL. Supramolecular Chemistry. John Wiley and Sons; 2000, 2009.
Bradshaw JS, Krakowiak KE, Izatt RM. Aza-Crown Macrocycles. Chemistry of Heterocyclic Compounds: A Series of Monographs. New York: Wiley-Blackwell; 1993.

References

1. Down JL, Lewis J, Moore B, Wilkinson G. The solubility of alkali metals in ethers. *J Chem Soc.* 1959; 3767–3773.
2. Dye JL. Electrons as anions. *Science.* 2003; **301**: 607–608.
3. Pedersen CJ. The discovery of crown ethers. *Nobel Prize Lecture, Angew Chem Int Ed.* 1988; **27**: 1021–1027.

4. Pedersen CJ. Cyclic polyethers and their complexes with metal salts. *J Am Chem Soc.* 1967; **89**: 7017–7036.

5. Pedersen CJ. Macrocyclic polyethers: dibenzo-18-crown-6 polyether and dicyclohexyl-18-crown-6 polyether. *Org Syn.* 1988; **50**: 395–400.

6. Gokel GW, Cram DJ, Liotta CL, Harris HP, Cook FL. 18-crown-6. *Org Synth.* 1988; **6**: 301–302.

7. Bush MA, Truter MR. The crystal structures of three alkali-metal complexes with cyclic polyethers. *J Chem Soc.* 1970; **D**: 1439–1440.

8. Bush MA, Truter MR. Crystal structures of complexes between alkali-metal salts and cyclic polyethers. Part IV. The crystal structures of dibenzo-30-crown-10 (2,3:17,18-dibenzo-1,4,7,10,13,16,19,22,25,28-decaoxacyclotriaconta-2,17-diene) and of its complex with potassium iodide. *J Chem Soc. Perkin Trans.* 1972; **2**: 345–350.

9. Fenton DE, Mercer M, Poonia NS, Truter MR. Preparation and crystal structure of a binuclear complex of potassium with one molecule of cyclic polyether: bis(potassium thiocyanate)dibenzo-24-crown-8. *J Chem Soc Chem Commun.* 1972; 66–67.

10. Mercer M, Truter MR. Crystal structures of complexes between alkali-metal salts and cyclic polyethers. Part VII. Complex formed between dibenzo-24-crown-8 (6,7,9,10,12,13,20,21,23,24,26,27-dodecahydrodibenzo[b,n]-1,4,7,10,13,16, 19,22-octaoxacyclotetracosin) and two molecules of potassium isothiocyanate. *J Chem Soc. Dalton Trans.* 1973; 2469–2473.

11. Mallinson PR, Truter MR. Crystal structures of complexes between alkali-metal salts and cyclic polyethers. Part V. The 1 : 2 complex formed between potassium iodide and 2,3,5,6,8,9,11,12-octahydro-1,4,7,10,13-benzopentaoxacyclopentadecin (benzo-15-crown-5). *J Chem Soc. Perkin Trans.* 1972; **2**: 1818–1823.

12. Izatt RM, Bradshaw JS, Nielsen SA, Lamb JD, Christensen JJ. Thermodynamic and kinetic data for cation–macrocycle interaction. *Chem Rev.* 1985; **85**: 271–339.

13. Izatt RM, Lamb JD, Christensen JJ, Haymore BL. Anomalous stability sequence of lanthanide(III) chloride complexes with 18-crown-6 in methanol. Abrupt decrease to zero from Gd3+ to Tb3+. *J Am Chem Soc.* 1977; **99**: 8344–8346.

14. Kyba EP, Helgeson RC, Madan K, Gokel GW, Tarnowski TL, Moore SS, Cram DJ. Host–guest complexation. 1. Concept and illustration. *J Am Chem Soc.* 1977; **99**: 2564–2571.

15. Belamril B, Bavoux C, Thozet A. Crystal structure of the 1 : 1 : 6 complex between 18-crown-6, hydroquinone and water. *J Incl Phenom Macro Chem.* 1990; **8**: 383–388.

16. Vögtle F, Weber E. Multidentate acyclic neutral ligands and their complexation. *Angew Chem Int Ed.* 1979; **18**: 753–776.

17. Knobler CB, Maverick EF, Parker KM, Trueblood KN, Weiss RL, Cram DJ, Helgeson RC. Structures of 4,5,15,16-tetraacetoxy[2.2]paracyclophane and two bis(crown-6)[2.2]paracyclophanes. *Acta Cryst.* 1986; **C42**: 1862–1868.

18. Newcomb M, Timko JM, Walba DM, Cram DJ. Host–guest complexation. 3. Organization of pyridyl binding sites. *J Am Chem Soc.* 1977; **99**: 6392–6398.

19. Moore SS, Tarnowski TL, Newcomb M, Cram DJ. Host–guest complexation. 4. Remote substituent effects on macrocyclic polyether binding to metal and ammonium ions. *J Am Chem Soc.* 1977; **99**: 6398–6405.

20. Newcomb M, Moore SS, Cram DJ. Host–guest complexation. 5. Convergent functional groups in macrocyclic polyethers. *J Am Chem Soc.* 1977; **99**: 6405–6410.

21. Kyba EP, Gokel GW, De Jong F, Koga K, Sousa LR, Siegel MG, Kaplan L, Sogah GDY, Cram DJ. Host–guest complexation. 7. The binaphthyl structural unit in host compounds. *J Org Chem.* 1977; **42**: 4173–4184.

22. Sousa LR, Sogah GDY, Hoffman DH, Cram DJ. Host–guest complexation. 12. Total optical resolution of amine and amino ester salts by chromatography. *J Am Chem Soc.* 1978; **100**: 4569–4576.

23. Peacock SC, Domeier LA, Gaeta FCA, Helgeson RC, Timko JM, Cram DJ. Host–guest complexation. 13. High chiral recognition of amino esters by dilocular hosts containing extended steric barriers. *J Am Chem Soc.* 1978; **100**: 8190–8202.

24. Lockhart JC, Robson AC, Thompson ME, Furtado SD, Kaura CK, Allan AR. Preparation of some nitrogen-containing polyether crown compounds. *J Chem Soc. Perkin Trans.* 1973; **1**: 577–581.

25. Gokel GW, Dishong DM, Schultz RA, Gatto VJ. Syntheses of aliphatic azacrown compounds. *Synthesis*. 1982; 997–1013.

26. Dietrich B, Lehn JM, Sauvage JP. Diaza-polyoxa-macrocycles et macrobicycles. *Tet Lett*. 1969; **10**: 2885–2888.

27. Elwahy AHM. New trends in the chemistry of condensed heteromacrocycles. Part A. Condensed azacrown ethers and azathiacrown ethers. *J Heterocyclic Chem*. 2003; **40**: 1–23.

28. Pedersen CJ, Frensdorff HK. Macrocyclic polyethers and their complexes. *Angew Chem Int Ed*. 1972; **11**: 16–25.

29. Richman JE, Atkins TJ. Nitrogen analogs of crown ethers. *J Am Chem Soc*. 1974; **96**: 2268–2270.

30. Atkins TJ, Richman JE, Oettle WF. 1,4,7,10,13,16-hexaazacyclooctadecane. *Org Synth Coll*. 1988; **6**: 652–662.

31. Reed DP, Weisman GR. 1,4,7,10-tetraazacyclododecane. *Org Synth Coll*. 2004; **10**: 667.

32. Suchy M, Hudson RHE. Synthetic strategies toward n-functionalized cyclens. *Eur J Org Chem*. 2008; **29**: 4847–4865.

33. Thuery P, Keller N, Lance M, Vigner J-D, Nierlich M. (H4-hexaaza-18-crown-6)$^{4+}$·4CF$_3$SO$_3{}^-$ and its hydrated form. *Acta Cryst*. 1995; **C51**: 1407–1411.

34. Bencini A, Bianchi A, Dapporto P, Garcia-España E, Micheloni M, Paoletti P, Paoli P. (PdCl$_4$)$^{2-}$ inclusion into the deca-charged polyammonium receptor (H10 [30]aneN10)$^{10+}$ ([30]aneN10 = 1,4,7,10,13,16,19,22,25,28-deca-azacyclotriacontane). *J Chem Soc Chem Commun*. 1990; 753–755.

35. Iitaka Y, Shina M, Kimura E. Crystal structure of dinitro(1,4,7,10-tetraazacyclododecane)cobalt(III) chloride. *Inorg Chem*. 1974; **13**: 2886–2891.

36. Schrodt A, Neubrand A, van Eldik R. Fixation of CO2 by zinc(II) chelates in alcoholic medium. X-ray structures of {[Zn(cyclen)]3(μ3-CO$_3$)}(ClO$_4$)$_4$ and [Zn(cyclen)EtOH](ClO$_4$)$_2$. *Inorg Chem*. 1997; **36**: 4579–4584.

37. El Ghachtouli S, Cadiou C, Déchamps-Olivier I, Chuburu F, Aplincourt M, Turcry C, Le Baccon M, Handel H. Spectroscopy and redox behaviour of dicopper(II) and dinickel(II) complexes of bis(cyclen) and bis(cyclam) ligands. *Eur J Inorg Chem*. 2005; **13**: 2658–2668.

38. Arpadjan S, Mitewa M, Bontchev PR. Liquid–liquid extraction of metal ions by the 6-membered N-containing macrocycle hexacyclen. *Talanta*. 1987; **34**: 953–956.

39. Royer DJ, Grant GJ, Van Derveer DG, Castillo MJ. Chiral cobalt(III) complexes of symmetrical hexaaza macrocycles. *Inorg Chem*. 1982; **21**: 1902–1908.

40. Chandrasekhar S, Fortier DG, McAuley A. Syntheses of chromium and copper complexes of hexaaza-macrocycles. Crystal structures of chromium(III) complexes of 1,4,7,10,13,16-hexaazacyclooctadecane and 1,4,7,11,14,17-hexaazacycloeicosane. *Inorg Chem*. 1993; 1424–1429.

41. Carrondo MAAF de CT, Félix V, Duarte MT, Santos MA. Synthesis, electrochemical behaviour and structural characterization of the mercury complex [Hg([18]aneN6)]·(HgCl$_4$). *Polyhedron*. 1993; **12**: 931–937.

42. Bu X-H, Lu S-L, Zhang R-H, Wang H-G, Yao X-K. Crystal structure and properties of neodymium complex with hexaazacyclooctadecane ligand. *Polyhedron*. 1997; **16**: 3247–3251.

43. Nierlich M, Sabattie J-M, Keller N, Lance M, Vigner J-D. Inclusion complex between uranyl and an azacrown; structure of [UO$_2$(18-azacrown-6)]$^{2+}$·2CF$_3$SO$_3{}^-$. *Acta Cryst*. 1994; **C50**: 52–54.

44. Barker JE, Liu Y, Martin ND, Ren T. Dicopper-[18]ane-N6 complex as the platform for phosphate monoester binding. *J Am Chem Soc*. 2003; **125**: 13332–13333.

45. Curtis NF, Curtis YM, Powell HKJ. Transition-metal complexes with aliphatic Schiff bases. Part VIII. Isomeric hexamethyl-1,4,8,11-tetra-azacyclotetradecadienenickel(II) complexes formed by reaction of trisdiaminoetha-nenickel(II) with acetone. *J Chem Soc A*. 1966; 1015–1018.

46. Curry JD, Busch DH. The reactions of coordinated ligands. VII. Metal ion control in the synthesis of chelate compounds containing pentadentate and sexadentate macrocyclic ligands. *J Am Chem Soc*. 1964; **86**: 592–594.

47. Radecka-Paryzeka W, Patroniaka V, Lisowski J. Metal complexes of polyaza and polyoxaaza Schiff base macrocycles. *Coord Chem Rev*. 2005; **249**: 2156–2175.

48. Busch DH. Distinctive coordination chemistry and biological significance of complexes with macrocyclic ligands. *Acc Chem Res*. 1978; **11**: 392–400.

49. Barefield EK, Wagner F, Herlinger AW, Dahl AR, Holt S. 1,4,8,11-tetraazacyclotetradecane)nickel(II) perchlorate and 1,4,8,11-tetraazacyclotetradecane. *Inorg Synth*. 1976; **16**: 220–225.

50. Cabbiness DK, Margerum DW. Macrocyclic effect on the stability of copper(II) tetramine complexes. *J Am Chem Soc.* 1969; **91**: 6540–6541.

51. Bakaj M, Zimmer M. Conformational analysis of copper(II) 1,4,8,11-tetraazacyclotetradecane macrocyclic systems. *J Mol Struct.* 1999; **508**: 59–72.

52. Tyson TA, Hodgson KO, Hedman B, Clark GR. Structure of Zn(cyclam)(ClO$_4$)$_2$. *Acta Cryst.* 1990; **C46**: 1638–1640

53. Endicott JF, Lilie J, Kuszaj JM, Ramaswamy BS, Schmonsees WG, Simic MG, Glick MD, Rillema DP. The trans-influence and axial interactions in low spin, tetragonal cobalt(II) complexes containing macrocyclic and/or cyano ligands. Pulse radiolytic studies in fluid solution, electron paramagnetic resonance spectra at 77 K, and single-crystal X-ray structures. *J Am Chem Soc.* 1977; **99**: 429–439.

54. Prasad L, Nyburg SC, McAuley A. The structure of Ni(cyclam)(ClO$_4$)$_2$. *Acta Cryst.* 1987; **C43**: 1038–1042.

55. Toriumi K, Yamashita M, Ito H, Ito T. Structures of one-dimensional PdII–PdIV mixed-valence complexes, [PdIIL][PdIVCl$_2$L]Y$_4$ (Y = ClO$_4$ and PF$_6$), and their parent PdII and PdIV complexes, [PdIIL](ClO$_4$)$_2$ and [PdIVCl$_2$L](NO$_3$)$_2$.HNO$_3$.H$_2$O, with 1,4,8,11-tetraazacyclotetradecane. *Acta Cryst.* 1986; **C42**: 963–968.

56. Ito T, Ito H, Toriumi K. The structures of both the kinetic and the thermodynamic isomers of 1,4,8,11-tetraazacyclotetradecanesilver(II) perchlorate as determined by X-ray analyses. *Chem Lett.* 1981; **10**: 1101–1104.

57. Alcock NW, Herron N, Moore P. Structural and dynamic behaviour of complexes of lead(II) with two tetra-aza macrocyclic ligands as studied by X-ray crystallography and natural-abundance carbon-13 and nitrogen-15 nuclear magnetic resonance spectroscopy. *J Chem Soc. Dalton Trans.* 1979; 1486–1491.

58. Alcock NW, Curson EH, Herron N, Moore P. Structural and dynamic behaviour of cadmium(II) and mercury(II) complexes of 1,4,8,11-tetra-azacyclotetradecane and 1,4,8,11-tetramethyl-1,4,8,11-tetra-azacyclotetradecane. *J Chem Soc. Dalton Trans.* 1979; 1987–1993.

59. Nakani BS, Welsh JJB, Hancock RD. Formation constants of some complexes of tetramethylcyclam. *Inorg Chem.* 1983; **22**: 2956–2958.

60. Liang X, Parkinson JA, Parsons S, Weishäup M, Sadler PJ. Cadmium cyclam complexes: interconversion of cis and trans configurations and fixation of CO$_2$. *Inorg Chem.* 2002; **41**: 4539–4547.

61. Pedersen CJ. Macrocyclic polyether sulphides. *J Org Chem.* 1971; **36**: 254–257.

62. Ochrymowycz LA, Mak C-P, Michna JD. Synthesis of macrocyclic polythiaethers. *J Org Chem.* 1974; **39**: 2079–2084.

63. Wolf RE Jr, Hartman JR, Storey JME, Foxman BM, Cooper SR. Crown thioether chemistry: structural and conformational studies of tetrathia-12-crown-4, pentathia-15-crown-5, and hexathia-18-crown-6. Implications for ligand design. *J Am Chem Soc.* 1987; **109**: 4328–4335.

64. Hintsa EJ, Hartman JR, Cooper SR. Crown thioether chemistry. The nickel(II) complex of 1,4,7,10,13,16-hexathiacyclooctadecane, the hexathia analogue of 18-crown-6. *J Am Chem Soc.* 1983; **105**: 3738–3739.

65. Grant GJ, Rogers KE, Setzer WN, VanDerveer DG. Crown thioether complexes of trivalent transition metal ions. The crystal structure of [Cr(18S6)Cl$_3$]. *Inorg Chim Acta.* 1995; **234**: 35–45.

66. Alcock NW, Herron N, Moore P. Comparison of the different modes of bonding of the macrocycle in μ-(1,4,8,11-tetrathiacyclotetradecane-S1S4;S8S11)-bis[dichloromercury-(II)] and aqua(1,4,8,11-tetrathiacyclotetradecane)mercury(II) perchlorate by X-ray structural analysis. *J Chem Soc. Dalton Trans.* 1978; 394–399.

67. Batchelor RJ, Einstein FWB, Gay ID, Gu J-H, Johnston BD, Pinto BM. Selenium coronands: synthesis and conformational analysis. *J Am Chem Soc.* 1989; **111**: 6582–6591.

68. Batchelor RJ, Einstein FWB, Gay ID, Gu J-H, Mehta S, Pinto BM, Zhou X-M. Synthesis, characterization, and redox behavior of new selenium coronands and of copper(I) and copper(II) complexes of selenium coronands. *Inorg Chem.* 2000; **39**: 2558–2571.

69. Batchelor RJ, Einstein FWB, Gay ID, Gu J-H, Pinto BM, Zhou X-M. Stereochemical analysis of palladium(II) complexes of the selenium coronands 1,5,9,13-tetraselenacyclohexadecane and 1,5,9,13,17,21-hexaselenacyclotetracosane. *Inorg Chem.* 1996; **35**: 3667–3674.

70. Hansheng X, Weiping L, Xiufang L, Lianfang S, Xian M, Ming L. Syntheses of telluracrown ethers and a platinum complex of one of them. *Acta Chimica Sinica.* 1994; **52**: 386–390.

71. Mealli C, Sabat M, Zanobini F, Ciampolini M, Nardi N. Macrocyclic polyphosphane ligands. Iron(II), cobalt(II), and nickel(II) complexes of (4RS,7RS,13SR,16SR)-tetraphenyl-1,10-dipropyl-1,10-diaza-4,7,13,16-tetraphosphacyclo-octadecane: crystal structures of their tetraphenylborate derivatives. *J Chem Soc. Dalton Trans.* 1985; 479–485.

72. Gokel GW. Lariat ethers: from simple sidearms to supramolecular systems. *Chem Soc Rev.* 1992; **21**: 39–47.

73. Steed JW, Atwood JL. Supramolecular Chemistry. Chichester: J.W. Wiley and Sons; 2000.

74. Tsukube H. Double armed crown ethers and armed macrocycles as a new series of metal-selective reagents: A review. *Talanta.* 1993; **40**: 1313–1324.

75. Fronczek FR, Gatto VJ, Schultz RA, Jungk SJ, Colucci WJ, Gandour RD, Gokel GW. Unequivocal evidence for sidearm participation in crystalline lariat ether complexes. *J Am Chem Soc.* 1983; **105**: 6717–6718.

76. Gandour RD, Fronczek FR, Gatto VJ, Minganti C, Schultz RA, White BD, Arnold KA, Mazzocchi D, Miller R, Gokel GW. Solid-state structural chemistry of lariat ether and BiBLE cation complexes: metal-ion identity and coordination number determine cavity size. *J Am Chem Soc.* 1986; **108**: 4078–4088.

77. Tsukube H, Yamashita K, Iwachido T, Zenki M. Pyridine-armed diaza-crown ethers: molecular design of effective synthetic ionophores. *J Org Chem.* 1991; **56**: 268–272.

78. Tsukube H, Takagi K, Higashiyama T, Iwachido T, Hayama N. Cation-binding properties of new armed macro-cyclic host molecules and their applications to phase-transfer reactions and cation membrane transport. *Chem Soc. Perkin Trans.* 1986; **1**: 1033–1037.

79. Dietrich B, Lehn JM, Sauvage JP. Les cryptates. *Tet Lett.* 1969; **10**: 2889–2892.

80. Lehn JM, Sauvage JP. Cryptates. XVI.[2]-Cryptates. Stability and selectivity of alkali and alkaline-earth macrobicyclic complexes. *J Am Chem Soc.* 1975; **97**: 6700–6707.

81. Dietrich B, Lehn JM, Sauvage JP. Cryptates: control over bivalent/monovalent cation selectivity. *J Chem Soc Chem Commun.* 1973; 15–16.

82. Metz B, Moras D, Weiss R. The crystal structure of a rubidium cryptate $[RbC_{18}H_{36}N_2O_6]SCN \cdot H_2O$. *J Chem Soc Chem Commun.* 1970; 217–218.

83. Metz B, Moras D, Weiss R. Crystal structures of three alkali-metal complexes with a macrobicyclic ligand. *J Chem Soc Chem Commun.* 1970; 444–445.

84. Mathieu F, Metz B, Moras D, Weiss R. Cavities in macrobicyclic ligands and complexation selectivity. Crystal structures of two cryptates, |Na$^+$·cntnd·221|·cntdot·SCN$^-$ and |K$^+$·cntnd. 221|·cntdot·SCN$^-$2. *J Am Chem Soc.* 1978; **100**: 4412–4416.

85. Lehn JM, Sonveaux E, Willard AK. Molecular recognition. Anion cryptates of a macrobicyclic receptor molecule for linear triatomic species. *J Am Chem Soc.* 1978; **100**: 4914–4916.

86. Dietrich B, Lehn JM, Sauvage JP. Oxathia-macrobicyclic diarnines and their cryptates. *J Chem Soc Chem Commun.* 1970; 1055–1056.

87. Ragunathan KG, Bharadwaj PK. Template synthesis of a macrobicyclic cryptand having mixed donors via [2 + 3] Schiff base condensation. *Proc Ind Acad Sci (Chem Sci).* 1993; **105**: 215–217.

88. Graf E, Lehn JM. Cryptates. XVII. Synthesis and cryptate complexes of a spheroidal macrotricyclic ligand with octahedrotetrahedral coordination. *J Am Chem Soc.* 1975; **97**: 5022–5024.

89. Graf E, Lehn JM. Anion cryptates: highly stable and selective macrotricyclic anion inclusion complexes. *J Am Chem Soc.* 1976; **98**: 6403–6405.

90. Metz B, Rosalky JM, Weiss R. [3] Cryptates: X-ray crystal structures of the chloride and ammonium ion complexes of a spheroidal macrotricyclic ligand. *J Chem Soc Chem Commun.* 1976; 533–534.

91. Cram DJ, Kaneda T, Helgeson RC, Lein GM. Spherands: ligands whose binding of cations relieves enforced electron–electron repulsions. *J Am Chem Soc.* 1979; **101**: 6752–6754.

92. Cram DJ, deGrandpre M, Knobler CB, Trueblood KN. Host–guest complexation. 29. Expanded hemispherands. *J Am Chem Soc.* 1984; **106**: 3286–3292.

93. Trueblood KN, Knobler CB, Maverick E, Helgeson RC, Brown SB, Cram DJ. Spherands, the first ligand systems fully organized during synthesis rather than during complexation. *J Am Chem Soc.* 1981; **103**: 5594–5596.

94. Cram DJ, Carmack A, de Grandpre MP, Lein GM, Goldberg I, Knobler CB, Maverick EF, Trueblood KN. Host–guest complexation. 44. Cavitands and caviplexes composed of eight anisyl groups. *J Am Chem Soc.* 1987; **109**: 7068–7073.

95. Paek K, Knobler CB, Maverick EF, Cram DJ. Host–guest complexation. 51. Cyanospherands, a new type of salt binder. *J Am Chem Soc.* 1989; **111**: 8662–8671.

96. Helgeson RC, Mazaleyrat J-P, Cram DJ. Synthesis and complexing properties of chiral macrocycles containing enforced cavities. *J Am Chem Soc.* 1981; **103**: 3929–3931.

97. Lein GM, Cram DJ. Host–guest complexation. 34. Bridged hemispherands. *J Am Chem Soc.* 1985; **107**: 448–455.

98. Cram DJ, Ho SP, Knobler CB, Maverick E, Trueblood KN. Host–guest complexation. 38. Cryptahemispherands and their complexes. *J Am Chem Soc.* 1986; **108**: 2989–2998.

99. Cram DJ, Ho SP. Host–guest complexation. 39. Cryptahemispherands are highly selective and strongly binding hosts for alkali metal ions. *J Am Chem Soc.* 1986; **108**: 2998–3005.

100. Kimura K, Maeda T, Shono TI. Extraction of alkali metal picrates with poly- and bis(crown ether)s. *Talanta.* 1979; **26**: 945–949.

101. Shinkai S, Ogawa T, Nakaji T, Manabe O. Light-driven ion-transport mediated by a photo-responsive bis(crown ether). *J Chem Soc Chem Commun.* 1980; 375–377.

102. Shinkai S, Nakaji T, Ogawa T, Shigematsu K, Manabe O. Photoresponsive crown ethers. 2. Photocontrol of ion extraction and ion transport by a bis(crown ether) with a butterfly-like motion. *J Am Chem Soc.* 1981; **103**: 111–115.

103. Yamamoto K, Yumioka H, Okamoto Y, Chikamatsu H. Synthesis and chiral recognition of an optically active bis-crown ether incorporating a diphenanthrylnaphthalene moiety as the chiral centre. *J Chem Soc Chem Commun.* 1987; 168–169.

104. Beer PD, Sikanyika H, Slawin AMZ, Williams DJ. The synthesis, coordination and electrochemical studies of metallocene bis(crown ether) receptor molecules. Single-crystal C-ray structure of a ferrocene bis(crown ether) potassium complex. *Polyhedron.* 1989; **8**: 879–886.

105. Grossel MC, Goldspink MR, Knychala JP, Cheetham AK, Hriljac JA. Metallocene-bridged cryptands. I. X-ray structural study of 1,1″:1′, -bis(1,4,10,13-tetraoxa-7,16-diazacyclooctadecane-7,16-diyldicarbonyl)bisferrocene. *J Organomet Chem.* 1988; **352**: C13–C16.

106. Hall CD, Danks IP, Beer PD, Chu SYF, Nyburg SC. The reaction of 1,1′-bis-(chlorocarbonyl) ferrocene with tetrathia-diaza-18-crown-6. *J Organomet Chem.* 1994; **468**: 193–198.

107. Kotzyba-Hibert F, Lehn JM, Vierling P. Multisite molecular receptors and co-systems ammonium cryptates of macrotricyclic structures. *Tet Lett.* 1980; **21**: 941–944.

108. Kotzyba-Hibert, Lehn JM, Saigo K. Synthesis and ammonium cryptates of triply bridged cylindrical macrotetracycles. *J Am Chem Soc.* 1981; **103**: 4266–4268.

109. Pascard C, Riche C, Cesario M, Kotzyba-Hibert F, Lehn JM. Coreceptor–substrate binding. Crystal structures of a macrotricyclic ligand and of its molecular cryptate with the cadaverine dication. *J Chem Soc Chem Commun.* 1982; 557–560.

110. Lehn JM, Simon J, Moradpour A. Synthesis and properties of chiral macrotricyclic ligands. Complexation and transport of chiral molecular cations and anions. *Helv Chim Acta.* 1978; **61**: 2407–2418.

111. Zaitsev YS, Zarudnaya EN, Möbius D, Bondarenko CV, Maksimov VI, Zaitsev IS, Ushakov EN, Lobova NA, Vedernikov AI, Gromov SP, Alfimov MV. Ultrathin chemosensing films with a photosensitive bis(crown ether) derivative. *Mendeleev Commun.* 2008; **18**: 270–272.

112. Huang ZB, Chang SH. Synthesis and complexation behavior studies of novel bis-crown ethers. *J Inclus Phenom Macrocyc Chem.* 2006; **55**: 341–346.

113. Pigge C, Dighe MK, Houtman JCD. Mono-, bis-, and tris(crown ether)s assembled around 1,3,5-triaroylbenzene scaffolds. *J Org Chem.* 2008; **73**: 2760–2767.

114. Huang ZB, Kang TJ, Chang SH. Synthesis, characterization and complexation behavior investigations of novel starburst-like tris-crown ethers. *New J Chem.* 2005; **29**: 1616–1620.

115. Thanabal V, Krishnan V. Porphyrins with multiple crown ether voids: novel systems for cation complexation studies. *J Am Chem Soc.* 1982; **104**: 3643–3650.

116. Maiya GB, Krishnan V. Electrochemical redox properties and spectral features of supermolecular porphyrins. *Inorg Chem.* 1985; **24**: 3253–3257.

117. Richardson NM, Sutherland IO, Camilleri P, Page JA. Cation binding by a crown-capped porphyrin. *Tet Lett.* 1985; **26**: 3739–3742.
118. Caia X, Sheng N, Zhang Y, Qi D, Jiang J. Structure and spectroscopic properties of phthalocyaninato zinc(II) complexes fused with different number of 15-crown-5 moieties. *Spectrochimica Acta A.* 2009; **72**: 627–635.
119. Michel SLJ, Barrett AGM, Hoffman BM. Peripheral metal-ion binding to tris(thia-oxo crown) porphyrazines. *Inorg Chem.* 2003; **42**: 814–820.
120. Wong L-H, Smid J. Binding of organic solutes to poly(crown ethers) in water. *J Am Chem Soc.* 1977; **99**: 5637–5642.
121. Kimura K, Maeda T, Tamura H, Shono T. Potassium-selective PVC membrane electrodes based on bis- and poly(crown ether)s. *J Electroanal Chem.* 1979; **95**: 91–101.
122. Nakajima M, Kimura K, Shono T. Ion-chromatographic behavior of silica gels modified by poly- and bis(crown ether)s of benzo-18-crown-6. *Bull Chem Soc Jpn.* 1983; **56**: 3052–3056.
123. Peramunage D, Fernandez JE, Garcia-Rubio LH. Poly(crown ether): a potential candidate for solid-state electrolytes. *Macromolecules.* 1989; **22**: 2845–2849.
124. Shirai M, Moriuma H, Tanaka M. Photoinduced release of alkali picrates using photoreactive poly(crown ether)s. *Macromolecules.* 1989; **22**: 3184–3186.
125. Ohno H, Yamazakia H, Tsukube H. Selective conduction of alkali metal ions in poly(diaza-crown ether). *Polymer.* 1993; **34**: 1533–1534.
126. Ohno H, Inoue Y, Tsukube H. Conduction of alkali metal cations in poly(diaza-crown ether)s having hydroxyethyl side arms. *Polymer.* 1994; **35**: 5753–5757.
127. Ion AC, Mouteta J-C, Pailleret A, Popescu1 A, Saint-Aman E, Siebert E, Ungureanu EM. Electrochemical recognition of metal cations by poly(crown ether ferrocene) films investigated by cyclic voltammetry and electrochemical impedance spectroscopy. *J Electroanal Chem.* 1999; **464**: 24–30.
128. Ma Y, Galal A, Lunsford SK, Zimmer H, Mark HB Jr, Huang Z, Bishop PB. Poly(binaphthyl-20-crown-6) as receptor based molecular selective potentiometric electrodes for catecholamines and other 1,2-dihydroxybenzene derivatives. *Bios Bioelec.* 1995; **8**: 705–715.
129. Chandra S, Buschbeck R, Lang H. A 15-crown-5-functionalized carbosilane dendrimer as ionophore for ammonium selective electrodes. *Talanta.* 2006; **70**: 1087–1093.
130. Montanari F, Tundo P. Hydroxymethyl derivatives of 18-crown-6 and [2.2.2]cryptand: versatile intermediates for the synthesis of lipophilic and polymer-bonded macrocyclic ligands. *J Org Chem.* 1982; **47**: 1298–1302.
131. Sugiyama H, Enokida Y, Yamamoto I. Nitrogen isotope separation with displacement chromatography using cryptand polymer. *J Nuc Sci Tech.* 2002; **39**: 442–446.
132. Liang X, Sadler PJ. Cyclam complexes and their applications in medicine. *Chem Soc Rev.* 2004; **33**: 246–266.

4

Calixarenes

4.1 Introduction

The calixarenes and the related resorcinarenes represent an actively researched group of compounds. Calixarenes are cyclic oligomers formed by the reaction of substituted phenols with formaldehyde and have the general structure shown in Figure 4.1a. They come in a variety of ring sizes but the compounds containing four, six or eight phenol rings within the macrocycle are the easiest to synthesise and the most widely studied. Resorcinarenes have the general formula shown in Figure 4.1b and are synthesised by the condensation of resorcinol with a variety of aldehydes, usually to form the cyclic tetramer, although other rings sizes are known. Both groups of compounds are examples of cyclophanes and a calix-4-arene for instance can in this context be thought of as a [1,1,1,1] metacyclophane. Much of the work that will be summarised within these chapters has been extensively reviewed, the majority of it within the series of monographs by David Gutsche (one of the pioneers within this field), along with several other titles which are all given in the Bibliography.

As can be seen from the three-dimensional molecular models (Figure 4.1a,b), both calixarenes and resorcinarenes generally adopt a bowl- or vase-shaped conformation. It is this structural shape that has led to the name 'calixarene' and it acts as a major contributor to the properties and applications of these compounds. The bowl shapes and the presence of the hydroxyl groups allow this family of compounds to interact with a wide variety of guests by a combination of hydrogen bonding and aromatic ring-based interactions. Calixarenes are usually thermally and chemically robust, can be made in large quantities from inexpensive precursors and can be easily derivatised at the hydroxy groups or on the aromatic rings to give a wide range of materials with highly specific purposes and potential commercial applications. Within this chapter we will attempt to present an overview of the history, synthesis, structure, chemistry and applications of these compounds.

4.2 History

The history of the calixarenes has been detailed elsewhere, especially within the series of monographs by Gutsche (see the Bibliography), so only a brief review will be given here. One of the major hurdles of the

Macrocycles: Construction, Chemistry and Nanotechnology Applications, First Edition.   Frank Davis and Séamus Higson.
© 2011 John Wiley & Sons, Ltd. Published 2011 by John Wiley & Sons, Ltd.

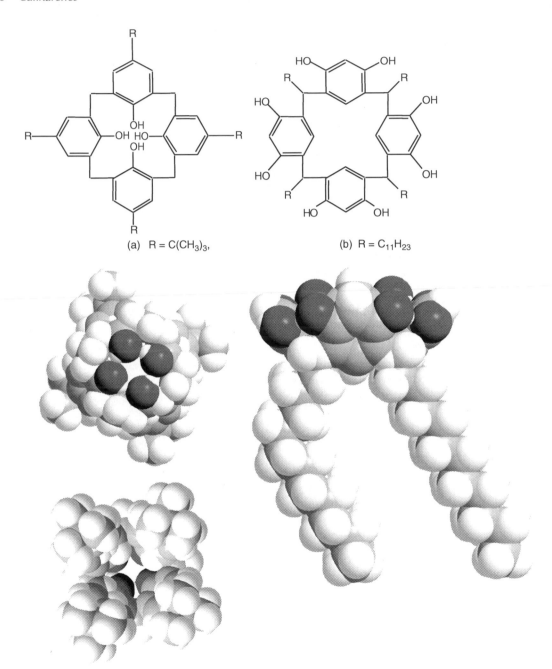

(a)   R = C(CH₃)₃,

(b) R = C₁₁H₂₃

Figure 4.1 *Structures and three-dimensional models of (a) t-butylcalix-4-arene and (b) tetraundecylcalix-4-resorcinarene*

early investigations of calixarenes was the difficulty in obtaining pure, crystalline materials. As far back as 1872 Baeyer reported the reaction between benzaldehyde and pyrogallol[1] to give a resinous product. He reported later the same year[2] that formaldehyde also gave resinous material when reacted with phenols. The field of phenol-formaldehyde chemistry was greatly advanced by the work of Leo Baekeland; at the age of 37 Baekeland sold the rights to a photographic process, Velox, to George Eastman for one million dollars (this in 1900). Baekeland obviously had a love for science because he used his wealth to set up a home laboratory and hire assistants. One of the projects that he studied was the reaction of phenol with formaldehyde, which could be catalysed by a small amount of alkali. This eventually led to the production of a resinous material which was patented and named 'Bakelite'. Bakelite is a crosslinked copolymer of phenol and formaldehyde with a complicated structure, part of which is shown in Figure 4.2. The formation of a complex network polymer is due to the fact that phenol has three reactive positions, two *ortho* to the hydroxy group and one *para*.

In an attempt to simplify the reaction, Alois Zinke used para(t-butyl) phenol instead of phenol, reasoning that the blocking of one of the positions would lead to the formation of a simple linear condensation polymer which might prove more tractable than the crosslinked resin. Instead he obtained a series of crystalline products for a series of phenols, detailed within his paper; based on molecular weights for these materials he proposed a cyclic tetrameric structure,[3] although it is likely that in the case of his early work with para(t-butyl) phenol the cyclic octamer was the actual product synthesised. Within Baeyer's original work he described the formation of high-melting and crystalline products from the reaction of resorcinol with a variety of aldehydes.[1] Others advanced this work and in 1943 Niederl and Vogel[4] published a series of syntheses of aldehyde resorcinol products (resorcinarenes) and proposed a cyclic tetrameric structure.

The calixarenes still proved difficult to characterise and their syntheses proved difficult to reproduce. This became a problem when the substituted (t-butyl)phenol-formaldehyde resins were utilised as deemulsifiers

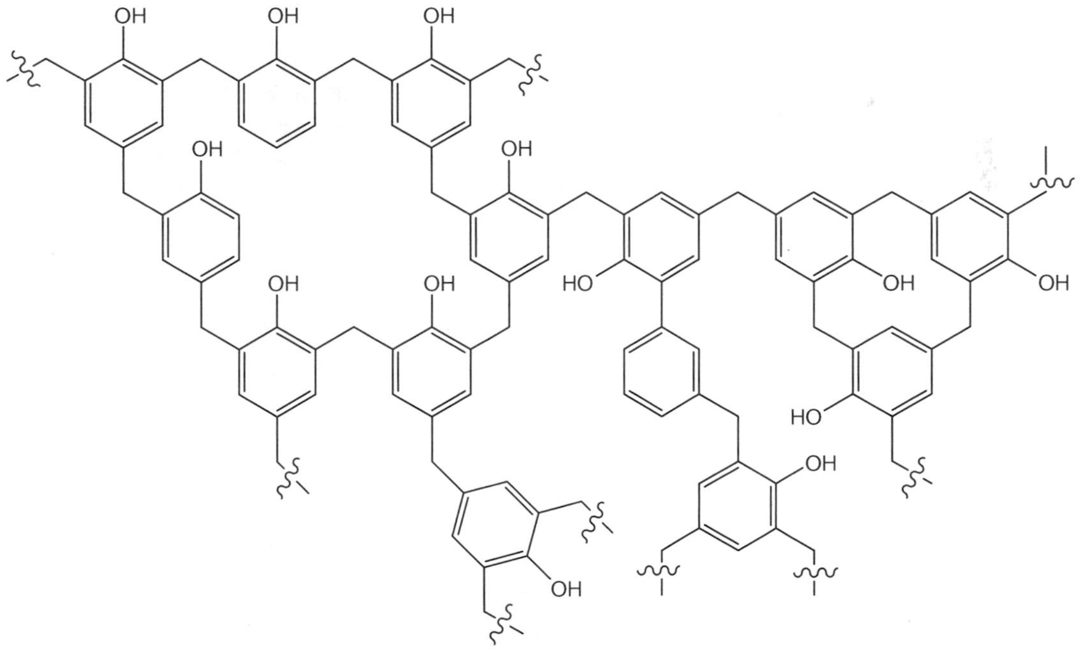

Figure 4.2 *Proposed partial structure of Bakelite*

for oil, the so-called Petrolite products. This problem was addressed by the group of Gutsche, who varied the solvents, reaction conditions and so on of the Petrolite procedure and finally formulated a series of protocols for the synthesis of the cyclic tetramer,[5] hexamer[6] and octamer.[7] They also coined the name 'calixarene' for this family of compounds, because of the resemblance between the shape of a calixarene and a vase, the Greek '*calix*' meaning 'vase'. Other oligomers such as the cyclic pentamer[8] and heptamer[9] have also been successfully synthesised. Finally, in an acid-catalysed reaction of trioxane with t-butylphenol, a series of cyclic oligomers with up to 20 phenol units[10] were isolated. Besides these one-pot reactions, there have also been a wide variety of synthetic schemes, some detailed in the monographs by Gutsche, where the calixarenes are assembled 'piece by piece' in a series of condensation reactions. Although this process is much more complex than the simple one-pot synthesis of calixarenes, it does enable the formation of more complex asymmetric calixarenes containing different rings and bridging groups.

4.3   Structures of Calixarenes

A variety of potential conformations are known; these are shown in Figure 4.3 for a calix-4-arene. They are generally labeled as 'cone' (where all the rings are in what Gutsche terms the 'up' conformation, with the OH below the plane of the macrocycle and the aromatic rings above), 'partial cone' (uuud), '1,3-alternate' (udud) and '1,2-alternate' (uudd), respectively. Although the basic shape of the majority of parent calixarenes is a cone, stabilised with so-called 'cyclic' hydrogen bonding between the —OH groups, substituted calixarenes have been synthesised in all the possible conformers.

The larger calixarenes are even more complex, with eight possible up–down conformers for calix-6-arenes and sixteen for calix-8-arenes. X-ray studies of calix-4-arenes with t-butyl,[11] t-octyl[12] or phenyl[13] *para* substituents have all shown the macrocycles to exist in the cone conformation in the solid state. Calix-6-arenes have a more complex behaviour, as shown by X-ray studies, in which crystallisation of t-butylcalix-6-arene from benzene gives a pinched cone conformation,[14] whereas crystallisation from a strongly H-bond-disrupting solvent like DMSO, dioxane or acetone, may isolate a 1,2,3-alternate conformation with three —OH groups above and three below the ring of the macrocycle. The earliest X-ray structure of t-butylcalix-8-arene crystallised from pyridine displayed a structure known as a pleated loop;[15] other workers[16] have also isolated a calix-8-arene–pyridine complex and showed that the macrocycle existed in an open chair-like conformation with virtually no central cavity. Other larger calixarenes have also been crystallised, such as for example the calix-10-arenes (Figure 4.4), which can exist in either pleated-loop or pinched-cone conformations (where the calix is essentially attempting to form several cavities within the macrocycle) depending on the solvent of crystallisation,[17] whereas the t-butylcalix-12-arene–pyridine complex[18] adopts a type of pleated-loop conformation. The largest calixarene structure successfully determined so far is the t-butylcalix-16-arene,[19] which combines pleated-loop and cone-like sections.

X-ray crystallography gives the structures of these molecules in the solid state. The parent hydroxy calixarenes are all conformationally mobile in solution, as proved by NMR studies. The aromatic groups of calix-4-arenes can rotate slowly at room temperature about the plane of the macrocycle, allowing transitions between cone, partial-cone and the other conformers, with the cone conformer dominating (as well as rotating faster at higher temperatures). Larger calixarenes tend to have lower energies of interconversion and exist very much in a fluxional state. This interconversion only occurs when the aromatic rings can freely rotate, but if substituents that are attached to the —OH groups are too bulky to pass through the centre of the macrocycle, it becomes possible to synthesise and isolate calixarenes that are frozen into one or more possible conformations.

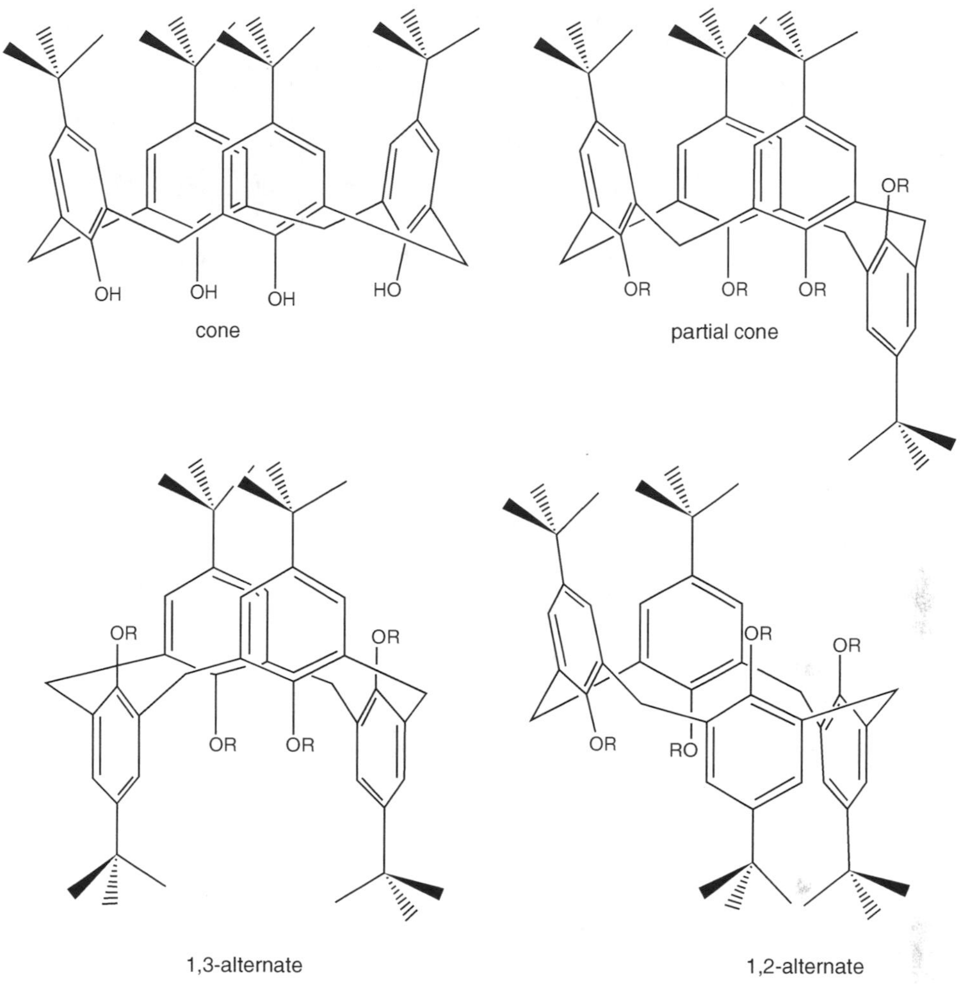

cone

partial cone

1,3-alternate

1,2-alternate

Figure 4.3   *Conformers of t-butylcalix-4-arene and substituted variants*

The tetra-acetates of calix-4-arenes and higher esters are all conformationally fixed, but calix-4-arenes with four methoxy or even four ethoxy substituents (Figure 4.5a,b) can interconvert, as shown by NMR measurements,[20] although inspection of molecular models implies this should not occur, indicating that sometimes molecules are more flexible than their models suggest. Tetramethoxy t-butylcalix-4-arene exists mainly as a partial cone, as does tetraethoxy t-butylcalix-4-arene, although at high temperatures the 1,2-alternate conformer is the major species.[20] When longer substituents such as propyl and butyl (Figure 4.5,c,d) are synthesised most of the product is a mixture of cone and partial-cone conformers, which can be isolated because the alkyl groups are now too large to allow interconversion.

Calix-5-arenes display more flexibility, with up to the butyl ether being flexible; for t-butylcalix-6-arene even the presence of large —O substituents such as p-phenylbenzyl[22] does not prevent interconversion, since the macrocycle is now large enough for the t-butyl group to pass through the central cavity, allowing the rings to rotate. For the larger calixarenes, passage of the t-butyl group through the ring is even easier.

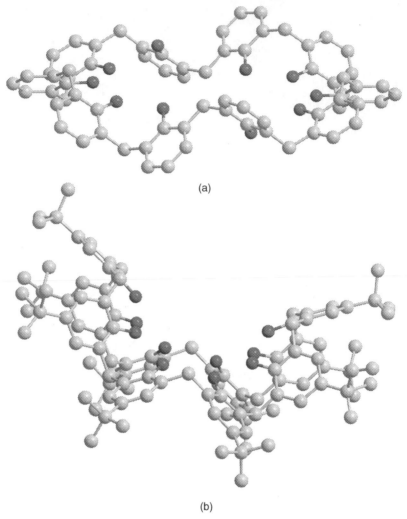

(a)

(b)

Figure 4.4   *Crystal structures of (a) the acetone and (b) the toluene complexes of t-butylcalix-10-arene (guest molecules not shown), showing pleated-loop and pinched-cone conformations of the macrocycle*

4.4   Chemical Modification of Calixarenes

There are a wide variety of methods by which calixarenes can be modified. If we begin with a simple t-butylcalixarene then several sites can be 'attacked', such as the —OH groups, the t-butyl groups, the methylene-bridging groups and the aromatic rings themselves.

4.4.1   Modification of the hydroxyls

The hydroxy groups are active species and can undergo reaction with a number of reagents. Early work, for example by the Petrolite process, involved reacting the calixarenes with ethylene oxide under basic

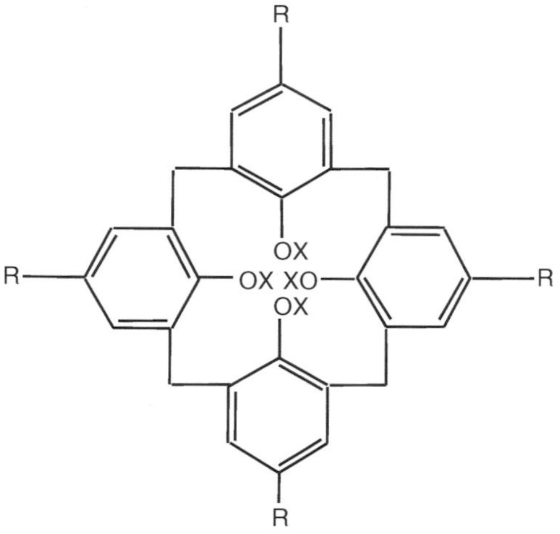

(a) R = C(CH$_3$)$_3$, X = CH$_3$
(b) R = C(CH$_3$)$_3$, X = CH$_2$CH$_3$
(c) R = C(CH$_3$)$_3$, X = CH$_2$CH$_2$CH$_3$
(d) R = C(CH$_3$)$_3$, X = CH$_2$CH$_2$CH$_2$CH$_3$

Figure 4.5 *Structures of t-butylcalix-4-arene ethers*

conditions to form calixarenes substituted with polyethylene oxide chains. The use of chemical substituents sometimes made the calixarenes easier to crystallise or to separate using chromatographic methods and indeed aided the characterisation of their structures, including for example the isolation of calixarenes via conversion to the trimethylsilyl derivative, followed by purification via chromatography and hydrolysis back to the parent calixarene.[23]

The calixarenes have been substituted with an immense number of different ester and ether groups, so only those of most interest will be given. When the substituent groups are large enough to prevent interconversion, different conformers can be isolated. The ratios of the conformers are dependent on a wide range of parameters, such as the temperature of the reaction, the nature of any base used, the size and steric constraints of the alkyl group, and the nature of the leaving groups. For example, the reaction of p-t-butylcalix-4-arene with 1-bromopropane gives a mixture of 42% cone, 55% partial cone and 3% 1,3-alternate conformers when sodium hydride is used as the base; when caesium carbonate is used as the base, the major conformer is the 1,3-alternate (57%), followed by the partial cone (34%) and the 1,2-alternate (9%), with the cone conformer being entirely absent.[20] Several other examples of this are discussed within Gutsche's 2008 monograph (see Bibliography). Studies have also been carried out on larger calixarenes; for example, not only has calix-8-arene been converted to its octamethylether but methods have been developed for the synthesis of the mono- and various dimethyl, trimethyl, tetramethyl and heptamethyl ethers.[24,25] Variation of conditions and alkylating agents had large effects and in this context the 1,2,3,4 and 1,3,5,7 substitution patterns could, for example, be obtained.[24,25]

So far in this section we have discussed the total substitution of calix-4-arenes, but it is also possible to perform these substitutions in a stepwise manner: t-butylcalix-4-arene for example can be mono-, 1,3-di-, tri- and tetraalkylated by alkyl halides[20] catalysed with NaH. When calix-4-arenes are reacted with excess acetyl, propionyl, butyryl or isobutyrl chloride with NaH as the base,[26] the fully esterified products (Figure 4.6a–d) are obtained in a variety of conformers.

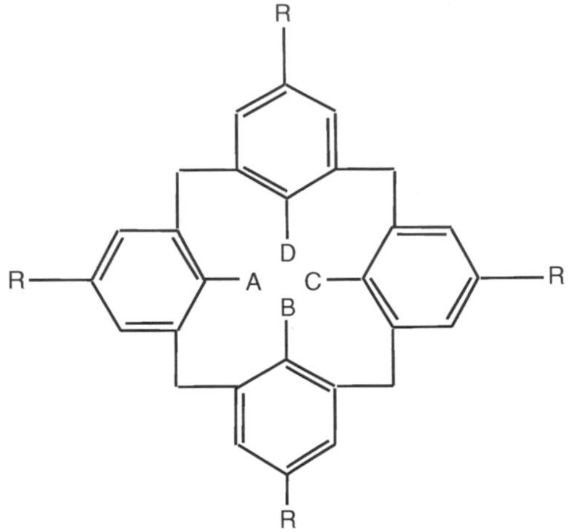

(a) R = C(CH$_3$)$_3$, A=B=C=D= OCOCH$_3$

(b) R = C(CH$_3$)$_3$, A=B=C=D= OCOCH$_2$CH$_3$

(c) R = C(CH$_3$)$_3$, A=B=C=D= OCOCH$_2$CH$_2$CH$_3$

(d) R = C(CH$_3$)$_3$, A=B=C=D= OCH$_2$CH(CH$_2$)$_2$

(e) R = C(CH$_3$)$_3$, A=H, B=C=D= OCH$_2$Ph

(f) R = C(CH$_3$)$_3$, A=C=H, B=D= OCH$_2$Ph

(g) R = C(CH$_3$)$_3$, A=B=C=D= OPPh$_2$

Figure 4.6   *Structures of t-butylcalix-4-arene esters*

When calix-4-arene is reacted with benzoyl chloride[27] in pyridine, the tribenzoate (Figure 4.6e) can be formed, and when refluxed in toluene/NaH the result is the A,C-diester[28] (Figure 4.6f) in an *anti*-conformation, with one benzyl ring on each side of the macrocycle. An alternative preparation using ice-cold THF/NaH gives the *syn*-conformer. A more complex situation occurs with dinitrobenzoyl chloride, where by careful selection of reaction conditions it is possible to synthesise the monoester, three different diester conformers or two triester conformers.[29] Phosphorus-containing moieties[30] can also be attached to these —OH groups, for example by reaction with Ph$_2$PCl (Figure 4.6g). A wide variety of other ester and ether functionalities have been attached to calixarene frameworks and the reader is advised to consult the Bibliography for further information.

Substitution of the —OH groups was utilised to build a new species of macrocycle combining calixarenes with crown ethers. These so-called 'calixcrowns' are made by reacting calixarenes with oligoethers substituted with leaving groups. Examples include systems such as those shown in Figure 4.7a,b, where the calix-4-arenes in the 1,3-conformer is bridged by reacting with oligoethyleneglycol ditosylate.[31,32] This appears to combine the ion-binding properties of crown ethers with those that will be discussed for calixarenes later in this chapter, with the crown(6) derivatives showing strong binding of caesium[31] and the crown(5) derivatives showing strong binding of potassium.[32] Multiple bridges are also possible, such as the doubly bridged calix-4-arenes (Figure 4.7c,d), where two crown ether chains bridge opposite sides of the ring.[33]

Larger calixarenes have also been studied: an example of even higher levels of bridging is the triply bridged calix-6-arene[34] shown in Figure 4.8a. Similar approaches have also been used to bridge

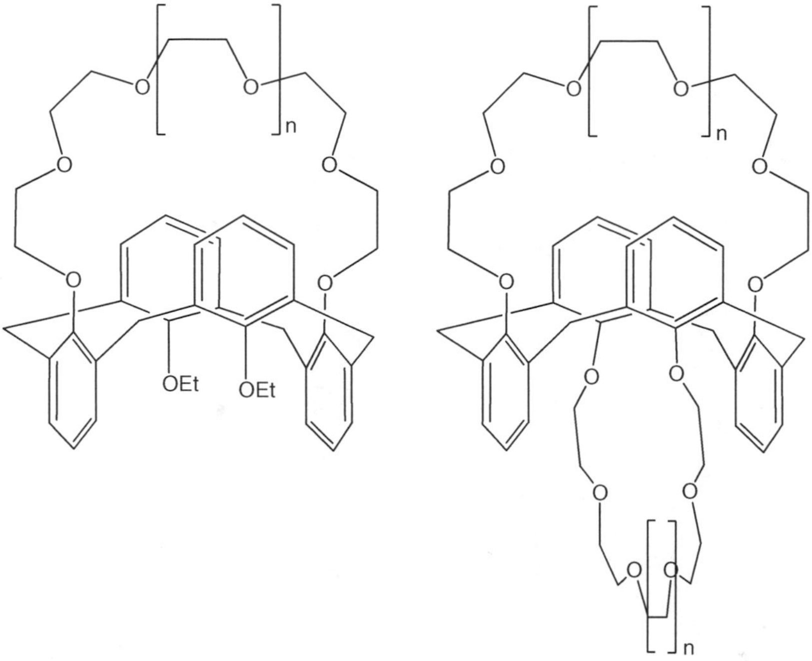

(a) n = 0
(b) n = 1

(c) n = 0
(d) n = 1

Figure 4.7 *Structures of (a,b) calixcrowns and (c,d) calix-bis-crowns*

calix-8-arenes. The doubly[35] bridged calix-8arene (Figure 4.8b) has been synthesised and shown to be chiral due to conformational immobility and can be resolved by chiral HPLC. Other examples of the numerous-bridged calix-8-arenes are doubly bridged by durylene units (Figure 4.8c) or alternatively contain a single durylene bridge (Figure 4.8d) attached via four of the —OH groups.[36] The calix-8-arenes are forced into an immobile pseudo pleated-loop conformation by the bridging.

Besides the calixcrowns, many other species have been utilised to bridge across calixarene skeletons. Many of course are simple bridges such as alkyl or xylenyl, for example those discussed in Chapter 2. Just a few examples will be given as an introduction to the types of more exotic bridging units, such as those obtained by reacting 1,10-dibromomethylphenanthroline with t-butylcalix-6-arene in THF with potassium carbonate catalysis, giving the product[37] shown in Figure 4.9a. This material adopts an unusual conformation in the solid state, where one of the rings *ortho* to one of the substituted aromatic rings adopts an inverted orientation respective to the other aromatic units. Diglycosyl units could also be incorporated to give a 'calixsugar',[38] as shown in Figure 4.9b. Dialkyldichlorosilanes were reacted with t-butylcalix-5-arene to give a mixture of mono- and disubstituted bridged structures and allow isolation of all four conformers.[39] In an elegant synthesis, t-butylcalix-9-arene was reacted with phosphorous pentachloride then underwent hydrolysis to give a product (Figure 4.9c) in which all nine of the hydroxy groups were substituted onto three phosphorus-bridging atoms. The substitution pattern with a central phosphorus atom and two 'outer' bridges, rather than the structure shown in Figure 4.9d, is held to be evidence of a pleated-loop conformation in solution.[40]

Since the combination of calixarene and crown ethers worked so well, studies have also been carried out on bridging calixarenes with hemispherands.[41] Metallocenes are suitable for use as bridging units with

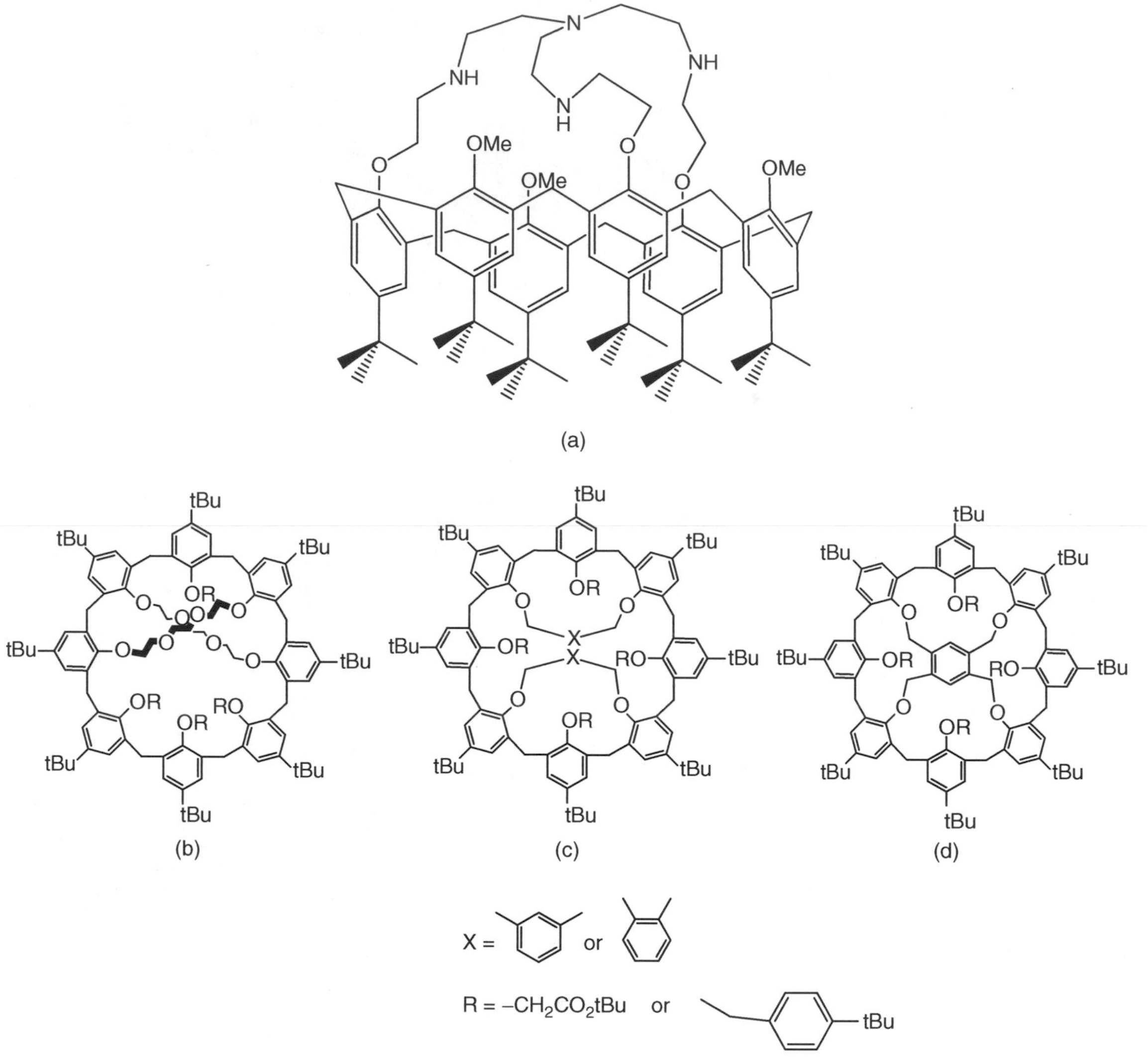

(a)

(b)   (c)   (d)

X = [structure] or [structure]

R = –CH$_2$CO$_2$tBu   or   [structure] –tBu

Figure 4.8 *Structures of (a) a bridged calix-6-arene, (b,c) doubly bridged calix-8-arenes and (d) four-point bridging*

the reaction of ferrocene dicarbonyl chloride or the corresponding ruthenium compound with t-butylcalix-4-arene, leading to formation of the 1,3-bridged ester.[42] Inorganic groups have also been used as bridging agents, such as those found when calix-4-arenes are reacted with WOCl$_4$ to form a complex in which the tungsten atom is coordinated to all four oxygens.[43] Calixarenes have actually been used to bridge other calixarenes to give structures such as calixtubes, which will be discussed later in this chapter.

4.4.2   Replacement of the hydroxyls

As an alternative to substituting the hydroxyl groups, it has proved possible to change them for another group entirely. Reaction of t-butyl-calix-4-arene with ClP(O)(OEt)$_2$ can give either the 1,3-diester or the

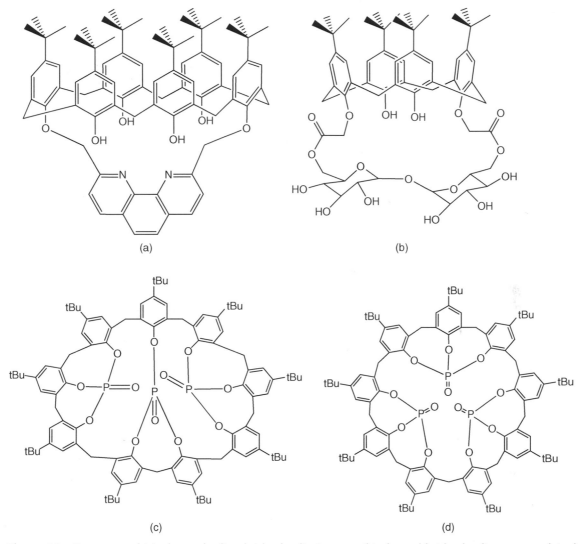

Figure 4.9 *Structures of (a) phenanthroline-bridged calix-6-arene, (b) glycosyl-bridged calixarenes and (c,d) phosphorus-bridged calix-9-arenes*

fully substituted phosphate ester,[43] which can then be reduced by potassium metal in liquid ammonia to give either the dihydroxy[45] (Figure 4.10a) or the fully dehydroxylated[44] (Figure 4.10b) calixarenes. The fully dehydroxylated species is a flexible macrocycle but in solution exists mainly in the 1,3 conformation. Similar chemistry has been used to synthesise the fully dehydroxylated[46] t-butylcalix-6-arene and t-butylcalix-8-arene.[44] The parent [1,1,1,1] metacyclophane (i.e. the completely dehydrogenated calix-4-arene; see Figure 4.10c) has been isolated and shown to be extremely flexible in solution, and adopts an unusual chair conformation in the solid state.[47]

Using a similar method, t-butylcalix-4-arene diphosphate ester was reacted with potassium amide in liquid ammonia to give a mixture of monoamine (Figure 4.10d) and diamine (Figure 4.10e), both of which were shown by NMR to be more flexible than the parent calixarene.[48] Reaction of t-butylcalix-4-arene

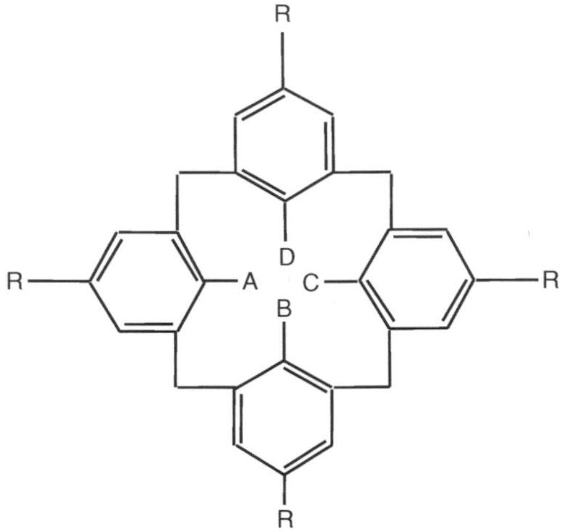

(a) R = C(CH$_3$)$_3$, A=C=H, B=D= OH

(b) R = C(CH$_3$)$_3$, A=B=C=D= H

(c) R = H, A=B=C=D= H

(d) R = C(CH$_3$)$_3$, A=C= OH, B= H, D= NH$_2$

(e) R = C(CH$_3$)$_3$, A=C= OH, B=D= NH$_2$

(f) R = C(CH$_3$)$_3$, A=B=C= OH, D= SH

(g) R = C(CH$_3$)$_3$, A=C= OH, B=D= SH

(h) R = C(CH$_3$)$_3$, A=B=C= SH, D= OH

(i) R = C(CH$_3$)$_3$, A=B=C=D= SH

Figure 4.10 *Structures of t-butylcalix-4-arenes with replacement of hydroxy groups*

(ArOH) with Me$_2$NC(S)Cl gives rise to substitution with Ar-OC(S)NMe$_2$ groups, which upon heating react via the Newman–Kwart rearrangement to Ar-SC(O)NMe$_2$ groups. These can then be hydrolysed to —SH groups. Using this method, the t-butylcalix-4-arene has been converted to the mono-, 1,3-di-, tri- and tetrathio derivatives.[49] X-ray studies show that the monothioderivative (Figure 4.10f) is a cone, the dithiol (Figure 4.10g) a flattened cone, the trithiol (Figure 4.10h) a partial cone and the tetrathiol (Figure 4.10i) a 1,3-alternate conformer.[49] NMR studies indicate that all the calixarenes are mobile in solution except for the tetrathio compound.

4.4.3   Reaction of the rings

Calixarenes are capable of being both oxidised and reduced. Chlorine dioxide and thallium trifluoroacetate have been successfully used to oxidise calixarenes into calixquinone (Figure 4.11a). Chlorine dioxide[50] is capable of oxidising the four-, five- and six-unit calixarenes to their corresponding quinones. Thallium acetate also oxidises the t-butylcalix-4-arene to its corresponding quinone and has been used to synthesise mono-, di- and triquinones when calixarenes with protected —OH groups are used as the starting material. Further reactions of the calixquinones have also been investigated,[51] with milder oxidation conditions leading to the formation of spirodienones (Figure 4.11b), which can then undergo further chemical modifications.[52]

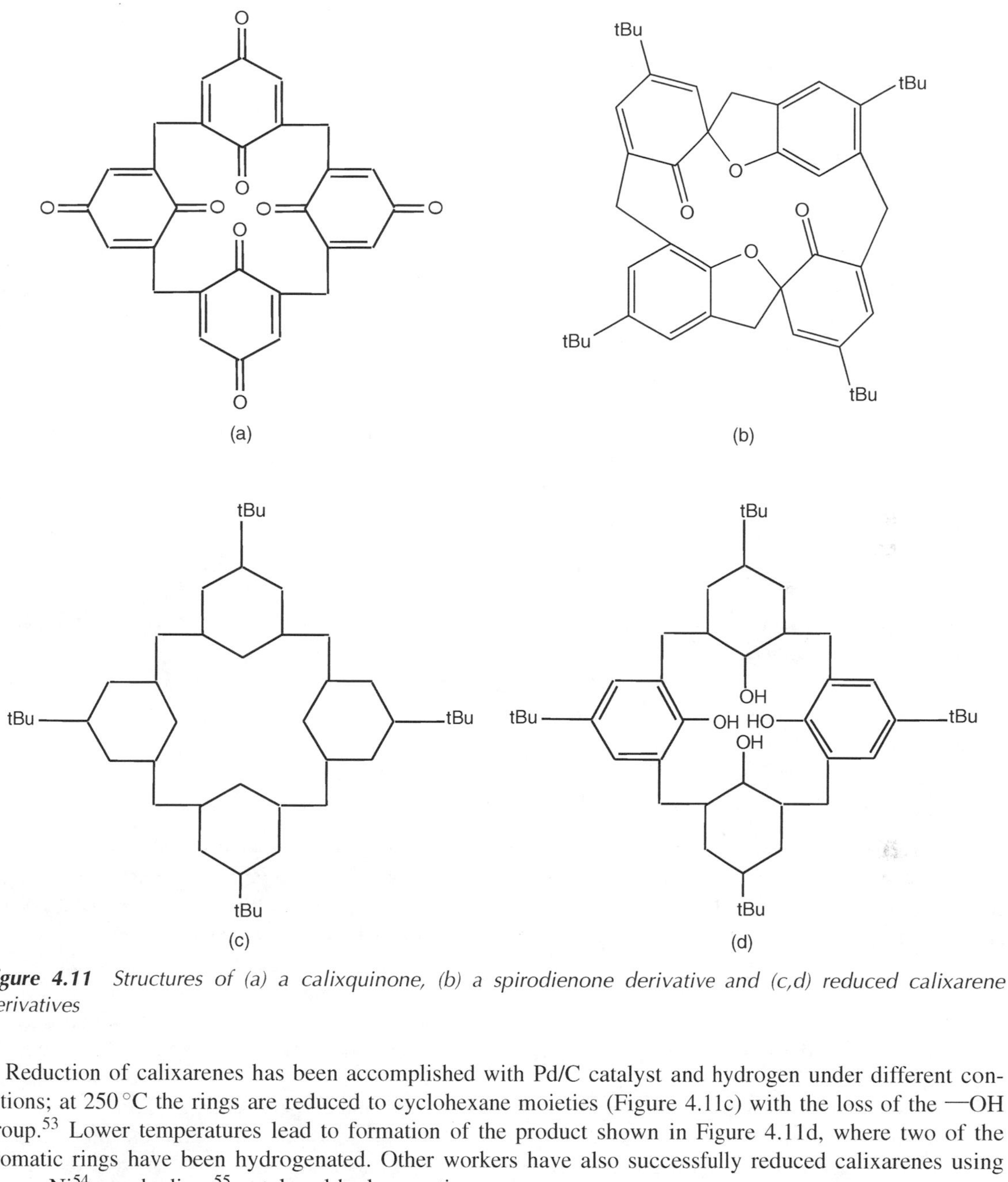

Figure 4.11 *Structures of (a) a calixquinone, (b) a spirodienone derivative and (c,d) reduced calixarene derivatives*

Reduction of calixarenes has been accomplished with Pd/C catalyst and hydrogen under different conditions; at 250 °C the rings are reduced to cyclohexane moieties (Figure 4.11c) with the loss of the —OH group.[53] Lower temperatures lead to formation of the product shown in Figure 4.11d, where two of the aromatic rings have been hydrogenated. Other workers have also successfully reduced calixarenes using Raney Ni[54]- or rhodium[55]-catalysed hydrogenation.

4.4.4   Reactions of the bridging groups

The bridging groups have also been the targets for some chemical modifications. The methylene units can be converted to carbonyl groups (Figure 4.12a) by oxidation of the tetraacetate of t-butylcalix-4-arene[56]

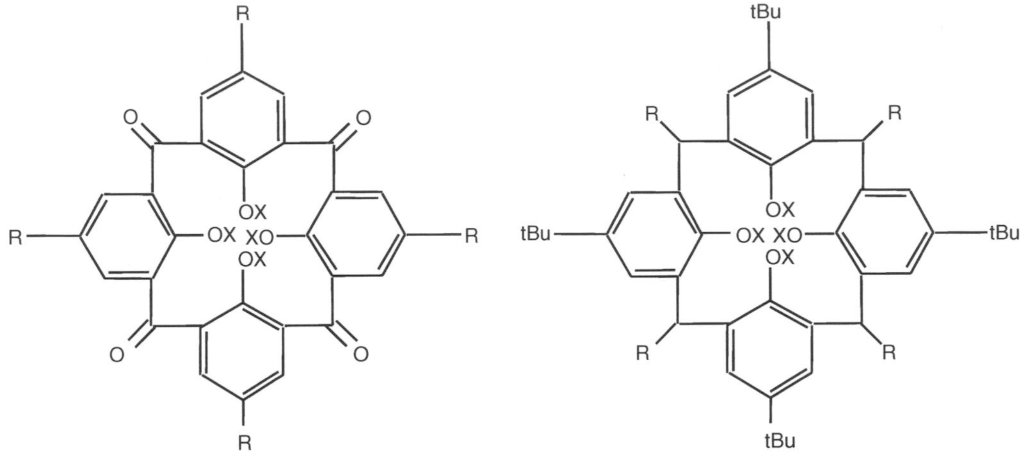

(a) R = C(CH₃)₃, X = COCH₃

(b) R = Li, X = CH₃
(c) R = COOH, X = CH₃
(d) R = Br, X = H

Figure 4.12   *Structures of methylene-substituted calix-4-arenes*

and the resulting carbonyl converted to alcohols by reduction with sodium borohydride. Lithiation of t-butylcalix-4-arene (with the —OHs converted to the methyl ether) by BuLi gave the tetralithyl derivative (Figure 4.12b), which could then be reacted with alkyl halides to give tetra-alkyl calixarene or carbon dioxide to give the acid derivative (Figure 4.12c).[57] N-bromosuccinimide can also be used to brominate (Figure 4.12d) the methylene groups.[58]

4.4.5   Modification of the upper rim

One of the easiest syntheses of calixarenes is from *para*-t-butylphenol, to give products which can be obtained in moderate to excellent yields in a variety of ring sizes and at high purity. Although phenol itself cannot be used directly in calixarene synthesis due to the formation of Bakelite-type materials, a phenol calixrarene with just a hydrogen atom in the *para* position can be obtained from the t-butylcalixarene by a simple Friedel–Crafts-type reaction with AlCl₃. What is also extremely synthetically useful is that this reaction is selective; for example, when the —OH groups are substituted, the rate of dealkylation *para* to those groups is greatly diminished. The 1,3-dimethyl ether of t-butylcalix-4-arene (Figure 4.13a) can therefore have the t-butyl groups *para* to the unsubstituted group selectively removed[59] to give the structure shown in Figure 4.13b.

Once the calixarene has been partially or totally dealkylated, there opens a large number of possible reactions, of which only a few examples can be given; many more are detailed within the Bibliography. The *para* hydrogen can be easily substituted by a wide range of reactions, some of which are listed below.

Early work[60] by David Gutsche's group utilised N-bromosuccinimide to brominate the *para* positions of calix-4-arene tetramethyl ether (Figure 4.13c) to give the fully brominated compound (Figure 4.13d). This was then used as a feedstock for a variety of reactions. Reaction with t-butyl lithium gave the tetralithio compound (Figure 4.13e), which could then be reacted with carbon dioxide to give the tetracarboxylic acid (Figure 4.13f); this could be converted into the methyl ester, while reaction of the brominated calixarene with CuCN gave the tetracyano compound (Figure 4.13g).

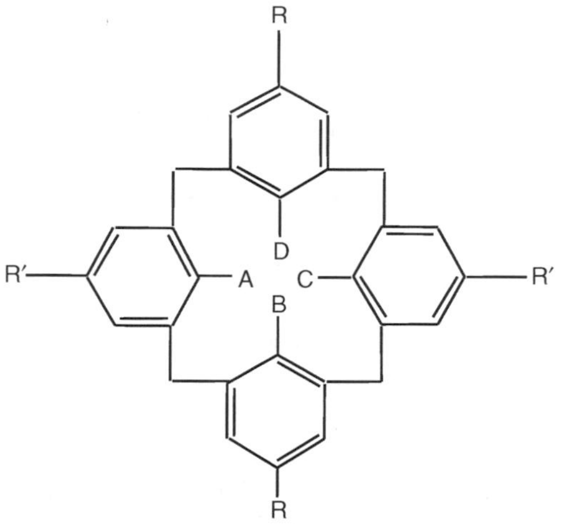

(a) R=R'= C(CH$_3$)$_3$, A=C= OH, B=D= OCH$_3$

(b) R= C(CH$_3$)$_3$, R'= H, A=C= OH, B=D= OCH$_3$

(c) R= R'= H, A=B=C=D= OCH$_3$

(d) R= R'= Br, A=B=C=D= OCH$_3$

(e) R= R'= Li, A=B=C=D= OCH$_3$

(f) R= R'= COOH, A=B=C=D= OCH$_3$

(g) R= R'= CN, A=B=C=D= OCH$_3$

(h) R= R'= H, A=B=C=D= OCH$_2$CH=CH$_2$

(i) R= R'= CH$_2$CH=CH$_2$, A=B=C=D= OH

(j) R= R'= CH$_2$CH$_2$ OH, A=B=C=D=OSO$_2$Tos

(k) R= R'= CH$_2$CH$_2$ Br, A=B=C=D=OSO$_2$Tos

(l) R= R'= CH$_2$CH$_2$ N$_3$, A=B=C=D=OSO$_2$Tos

(m) R= R'= CH$_2$CH$_2$ NH$_2$, A=B=C=D=OSO$_2$Tos

(n) R= R'= CH=CHCH$_3$, A=B=C=D=OSO$_2$Tos

(o) R= R'= CHO, A=B=C=D=OSO$_2$Tos

(p) R= R'= CH=NOH, A=B=C=D=OSO$_2$Tos

Figure 4.13   *Dealkylation and substitution of t-butylcalix-4-arenes*

Simultaneously, the same group[61] investigated calixarene modification via a Claisen rearrangement route. Calix-4-arene was converted to the tetraallyl ether (Figure 4.13h) and then refluxed in dimethylaniline to give the rearranged product shown in Figure 4.13i. After protection of the —OH groups with p-toluenesulfonyl groups, the *para* allyl group could undergo further modification. The allyl group could be reacted with ozone and then reduced with sodium borohydride to give a 2-hydroxyethyl group on the *para* position (Figure 4.13j) – or could be brominated to give the 2-bromoethyl-substituted calix-4-arene (Figure 4.13k). The bromide could be converted to an azide (Figure 4.13l), which could then be reduced to give the 2-aminoethyl group (Figure 4.13m). The allyl group could also be rearranged to give the 2-propenyl substituent (Figure 4.13n) and this could be converted to a formyl group (Figure 4.13o) and then to an oxime (Figure 4.13p). Further work by the same group demonstrated an alternative to tosylation, a method utilising silyl ether-protecting groups,[62] which has proved to be successful in introducing selectivity into this reaction protocol, making it possible, for instance, to synthesise calix-5-arenes with between one and five p-allyl groups.

Other reactions at the *para* position include utilising Friedel–Crafts-type conditions to alkylate or acylate the *para* positions. Chloromethylation is one of the most investigated, since the introduction of a chloromethyl group allows many further reactions. Calixarenes could be reacted with chloromethyl octyl ether with $SnCl_4$ catalysis to introduce chloromethyl groups in the *para* positions;[63] these could then be converted to phosphate esters and hydrolysed to give water-soluble calixarene phosphonic acids. The chloromethyl group is also amenable to substitution reactions such as those described earlier in this chapter to give alcohols, ethers, amines, quaternary ammonium salts, thiols and so on. It can also be coupled with alkyl or aryl halides to give p-ethyl or p-benzyl calixarenes.[64] Introduction of acyl groups is also possible, with for example calixarenes having been reacted with hexanoyl chloride with $AlCl_3$ catalysis[65] to give p-hexanoyl calixarenes, which can then be reduced to p-hexyl calixarenes, thereby giving a route to p-alkyl calixarenes, which generally cannot be synthesised directly from the corresponding phenols. Other groups such as acetyl or benzoyl have also been successfully incorporated.

One field of interest is the synthesis of so-called 'deep-cavity' calixarenes, in which the benzene rings are replaced or extended by further reactions to give much larger aromatic systems. Although some direct syntheses such as the condensation of p-phenyl phenol with formaldehyde have been reported, these tend to give rather intractable mixtures of ring sizes. A better method is to take a preformed calixarene and chemically extend the system. Several methods exist to do this, with the Suzuki method being one, which in this example[66] involves the use of calix-4-arene, which first has its hydroxy groups protected by conversion to the benzyl ether, is then brominated in the *para* positions, and is finally coupled with 3-benzyloxyphenyl boronic acid, with the protecting groups being removed to give the structure shown in Figure 4.14a. Alternative methods also exist, such as coupling p-lithiocalixarenes with aryl iodides catalysed by zinc chloride to give p-phenyl calixarenes.[67] Formyl-substituted calixarenes have also been used, allowing the synthesis of calixarenes with p-phenanthryl[68] groups linked by imidazole units (Figure 4.14b) – or the structure shown in Figure 4.14c, in which a p-formyl calixarene was converted to the tetraethyne-substituted compound; the ethyne groups can then reacted with tetraphenylcyclopentadienone in a Diels–Alder reaction to give the highly substituted calixarene, as shown.[69]

Calixarenes are also easily substituted via the Mannich reaction with formaldehyde and a dialkylamine to form the dialkylaminomethyl calixarene (Figure 4.15a shows a typical structure). Reaction with methyl iodide gives a quaternary ammonium salt, which can then undergo nucleophilic substitution with a range of nucleophiles to give calixarenes with *para* substituents of the formula —CH_2X, where X can be cyanide, an ether, azide or diethyl malonate, to name a few.

A series of other reactions have also been used to substitute the aromatic units of calixarenes. Two of the most widely investigated are sulfonation and nitration. Sulfonation was studied because the presence of sulfonate salts on the calixarene renders it extremely water soluble. Initial studies utilised p-H-calixarenes,[70] which could be reacted with concentrated sulfuric acid to give the types of structure shown in Figure 4.15b. Further reaction of these compounds with concentrated nitric acid[70] gave the p-nitro compounds (Figure 4.15c). Direct *ipso*-substitution of the easily available t-butylcalixarenes has since been demonstrated for both sulfonation[71] and nitration.[72] This form of nitration also has the advantage of being selective, depending both on the nature of the group being removed and whether or not there is an —OH or —OR group on the ring. Nitro groups are synthetically useful since they can be reduced to amines, which can then serve as a base to attach a wide range of other groups.

Selective iodination of calixarenes has been demonstrated and the iodo groups converted into amines and methoxy groups.[73] Diazotisation[74] with p-nitrobenzenediazonium tetrafluoroborate has also proved a successful method of derivatising calixarenes, with the interesting effect that the reaction appears to be

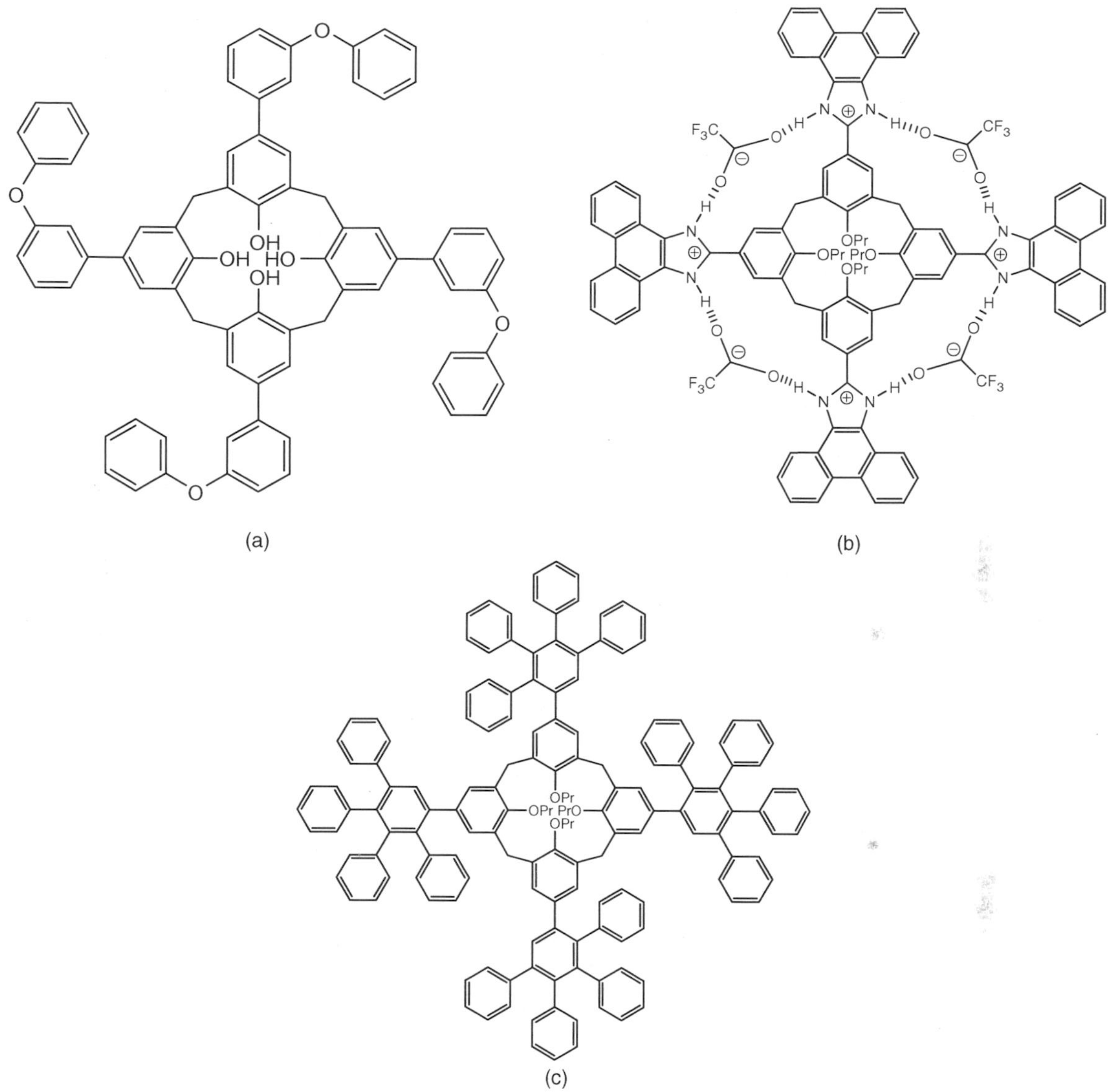

(a)

(b)

(c)

Figure 4.14   *Deep-cavity calixarenes*

autoaccelerating. If the amount of the diazonium salt is less than that required for full substitution, a mixture of tetra and unsubstituted calixarene (Figure 4.15d) is recovered, with almost none of the mono-, di- and tri-substituted products being formed. Calixarenes derivatised by this method are highly coloured due to the presence of azobenzene units and have been shown to act as chromogenic sensors[75] for species such as Li$^+$.

Bridging reactions of calixarenes have been studied not just at the lower rim, as demonstrated earlier in this chapter, but also at the upper rim. The earliest examples were synthesised by the reaction[76] of alkyl-*bis*-phenols with di(bromomethyl)phenols (Figure 4.16a,b). X-ray crystallographic studies[76] showed the calixarenes were in the cone conformer, although in the case of the calixarene with a five-carbon

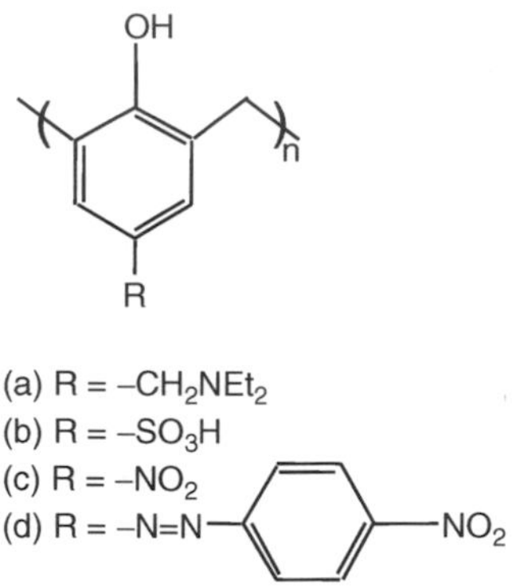

(a) R = –CH$_2$NEt$_2$
(b) R = –SO$_3$H
(c) R = –NO$_2$
(d) R = –N=N–⟨benzene⟩–NO$_2$

Figure 4.15   *Structures of p-substituted calixarenes*

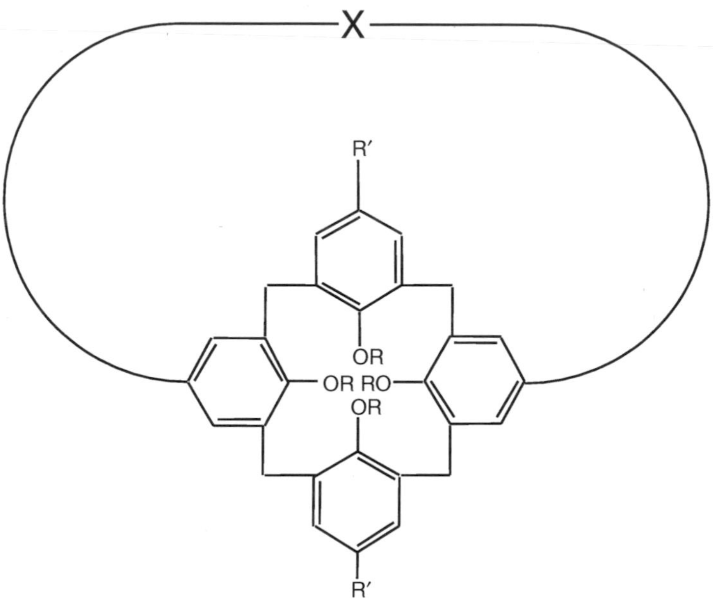

(a) R= OH, R'= CH$_3$, X= –(CH$_2$)$_5$–
(b) R= OH, R'= CH$_3$, X= –(CH$_2$)$_7$–

Figure 4.16   *Upper-rim-bridged calix-4-arenes*

bridge, the strain led to distortion of the cone. These types of compound have been named 'arrichoarenes', from the Greek '*arriches*', meaning a basket with a handle. Calixarenes bridged at the upper rim are much less common than lower-rim-bridged calixarenes and are usually made by reacting a bridging group with an upper-rim-disubstituted calixarene.

4.5 Complexes with Calixarenes

The calixarene family of macrocycles has been shown to form complexes with metal ions, common solvents and a wide range of organic compounds. Some examples will be given for all of these, but for more detailed descriptions of a much wider range of host–guest complexes the reader is once again referred to the Bibliography.

The first crystal structure obtained for t-butylcalix-4-arene[11] clearly showed inclusion of a toluene molecule within the calixarene cavity. These types of complex can often be very stable, with the calixarenes clinging on tightly to their guests, although larger calixarenes do tend to lose bound solvent more easily. Other examples of guest compounds include acetone,[77] which can form a 1 : 1 or 3 : 1 complex with calix-4-arene, and anisole,[78] which forms a complex in which the anisole is inside the cavity between two calixarene molecules. Different calixarene sizes and substituents have been examined, such as t-butylcalix-4-arene tetracarbonate and tetraacetate, which form 1 : 1 complexes with acetonitrile and acetic acid respectively. Interestingly, the X-ray crystal structures show that the tetracarbonate adopts a cone conformer,[79] whereas the tetraacetate is in the 1 : 3 alternate conformation.[80] Many other calixarene/solvent combinations have been isolated.

Similarly to the crown ethers, there has been a large volume of work concerning the binding of metal ions by calixarenes. Many calixarenes have been shown to selectively bind metal ions in both solution and solid state. The parent calixarenes themselves are not especially strong ion-binding materials, although in strongly basic solution the phenols can ionise and allow transport of alkali metal ions through chloroform,[81] with caesium being the preferred ion for transport in mixed systems. A crystal structure[82] has also been obtained for the 1 : 1 complex formed between caesium and t-butylcalix-4-arene, which clearly shows the caesium ion to be inside the cavity rather than associated with the oxygen atoms.

More effective binding has been obtained by substituting the hydroxy groups. A crystal structure has been obtained for the tetramethyl ether of t-butylcalix-4-arene (Figure 4.17a), which shows double guest inclusion, with a sodium ion being bound by the ether groups and a molecule of toluene within the calixarene cavity.[83] More effective than the ethers though are the calixarene esters, an example of which is shown in Figure 4.17b, and the similar ketones (Figure 4.17c). Extraction of alkali metals with an extensive range of calixarenes of various sizes and substituents was attempted,[84] with the tetramer showing good affinity for sodium and the hexamer for rubidium and caesium, and larger calixarenes displaying very poor binding. The ketones generally showed stronger binding than the esters and variation of the ester group also affected the binding capabilities.

Even more effective binding was found for a range of amide-substituted calixarenes, such as that shown within Figure 4.17d. This calixarene[85] was shown to bind potassium ions with the calix in the cone conformation and the cation encapsulated by the polar headgroups. Also, unlike the previous calixarenes, besides complexing alkali metals these materials also form strong complexes with alkaline earths.[85] The larger alkaline earths are more strongly bound, the calixarene shown in Figure 4.17d for example has a 10^8-fold selectivity for calcium over magnesium. Larger calixarenes of this type also show preference for strontium and barium, with for example strontium being favoured by the calix-6-arene. Other divalent metals efficiently encapsulated[86] include iron, nickel, copper, zinc and lead. Crystal structures for all of these complexes can be obtained. The amides also act as effective hosts for trivalent metals[87] such as lanthanum, which is strongly bound by calixarene tetraamides synthesised in the partial-cone conformation.

Acid-substituted calixarenes can form complexes with both alkali metal ions and display even stronger complexation of the alkaline earths, with the calixarene shown in Figure 4.17e for example displaying a very strong preference for calcium over other ions.[88] Even better preference occurs with the calixarene containing two acid and two amide chains (Figure 4.17f), for which the calcium/magnesium selectivity is

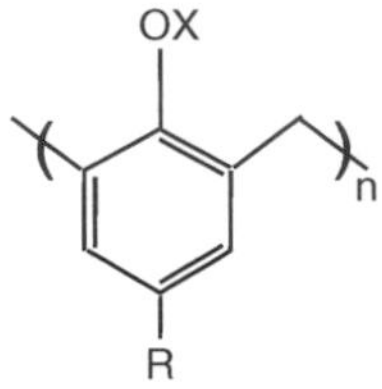

(a) R = C(CH₃)₃, X= CH₃

(a) R = C(CH_3)_3, X= CH_3
(b) R = C(CH_3)_3, X= CH_2CO_2CH_2CH_3
(c) R = C(CH_3)_3, X= CH_2COCH_3
(d) R = C(CH_3)_3, X= CH_2CON(CH_2CH_3)_2
(e) R = C(CH_3)_3, X= CH_2COOH
(f) R = C(CH_3)_3, X= CH_2COOH (1,3) +
        CH_2CON(CH_2CH_3)_2 (2,4 positions)
(g) R = C(CH_3)_3, X= PPh_2
(h) R = C(CH_3)_3, X= CH_2CSN(CH_2CH_3)_2
(i) R = SO_3H, X= CH_2COOH
(j) R = C(CH_3)_3, X= CH_2CONHOH

Figure 4.17   *Examples of metal ion-binding calixarenes*

essentially infinite.[89] Even stronger binding has been observed for lanthanide and actinide ions, especially when the calixarenes are ionised. The hexamer has been shown to bind more strongly than the tetramer,[90] with neodynium and europium being the most strongly bound.

Other groups that have been utilised include phosphorus-containing groups such as that shown in Figure 4.17g, which has a high affinity for copper(II);[91] similar calixarenes have been shown to bind palladium(II).[92] Further work utilised mono- to tetra-substituted calixarenes of the same family to design materials which bound palladium(II) and platinum(II).[93] Other phosphorus-containing calixarenes bind lanthanides and actinides.[94] Sulfur-containing thioamides, such as the structure shown in Figure 4.17h, are effective binders for many heavy metals. The tetramer is an effective host for lead(II), the pentamer a host for cadmium(II) and the related hexamer a host for for silver(I).[95]

Other calixarenes that have very high affinities for uranyl ions have been devised. The water-soluble calixarenes containing p-sulfonato groups have very high affinities for uranyl cation $(UO_2)^{2+}$. A range of calixarenes containing sulfonato and carboxylic acid groups (Figure 4.17i) have been synthesised, with the pentamer and especially the hexamer being shown to have exceptional affinities for uranyl cation.[96] The related hydroxamic acid hexamer (Figure 4.17j) has been shown to be even more efficient in this respect than the corresponding carboxylic acid.[97]

As demonstrated in this and the previous chapter, both calixarenes and crown ethers are capable of binding metal ions, and so combined structures with high affinities would be expected to be possible. A wide range of these compounds have been synthesised and tested, many aiming at the selective extraction of caesium. Cs(137) is a radioactive isotope of caesium found in nuclear waste and displays high mobility in the environment. Calixcrowns have been studied as potential hosts for caesium. A calixarene bridged with an 18-crown-6 ether chain (Figure 4.7a) normally adopts the cone conformation, but adopts the 1,3-alternate conformation when binding caesium to maximise cation-π interactions.[98] Similar work utilised benzocrownethers to bridge the calixarene ring and again shown a very high selectivity for caesium.[99]

Anion-selective binding by calixarenes is much less developed than that of its cation counterpart. This may be related to the various differences that exist between anions/cations,[100] including ionic size (anions are generally larger and require larger binding sites, for example Cl^- has an ionic radius of 0.167 nm, K^+ 0.133 nm, the same as the smallest anion F^-), shape (metals are generally spherical whereas sulfate is tetrahedral and nitrate trigonal planar) and ionisation state (which can be pH-dependent, e.g. carbonate vs bicarbonate). Anions are also relatively strongly hydrated in comparison to similar-sized cations; however, with suitable substitution, calixarenes suitable as receptors for various anions have been developed.

One of the earliest approaches[101,102] was to bridge the calixarene rings with metallocenes (Figure 4.18a) in order to give calixarenes capable of binding a variety of anions with especially high affinity for dihydrogen phosphate.[102] Direct attachment of aryl ruthenium species[103] to the aromatic rings of calix-4-arenes gave species that bound halides in solution. A complex with tetrafluoroborate ion could be crystallised and X-ray studies showed the BF_4^- unit to be located inside the calixarene cavity. Calixarenes substituted with sulfonamide groups on the upper rim[104] have been shown to complex a variety of anions, with the structure shown in Figure 4.18b having a high selectivity for hydrogen sulfate over chloride and nitrate. Calix-6-arenes containing three urea or thiourea groups[105] on the upper rim (see for example Figure 4.18c,d) have been shown to selectively bind bromide over chloride and have an even stronger affinity for tricarboxylate anions. Other work used amide- and urea-substituted calix-4-arenes, which could simultaneously stabilise both the anion and the cation, as well as showing a high affinity for sodium carboxylates.[106] Much of the work on anion-binding calixarenes up to 2005 is reviewed by Matthews and Beer.[107]

We have already described the encapsulation of solvents within calixarenes, but larger neutral molecules have also been encapsulated. Calixarenes substituted at the upper rim with dimethylaminomethyl or carboxyethyl groups are soluble in acidic and basic solutions respectively. These can be used to solubilise organic guests in water.[108] Calix-4-arenes cannot even solubilise durene; while calix-5 and 6-arenes can solubilise larger species such as naphthalene and anthracene, calixarenes with seven or more units can solubilise pyrene and perylene. Fullerene C_{60} has also been shown to complex strongly with t-butylcalix-8-arene[109] to give a soluble complex, and this can actually be used as a method of separating C_{60} and C_{70}.

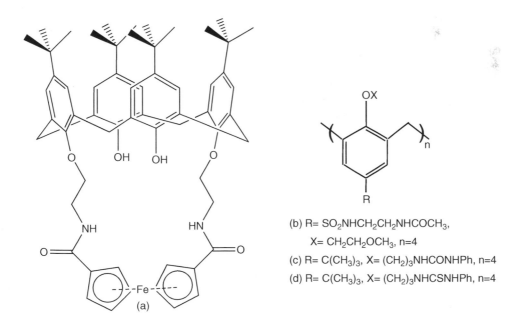

(b) R= $SO_2NHCH_2CH_2NHCOCH_3$,
    X= $CH_2CH_2OCH_3$, n=4
(c) R= $C(CH_3)_3$, X= $(CH_2)_3NHCONHPh$, n=4
(d) R= $C(CH_3)_3$, X= $(CH_2)_3NHCSNHPh$, n=4

Figure 4.18   *Examples of anion-binding calixarenes*

The p-sulfanatocalix-6-arene has been shown to solubilise lipophilic dyes such as orange OT in water.[110] The corresponding tetramers have also been shown to form complexes with a variety of amino acids,[111] with the amino acids being located in the hydrophobic cavity of the calixarene.

4.6 *Bis* and Multicalixarenes

One common theme within this work is of increasing complexity. Simple ring compounds such as benzene have been combined to form multiple-ring compounds or cyclophanes. After the basic calixarene chemistries had been established, researchers similarly began to look at ways to combine calixarenes in more complex architectures. Calixarenes have been bridged with crown ethers, linked to other functionalities such as metallocenes, and in this section the linking together of two or more calixarenes in multicalixarene assemblies will be discussed.

There exist several ways to link together calixarenes. The most obvious is via covalent bonding through one or more active groups, which can be on the upper or lower rims. Calixarenes can be incorporated into polymeric systems. Finally, noncovalent interactions can be utilised, such as hydrogen bonding, electrostatic interactions and other noncovalent interactions.

As an alternative to the bridged calixs and calixcrowns, where two hydroxyl groups on the calixarene are bridged by a crown ether or other chain, it is possible to bridge between different calixarenes. One of the earliest examples came from the Gutsche group, which bridged calixarenes via the lower rims. For example, the pentamethyl ether of t-butylcalix-6-arene can be bridged by reacting with 1,4-dibromomethylbenzene and sodium hydride to give the *bis*-calixarene[112] with a p-xylene-bridging group (Figure 4.19b). Similarly, the monopropargyl pentamethyl ether of t-butylcalix-6-arene can be coupled together[112] to give a diacetylene-bridged dimer (Figure 4.19b). In both compounds the calixs were shown to be in the cone conformer. The bridging is not limited to single bridges, as can be seen in Figure 4.19c, where four bridges have been used to link together two t-butylcalix-4-arenes.[113] This format has been named a 'calix-tube', with the compound shown displaying a strong selective affinity for potassium, which can be tailored by modification of the upper-rim groups. Extended versions of this can be obtained by using calix-4-arenes in the 1,3-conformation,[114] such as the structures in Figure 4.19d,e, which were shown to complex potassium and caesium ions. Esterification reactions have also been effective, with reactions of t-butylcalix-4-arene with aromatic diacid or disulfonyl chlorides leading to substitution at the 1,3-positions and the isolation of double and in one case triple calixarenes.[115] Similar reactions also occurred when 1,1'-metallocene dicarbonyl chloride was used[116] to give *bis*-calix-4-arenes, with a crystal structure of the ferrocene-containing dimer being obtained, which showed the calixarenes maintained their con conformations. A trimer was also obtained for the ferrocene-substituted calix-4-arene.

An interesting variation on this theme is where one or two bridges can be introduced by reacting calix-4, 5 or 6-arene with 1,4-dibromobut-2-ene to give the mono- or dibridged calixarenes.[117] An example of the doubly bridged *bis*-calix-6-arene is shown in Figure 4.20a. What is especially interesting about this series of compounds is that they undergo a tandem Claisen rearrangement to give reasonable yields of the corresponding calixarenes bridged at the upper rather than the lower rim (Figure 4.20b). The *bis*-calix-5 and 6-arenes with a single bridge both bound C_{60} 1:1 more tightly than the single calixarenes; however, none of the calix-4-arenes or the doubly bridged *bis*-calixarenes could bind the fullerene.

Direct methods have also been used to bridge calixarenes at the upper rim. One of the simplest is direct oxidative coupling[118] of two calix-4, 5, 6 or 8-arenes with ferric chloride to form a C—C bond between two calixarenes (Figure 4.21a). A second method is to take the calix-4-arene with amine substituents on the 1,3-upper-rim positions[119] and utilise various condensations such as with thiophene-2,5-dicarboxaldehyde to synthesise a double calixarene capable of binding viologen molecules along with other aldehydes to

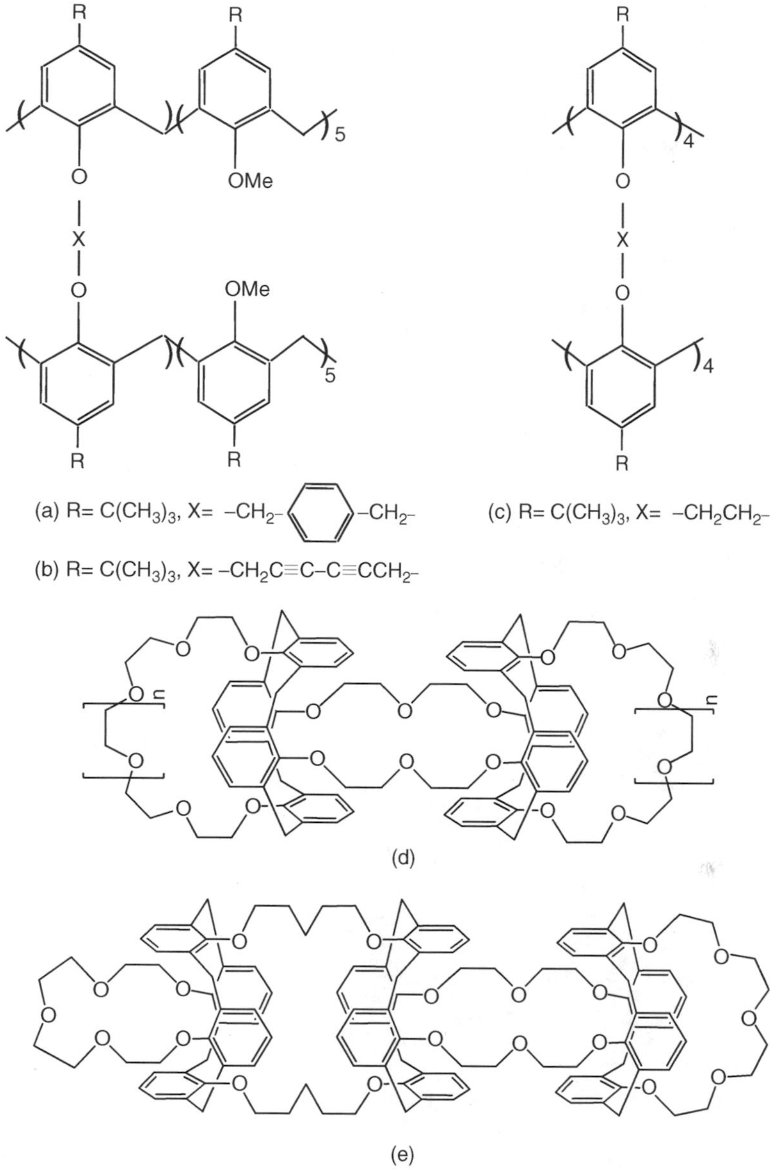

(a) R= C(CH₃)₃, X= –CH₂–⟨ ⟩–CH₂– (c) R= C(CH₃)₃, X= –CH₂CH₂–

(b) R= C(CH₃)₃, X= –CH₂C≡C–C≡CCH₂–

(d)

(e)

Figure 4.19 *Examples of multicalixarenes*

create a wide variety of double calixarenes. Acylation reactions, for example with aromatic disulfonyl chlorides, also gave double and in one case quadruple calixarenes.[119]

The reaction of t-butylcalix-8-arene with the tetramethoxy ether of p-chloromethyl calix-4-arene gave a linked system in which the lower rim of the octamer was bound by four —OCH₂- to the upper rim of the calix-4-arene (Figure 4.21b), giving a calixarene dimer with a very deep cavity.[120]

Calixarenes have also been incorporated into polymeric systems. One of the simplest methods is simply substituting a calixarene with a reactive group and allowing it to react with a preformed polymer, such

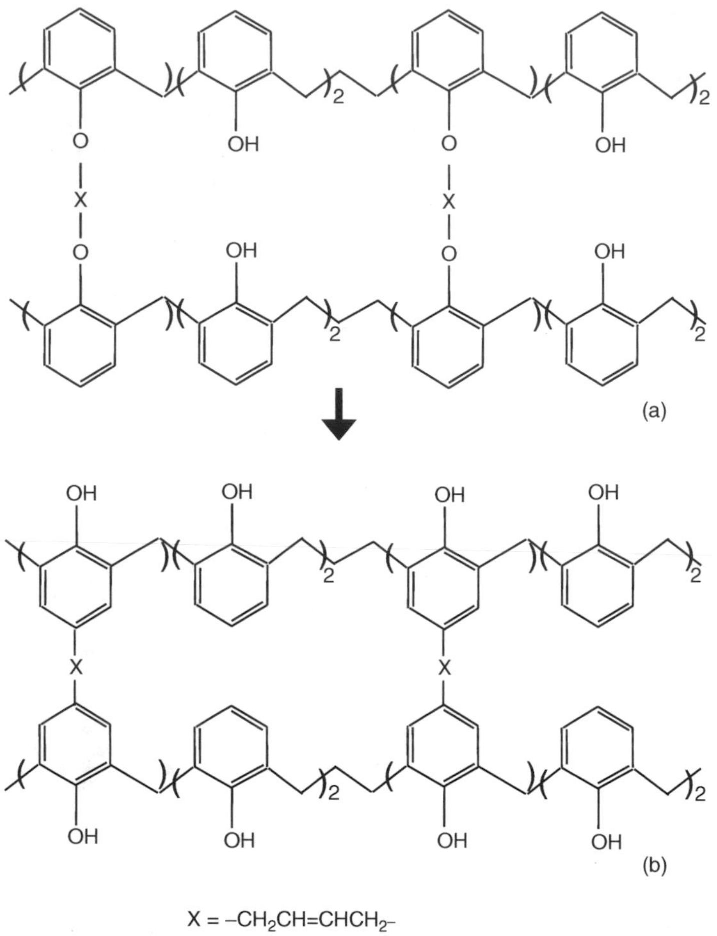

Figure 4.20   *Claisen rearrangement of a* bis-*bridged calix-6-arene*

as the reaction of polyethylenimine with a calix-4-arene substituted with a single upper-rim bromopropyl group[121] to give a water-soluble polymer or with tetrachloromethyl calix-4-arene to give an insoluble resin. Alternatively, it is possible to synthesise polymers by free-radical methods, by substituting the calixarene with a methacrylate on the lower rim and polymerising this to give a chloroform-soluble calixarene polymer[122] which can form stable sodium complexes. Methacryl groups can also be added on the upper rim and the resultant calixarene[123] can be copolymerised with a second sulfonate-containing monomer, which gives water-soluble polymers with calixarene units capable of solubilising perylene. Condensation polymers can also be synthesised by the reaction of 1,3-disubstituted calix-4-crowns with difunctional monomers such as diacid chlorides, diamines and so on, to give low-molecular-weight polymers.[124] Copolymers of calixarenes with bisphenol-A[125] can also be obtained, which display a 100-fold enhanced affinity for silver ion compared with the monomeric calixarene. Dendrimers containing multiple calixarenes have also been reported, such as the structure[126] in Figure 4.22, synthesised by amidation reactions.

   Besides covalent binding, other forces have been used to assemble di- and multicalixarenes. This leads to the field of self-assembly, where simply dissolving a calixarene in a suitable solvent or mixing together

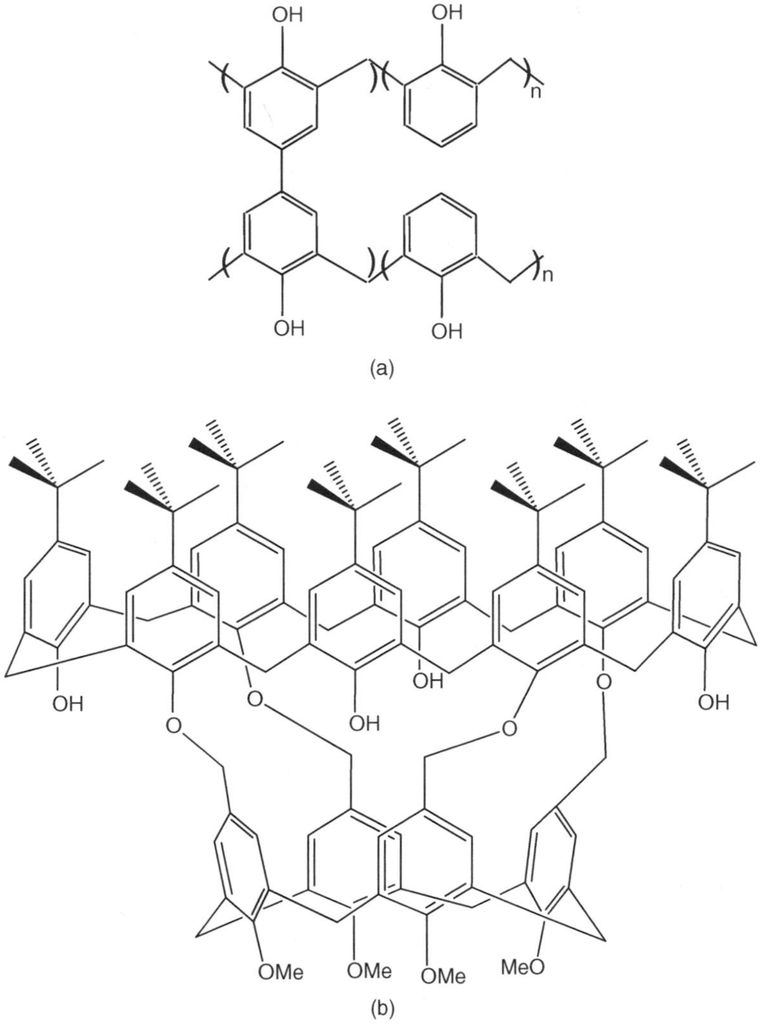

(a)

(b)

Figure 4.21   *Upper-rim-bridged calixarenes*

two calixarenes can lead to spontaneous formation of highly organised systems. Again, a few examples of what is a wide-ranging field will be given. Hydrogen-bonding interactions in particular have been utilised to construct some of the most intricate structures. The earliest examples came from calix-4-arenes with two pyridinone groups[127] attached to the upper rim (Figure 4.23a), which formed aggregates in chloroform. Two different calixarenes containing either carboxylic acids or pyridine moieties on their upper rims (Figure 4.23b,c) were shown to form a heterdimer in chloroform.[128] Substituting the upper rims with urea groups led to a productive series of materials, such as the calixarene shown in Figure 4.23d, which spontaneously forms dimers in solution or when crystallised from a solution containing triethylammonium ion, with the crystal structure clearly showing encapsulation of the ammonium ion within the cavity of the calixarene dimer.[129] Other papers from this group have shown that often the presence of a guest can aid in dimer formation. It is also possible to assemble these systems and then perform chemical reactions so as to 'lock' the two calixarenes permanently together.[130]

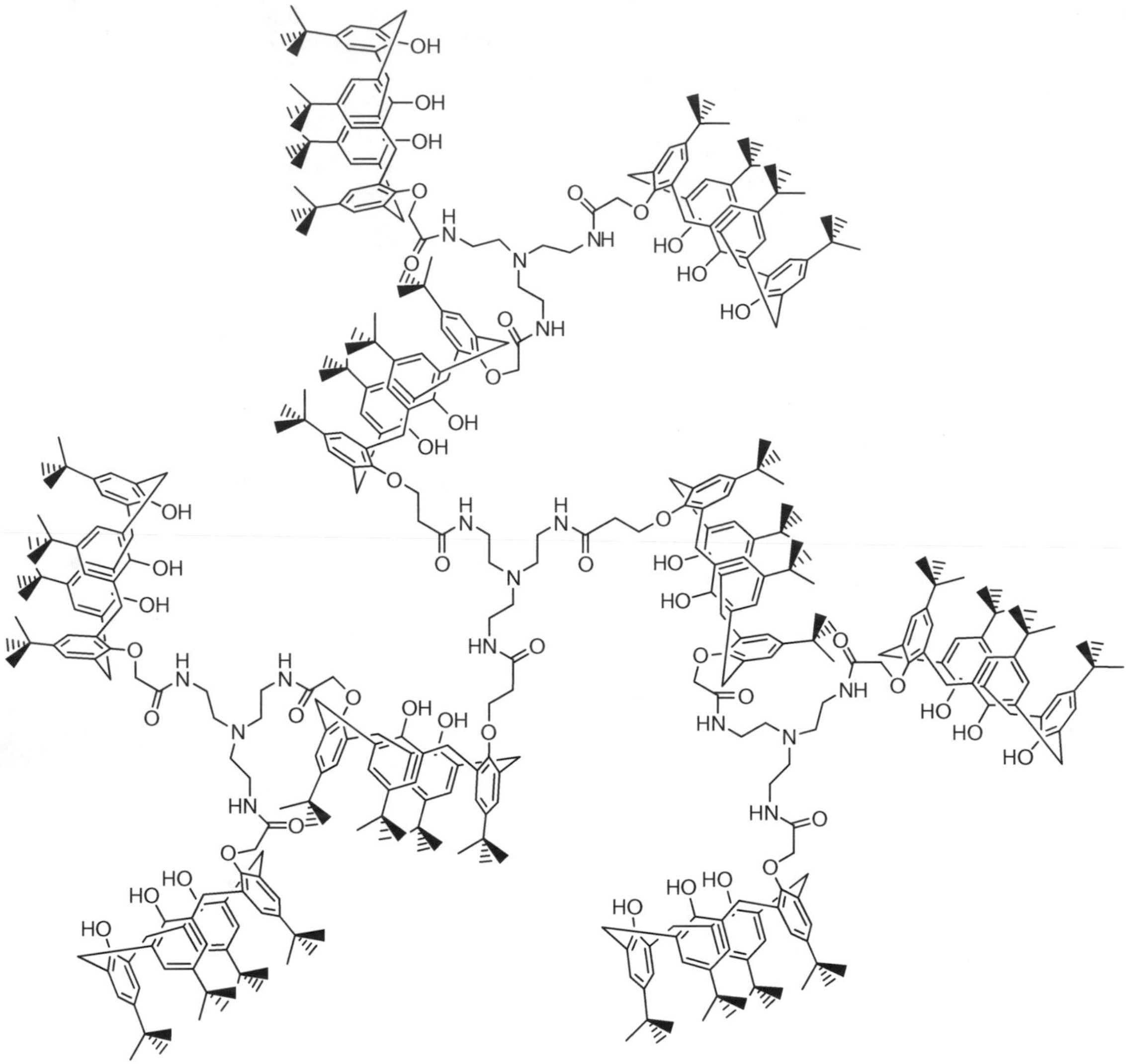

Figure 4.22   *Dendrimer calixarenes*

Other hydrogen-bonding moieties have also been attached, such as barbituric acid groups, uracil groups and melamine groups. One very elegant procedure is the synthesis of calixarenes bearing a single melamine group, which in solution with diethylbarbituric acid forms the hydrogen-bonded trimer[131] (Figure 4.24a). When two calixarenes containing either melamine or isocyanuric acid groups were mixed, a cyclic hexamer was formed[131] (Figure 4.24b). Electrostatic forces have also been used, and a mixture of calix-4-arenes with either sulfonato or quaternary ammonium salts at the upper rim spontaneously formed heterodimers which could complex guests such as acetylcholine.[132]

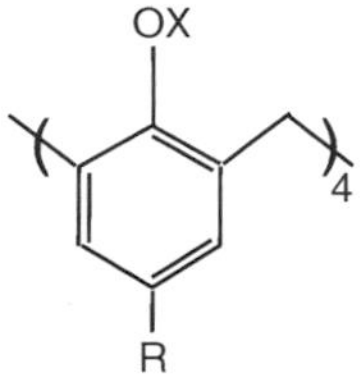

(a) R= H (1,3 positions) or –NHCO— [pyridone structure]

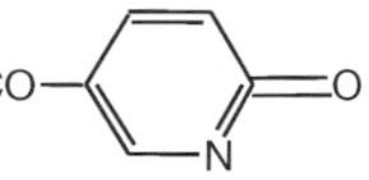

(2,4 positions), X= $CH_2OCH_2CH_2OCH_3$
(b) R= COOH, X= $CH_2CH_2CH_3$
(c) R= CH=CH–4–Pyridyl, X= $CH_2CH_2CH_3$
(d) R= $NHCONHC_6H_4CH_3$, X= C_5H_{11}
(e) R= $C(CH_3)_3$, X= CH_2COOH

Figure 4.23   *Hydrogen-bonded calixarenes*

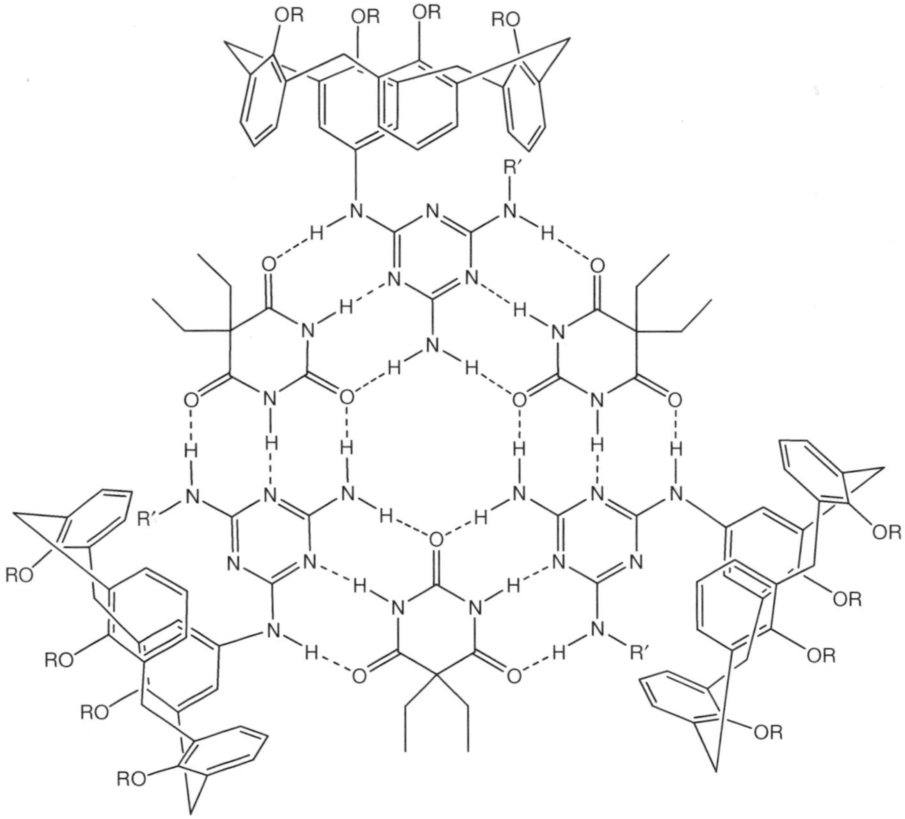

(a)

Figure 4.24   *Hydrogen-bonded calixarenes*

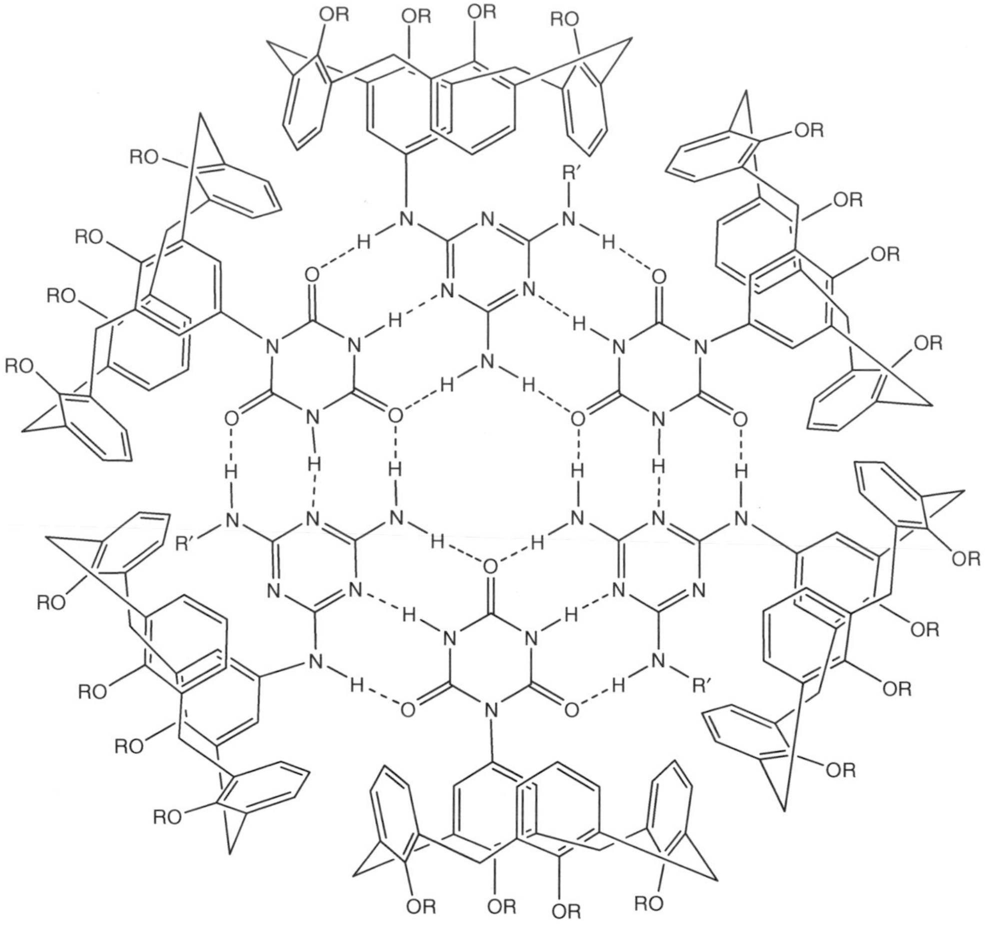

(b)

Figure 4.24   (continued)

4.7   Oxacalixarenes, Azacalixarenes and Thiacalixarenes

As an alternative to the methylene groups between the rings, a series of other groups containing oxygen and nitrogen atoms have been utilised. During the standard Petrolite procedure for the synthesis of t-butylcalix-4-arene, besides condensation of phenol with formaldehyde, an alternative reaction may occur where two hydroxymethyl groups undergo a dehydration reaction[133] to give a calixarene in which one of the linking groups is not —CH$_2$— but rather —CH$_2$—O—CH$_2$—. This compound proved difficult to isolate from the mixture however and if the reaction is driven to completion, is not found at all. It can be synthesised by taking the linear tetramer substituted with hydroxymethyl groups and cyclising that. Using similar methods, tetramers and trimers containing oxygen atoms can be synthesised, as shown in Figure 4.25b,c.[133]

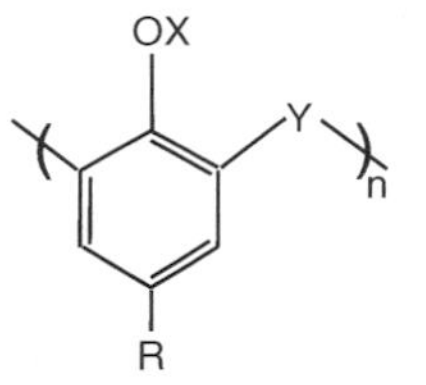

(a) R= C(CH$_3$)$_3$ X=H, n=4, Y= –CH$_2$– (1,2,3
       positions) or –CH$_2$OCH$_2$– (4 position)
(b) R= C(CH$_3$)$_3$, X=H, n=4, Y= –CH$_2$– (1, 3
       positions) or –CH$_2$OCH$_2$– (2,4 positions)
(c) R= C(CH$_3$)$_3$, X=H, n=3, Y= –CH$_2$OCH$_2$–
(d) R= CH$_3$, X=H, n=3, Y= –CH$_2$–NPh–CH$_2$–
(e) R= C(CH$_3$)$_3$ X=H, n=4, Y= S
(f) R= C(CH$_3$)$_3$ X=H, n=4, Y= SO$_2$
(g) R= C(CH$_3$)$_3$ X=H, n=4, Y= SO

Figure 4.25   *Oxa-, aza- and thiacalixarenes*

Azacalixarenes have also proved amenable to synthesis, such as for example by condensing a *bis*-hydroxymethyl phenol with an amine such as benzylamine (Figure 4.25d) to give good yields of calix-3-arenes with bridging —CH$_2$—NR—CH$_2$— groups.[134] This system was shown to selectively complex potassium, barium and uranyl ions. Further work by the same group utilising phenol dimers and trimers in a similar manner to the work above on oxacalixarenes gave mixed systems with both —CH$_2$—NR—CH$_2$— and —CH$_2$— bridging groups.[135] Other workers have also demonstrated the ability of these systems to bind lanthanides.[136]

Thiacalixarenes are relatively recent additions to the field. They were serendipitously discovered by a group working on synthesising polyphenols by a condensation reaction with sulfur. Much of the history and properties of these materials is extensively reviewed by Morohashi *et al.*[137] We will only present a short study of their chemistry, since much of it is very similar to the carbon-linked calixarenes – but will instead highlight the differences between the two systems.

Thiacalixarenes are synthesised by a condensation reaction similar to that of the calixarenes. The initial report described how t-butyl phenol, sulfur and sodium hydroxide were heated to 230 °C and then purified to give reasonable yields of t-butylthiacalix-4-arene[138] (Figure 4.25e). This compound had actually been synthesised earlier by a stepwise procedure,[139] along with calixarenes with mixed sulfur and methylene bridges, although yields were rather poor. Use of a *bis*-t-butylphenol sulfur dimer improved the yield of the tetramer and gave isolatable amount of hexamer and octamer.[140]

Crystal structures of the tetramer[141] and octamer[142] have shown that the thiacalixarenes adopt similar conformations (cone for tetramer, pleated loop for the octamer) to those of the corresponding calixarenes. However, the greater length of the C—S bond compared to the C—C bond (by about 15%) does mean that thiacalixarenes are somewhat larger than their all-carbon analogues.[137] This had led to differences in their inclusion behaviour towards some solvents compared to the methylene-bridged calixarenes, a subject again reviewed by Morohashi *et al.*[137]

Much of the physics and chemistry of the thiacalixarenes are similar to those of their analogues; for example, solvents and other molecules can be complexed within the cavity; removal of t-butyl groups allows reactions at the upper rim, with the lower rim being capable of being esterified, etherified and undergoing many other reactions.[137] The thiacalixarenes easily interconvert in solution and are thought to be somewhat more flexible than the calixarenes.[137] The effect of the larger cavity can be seen for the tetrapropyl ether of thiacalixarene, which still shows conformational mobility at higher temperatures,[143] unlike the calixarene analogue.[20] Heating the thiacalixarene solution and cooling to room temperature allows isolation of the conformers.

The presence of the sulfur-bridging groups allows novel chemistry and binding behaviour not seen in calixarenes. The t-butylthiacalix-4-arene can easily be oxidised by hydrogen peroxide[144] to give the tetrasulfone (Figure 4.25f), with similar oxidations capable of being performed on the larger thiacalixarenes. X-ray crystallography showed that the tetrasulfones preferred the 1,3-alternate conformation[144] to the cone of a methylene-bridged calixarene, which is probably due to hydrogen-bonding interactions between the hydroxyl and sulfone groups.

The sulfinyl calixarenes, in which the S atom has been oxidised to give an S=O group, have also been synthesised. What is of interest is that because of the tetrahedral orientation around the sulfur atom, there are four isomers, since the S=O bond can be orientated with the oxygen pointed towards or away from the hydroxyl groups, as shown in Figure 4.26a. The *rtct* and *rctt* isomers can both be isolated from a simple oxidation of the parent thiacalixarene, whereas synthesis of the tetrabenzyl ester in different conformations and oxidation followed by debenzylation lead to the production of the other two isomers.[145] Conversion to the methyl ethers and X-ray crystallography confirmed the conformations of these isomers.

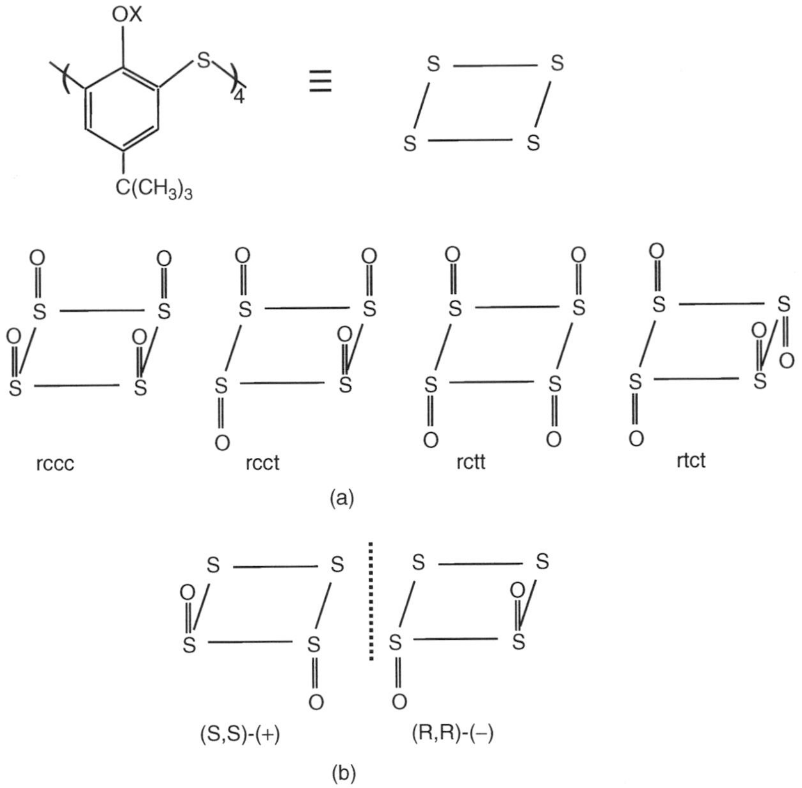

Figure 4.26   *Stereoisomer of sulphinyl calixarenes*

Besides complete oxidation of the sulfur atoms, it has also been possible to partially oxidise the thiacalixarenes. The tetramethyl ether of t-butylthiacalix-4-arene could be oxidised by two equivalents of sodium borate to give a mixture of monosulfinyl and two isomers of the disulfinyl thiacalixarenes.[146] The disulfinyl shown in Figure 4.26b is of especial interest since it has no centre of symmetry and is inherently chiral. The isomers shown could be separated by chiral HPLC, while their structures were confirmed by X-ray crystallography.

Presence of the sulfur atoms also modifies the metal-binding ability of thiacalixarenes compared to calixarenes. Examples of this include the thiacalix-4-crowns showing better extraction of silver ions than alkali metals, which are preferred by the calix-4-crowns.[147] Similar thiacalixcrowns have been shown to strongly complex $^{226}Ra^{2+}$, which occurs naturally and is found in mining waste streams.[148] What is of even more interest however is that whereas the unsubstituted calixarenes are relatively poor ion binders and transporters, the thiacalixarenes bind a much wider range of metal ions.[137] Thia, sulfinyl and sulfonylcalixarenes have been shown to transport ions including nickel, cobalt, copper, zinc and alkaline earths much more effectively than calixarenes.[149,150] The tendency is for the thiacalixarenes to transport 'soft' ions and the sulfonylcalixarenes 'hard' ions, with the sulfinyl being intermediate and transporting both types.[151] It will be interesting to see whether these and other properties will allow thiacalixarenes to rival calixarenes as objects of research.

4.8   Resorcinarenes: Synthesis and Structure

The condensation of resorcinol and related compounds with aldehydes has been known since the days of Baeyer,[1] who reacted benzaldehyde with pyrogallol but was unable to fully characterise the product. The syntheses of several of these compounds were described by Niederl and Vogel,[4] who also proposed the cyclic tetrameric structure. The resorcinarenes, although like the calixarenes in some ways, have some basic differences. Whereas calixarenes tend to be made from formaldehyde, a much wider range of aldehydes are suitable for the synthesis of resorcinarenes, which adds a degree of complexity to the stereochemistry of these materials since the substituents restrict rotation. In a calixarene in the cone formation, the hydroxyls are attached to the 'narrow' lower rim of the calix and take part in cyclic hydrogen bonding, whereas in resorcinarenes the hydroxyls are on the upper rim, as shown in Figure 4.1. This makes it impossible to achieve a cyclic hydrogen and instead the hydroxyls tend to exist as hydrogen-bonded pairs. Much of the earlier work on resorcinarenes has been extensively reviewed.[152] These compounds have been variously named 'resorcinarenes', 'calix-n-resorcinarenes' and 'resorcarenes' and are usually but not always tetramers. We will mainly refer to them as resorcinarenes throughout this work unless we wish to specify the ring size.

The resorcinarenes are usually synthesised by acid-catalysed reactions between resorcinol and various aldehydes, apart from formaldehyde, which although it undergoes ready condensation with resorcinol under both acidic and basic conditions, results in materials that are polymeric in nature and widely used within the adhesives industry. Much of the groundbreaking work on the syntheses of resorcinarenes was carried out by Hogberg, who reacted a range of aldehydes including acetaldehyde[153] and benzaldehyde[154] with resorcinol under different acidic conditions. There are a range of possible conformations for the resorcinarenes (Figure 4.27a), known as crown (or cone), saddle, boat, diamond and chair. The bridging substituents can also exist is a variety of *cis* and *trans* conformations (Figure 4.27b). From this it can be deduced that there are a large range of possible isomers for a single resorcinarene. Using the methods described by Hogberg to synthesise and esterify these materials, it was found that it was possible to isolate two separate conformers whose relative yields varied with differing reaction conditions and times. What is important is that these studies showed that the reaction is in constant equilibrium and whereas

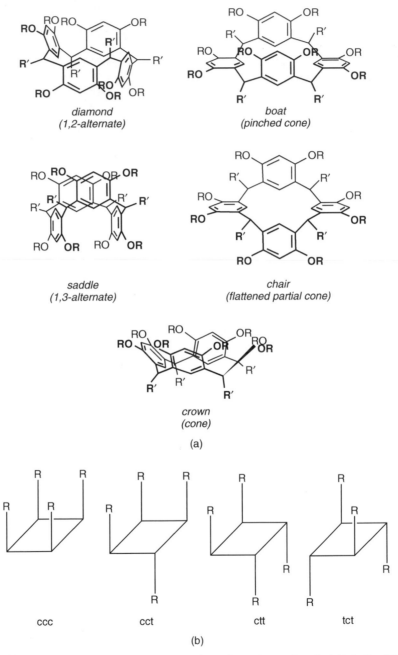

Figure 4.27   *Conformations of resorcinarenes (Copyright Wiley-VCH Verlag GmbH & Co. KGaA. Reproduced with permission[199])*

one product is favoured kinetically and is the major constituent after a short reaction time, longer times lead to the formation of the second isomer as it is thermodynamically more stable. Mass spectra of the esters of these materials proved the tetrameric structure and from NMR it was possible to deduce the first product to be the chair conformer and the second the boat. Other work also provided the first X-ray crystallographic proof of a tetramer structure – that of the octabutyrate of the reaction product of resorcinol and p-bromobenzaldehyde.[155] A more recent study showed that in the early part of the reaction a wide range of oligomeric products are produced,[156] but since the condensation reaction is reversible, as the reaction proceeds, the thermodynamically most stable products dominate.

Further work was focussed towards the reaction of a range of aldehydes[157] and substituted resorcinols. Simple aliphatic aldehydes were shown to give just the crown conformer, whereas with *para*-substituted benzaldehydes the composition depended on the substituent but was usually a mixture of crown and boat – although the presence of some substituents such as nitro prevented condensation from occurring. A substituted resorcinol, 2-methyl resorcinol, also formed tetramers, although no product was isolated for the corresponding nitro and bromo compounds. Crystal structures of two of the compounds showed them to be in the crown conformation and include solvent (acetonitrile). More recently, dodecanal tetramers with resorcinol[158] and pyrogallol[159] have been crystallised from ethanol and have been shown to adopt the crown conformation with bowl-to-bowl hydrogen bonding and interdigitation of the long side chains (Figure 4.1b). Resorcinols that have been blocked at the 2-position, such as pyrogallol and 2-methyl resorcinol,[160] can form cyclic compounds with formaldehyde. A detailed list of the aldehydes that have been utilised is given by Timmerman *et al*[152] and as can be seen, most aldehydes can be used unless they are extremely sterically hindered or contain active groups very close to the aldehyde such as glucose or $ClCH_2CHO$.

There are variations on these syntheses: resorcinarenes can be synthesised by simply combining aldehyde, resorcinol and a catalytic amount of toluene sulfonic acid and then grinding together in a mortar and pestle for just a few minutes to give high yields of the pure resorcinarene[161] under solvent-free conditions. Condensation of 2-methyl, propyl or hexyl resorcinol with trioxane under acid conditions[162] gave a mixture of the cyclic tetramers and hexamers (prolonged reaction times led to formation of the tetramer only) and NMR studies showed the macrocyclic ring to be highly flexible in solution. Condensation of 2,4-allyloxybenzyl alcohol using a Lewis acid catalyst, scandium trifluoromethyl sulfonate,[163] gave a cyclic tetramer (Figure 4.28a) which could have the protective allyl groups removed to give the parent resorcinarene (Figure 4.28b), which again was shown to have high conformational flexibility. Other reactions include the Lewis acid-catalysed condensation of 2,4-dimethoxy cinnamates, with for instance the reaction of 2,4-dimethoxy cinnamic acid, methyl ester (Figure 4.28c), with BF_3/diethyl ether[164] giving the calix-4-resorcinarene octamethyl ether (Figure 4.28d) in a mixture of diamond and boat conformers. A wide variety of other esters can be used and the parent resorcinarenes generated by removal of the methyl groups with BBr_3. One interesting variation on this condensation reaction is to utilise the monomethyl ether of resorcinol to form a resorcinarene to give 95% of a product, which we would initially expect to be a highly complex mixture due to the fact that the resorcinol units could be incorporated into the macrocycle in different orientations respective to each other. However, it was found that the hydroxy group of one resorcinol unit is always adjacent to the methoxy of the adjoining unit, as shown in Figure 4.28e,[165] and this structure is probably thermodynamically favoured since it maximises hydrogen bonding. This macrocycle does not have a centre of symmetry, making it inherently chiral.

Many of the reactions described previously for calixarenes work equally effectively for resorcinarenes. Resorcinarenes can be alkylated to give ethers, and silylated, phosphorylated or esterified with acid anhydrides or chlorides. Many of the calixarene reactions depend on removing the group *para* to the phenols, followed by reactions at the activated *para* position. This approach cannot be followed with resorcinarenes since the hydroxy groups are *para* to the bridging groups and reactions at these positions would tear apart

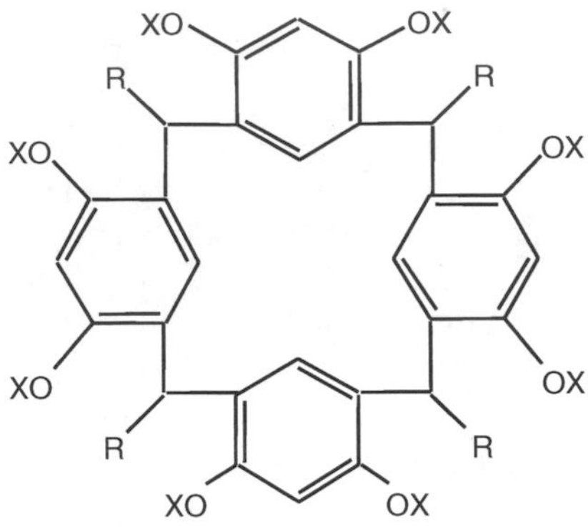

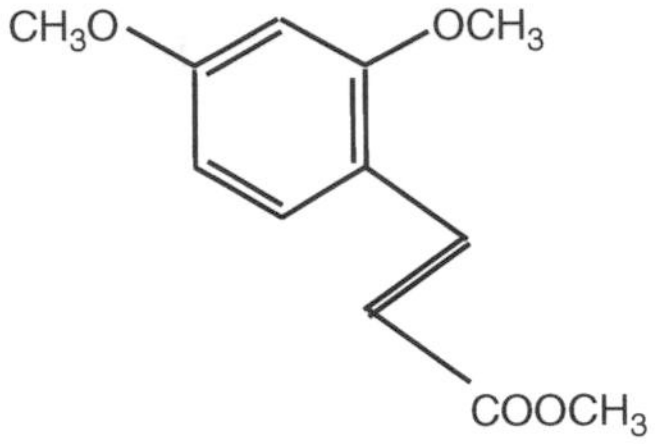

(a) R=Y= H, X = CH₂CH=CH₂

(b) R=X= H

(d) R= CH₂COOCH₃, X=CH₃

(c)

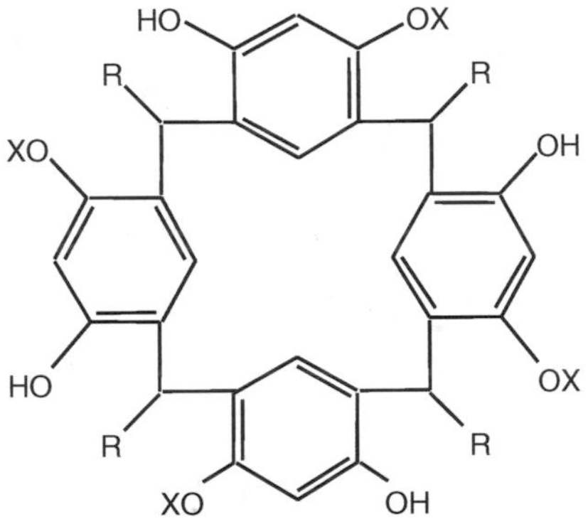

(e) R= –C₅H₁₁, X=CH₃

Figure 4.28   *Structures of resorcinarenes*

the macrocycle. However, the presence of two electron-releasing hydroxyl groups *ortho* to the 2-position means that it is highly activated and can readily take part in substitution reactions.

Bromination with n-bromosuccinimide[166] gives the tetrabromo compounds such as that shown in Figure 4.29a. In a similar manner, a water-soluble resorcinarene can be synthesised by diazo coupling (Figure 4.29b), and this is capable of solubilising large aromatic molecules such as coronene in water.

A variety of Mannich reactions can also take place, with for example the reaction with formaldehyde and sodium sulfite[167] giving rise to water-soluble resorcinarenes of the type shown in Figure 4.29c. Reactions with amines have also been utilised, when a secondary amine is used, such as the reaction with formaldehyde and diethylamine,[168] the product having the structure shown in Figure 4.29d. When a primary amine is used with one equivalent of formaldehyde, a complex mixture is obtained; however, when two equivalents of formaldehyde are utilised,[168] the final product has the structure shown in Figure 4.29e. What probably happens is an initial Mannich reaction with the amine at the 2-position, followed by further condensation to give the oxazine rings; of interest is that the resorcinarene again exists as one isomer

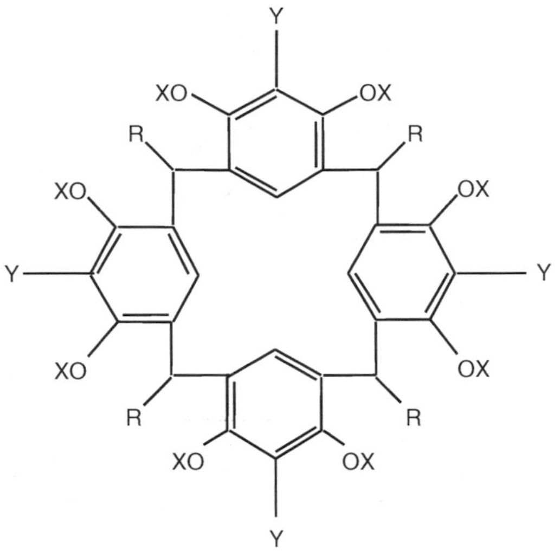

(a) R= CH$_3$, X = H, Y = Br

(b) R= CH$_3$, X= H, Y = –N=N– ⬡ –SO$_3$Na

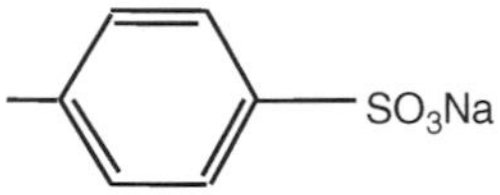

(c) R= CH$_3$, X= H, Y= SO$_3$Na

(d) R= CH$_3$, X= H, Y= CH$_2$N(CH$_2$CH$_3$)$_2$

(f) R= CH$_2$CH$_2$SO$_3$Na, X= H, Y = H or OH

(g) R= (CH$_2$)$_{10}$SH, X= H, Y = H

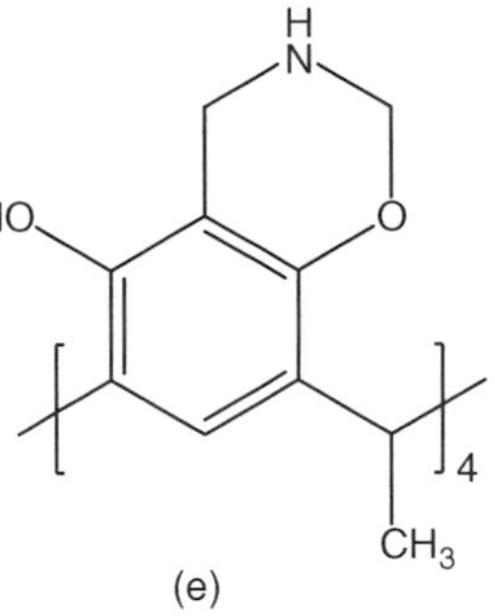

(e)

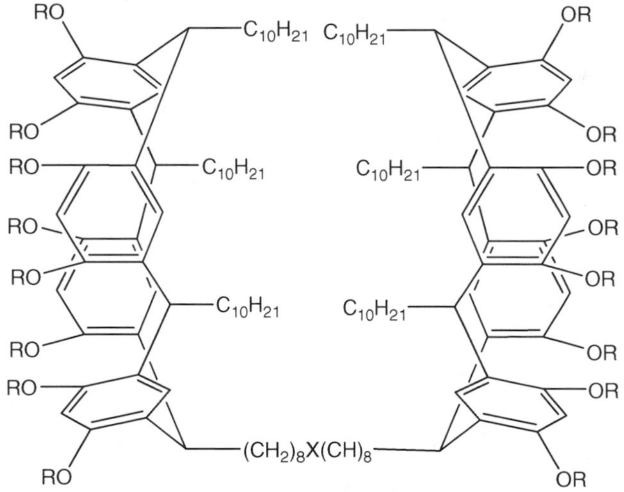

(f) R = pivalate, X = –CH$_2$CH=CHCH$_2$– or –(CH$_2$)$_4$–

Figure 4.29   *Substituted resorcinarenes*

and is inherently chiral. Diasterioisomers can be obtained when a chiral amine (2-aminobutane) is used; however, only one diasterioisomer is formed for each amine enantiomer, indicating that the chirality of the amine has a high affect on the structure and chirality of the macrocycle.[169]

There has been a great deal of work concerning bridging of resorcinarenes, most of which will be dealt with later in this chapter, via reactions of the hydroxy groups or the 2-positions. However, other methods do exist, with for example resorcinol being condensed with a mixture of undecanal and 10-undecenal to give a mixture of resorcinarenes with different ratios of the two side chains.[170] Protection of the hydroxyls with pivalate groups and careful chromatography allows isolation of resorcinarenes bearing one unsaturated side chain. The double bonds can be converted to give a wide variety of monofunctionalised resorcinarenes – or they can be linked together with a Grubbs catalyst, which allows synthesis of *bis*-resorcinarenes linked by a single side chain, as shown in Figure 4.29f.

Like their counterparts the calixarenes, the complex-forming properties of the resorcinarenes have been widely investigated. Much of the complexation chemistry of resorcinarenes has been reviewed[152] and several groups of compounds have been studied. One of the earliest investigations was of the solubilisation of various sugars in chloroform by a resorcinarene bearing four undecyl sidechains (Figure 4.1b). This compound was shown to form 1 : 1 complexes with sugars with a great deal of selectivity,[171] and was found for example to form strong complexes with fucose and xylose, although only minimal amounts of glucose, galactose and mannose were extracted. This was thought to be due to the stereochemistry of the sugars and only those that could bind via a two-point attachment would be strongly extracted. In the case of ribose this was confirmed by the fact that the sugar only extracted in its pyranose form. This rule also held true for simple cyclohexanediols, with the *cis*-1,4-isomer being most strongly extracted – showing high selectivity over the *trans* isomer. Glycerol and water were also extracted into the chloroform phase but only via single-point binding and as a 4 : 1 complex. No transfer was observed when 4-dodecyl resorcinol or the octa-acetate of the resorcinarene was used. Binding was also observed in aqueous systems using a water-soluble resorcinarene[172] (Figure 4.29f) or in monolayers of resorcinarenes on water,[173] although with different selectivities. Other compounds that were found to form complexes were dicarboxylic acids, again with strong selectivity and with glutaric acid being preferred over pimellic and malonic.[173]

Complexation of water-soluble resorcinarenes (Figure 4.29f) with amino acids has also been studied. It was found that only amino acids with a hydrophobic residue bound, whereas there was no affinity for hydrophilic compounds such as serine.[174] It is also possible to dissolve resorcinarenes in basic solutions to give the tetraanion in which four of the phenolic groups have ionised, with ionisation of the other hydroxyls requiring much harsher conditions. The tetraanion binds ammonium compounds very well via an electrostatic interaction, with large side chains on the ammonium salt reducing the binding constant.[175] Neutral resorcinarenes have also been shown to form complexes with ammonium cations in ethanol/water, and in the case of triethylammonium sulfate a 1 : 1 complex was crystallised, with X-ray diffraction showing the resorcinarene to be in the cone conformation[176] (unlike the uncomplexed resorcinarene, which is a mixture of boat and chair[153]). Pyridinium compounds also complex with resorcinarene, with for example a fluorescent pyrene–pyridinium compound forming a complex with resorcinarene tetraanion, with the pyrene fluorescence being quenched. Addition of the neurotransmitter acetyl choline displaces the pyridinium and restores fluorescence.[177]

The facile substitution of the side chains has also led to investigation of these systems as immobilised monolayers. Immobilisation of a thiol-substituted resorcinarene (Figure 4.29g) onto a gold surface gave a system which also showed high selectivity for some polyhydroxy compounds, such as ascorbic acid, over others.[157–179] Sulfide-substituted resorcinarenes have also been adsorbed onto gold and shown to bind aromatic molecules from the aqueous phase[180] or chlorinated solvents from the vapour phase.[181]

4.9   Cavitands and Carcerands

The proximity of the hydroxyls of adjacent resorcinol units makes bridging an attractive method for fixing the cavity into a cone shape and also for extending the cavity. These fixed-cone-type molecules have been extensively described in the Crams' book (see Bibliography); highlights will be given here. Resorcinarenes are particularly attractive because a wide variety of functional groups can be introduced on the lower rim, thereby allowing manipulation of solubility or attachment of the resorcinarenes to solid supports without affecting the structure and chemistry of the upper rim. One of the simplest bridges is a methylene bridge, as shown in Figure 4.30a, where the resorcinarene in Figure 4.29a is reacted with CH_2BrCl and base.[166] Bridges containing two or three carbon atoms could also be incorporated. X-ray crystallographic studies proved the bridged resorcinarenes to be in the cone conformation, as well as including guest molecules of the various crystallisation solvents. Variation of the ratios of resorcinarene and bridging compound allowed the synthesis of tribridged, dibridged and monobridged resorcinarenes with unreacted —OH groups.[182] Both A,B and A,C dibridged resorcinarenes have been synthesised.[182] More complex bridging ligands have also been utilised, some of which will be described here. When 2,3-dichloroquinoxaline is utilised, a good yield of a resorcinarene (Figure 4.30b) with an extended cavity is obtained.[183] NMR studies show that the compound can exist in a mixture of 'vase' form, where the quinoxaline rings are essentially perpendicular to the plane of the resorcinarene macrocycle, and 'kite' form, where they are parallel, depending on the temperature and solvent. The 'vase' form has been proved by X-ray crystallography[183] and has been shown to contain a cavity approximately 0.7 nm in diameter and 0.8 nm deep. Again, partially substituted derivatives have also been synthesised. Quinone-bridging groups have also been used, such as for example the structure[184] in Figure 4.30c, which is formed by the reaction of a resorcinarene with 2,3-dichloronaphthoquinone. Other bridging groups include dialkylsilyl,[185] such as that shown in Figure 4.30d, which selectively complexes linear guests such as O_2 or CS_2, and phosphoryl[186] hyphen;bridged compounds (Figure 4.30e), which strongly bind copper or gold ions.

The next stage after synthesis of these cavitands is to attempt to join them together to form carcerands, where two resorcinarene bowls are linked together to form an approximately spherical host molecule (the carcerand), which may contain a trapped guest, unable to leave without breaking covalent bonds in the host. One of the first of these was obtained[187] by bridging a resorcinarene with methylene groups and then substituting the 2-position to give the two compounds in Figure 4.30f,g. These could then be coupled together to give a double resorcinarene with the two macrocycles linked together by four —CH_2SCH_2— linkers. Analysis showed that the carcerand contained solvent (DMF), water and caesium ions, and even a small amount of the inert atmosphere (Ar) atoms was trapped. Many of these species could only be released by hydrolysing the carcerand with trifluoroacetic acid.

Although the carcerand is a molecule of interest, it is limited by the fact it can only capture the guest during the closure process and cannot release it. The hemicarcerands, which are molecules that can capture and release a guest, have been objects of wider studies. There are two ways to achieve these. The first is to use less linkages, such as the compound shown in Figure 4.31a, which has only three methylene-bridging groups and contains a 'portal' through which small molecules can enter and leave.[188] Alternatively, the compounds described by Cram *et al.*[189] may be used, where groups have been used not only to bridge between the hydroxyls of a resorcinarene but to form one or more bridges between two different resorcinarene units, by using tetra-substituted species such as tetrafluorobenzoquinone or similar quinoxalines. Another approach involves using four-fold bridging; if the bridges are flexible enough, guests may enter and leave the cavity. Typical bridges include Schiff-base bridges, made by synthesising the resorcinarene with four aldehyde groups (Figure 4.30h), which can be bridged using either 1,2- or

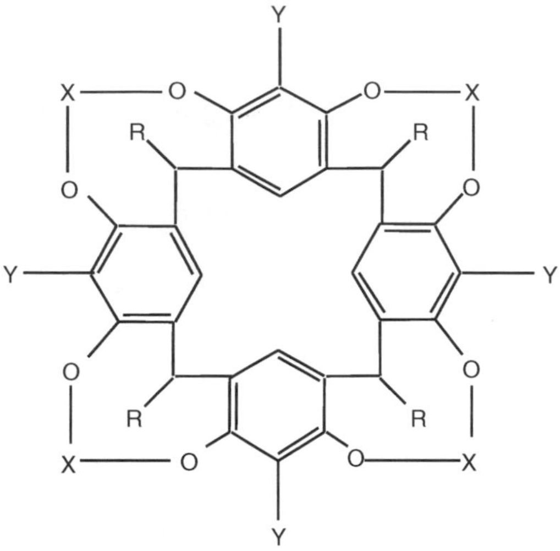

(a) R= CH$_3$, Y = Br, X = –CH$_2$–

(b) R= CH$_3$, Y= H, X =

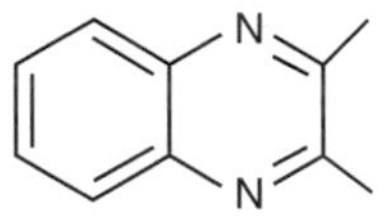

(c) R= CH$_3$, Y= H, X=

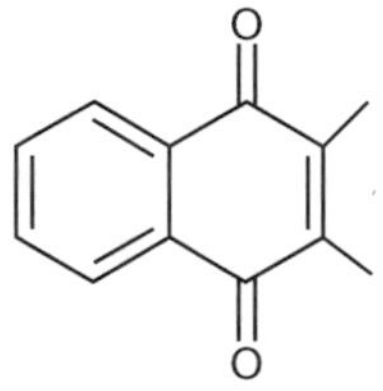

(d) R= CH$_2$CH$_2$Ph, Y= H, Y= Si(CH$_3$)$_2$
(e) R= CH$_2$CH$_2$Ph, Y= H, X= PPh
(f) R= CH$_3$, Y = CH$_2$Cl, X = –CH$_2$–
(g) R= CH$_3$, Y = CH$_2$SH, X = –CH$_2$–
(h) R= CH$_2$CH$_2$Ph, Y= CHO, X= –CH$_2$–

Figure 4.30 *Bridged resorcinarenes*

1,3-diaminobenzene.[190] This hemicarcerand has been shown to be capable of including large guests such as ferrocene, [2,2']paracyclophane and adamantane. Another approach was to utilise Mannich reactions to join together resorcinarenes;[191] however, these compounds (Figure 4.31c) tended to collapse together and no guest incorporation could be detected. Many of these compounds are extensively reviewed in the Crams' book (see Bibliography).

One interesting aspect of these systems is the potential to utilise them as nanoreactors. Compounds that are unstable in solution may well be protected by encapsulation within a hemicarcerand. An example of this is the room-temperature synthesis of cyclobutadiene;[188] cyclobutadiene can be photochemically

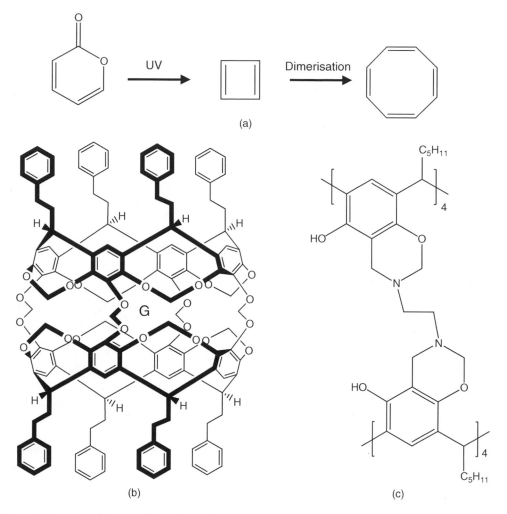

Figure 4.31 *(a) Synthesis of butadiene, (b) the hemicarcerand and (c) Mannich-bridged carcerands*

synthesised from α-pyrone (Figure 4.31a), but this compound is highly unstable, readily reacting with oxygen or solvents or dimerising to form cyclooctatetraene. Refluxing solutions of the hemicarcerand with α-pyrone was found to cause the pyrone to become entrapped, and upon irradiation this reacted to form cyclobutadiene within the cavity. This species could be detected by NMR and by trapping experiments with oxygen. Heating the hemicarcerand solution caused loss of butadiene from the cavity, with the formation of cyclooctatetraene.

The joining together of two resorcinarenes usually (but not always) gives a symmetric carcerand or hemicarcerand. As an alternative system, the combination of resorcinarenes with calixarenes has been studied. The synthesis of 1:1, 1:2, 2:1 and 2:2 compounds of resorcinarenes and calix-4-arenes has been reported,[192] with the 2:2 adduct being shown by NMR and modelling studies to be rigid and to contain a pore approximately 1 nm³ in volume. Similarly, a 2:1 calix-4-arene:resorcinarene adduct was synthesised and shown to selectively form complexes with steroids such as prednisolone-21-acetate.[193]

A 1:1 adduct of a resorcinarene with a calix-4-arene was shown to include dimethylacetamide as a guest,[194] and because of the size and rigidity of the system, at low temperatures the guest could adopt two different orientations within the cavity rather than rotating freely. This raises the possibility of using such systems as molecular switches, while a sulfide-substituted version of the carcerand has been synthesised and immobilised on a gold surface.[195]

The multiple systems described above are all linked by covalent bonds. However, multiple resorcinarene systems have been shown to assemble in solution. C-undecyl resorcinarene (molecular weight 1104) has shown by vapour pressure osmometry to have apparent molecular weights of between 3000 and 7000 in nonhydrogen-bonding solvents,[196] but of 1100 in ethanol, indicating that the resorcinarenes associate in some solutions. More detailed work showed that the C-undecylresorcinarene associates in benzene solution, along with adventitious water, to form a hexameric species,[197] and in the case of the C-methylresorcinarene this has been crystallised and shown by X-ray crystallography to have a snub cube geometry. A resorcinarene with fluorinated side chains has been shown to form a hexamer in wet fluorinated solvents.[198]

4.10   Conclusions

A variety of uses have been proposed for calixarenes and their related systems. The advantages of these materials includes their ability to be made organic or water-soluble, their ability to be easily immobilised, their chemical and thermal stability compared to many other macrocyclic systems, their ability to be exquisitely tailored and the fact that many can be synthesised in bulk from inexpensive materials.

The ion-binding properties of these materials are of great interest to the nuclear industry as they have the potential to be used as extractants for the removal of radioactive material from waste streams. There has also been interest in using them as potential extractants for uranium from seawater. The high selectivity of calixarenes has also led to their commercialisation as ion binders for use in ion-selective electrodes, with three of the Selectophore range of ionophores, namely amine ionophore I, caesium ionophore I and calcium ionophore VI, being calixarene-based.

A variety of other uses have been proposed for these compounds and in his most recent book Gutsche gives details of some of the 200+ patents listed for them (see Bibliography); they have been investigated for uses including the active components of sensors, catalysts and stationary phases for HPLC. It would be fitting if calixarenes were to prove suitable for commercial uses since they were originally discovered as a by-product of industrial products such as Bakelite and the Petrolite series of compounds.

Bibliography

Gutsche CD. Calixarenes. Monographs in Supramolecular Chemistry. Royal Society of Chemistry; 1989.

Gutsche CD. Calixarenes Revisited. Monographs in Supramolecular Chemistry. Royal Society of Chemistry; 1998.

Gutsche CD. Calixarenes: An Introduction. Monographs in Supramolecular Chemistry. Royal Society of Chemistry; 2008.

Vicens J, Böhmer V. Calixarenes: A Versatile Class of Macrocyclic Compounds. Topics in Inclusion Science. Springer; 1990.

Cram DJ, Cram JM. Container Molecules and Their Guests. Monographs in Supramolecular Chemistry S. Royal Society of Chemistry; 1994.

Mandolini L, Ungaro R. Calixarenes in Action. Imperial College Press; 2000.

Asfari M-Z, Böhmer V, Harrowfield J, Vicens J. Calixarenes 2001. Springer; 2001.

Baklouit L, Vicens J, Harrowfield J. Calixarenes in the Nanoworld. Springer; 2006.

Sliwa W, Kozlowski C. Calixarenes and Resorcinarenes: Synthesis, Properties and Applications. Wiley VCH; 2009.

References

1. Baeyer A. Ueber die Verbindungen der Aldehyde mit den Phenolen [About the binding of aldehydes with phenols]. *Chem Ber.* 1872; **5**: 280–282.
2. Baeyer A. Ueber die Verbindungen der Aldehyde mit den Phenolen und aromatischen Kohlenwasserstoffen [About the binding of aldehydes with phenols and aromatic carbohydrates]. *Chem Ber.* 1872; **5**: 1094–1100.
3. Zinke A, Kretz R, Leggewie E, Hossinger K. Zur Kenntnis des Härtungsprozesses von Phenol-Formaldehyd-Harzen [About the knowledge of the hardening process of phenol-formaldehyde resins]. *Monatsh Chem.* 1952; **83**: 1213–1227.
4. Niederl JB, Vogel HJ. Aldehyde–resorcinol condensations. *J Am Chem Soc.* 1940; **62**: 2512–2514.
5. Gutsche CD, Iqbal M, Stewart D. Calixarenes. 19. Syntheses procedures for p-tert-butylcalix[4]arene. *J Org Chem.* 1986; **51**: 742–745.
6. Gutsche CD, Dhawan B, Leonis M. p-tert-butylcalix[6]arene. *Org Synth.* 1990; **68**: 238–242.
7. Munch JH, Gutsche CD. p-tert-butylcalix[8]arene. *Org Synth.* 1990; **68**: 243–246.
8. Stewart DR, Gutsche CD. The one-step synthesis of p-tert-Butylcalix[5]arene. *Org Prep Proc Inc.* 1993; **25**: 137–140.
9. Vocanson F, Lamartine R, Lanteri P, Longeray R, Gauvrit I. A one-step synthesis of p-tert-butylcalix[7]arene is proposed–optimization is carried out by the method of experimental plants. *New J Chem.* 1995; **19**: 825–829.
10. Stewart DR, Gutsche CD. Isolation, characterization, and conformational characteristics of p-tert-Butylcalix[9–20]arenes. *J Am Chem Soc.* 1999; **121**: 4136–4146.
11. Andreetti GD, Ungaro R, Pochini A. Crystal and molecular structure of cyclo{quater[(5-t-butyl-2-hydroxy-1,3-phenylene)methylene]} toluene (1 : 1)clathrate. *J Chem Soc Chem Commun.* 1979; 1005–1007.
12. Andreetti GD, Pochini A, Ungaro R. Molecular inclusion in functionalized macrocycles. Part 6. The crystal and molecular structures of the calix[4]arene from p-(1,1,3,3-tetramethylbutyl)phenol and its 1 : 1 complex with toluene. *J Chem Soc. Perkin Trans.* 1983; **2**: 1773–1779.
13. Juneja RK, Robinson KD, Johnson CP, Atwood JL. Synthesis and characterization of rigid, deep-cavity calix[4]arenes. *J Am Chem Soc.* 1993; **115**: 3818–3819.
14. Wolfgong WJ, Talafuse LK, Smith JM, Adams MJ, Adeogba F, Valenzuela M, Rodriguez E, Contreras K, Carter DM, Bacchus A, McGuffey AR, Bott SG. The influence of solvent of crystallization upon the solid-state conformation of calix[6]arenes. *Supramol Chem.* 1996; **7**: 67–78.
15. Gutsche CD, Gutsche AE, Karaulov AI. Calixarenes 11. Crystal and molecular structure of p-tert-butylcalix[8]arene. *J Incl Phenom Macro Chem.* 1985; **3**: 447–451.
16. Czugler M, Tisza S, Speier G. Versatility in inclusion hosts. Unusual conformation in the crystal structure of the p-t-butylcalix[8]arene: pyridine (1 : 8) clathrate. *J Incl Phenom Macro Chem.* 1991; **11**: 323–331.
17. Perrin M, Ehlinger N, Viola-Motta L, Lecocq S, Dumazet I, Bouoit-Montesino S, Lamartine R. Crystal structures of two calix[10]arenes complexed with neutral molecules. *J Incl Phenom Macro Chem.* 2001; **39**: 273–276.
18. Leverd PC, Nierlich M, Dumazet-Bonnamour I, Lamartine R. Using a large calixarene as a polyalkoxide ligand: tert-butylcalix[12]arene and its complex with the uranyl cation. *J Chem Soc Chem Commun.* 2000; 493–494.
19. Bavoux C, Baudry R, Dumazet-Bonnamour I, Lamartine R, Perrin M. Large calixarenes: structure and conformation of a calix[16]arene complexed with neutral molecules. *J Incl Phenom Macro Chem.* 2001; **40**: 221–224.
20. Iwamoto K, Araki K, Shinkai S. Conformations and structures of tetra-O-alkyl-p-tert-butylcalix[4]arenes. How is the conformation of calix[4]arenes immobilized? *J Org Chem.* 1991; **56**: 4955–4962.
21. Ferguson G, Notti A, Pappalardo S, Parisi MF, Spek AL. Influence of the size of upper and lower rim substituents on the fluxional and complexation behaviour of calix[5]arenes. *Tet Lett.* 1998; **39**: 1965–1968.
22. van Duynhoven JPM, Janssen RG, Verboom W, Franken SM, Casnati A, Pochini A, Ungaro R, de Mendoza J, Nieto PM. Control of calix[6]arene conformations by self-inclusion of 1,3,5-tri-O-alkyl substituents: synthesis and NMR studies. *J Am Chem Soc.* 1994; **166**: 5814–5822.
23. Gutsche CD, Muthukrishnan R. Calixarenes. 1. Analysis of the product mixtures produced by the base-catalyzed condensation of formaldehyde with para-substituted phenols. *J Org Chem.* 1978; **43**: 4905–4906.

24. Neri P, Battocolo E, Cunsolo F, Geraci C, Piattelli M. Study on the alkylation of p-tert-butylcalix[8]arene. Partially O-alkylated calix[8]arenes. *J Org Chem.* 1994; **59**: 3880–3889.

25. Neri P, Consoli GML, Cunsolo F, Rocco C, Piattelli M. Alkylation products of a calix[8]arene trianion. effect of charge redistribution in intermediates. *J Org Chem.* 1997; **62**: 4236–4239.

26. No K, Koo HJ. Conformations of the acyl esters of p-tert-butylcalix[4]arene and calix[4]arene. *Bull Kor Chem Soc.* 1994; **15**: 483–488.

27. Gutsche CD, Lin L-G. Calixarenes. 12. *The synthesis of functionalized calixarenes' tetrahedron.* 1986; **42**: 1633–1640.

28. Shu C-M, Liu W-C, Ku M-C, Tang F-S, Yeh M-L, Lin L-G. 25,27-bis(benzoyloxy)calix[4]arenes: synthesis and structure elucidation of syn and anti isomers. *J Org Chem.* 1994; **59**: 3730–3733.

29. See KA, Fronczek FR, Watson WH, Kashyap RP, Gutsche CD. Calixarenes. 26. Selective esterification and selective ester cleavage of calix[4]arenes. *J Org Chem.* 1991; **56**: 7256–7268.

30. Moran JK, Roundhill DM. Introduction of diphenylphosphinite functional groups onto selected positions on the lower rim of calix[4]arenes and calix[6]arenes. *Inorg Chem.* 1992; **31**: 4213–4215.

31. Casnati A, Pochini A, Ungaro R, Ugozzoli F, Arnaud F, Fanni F, Schwing M-J, Egberink RJM, de Jong F, Reinhoudt DN. synthesis, complexation, and membrane transport studies of 1,3-alternate calix[4]arene-crown-6 conformers: a new class of cesium selective ionophores. *J Am Chem Soc.* 1995; **117**: 2767–2777.

32. Ghidini E, Ugozzoli F, Ungaro R, Harkema S, El-Fadl AA, Reinhoudt DN. Complexation of alkali metal cations by conformationally rigid, stereoisomeric calix[4]arene crown ethers: a quantitative evaluation of preorganization. *J Am Chem Soc.* 1990; **112**: 6979–6985.

33. Asfari Z, Pappalardo S, Vicens J. New preorganized calix[4]arenes. Part I. A doubly-crowned calix and a double-calixcrown derived from 4,6,10,12,16,18,22,24,25,26,27,28-Dodecamethyl-5,11,17,23-tetrahydroxycalix[4]arene. *J Mol Inclus Chem.* 1992; **14**: 189–192.

34. Jabin I, Reinaud O. First C_{3v}-symmetrical calix[6](aza)crown. *J Org Chem.* 2003; **68**: 3416–3419.

35. Caccamese S, Principato GA, Geraci CA, Neri P. Resolution of inherently chiral 1,4-2,5-calix[8]bis-crown-4 derivatives by enantioselective HPLC. *Tet Asymm.* 1997; **8**: 1169–1173.

36. Cunsolo F, Consoli GML, Piattelli M, Neri P. Double-two-point and four-point intramolecular bridging of p-tert-butylcalix[8]arene. *Tet Lett.* 1996; **39**: 715–718.

37. Eggert JPW, Harrowfield J, Lüning U, Skelton BW, White AH, Löffler F, Konrad S. Improved synthesis and conformational analysis of an A,D-1,10-phenanthroline-bridged calix[6]arene. *Eur J Org Chem.* 2005; 1348–1353.

38. Dondoni A, Hu X, Marra A, Banks HD. First synthesis of bridged and double calixsugars. *Tet Lett.* 2001; **42**: 3295–3298.

39. Sood P, Zhang H, Lattman M. Isolation of All four conformations of p-tert-butylcalix[5]arene using bridging silyl groups. *Organometallics.* 2002; **21**: 4442–4447.

40. Gloede J, Ozegowski S, Costisella B, Gutsche CD. Threefold bridged p-tert-butylcalix[9]arene triphosphate. *Eur J Org Chem.* 2003; 4870–4873.

41. Reinhoudt DN, Dijkstra PJ, In't Veld PJA, Bugge KE, Harkema S, Ungaro R, Ghidini E. Kinetically stable complexes of alkali cations with rigidified calix[4]arenes. X-ray structure of a calixspherand sodium picrate complex. *J Am Chem Soc.* 1987; **109**: 4761–4762.

42. Beer PD, Keefe AD. The synthesis of metallocene calix[4]arenes. *J Incl Phenom.* 1987; **5**: 499–504.

43. Corazza F, Floriani C, Chiesi-Villa A, Rizzoli C. Mononuclear tungsten(VI) calix[4]arene complexes. *Inorg Chem.* 1991; **30**: 4465–4468.

44. Goren Z, Biali SE. Multiple reductive cleavage of calixarene diethyl phosphate esters: a route to [1n]metacyclophanes. *J Chem Soc. Perkin Trans.* 1990; **1**: 1484–1487.

45. Ting Y, Verboom W, Groenen LC, van Loon J-D, Reinhoudt DN. Selectively dehydroxylated calix[4]arenes and 1,3-dithiocalix[4]arenes; novel classes of calix[4]arenes. *J Chem Soc Chem Commun.* 1990; 1432–1433.

46. Regnouf de Vains J-B, Pellet-Rostaing S, Lamartine R. Synthesis of a fully OH-depleted p-tert-butyl-calix[6]arene. *Tet Lett.* 1994; **35**: 8147–8150.

47. McMurry JE, Phelan JC. Synthesis and conformation of unsubstituted calix[4]arene. *Tet Lett.* 1991; **32**: 5655–5658.

48. Ohseto F, Murakami H, Araki K, Shinkai S. Substitution of OH with NH_2 calix[4]arenes: An approach to the synthesis of aminocalixarenes. *Tet Lett.* 1992; **33**: 1217–1220.

49. Gibbs CG, Sujeeth PK, Rogers JS, Stanley GG, Krawiec M, Watson WH, Gutsche CD. Syntheses and conformations of the p-tert-butylcalix[4]arenethiols. *J Org Chem.* 1995; **60**: 8394–8402.

50. Reddy PA, Kashyap R, Watson WH, Gutsche CD. Calixarenes. 30. Calixquinones. *Isr J Chem.* 1992; **32**: 89–96.

51. Reddy PA, Gutsche CD. Calixarenes. 32. Reactions of calixquinones. *J Org Chem.* 1993; **58**: 3245–3251.

52. Litwak AM, Grynszpan F, Aleksiuk O, Cohen S, Biali SE. Preparation, stereochemistry, and reactions of the bis(spirodienone) derivatives of p-tert-butylcalix[4]arene. *J Org Chem.* 1993; 393–402.

53. Columbus I, Biali SE. Totally and partially saturated calixarene analogues. *J Am Chem Soc.* 1998; **120**: 3060–3067.

54. Grynszpan F, Biali SE. From calixarenes to macrocyclic polyethers. *J Chem Soc Chem Commun.* 1996; 195–196.

55. Columbus I, Haj-Zaroubi M, Biali SE. Fully hydrogenated calixarene derivatives: calix[4]cyclohexanone and calix[4]cyclohexanol. *J Am Chem Soc.* 1998; **120**: 11806–11807.

56. Görmar G, Seiffarth K, Schulz M, Zimmermann J, Flämig G. Synthese und Reduktion des Tetra-tert-butyltetraoxocalix[4]arens. *Makromol Chem.* 1990; **191**: 81–87.

57. Scully PA, Hamilton TM, Bennett JL. Synthesis of 2-alkyl- and 2-carboxy-p-tert-butylcalix[4]arenes via the lithiation of tetramethoxy-p-tert-butylcalix[4]arene. *Org Lett.* 2001; **3**: 2741–2744.

58. Kumar S, Chawla HM, Varadarajan R. One step facile synthesis of bromo calix[n]arenes. *Tet Lett.* 2002; **43**: 7073–7075.

59. van Loon J-D, Arduini A, Verboom W, Ungaro R, van Hummel GJ, Harkema S, Reinhoudt DN. Selective functionalization of calix[4]arenes at the upper rim. *Tet Lett.* 1989; **30**: 2681–2684.

60. Gutsche CD, Pagoria PF. Calixarenes. 16. Functionalized calixarenes: the direct substitution route. *J Org Chem.* 1985; **50**: 5795–5802.

61. Gutsche CD, Levine JA, Sujeeth PK. Calixarenes. 17. Functionalized calixarenes: the claisen rearrangement route. *J Org Chem.* 1985; **50**: 5802–5806.

62. Gibbs CG, Wang J-S, Gutsche CD. Calixarenes for separations. *ACS Symposium Series* 2002; **757**: 313.

63. Almi M, Arduini A, Casnati A, Pochini A, Ungaro R. Chloromethylation of calixarenes and synthesis of new water soluble macrocyclic hosts. *Tetrahedron.* 1989; **45**: 2177–2182.

64. Arduini A, Pochini A, Rizzi A, Sicuri AR, Ugozzoli F, Ungaro R. Extension of the hydrophobic cavity of calix[4]arene by 'upper rim' functionalization. *Tetrahedron.* 1992; **48**: 905–912.

65. Shinkai S, Nagasaki T, Iwamoto K, Ikeda A, He G-X, Matsuda T, Iwamoto M. New syntheses and physical properties of p-alkylcalix[n]arenes. *Bull Chem Soc Jpn.* 1991; **64**: 381–386.

66. Gleave CA, Sutherland IO. A new calix[4]arene binding site. Strong cooperativity in cation binding by a two site receptor. *J Chem Soc Chem Commun.* 1994; 1873–1874.

67. Larsen M, Jørgensen M. Negishi type biaryl cross-coupling reactions on calix[4]arenes. *J Org Chem.* 1997; **62**: 4171–4173.

68. Botana E, Nättinen K, Prados P, Rissanen K, de Mendoza J. p-(1H-phenanthro[9,10-d]imidazol-2-yl)-substituted calix[4]arene, a deep cavity for guest inclusion. *Org Lett.* 2004; **6**: 1091–1094.

69. Mastalerz M, Dyker G, Flörke U, Henkel G, Oppel IM, Merz K. Oligophenylcalix[4]arenes as potential precursors for funnelenes and calix[4]triphenylenes: syntheses and preliminary cyclodehydration studies. *Eur J Org Chem.* 2006; 4951–4962.

70. Shinkai S, Araki K, Tsubaki T, Arimura T, Manabe O. New syntheses of calixarene-p-sulphonates and p-nitrocalixarenes. *J Chem Soc. Perkin Trans.* 1987; **1**: 2297–2299.

71. Shinkai S, Kawaguchi H, Manabe O. Selective adsorption of uo22+ to a polymer resin immobilizing calixarene-based uranophiles. *J Polym Sci C-Polym Lett.* 1988; **26**: 391–396.

72. Kumar S, Varadarajan R, Chawla HM, Hundal G, Hunda MS. Preparation of p-nitrocalix[n]arene methyl ethers via ipso-nitration and crystal structure of tetramethoxytetra-p-nitrocalix[4]arene. *Tetrahedron.* 2004; **60**: 1001–1005.

73. Timmerman P, Verboom W, Reinhoudt DN, Arduini A, Grandi S, Sicuri AR, Pochini A, Ungaro RO. Novel routes for the synthesis of upper rim amino and methoxycarbonyl functionalized calix[4]arenes carrying other types of functional groups. *Synthesis.* 1994; 185–189.

74. Shinkai S, Araki K, Shibata J, Manabe O. Autoaccelerative diazo coupling with calix[4]arene: unusual co-operativity of the calixarene hydroxy groups. *J Chem Soc. Perkin Trans.* 1989; **1**: 195–196.

75. Shinkai S, Araki K, Shibata J, Tsugawa D, Manabe O. Autoaccelerative diazo coupling with calix[4]arene: substituent effects on the unusual co-operativity of the OH groups. *J Chem Soc. Perkin Trans.* 1990; **1**: 3333–3337.

76. Böhmer V, Goldmann H, Vogt W, Paulus EF, Tobiason FL, Thielman MJ. Bridged calix[4]arenes: X-ray crystal and molecular structures and spectroscopic studies. *J Chem Soc. Perkin Trans.* 1990; **2**: 1769–1775.

77. Ungaro R, Pochini A, Andreetti GD, Sangermano V. Molecular inclusion in functionalized macrocycles. Part 8. The crystal and molecular structure of calix[4]arene from phenol and its (1 : 1) and (3 : 1) acetone clathrates. *J Chem Soc. Perkin Trans.* 1984; **2**: 1979–1985.

78. Ungaro R, Pochini A, Andreetti GD, Domiano P. Molecular inclusion in functionalized macrocycles. Part 9. The crystal and molecular structure of p-t-butylcalix[4]arene–anisole (2 : 1) complex: a new type of cage inclusion compound. *J Chem Soc. Perkin Trans.* 1985; **2**: 197–201.

79. McKervey MA, Seward EM, Ferguson G, Ruhl BL. Molecular receptors. Synthesis and x-ray crystal structure of a calix[4] arene tetracarbonate-acetonitrile (1 : 1) clathrate. *J Org Chem.* 1986; **51**: 3581–3584.

80. Rizzoli C, Andreett GD, Ungaro R, Pochini A. Molecular inclusion in functionalized macrocycles 4.the crystal and molecular structure of the cyclo {tetrakis[(5-t-butyl-2-acetoxy-1,3-phenylene)methylene]}-acetic acid (1 : 1) clathrate. *J Mol Struct.* 1982; **82**: 133–141.

81. Izatt SR, Hawkins RT, Christensen JJ, Izatt RM. Cation transport from multiple alkali cation mixtures using a liquid membrane system containing a series of calixarene carriers. *J Am Chem Soc.* 1985; **107**: 63–66.

82. Harrowfield JM, Ogden MI, Richmond WR, White AH. Calixarene-cupped caesium: a coordination conundrum? *J Chem Soc Chem Commun.* 1991; 1159–1161.

83. Bott SG, Coleman AW, Atwood JL. Inclusion of both cation and neutral molecule by a calixarene. Structure of the [p-tert-butylmethoxycalix[4]arene-sodium-toluene]+ cation. *J Am Chem Soc.* 1986; **108**: 1709–1710.

84. Arnaud-Neu F, Collins EM, Deasy M, Ferguson G, Harris SJ, Kaitner B, Lough AJ, McKervey MA, Marques E. Synthesis, x-ray crystal structures, and cation-binding properties of alkyl calixaryl esters and ketones, a new family of macrocyclic molecular receptors. *J Am Chem Soc.* 1989; **111**: 8681–8691.

85. Arduini A, Ghidini E, Pochini A, Ungaro R, Andreetti GD, Calestani G, Ugozzoli F. p-t-butylcalix[4]arene tetra-acetamide: a new strong receptor for alkali cations. *J Mol Inclus Macro Chem.* 1988; **6**: 119–134.

86. Beer PD, Drew MGB, Leeson PB, Ogden MI. Versatile cation complexation by a calix[4]arene tetraamide (L). Synthesis and crystal structure of [ML][ClO$_4$]$_2$ · nMeCN (M = Fe^{II}, Ni^{II}, Cu^{II}, Zn^{II} or Pb^{II}). *J Chem Soc. Dalton Trans.* 1995; 1273–1283.

87. Beer PD, Drew MGB, Ogden MI. First- and second-sphere co-ordination of a lanthanum cation by a calix[4]arene tetraamide in the partial-cone conformation. *J Chem Soc. Dalton Trans.* 1997; 1489–1492.

88. Ungaro R, Pochini A, Andreetti GD. New ionizable ligands from p. t-butylcalix[4]arene. *J Inclus Phenom Macro Chem.* 1984; **2**: 199–206.

89. Ogata M, Fujimoto K, Shinkai S. Molecular design of calix[4]arene-based extractants which show high Ca^{2+} selectivity. *J Am Chem Soc.* 1994; **116**: 4505–4506.

90. Ludwig R, Inoue K, Yamato T. Solvent extraction behavior of calixarene-type cyclophanes towards trivalent La, Nd, Eu, Er and Yb. *Solv Extr Ion Exch.* 1993; **11**: 311–320.

91. Floriani C, Jacoby D, Chiesi-Villa A, Guastini C. Aggregation of metal ions with functionalized, calixarenes: synthesis and structure of an, octanuclear copper(I) chloride complex. *Angew Chem Int Ed.* 1989; **28**: 1376–1377.

92. Dieleman C, Loeber C, Matt D, De Cian A, Fischer J. Facile synthetic route to cone-shaped phosphorylated [CH$_2$P(O)Ph$_2$] calix[4]arenes. *J Chem Soc. Dalton Trans.* 1995; 3097–3100.

93. Loeber C, Matt D, Briard P, Grandjean D. Transition-metal complexation by calix[4]arene-derived phosphinites. *J Chem Soc. Dalton Trans.* 1996; **4**: 513–524.

94. Arnaud-Neu F, Böhmer V, Dozol J-F, Grüttner C,. Jakobi RA, Kraft D, Mauprivez O, Rouquette H, Schwing-Weill M-K, Simon N, Vogt W. Calixarenes with diphenylphosphoryl acetamide functions at the upper rim. A new class of highly efficient extractants for lanthanides and actinides. *J Chem Soc. Perkin Trans.* 1996; **2**: 1175–1182.

95. Arnaud-Neu F, Barrett G, Corry D, Cremin S, Ferguson G, Gallagher JF, Harris SJ, McKervey MA, SchwingWeill M-J. Cation complexation by chemically modified calixarenes. Part 10. Thioamide derivatives of p-tert-butylcalix[4]-, [5]- and [6]-arenes with selectivity for copper, silver, cadmium and lead. X-Ray molecular structures of calix[4]arene thioamide–lead(II) and calix[4]arene amide–copper(II) complexes. *J Chem Soc. Perkin Trans.* 1997; **2**: 575–580.

96. Shinkai S, Koreishi H, Ueda K, Manabe O. A new hexacarboxylate uranophile derived from calix[6]arene. *J Chem Soc Chem Commun.* 1991; 233–234.

97. Nagasak T, Shinkai S. Synthesis and solvent extraction studies of novel calixarene-based uranophiles bearing hydroxamic groups. *J Chem Soc. Perkin Trans.* 1991; **2**: 1063–1066.

98. Casnati A, Pochini AA, Ungaro R, Ugozzoli F, Arnaud F, Fanni S, Schwing M-J, Egberink RJM, de Jong F, Reinhoudt DN. synthesis, complexation, and membrane transport studies of 1,3-alternate calix[4]arene-crown-6 conformers: a new class of cesium selective ionophores. *J Am Chem Soc.* 1995; **117**: 2767–2777.

99. Kim JS, Pang JH, Yu IY, Lee WK, Suh IH, Kim JGK, Cho MH, Kim ET, Ra DY. Calix[4]arene dibenzocrown ethers as caesium selective extractants. *J Chem Soc. Perkin Trans.* 1999; **2**: 837–846.

100. Davis F, Collyer SD, Higson SPJ. The construction and operation of anion sensors: current status and future perspectives. *Top Curr Chem.* 2005; **255**: 97–124.

101. Beer PLD, Drew MGB, Hesek D, Chun Nam K. A new carboxylate anion selective cobaltocenium calix[4]arene receptor. *Chem Commun.* 1997; **1**: 107–108.

102. Beer PD, Chen Z, Goulden AJ, Graydon A,. Stokes SE, Wear T. Selective electrochemical recognition of the dihydrogen phosphate anion in the presence of hydrogen sulfate and chloride ions by new neutral ferrocene anion receptors. *J Chem Soc Chem Commun.* 1993; 1834–1836.

103. Staffilani MA, Hancock KSB, Steed JW, Holman KT, Atwood JL, Juneja RK, Burkhalter RS. Anion binding within the cavity of π-metalated calixarenes. *J Am Chem Soc.* 1997; **119**: 6324–6335.

104. Morzherin Y, Rudkevich DM, Verboom W, Reinhoudt DN. Chlorosulfonylated calix[4]arenes: precursors for neutral anion receptors with a selectivity for hydrogen sulfate. *J Org Chem.* 1993; **58**: 7602–7605.

105. Scheerder J, Engbersen JFJ, Casnati A, Ungaro R, Reinhoudt DN. Complexation of halide anions and tricar-boxylate anions by neutral urea-derivatized p-tert-butylcalix[6]arenes. *J Org Chem.* 1995; **60**: 6448–6454.

106. Pelizzi N, Casnati A, Friggeri A, Ungaro R. Synthesis and properties of new calixarene-based ditopic receptors for the simultaneous complexation of cations and carboxylate anions. *J Chem Soc. Perkin Trans.* 1998; **2**: 1307–1311.

107. Matthews SE, Beer PD. Calixarene-based anion receptors. *Supramol Chem.* 2005; **17**: 411–435.

108. Gutsche CD, Alam I. Calixarenes. 23. The complexation and catalytic properties of water soluble calixarenes. *Tetrahedron.* 1988; **44**: 4689–4694.

109. Atwood JL, Koutsantonis GA, Raston CL. Purification of C_{60} and C_{70} by selective complexation with cal-ixarenes. *Nature.* 1994; **368**: 229–231.

110. Shinkai S, Mori S, Koreishi H, Tsubaki T, Manabe O. Hexasulfonated calix[6]arene derivatives: a new class of catalysts, surfactants, and host molecules. *J Am Chem Soc.* 1986; **108**: 2409–2416.

111. Arena G, Contino A, Gulino FG, Magrì A, Sansone F, Sciotto D, Ungaro R. Complexation of native L-α-aminoacids by water soluble calix[4]arenes. *Tet Lett.* 1999; **40**: 1597–1600.

112. Kanamathareddy S, Gutsche CD. Conformational characteristics of p-tert-butylcalix[6]arene ethers. *J Org Chem.* 1994; **59**: 3871–3879.

113. Matthews SE, Schmitt P, Felix V, Drew MGB, Beer PD. Calix[4]tubes: a new class of potassium-selective ionophore. *J Am Chem Soc.* 2002; **124**: 1341–1353.

114. Kim SK, Vicens J, Park K-M, Lee SS, Kim JS. Complexation chemistry. Double- and multi-1,3-alternate-calixcrowns. *Tet Lett.* 2003; **44**: 993–997.

115. Kraft D, van Loon J-D, Owens M, Verboom W, Vogt W, McKervey MA, Böhmer V, Reinhoudt DN. Double and triple calix[4]arenes connected via the oxygen functions. *Tet Lett.* 1990; **31**: 4941.

116. Beer D, Keefe AD, Slawin AMZ, Williams DJ. Dimer and trimer calix[4]arenes containing multiple metallocene redox-active centres. Single-crystal X-ray structure of a bis(ferrocene)-bis(p-t-butylcalix[4]-arene) hydrophobic host molecule. *J Chem Soc. Dalton Trans*. 1990; 3675–3682.

117. Wang J, Gutsche CD. Complexation of fullerenes with bis-calix[n]arenes synthesized by tandem claisen rearrangement. *J Am Chem Soc*. 1998; **120**: 12226–12231.

118. Bottino A, Cunsolo F, Piattelli M, Garozzo D, Neri P. Synthesis of 5,5'-bicalix[6]arene and 5,5'-bicalix[8]arene systems. *J Org Chem*. 1999; **64**: 8018–8020.

119. Hwang GT, Kim BH. Synthesis and binding studies of multiple calix[4]arenes. *Tet Lett*. 2002; **58**: 9019–9028.

120. Arduini A, Pochini A, Secchi A, Ungaro R. A new macrocavitand from the head to tail four-point capping of p-tert-butylcalix[8]arene with a calix[4]arene. *J Chem Soc Chem Commun*. 1995; 879–880.

121. Georgiev EM, Troev K, Roundhill DM. Formation of ionic polymers by the alkylation of polyethyleneimine with tetra-chloromethyl and mono-bromopropyl substituted calix[4]arenes. *Supramol Chem*. 1993; **2**: 61–64.

122. Harris SJ, Barrett G, McKervey MA. Polymeric calixarenes. Synthesis, polymerisation and Na^+ complexation of a calix[4]arene methacrylate. *J Chem Soc Chem Commun*. 1991; 1224–1225.

123. Gravett DM, Guillet JE. Synthesis and photophysical properties of a novel water-soluble, calixarene-containing polymer. *Macromolecules*. 1996; **29**: 617–624.

124. Zhong Z-L, Tang C-P, Wu C-Y, Chen Y-Y. Synthesis and properties of calixcrown telomers. *J Chem Soc Chem Commun*. 1997; 1737–1738.

125. Dondoni A, Ghiglione C, Marra A, Scoponi M. Synthesis and receptor properties of calix[4]arene–bisphenol-A copolymers. *Chem Commun*. 1997; 1673–1674.

126. Cheriaa N, Abidi R, Vicens J. Calixarene-based dendrimers. Second generation of a calix[4]-dendrimer with a 'tren' as core. *Tet Lett*. 2005; **46**: 1533–1536.

127. van Loon J-D, Janssen RG, Verboom W, Reinhoudt DN. Hydrogen bonded calix[4]arene aggregates. *Tet Lett*. 1992; **33**: 5125–5128.

128. Koh K, Araki K, Shinkai S. Self-assembled molecular capsule based on the hydrogen-bonding interaction between two different calix[4]arenes. *Tet Lett*. 1994; **35**: 8255–8258.

129. Thondorf I, Broda F, Rissanen K, Vysotsky M, Böhmer V. Dimeric capsules of tetraurea calix[4]arenes. MD simulations and X-ray structure, a comparison. *J Chem Soc. Perkin Trans*. 2002; **11**: 1796–1800.

130. Vysotsky MO, Bolte M, Thondorf I, Böhmer V. New molecular topologies by fourfold metathesis reactions within a hydrogen-bonded calix[4]arene dimer. *Chem Eur J*. 2003; **9**: 3375–3382.

131. Vreekamp RH, van Duynhoven JPM, Hubert M, Verboom W, Reinhoudt DN. Molecular boxes based on calix[4]arene double rosettes. *Angew Chem Int Ed*. 1996; **35**: 1215–1218.

132. Corbellini F, Fiammengo R, Timmerman P, Crego-Calama M, Versluis K, Heck AJR, Luyten I, Reinhoudt DN. Guest encapsulation and self-assembly of molecular capsules in polar solvents via multiple ionic interactions. *J Am Chem Soc*. 2002; **12**: 6569–6575.

133. Dhawan B, Gutsche CD. Calixarenes. 10. Oxacalixarenes. *J Org Chem*. 1983; **48**: 1536–1539.

134. Takemura H, Yoshimura K, Khan IU, Shinmyozu T, Inazu T. The first synthesis and properties of hexahomo-triazacalix[3]arene. *Tet Lett*. 1992; **33**: 5575–5578.

135. Khan IU, Takemura H, Suenaga M, Shinmyozu T, Inazu T. Azacalixarenes: new macrocycles with dimethyleneaza-bridged calix[4]arene systems'. *J Org Chem*. 1993; **58**: 3158–3161.

136. Thuéry P, Nierlich M, Vicens JS, Takemura H. Crystal structure of p-chloro-N-benzylhexahomotriazacalix[3]arene and of the complex of its zwitterionic form with neodymium(III) nitrate. *J Chem Soc. Dalton Trans*. 2000; 279–283.

137. Morohashi N, Narumi F, Iki N, Hattori T, Miyano S. Thiacalixarenes *Chem Rev*. 2006; **106**: 5291–5316.

138. Kumagai H, Hasegawa M, Miyanari S, Sugawa Y, Sato Y, Hori TI, Ueda S, Kamiyama H, Miyano S. Facile synthesis of p-tert-butylthiacalix[4]arene by the reaction of p-tert-butylphenol with elemental sulfur in the presence of a base. *Tet Lett*. 1997; **38**: 3971–3972.

139. Sone T, Ohba Y, Moriya K, Kumada H, Ito K. Synthesis and properties of sulfur-bridged analogs of p-tert-butylcalix[4]arene. *Tetrahedron*. 1997; **53**: 10689–10698.

140. Kon N, Iki N, Miyano S. Synthesis of p-tert-butylthiacalix[n]arenes (n = 4, 6, and 8) from a sulfur-bridged acyclic dimer of p-tert-butylphenol. *Tet Lett*. 2002; **33**: 2231–2234.

141. Akdas H, Bringel L, Graf E, Hosseini MW, Mislin G, Pansanel J, De Cian A, Fischer J. Thiacalixarenes: synthesis and structural analysis of thiacalix[4]arene and of p-tert-butylthiacalix[4]arene. *Tet Lett.* 1985; **39**: 2311–2314.

142. Kondo Y, Endo K, Iki N, Miyano S, Hamada F. Synthesis and crystal structure of p-tert-butylthiacalix[8]arene: a new member of thiacalixarenes. *J Inclus Phenom Macro Chem.* 2005; **52**: 45–49.

143. Lang J, Vlach J, Dvoáková H, Lhoták P, Himl M, Hrabal R, Stibor I. Thermal isomerisation of 25,26,27,28-tetrapropoxy-2,8,14,20-tetrathiacalix[4]arene: isolation of all four conformers. *J Chem Soc. Perkin Trans.* 2001; **2**: 576–580.

144. Mislin G, Graf E, Hosseini MW, Mislin G, De Cian A, Fischer J. Sulfone-calixarenes: a new class of molecular building block. *J Chem Soc Chem Commun.* 1998; 1345–1346.

145. Morohashi N, Katagiri H, Iki N, Yamane Y, Kabuto C, Hattori T, Miyano S. Synthesis of all stereoisomers of sulfinylcalix[4]arenes. *J Org Chem.* 2003; **68**: 2324–2333.

146. Morohashi N, Iki N, Onodera T, Kabuto C, Miyano S. Novel molecular chirality in the calixarene family: formation of chiral disulfinyldithiacalix[4]arenes via partial oxidation of two adjacent sulfides of tetrathiacalix[4]arene. *Tet Lett.* 2000; **41**: 5093–5097.

147. van Leeuwen FWB, Beijleveld H, Kooijman H, Spek AL, Verboom W, Reinhoudt DN. Cation control on the synthesis of p-t-butylthiacalix[4](bis)crown ethers. *Tet Lett.* 2002; **43**: 9675–9678.

148. van Leeuwen FWB, Beijleveld H, Miermans CJH, Huskens J, Verboom W, Reinhoudt DN. Ionizable (thia)calix[4]crowns as highly selective $^{226}Ra^{2+}$ ionophores. *Anal Chem.* 2005; **77**: 4611–4617.

149. Iki N, Kumagai H, Morohashi N, Ejima KI, Hasegawa M, Miyanari S, Miyano S. Selective oxidation of thiacalix[4]arenes to the sulfinyl- and sulfonylcalix[4]arenes and their coordination ability to metal ions. *Tet Lett.* 1998; **39**: 7559–7562.

150. Iki N, Morohashi N, Narumi F, Miyano S. High complexation ability of thiacalixarene with transition metal ions. the effects of replacing methylene bridges of tetra(p-t-butyl)calix[4]arenetetrol by epithio groups. *Bull Chem Soc Jpn.* 1998; **71**: 1597–1603.

151. Morohashi N, Iki N, Sugawara A, Miyano S. Selective oxidation of thiacalix[4]arenes to the sulfinyl and sulfonyl counterparts and their complexation abilities toward metal ions as studied by solvent extraction. *Tetrahedron.* 2002; **57**: 5557–5563.

152. Timmerman P, Verboom W, Reinhoudt DN. Resorcinarenes. *Tetrahedron.* 1996; **52**: 2663–2704.

153. Hoegberg AGS. Two stereoisomeric macrocyclic resorcinol-acetaldehyde condensation products. *J Org Chem.* 1980; **45**: 4498–4500.

154. Hoegberg AGS. Cyclooligomeric phenol-aldehyde condensation products. 2. Stereoselective synthesis and DNMR study of two 1,8,15,22-tetraphenyl[14]metacyclophan-3,5,10,12,17,19,24,26-octols. *J Am Chem Soc.* 1980; **102**: 6046–6050.

155. Erdmann H, Hoegberg AGS. Cyclooligomeric phenol-aldehyde condensation products. I. *Tet Lett.* 1968; **9**: 1679–1682.

156. Weinelt F, Schneider HJ. Host–guest chemistry. 27. Mechanisms of macrocycle genesis. The condensation of resorcinol with aldehydes. *J Org Chem.* 1991; **56**: 5527–5535.

157. Tunstad LM, Tucker JA, Dalcanale E, Weiser J, Bryant JA, Sherman JC, Helgeson RC, Knobler CB, Cram DJ. Host–guest complexation. 48. Octol building blocks for cavitands and carcerands. *J Org Chem.* 1989; **54**: 1305–1312.

158. Adams H, Davis F, Stirling CJM. Selective adsorption in gold-thiol monolayers of calix-4-resorcinarenes. *J Chem Soc Chem Commun.* 1994; 2527–2529.

159. Adams H, Davis F, Hibbs D, Hursthouse MB, Abdul-Malik KM, Stirling CJM. Tetraundecylpentacyclooctacosadodecenooctol and tetraundecylpentacycloocta cosa dodecenododecol. *Acta Cryst.* 1998; **C54**: 987.

160. Cometti G, Dalcanale E, Du Vosel A, Levelut A-M. A new, conformationally mobile macrocyclic core for bowl-shaped columnar liquid crystals. *Liquid Crystals.* 1992; **11**: 93–100.

161. Roberts BA, Cave GWV, Raston CL, Scott JL. Solvent-free synthesis of calix[4]resorcinarenes. *Green Chem.* 2001; **3**: 280–284.

162. Konishi H, Ohata K, Morikawa O, Kobayashi K. Calix[6]resorcinarenes: the first examples of [16]metacyclophanes derived from resorcinols. *J Chem Soc Chem Commun.* 1995; 309–310.

163. Konishi H, Sakakibara H, Kobayashi K, Morikawa O. Synthesis of the parent resorcin[4]arene. *J Chem Soc. Perkin Trans.* 1999; **1**: 2583–2584.

164. Botta B, Di Giovanni MC, Monache GD, De Rosa MC, Gacs-Baitz E, Botta M, Corelli F, Tafi A, Santini A. A novel route to calix[4]arenes. 2. Solution- and solid-state structural analyses and molecular modeling studies. *J Org Chem.* 1994; **59**: 1532–1541.

165. Boxhall JY, Page PCB, Elsegood MRJ, Chan Y, Heaney H, Holmes KE, McGrath MJ. The synthesis of axially chiral resorcinarenes from resorcinol monoalkyl ethers and aldehyde dimethylacetals. *Synlett.* 2003; **7**: 1002–1006.

166. Cram DJ, Karbach S, Kim HE, Knobler CB, Maverick EF, Ericson JL, Helgeson RC. Host–guest complexation. 46. Cavitands as open molecular vessels form solvates. *J Am Chem Soc.* 1988; **110**: 2229–2237.

167. Kazakova EK, Makarova NA, Ziganshina AU, Muslinkina LA, Muslinkin AA, Habicher WD. Novel water-soluble tetrasulfonatomethylcalix[4]resorcinarenes. *Tet Lett.* 2000; **41**: 10111–10115.

168. Matsushita Y, Matsui T. Synthesis of aminomethylated calix[4]resorcinarenes. *Tet Lett.* 1993; **43**: 7433–7436.

169. El Gihani MT, Heaney H, Slawin AMZ. Highly diastereoselective functionalisation of calix[4]resorcinarene derivatives and acid catalysed epimerisation reactions. *Tet Lett.* 1995; **36**: 4905–4908.

170. Fairful-Smith K, Redon PMJ, Haycock JW, Williams NH. Monofunctionalised resorcinarenes. *Tet Lett.* 2007; **48**: 1317–1319.

171. Aoyama Y, Tanaka Y, Sugahara S. Molecular recognition. 5. Molecular recognition of sugars via hydrogen-bonding interaction with a synthetic polyhydroxy macrocycle. *J Am Chem Soc.* 1989; **111**: 5397–5404.

172. Kobayashi K, Asakawa Y, Kato Y, Aoyama Y. Complexation of hydrophobic sugars and nucleosides in water with tetrasulfonate derivatives of resorcinol cyclic tetramer having a polyhydroxy aromatic cavity: importance of guest–host CH-pi interaction. *J Am Chem Soc.* 1992; **114**: 10307–10313.

173. Kurihara K, Ohto K, Tanaka Y, Aoyama Y, Kunitake T. Binding of sugars and water-soluble polymers to a monolayer of cyclic resorcinol tetramer at the air–water interface. *Thin Solid Films.* 1989; **179**: 21–26.

174. Kobayashi K, Tominaga M, Asakawa Y, Aoyama Y. Binding of amino acids in water to a highly electron-rich aromatic cavity of pyrogallol or resorcinol cyclic tetramer as π-base. *Tet Lett.* 1993; **34**: 5121–5124.

175. Schneider HJ, Guettes D, Schneider U. Host–guest chemistry. 15. Host–guest complexes with water-soluble macrocyclic polyphenolates including induced fit and simple elements of a proton pump. *J Am Chem Soc.* 1988; **110**: 6449–6454.

176. Lippmann T, Wilde H, Pink M, Schäfer A, Hesse M, Mann G. Host–guest complexes between calix[4]arenes derived from resorcinol and alkylammonium ions. *Angew Chem Int Ed.* 1993; **32**: 1195–1197.

177. Inouye M, Hashimoto K, Isagawa K. Nondestructive detection of acetylcholine in protic media: artificial-signaling acetylcholine receptors. *J Am Chem Soc.* 1994; **116**: 5517–5518.

178. Davis F, Stirling CJM. Calix-4-resorcinarene monolayers and multilayers: formation, structure, and differential adsorption. *Langmuir.* 1996; **12**: 5365–5374.

179. Faull JD, Gupta VK. Selective guest–host association on self-assembled monolayers of calix[4]resorcinarene. *Langmuir.* 2001; **17**: 1470–1476.

180. Friggeri A, van Veggel FCJM, Reinhoudt DN. Self-assembled monolayers of cavitand receptors for the binding of neutral molecules in water. *Langmuir.* 1998; **14**: 5457–5463.

181. Huisman B-H, Kooyman RPH, van Veggel FCJM, Reinhoudt DN. Molecular recognition by self-assembled monolayers detected with surface plasmon resonance. *Adv Mat.* 1996; **8**: 561–564.

182. Cram DJ, Tunstad LM, Knobler CB. Host–guest complexation. 61. C- and Z-shaped ditopic cavitands, their binding characteristics, and monotopic relatives. *J Org Chem.* 1992; **57**: 528–535.

183. Moran JR, Ericson JL, Dalcanale E, Bryant JA, Knobler CB, Cram DJ. Vases and kites as cavitands. *J Am Chem Soc.* 1991; **113**: 5707–5714.

184. Davis F, Lucke A, Smith KA, Stirling CJM. Order and structure in Langmuir–Blodgett mono- and multilayers of resorcarenes. *Langmuir.* 1998; **14**: 4180–4185.

185. Cram DJ, Stewart KD, Goldberg I, Trueblood KN. Complementary solutes enter nonpolar preorganized cavities in lipophilic noncomplementary media. *J Am Chem Soc.* 1985; **107**: 2574–2575.

186. Xu W, Rourke JP, Vittal JJ, Puddephatt RJ. Anion inclusion by a calix[4]arene complex: a contrast between tetranuclear gold(I) and copper(I) complexes. *J Chem Soc Chem Commun.* 1993; 145–147.

187. Cram DJ, Karbach S, Kim YH, Baczynskyj L, Kallemeyn GW. Shell closure of two cavitands forms carcerand complexes with components of the medium as permanent guests. *J Am Chem Soc.* 1985; **107**: 2575–2576.

188. Cram DJ, Tanner ME, Thomas R. The taming of cyclobutadiene. *Angew Chem Int Ed Engl.* 1991; **30**: 1024–1027.

189. Cram DJ, Tunstad LM, Knobler CB. Host–guest complexation. 61. C- and Z-shaped ditopic cavitands, their binding characteristics, and monotopic relatives. *J Org Chem.* 1992; **57**: 528–535.

190. Quan MLC, Cram DJ. Constrictive binding of large guests by a hemicarcerand containing four portals. *J Am Chem Soc.* 1991; **113**: 2754–2755.

191. Schmidt C, Thondorf I, Kolehmainen E, Bohmer V, Vogt W, Rissanen K. One-step synthesis of resorcarene dimers composed of two tetra-benzoxazine units. *Tet Lett.* 1998; **39**: 8833–8836.

192. Timmerman P, Verboom W, van Veggel FCJM, van Hoorn WP, Reinhoudt DN. An organic molecule with a rigid cavity of nanosize dimensions. *Angew Chem Int Ed.* 1994; **33**: 1292–1295.

193. Timmerman P, Brinks EA, Verboom W, Reinhoudt DN. Synthetic receptors with preorganized cavities that complex prednisolone-21–acetate. *J Chem Soc Chem Commun.* 1995; 417–418.

194. Timmerman P, Verboom W, van Veggel FCJM, van Duynhoven JPM, Reinhoudt DN. A novel type of stereoisomerism in calixI4larene-based carceplexes. *Angew Chem Int Ed.* 1994; **33**: 2345–2348.

195. Huisman B-H, Rudkevich DM, van Veggel FCJM, Reinhoudt DN. Self-assembled monolayers of carceplexes on gold. *J Am Chem Soc.* 1996; **118**: 3523–3524.

196. Davis F, Stirling CJM. Spontaneous multilayering of calix-4-resorcinarenes. *J Amer Chem Soc.* 1995; **117**: 10385–10386.

197. MacGillivray LR, Atwood JL. Achiral sphericalmolecular assembly held together by 60 hydrogen bonds. *Nature.* 1997; **389**: 469–472.

198. Shimizu S, Kiuchi T, Pan N. A 'teflon-footed' resorcinarene: a hexameric capsule in fluorous solvents and fluorophobic effects on molecular encapsulation. *Angew Chem.* 2007; **119**: 6562–6565.

199. Dvoáková H, Stursa J, Cajan M, Moravcová J. Synthesis and conformational properties of partially alkylated methylene-bridged resorc[4]arenes: study of the flip-flop inversion. *Eur J Org Chem.* 2006; 4519–4527.

5

Heterocalixarenes and Calixnaphthalenes

5.1 Introduction

The calixarenes and the related resorcinarenes described in the previous chapter represent an actively researched group of compounds. Calixarenes are cyclic oligomers formed by the condensation of substituted phenols with formaldehyde, although phenols are not the only compounds capable of undergoing these types of reaction. The larger naphthol aromatics have been shown to undergo almost identical condensation reactions with formaldehyde. Similarly, a range of heteroaromatic compounds such as furan, pyrrole and thiophenes can also form these types of macrocycle. Within this chapter we will give an overview of this extended family of compounds and discuss their structures and binding capabilities. The use of a range of aromatic compounds instead of just phenols should greatly enhance the structural and functional diversity seen so far.

5.2 Calixnaphthalenes

Once the condensation of phenols with formaldehyde was successfully attained, attention turned towards other hydroxyl-substituted aromatic compounds. Much of this work was pursued by the group of Paris Georghiou. Naphthols have also been shown to undergo condensation reactions; 2-naphthol, for example, condenses with formaldehyde under acidic or basic conditions to give the dimeric species shown in Figure 5.1a. With the corresponding 1-naphthol, however, the base-catalysed reaction with formaldehyde gives a mixture of products. When the condensation reaction was catalysed by potassium carbonate in DMF, a complex mixture was obtained, from which three crystalline products could be taken.[1] These had the ratio 1.0 : 2.2 : 3.0 in the crude mixture (in order of crystallisation) and were all shown by mass spectrometry to have a mass of 624, corresponding to the cyclic tetramer analogous to calix[4]arenes. However, there are four possible isomers, as shown in Figure 5.2. Studies of the NMR spectrum indicated that the first product had the structure shown in Figure 5.2a (9.4% final yield), the second that of Figure 5.2c (16.0% final yield) and the third that of Figure 5.2d, but the latter proved difficult to crystallise and could only be

Macrocycles: Construction, Chemistry and Nanotechnology Applications, First Edition.   Frank Davis and Séamus Higson.
© 2011 John Wiley & Sons, Ltd. Published 2011 by John Wiley & Sons, Ltd.

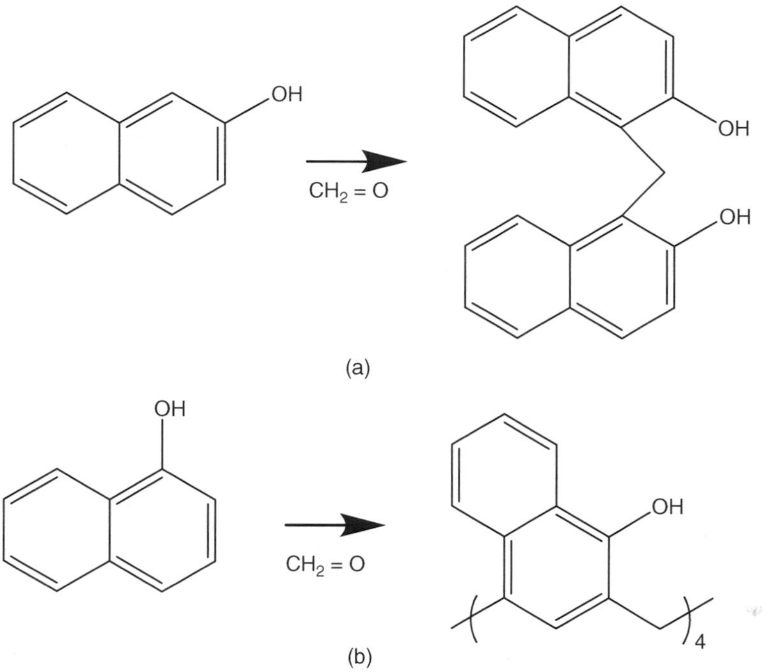

Figure 5.1 Condensation products of formaldehyde with (a) 2-naphthol and (b) 1-naphthol

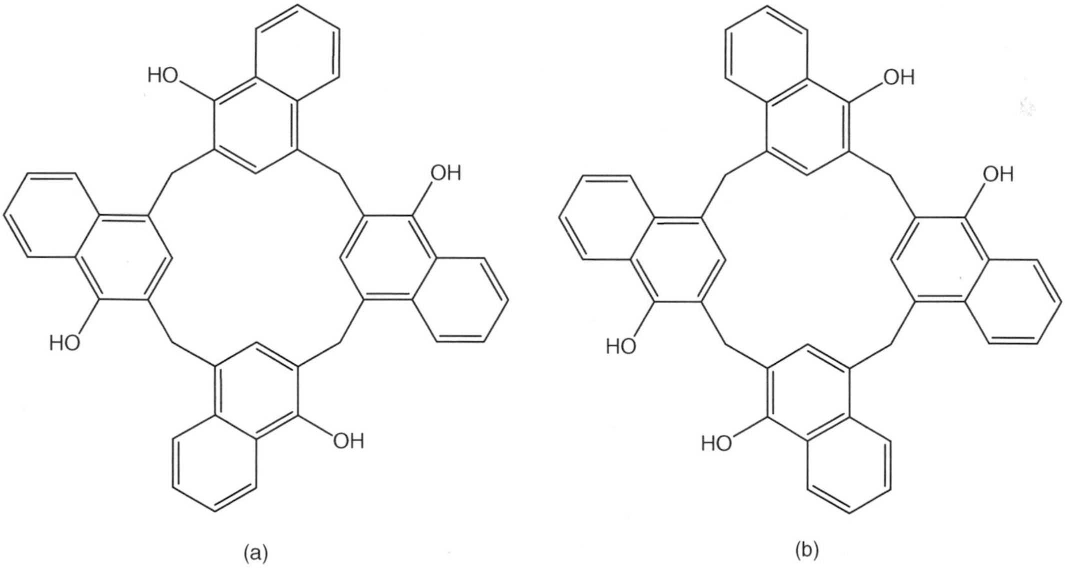

Figure 5.2 Isomers of condensation products of formaldehyde with 1-naphthol

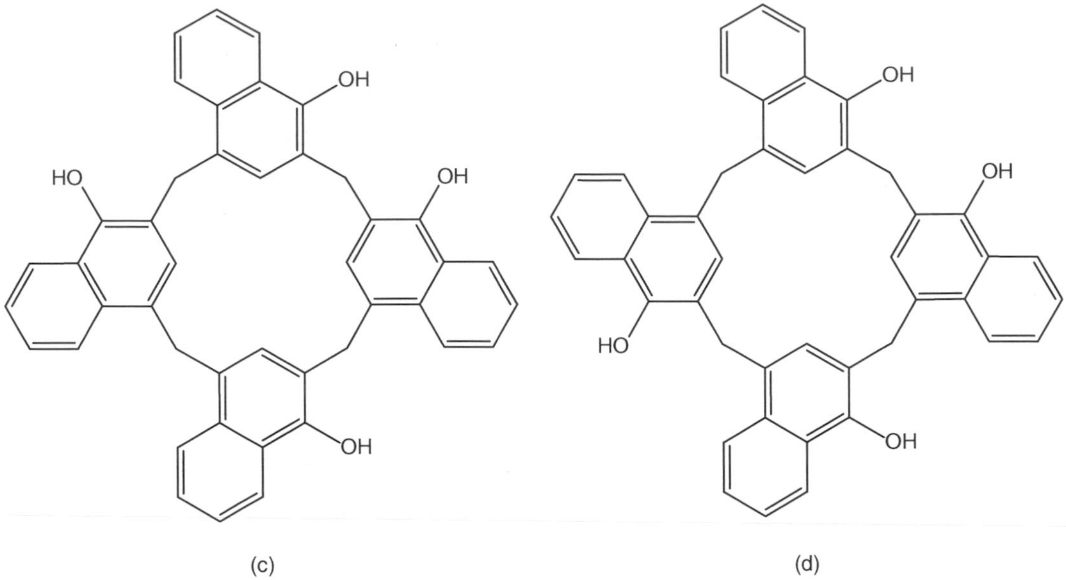

(c)   (d)

Figure 5.2   *(continued)*

obtained in a 5.0% yield. Further analysis showed the isomers were all conformationally mobile, although substitution to give the benzoate esters restricted this.[2] Much of the work up to 2005 on calixnaphthalenes has been reviewed by Georghiou *et al.*[3]

All four isomers of the calix[4]naphthalene could be obtained by stepwise synthesis.[4] These structures show one major difference from the classic calixarenes, in that the —OH groups are positioned on the 'outside' of the macrocycle (similar to the resorcinarenes) rather than 'inside' the main ring. A different series of macrocycles were synthesised, again using a stepwise process rather than a simple condensation, to give dihomocalix[4]naphthalenes,[5] the structures of which are shown in Figure 5.3a, with the methoxy groups inside the macrocycle. X-ray studies showed them to have a 1,2-alternate structure in the crystal (Figure 5.3b). A structure with groups both inside and outside the macrocycle could also be synthesised (Figure 5.3c). Both these materials were again shown to be conformationally flexible in solution[5] and a cone-like structure was not adopted. A variety of hydroxyl- and alkyl-substituted versions of this material have been described.[6] It has also proved possible to synthesise the hexahomotrioxacalix[3]naphthalenes (Figure 5.4a,b) in symmetrical and unsymmetrical forms.[7]

One of the reasons for the study of these compounds is that a calix[4]naphthalene which adopted a cone conformation would have a deeper and more electron-rich cavity than a simple calix[4]arene. A calix[4]naphthalene containing hydroxyl groups in *endo* positions[8] has been synthesised (Figure 5.4c) and NMR studies[9] showed it adopts a cone conformation at low temperatures, while X-ray crystal studies showed it adopts a pinched-cone conformation in the solid state. This compound was initially thought to form strongly associated complex with fullerenes,[9] but later work indicated that calix[4]naphthalenes substituted with *tert*-butyl groups gave much better complexation with the C_{60} molecule deeply inserted into the cavity.[10] Complexes have also been formed with the hexahomotrioxacalix[3]naphthalenes,[11] in which X-ray studies have shown the fullerene to be encased in a 'capsule' formed by two macrocycles (Figure 5.4d). This compound has also proven amenable to oxidation to form spirodienones[8] in the same manner as some calixarenes, a reaction which does occur with the calix[4]naphthalenes with *exo* hydroxyl

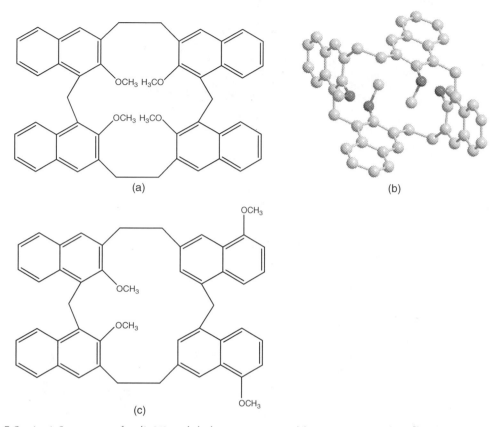

(a)

(b)

(c)

Figure 5.3 *(a,c) Structures of calix[4]naphthalenes constructed by stepwise synthesis and (b) X-ray structure of 5.3a*

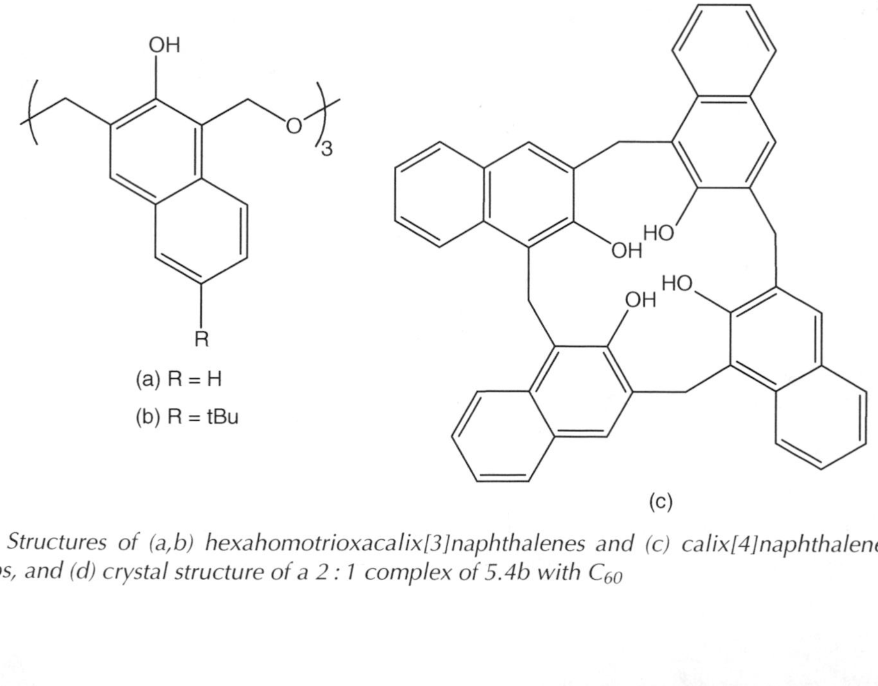

(a) R = H
(b) R = tBu

(c)

Figure 5.4 *Structures of (a,b) hexahomotrioxacalix[3]naphthalenes and (c) calix[4]naphthalenes with endo* —OH *groups, and (d) crystal structure of a 2 : 1 complex of 5.4b with C*$_{60}$

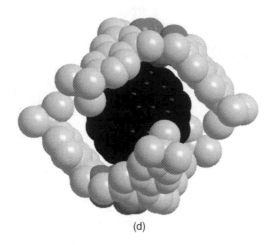

(d)

Figure 5.4　*(continued)*

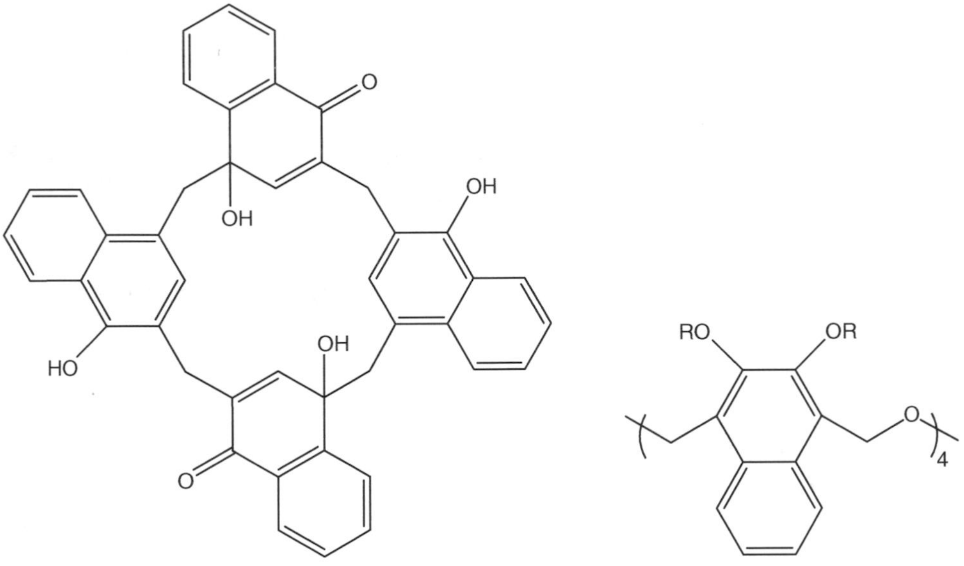

(a)

(b) R = H, Me, Et, nPr, nBu

Figure 5.5　*Structures of (a) oxidised calixnaphthalene and (b) 'Zorbarene'*

groups. Oxidation has also been shown to occur for the compound in Figure 5.1a, which can be oxidised aerobically to give the compound shown in Figure 5.5a, this being confirmed by X-ray crystallography, although the results did not give enough data for a full crystal refinement.[12] A series of 2,3-disubstituted naphthalenes have also been linked to form cyclic tetramers,[13] which the authors named 'Zorbarenes' (Figure 5.5) and which formed complexes with tetramethyl ammonium ions.

Substituted naphthalenes have also been incorporated into macrocyclic structures. For example, 1,8-naphthalene sultone[12] can be condensed with formalin to give the water-soluble cyclic tetramer shown

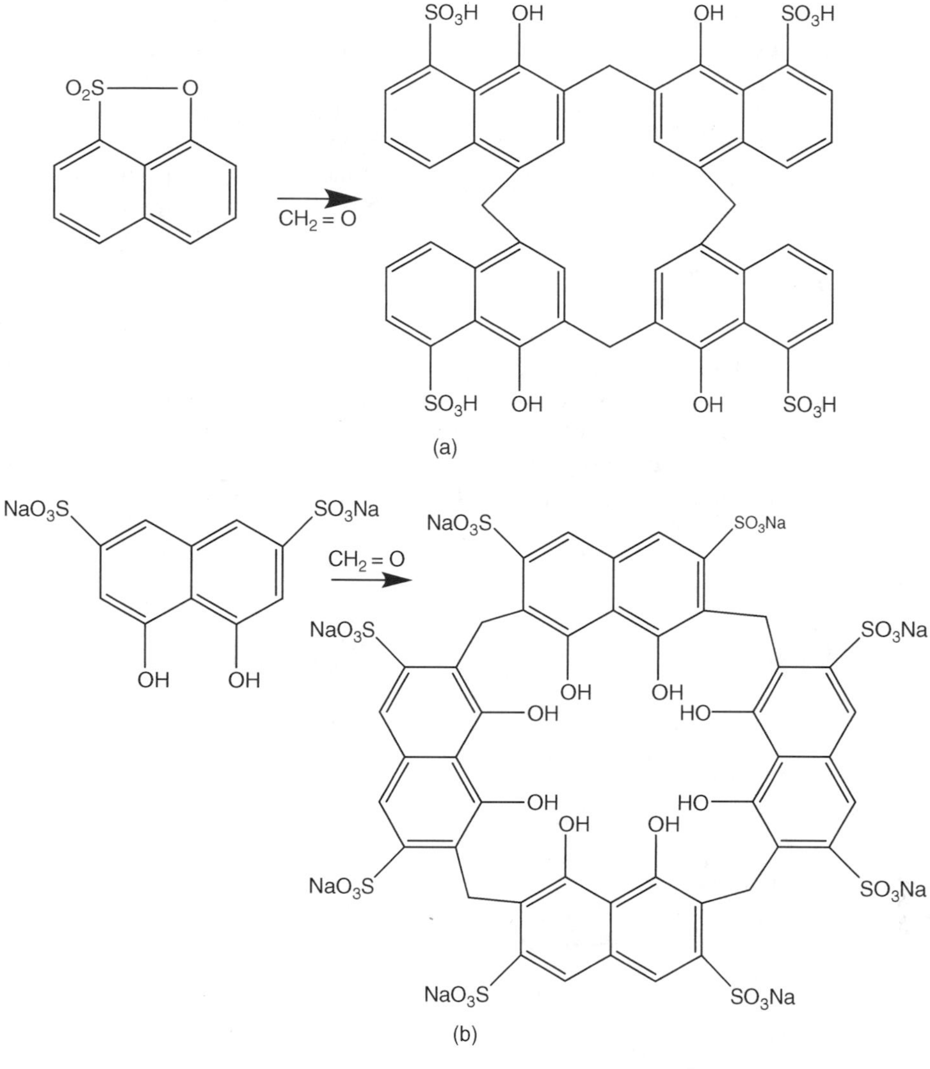

Figure 5.6 *Condensation products of formaldehyde with (a) 1,8-naphthalene sultone and (b) chromotropic acid*

in Figure 5.6a in 15% yield. NMR studies confirmed the structure and indicated that the macrocycle is conformationally mobile. A more extensive study has been made of the condensation products between chromotropic acid and formaldehyde by the group of Poh.[14] When chromotropic acid and a fivefold excess of formaldehyde were combined in aqueous solution for a week, a dark-red water-soluble product could be obtained.[14] Mass spectral and NMR studies indicated that the product was the cyclic tetramer shown in Figure 5.6b, which is conformationally mobile in solution. Earlier work proposed a linear polymeric structure[15] for this molecule and although much work has been published by Poh's group on this compound, as yet no X-ray structure has been successfully obtained to prove the cyclic tetramer structure.

The same group studied the interactions of the 'chromotropylene' tetramer with a wide variety of substrates. A range of aromatic compounds could be solubilised in water by the tetramer and the association constants were shown to increase with the number of aromatic rings in the guest.[16] Metal ions such as Ni²⁺ and Co²⁺ can be included in a 1:1 ratio.[17] Also, when the macrocycle is deprotonated, amines[18] can be included as guests. Other guests also bound in a 1:1 ratio include phenols,[19] alcohols[20] and sugars;[20] in the case of the latter two species, the binding strength appears to be determined by the strength of the CH–π interaction. Biologically significant guests such as amino acids[21] have also been shown to form complexes in aqueous solutions; in the case of these guests, aliphatic CH–π interactions were shown to be the major factor in determining binding, while in the case of aromatic amino acids, the major factor was π–π interactions. Nucleosides[22] also bind to the macrocycle in aqueous solution (1:1 ratio), with NMR studies showing the base unit is included in the cavity, and the same macrocycle also forms 1:1 complexes with paraquat,[23] a widely used herbicide, with the aromatic units of the guest being included within the cavity. Other macrocyclic ring compounds have also been shown to form complexes with cyclotetrachromotropylene. Crown ethers, for example, could be partially included in the cavity of the macrocycle to form 1:1 complexes in water,[24] with higher binding constants being observed for dibenzo-18-crown-6 than for the parent crowns, again indicating π–π interactions. Another group of macrocycles, the cyclodextrins (cyclic polysugars, which will be discussed in detail in Chapter 6), also formed complexes[25] with cyclotetrachromotropylene. This is the first case of cyclodextrins serving as guests

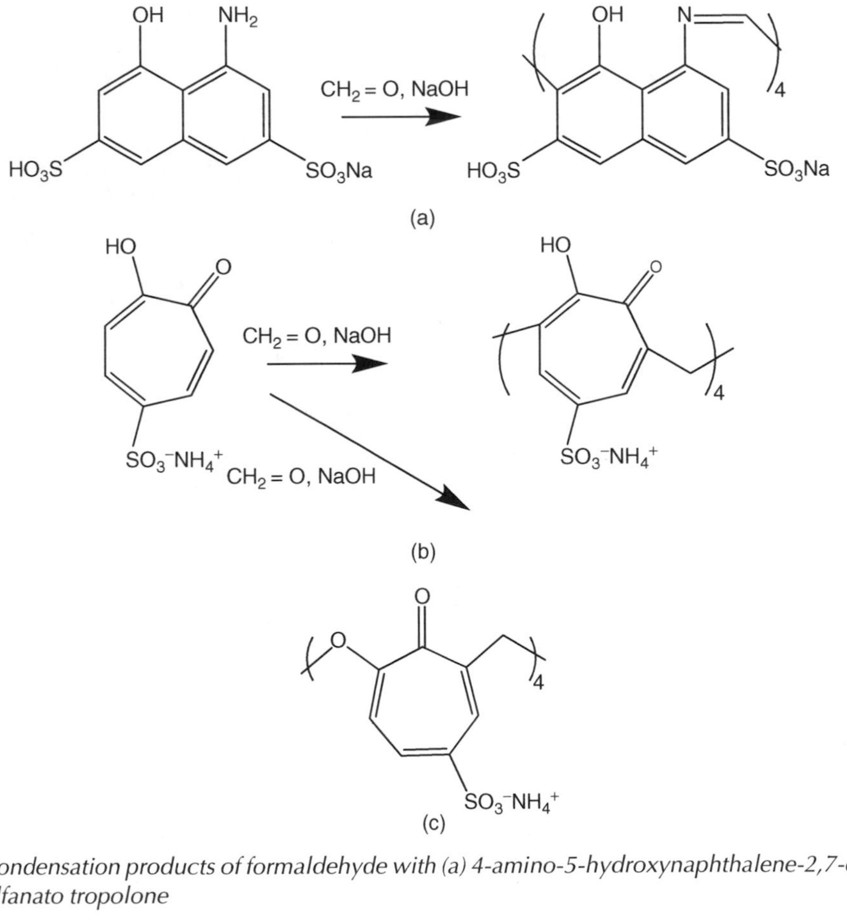

Figure 5.7 *Condensation products of formaldehyde with (a) 4-amino-5-hydroxynaphthalene-2,7-disulfonic acid and (b,c) 5-sulfanato tropolone*

rather than hosts and again the guest is only partially included in the cavity since neither macrocycle is of a size to completely encapsulate the other. An oxidised version of the tetramer has also been synthesised by reaction with dichromate[26] and has been shown to complex a variety of amines, tetraalkyl ammonium salts and amino acids as both 1:1 and 1:2 host–guest complexes. A cyclic tetrameric structure has also been claimed for the condensation product of 4-amino[5]hydroxynaphthalene-2,7-disulfonic acid with formaldehyde,[27] and this product (Figure 5.7a) has been shown to form complexes with polycyclic aromatic compounds in water, with complex strength increasing with the number of rings in the guest.

   Expanded calixnaphthalenes have been synthesised by the reaction of various derivatives of 2,7-dihydroxynaphthalenes.[28] When a 2,7-dihydroxynaphthalene is reacted with formaldehyde it will react at the C1 position to form dimers rather than a cyclic structure, so substitution of this system is required. When the derivative shown in Figure 5.8a is condensed using triflic acid as a catalyst, a cyclic trimer (Figure 5.8b) is formed in 23% yield, with mass spectral studies showing the existence of the tetramer as a minor product. Calculations indicate that this material will have a twisted propeller-like conformation. Likewise, a naphthalene substituted with cyclic ether units (Figure 5.8c) was found to cyclise to form a mixture of cyclic oligomers with three to six naphthalene units.[28] The monomers could also be condensed with formaldehyde to form a dimer and then reacted with hexanal to give the cyclic tetramer (Figure 5.8d) as a mixture of *cis* and *trans* isomers, the latter of which were shown in X-ray studies to have the structure in Figure 5.8e. This compound displays a flattened partial-cone structure, with two opposing naphthalenes being approximately planar and the other naphthalenes perpendicular to this plane.

5.3   Tropolone-Based Macrocycles

The group that investigated the chromotropylenes has also investigated other water-soluble macrocycles. A sulfonated tropolone derivative (Figure 5.7b) underwent condensation reactions with formaldehyde to form – based on NMR and MALDI mass spectral results – a cyclic tetramer.[29] NMR investigations

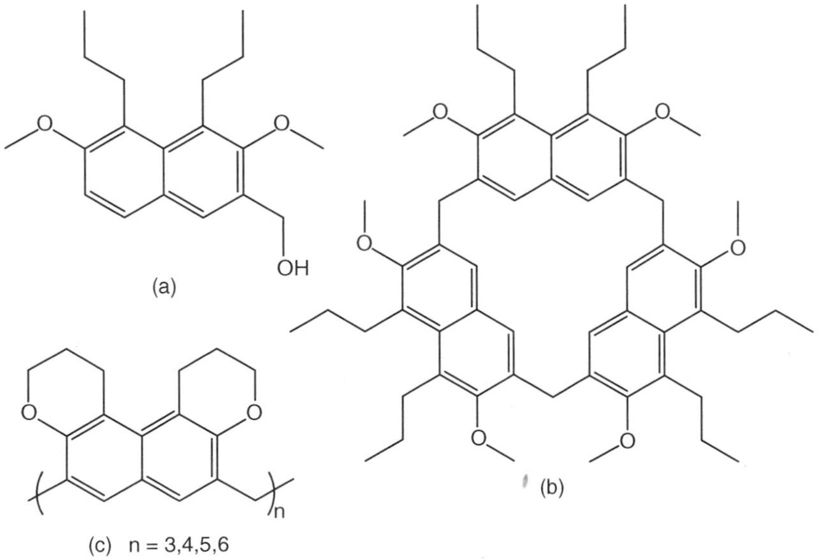

(a)

(c)   n = 3,4,5,6

(b)

Figure 5.8   *Expanded calix[n]naphthalenes*

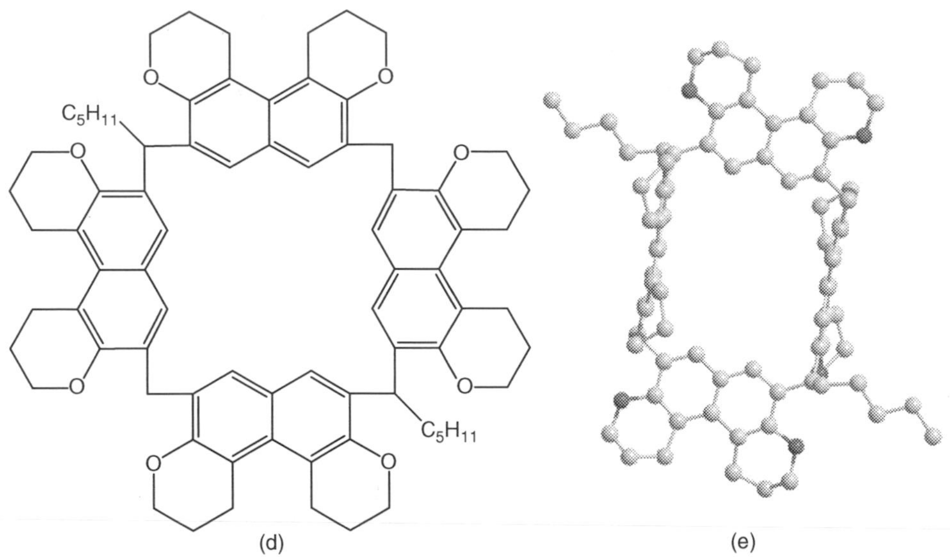

(d)   (e)

Figure 5.8   *(continued)*

showed this to exist in aqueous solution as a mixture of cone, partial-cone, 1,2- and 1,3-alternate configurations. The tetramer was shown to form weakly-bound complexes with alcohols in aqueous solution and more strongly-bound ones with aromatic species.[30] Again, the tropolone-derived tetramer[31] has been shown to form complexes with polycyclic aromatic compounds in water, with complex strength increasing with the number of rings in the guest. A second product (Figure 5.7c) with an oxacalixarene-type structure could also be obtained in low yield and was shown to have a 1,3-alternate structure in aqueous solution.[32]

5.4   Calixfurans

It is not only benzene rings which undergo condensation reactions with formaldehyde and other aldehydes and ketones – many heterocyclic compounds also undergo these reactions. One example is furan, which can form cyclic oligomers with a variety of aldehydes and ketones. A number of ketones were condensed with furan under acidic conditions in reaction media containing lithium perchlorate, which is thought to act as a templating agent.[33] Cyclic tetramers of acetone with furan (Figure 5.9a) could be obtained in 35% yield and other linear and cyclic ketones also gave cyclic tetramers, although in lower yields. These have been named 'calixfurans'. Reduction with hydrogen and Raney nickel of the acetone/furan tetramer was found to give good yields of the fully reduced compound (Figure 5.9c) as well as a by-product compound in which two adjacent furan rings have been reduced (Figure 5.9b). Varying the metal salts affected the yield of product, with lithium perchlorate giving the best yields.[33] However, no isolatable complexes could be obtained for the calixfurans with any of the metal salts used, although the tetrahydrofuran macrocycle formed isolatable complexes with species such as LiSCN and Ni(ClO$_4$)$_2$. When a linear acetone furan tetramer was reacted with acetone, a mixture of the cyclic tetramer and octamer was formed, which could be separated, and when a linear hexamer was used, the cyclic hexamer could be obtained as a pure compound. Later work utilised linear trimers and showed the synthesis of both the cyclic hexamer and

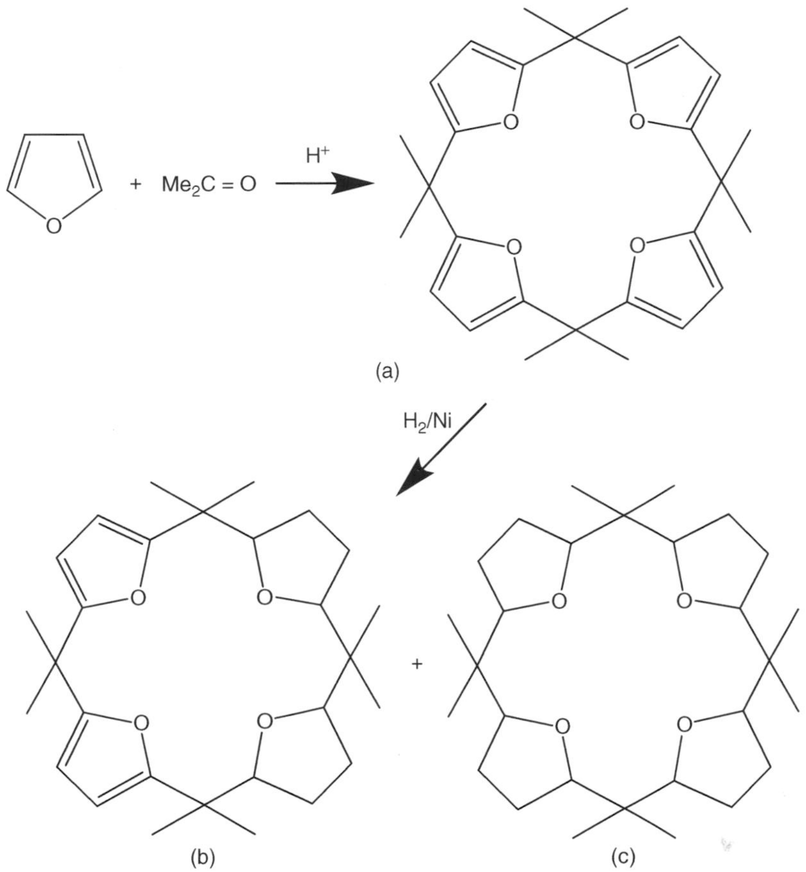

Figure 5.9 (a) Condensation products of furan with acetone; (b) partially reduced and (c) fully reduced

the linear product containing nine furan residues.[34] When this linear nonamer was further reacted with acetone, a cyclic nonamer was obtained in reasonable yield.

Direct reaction of formaldehyde with furan does not give appreciable yields of cyclic compounds, but when *bis-* and *tris*-furans were reacted with dimethoxyethane (Figure 5.10), reasonable quantities of the cyclic tetramer could be obtained, along with small amounts of the pentamer, hexamer and octamer.[35] Attempts were made to find template species that would increase the yields of these materials, but these were unsuccessful. These compounds were however shown to form gas-phase complexes with ammonium ions by mass spectrometry.[36] Less symmetric calixfurans could be synthesised by reacting linear *bis-* and *tris*-furans (formed by reacting furan with various aldehydes and ketones) with acetone or other compounds to give cyclic tetramers and larger oligomers.[37] These again could be reduced to give the tetrahydrofuran macrocycles, which were found to be capable of transporting a range of alkali metals, ammonium and silver ions through an organic medium.

X-ray studies have been carried out on these macrocycles. The cyclic tetramers usually adopt a 1,3-alternate configuration, as shown in Figure 5.11a for the acetone tetramer.[38] The cyclic hexamer has also been studied;[34,39] it is shown in Figure 5.11b. The hexamer has a structure in which two of the furan rings are in the plane of the macrocycle and the remaining four are in an up-down-up-down orientation.

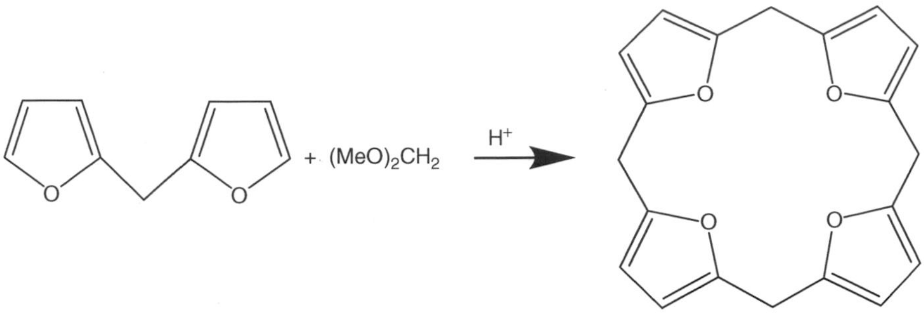

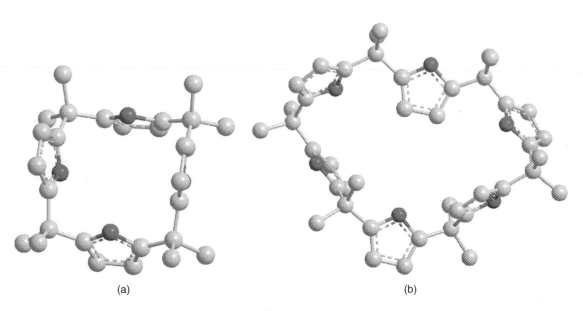

Figure 5.10   *Condensation of bis-furans with dimethoxymethane*

Figure 5.11   *Crystal structures of acetone-furan (a) tetramer and (b) hexamer*

   Because synthetic schemes exist to make a wide variety of calixfurans, one field of investigation has been their ring-opening reactions. For example, the cyclic tetramer (Figure 5.9a) can be oxidised[40] to convert the furan units into unsaturated diketones such as that shown in Figure 5.12a. This can then be hydrogenated to give the cyclic octaketone shown in Figure 5.12b. The cyclic hexamer can also be oxidised in the same manner and hydrogenated to give a cyclic dodecaketone. In the same work, treatment with bromine followed by hydrolysis gave rise to enedione species such as that shown in Figure 5.12c, and by careful control of the reaction conditions partially oxidised products such as that in Figure 5.12d can be obtained. The diketone compounds are of interest since they can easily be cyclised to give furan, pyrrole or thiophene units, thereby offering a route to compounds which contain a mixture of heterocyclic groups within a single macrocycle;[41] this will be discussed in more detail later.

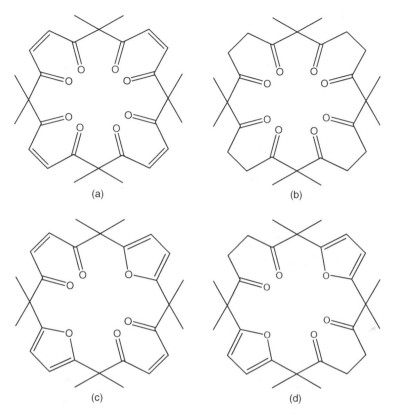

Figure 5.12   *Oxidation products of acetone-furan tetramers*

Furans also readily undergo Diels–Alder reactions, and the cyclic furan-acetone oligomers have been studied as a substrate for this type of chemistry. The cyclic tetramers[42] undergo Diels–Alder reactions with species such as benzyne to give products such as those shown in Figure 5.13a. These could be converted into the hydrogenated derivatives (Figure 5.13b), but attempts to convert either of the two macrocycles into the corresponding naphthalene tetramers failed. Only one isomer of the tetramer was obtained, which was shown by X-ray crystallography (Figure 5.13c,d) to again have a 1,3-alternate-type structure, in which the bridging oxygen atoms are alternately above and below the mean plane of the macrocycle. When crystallised from a mixture of *o*-, *m*- and *p*-xylenes, only *p*-xylene was found to be incorporated in the lattice, demonstrating molecular recognition by the tetramer. Within this work the synthesis of mono- di- and tri-adducts of benzyne with the calixfuran are also described, along with their hydrogenated derivatives. In the case of the 1 : 1 adduct, it was found to be possible to dehydrate the epoxynaphthalene compound to give the calix[3]furan[1]naphthalene system, the crystal structure of which is shown in Figure 5.14, demonstrating a more cone-like structure with a central cavity.

The similar hexamer could also undergo these types of reaction.[39] It proved possible to monosubstitute the calix[6]furan with benzyne and convert the product into the calix[5]furan[1]naphthalene ring system, and to also add two benzyne units to the hexamer and convert that to the calix[4]furan[2]naphthalene

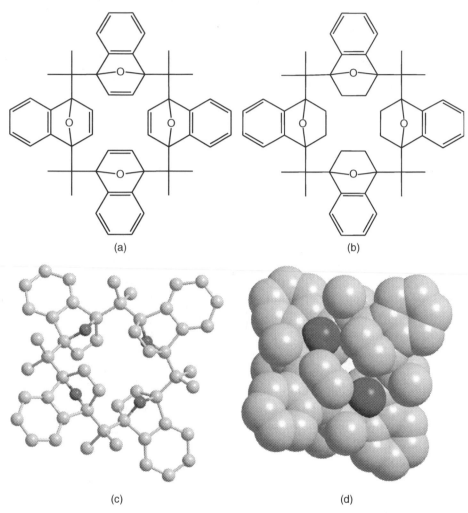

Figure 5.13 *(a) Diels–Alder adducts formed by reaction with benzyne and (b) after reduction with H_2, Pd/C; (c) the crystal structure and (d) the space-filling model of 5.13a*

macrocycle (with the two naphthalenes at the 1,3 or 1,4 positions of the macrocycle (Figure 5.15)). Several other adducts could also be synthesised and their crystal structures determined. Dimethylacetylene dicarboxylate also underwent Diels–Alder additions to the hexamer, allowing the synthesis of calixfurans substituted with carboxylic acid groups.

5.5   Calixpyrroles

5.5.1   Introduction

We have already discussed (in Chapter 1) the porphyrin series of molecules. These can by synthesised via the condensation of aldehydes with pyrroles to give a nonaromatic porphrinogen tetramer (Figure 1.7b),

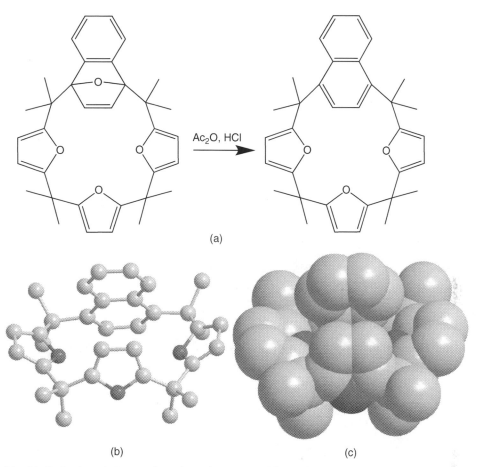

Figure 5.14 *(a) Dehydrated 1:1 calix[4]furan:benzyne adduct, (b) its crystal structure and (c) its space-filling model*

which undergoes oxidation to give the aromatic porphyrin molecule. A similar reaction occurs when ketones are condensed with pyrrole, but in this case the porphrinogen tetramer is stable and can be isolated as a crystalline compound. These classes of molecules have been termed the 'calixpyrroles'.

The earliest reports in this field come from 1886, when the great German chemist Adolf von Baeyer, who was mentioned in Chapter 4 for his early work on phenol/aldehyde condensation and won the 1905 Nobel Prize in Chemistry, utilised an acid-catalysed condensation of acetone and pyrrole[43] to synthesise an octamethyl-substituted calix[4]pyrrole (Figure 5.16a; R = —CH$_3$). Although it appears very similar to a porphyrin, the aromatic pyrrole units are not conjugated together to form an extended delocalised system. This led to major differences in the chemistry and structure of the porphyrins and calixpyrroles. For example, porphyrins are usually highly coloured due to the extended pi system, whereas calixpyrroles are usually white or pale yellow in colour unless substituted with other chromogenic groups. Furthermore, porphyrins are flat, to allow extended interaction of the p-orbitals, whereas calixpyrroles are nonplanar in structure, as will be discussed later. There have been several reviews of calixpyrroles, which provide much more detailed descriptions than will be given in this chapter.[44,45]

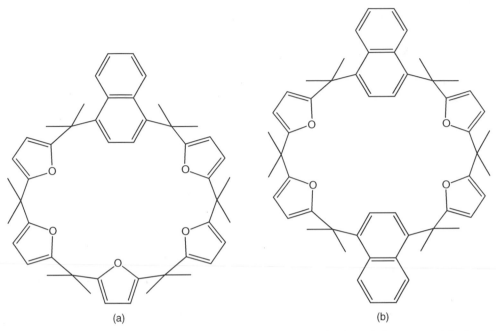

Figure 5.15   *(a) 1 : 1 and (b) 2 : 1 Diels–Alder adducts of calix[6]furan with benzyne, dehydrated to give naphthalenes*

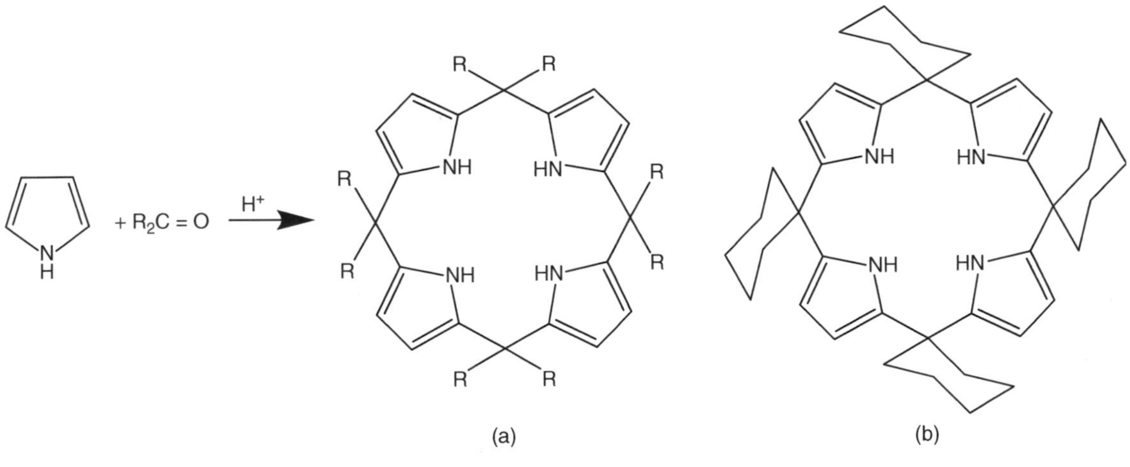

Figure 5.16   *Synthesis of calix[4]pyrroles with (a) linear ketones and (b) cyclohexanone*

A wide variety of ketones have been reacted with pyrrole to give the cyclic tetramers. Two of the most widely used are acetone (R = Me) and cyclohexanone, which give the compounds shown in Figure 5.16. X-ray crystallographic studies of these compounds have been carried out; they demonstrated the non-planarity of the structures.[46] The calixpyrroles instead usually adopt a 1,3-alternate conformation of the pyrrole rings, as shown in Figure 5.17a,b. This conformation is the most common for calixpyrroles, but

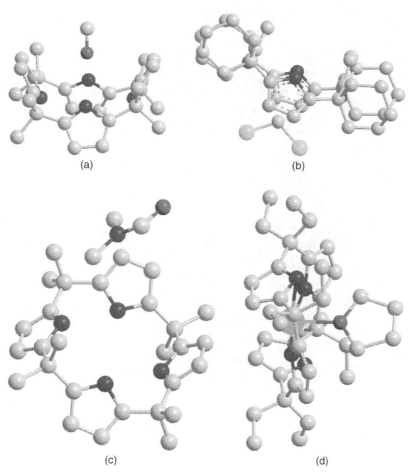

(a)

(b)

(c)

(d)

Figure 5.17 *Crystal structure of calix[4]pyrroles synthesised with acetone crystallised from (a) methanol, (b) cyclohexanone calix[4]pyrrole/dichloromethane, (c) DMF and (d) octaethylcalix[4]pyrrole complex with Zr(IV)*

complexation can have a major effect on the structure and when for example the acetone-pyrrole tetramer is crystallised from DMF, it instead adopts a 1,2-alternate conformation,[47] as shown in Figure 5.17c. Studies were carried out on other solvent/calixpyrrole combinations both in solution and in the crystalline state; interestingly, the calixpyrroles tended to form 1 : 1 complexes with hydrophilic solvents in benzene but 1 : 2 host–guest complexes in the crystal.[47]

We have already described how porphyrins can bind a wide variety of metals. However, for simple calixpyrroles, metal binding does not appear to occur to any great extent. Some success has been achieved by deprotonating the pyrrolic N—H groups with BuLi and then utilising these species to form complexes with iron and molybdenum species.[48] In these complexes the calixpyrrole adopts a cone-like conformation, as shown by X-ray crystallography, with all of the nitrogens complexing to the metal. The zirconium complex[49] displays quite different behaviour, with two of the pyrrole rings forming σ-bonds to the metal via the nitrogen atom and the other two showing π-interactions via the aromatic system rather than via the nitrogen.

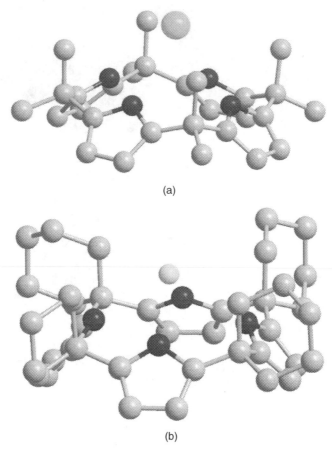

(a)

(b)

Figure 5.18 *Crystal structure of (a) octamethylcalixpyrrole complex with chloride and (b) octacyclohexylcalixpyrrole complex with fluoride*

One field of calixpyrrole chemistry that has excited much interest is their anion-binding capability. For example, the simple acetone-pyrrole tetramer can be mixed in dichloromethane with tetrabutylammonium chloride or fluoride and slowly evaporated to give crystalline complexes.[46] X-ray studies show that the calixpyrrole adopts a cone conformation, allowing all four N—H protons to hydrogen-bond to the anion (Figure 5.18a). The two complexes are similar, but the fluoride ion is 0.15 nm above the plane of the nitrogen atoms, while the larger chloride ion is 0.23 nm above the plane. Stability constants showed stronger binding for fluoride (17170 M^{-1}) than chloride (350 M^{-1}); they also showed some binding to dihydrogen phosphate but little or no binding for bromide, iodide and hydrogen sulfate. Binding of fluoride and chloride was also observed for the cyclohexanone-based tetramer (Figure 5.16b), although it was somewhat weaker.

Most calixpyrroles are cyclic tetramers but other ring sizes are known. One problem is that the calix[4]pyrroles are the most ready to form, and although other ring sizes can be detected in reaction mixtures, they cannot easily be isolated and tend to re-equilibrate to give the tetramer. Calix[4]pyrroles can be synthesised directly from fluorinated pyrroles such as 3,4-difluoropyrrole[50] by a methane sulfonic acid-catalysed reaction with acetone (Figure 5.19). The fluorinated derivatives show increased affinities to

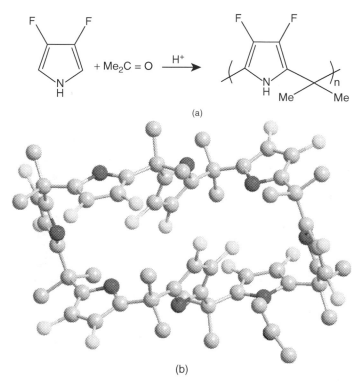

(a)

(b)

Figure 5.19 *(a) Synthesis of fluorinated calixpyrroles and (b) crystal structure of the octamer*

chloride and especially $H_2PO_4^-$ ions (increase in binding constant of >1000). In the case of $H_2PO_4^-$, binding is accompanied by a visual colour change from yellow to orange. The solid-state structures of these compounds are very complex and depending on the presence of various ions or solvents, cone, partial-cone, 1,3- and 1,2-alternate structures can all be determined.

The reaction of fluorinated pyrrole is noticeably slower than the simple pyrrole-acetone condensation and it was found that by careful optimisation of the reaction conditions, both the cyclic pentamer and the octamer could be isolated from the reaction mixture in yields of 23% and 14% respectively.[51] X-ray crystallography studies showed that the calixpyrrole cores adopt what can best be described as distorted alternate conformations. In both cases the macrocycles appear to contain a large cavity with (in the fluoride complex) many but not all of the pyrrolic NH moieties pointing into it and displaying some NH-F hydrogen bonding. Further work[52] succeeded in isolating the cyclic hexamer and studied the interactions of this series of calixpyrroles with anions. Strong binding was found to occur between the cyclic tetramer and a range of anions such as chloride, bromide, acetate, benzoate and phosphate in acetonitrile and DMSO. It appears that the strongly electron-withdrawing fluoride groups increase the affinity of the calixpyrrole for anions compared to the unsubstituted tetramers; also, the fluorinated tetramer generally binds this group of anions more strongly than the higher analogues. X-ray studies of the complex of the tetramer with chloride showed a conical form for the calixpyrrole, stabilised by hydrogen bonding to the anion, as found for the unsubstituted tetramer. A recent report has also described the condensation of pyrrole and acetone in the presence of bismuth nitrate; using chromatographic methods the cyclic pentamer was obtained in moderate yields.[53]

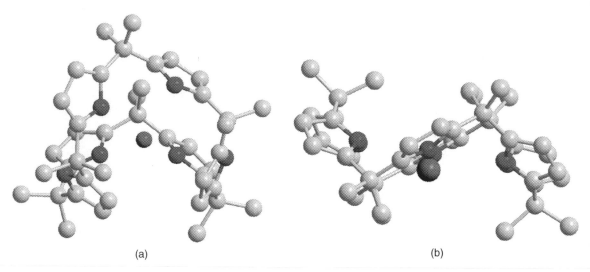

(a)   (b)

Figure 5.20   *(a) Crystal structure of calix[6]pyrrole/water and (b) its complex with bromide ion*

We have already discussed the synthesis of calix[6]furans and their oxidation to unsaturated dodecaketones.[40] These can be reacted with ammonium acetate to give the corresponding calix[6]pyrrole in 42% yield.[54] X-ray crystallographic studies of the 1 : 1 complex with water (Figure 5.20a) showed it to have a structure described as being like a 'tennis-ball seam'. Complexation experiments with a variety of ions were undertaken, and under the conditions used in this work, calix[6]pyrrole bound fluoride much more weakly and chloride much more strongly than calix[4]pyrrole. The crystal structure of the complex of the hexamer with bromide (Figure 5.20b) shows considerable change from the water complex, with a cone-like shape and all six N—H groups directed towards the central anion.

Dipyrrolyl compounds have also been used in the synthesis of calix[6]pyrroles. Reacting pyrrole with 4-nitroacetophenone gave a dipyrrole species which could be condensed with acetone[55] to give two calix[4]pyrrole isomers, one with the nitrophenyl groups on the same side of the macrocycle and the other with substituents on opposite sides, along with a calix[6]pyrrole with two nitrophenyl groups on one side and one on the other (Figure 5.21). The calix[6]pyrrole with all three nitrophenyl groups on the same side of the macrocycle was not detected. Crystal structures were obtained for both tetramers. A range of anion-binding experiments were performed, which showed that both ring size and conformation affected binding strength. Under dry conditions, fluoride bound to the tetramer in Figure 5.21a and the hexamer in Figure 5.21c with approximately the same binding coefficient, and more weakly to the tetramer in Figure 5.21b, although chloride was found to bind much more strongly to the hexamer. It should be noted that solvent and especially adventitious water had large effects on the binding constants of anions to calixpyrroles and therefore it is very difficult to compare results obtained by different groups under different conditions.

The reactions of pyrroles with bulky ketones such as benzophenone did not progress to give calix-pyrroles; instead dipyrrolyl species were isolated,[56] probably due to the effect of steric hindrance preventing cyclisation. These could however be condensed with acetone using a methane sulfonic acid catalyst to give asymmetric calix[4]pyrroles (Figure 5.22). However, when trifluoroacetic acid was used as the catalyst, calix[6]pyrroles were obtained. X-ray studies of the calix[6]pyrrole synthesised from benzophenone showed that the pyrrole rings adopted a 1,3,5-alternate structure with the diphenylmethylene groups all on

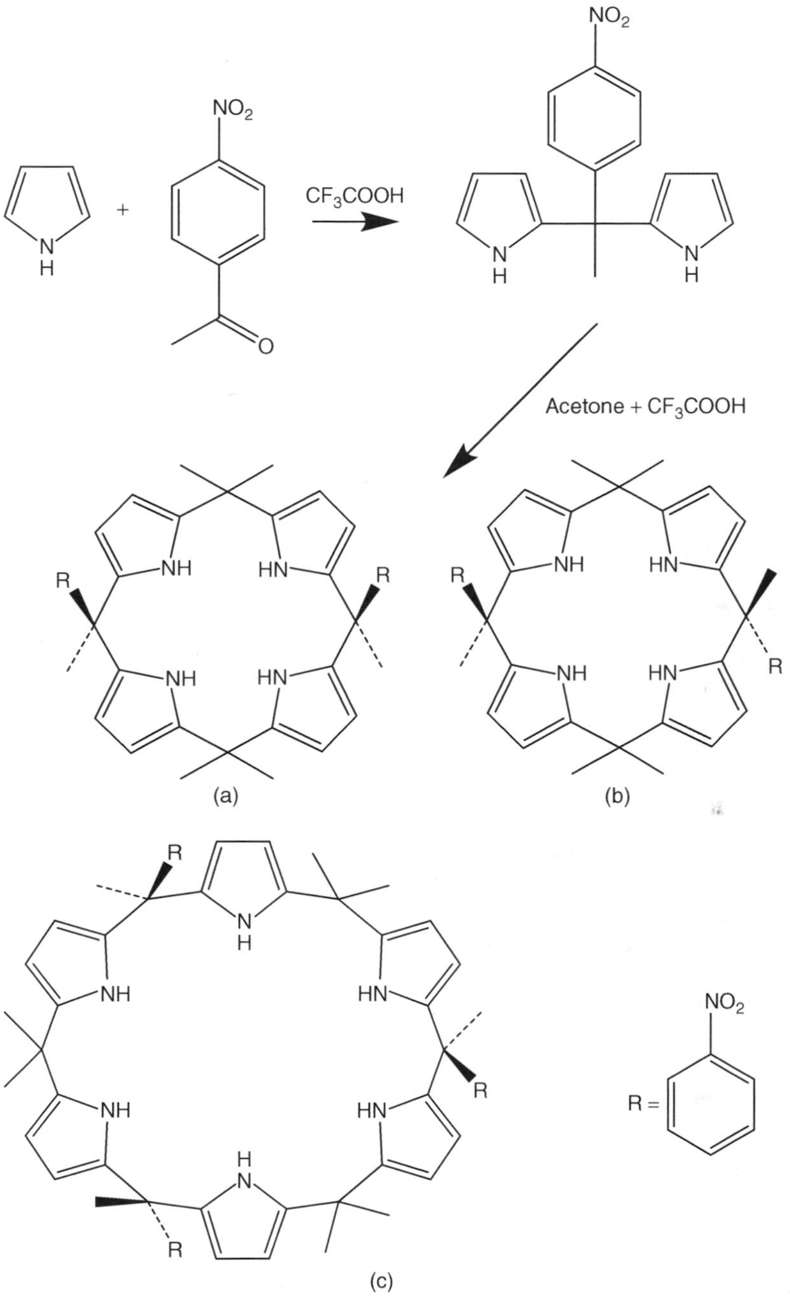

Figure 5.21 *Synthesis of nitrophenyl-substituted calix[n]pyrroles: (a) tetramer (cis isomer), (b) tetramer (trans isomer) and (c) hexamer (up-up-down isomer)*

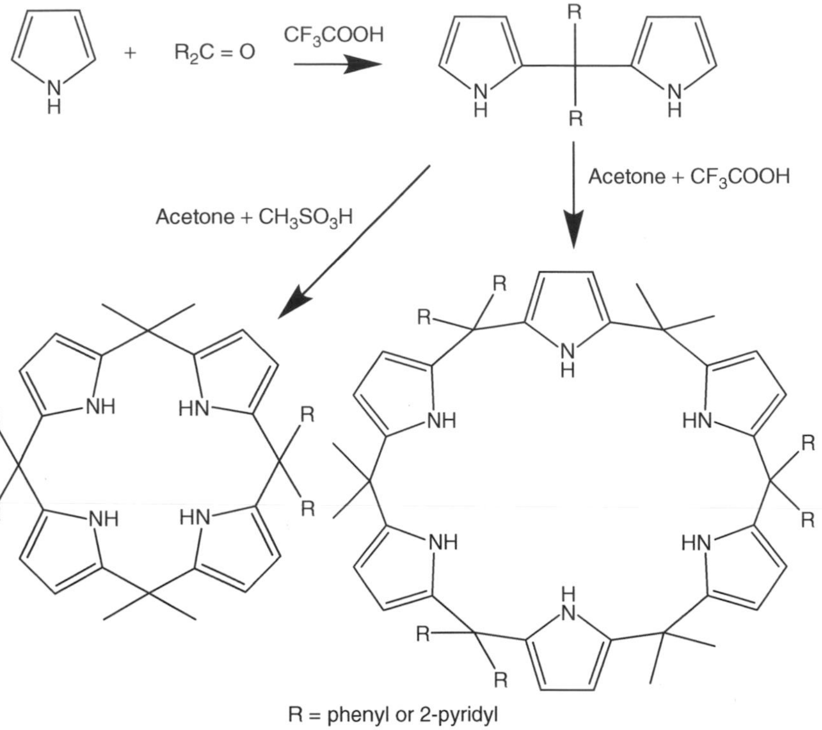

R = phenyl or 2-pyridyl

Figure 5.22   *Synthesis of calix[4]pyrroles and calix[6]pyrroles*

one side of the macrocycle and the dimethylmethylene groups all on the other. The overall shape of the molecule is an asymmetric cone, with three N—H bonds perpendicular to the macrocycle on the side of the dimethylmethylene groups and the other three pointing towards the centre of the cavity.

Synthesis of a substituted dipyrrole species and linkage to 1,3,5-trihydroxy benzene gave moieties such as that shown in Figure 5.23, which could then be reacted with acetone to give the capped calix[6]pyrrole,[57] albeit in only 2–3% yield. A similar species capped with a benzene tricarboxylate unit was also synthesised.

5.5.2   Chemical modification of the calixpyrroles

There appear to be three sites at which a calixpyrrole can be modified, namely the linking groups between the pyrrole rings, the pyrrole N—H groups and the pyrrole 3-positions. The first, modification of the linking groups, has not been widely studied since it is usually simpler to incorporate the required substituent into the ketone used to assemble the calixpyrrole. If the desired substituent would either interfere with the cyclisation reaction or be degraded by it then in some cases a protecting group can be utilised. For instance, the Cbz-protected 3-aminoacetophenone could be reacted with pyrrole and 3-pentanone with boron trifluoride catalysis and the resultant mixture of products separated by flash chromatography to give the mono-substituted calix[4]pyrrole; the protecting group could then be removed to give a calix[4]pyrrole bearing one aminophenyl side group[58] and this side chain could be substituted with a number of dye groups to give highly fluorescent anion sensors. These sensors showed high affinity for a number of anions, especially for phosphate and pyrophosphate ions.

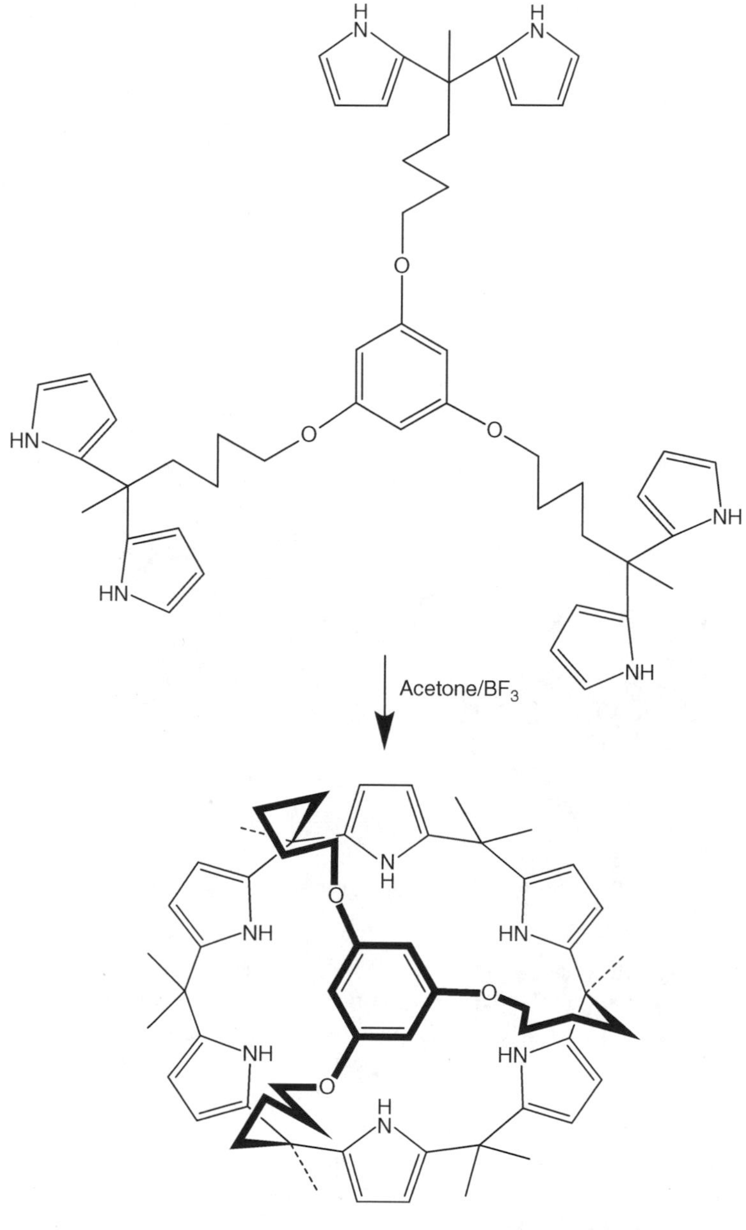

Figure 5.23 *Synthesis of a trihydroxybenzyl-capped calix[6]pyrrole*

Replacement of the pyrrole 3,4-hydrogens by a variety of methods has been studied. The simplest method is to utilise a substituted pyrrole. We have already mentioned for example the use of 3,4-difluoropyrrole. Another compound that has been successfully utilised is 3,4-dimethoxypyrrole, which could be condensed with cyclohexanone to give the material shown in Figure 5.24a, although with only an 8% yield after chromatographic purification.[59] In the same work a calix[4]pyrrole could be reacted with N-bromo succinimide

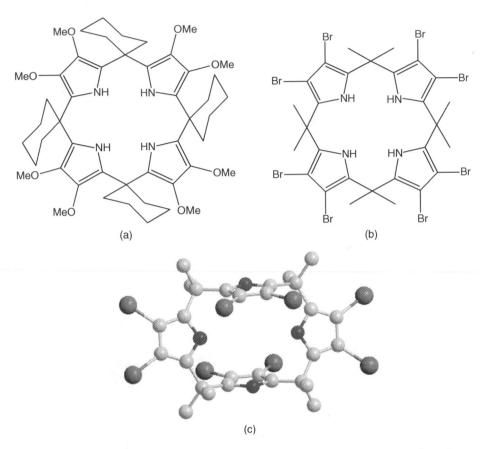

Figure 5.24 *(a,b) C-substituted calix[4]pyrroles and (c) crystal structure of 5.24b*

to give the octabromo derivative shown in Figure 5.24b, which in the crystal form adopts an unusual flattened 1,2-alternate configuration. The hydrogens at the 3-positions of calix[4]pyrroles are acidic enough to react with strong bases such as BuLi and the resultant carbanions react with electrophiles. Addition of BuLi to octamethylcalix[4]pyrrole followed by reaction with ethyl bromoacetate[59] leads to the formation of both the mono-substituted calixpyrrole (26%) and a small amount (3%) of a single isomer of diester (Figure 5.25). Complexation studies showed that electron-donating methoxy groups lowered anion-binding strength, while electron-withdrawing bromo groups increased it. This synthetic methodology was expanded to include a wider range of electrophiles and calix[4]pyrroles monosubstituted with carboxylic acid, formyl, ethyl alcohol, ethyl piperidine and iodo groups have all been obtained.[60] Small amounts of di-substituted products could also be observed. X-ray crystal studies showed that in several cases the side groups underwent interactions with the N—H groups of adjacent pyrroles, forming dimeric species.

An alternative synthetic procedure[61] involved reaction of iodine-[bis(trifluoroacetoxy)iodo] benzene with calix[4]pyrrole to give the monoiodinated product shown in Figure 5.25c. This could then be reacted with excess trimethylsilyl acetylene to give the structure shown in Figure 5.25d, which could be deprotected to give that shown in Figure 5.25e. Finally, this could be coupled with aromatic iodides using palladium catalysis to give structures of the type shown in Figure 5.25f (where Ar = tolyl, nitrophenyl, dinitrophenyl, *p*-dimethylamino azobenzene, phenanthrene and other aromatic moieties) in good yields. These compounds

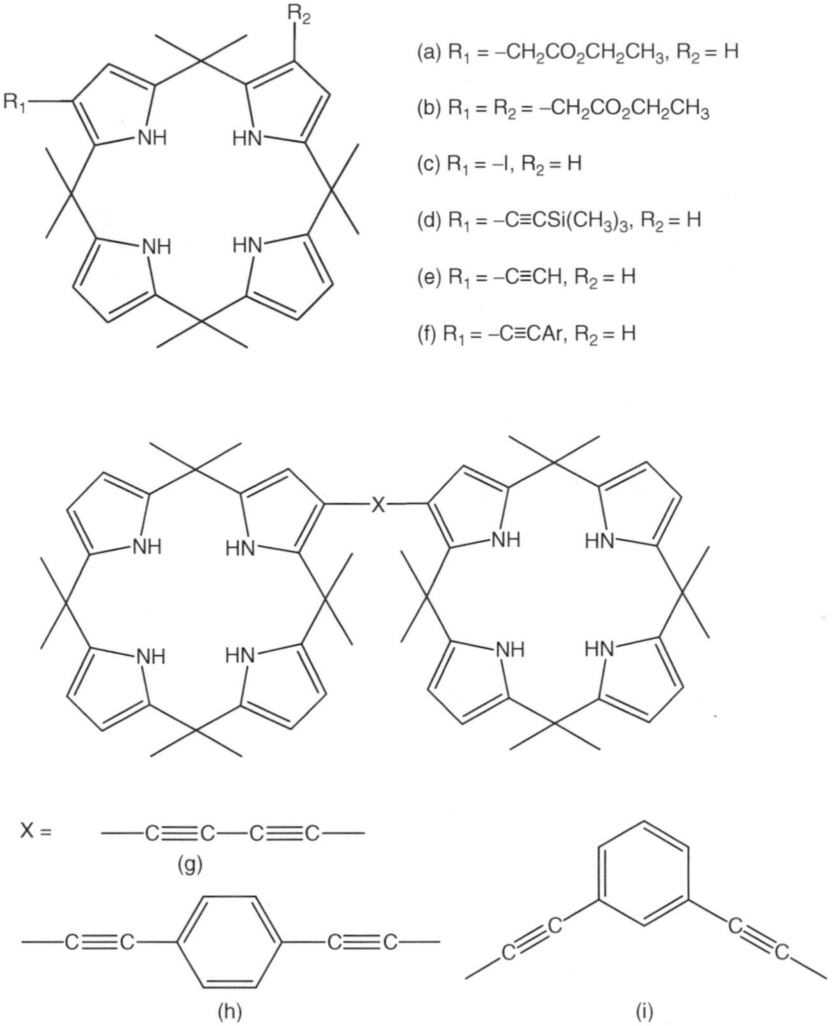

(a) $R_1 = -CH_2CO_2CH_2CH_3$, $R_2 = H$

(b) $R_1 = R_2 = -CH_2CO_2CH_2CH_3$

(c) $R_1 = -I$, $R_2 = H$

(d) $R_1 = -C\equiv CSi(CH_3)_3$, $R_2 = H$

(e) $R_1 = -C\equiv CH$, $R_2 = H$

(f) $R_1 = -C\equiv CAr$, $R_2 = H$

$X =$

—C≡C—C≡C—
(g)

—C≡C——C≡C—
(h)

(i)

Figure 5.25 *(a–f) Mono- and di-C-substituted calix[4]pyrroles and (g–i) structures of calix[4]pyrrole dimers*

could serve as colorimetric or fluorimetric sensor materials for anions. Alternatively, two calix[4]pyrrole units could be coupled together directly or via a di-iodo benzene[62] to give the dimeric species shown in Figure 5.25g. These structures open up the possibility of cooperative binding between two calix[4]pyrrole units. The binding constants of the structure in Figure 5.25g were determined for complexation to benzoate, phthalate and isophthalate ions were determined, and in the case of benzoate and phthalate, 1 : 2 host–guest complexes were formed with association constants three to four times lower than those with the parent calix[4]pyrrole (which formed 1 : 1 complexes). The bent isophthalate was found to form a 1 : 1 complex with the structure in Figure 5.25g, with an association constant six times that of the complex with the parent compound, indicating cooperative binding of the two carboxylate groups to the two pyrrole rings.

Usually reactions with strong bases and electrophiles give rise to C-substitution but N-substitution is also known to occur. When octaethylcalix[4]pyrrole was treated with sodium hydride, 18-crown-6 and methyl iodide, a series of products could be obtained.[63] These were the mono-N-methylated, 1,2-di-N-methylated (in very low <1% yield), 1,3-di-N-methylated, tri-N-methylated and fully N-methylated products. The major product (1,3-di-N-methylated) could be crystallised and was shown to have a 1,3-alternate configuration in the crystal. When ethyl iodide was used, only the mono-N-ethylated product was obtained. Direct condensation of ketones with N-methyl pyrrole has been attempted[63] but as yet macrocyclic products have not been obtained.

Attempts have been made to extend the cavities of calix[4]pyrroles, often by the use of functionalised ketones. For example, 4-hydroxyacetophenone can be condensed with pyrrole to give cyclic tetramers as a mixture of four configurational isomers (Figure 5.26a–d), which can be separated chromatographically.[64] What is interesting is that the major product (45%) is the isomer with all four hydroxyphenyl units

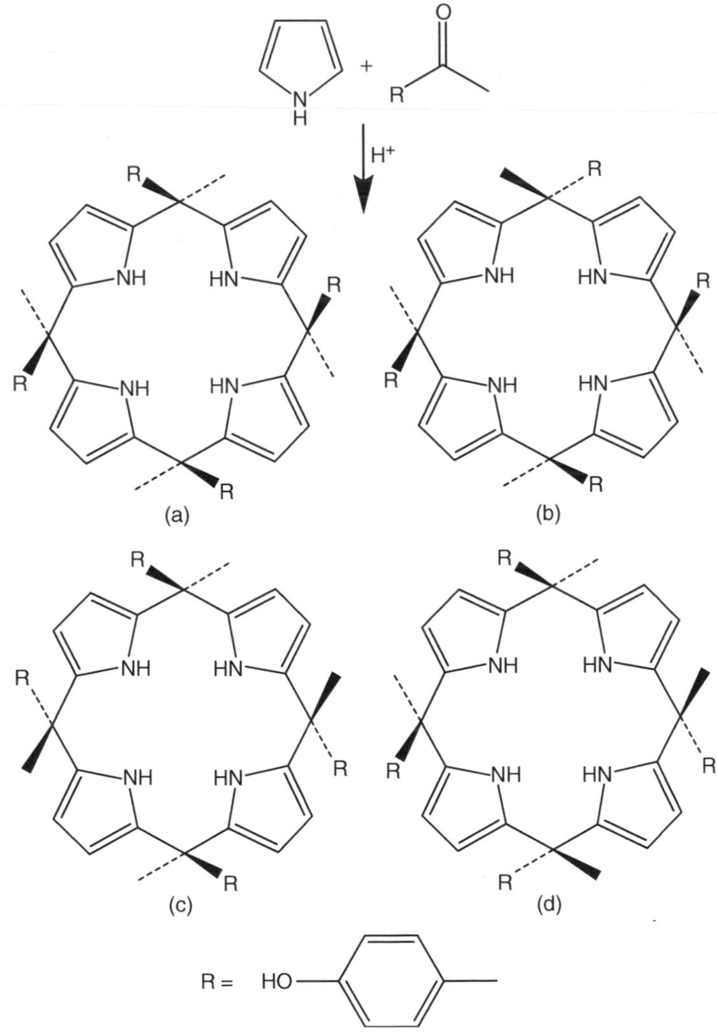

Figure 5.26   *Isomers of tetra-substituted calix[4]pyrroles*

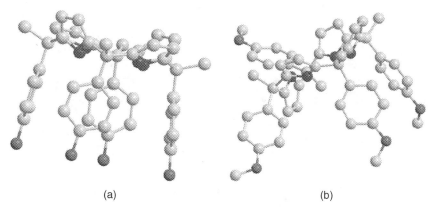

(a)   (b)

Figure 5.27 *Crystal structures of (a) αααα 4-hydroxyphenyl calix[4]pyrrole and (b) the 4-methoxyphenyl αααβ isomer*

on the same side of the macrocycle (αααα; Figure 5.26a), while the isomer with three units on the same side (αααβ; Figure 5.26b) is synthesised in 30% yield. The other two isomers are the system with two adjacent hydroxyphenyl units above and below the macrocycle (ααββ, 25%; Figure 5.26c) and the isomer with the αβαβ conformation (<5%; Figure 5.26d). This last isomer is the least sterically hindered, so a simplistic model would predict it would be the most common, and the αααα the least. From this it appears that there must be favourable interactions between adjacent hydroxyphenyl units, possibly hydrogen bonding, π–π interactions or a combination, similar to the cooperative binding observed in the formation of calixarenes and related species. Reaction with methyl iodide gave the methoxyphenyl-substituted compounds.

The crystal structures of the αααα isomer of the hydroxyphenyl calix[4]pyrrole and the αααβ isomer of the methoxyphenyl counterpart demonstrate how a deeper, extended cavity is formed by the hydroxyphenyl units. Each contains a molecule of ethanol (Figure 5.27a) or acetonitrile (Figure 5.27b). Studies of their interactions with fluoride, chloride and phosphate ions have been made. Generally stronger binding is seen for the hydroxyphenyl systems, with the ααββ isomer showing the strongest binding for chloride and with the αααα binding more strongly that the αααβ system. None of these bind as strongly as the parent octamethylcalix[4]pyrrole, although their relative selectivity for fluoride appears to be better.

A similar group of compounds using 3-hydroxyacetophenone was also studied by different workers.[65] Again three isomers were obtained, in the same order of occurrence. These occurred in a variety of crystal forms, as determined by X-ray crystallography, with for example the αααα or cone isomer crystallised as a dimer from DMF (Figure 5.28a) and as a trimer from acetic acid (Figure 5.28b). The arrangement of the hydroxyphenyl units was likened by the authors to that occurring in calixarenes. The cavity could be extended even further by converting the hydroxyl groups into N-phenyl carbamate, the crystal structure of which is shown in Figure 5.28c, which demonstrates binding of the solvent of crystallisation (acetone), although this is lost on gentle heating.

Even though the *para*-substituents of the compounds shown in Figure 5.26 are some distance from the pyrrolic N—H groups which compose the anion-binding site, they obviously affect binding, as was shown by the lower binding coefficients for methoxy- than for hydroxyl-substituted systems. Other reactions were used to introduce ester (—OCH$_2$CO$_2$Et) and amide (—OCH$_2$CONEt$_2$) groups in these positions,[66] and the resultant calix[4]pyrroles were shown to be highly selective to fluoride ion, binding fluoride from DMSO solution but showing no affinity at all for other halides, dihydrogen phosphate or hydrogen sulfate ions.

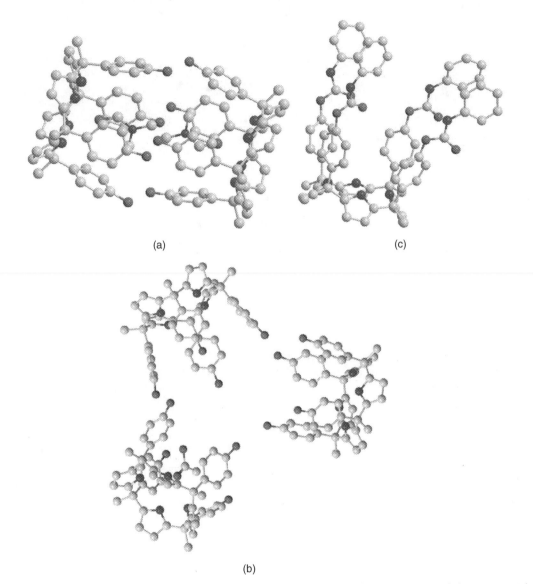

(a)   (c)

(b)

Figure 5.28   *Crystal structures of αααα 3-hydroxyphenyl calix[4]pyrrole from (a) DMF and (b) acetic acid, and (c) the phenylcarbamate product from acetone*

5.5.3   Confused calixpyrroles

Although pyrroles usually react via the 2,5-positions to form calixpyrroles, they can also undergo reaction via the 2,4-positions to give N-confused calixpyrroles. These are much less common, with only a few syntheses being known. One example involved pyrrole and cyclohexanone being reacted under a wide range of conditions (solvent, catalyst) to give, besides the expected pyrrole, the single N-confused pyrrole (Figure 5.29) – in some cases as the major isomer.[67] In the same work, other reaction conditions gave mixtures of the isomers of calix[4]pyrroles with two N-confused rings in >30% yield. The confused

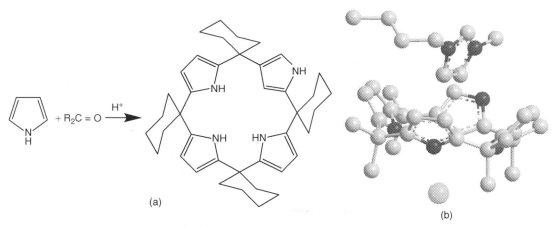

Figure 5.29 (a) Synthesis of N-confused calix[4]pyrroles and (b) crystal structures of a complex with butyl methyl imidazolium chloride

calixpyrroles were much less soluble than the usual isomer. Crystal structures for N-confused octamethyl calix[4]pyrrole with imidazolium chloride complexes[68] have been obtained and demonstrate that in the solid phase the chloride ion is hydrogen-bonded by the three 'unconfused' N—H groups, with the imidazole unit being included in the calix[4]pyrrole cup (Figure 5.29b).

5.5.4 Substituted calixpyrroles

Recent years have seen a great increase in interest in the calixpyrroles and a brief review of this work will be given here. Numerous other side chains have been attached to calixpyrroles – and calixpyrroles can be bridged in the same way as other macrocyclic systems such as calixarenes to give strapped pyrroles. A pyrrole substituted with a tetrathiafulvalene unit[69] could be condensed with acetone to give the cyclic tetramer shown in Figure 5.30a in 18% yield. This compound was found to form host–guest complexes with electron-deficient materials such as 1,3,5-trinitrobenzene in dichloromethane solution, where each calixpyrrole adopted a 1,3-conformation with a guest sandwiched between opposing tetrathaifulvalenes. In the case of 1,3,5-trinitrobenzene, X-ray crystal structures demonstrated this binding. Other guests included quinones and binding was accompanied by visible colour changes. Interestingly, the host–guest complex could be dissociated by the addition of chloride, which was thought to form a complex, thereby pulling the calixpyrrole into the cone conformation and disrupting binding of the neutral guest.[69] Further work[70] showed that the conical chloride complex was capable of binding both fullerene (but not the uncomplexed calixpyrrole) and a fullerene substituted with an electron-deficient fluorine moiety that could be bound by either the chloride complex or uncomplexed calixpyrrole, albeit with very different binding modes.

As an alternative to substituting the pyrrole unit, acetyl ferrocene could be condensed with cyclohexanone and pyrrole[71] to give a cyclic tetramer in 30% yield containing a single ferrocene unit (Figure 5.30b), the structure of which was proved by NMR and X-ray crystallography. Cyclic voltammetry showed that the material bound halide and dihydrogen phosphate ions with resultant shifts in cathodic potential of the ferrocene unit. This material could be immobilised onto carbon paste electrodes[72] to give electrochemical detection of fluoride and dihydrogen phosphate ions. Calix[4]pyrroles bearing ferrocene amide groups could also be synthesised,[73] and demonstrated sensitivity to fluoride and acetate ions when studied by NMR and electrochemically.

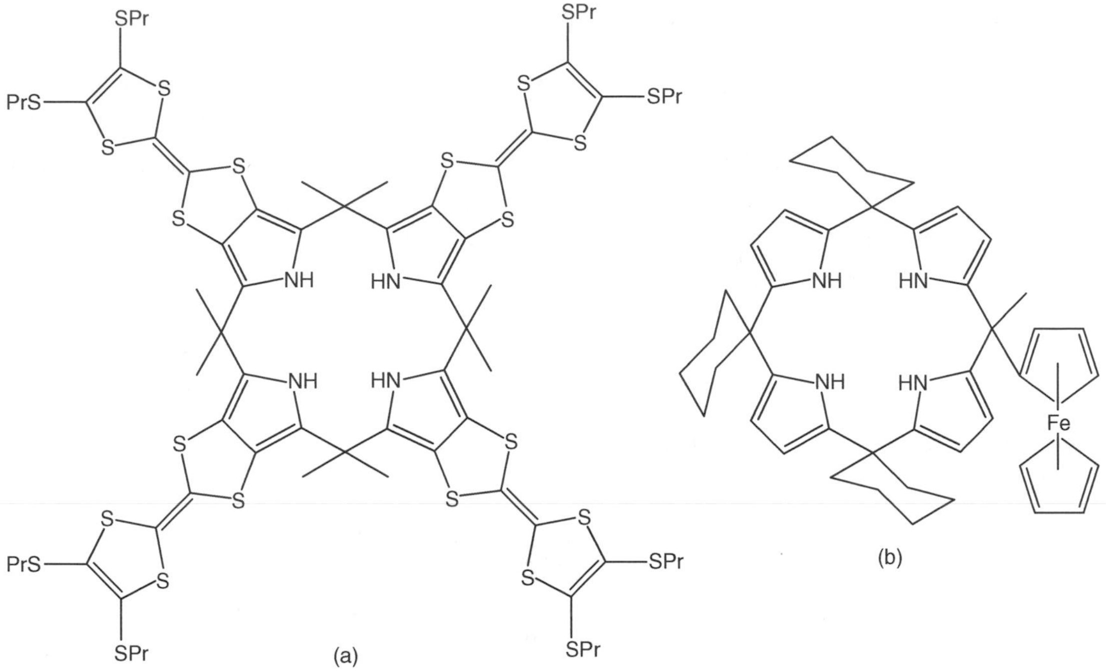

Figure 5.30   *(a) Tetrathiafulvalene-substituted calix[4]pyrrole and (b) ferrocene-substituted calix[4]pyrrole*

Calix[4]pyrroles substituted at the bridging positions by a variety of *p*-phenoxy ethers could be synthesised,[74] and in the case of the material substituted with thioether groups (Figure 5.31a) were shown to be an efficient extractant for both fluoride and mercury ions. Anion transport through dichloromethane has also been demonstrated by the octafluoro calixpyrroles described earlier.[75] Extended-cavity calix[4]pyrroles could also be obtained, containing phenyl urea units[76] (Figure 5.31b), and were shown by NMR studies to selectively form 1:2 complexes with fluoride, where one anion binds to the macrocycle and one to the phenyl urea groups, or 1:1 complexes with dihydrogen phosphate.

Chromogenic derivatives of normal calix[4]pyrroles substituted with azobenzyl and tricyanoethylene moieties (Figure 5.31c,d) have also been synthesised, along with their N-confused isomers,[77] with their structures being determined by X-ray crystal studies. These were shown to undergo colour changes in solution on exposure to various anions. Interestingly, the confused calixpyrrole showed stronger colour changes and different selectivities than the normal analogue. Also, the N-confused analogues of Figure 5.31d underwent a tautomerisation of the side group with the pyrrole moiety to give a cyclic pyrrolizine-type group. Immobilisation of the calixpyrroles in polyurethane matrices was found to give materials which gave analyte-specific colour changes. The behaviour of these materials and their utilisation in sensor arrays for anions has also been reviewed.[78]

Condensation of acenaphthene quinone with a dipyrrole compound gave the calix[4]pyrrole shown in Figure 5.32a, which was demonstrated to have a 1,3-alternate conformation in the crystal (Figure 5.32b), with the naphthalene rings almost perpendicular to the plane of the calix[4]pyrrole.[79] Binding of fluoride in solution affected the UV adsorption of the naphthalene units and also led to the adoption of a cone conformation in the crystal (Figure 5.32c).

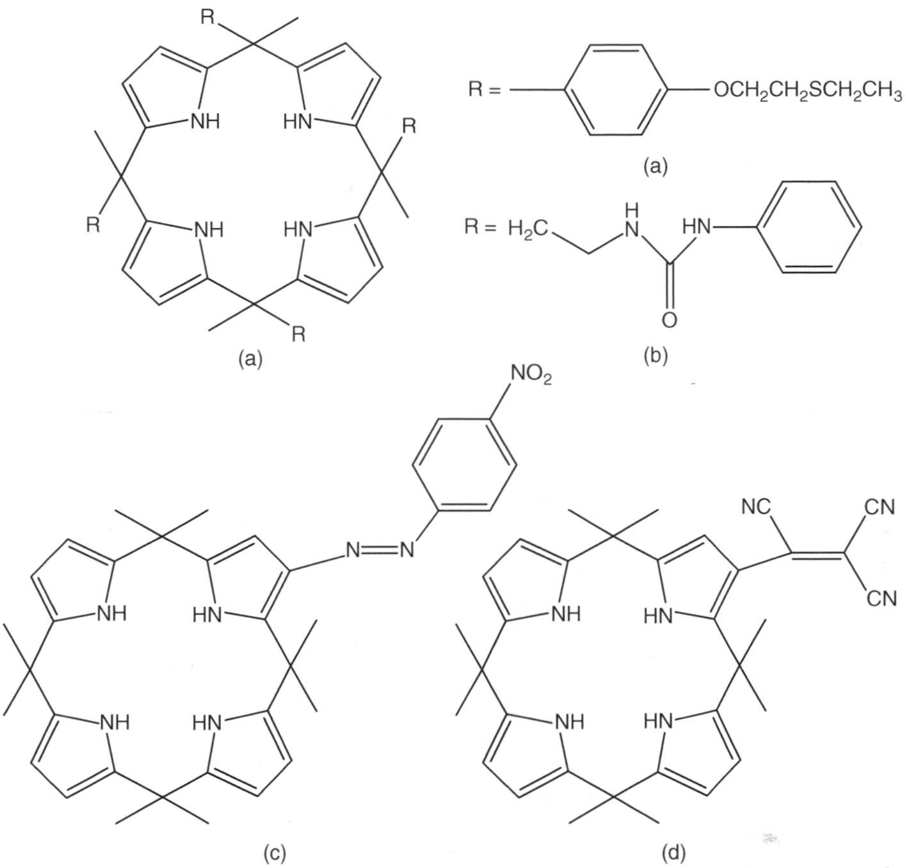

Figure 5.31   (a) Thioether-substituted calix[4]pyrrole, (b) phenylurea substituted calix[4]pyrrole, (c) azobenzene-substituted calix[4]pyrrole and (d) tricyanoethylene-substituted calix[4]pyrrole

Selective functionalisation of calixpyrroles has also been studied, with for example the compound shown in Figure 5.33a bearing two *p*-nitrophenyl groups on the same side of a calix[4]pyrrole macrocycle.[80] This material is U-shaped and can be thought of as acting like a pair of molecular tweezers, which explains its ability to form complexes with various hydroxybenzoic acids and benzene dicarboxylic acids. The macrocycle formed complexes with many of these species; for example, it formed a 1 : 1 complex with 3-hydroxybenzoic acid, where the structure was pH-dependent: at lower pH binding of the macrocycle was to the acid group but at higher pH the hydroxy group deprotonated and binding was to the phenoxide. Other substrates, such as that of 1,3-benzene dicarboxylic acid, displayed a structure in which two macrocycles formed a capsule around a single guest. This behaviour was also found for 4-hydroxybenzoic acid. Further studies were carried out on these types of macrocycle, such as for example the compound shown in Figure 5.33b, which has the two *p*-nitrophenyl groups on opposing sides of the macrocycle and has been shown to catalyse Diels–Alder reactions, possibly due to its potential to act as a hydrogen-bond donor.[81] However, the macrocycle with both *p*-nitrophenyl groups on the same side was totally ineffective as a catalyst, as were simple dipyrrole species. Recent work however has demonstrated the potential of this *cis* compound to bind neutral aromatic species and to act as a protecting agent and allow regioselective

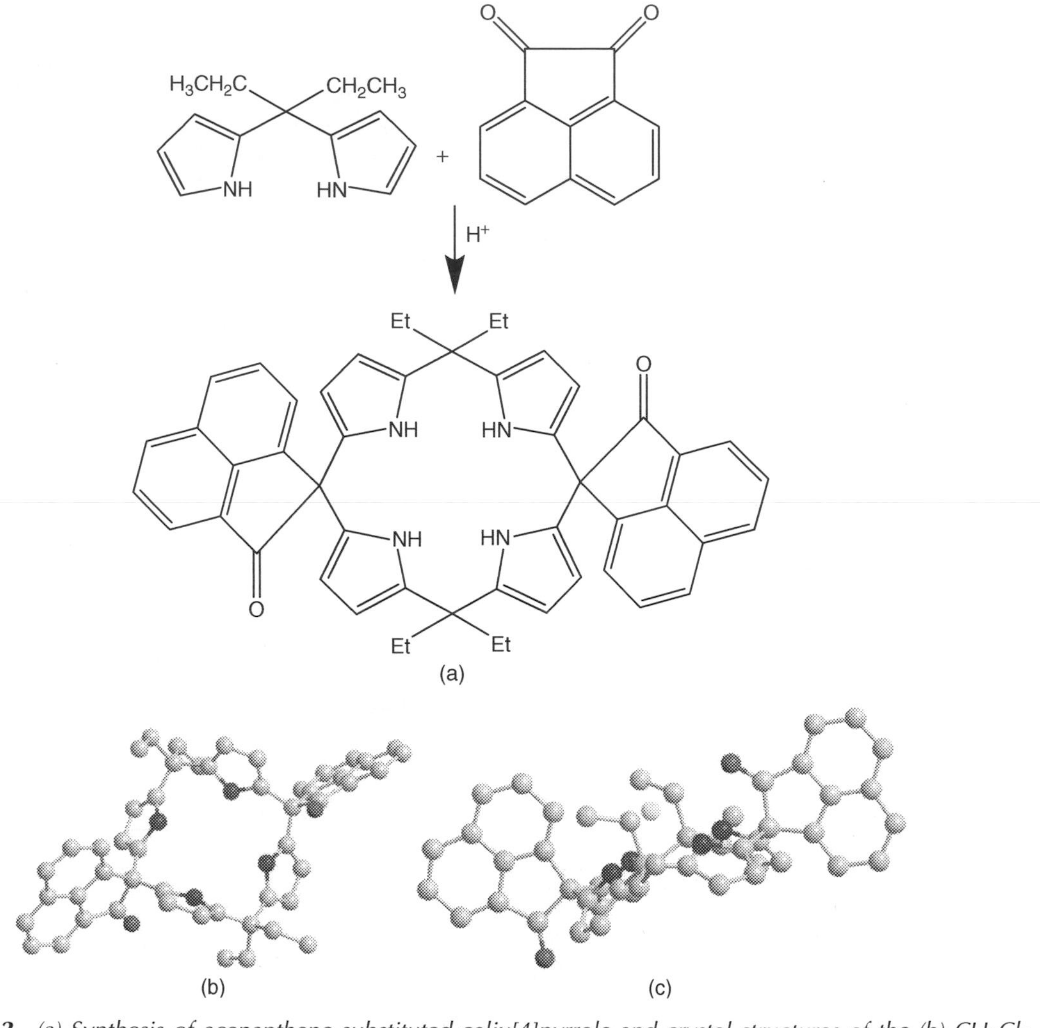

Figure 5.32 *(a) Synthesis of acenapthene-substituted calix[4]pyrrole and crystal structures of the (b) CH₂Cl₂* *and (c) F⁻ complexes*

alkylation of several dihydroxynaphthalene compounds;[82] as an example, reaction of benzyl bromide with 1,6-dihydroxynaphthalene and base led to formation (in the presence of the structure in Figure 5.33a) of the 1-monobenzyl ether – whereas in the absence of the macrocycle a mixture of mono- and dibenzyl ethers was obtained.

An effective microwave synthetic strategy has been developed for the synthesis of calix[4]pyrroles bearing substituents at the bridging positions[83] which requires reaction times of minutes rather than hours. This has allowed the syntheses of a variety of tetrasubstituted calix[4]pyrroles with a variety of groups. One includes the *p*-nitrophenyl group (Figure 5.34a), the nitro group of which could then be reduced to give the hydroxyamine (Figure 5.34b), which was then converted to a variety of hydroxamic acid derivatives. One of these (Figure 5.34c) was found to be capable of forming a complex with two VO_2^+ ions where the hydroxamic acid units are thought to complex the vanadium atoms and the pyrroles

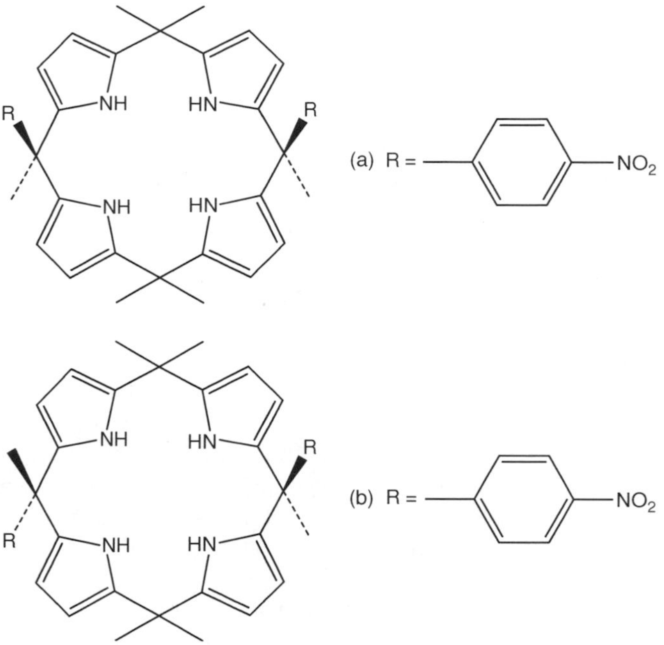

Figure 5.33 *Structures of the (a)* cis *and (b)* trans *calix[4]pyrroles*

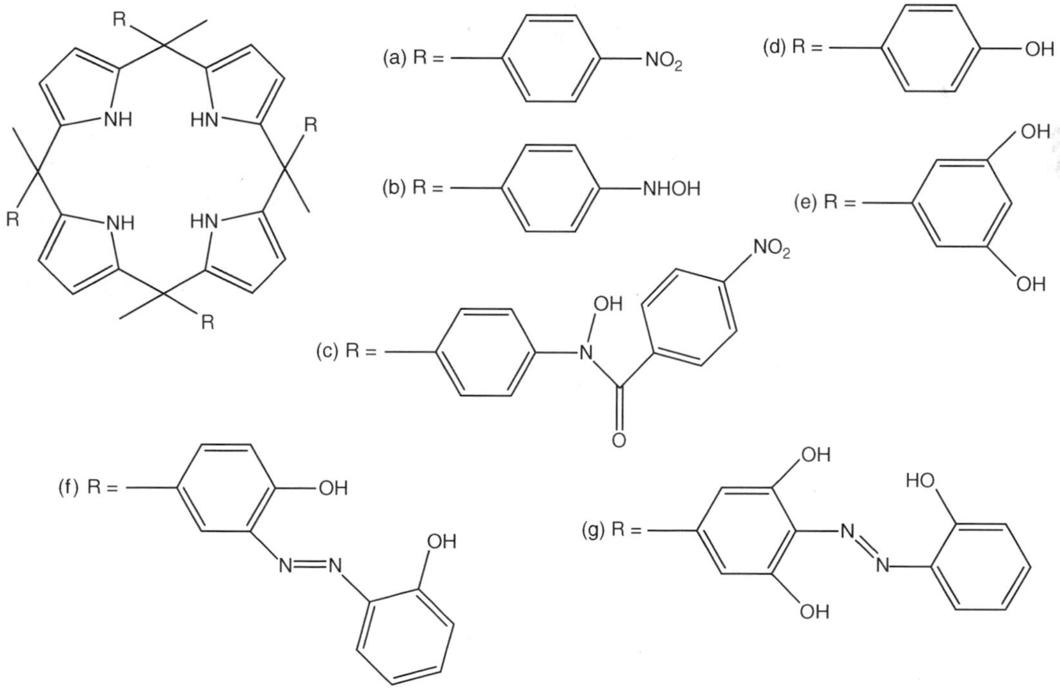

Figure 5.34 *Structures of the tetra (a) p-nitrophenyl, (b) hydroxyamino, (c) hydroxamic acid, (d) phenolic, (e) resorcinol and (f,g) azobenzene derivatives*

hydrogen-bond to one of the oxygens. A similar method was used to make calix[4]pyrroles with meso phenolic (Figure 5.34d) and resorcinol (Figure 5.34e) substituents.[84] These could then be further reacted with diazonium salts to make a series of azobenzene-derived calix[4]pyrroles; examples of two of these are shown in Figure 5.34f,g. Early studies showed that these azo compounds could bind copper(II) ions in ethanol solution with a concurrent colour change from yellow to red. The calix[4]pyrrole shown in Figure 5.34e can be isolated as its αααα isomer, which has all the resorcinol units on the same side of the macrocycle[85] and has a structure somewhat analogous to that of a resorcinarene. This compound, when crystallised with tetraalkyl ammonium chlorides, shows a columnar structure with the chloride ions bound within the calix[4]pyrrole unit. However, when a large excess of chloride is used, the complex instead assembles into a hexameric cage compound with six calix[4]pyrroles assembled into an octahedral arrangement.

5.5.5   Strapped calixpyrroles

In earlier chapters we have described how macrocycles can be bridged to give compounds such as cryptands or bridged calixarenes, which affects their properties and binding ability. Similar work has been performed using calixpyrroles; this is reviewed by Lee *et al.*,[86] and we will concentrate here on the most recent studies. Calix[4]pyrroles that were *cis* and *trans* strapped could be synthesised with phthalamide linkers.[87] Phthalamides were synthesised with side chains substituted with ketone groups, these being condensed with pyrrole to give dipyrrolic end groups, which were then cyclised with acetone (Figure 5.35a), where the ends of the 'strap' could all be on one side of the calixpyrrole unit or be on opposing sides. These materials demonstrated enhanced binding of halides in comparison to unstrapped calixpyrroles but did not show enhancements in selectivity. Formylation of a calix[4]pyrrole strapped with a phthalyl ester unit and condensation of the product with 1,3-indanedione[88] gave two isomeric products, one of which is shown in Figure 5.35b. Both of these materials interacted with anions; however, a specific bleaching of the macrocycles was observed for cyanide ions, even in the presence of other ions, indicating the possibility of using this material as a specific colourimetric sensor.

Dipyrrolylquinoxalines units could also be incorporated into the bridge to give chromogenic strapped calix[4]pyrroles[89] (Figure 5.35c), which displayed colour changes when complexed with various anions. Colour changes were particularly observed for fluoride and dihydrogen phosphate ions, with the affinities being enhanced compared to unstrapped macrocycles. Pyrrole units could also be incorporated into the strap to give a structure like that shown in Figure 5.35d, which allowed real-time determination of chloride ion in DMSO.[90] A triazole-bridged calix[4]pyrrole has been synthesised which shows high affinity for chloride ion and allows transport of the anion through a lipid bilayer.[91]

Other macrocycles have also been used to bridge calixpyrroles; a calix[4]arene can for example be synthesised bearing four ketone groups and then reacted with four equivalents of pyrrole to give a hybrid calix[4]arene/calix[4]pyrrole dimer in 32% yield.[92] The resultant compound is cylindrical in nature (Figure 5.36a) and its structure has been proved by X-ray crystallography. What is unusual is that combining a tetraketone with pyrrole would be expected to give a highly crosslinked polymeric material. The isolation of a pure compound in such good yields indicates that the conical calix[4]arene must help template the formation of the calix[4]pyrrole and it has been shown by NMR that there is hydrogen bonding between the pyrrole N—H groups and the carbonyls of the calix[4]arene. This idea has been extended to the use of a calix[5]arene with identical substituents; when this was combined with pyrrole under acid catalysis, a similar double macrocyclic compound was formed, except that in this case a calix[5]pyrrole was formed, the first synthesis of a compound of this nature.[93] Although in this case the yield was poorer (10%), the formation of this 'unnatural' macrocycle again confirms the existence of a template effect.

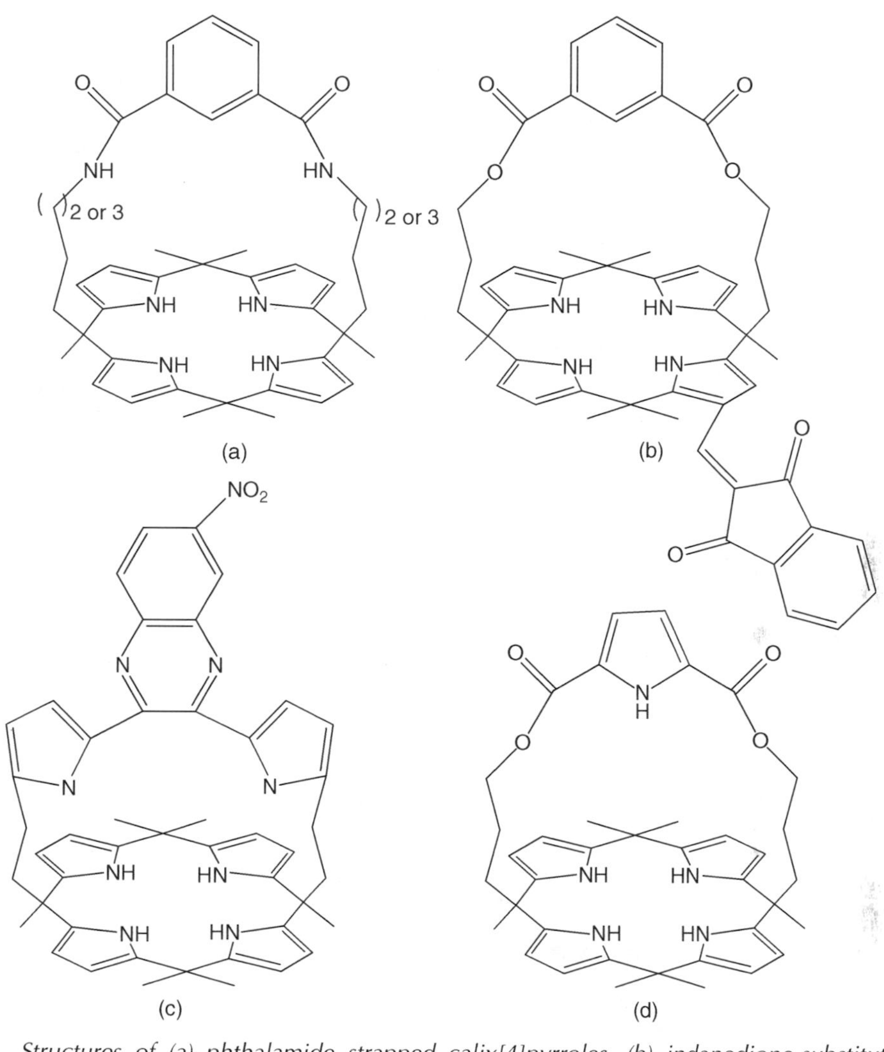

Figure 5.35 *Structures of (a) phthalamide strapped calix[4]pyrroles, (b) indanedione-substituted strapped calix[4]pyrrole, (c) dipyrrolylquinoxaline strapped calix[4]pyrrole and (d) pyrrole strapped calix[4]pyr*

An even more complex structure contained three types of ring system (Figure 5.36c), a calix[4]pyrrole being bridged by a calix[4]arene, which was in turn bridged by a 18-crown-6-ether.[94] This species formed a strong 1 : 1 complex with CsF, where X-ray crystallography showed complexation of the caesium by the calixcrown moiety and hydrogen-bonding between the calix[4]pyrrole ring and the fluoride ion. With caesium perchlorate, only binding of the metal ion was observed, and no binding at all occurred with tetrabutyl ammonium fluoride; when both salts were used together, CsF was found to be incorporated. A series of Ni(II) porphyrins were also synthesised bearing two side chains terminated with dipyrrollic units and then cyclised[95] to give porphyrin-bridged calix[4]pyrroles such as that shown in Figure 5.36e. These were found to bind fluoride as a 1 : 1 complex (although not other halide ions), with the anion being shown to reside inside the cavity formed by the two macrocycles.

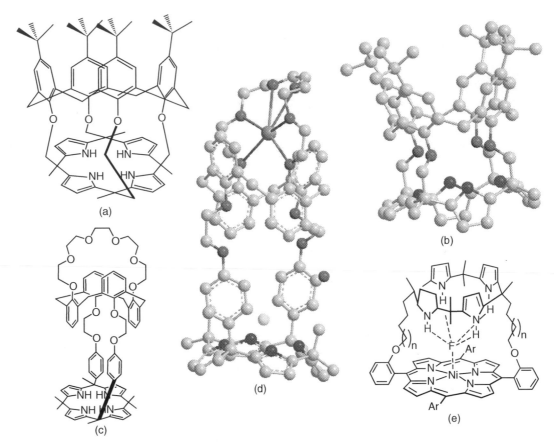

Figure 5.36 *Chemical and X-ray structures of (a,b) calix[4]arene-bridged calix[4]pyrrole, (c,d) crown/calix[4]arene-bridged calix[4]pyrrole and (e) porphyrin-capped calix[4]pyrrole with fluoride ion (Reprinted with permission from[95]. Copyright 2005 American Chemical Society)*

5.5.6 Modification of the pyrrole units

The aromatic rings of calix[4]pyrroles can also undergo other reactions. Recently the reduction of the pyrrole rings using hydrogen and palladium/carbon catalysis[96] in a solution of octamethyl calix[4]pyrrole in acetic acid gave a 30% yield of the structure shown in Figure 5.37a, where two of the rings have been reduced to pyrrolidine units. This compound is soluble when protonated and the X-ray structure (Figure 5.37b) is quite different to that of the parent compound in that the two pyrrole rings are almost planar, with the N—H bonds pointing to the centre of the macrocycle, and the pyrrolidine rings are almost perpendicular. Optimisation of the reaction also allowed the isolation of about 7% of the fully reduced compound (Figure 5.37c). This compound formed complexes with Cu(II), Ni(II) and Pd(II) ions, and X-ray studies (shown for the copper complex in Figure 5.37d) demonstrated the four nitrogen atoms forming a planar structure with the chloride ions assuming the axial positions of an octahedral coordination arrangement around the copper ion. Similar structures were found for the nickel and palladium complexes.

Direct reaction of a calix[4]pyrrole with an organoruthenium compound allowed the formation of the metallated compound[97] shown in Figure 5.38a in 91% yield. Attachment of the organometallic moieties

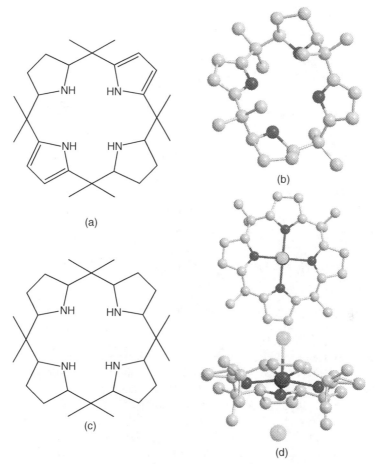

Figure 5.37 *(a) Chemical and (b) crystal structure of partially reduced calixpyrrole; (c) the wholly reduced calixpyrrole and (d) the crystal structure of its Cu(II) complex*

to the calix[4]pyrrole skeleton causes great changes in its nature. First the acidity of the N—H protons is greatly increased, leading to loss of two of the macrocycle protons. Also, the pyrrole macrocycle becomes a much better host for cation binding; a Cu(II) complex with the calixpyrrole could be isolated and characterised by X-ray crystallography (Figure 5.38b), again with the nitrogen atoms assuming a square-planar configuration about the central copper ion.

5.5.7 Sensing and other applications of calix[4]pyrroles

We have already discussed a large number of calixpyrroles and their behaviour towards many anions. We will however mention here a few more examples of selective sensing with these compounds. A series of chromogenic-substituted calix[4]pyrroles with tricyanoethylene substituents (Figure 5.31d) and a similar compound bearing an anthracene moiety were screened against a range of anions and were shown to undergo specific colour changes,[98] especially to pyrophosphate, giving a bright-blue colouration. Encapsulation of these compounds inside polyurethane membranes prevented access of these anions, however, but

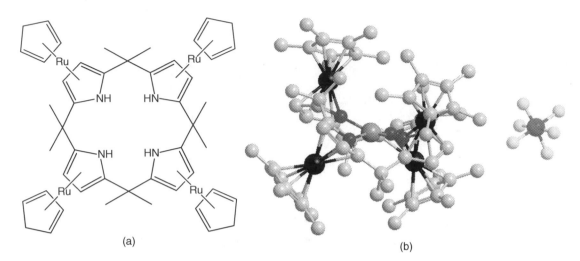

(a)   (b)

Figure 5.38 *(a) Chemical structure of metallated calixpyrrole and (b) crystal structure of its copper complex*

allowed the access of relatively hydrophobic carboxylate anions such as ibuprofen and salicylate, allowing detection of these species in environments that contained chloride and hydrogen phosphate, such as in blood plasma.

Calixpyrroles have also been used in chromatographic applications. Calix[4]pyrroles derived from acetone or cyclohexanone were derivatised via reaction at one of the pyrrole 3-positions and this derivative was used to tether the calix[4]pyrrole onto silica gel.[99] These materials could be used as packing materials in HPLC columns and successfully separated mixtures of anions. Within the same work, other experiments showed separations of mixtures of either amino acids, fluorinated aromatic compounds or nucleotides. Later work with similar systems[100] succeeded in separating mixtures of other compounds, with the families of compounds that could be separated including amino acids, phenols, benzene dicarboxylic acids and some medicines.

We have discussed the complexation of calix[4]pyrroles with anions, but they can also complex with some cations, such as the soft, polarisable silver ions. Silver has been shown to be capable of being transported through supported liquid membranes containing a variety of calix[4]pyrroles, with high selectivity over a number of other metal ions.[101] Calixpyrroles are also capable of transporting a range of salts through bilayer lipid membranes.[91,102] Other workers have studied complex formation with tetraalkyl ammonium salts and showed that although complexation of the anion is important, it is not the sole factor in determining binding. For example, methyl, ethyl and n-butyl ammonium chloride have binding coefficients in dichloromethane of 100 000, 10 000 and 100 M^{-1} respectively towards the simple octamethylcalix[4]pyrrole.[103] However, in other solvents such as acetonitrile, this difference in binding strength is minimal. From this we can infer that interactions between the ammonium ions and the calix[4]pyrrole bowl are also important in determining binding strength in certain solvents. Investigation of a range of ammonium salts showed that the chloride ion causes the calix[4]pyrrole to adopt the cone conformation, which then binds the ammonium cations. X-ray crystal studies confirm that the cations are located in the calix[4]pyrrole cones. Other workers studied binding phenomena using UV spectroscopy and showed that binding was dependent on the nature of the alkyl groups in the meso positions, the nature of the anion and the nature of the tetraalkyl ammonium cation.[104] It has also been shown that the nature

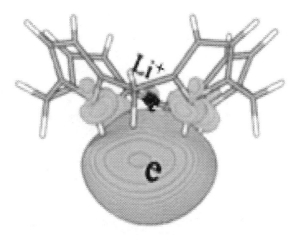

Figure 5.39   *Theoretical HOMO structure of Li@calix[4]pyrrole (Reprinted with permission from[107]. Copyright 2005 American Chemical Society)*

of the meso substituents has a major effect; for example, tetraphenyl-substituted calix[4]pyrroles with a variety of substituents (acid, amino, ester) on the phenyl groups were shown to give highly stable 1:1 complexes in water with a number of pyridine-N-oxides.[105] The guests are bound via H-bonding of the pyrrole N—H groups with the oxygen of the N-oxide, along with a combination of CH$-\pi$, $\pi-\pi$ and hydrophobic interactions.

Metal complexes with calix[4]pyrroles are also provoking considerable interest. We mentioned above how calix[4]pyrroles can form zirconium complexes[49] and recent work has demonstrated the synthesis of *mono-* and *bis*-zirconium complexes of octamethyl calix[4]pyrrole.[106] The *bis*-zirconium species in particular displayed a high catalytic activity towards the polymerisation of ethylene. The potential of calix[4]pyrroles for the stabilisation of electride-type materials has also been investigated, for example in theoretical investigations of the Li@calix[4]pyrrole complex, where the lithium forms a Li$^+$ cation and a free-electron anion may be a stable electride at room temperature and display large nonlinear optical properties.[107] Figure 5.39 shows the theoretical highest occupied molecular orbital (HOMO) of this material, with the lithium cation bound within the calix[4]pyrrole cup and the free electron beneath it. The same group has also made theoretical studies of the alkalide compounds Li$^+$ (calix[4]pyrrole)M$^-$ (M = Li, Na, and K) and showed that these have even higher nonlinear properties than the electride and that the hyperpolarisability increases with the atomic number of M.[108]

5.5.8   Calixpyrroles including nonpyrrole units

There has been a great deal of interest in the development of mixed heteroaromatic systems, where within the same macrocycle a number of heteroaromatic units are incorporated. These can be assembled in two ways, either via stepwise incorporation of various heteroaromatic units to form a mixed macrocycle, or alternatively by synthesising a macrocycle containing heteroaromatic units and then modifying certain of the units. For example, a carbazole unit could be substituted with two pyrrole units[109] to give the structure in Figure 5.40a, reaction of which with acetone gives the expanded calixpyrrole shown. This material forms complexes with anions such as acetate or benzoate but not bromide and nitrate. X-ray studies of various complexes shows that the compound adopts a winged conformation like that seen in the benzoate complex,

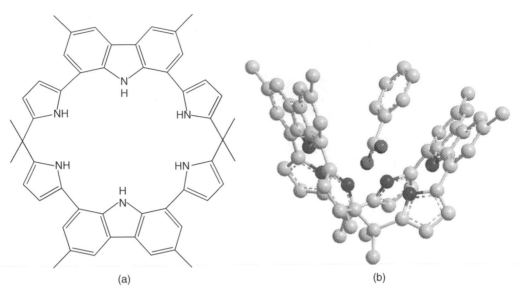

(a)   (b)

Figure 5.40   *(a) Structure of the carbazole expanded calixpyrrole and (b) crystal structure of the complex with benzoate*

where the benzoate anion is encapsulated within the macrocycle structure (Figure 5.40b). Another method utilised a base-catalysed condensation of either furan or thiophene with acetone to give the disubstituted heterocyclics shown in Figure 5.41. These could then be condensed with pyrrole to give the linear trimers shown. These trimer could then be condensed with the disubstituted thiophenes or furans to give cyclic tetramers as the major product, with cyclic octameric compounds as products in lower yields, along with small amounts of cyclic hexamers.[110] All these products could be separated by chromatographic techniques. The reaction mechanisms and yields are shown in Figure 5.41 and were found to be relatively insensitive to reaction temperatures and to be unaffected by the presence of inorganic ions, which were utilised in a failed attempt to exploit template effects. The X-ray structure of the calix[2]furan[2]pyrrole was obtained and demonstrated a typical 1,3-alternate conformation for this macrocycle.

A similar method was expanded to give a range of hybrid macrocycles[111] which contained benzene rings. Disubstituted benzene derivatives analogous to the furan and thiophene systems described above could be condensed with pyrrole to give calix[2]benzene[2]pyrroles directly (Figure 5.42a) or could instead be condensed to give trimers as before, which could then be cyclised with more pyrrole to give calix[1]benzene[3]pyrroles (Figure 5.42b). Either unsubstituted or methoxybenzene rings could be incorporated. X-ray studies of these systems showed them to have 1,3-alternate conformations as expected, and binding of fluoride and chloride was observed in the case of the calix[1]benzene[3]pyrroles, albeit with lesser binding constants than the calix[4]pyrroles. Also within this work, the reaction of calix[2]benzene[2]pyrroles and calix[4]pyrroles with dichlorocarbene was described. Dichlorocarbene inserts into pyrrole rings to expand the ring to six and thereby generates calix[2]benzene[1]pyridine[1]pyrroles and calix[1]pyridine[3]pyrroles respectively (Figure 5.42c,d). These structures could also be examined by X-ray crystallography and were shown to form flattened partial cones.

A novel electroactive *bis*(tetrathiafulvalene)calix[2]thiophene[2]pyrrole (Figure 5.43a) has been synthesised[112] and has been shown to adopt an unusual 1,2-alternate configuration in the crystal

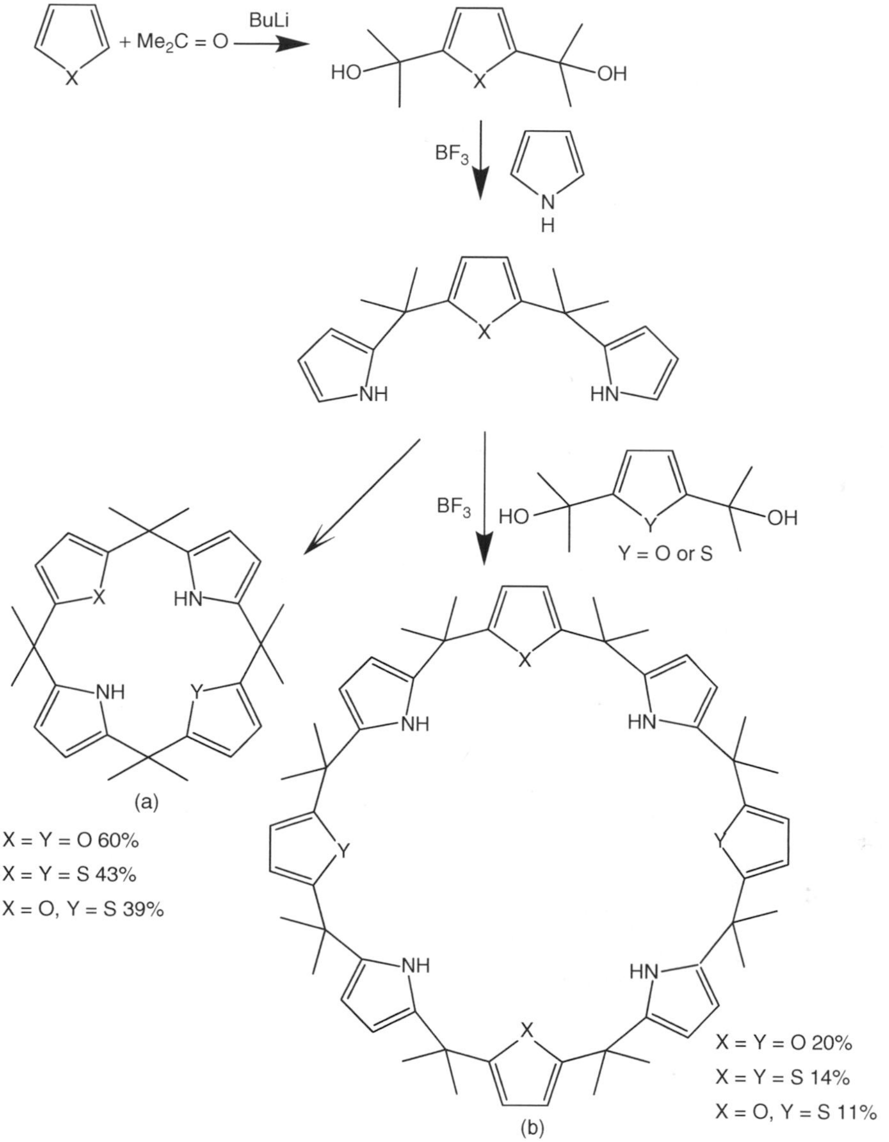

Figure 5.41 *Structures of (a) tetramers and (b) octamers containing pyrrole, furan and thiophene units*

form (Figure 5.43b). This macrocycle forms a strong charge-transfer complex with two molecules of tetracyanoquinodimethane (TCNQ), as shown by UV/Vis and ESR spectroscopy. In the crystalline form the macrocycle now adopts the expected 1,3-alternate conformation with one TCNQ inside the cavity and the other between tetrathiafulvalene units of adjacent macrocycles (Figure 5.43c).

A stepwise synthesis method[113] was also employed to generate cyclic tetramers, where a variety of heterocyclic dimers were synthesised and then condensed with dipyrrolylmethane units to give the structures

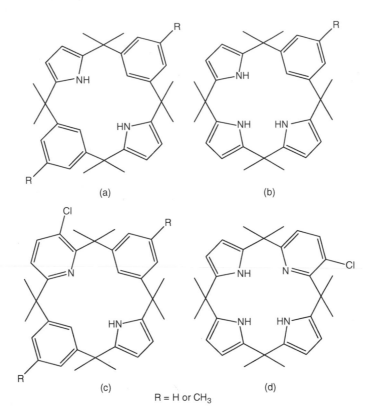

(a)   (b)

(c)   (d)

R = H or CH₃

Figure 5.42   *Mixed heterocalixarenes*

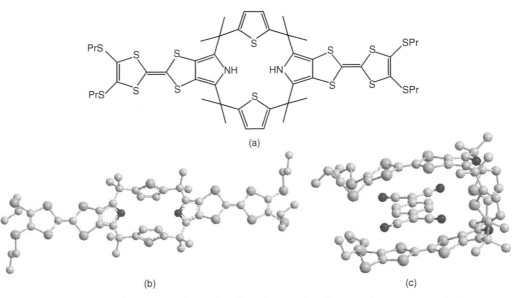

(a)

(b)   (c)

Figure 5.43   *(a) Structure of bis(TTF)calix[2]thiophene[2]pyrrole, (b) crystal structure and (c) as a complex with TCNQ*

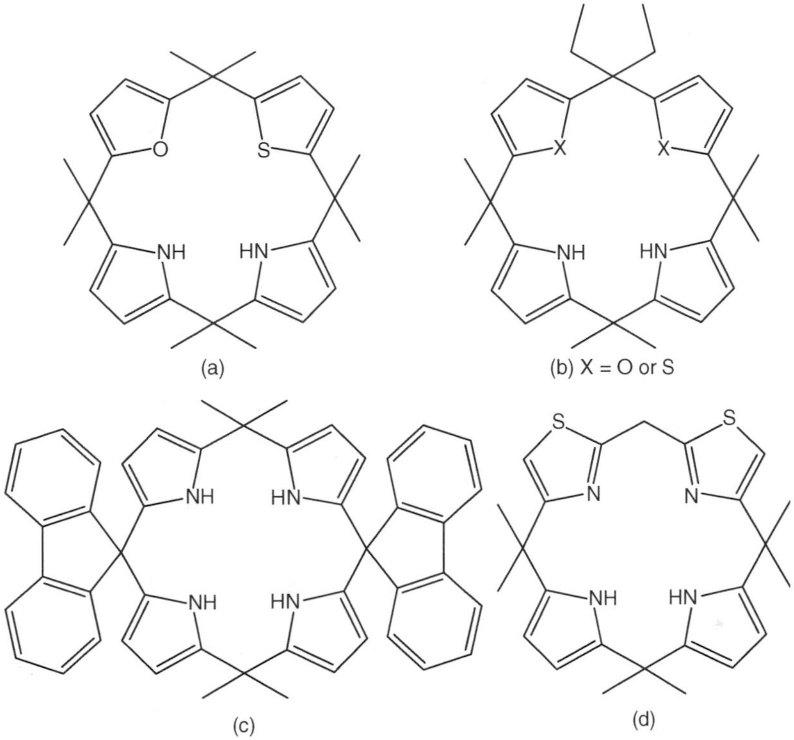

(a)

(b) X = O or S

(c)

(d)

Figure 5.44 *Mixed heterocalixarenes*

show in Figure 5.44a–d, where furan, thiophenes, fluorenes and thiazoles can all be incorporated into the macrocycle. The compound shown in Figure 5.44c is fluorescent and its emission can be quenched by the binding of chloride, although the binding coefficient is much less than that observed for calix[4]pyrroles. The other macrocycles showed no obvious affinity for fluoride or chloride. A disubstituted phosphole could be reacted with two pyrrole moieties[114] as before and this system could then be ring-closed with a furan or thiophene unit (Figure 5.45a,b). The furan-containing compound adopts a cone conformation in the crystal (Figure 5.45c), whereas the thiophene compound adopts a partial cone (Figure 5.45d). In the case of the thiophene compound, the P=S bond could be removed and the resultant compound used to synthesise a platinum(II) complex, where the Pt ion coordinates to the phosphorus and the N—H bonds are pointed towards one of the chloride counter-ions (Figured 5.45e).

It should be possible to construct calix[4]pyrroles by reaction of aldehydes with pyrrole, but the resultant tetramers are easily oxidised to the corresponding porphyrin. When adamantane carbaldehyde was reacted with pyrrole, a stable calix[4]pyrrole (Figure 5.46a) could however be obtained, albeit in only 23% yield.[115] It also proved possible to condense 2-adamantone with pyrrole to give cyclic tetramers, the structures and X-ray crystallography studies of which are shown in Figure 5.46b,c. This compound, when ground with chloride salts, formed a complex with the chloride ion. Phyrins are macrocycles that can be thought to be intermediate between a porphyrin and a calixpyrrole. The stability of the adamantane carbaldehyde materials meant that they could be reacted to give a dipyrroyl-substituted compound, which could then be reacted with benzaldehyde to give a cyclic tetramer.[115] This could then be oxidised at the benzaldehyde-derived bridges to give the calix[4]phyrin (Figure 5.47a), the structure of which could be proved by X-ray

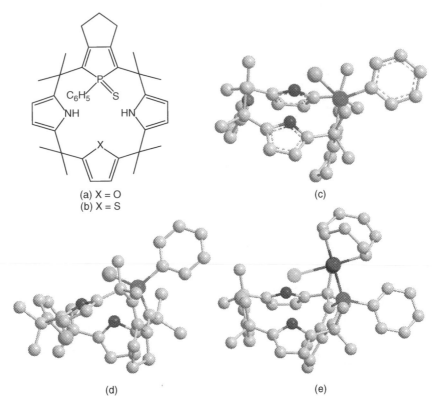

Figure 5.45   *(a,b) Phosphorus-containing heterocalixarenes, (c,d) crystal structures of the heterocalixarenes and (e) the platinum complex*

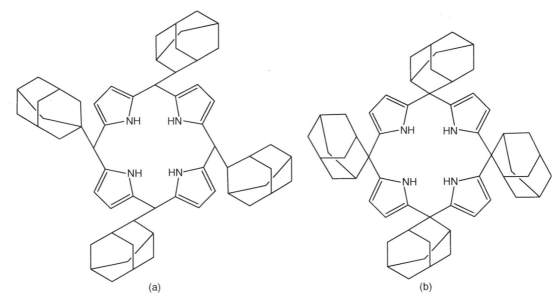

Figure 5.46   *Calixpyrroles with (a) adamantane carbaldehyde and (b) 2-adamantone; (c) the crystal structure of 5.46b*

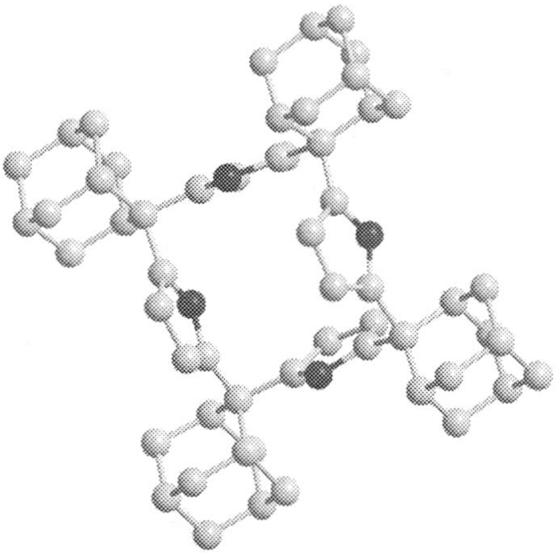

(c)

Figure 5.46 *(continued)*

crystallography (Figure 5.47b). No binding of anions in solution was observed for this material. In other studies, workers[116] first condensed pyrrole with acetone to give the dipyrrolyl species and then reacted it with a wide range of suitable aromatic aldehydes and oxidised it to give the calix[4]phyrins (Figure 5.47c). These were characterised by X-ray crystallography and NMR. The conformations of these macrocycles were found to be dependent on their substituents and in one case a doubly confused macrocycle was isolated (Figure 5.47d)

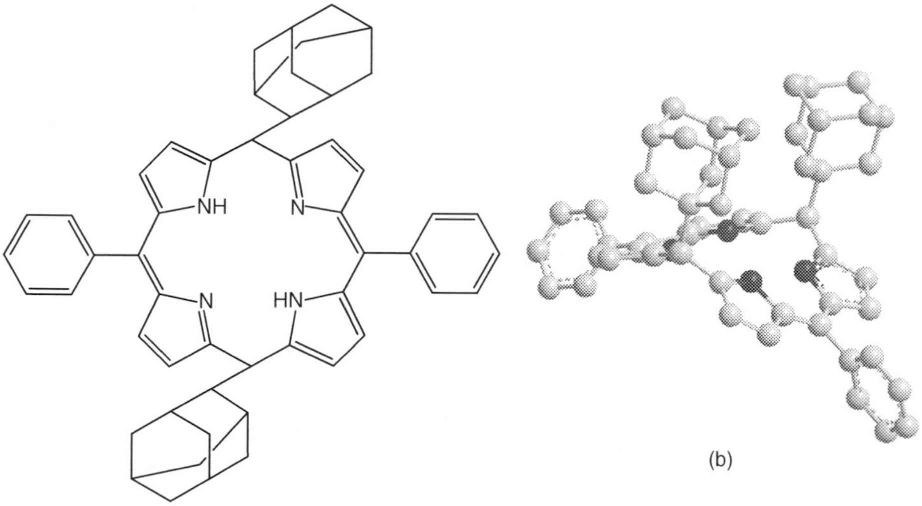

(a)

(b)

Figure 5.47 *(a) Calix[4]phyrin with adamantane carbaldehyde, (b) its crystal structure; (c) 'normal' and (d) 'confused' calix[4]phyrins*

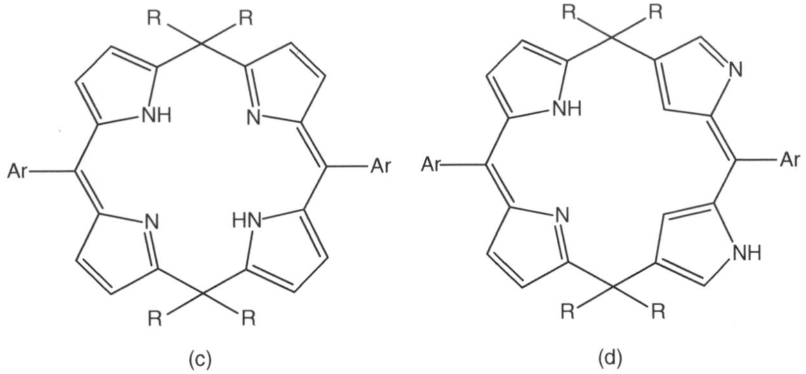

(c)   (d)

Figure 5.47   *(continued)*

An unusual calix[3]dipyrrin has also obtained as a by-product of the synthesis of Ni(II)porphyrin.[117] As can be seen from the structure (Figure 5.48a), a cyclic hexamer has formed, which is only 50% oxidised and binds three nickel atoms. In the crystal (Figure 5.48b) the complex has the nickel atoms forming an equilateral triangle and the macrocycle consists of three planar dipyrrin units, which tilt to form a bowl-shaped macrocycle. The complex could be demetallated with trifluoracetic acid and then treated with copper salts to form a *tris*-copper complex.

We have already discussed how a calix[6]furan can be converted into a calix[6]pyrrole. Within this work other mixed heteroaromatic systems apart from the calixpyrrole were synthesised.[41] By careful control of reaction conditions and the amounts of oxidation reagents, it proved possible to synthesise calix[3]furan[3]pyrrole and calix[2]furan[4]pyrrole (Figure 5.49a,b), along with the calix[1]furan[5]-pyrrole.[54] X-ray crystallographic studies of these mixed pyrroles showed them to have a 'tennis-ball

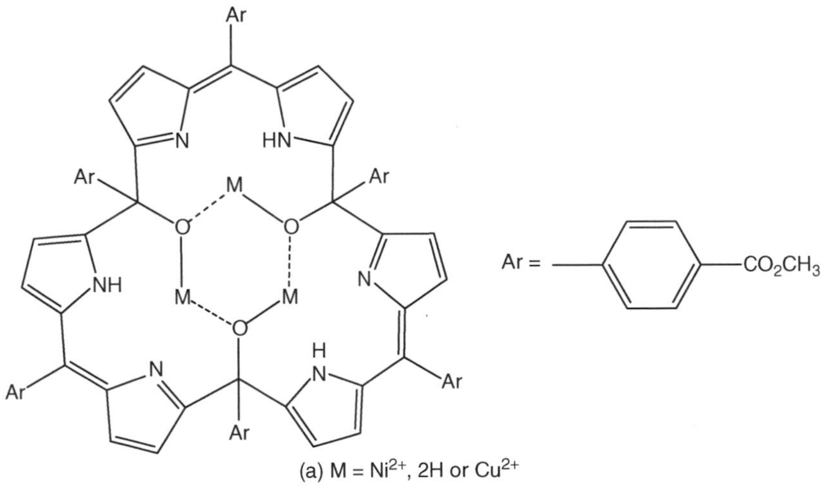

$Ar = $ —⟨benzene⟩—CO_2CH_3

(a) M = Ni^{2+}, 2H or Cu^{2+}

Figure 5.48   *(a) Calix[3]dipyrrin and (b) its crystal structure (Ni complex)*

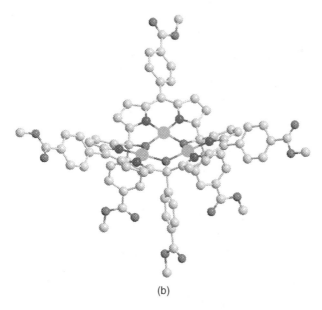

(b)

Figure 5.48   *(continued)*

seam'-like structure, similar to the calix[6]pyrrole.[54] Binding studies have also been made on these four hexamers, with calix[3]furan[3]pyrrole basically not showing any binding of anions at all, although the other compounds did bind to anions, with binding strength being found to increase with the number of pyrrole units within the macrocycle. Another extensive paper studied a wide range of macrocyclic tetramers and an octamer with various combinations of pyrrole, furan, thiophene and bipyrrole as well as their complexes with a variety of species.[117,118] Many of these species showed weakened binding of anions compared to calix[4]pyrroles. However, species containing bipyrrole units (Figure 5.49c–e) were good receptors for carboxylate anions and bound these with high selectivity. This is thought to be in part due to the effective matching of geometry between the host and guest, as supported by X-ray studies, with for example a close matching being seen between the host shown in Figure 5.49d and the benzoate anion in Figure 5.49f.

   In a reaction analogous to calix[4]pyrrole synthesis, bipyrrole (Figure 5.50a) can be reacted with acetone to form calix[3]bipyrrole and calix[4]bipyrrole in reasonable yields after chromatographic separation.[119] These can be thought of as larger analogues to calix[4]pyrroles. The calix[3]bipyrrole did not give crystals of sufficient quality to be suitable for X-ray analysis – although the chloride complex did – and showed the macrocycle to adopt a cone-like structure with all six N—H bonds binding to the anion (Figure 5.50b). The calix[4]bipyrrole could be crystallised from THF and gave a 1,3-alternate-type structure (Figure 5.50c). Of the two compounds, the cyclic trimer was the most efficient at binding anions, binding chloride approximately as strongly as the calix[4]pyrrole and having much larger affinities for bromide and iodide, probably due to its larger size. The calix[4]bipyrrole, because of its large size and flexible nature, was thought to be unsuitable for use as an anion-binding agent, however studies have shown[120] that it forms V-shaped complexes with chloride (Figure 5.50d) and bromide in the solid state, with the anions encapsulated within the cavity of the macrocycle. In acetonitrile the tetramer had a binding constant for chloride over 20 times that of the trimer or calix[4]pyrrole, although in DMSO no binding of anion occurs at all.

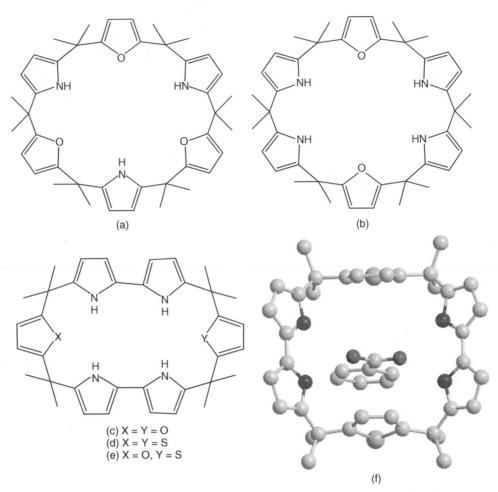

Figure 5.49 *(a) Calix[3]furan[3]pyrrole, (b) calix[2]furan[4]pyrrole, (c–e) bipyrrole based macrocycles, (f) crystal structure of 5.47d (benzoate complex)*

5.6   Calixindoles, Calixpyridines and Calixthiophenes

Indole was first shown to form calixarene-type structures in 1989 when it was demonstrated that aryl aldehydes could be condensed with 4,6-dimethoxy-3-methylindole and catalysed by phosphoryl chloride[121] to give macrocycles with arylmethine links, an example of which is shown in Figure 5.51a. Later work[122] utilised a series of 7- or 2-hydroxymethyl indoles to synthesise the methylene-bridged compounds such as that shown in Figure 5.51b and also demonstrated the synthesis of the calix[3]indoles from other substituted indoles. Another synthetic method utilised antimony chloride as the catalyst for this reaction.[123] Reaction of an indolylglyoxylamide gave the compound shown in Figure 5.50c, which gave crystals suitable for X-ray crystallography and showed the calix[3]indole to have a cone-shaped conformation (Figure 5.51d) as well as binding a molecule of ethanol within the cup.[124] The conformation is thought to be stabilised by hydrogen

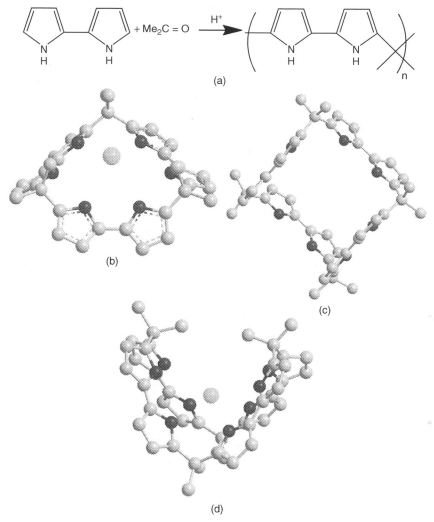

Figure 5.50 *(a) Synthesis of calix[3 and 4]bipyrrole and crystal structures of (b) calix[3]bipyrrole/Cl⁻, (c) calix[4]bipyrrole and (d) calix[4]bipyrrole/Cl⁻*

bonding between the indole N—H groups and the linkage carbonyl groups. Later work[125] also succeeded in isolating a calix[4]indole (Figure 5.52a), which was shown by NMR studies to be strikingly more rigid than its corresponding trimer. X-ray data (Figure 5.52b) showed it to have a 1,3-alternate structure.

Calixpyridines have also been reported; these have so far not been synthesised directly but instead may be formed from a ring expansion reaction of calixpyrroles. Attempts were made to condense pyridine-N-oxide in the same manner as phenols, but these were unsuccessful.[126] Reaction of octamethyl calix[4]pyrrole with dichlorocarbene led to insertion of the carbene species into pyrroles to give chloropyridines. By varying the reaction conditions, between one and all four of the pyrrole rings could be converted to

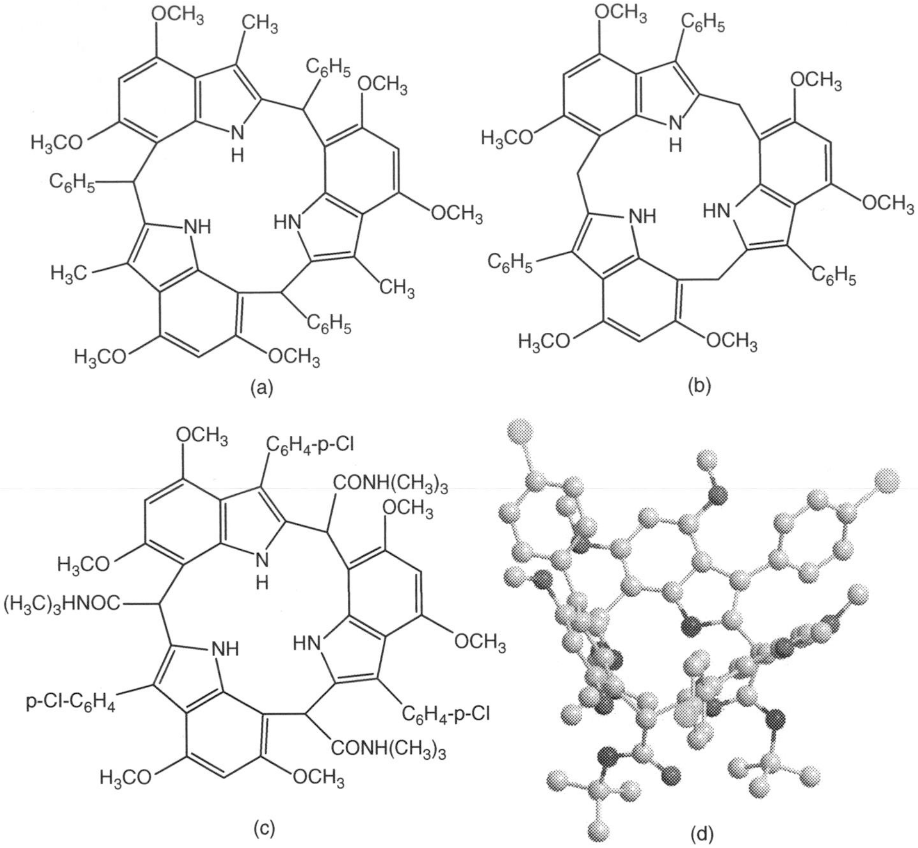

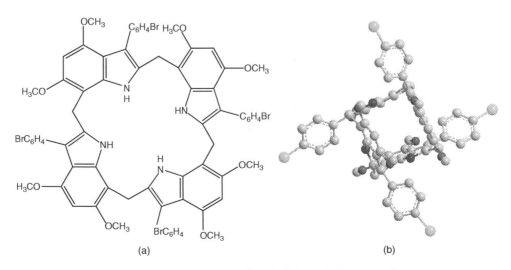

Figure 5.51   *(a–c) Structures of various calix[3]indoles and (d) crystal structure of 5.51c*

Figure 5.52   *(a) Structure of a calix[4]indole and (b) its crystal structure*

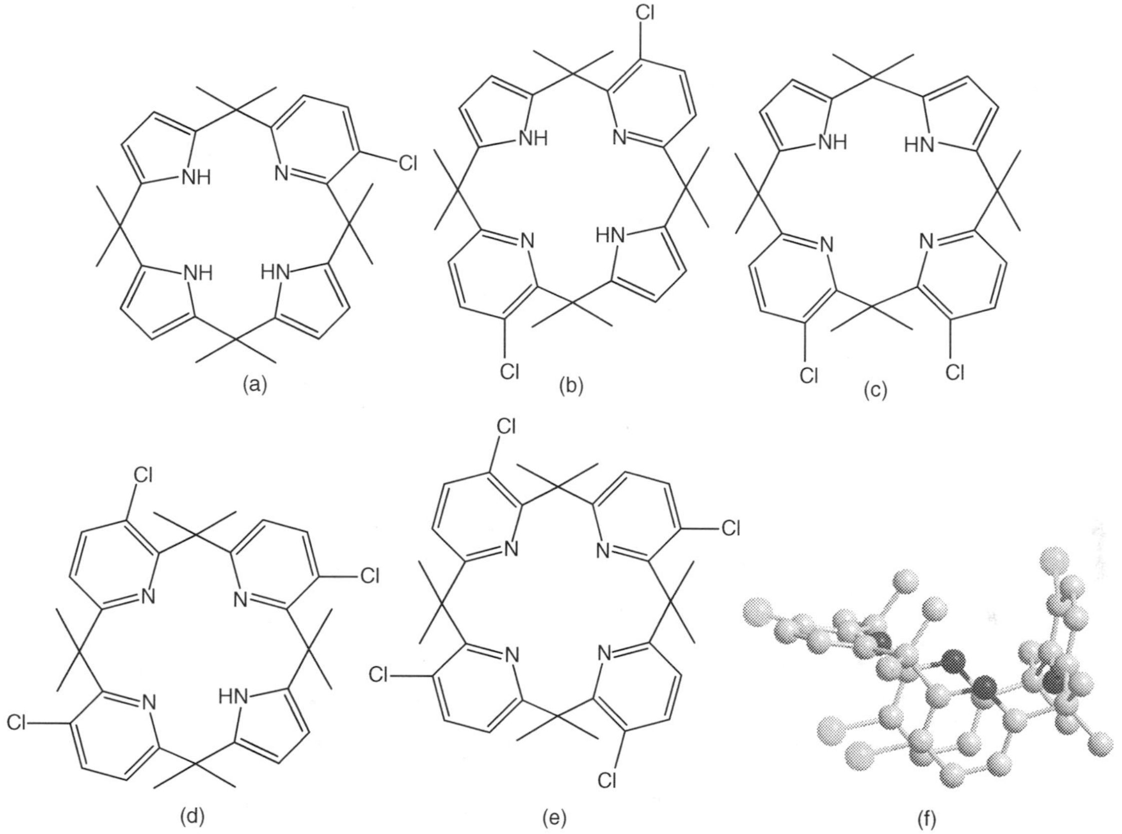

Figure 5.53   (a–e) Structures of calixpyridines/pyrroles and (f) crystal structure of one isomer of 5.53e

chloropyridines.[126] Structures of these materials are shown in Figure 5.53a–e and it should be noted that since the carbene can attack either pyrrole double bond, the chlorine atoms can be on either side of the pyridine ring. This means that isomeric mixtures are often obtained (we have only shown one isomer for each compound). However, these have been separated and in some cases examined by X-ray crystallography; the tetrachloro calix[4]pyridine, for example, was found to have a structure with two rings parallel to the plane of the macrocycle (Figure 5.53f) and two perpendicular. A similar structure occurs for the calix[1]pyridine[3]pyrrole, whereas the compounds with two and three pyridine rings both adopt two different conformations in the crystal, while the *bis*-substituted product adopts conical forms to maximise N—H to N hydrogen-bonding and the compound with three pyridine rings adopts a conformation somewhat similar to that of the calix[4]pyridine.

A stepwise building approach has been used to synthesise azacalixpyridines, where nitrogen atoms rather than carbons bridge the rings.[127] Both benzene and pyridine rings could be utilised to build the macrocycle, with both cyclic tetramers and octamers being isolated (Figure 5.54a–d) along with linear species. The azacalix[4]pyridine has a 1,3-alternate structure with all eight of its nitrogen atoms almost in the same plane (Figure 5.44e). The tetramer containing two pyridine and two benzene units adopts a twisted 1,3-alternate structure and the corresponding octamer (Figure 5.54d) adopts a structure described as a 'double-ended spoon', whereas the azacalix[8]pyridine has a pleated-loop structure (Figure 5.54f).

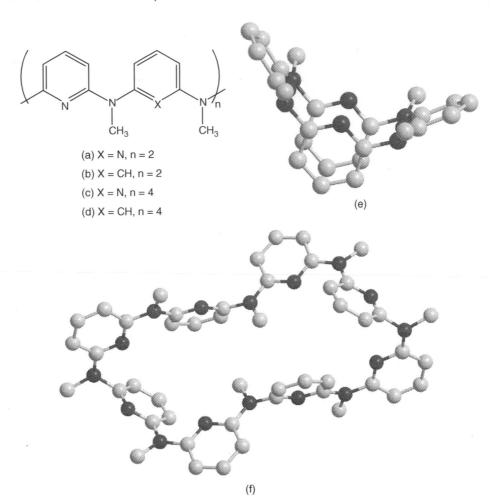

(a) X = N, n = 2
(b) X = CH, n = 2
(c) X = N, n = 4
(d) X = CH, n = 4

(e)

(f)

Figure 5.54 *(a–d) Structures of azacalixpyridines and crystal structures of (e) azacalix[4]pyridine and (f) azacalix[8]pyridine*

X-ray studies of protonated versions of these compounds were also obtained. Complexes were formed with solvents and anions, and in the case of the octamers, complexes have been formed with fullerenes with binding constants higher than those of calixarenes. Octamer 5.54d bound more strongly than 5.54c and both species bound C_{70} more strongly than C_{60}. The tetramers did not form complexes with fullerenes.

Pyridine-containing versions of resorcinarenes have also been synthesised. In a similar manner to resorcinarene formation, 2,6-dihydroxypyridine could be condensed with isobutanal to give a mixture of cyclic tetramers along with some linear products.[128] The macrocyclic compounds were poorly soluble and could not be purified so were converted into their benzyl esters and then separated by column chromatography and shown to be cyclic tetramers with different conformations. However, using glycol monomethyl ether as the solvent allowed reflux at higher temperatures and prolonged heating of the reagents gave the rccc isomer in 72% yield (Figure 5.55a). X-ray crystallographic studies proved the cone conformation and showed that the macrocycle exists as a head-to-head dimer in the crystal (Figure 5.55b), although

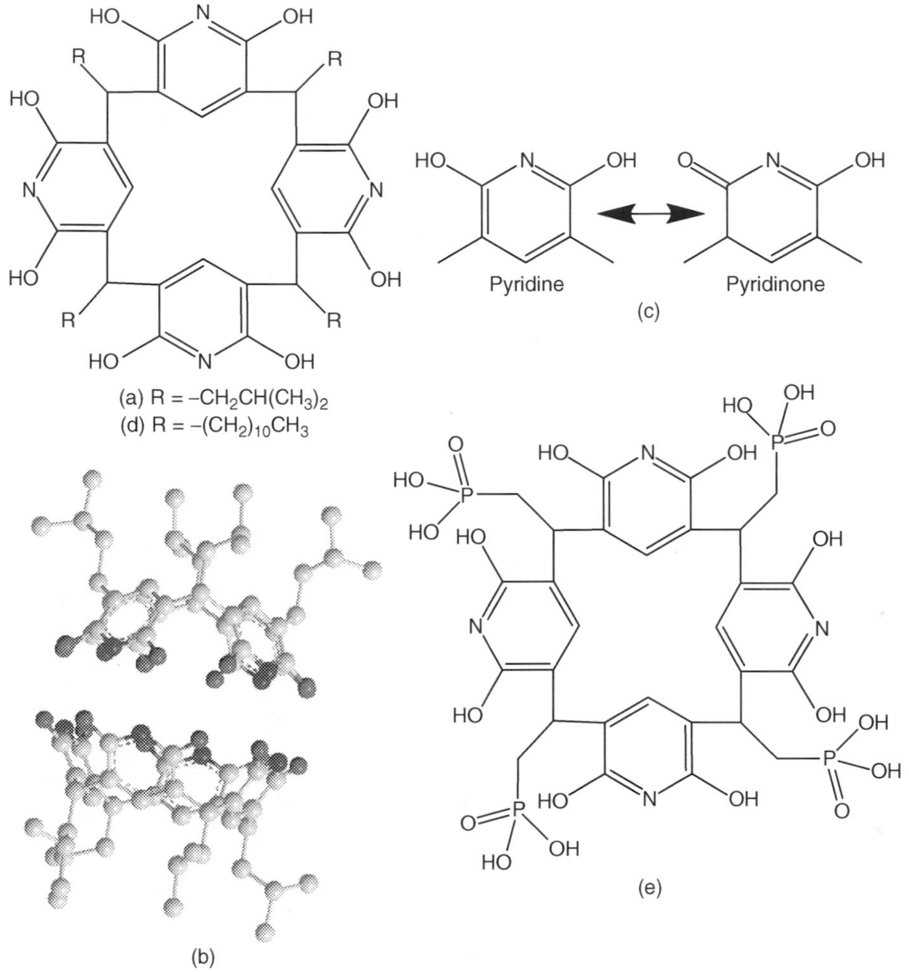

(a) R = –CH$_2$CH(CH$_3$)$_2$
(d) R = –(CH$_2$)$_{10}$CH$_3$

Pyridine   Pyridinone

(c)

(b)

(e)

Figure 5.55   *(a,d) Structure of pyridine-containing resorcinarenes, (b) crystal structure of the dimer, (c) tautomers of the hydroxypyridines and (e) phosphorylated resorcinarene*

this dimer appears to dissociate in THF. The X-ray and other spectrographic studies indicated that the structure is highly dominated by the pyridinone tautomer (this equilibration is shown in Figure 5.55c). A range of other aldehydes were found to form tetramers but formaldehyde gave amorphous solids and many aromatic aldehydes gave either insoluble products that could not be characterised or highly coloured quinone-type products.

Further work using these materials solubilised by using long alkyl side chains (Figure 5.55d) again showed that these materials formed dimers and were capable of encapsulating various acids and amides in chloroform solution, as shown by mass spectrometry.[129] The cavity appears to be quite small, with molecules larger than propionic acid not being encapsulated. Other work also showed that, like resorcinarenes and pyrogallenes, these species can form hexameric capsules in solution as well as dimers.[130] Condensation of a phosphoryl-substituted acetal with 2,6-dihydroxypyridine[131] gave the corresponding resorcinarene-like compound (Figure 5.55e).

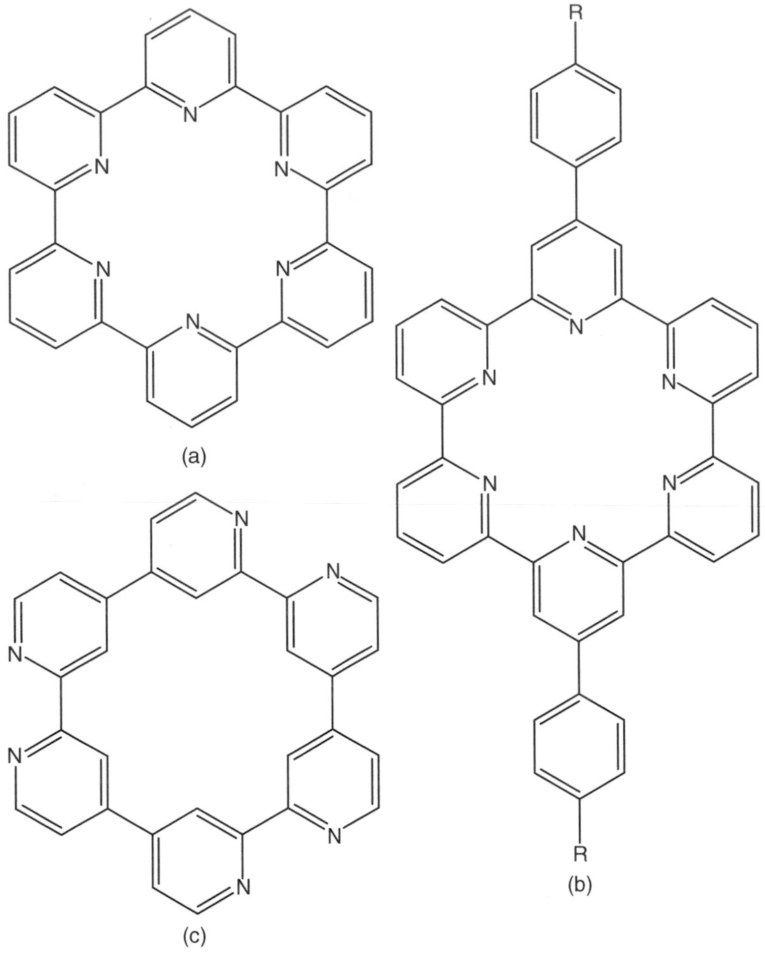

Figure 5.56 *(a–c) Structure of sexipyridine and analogues*

A pyridine analogue of a spherand, sexipyridine (Figure 5.56a), was first successfully synthesised[132] in 1983. Other workers also synthesised similar compounds, such as that shown in Figure 5.56b, which is capable of forming 1 : 1 complexes with sodium ions.[133] A version of this compound with the nitrogen atoms on the outside of the macrocycle (Figure 5.56c) has also been reported.[134]

Condensation of 2,6-dibromopyridine with sodium hydrosulfide can give thiacalixpyridines such as those shown in Figure 5.57a. Using a one-pot reaction the cyclic trimers, tetramers and hexameres were prepared in yields of 40%, 8% and 4% respectively and separated.[135] Both nitrogen and sulfur atoms can participate in metal binding, for instance when crystallised from dichloromethane the cyclic trimer can form a 2 : 2 complex with copper(I), as shown in Figure 5.57b. However, when crystallised from acetonitrile, no Cu—S bonding is observed; a 1 : 1 complex is formed, with acetonitrile binding to the copper (Figure 5.57c). The tetramer, which when crystallised without any guest adopts a 1,3-alternate conformation, can also bind copper(I) and the crystal structure again shows both Cu—N and Cu—S intercations (Figure 5.57d).

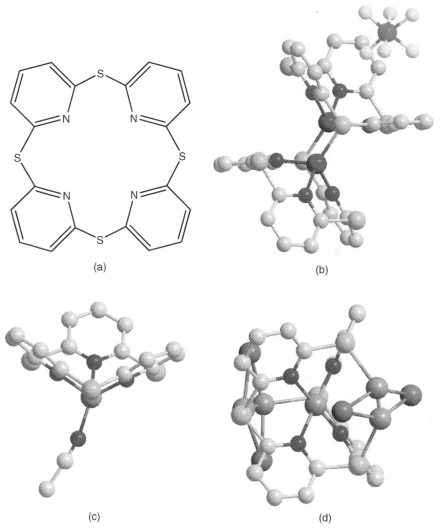

(a)   (b)

(c)   (d)

Figure 5.57   *(a) Structure of thia-calix[4]pyridine and crystal structures of (b) $Cu(I)_2(Py_3S_3)_2(PF_6)_2$, (c) Cu(I)-$(Py_3S_3)(CH_3CN)(PF_6)$ and (d) $[Cu(I)(Py_4S_4)(CuBr_2)_2]_n$*

Calixthiophenes have also been of some interest and we have already discussed earlier in this chapter how thiophene units can be included along with other aromatic moieties in macrocyclic compounds[110,112–114]. Other workers have also attempted the direct synthesis of calixthiophene. Condensation of dithienyl species gave the cyclic tetramers shown in Figure 5.58a,b, albeit in very poor yields.[136] Calixthiophenes linked by sulfur bridges have been synthesised[137] by condensation of linear thiophene oligomers with sodium sulfide to give cyclic tetramers and pentamers (Figure 5.58c,d) and shown to form monolayers on gold electrodes, which block the access of large ions such as ferricyanide to the electrodes but allow access of smaller ions such as silver.[138] Homooxacalix[n]thiophenes[139] were prepared by dehydrating thiophene-2,5-dimethanol to give trimers and tetramers (Figure 5.59a), which could be easily isolated, plus smaller

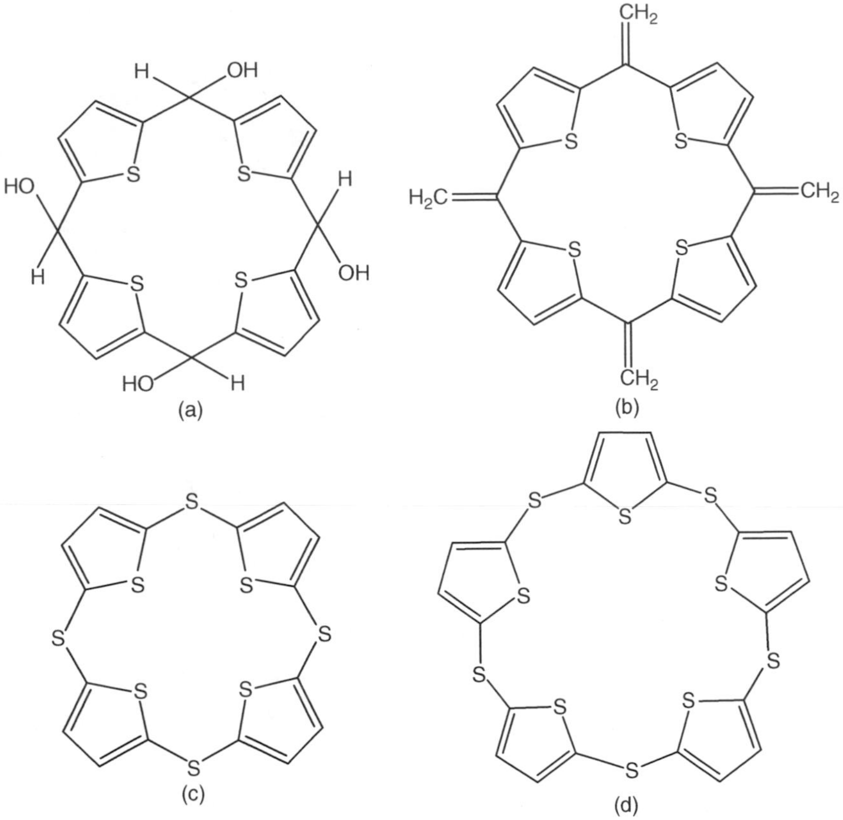

Figure 5.58   *(a,b) Structure of calix[4]thiophenes and (c,d) sulfur-linked calix[4]thiophene and calix[5]thiophene*

amounts of n = 5, 6 and 7. X-ray studies revealed that trimer has helical chirality, while the tetramer adopts a 1,2-alternate conformation in the solid state (Figure 5.59c,d). A similar dehydration route was utilised with 3,4-diethylthiophene methanol[140] to give the calixthiophenes shown in Figure 5.60. Oxidation of these species gave compounds such as the octaethyltetrathiaporphyrin dication, a sulfur analogue of a porphyrin which could be crystallised to give metallic-looking blue crystals and was shown to have an approximately planar structure (Figure 5.60c). The large sulfur atoms lead to small deviations from planarity, but the NMR spectra also indicate an aromatic nature for this species.

A cyclic tetramer containing uracil moieties has also been synthesised (Figure 5.61); it forms a complex with silver.[141] The crystal structure shows it adopts a pinched-cone conformation in the solid state.

5.7   Conclusions

We have attempted to demonstrate the richness of chemistry and structural versatility of the heterocal-ixarenes. Many of these species are relatively new compared to the earlier crowns, cyclodextrins and other heterocyclics. The use of heterocyclic compounds greatly expands the richness and diversity of these

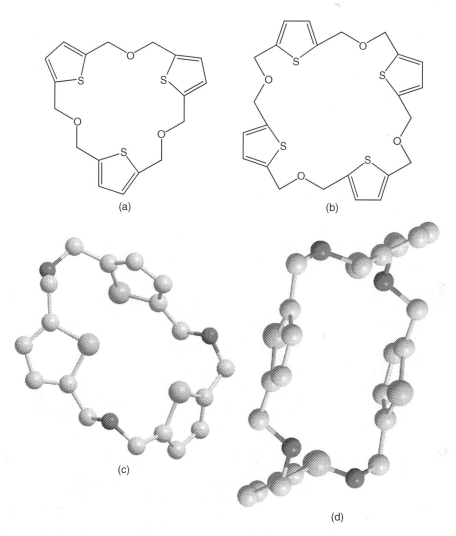

Figure 5.59 *(a,b) Structures of homooxacalix[n]thiophenes and (c,d) their crystal structures*

macrocyclic systems. The calixnaphthalenes offer deep, electron-rich cavities[3] with potential novel binding behaviours. This has been seen especially in the cyclochromotropylenes and related species, which display a high water-solubility. Although their crystal structure is still not elucidated, they have been shown to complex a wide variety of materials in water[17-26] and even to solubilise large, hydrophobic molecules such as polycyclic aromatic hydrocarbons.[16,27] The presence of the multiple negative charges on these species has allowed their incorporation into self-assembled polyelectrolyte multilayers[142] and these have been shown to act as optical sensors for ammonia as well as being able to be combined in the multilayer with urease to give a urea biosensor.

The other heterocalixarenes also have potential applications; for example, the calixfurans can be easily oxidised and then chemically modified to allow easy synthesis of a wide range of heterocalixarenes, as

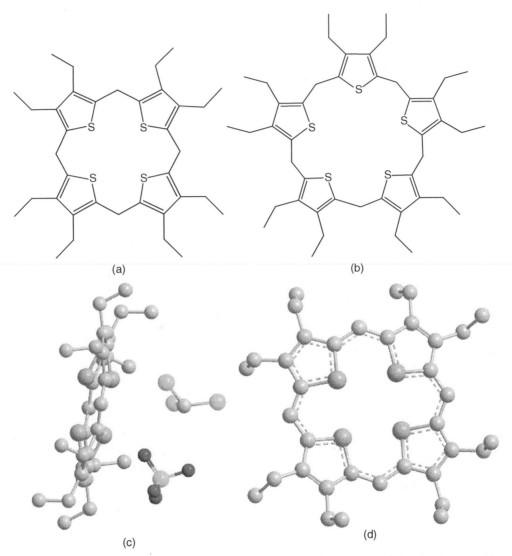

Figure 5.60 *(a,b) Structures of calix[n]thiophenes and the crystal structures of the dication (c) side-on with counter-ions and (d) face-on*

well as being able to be modified using Diels–Alder type reactions to give expanded ring systems. The calixpyrroles have been widely studied as both colourimetric[58,61,78,81,88,98] and fluorimetric[61] sensors for a variety of anions, for example as selective sensors for cyanide.[81] Compounds have been developed to display optical or electrochemical changes in response to the presence of various species. There are few specific sensors for anions and the calixpyrroles nicely fill this niche, especially since the crown ethers and calixarenes complex cations so well. They can be incorporated into polymer films to give selective colourimetric anion sensors,[78] which have even been shown to function in plasma.[98]

We can conclude from this that the heterocalixarenes are a series of molecules that will continue to be studied widely and will contribute greatly to the depth and richness of heterocyclic chemistry. The

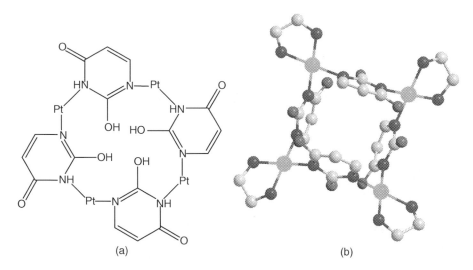

(a)   (b)

Figure 5.61   *Structure of a tetra-uracil calixarene analogue*

development and increased availability of various modified, funtionalised and deep-cavity heterocalixarenes will lead to the development of compounds with improved affinities and selectivities.

Bibliography

Gutsche CD. Calixarenes. Monographs in Supramolecular Chemistry. Royal Society of Chemistry; 1989.
Gutsche CD. Calixarenes Revisited. Monographs in Supramolecular Chemistry. Royal Society of Chemistry; 1998.
Gutsche CD. Calixarenes: An Introduction. Monographs in Supramolecular Chemistry. Royal Society of Chemistry; 2008.

References

1. Georghiou PE, Li Z. Calix[4]naphthalenes: cyclic tetramers of 1-naphthol and formaldehyde. *Tet Lett.* 1993; **34**: 2887–2890.
2. Georghiou PE, Li Z. Conformational properties of the calix[4]naphthalenes. *J Inclus Phenom Mol Recog.* 1994; **19**: 55–66.
3. Georghiou PE, Li ZP, Ashram M, Chowdhury S, Mizyed S, Tran AH, Al-Saraierh H, Miller DO. Calixnaphthalenes: deep, electron-rich naphthalene ring-containing calixarenes. The first decade. *Synlett.* 2005; 879–891.
4. Georghiou PE, Ashram M, Li Z, Chaulk G. Syntheses of calix[4]naphthalenes derived from 1-naphthol. *J Org Chem.* 1995; **60**: 7284–7289.
5. Georghiou PE, Li Z, Ashram M, Miller DO. Synthesis of dihomocalix[4]naphthalenes: first members of a new class of [1.2.1.2](1,3)naphthalenophanes. *J Org Chem.* 1996; **61**: 3865–3869.
6. Chowdbury S, Georghiou PE. Synthesis and properties of a new member of the calixnaphthalene family: a C2-symmetrical endo-calix[4]naphthalene. *J Org Chem.* 2002; **67**: 6808–6811.
7. Mizyed S, Georghiou PE. Synthesis of hexahomotrioxacalix[3]naphthalenes and a study of their alkali-metal cation binding properties. *J Org Chem.* 2001; **66**: 1473–1479.
8. Georghiou PE, Ashram M, Clase HJ, Bridson JN. Spirodienone and bis(spirodienone) derivatives of calix[4]naphthalenes. *J Org Chem.* 1998; **63**: 1819–1826.

9.   Georghiou PE, Mizyed S, Chowdhury S. Complexes formed from [60]fullerene and calix[4]naphthalenes. *Tet Lett.* 1999; **40**: 611–614.

10. Georghiou PE, Tran AH, Stroud SS, Thompson DW. Supramolecular complexation studies of [60]fullerene with calix[4]naphthalenes: a reinvestigation. *Tetrahedron.* 2006; **62**: 2036–2044.

11. Mizyed S, Ashram M, Miller DO, Georghiou PE. Supramolecular complexation of [60]fullerene with hexaho-motrioxacalix[3]naphthalenes: a new class of naphthalene-based calixarenes. *J Chem Soc. Perkin Trans.* 2001; **2**: 1916–1919.

12. Georghiou PE, Li Z, Ashram M. Chemistry of 1,8-naphthalenesultone: synthesis of a water-soluble tetrasulfonated C2v calixnaphthalene. *J Org Chem.* 1998; **63**: 3748–3752.

13. Tran AH, Miller DO, Georghiou PE. Synthesis and complexation properties of 'zorbarene': a new naphthalene ring-based molecular receptor. *J Org Chem.* 2005; **70**: 1115–1121.

14. Poh B-L, Lim CS, Khoo KS. A water-soluble cyclictetramer from reacting chromotropic acid with formaldehyde. *Tet Lett.* 1989; **30**: 1005–1008.

15. Georghiou PE, Ho CK. The chemistry of the chromotropic acid method for the analysis of formaldehyde. *Can J Chem.* 1989; **67**: 871–876.

16. Poh B-L, Koay L-S. Complexation of aromatic hydrocarbons with cyclotetrachromotropylene in aqueous solution. *Tet Lett.* 1990; **31**: 1911–1914.

17. Poh B-L, Seah LH, Lim CS. Complexations of metal cations with ccyclotetrachromotropylene in water and methanol. *Tetrahedron.* 1990; **46**: 4379–4386.

18. Poh B-L, Lim C. S. Complexations of amines with water-soluble cyclotetrachromotropylene. *Tetrahedron.* 1990; **46**: 3651–3658.

19. Poh B-L, Lim CH, Tan CM, Wong WM. 1H NMR study on the complexation of phenols with cyclotetrachromotropylene in aqueous solution. *Tetrahedron.* 1994; **49**: 7259–7266.

20. Poh B-L, Tan CM. Contribution of guest–host CH–π interaction to the stability of complexes formed from cyclotetrachromotropylene as host and alcohols and sugars as guests in water. *Tetrahedron.* 1993; **49**: 9581–9592.

21. Poh B-L, Tan CM. Complexation of amino acids by cyclotetrachromotropylene in aqueous solution: importance of CH–π and π–π interactions. *Tetrahedron.* 1994; **50**: 3453–3462.

22. Poh B-L, Tan CM. Complexation of nucleotides in water with cyclotetrachromotropylene. *J Inclus Phenomn Macrocy Chem.* 2000; **38**: 69–74.

23. Poh B-L, Tan CM, Loh C. L. ¹H NMR and visible spectroscopic studies on the complexation of Paraquat dication with cyclotetrachromotropylene in water. *Tetrahedron.* 1993; **49**: 3849–3856.

24. Poh B-L, Tan CM. Crown ethers as guests of cyclotetrachromotropylene in water. *Tetrahedron.* 1995; **51**: 953–958.

25. Poh B-L, Tan CM. Cyclotetrachromotropylene plays host to α-, β- and γ-cyclodextrin in water. *Tet Lett.* 1994; **34**: 6397–6390.

26. Poh B-L, Teem CM. A ¹H NMR study on the complexation of tetraalkylammonium cations, mono and diprotonated amines, and amino acids with a derivatized cyclotetrachromotropylene in an aqueous solution. *Tetrahedron.* 2005; **61**: 5123–5129.

27. Poh B-L, Chin LY, Lee CW. A cyclic tetramer from reacting 4-amino[5]hydroxynaphthalene-2,7-disulfonic acid with formaldehyde and its complexation with polyaromatic hydrocarbons in water. *Tet Lett.* 1995; **36**: 3877–3880.

28. Shorthill BJ, Granucci G, Powell DR, Glass TE. Synthesis of 3,5- and 3,6-linked calix[n]naphthalenes. *J Org Chem.* 2002; **67**: 904–909.

29. Poh B-L, Ng YY. Synthesis and conformations of a water-soluble cyclic tetramer obtained from reacting 5-sulfonatotropolone with formaldehyde. *Tetrahedron.* 1997; **53**: 8635–8642.

30. Poh B-L, Ng Y. Y. Proton NMR study on the complexation of organic molecules with cyclotetra[5]sulfonatotropolonylene in aqueous solution. *Tetrahedron.* 1998; **54**: 129–134.

31. Poh B-L, Ng YY. Complexation of aromatic hydrocarbons with a macrocycle containing four tropolone units in water. *Tetrahedron.* 1997; **53**: 11913–11918.

32. Poh B-L, Yue CS. A cyclic tetramer that exists solely in the 1,3-alternate conformation in water. *Tetrahedron.* 1999; **55**: 5515–5518.

33. de Sousa Healy M, Rest AJ. Role of metal salts in the synthesis of furan–ketone condensation macrocycles: an apparent metal template effect. *J Chem Soc. Perkin Trans.* 1985; **1**: 973–982.

34. Kohnke FH, La Torte GL, Parisi MF. Large cyclic oligomers of furan and acetone. X-ray crystal structure of the hexamer and first synthesis of the nonamer. *Tet Lett.* 1996; **37**: 4593–4596.

35. Musau RM, Whiting A. The synthesis of furan-derived calixarenes. *J Chem Soc Chem Commun.* 1993; 1029–1031.

36. Musau RM, Whiting A. Synthesis of calixfuran macrocycles and evidence for gas-phase ammonium ion complexation. *J Chem Soc. Perkin Trans.* 1994; **I**: 2881–2888.

37. Kobuke Y, Hanji K, Horiguchi K, Asada M, Nakayama Y, Furukawa J. Macrocyclic ligands composed of tetrahydrofuran for selective transport of monovalent cations through liquid membranes. *J Am Chem Soc.* 1976; **98**: 7414–7419.

38. Hazell A. The structure of octamethyltetraoxaquaterene. *Acta Cryst.* 1989; **C45**: 137–140.

39. Cafeo G, Giannetto M, Kohnke FH, La Torre GL, Parisi MF, Menzer S, White AJP, Williams DJ. Chemical modifications of furan-based calixarenes by Diels–Alder reactions. *Chem Eur J.* 1999; **5**: 356–368.

40. Williams PD, LeGoff E. Enedione-functionalized macrocycles via oxidative ring opening of furans. *J Org Chem.* 1981; **46**: 4143–4147.

41. Cafeo G, Kohnke FH, La Torre GL, White AJP, Williams DJ. From large furan-based calixarenes to calixpyrroles and calix[m]furan[n]pyrroles: syntheses and structures. *Angew Chem Int Ed.* 2002; **39**: 1496–1498.

42. Kohnke FH, Parisi MF, Raymo FM, O'Neil PA, Williams DJ. Acenaphane derivatives from furan macrocycles. *Tetrahedron.* 1994; **50**: 9113–9124.

43. Baeyer A. Ueber ein Condensationsproduct von Pyrrol mit Aceton. *Ber Dtsch Chem Ges.* 1886; **19**: 2184–2185.

44. Gale PA, Sessler JL, Král V. Calixpyrroles. *J Chem Soc Chem Commun.* 1998; 1–8.

45. Gale PA, Anzenbacher P, Sessler JL. Calixpyrroles II. *Coord Chem Rev.* 2001; **222**: 57–102.

46. Gale PA, Sessler JL, Král V, Lynch V. Calix[4]pyrroles: old yet new anion-binding agents. *J Am Chem Soc.* 1996; **118**: 5140–5141.

47. Allen W. E, Gale P. A, Brown C. T, Lynch V. M, Sessler J. L. Binding of neutral substrates by calix[4]pyrroles. *J Am Chem Soc.* 1996; **118**: 12471–12472.

48. Jacoby D, Floriani C, Chiesi-Villa A, Rizzoli C. Meso-octamethyl-porphyrinogen metal complexes: an entry to high valent unsaturated metal centres. *J Chem Soc Chem Commun.* 1991; 220–222.

49. Jacoby D, Floriani C, Chiesi-Villa A, Rizzoli C. The π and σ bonding modes of meso-octaethylporphyrinogen to transition metals: the X-ray structure of a meso-octaethylporphyrinogen–zirconiuml(IV) complex and of the parent meso-octaethylporphyrinogen ligand. *J Chem Soc Chem Commun.* 1991; 790–792.

50. Anzenbacher P, Try A. C, Miyaji H, Jursíková K, Lynch VM, Marquez M, Sessler JL. Fluorinated calix[4]pyrrole and dipyrrolylquinoxaline: neutral anion receptors with augmented affinities and enhanced selectivities. *J Am Chem Soc.* 2000; **122**: 10268–10272.

51. Sessler JL, Anzenbacher P, Shriver JA, Jursikova K, Lynch VM, Marquez M. Direct synthesis of expanded fluorinated calix[n]pyrroles: decafluorocalix[5]pyrrole and hexadecafluorocalix[8]pyrrole. *J Am Chem Soc.* 2000; **122**: 12061–12062.

52. Sessler JL, Cho W-S, Gross DE, Shriver JA, Lynch VM, Marquez M. Anion binding studies of fluorinated expanded calixpyrroles. *J Org Chem.* 2005; **70**: 5982–5986.

53. Chacón-García L, Chávez L, Cacho DR, Altamirano-Hernández J. The first direct synthesis of β-unsubstituted meso-decamethylcalix[5]pyrrole. *Beilstein J Org Chem.* 2009; **5**: art. no. 2.

54. Cafeo G, Kohnke FH, La Torre GL, Parisi MF, Nascone RP, White AJP, Williams DJ. Calix[6]pyrrole and hybrid calix[n]furan[m]pyrroles (n + m = 6): syntheses and host ± guest chemistry. *Chem Eur J.* 2002; **8**: 3148–3156.

55. Bruno G, Cafeo G, Kohnkeb FH, Nicol F. Tuning the anion binding properties of calixpyrroles by means of p-nitrophenyl substituents at their meso-positions. *Tetrahedron.* 2007; **63**: 10003–10010.

56. Turner B, Botoshansky M, Eichen Y. Extended calixpyrroles: meso-substituted calix[6]pyrroles. *Angew Chem Int Ed.* 1998; **37**: 2475–2478.

57. Yoon D-W, Jeong S-D, Song M-Y, Lee C-H. Calix[6]pyrroles capped with 1,3,5-trisubstituted benzene. *Supramol Chem.* 2007; **19**: 265–270.

58. Anzenbacher P, Jursíková K, Sessler JL. Second generation calixpyrrole anion sensors. *J Am Chem Soc.* 2000; **122**: 9350–9351.

59. Gale PA, Sessler JL, Allen WE, Tvermoes NA, Lynch V. Calix[4]pyrroles: C-rim substitution and tunability of anion binding strength. *J Chem Soc Chem Commun.* 1997; 665–666.

60. Anzenbacher P, Jursíková K, Shriver JA, Miyaji H, Lynch VM, Sessler JL, Gale PA. Lithiation of meso-octamethylcalix[4]pyrrole: a general route to c-rim monosubstituted calix[4]pyrroles. *J Org Chem.* 2000; **65**: 7641–7645.

61. Miyaji H, Sato W, Sessler JL, Lynch VM. A 'building block' approach to functionalized calix[4]pyrroles. *Tet Lett.* 2000; **41**: 1369–1373.

62. Sato W, Miyaji H, Sessler JL. Calix[4]pyrrole dimers bearing rigid spacers: towards the synthesis of cooperative anion binding agents' *Tet Lett.* 2000; **41**: 6731–6736.

63. Furusho Y, Kawasaki H, Nakanishi S, Aida T, Takata T. Design of muitidentate pyrrolic ligands by n-modification: synthesis of n-monoethyl, dimethyl, trimethyl, and tetramethylporphyrinogens. *Tet Lett.* 1998; **39**: 3537–3540.

64. Anzenbacher P, Jursíková K, Lynch VM, Gale PA, Sessler JL. Calix[4]pyrroles containing deep cavities and fixed walls. Synthesis, structural studies, and anion binding properties of the isomeric products derived from the condensation of p-hydroxyacetophenone and pyrrole. *J Am Chem Soc.* 1999; **121**: 11020–11021.

65. Bonomo L, Solari E, Toraman G, Scopelliti R, Floriani C, Latronico M. A cylindrical cavity with two different hydrogen-binding boundaries: the calix[4]arene skeleton screwed onto the meso-positions of the calix[4]pyrrole. *J Chem Soc Chem Commun.* 1999; 2413–2414.

66. Camiolo S, Gale PA. Fluoride recognition in super-extended cavity calix[4]pyrroles. *Chem Commun.* 2000; 1129–1130.

67. Depraetere S, Smet M, Dehaen W. N-confused calix[4]pyrroles. *Angew Chem Int Ed.* 1999; **38**: 3359–3361.

68. Bates GW, Kostermans M, Dehaen W, Gale PA, Light ME. Organic salt inclusion: the first crystal structures of anion complexes of N-confused calix[4]pyrrole. *Cryst Eng Comm.* 2006; **8**: 444–447.

69. Nielsen KA, Cho W-S, Jeppesen JO, Lynch VM, Becher J, Sessler JL. Tetra-TTF Calix[4]pyrrole: a rationally designed receptor for electron-deficient neutral guests. *J Am Chem Soc.* 2004; **126**: 16296–16297.

70. Nielsen KA, Martin-Gomis L, Sarovac GH, Sanguinet L, Gross DE, Fernandez-Lazaro F, Stein PC, Levillain E, Sessler JL, Guldi DM, Sastre-Santos A, Jeppesen JO. Binding studies of tetrathiafulvalene-calix[4]pyrroles with electron-deficient guests. *Tetrahedron.* 2008; **64**: 8449–8463.

71. Gale PA, Hursthouse MB, Light ME, Sessler JL, Warriner CN, Zimmerman S. Ferrocene-substituted calix[4]pyrrole: a new electrochemical sensor for anions involving CHanion hydrogen bonds. *Tetrahedron Lett.* 2001; **42**: 6759–6762.

72. Szymanska I, Radecka H, Radecki J, Gale PA, Warriner CN. Ferrocene-substituted calix[4]pyrrole modified carbon paste electrodes for anion detection in water. *J Electroanal Chem.* 2006; **591**: 223–228.

73. Yang W, Yin Z, Wang C-H, Huang C, He J, Zhu X, Cheng J-P. New redox anion receptors based on calix[4]pyrrole bearing ferrocene amide. *Tetrahedron.* 2008; **64**: 9244–9252.

74. de Namor AFD, Khalife R. Calix[4]pyrrole derivative: recognition of fluoride and mercury ions and extracting properties of the receptor-based new material. *J Phys Chem B.* 2008; **112**: 15766–15774.

75. Cui R, Li Q, Gross DE, Meng X, Li B, Marquez M, Yang R, Sessler JL, Shao Y. Anion transfer at a micro-water/1,2-dichloroethane interface facilitated by octafluoro-meso-octamethylcalix[4]pyrrole. *J Am Chem Soc.* 2008; **130**: 14364–14365.

76. Danil de Namor AF, Shehab M. Double-cavity calix[4]pyrrole derivative with enhanced capacity for the fluoride anion. *J Phys Chem B.* 2005; **109**: 17440–17444.

77. Nishiyabu R, Palacios MA, Dehaen W, Anzenbacher P. Synthesis, structure, anion binding, and sensing by calix[4]pyrrole isomers. *J Am Chem Soc.* 2006; **128**: 11496–11505.

78. Anzenbacher P, Nishiyabu R, Palacios MA. N-confused calix[4]pyrroles. *Coord Chem Rev.* 2006; **250**: 2929–2938.

79. Yang W, Yin Z, Li Z, He J, Cheng J-P. A new acenaphthenequinone-based calix[4]pyrrole: Synthesis, structure and anion binding study. *J Mol Struct.* 2008; **889**: 279–285.

80. Cafeo G, Kohnke FH, Valenti L, White AJP. pH-controlled molecular switches and the substrate-directed self-assembly of molecular capsules with a calix[4]pyrrole derivative. *Chem Eur J.* 2008; **14**: 11593–11600.

81. Cafeo G, De Rosa M, Kohnke FH, Neri P, Soriente A, Valenti L. Efficient organocatalysis with a calix[4]pyrrole derivative. *Tet Lett.* 2008; **49**: 153–155.

82. Cafeo G, Kohnke FH, Valenti L. Regioselective O-alkylations and acylations of polyphenolic substrates using a calix[4]pyrrole derivative. *Tet Lett.* 2009; **50**: 4138–4140.

83. Jain VK, Mandalia HC, Suresh E. A facial microwave-assisted synthesis, spectroscopic characterization and preliminary complexation studies of calix[4]pyrroles containing the hydroxamic-acid moiety. *J Incl Phenom Macrocycl Chem.* 2008; **62**: 167–178.

84. Jain VK, Mandalia HC, Suresh E. Azocalix[4]pyrroles: one-pot microwave and one drop water assisted synthesis, spectroscopic characterization and preliminary investigation of its complexation with copper(II). *J Incl Phenom Macrocycl Chem.* 2009; **63**: 27–35.

85. Gil-Ramirez G, Benet-Buchholz J, Escudero-Ada EC, Ballester P. Solid-state self-assembly of a calix[4]pyrrole-resorcinarene hybrid into a hexameric cage. *J Am Chem Soc.* 2007; **129**: 3920–3921.

86. Lee C-H, Miyaji H, Yoon D-W, Sessler JL. Strapped and other topographically nonplanar calixpyrrole analogues. Improved anion receptors. *J Chem Soc Chem Commun.* 2008; 24–34.

87. Lee C-H, Lee J-S, Na H-K, Yoon D-W, Miyaji H, Cho W-S, Sessler JL. Cis- and trans-strapped calix[4]pyrroles bearing phthalamide linkers: synthesis and anion-binding properties. *J Org Chem.* 2005; **70**: 2067–2074.

88. Kim S-H, Hong S-J, Yoo J, Kim SK, Sessler JL, Lee C-H. Strapped calix[4]pyrroles bearing a 1,3-indanedione at a β-pyrrolic position: chemodosimeters for the cyanide anion. *Org Lett.* 2009; **11**: 3626–3629.

89. Yoo J, Kim M-S, Hong S-J, Sessler JL, Lee C-H. Selective sensing of anions with calix[4]pyrroles strapped with chromogenic dipyrrolylquinoxalines. *J Org Chem.* 2009; **74**: 1065–1069.

90. Yoon D-W, Gross DE, Lynch VM, Lee C-H, Bennett PC, Sessler JL. Real-time determination of chloride anion concentration in aqueous-DMSO using a pyrrole-strapped calixpyrrole anion receptor. *Chem Soc Chem Commun.* 2009; 1109–1111.

91. Fisher MG, Gale PA, Hiscock JR, Hursthouse MB, Light ME, Schmidtchen FP, Tong CC. 1,2,3-triazole-strapped calix[4]pyrrole: a new membrane transporter for chloride. *J Chem Soc Chem Commun.* 2009; 3017–3019.

92. Gale PA, Sessler JL, Lynch V, Sansom PI. Synthesis of a new cylindrical calix[4]arene-calix[4]pyrrole pseudo dimer. *Tet Lett.* 1996; **41**: 7881–7884.

93. Gale PA, Genge JW, Kral V, McKervey MA, Sessler JL, Walker A. First synthesis of an expanded calixpyrrole. *Tet Lett.* 1997; **38**: 8443–8444.

94. Sessler JL, Kim SK, Gross DE, Lee C-H, Kim JS, Lynch VM. Crown-6-calix[4]arene-capped calix[4]pyrrole: an ion-pair receptor for solvent-separated CsF ions. *J Am Chem Soc.* 2008; **130**: 13162–13166.

95. Panda PK, Lee C-H. Metalloporphyrin-capped calix[4]pyrroles: heteroditopic receptor models for anion recognition and ligand fixation. *J Org Chem.* 2005; **70**: 3148–3156.

96. Blangy V, Heiss C, Khlebnikov V, Letondor C, Stoeckli-Evans H, Neier R. Synthesis, structure, and complexation properties of partially and completely reduced meso-octamethylporphyrinogens (calix[4]pyrroles). *Angew Chem Int Ed.* 2009; **48**: 1688–1691.

97. Cuesta L, Gross D, Lynch VM, Ou Z, Kajonkijya W, Ohkubo K, Fukuzumi S, Kadish KM, Sessler JL. Design and synthesis of polymetallic complexes based on meso-calix[4]pyrrole: platforms for multielectron chemistry. *J Am Chem Soc.* 2007; **129**: 11696–11697.

98. Nishiyabu R, Anzenbacher P. Sensing of antipyretic carboxylates by simple chromogenic calix[4]pyrroles. *J Am Chem Soc.* 2005; **127**: 8270–8271.

99. Sessler JL, Gale PA, Genge JW. Calix[4]pyrroles: new solid-phase HPLC Supports for the separation of anions. *Chem Eur J.* 1998; **4**: 1095–1099.

100. Zhou C, Tang H, Shao S, Jiang S. Calix[4]pyrrole-bonded hplc stationary phase for the separation of phenols, benzenecarboxylic acids, and medicines. *J Liq Chromato Tech.* 2006; **29**: 1961–1978.

101. Amiri AA, Safavi A, Hasaninejad AR, Shrghi H, Shamsipur M. Highly selective transport of silver ion through a supported liquid membrane using calix[4]pyrroles as suitable ion carriers. *J Membrane Sci.* 2008; **325**: 295–300.

102. Tong CC, Quesada R, Sessler JL, Gale PA. meso-Octamethyl calix[4]pyrrole: an old yet new transmembrane ion-pair transporter. *J Che Soc Chem Commun.* 2008; 6321–6323.

103. Gross DE, Schmidtchen FP, Antonius W, Gale PA, Lynch VM, Sessler JL. Cooperative binding of calix[4]pyrrole–anion complexes and alkylammonium cations in halogenated solvents. *Chem Eur J.* 2008; **14**: 7822–7827.

104. Liu K, Xu J, Sun Y, Guo Y, Jiang S, Shao S. Spectroscopic study on anion recognition properties of calix[4]pyrroles: effects of tetraalkylammonium cations. *Spectrochim Acta A: Mol Biomol Spec.* 2008; **69**: 1201–1208.

105. Verdejo B, Gil-Ramirez G, Ballester P. Molecular recognition of pyridine N-oxides in water using calix[4]pyrrole receptors. *J Am Chem Soc.* 2009; **131**: 3178–3179.

106. Mohebbi S, Assoud J. Tetrapyrrole bizirconium complexes as active catalysts for ethylene polymerization. *Monatsh Chem.* 2008; **139**: 1163–1167.

107. Chen W, Li ZR, Wu D, Li Y, Sun CC, Gu FL. The structure and the large nonlinear optical proerties of Li@Calix[4]pyrrole. *J Am Chem Soc.* 2005; **127**: 10977–10981.

108. Chen W, Li Z-R, Wu D, Li Y, Sun C-C, Gu FL, Aoki Y. Nonlinear optical properties of alkalides Li+(calix[4]pyrrole)M- (M = Li, Na, and K): alkali anion atomic number dependence. *J Am Chem Soc.* 2006; **128**: 1072–1073.

109. Piatek P, Lynch VM, Sessler JL. Calix[4]pyrrole[2]carbazole: a new kind of expanded calixpyrrole. *J Am Chem Soc.* 2004; **126**: 16073–16076.

110. Jang Y-S, Kim H-J, Lee P-H, Lee CH. Synthesis of calix[n]furano[n]pyrroles and calix[n]thieno[n]pyrroles (n = 2,3,4) by '3 + 1' approach. *Tet Lett.* 2000; **41**: 2919–2923.

111. Sessler JL, Cho W-S, Lynch V, Kral V. Missing-link macrocycles: hybrid heterocalixarene analogues formed from several different building blocks. *Chem Eur J.* 2002; **8**: 1134–1143.

112. Poulsen T, Nielsen KA, Bond AD, Jeppesen JO. Bis(tetrathiafulvalene)-calix[2]pyrrole[2]-thiophene and its complexation with TCNQ. *Org Lett.* 2007; **9**: 5485–5488.

113. Song M-Y, Na H-K, Kim E-Y, Lee S-J, Kim K-I, Baek E-M, Kim H-S, Ana DK, Lee C-H. Hetero-calix[4]pyrroles: incorporation of furans, thiophenes, thiazoles or fluorenes as a part of the macrocycles. *Tet Lett.* 2004; **45**: 299–301.

114. Matano Y, Nakabuchi T, Miyajima T, Imahori H. Phosphole-containing hybrid calixpyrroles: new multifunctional macrocyclic ligands for platinum(II) ions. *Organometallics.* 2006; **25**: 3105–3107.

115. Aleskovic M, Halasz I, Basaric N, Mlinaric-Majerski K. Synthesis, structural characterization, and anion binding ability of sterically congested adamantane-calix[4]pyrroles and adamantane-calixphyrins. *Tetrahedron.* 2009; **65**: 2051–2058.

116. Dolensky B, Kroulik J, Kral V, Sessler JL, Dvorakova H, Bourÿ P, Bernatkova M, Bucher C, Lynch V. Calix[4]phyrins. Effect of peripheral substituents on conformational mobility and structure within a series of related systems. *J Am Chem Soc.* 2004; **126**: 13714–13722.

117. Inoue M, Ikeda C, Kawata Y, Venkatraman S, Furukawa K, Osuka A. Synthesis of calix[3]dipyrrins by a modified Lindsey protocol. *Angew Che Int Ed.* 2007; **46**: 2306–2309.

118. Sessler JL, An D, Cho W-S, Lynch V, Yoon D-W, Hong S-J, Lee C-H. Anion-binding behavior of hybrid calixpyrroles. *J Org Chem.* 2005; **70**: 1511–1517.

119. Sessler JL, An D, Cho W-S, Lynch V. Calix[n]bipyrroles: synthesis, characterization, and anion-binding studies. *Angew Chem Int Ed.* 2003; **42**: 2278–2281.

120. Sessler JL, An D, Cho W-S, Lynch V, Marquez M. Calix[4]bipyrrole: a big, flexible, yet effective chloride-selective anion receptor. *J Chem Soc Chem Commun.* 2005; 540–542.

121. Black DSC, Craig DC, Kumar N. Synthesis of a new class of indole-containing macrocycles. *J Chem Soc Chem Commun.* 1989; 425–426.

122. Black DSC, Bowyer MC, Kumar N, Mitchell PSR. Calix[3]indoles, new macrocyclic tris(indolylmethylene) compounds with 2,7-linkages. *J Chem Soc Chem Commun.* 1993; 819–821.

123. Black DSC, McConnell DB. Tri(indolyl)phosphine oxides: attempted formation of lower rim-capped cone conformers of calix[3]indoles. *Heteroatom Che.* 1996; **7**: 437–441.

124. Black DSC, Craig DC, Kumar N, McConnell DB. Synthesis and crystal structure of a calix[3]indole with cone conformation: a new molecular receptor. *Tet Lett.* 1996; **37**: 241–244.

125. Black DSC, Craig DC, Kumar N. Calix[4]indoles: new macrocyclic tetra(indolyimethylene) compounds with 2,7-linkages. *Tet Lett.* 1995; **36**: 8075–8078.

126. Kral V, Gale PA, Anzenbacher P, Jursikova K, Lynch V, Sessler JL. Calix[4]pyridine: a new arrival in the heterocalixarene family. *J Chem Soc Chem Commun.* 1998; 9–10.

127. Gong H-Y, Zhang X-H, Wang D-X, Ma H-W, Zheng Q-Y, Wang M-X. Methylazacalixpyridines: remarkable bridging nitrogen-tuned conformations and cavities with unique recognition properties. *Chem Eur J.* 2006; **12**: 9262–9275.

128. Gerkensmeier T, Mattay J, Näther C. A new type of calixarene: octahydroxypyridine[4]arenes. *Chem Eur J.* 2001; **7**: 465–474.

129. Letzel MC, Decker B, Rozhenko AB, Schoeller WW, Mattay J. Encapsulated guest molecules in the dimer of octahydroxypyridine[4]arene. *J Am Chem Soc.* 2004; **126**: 9669–9674.

130. Evan-Salem T, Cohen Y. Octahydroxypyridine[4]arene self-assembles spontaneously to form hexameric capsules and dimeric aggregates. *Chem Eur J.* 2007; **13**: 7659–7663.

131. Burilov AR, Knyazeva IR, Pudovik MA, Syakaev VV, Latypov SK, Baier I, Habicher WD, Konovalov AI. New phosphorus-containing analog of calix[4]resorcinarene based on 2,6-dihydroxypyridine. *Russ Chem Bull Int Ed.* 2006; **56**: 364–366.

132. Newkome GR, Lee H-W. 18[(2,6) pyridino$_6$coronand-6]: sexipyridine. *J Am Chem Soc.* 1983; **105**: 5956–5957.

133. Toner JL. Cyclosexipyridines. *Tet Lett.* 1983; **24**: 2707–2710.

134. Kelly TR, Lee Y-J, Mears RJ. Synthesis of cyclo-2,2':4.4":2'.2'":4".4'"":2'".2'"":4'"".4-sexipyridine. *J Org Chem.* 1997; **62**: 2774–2781.

135. Tanaka R, Yano T, Nishioka T, Nakajo K, Breedlove BK, Kimura K, Kinoshita I, Isobe K. Thia-calix[n]pyridines, synhesis and coordination to Cu(I,II) ions with both N and S donor atoms. *J Chem Soc Chem Commun.* 2002; 1686–1687.

136. Ahmed M, Meth-Cohn O. The preparation and properties of thiophen analogues of porphyrins and related systems. Part III. *J Chem Soc.* 1971; 2104–2111.

137. Katano N, Sugihara Y, Ishii A, Nakayama J. Synthesis and structure of sulfur-bridged [1·n](2,5)thiophenophanes (n = 4–6). *Bull Chem Soc Jpn.* 1998; **71**: 2695–2700.

138. Nakabayashi S, Fukushima E, Baba R, Katano N, Sugihara Y, Nakayama J. Stereo-electrochemistry by a self-assembled monolayer of sulfur-bridged calixthiophene on gold. *Electrochem Comm.* 1999; **1**: 550–553.

139. Komatsu N, Taniguchi A, Suzuki H. Homooxacalix[n]thiophenes: their one-pot serial synthesis and X-ray structures. *Tet Lett.* 1999; **40**: 3749–3752.

140. Vogel E, Pohl M, Herrmann A, Wiss T, Konig CH, Lex J, Gross M, Gisselbrecht JP. Porphyrinoid macrocycles based on thiophene: the octaethyltetrathiaporphyrin dication. *Angew Chem Int Ed.* 1996; **35**: 1520–1524.

141. Rauter H, Hillgeris EC, Erxleben A, Lippert B. [(en)Pt(uracilate)]$_4^{4+}$A: metal analogue of calix[4]arene. similarities and differences with classical calix[4]arenes. *J Am Chem Soc.* 1994; **116**: 616–624.

142. Nabok AV, Davis F, Hassan AK, Ray AK, Majeed R, Ghassemlooy Z. Polyelectrolyte self-assembled thin films containing cyclo-tetrachromotropylene for chemical and biosensing. *Mater Sci Eng.* 1999; **C8–C9**: 123–126.

6

Cyclodextrins

6.1   Introduction

Many of the compounds mentioned previously within this work are entirely synthetic in origin and as such do not occur naturally. The cyclodextrins are somewhat different however since they have a natural origin, being synthesised originally by bacterial fermentation and now via the use of enzymes from a natural feedstock. We have already discussed how small rings can be incorporated into macrocycles, such as in crown ethers and calixarenes. The cyclodextrins, also known as cycloamyloses, represent further examples of macrocyclic compounds formed by the joining of small rings to give rise to a much larger ring system. We intend to provide a general overview of cyclodextrin chemistry within this chapter; for further detail the reader is recommended to study some of the works listed in the Bibliography.

Amylose is a water-soluble polysaccharide which is made up of 1,4-linked glucose units. Together with amylopectin it is one of the major constituents of starch. Cyclodextrins are made up of a series of α-D-glucopyranoside sugars linked together to form a macrocycle and can be thought of as cyclic versions of amylose. The three most common members of this group occur naturally; these are the hexamer (α-cyclodextrin), heptamer (β-cyclodextrin) and octamer (γ-cyclodextrin), as shown in Figure 6.1. The cyclic pentamer has been successfully synthesised in the laboratory – as have larger members of the series.

Cyclodextrins were initially isolated by Villiers[1] as long ago as 1891; Villiers coined the name 'cellulosine' for these materials. He digested starch with *Bacillus amylobacter* to obtain amongst other things about 0.3% of a crystalline material. Elemental analysis showed it to be a polysugar and to have two distinct crystal forms. However, the techniques for in-depth separation and characterisation of these materials (thought to have been α- and β-cyclodextrin) were not then available. Later work by Schardinger,[2] again using bacterial digestion of starch, managed to produce materials similar to cellulosine in yields of up to 30% and to separate and characterise the natural α- and β-cyclodextrins, which were then named 'Schardinger sugars' or 'Schardinger dextrins'. The early history of these and the γ-cyclodextrins, the elucidation of their composition from glucose (they can also be thought of as consisting of maltose; for example, α-cyclodextrin can be thought of as consisting of six glucose or three maltose units) and their cyclic structure have all been reviewed.[3] In 1998, the journal *Chemical Reviews* published a special issue on cyclodextrins, to which the reader is referred in the Bibliography.

Macrocycles: Construction, Chemistry and Nanotechnology Applications, First Edition.   Frank Davis and Séamus Higson.
© 2011 John Wiley & Sons, Ltd. Published 2011 by John Wiley & Sons, Ltd.

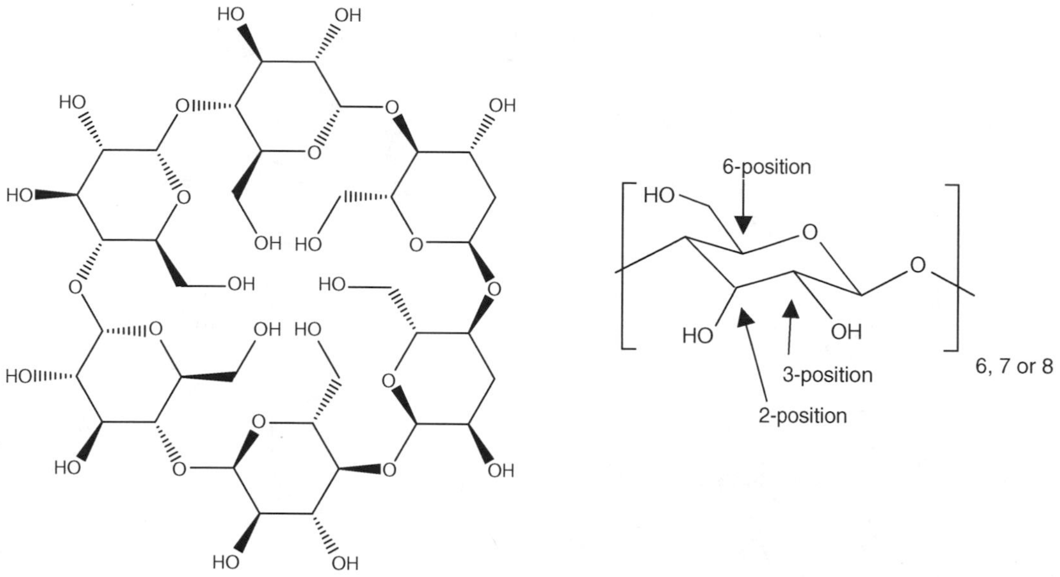

(a)

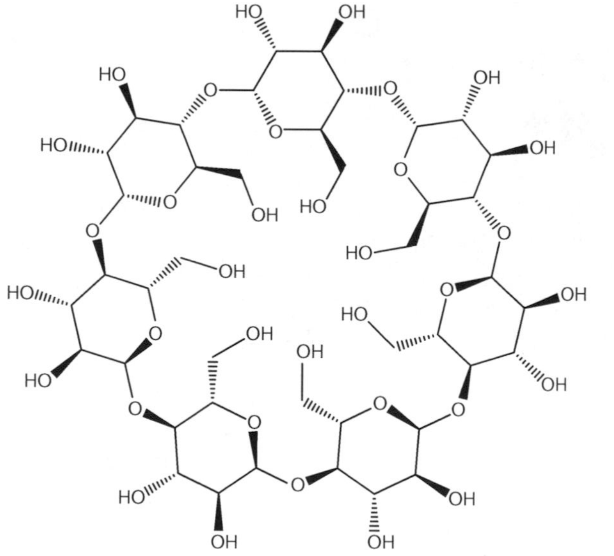

(b)

Figure 6.1   *Structures of (a) α-, (b) β- and (c) γ-cyclodextrin*

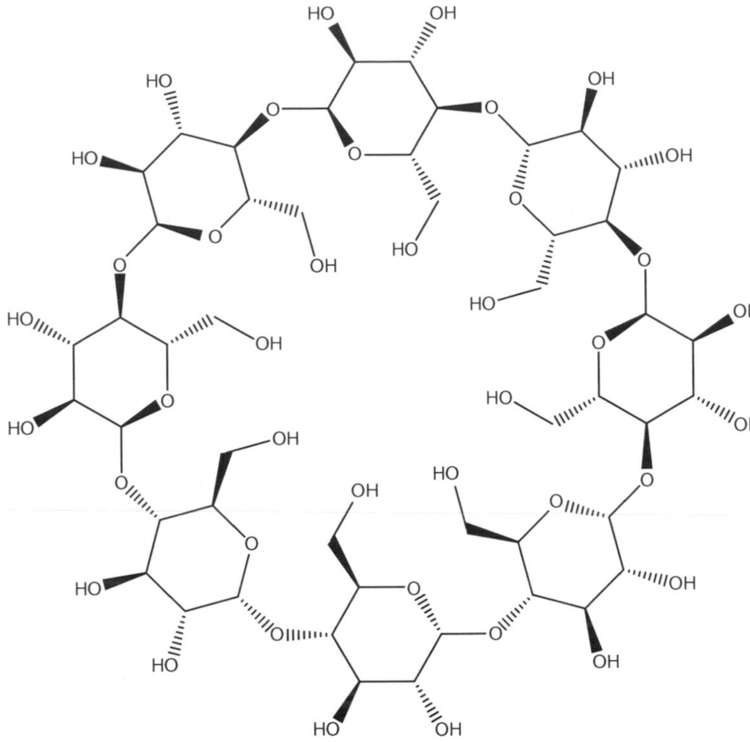

(c)

Figure 6.1   *(continued)*

The most widely used methods for the synthesis of cyclodextrins utilise starch as the feedstock and degrade it with any of a number of easily available enzymes, such as cyclodextrin glycosyltransferase. Starch is first degraded and solubilised either enzymatically by α-amylase or via heat treatment, then cyclodextrin glycosyltransferase is added, which catalyses the enzymatic conversion to cyclodextrins. There are a number of cyclodextrin glycosyltransferase enzymes available and each has its own characteristic α: β: γ synthesis ratio. The synthesis of cyclodextrins by enzymatic means has recently been reviewed.[4]

Once this mixture has been obtained, it remains to separate out the different compounds. The easiest of the family to separate is β-cyclodextrin, which is approximately an order of magnitude less soluble than the α and γ compounds and therefore can be crystallised from the mixture. Separation of the other two isomers normally requires chromatographic methods. However, yields can be improved and separation simplified by the addition of a 'complexing agent' during the enzymatic conversion step. As we will mention in detail later within this chapter, cyclodextrins have a high complexing ability towards organic molecules such as toluene. If such a species is added to the conversion mixture, it complexes with the cyclodextrins as they form and leads to precipitation. This has several benefits since precipitation simplifies separation of the cyclodextrin by filtration or centrifugication, and since we are removing the product from the reaction mixture, agents such as these drive the reaction towards the formation of cyclodextrins, thus improving their yields. Furthermore, by correct selection of guest and enzyme, it can in some instances prove possible to maximise the yield of one particular size of cyclodextrin. For example,[3] a linear alcohol such as 1-decanol favours the production of α-cyclodextrin, whereas if toluene is added, the toluene−β-cyclodextrin adduct precipitates very rapidly, making it the preferred isomer. Use of cyclohexadecenol leads to preferential formation of γ-cyclodextrin. Once the cyclodextrin complexes are isolated, the 'guest' can be removed to leave pure cyclodextrins.

Although many of the early reports only succeeded in isolating relatively small amounts of cyclodextrins, the common ones can now be produced at reasonable cost on an industrial scale, facilitating research on these series of molecules as well as allowing their use in numerous household, medical and industrial applications. The price of cyclodextrins varies with purity and size; the cheapest is usually β-cyclodextrin (due to ease of separation), which can cost as little as a few dollars per kilogram. Many derivatives such as permethylated and acetylated cyclodextrins, along with such derivatives as the hydroxypropyl ethers (Kleptose HPB) and 4-sulfonatobutyl ethers (Captisol), are also available.

The chemical structures of the α-, β- and γ-cyclodextrins (Figure 6.1) indicate that they might well form a macrocyclic structure with a 'hole' in the middle that would be suitable for binding guests. X-ray crystallographic measurements have been performed on many of these compounds. For example, α-cyclodextrin can be crystallised from water[5] and is shown to consist of six glucose units, each in the 'chair' conformation. The macrocycle is approximately doughnut-shaped, with a central cavity, and similar conformations exist for the macrocycles with seven[6], eight[7] and nine[8] sugar units that have been crystallised from water. For many of these species, differing crystal forms and hydrates have been isolated depending on crystallisation conditions; a detailed discussion of these is beyond the scope of this chapter. A three-dimensional representation of β-cyclodextrin is given in Figure 6.2. As can be seen from this structure, the molecule is a conical cylinder with a hydrophilic upper rim consisting of the secondary —OH groups in the 2,3-positions on the sugar unit and a lower ring consisting of primary —OH groups. The inside of the cavity appears to be relatively hydrophobic, consisting mainly of hydrogen atoms (from the C3 carbons) and the glycosidic oxygen bridges. The lone pairs of oxygen atoms are directed into the cavity, in some ways as in crown ethers. This dual nature is one of the factors that make cyclodextrins such effective complexing agents, and in fact cyclodextrins are rarely found with an empty cavity, containing either water, solvent or another guest.

There are some variations between the structures of the three common cyclodextrins. Intermolecular hydrogen-bonding occurs within these compounds between the 2-hydroxy group and the 3-hydroxy of the adjacent ring. In β-cyclodextrin this manifests itself as a complete belt of hydrogen bonds around the macrocycle, making it a rather rigid structure and probably explaining its relatively lower solubility. In the other two common cyclodextrins, steric effects prevent formation of a complete belt; in the hexamer one

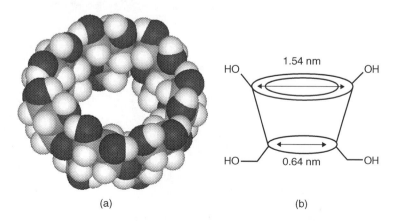

(a)   (b)

Figure 6.2   *(a) Three-dimensional model and (b) schematic of β-cyclodextrin*

of the glucose units is tilted relative to the others, whereas the octamer is quite flexible in solution – in both cases leading to increased solubility.[3]

The common cyclodextrins (α, β and γ) can be thought to be effectively cones, of height approximately 0.8 nm (Figure 6.2), all containing this central cavity. Quoted diameters in the literature differ due to different measuring techniques and what people consider to be the actual diameter (e.g. large rim, small rim); quotes range from 0.42–0.58 nm (α) up to 0.95 nm (γ-cyclodextrin). This is probably very dependent on the actual environment in which the cyclodextrin is located, and whether or not it contains a guest; we will show later that cyclodextrins are not necessarily fixed rigid species, but can in fact be quite flexible.

A simplistic view would be that the cavity size increases with the size of the cyclodextrin, but this is not the case. Larger cyclodextrins have been isolated and characterised but their shapes differ from the classic toroidal shape due to the large ring strain that would exist in a such structure. This means that the cyclodextrins can adopt 'pinched' structures, which can cause the cavity to collapse, restricting their use as complexing agents. The nine-unit (δ) cyclodextrin is boat-shaped,[8] as is the decamer;[9,10] the cyclodextrin with 14 sugar units[10,11] is bent into a saddle-like shape (Figure 6.3a) and the much larger 26-unit cyclodextrin[12] displays a structure containing two left-handed single helices connected in the form of a 'figure eight', as shown schematically in Figure 6.3b. Helical structures are known for amylase, which can exist as either double-helical A- and B-forms or alternatively as single-helical V-amylose.[12] A wide variety of other rings sizes have also been synthesised; these will be discussed below.

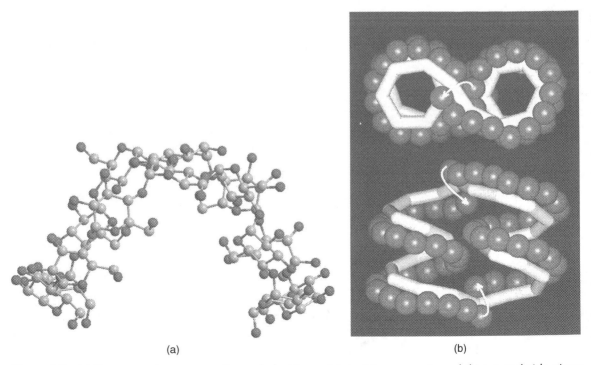

(a)   (b)

Figure 6.3   (a) X-ray crystal structures of a cyclodextrin containing 14 sugar units and (b) top and side views of a 26-unit cyclodextrin showing schematically the folding of the macrocycle into two left-handed single helices connected in the form of a 'figure eight'. (Reprinted from.[12] Copyright (1999) National Academy of Sciences, U.S.A)

6.2   Complex Formation by Cyclodextrins

Since their isolation and purification, the structure of the cyclodextrins has evoked much interest within the scientific world. The existence of molecules with a hydrophilic exterior and a relatively hydrophobic interior immediately led to investigations of their abilities to form complexes with other molecules in water or in the solid phase. Another major point of interest is that the cyclodextrins are all inherently chiral, since they are made up of chiral sugar molecules. In aqueous solution the interiors of the cyclodextrins are occupied by water molecules, as are those of many of the crystallised cyclodextrins.[5,6] However, the water–cyclodextrin interaction is relatively unfavourable and addition of a guest molecule leads to formation of a complex, with the 'driving force' being the removal of these unfavourable interactions. This process is aided if the guest is relatively nonpolar since unfavourable repulsive forces between the guest and water are also minimised. These guest–water interactions do not exist in the solid state and it is quite common that although a guest and cyclodextrin may associate very well in aqueous solution, when dried, the resultant complex is unstable and disassociates. Other forces that may also possibly contribute to complex formation are the relief of conformational strain of the cyclodextrin host, hydrophobic interactions, hydrogen-bonding and induction and dispersion forces. There is still debate in the literature about the importance of these various forces and a detailed discussion of these interactions has been published.[13]

The formation and stability of a wide variety of cyclodextrin complexes in solution have been extensively reviewed by other workers[13,14] and so only a general overview will be given here. The cyclodextrins possess an ability to form complexes with a wide range of guests. This is thought to be due in part to the fact that cyclodextrins are not necessarily rigid cones but instead possess a degree of flexibility which allows them to adapt their shape to accommodate a guest. NMR studies[15] have demonstrated that α-cyclodextrin is conformationally flexible in aqueous solution and when such a solution is frozen the macrocycle is 'trapped' in a range of conformations. This has been confirmed by molecular dynamics calculations,[16] which demonstrated that the crystal form is relatively static but the solvated form has much more mobility and displays a quite different hydrogen-bonding arrangement than in the solid phase. Molecular dynamics calculations have also showed that all three common cyclodextrins are conformationally mobile and that the symmetric structures often portrayed are actually only the time-averaged structures.[17]

There are many different reasons why various groups have studied complex formation with cyclodextrins. Complexation can be a way of getting a poorly water-soluble compound to dissolve in an aqueous environment. Also, incorporation into the cavity can modify the spectral properties, for example by reducing the quenching of fluorescence and enhancing lifetime. Cyclodextrin incorporation can also modify the chemical properties or protect the guest from undesirable reactions. Solid complexes can bestow improved properties; the volatility of the guest for instance can be greatly lowered, or release of the guest can occur upon utilising a desired stimulus such as heating. The interactions of solid or immobilised cyclodextrins have led to development of their use as solid-phase materials in various chromatographic applications. The inherent chirality of the cyclodextrins has led to their use in separating enantiomeric guests and chiral chromatography, as will be discussed later. We will also discuss the use of cyclodextrins as drug-delivery systems, due to their ability to form complexes with a wide range of therapeutic compounds.

Some of the earliest guests bound to cyclodextrins were in fact common solvents. As we have mentioned previously, such compounds as toluene can be used to modify the solubilities of certain cyclodextrins and cause their precipitation during their synthesis. Many other solvents have also been shown to bind within cyclodextrin cavities. We have already described some of the hydrates of these compounds, in which some of the water molecules are included in the cavity and others between the cyclodextrin molecules.

A wide variety of simple alkyl compounds have been shown to form complexes with cyclodextrins. Often these structures can be very complex, but there are some broad rules that are obeyed in most cases. First, the larger cyclodextrins can bind larger guests (until the cyclodextrins start to adopt nonconical shapes, as discussed earlier). Guests are thought to enter via the wider secondary alcohol rim as there are less steric problems there. The smaller α-cyclodextrin is thought to always enter this way, although it is believed that the larger cyclodextrins will permit entry from either rim if the guest is small enough. The guests tend to be accommodated with the hydrophobic portion of the molecule inside the cavity and any polar guest headgroups usually undergoing interactions with the cyclodextrin hydroxyl groups. In the case of longer guests, the polar groups will often protrude from the cavity and interact with a second cyclodextrin unit.

The relationship between the structures of the cyclodextrins and guests and the stability of their complexes in solution is a complex one. Connors[13] details many of the theories as to what are the major factors in determining binding constants. For example, in complexes between alcohols and cyclodextrins there are a number of effects. Binding constants in aqueous solution were determined for a range of linear, branched and cyclic alcohols with both α- and β-cyclodextrins.[18] It was found for simple linear alcohols the binding constant increased with chain length as far as the limits of the study (1-octanol). What is also of interest is the effect of the size of the cyclodextrin and that linear alcohols were bound more strongly by α-cyclodextrin, whereas bulkier branched and cyclic alcohols were bound more strongly by the β-cyclodextrin. This is probably because the bulkier branched systems are more capable of accommodation within the larger ring while the linear alcohols 'fit' better inside the smaller. This has been referred to as the 'Goldilocks' effect, where for a certain guest some cyclodextrins may be too small or too large but one will be just right.

Many other cyclodextrin–guest complexes are known. The most common seem to be simple 1 : 1 adducts in solution. Alkane sulfonates have been shown to associate with both α- and β-cyclodextrins,[19,20] with the degree of association increasing with alkyl chain length from 5 to 10 carbon atoms, above which a plateau is reached. Molecular modelling shows that only about six carbons can fit inside the hydrophobic cavity and some must protrude out of the opposing rim. A model for the structure of the complex is shown in Figure 6.4a. It was noted that association constants for the smaller compounds (pentyl or hexyl sulfonate) with α-cyclodextrin were more than double those with β-cyclodextrin, but as the chain length increased to octyl the association constants became equivalent and above this β-cyclodextrin was the superior host. It was also noted that variation of the headgroup of dodecyl surfactants had minimal effect on association constant apart from a strong increase in interaction between dodecyl benzene sulfonate and β-cyclodextrin.[20] Similar increases have been noted for alkane–cyclodextrin complexes,[21] where α-cyclodextrin forms 1 : 1 and 2 : 1 complexes with alkanes, with larger cyclodextrins just giving 1 : 1 complexes. Association constants tended

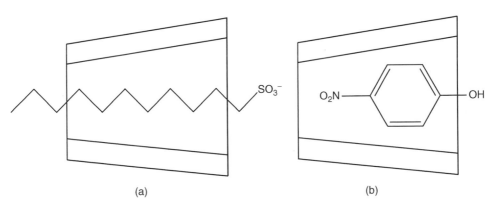

(a)   (b)

Figure 6.4   *Schematics of (a) the cyclodextrin–dodecyl sulphonate and (b) the cyclodextrin–p-nitrophenol complexes*

to be larger for β-cyclodextrin with short alkanes, while larger ones tended to have very similar association constants with all three cyclodextrins, which was unexpected since it was thought γ-cyclodextrin would be too large to suitably accept an alkane molecule. Other workers studied a range of alkyl compounds and concluded that entry into the cyclodextrin is via the 'wider' end of the cone and that there is interaction between polar headgroups and the secondary alcohols of the cyclodextrin.[22] Disubstituted species have also been studied;[23] α-cyclodextrin for example in this context was shown to form 1 : 1 complexes with α,ω-diols with chain lengths of three to seven carbon atoms, whereas for longer (8–10 carbons) diols, complexes containing one or two cyclodextrins per diol molecule could be observed.

As mentioned before, complex formation in solution does not guarantee that a solid-state complex can be isolated, but many stable adducts are known. Typical examples of solid-state cyclodextrin complexes with simple compounds include methanol and water with α-cyclodextrin, in which the methanol can be included either on one side of the cavity, bound to a glucose unit, or in the centre.[24] The structure of the cyclodextrin is also much closer to that of the six-fold symmetric cone shape than the twisted form found in the hydrate. Longer-chain alcohols such as propanol[25] also form a 1 : 1 adduct with α-cyclodextrin – as does propane sulfonate.[26] Both show a hexagonal conformation for the cyclodextrin, as shown for the propane sulfonate complex (Figure 6.5a), which has a centrally located guest with the primary hydroxyl groups hydrogen-bonding to it. The cyclodextrins stack together in a head-to-tail arrangement (Figure 6.5b), so the guests can be thought of as occupying a channel within the crystal.

When larger cyclodextrins are used, larger guests can be incorporated. 1-adamantane methanol for example can be incorporated into the β-cyclodextrin cavity, with the methanol unit protruding out of the primary alcohol end of the macrocycle.[27] In this structure the cyclodextrins adopt a symmetrical dimer structure, with the two secondary alcohol ends hydrogen-bonding to each other and the alcohol unit of the guest forming a strong hydrogen bond to the adjacent dimer's primary alcohol end. A similar structure was obtained for the adamantane carboxylic acid adduct.[28]

Disubstituted molecules have also been incorporated into these macrocycles. An X-ray study of the β-cyclodextrin–1,4-butanediol 1 : 1 complex,[27] for example, shows the diol is threaded through the cyclodextrin with one hydroxyl at each end of the cavity interacting with the cyclodextrin hydroxyls. Other polyfunctional alcohols such as ethylene glycol and glycerol have also been shown to form 1 : 1 complexes with β-cyclodextrin[30] and to have the guests bound within the cavity, but in a relatively disordered manner, with the alcohols spread between different sites within the cavity. Molecules with two different functional groups can also be incorporated.

The ready inclusion of aromatic species such as toluene and pyridine has led to widespread study of complex formation with many substituted aromatic rings, an overview of which will be given. The size of the cyclodextrins has a major effect on their binding of aromatic species. It would be impossible within this work to give a detailed description of the huge amount of work that has gone into determining the complexation behaviour of cyclodextrins with a wide variety of aromatic systems, so just a few examples will be given. Single-aromatic-ring compounds are easily bound by all the common cyclodextrins. Binding constants vary with structure, but molecules that are capable of association with the hydroxyl groups of the cyclodextrins tend to bind more strongly in aqueous solution. One type of system that has been widely studied is that of the nitrophenols.

NMR studies of the complexation of nitrophenols by α-cyclodextrin have shown that steric factors and hydrogen-bond formation both play a part in determining the association strength.[31] Simple *p*-nitrophenol binds to the cyclodextrin with a binding constant of 190 M^{-1}. The nitrophenol can be encapsulated with either its nitro group or phenol buried inside the cavity. When it is deprotonated to give the phenoxide ion, however, the binding constant increases by an order of magnitude and this is indicative of increased interaction between the phenoxide and the hydroxyl groups of the cyclodextrin, which can be thought of as being similar to the aliphatic compounds above in that the most hydrophilic group is located at

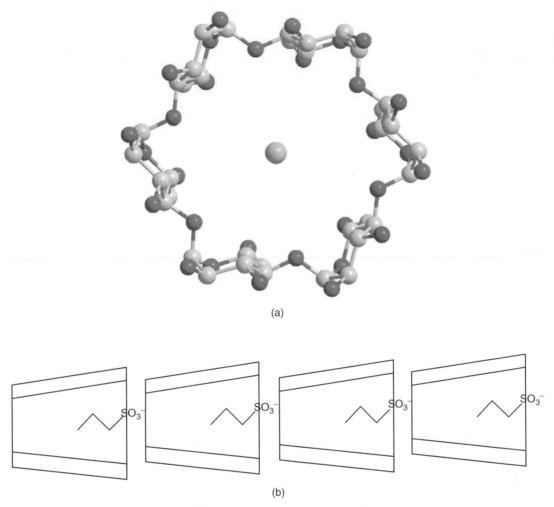

(a)

(b)

Figure 6.5   *(a) X-ray crystal structure of the α-cyclodextrin–propane sulfonate complex (only central sulfur atom can be resolved); (b) schematic of complex stacking*

the secondary hydroxyl ring. Indeed, this is confirmed by studies of the methyl-substituted compounds. Substituting the two positions *ortho* to the phenol lowers the binding constant by a factor of two for the phenoxide and three for the phenol. However, substituting a single *ortho* hydrogen with a methyl group prevents binding of the phenol completely and lowers that of the phenoxide by a factor of approximately 100, while methylating both positions *ortho* to the nitro group prevents binding completely.[31] The dramatic lowering of the binding constants by steric hindrance at the 'nitro end' of the guest proves it must orientate with the nitro group inside the cavity as shown in Figure 6.4b.

 The larger β-cyclodextrin also forms complexes with substituted phenols, but the binding forces are somewhat weaker; *p*-nitrophenol for example displays binding constants of 130 and 410 M^{-1} for the neutral and deprotonated forms.[32] Although the *ortho* and *meta* isomers display similar binding constants in their undissociated forms, there is a loss in binding strength upon loss of a proton rather than a marked enhancement, which is also found for phenol. This strong enhancement also has an effect on the actual dissociation constants of the phenols, with *p*-nitrophenol for example having a pKa of 6.90 in aqueous solution but 6.4 when complexed with β-cyclodextrin.[32,33] This is unusual and most other compounds show increases in pKa when complexed, indicating that the ionised form is less stable. The pKa of phenol increases from 9.82 to 10.75 when complexed with β-cyclodextrin, with similar trends being observed in the case of *ortho*- (7.05 to 7.21) and *meta*-nitrophenol (8.09 to 8.33).[33] Several explanations have been proposed for this effect, for example that the *p*-nitrophenyl ion is much more polarisable than the neutral compound and therefore increases induced dipole effects and binding strengths.[32] Anilines have also been shown to display pH-dependent binding,[32] with for example the binding constant for aniline being 56 M^{-1}, approximately 20 times that of the protonated form, with the same effect being observed for *p*-nitroaniline, which also binds strongly (300 M^{-1}) – three times as much as the protonated form. Other workers have performed similar studies[13,34] with a range of substituted benzoic acid compounds and have demonstrated that binding is stronger for the unionised substrates, indicating that the —COOH group is the primary binding site. The presence of electron-donating *para* substituents also favours the binding of the acid but hinders that of the anions.

Substituted *p*-nitro compounds have also been studied and shown to demonstrate interesting effects. Esters of *p*-nitrophenol are capable of being included inside α- and β-cyclodextrin and when these complexes are subjected to alkaline hydrolysis an interesting kinetic effect is observed.[35] Esters with short side chains such as *p*-nitrophenyl acetate hydrolyse more quickly when complexed; this is thought to be due to the cyclodextrin hydroxyls participating in the reaction.[36] The acetate or other ester group is transesterified onto a secondary hydroxyl group of one of the amylose units and then hydrolysed in a second reaction step. The hydrolysis of *p*-nitrophenyl acetate is accelerated when complexed with α-cyclodextrin, but as the length of the ester side chain increases to butyrate, this effect becomes less apparent, then it accelerates again for the hexanoate and even more for longer esters.[35] The acceleration is even more pronounced for β-cyclodextrin, dropping as chain length increases as far as hexanoate and then again accelerating as the side chain length increases. The explanation for this is that for the acetates, the *p*-nitrophenyl group is embedded in the cyclodextrin, allowing easy access of the hydroxyl groups to the ester unit. As the alkyl chain length increases there is more steric hindrance, lowering the catalytic effect. However, as the alkyl side chain becomes still longer it displaces the aromatic unit out of the cyclodextrin cavity, allowing access to the hydroxyl groups and thereby facilitating the catalytic effect once again.[35] The larger γ-cyclodextrin has much smaller acceleration effects,[36] probably due to less favoured complex formation.

Cyclodextrins can form complexes with compounds with two aromatic rings. One example is the complex of α-cyclodextrin with methyl orange[37] (Figure 6.6a), where a 2:1 complex is formed, with the dye molecules being incorporated into one of the macrocycles in a number of conformations. The cyclodextrins are in a head-to-tail arrangement with the dimethylamino group protruding from the 'narrow' end of the cyclodextrin (Figure 6.6b) and making contact with an adjacent cyclodextrin ring that is unoccupied. Similarly, sulfonato groups protrude from the other ends of the complex and make contact with an adjacent cyclodextrin. Schematics of the packing arrangement are shown in Figure 6.6b. Another compound that has been incorporated is biphenyl dicarboxylic acid (Figure 6.6c), where once again a 2:1 complex is formed with the α-cyclodextrins[38] in a structure in which each of the two cyclodextrins is occupied by one of the aromatic rings.

Complex formation can often be stronger with the larger cyclodextrins because the guest can penetrate deeper into the larger cavities. Again there is a Goldilocks effect, whereby larger cyclodextrins will often form complexes with guests that cannot be accommodated inside the smaller macrocycles. As has been shown, benzene rings can comfortably fit within the smaller cyclodextrins. Larger naphthalene rings can also fit within some of these macrocyclic structures, but the size of the cyclodextrin usually determines the binding conformation of the ring. Naphthalene has been shown to form 1:1 complexes in water[39] with all three of the major cyclodextrins, with binding constants of 630^{-1} M (β), 130^{-1} M (γ) and 83^{-1} M (α). In

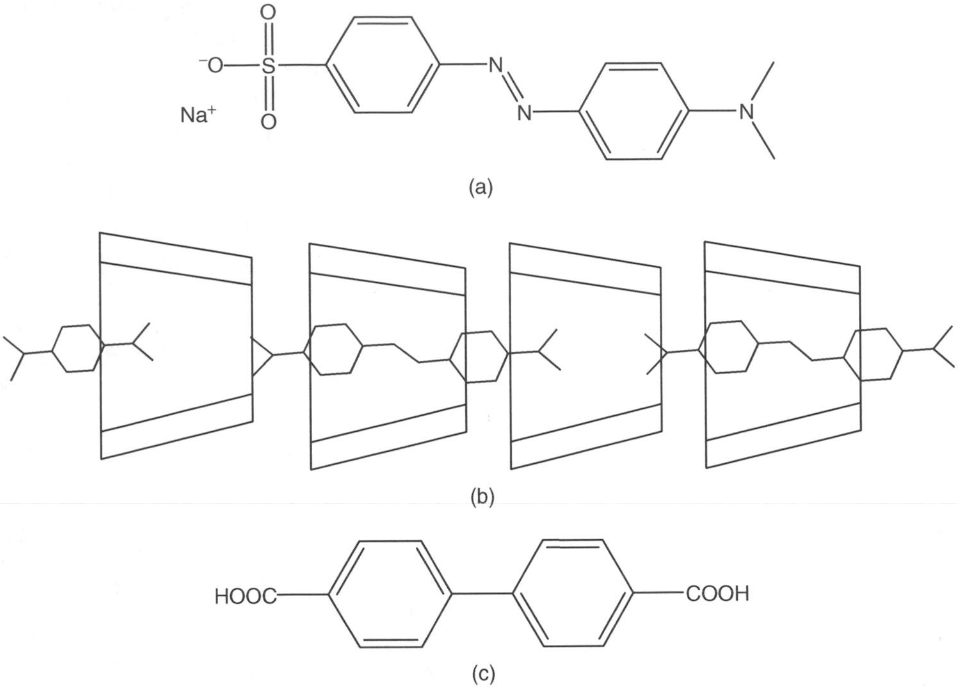

Figure 6.6 *(a) Methyl orange, (b) schematic of complex stacking and (c) biphenyl dicarboxylic acid*

addition, 1 : 2 naphthalene–α-cyclodextrin complexes have been observed. Methyl-substituted naphthalenes have also been studied and a number of models have been proposed for the structures of these complexes. It appears that α-cyclodextrin can only partially accommodate the aromatic molecule (i.e. the naphthalene is inserted via the long axis), whereas the γ- and β-cyclodextrins can bind the molecule completely.

Work on substituted naphthalenes has shown that different orientations can exist, with for example both the 1- and 2-naphthyl acetates having been complexed with cyclodextrins.[40] We have already seen with other aromatic esters that complexation with cyclodextrins can affect the rates of hydrolysis reactions. In the case of α-cyclodextrin, the 1-naphthylacetate was hydrolysed relatively faster than 2-naphthylacetate, indicating that the molecule is inserted with its long axis perpendicular to the plane of the hydroxyl groups. As is shown in Figure 6.7a, this arrangement brings the ester group in much closer contact with the cyclodextrin hydroxyls. In the case of the γ-cyclodextrin, the cavity is large enough to encapsulate the naphthalene in a parallel orientation (Figure 6.7b), thereby favouring hydrolysis of the 2-naphthyl ester. In the case of the β-cyclodextrin, no obvious preference is seen, perhaps indicating that there is a mixture of the two conformations. X-ray analysis has shown that β-cyclodextrin can encapsulate a naphthalene unit in the parallel orientation; 2,7-dihydroxynaphthalene for example adopts this conformation, with the two hydroxyl groups protruding from the rim of the host.[41] However, the cyclodextrin is distorted into an elliptical shape to accommodate this guest, again demonstrating the flexibility of these systems.

Systems containing both benzene and naphthalene rings, such as 1-anilino-8-naphthalene sulfonate (Figure 6.7c), have been complexed with the major cyclodextrins.[42] In aqueous solution the α-cyclodextrin was shown to complex weakly and bind the phenyl unit; the β-cyclodextrin bound the dye much more strongly but again via the phenyl unit, whereas the γ-cyclodextrin was found to bind either the phenyl or the

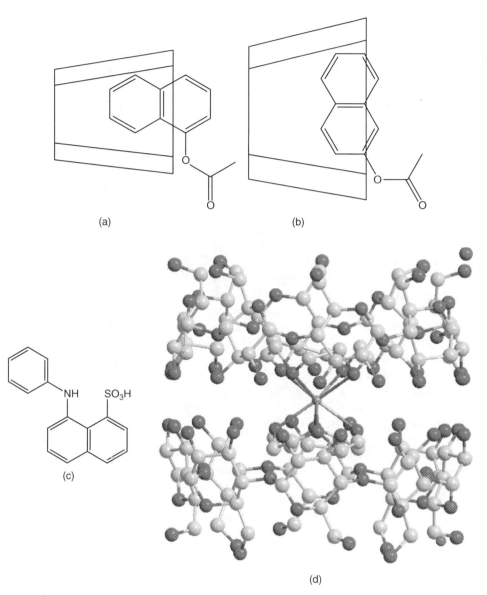

Figure 6.7 *Schematic of complex formation showing (a) α-cyclodextrin–1-naphthyl acetate, (b) γ-cyclodextrin–1-naphthyl acetate, (c) 1-anilino-8-naphthalene sulphonic acid, and (d) the crystal structure of the γ-cyclodextrin–12-crown-4/Li⁺ complex*

naphthyl unit, leading to two different complexes in solution. A series of substituted carboxylic acids with various cyclic groups such as adamantane (Figure 1.2b), which is essentially a spherical functional group, have also been studied[43] and shown to form complexes whose structures are dependent on both host–guest sizes and pH. Adamantane carboxylic acid forms complexes strongly with β-cyclodextrin, less strongly with γ-cyclodextrin and not at all with α-cyclodextrin – although this does complex smaller cyclic acids. Also of interest is that whereas β-cyclodextrin forms 1:1 complexes irrespective of pH, α-cyclodextrin forms 1 : 1 complexes at high pH, with the ionised forms of the acids forming 2 : 1 cyclodextrin:acid complexes at low pH.

Larger ring systems can also be included within cyclodextrins; 2-anthracene sulfonate for example forms a 2 : 2 complex with β-cyclodextrin[44] and when the larger γ-cyclodextrin is utilised as a host it can complex two anthracene molecules within its cavity. No complex formation was observed for α-cyclodextrin. When anthracene derivatives are irradiated they can react with each other and dimerise. The yields of this reaction are greatly enhanced when the anthracenes are irradiated whilst incorporated in cyclodextrin complexes compared to when they are in simple aqueous solution, indicating that the molecules are held in close proximity to each other. The larger pyrene system also forms 1 : 1 adducts with β-cyclodextrin[45] when the pyrene is only partially encapsulated by the macrocycle, and 1 : 2 complexes when a pyrene is almost completely encapsulated inside a head-to-head dimer. The larger γ-cyclodextrin usually forms a 1 : 1 complex in solution[45] but in the solid state has been shown to be capable of accommodating two pyrene units, as shown by luminescence measurements.[46] Even larger polycyclic aromatic hydrocarbons have been shown to form complexes with cyclodextrins, for example benzoperylene forms complexes in aqueous solution with β-cyclodextrin[47] and perylene has been shown to form a 1 : 2 complex in aqueous solution with γ-cyclodextrin,[48] as does the even larger coronene[49] in a water–methanol mixture.

Other large ring systems have been complexed with cyclodextrins, with for example cholesterol forming a complex with up to three β-cyclodextrin units.[50] This has led to one of the major commercial uses of cyclodextrins, namely the sequestration and removal of cholesterol, which will be discussed later. We have already discussed in detail the properties of crown ethers as hosts, but when 12-crown-4 is combined with γ-cyclodextrin[51] a complex is formed in which the crown is a guest inside the cyclodextrin ring. When lithium ions are also included, the crown can be thought of as host and guest. A complex crystal structure is formed (Figure 6.7d) in which the ion is complexed within the crown, which is complexed along with water within the cyclodextrin, an arrangement which has been called a 'Russian doll' system. Other workers have also isolated γ-cyclodextrin–[2,2,1]cryptand–Ca^{2+} complexes and proved their structure.[52]

We have already seen how long, thin guests can in effect be 'threaded' through the cyclodextrin rings, such as the alkane sulfonates mentioned earlier.[19,20] Studies performed on the compound shown in Figure 6.8a indicated that in aqueous solution the molecule threads itself through both α- and β-cyclodextrin[53] as shown, with the cyclodextrin encasing the alkyl chain spacer. We have already described many 2 : 1 cyclodextrin–guest complexes, usually as a result of encapsulation. Other workers investigated whether two or more cyclodextrin rings could be threaded onto a long guest, rather than simply having the guest located within a double cavity. Obvious candidates include long (possibly polymeric) hydrophobic compounds.

When polyethylene glycols[54] (Figure 6.8b) of molecular weight 400–10 000 were combined with α-cyclodextrin in water, the solutions became turbid and a crystalline precipitate was obtained, with the molecular weight of the polymer affecting precipitation, peaking at MW = 1000. Short-chain compounds such as ethylene glycol, diethylene glycol and triethylene glycol, did not appear to form complexes with α-cyclodextrin under these conditions. Heating the complexes reversed the complexation and caused resolution. No complexation was observed when polyethylene glycol which had its terminal hydroxyls substituted with bulky end groups was used; likewise no complexation occurred when polypropylene glycol (Figure 6.8c) was used. However, if the polymer–cyclodextrin complexes were treated to cap the end groups[55] and prevent the cyclodextrins 'slipping off', a stable complex was formed, which is an example of a polyrotoxane, a structure that the authors refer to as a 'molecular necklace'. What is interesting is that this material is much more soluble under alkaline conditions; studies indicate that this is because under neutral conditions the cyclodextrins are hydrogen-bonded together whereas under basic conditions they deprotonate and move away from each other. Many cyclodextrins have been included in rotaxane structures and these will be discussed in more detail in Chapter 9.

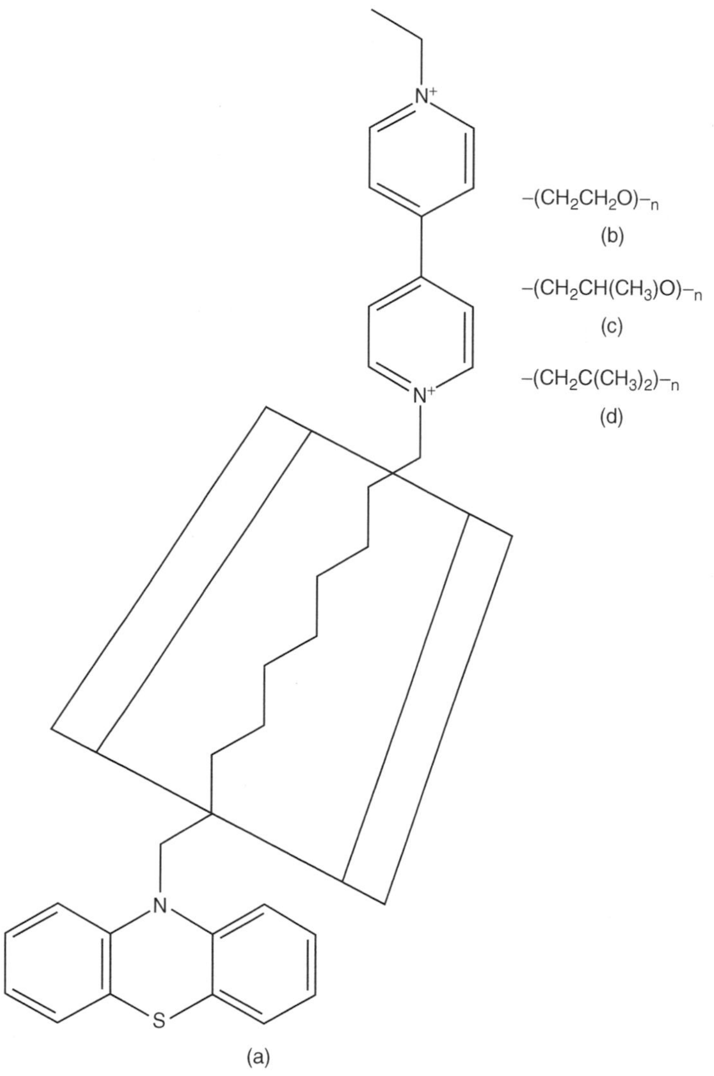

−(CH₂CH₂O)−ₙ

(b)

−(CH₂CH(CH₃)O)−ₙ

(c)

−(CH₂C(CH₃)₂)−ₙ

(d)

(a)

Figure 6.8 *(a) Structure of a cyclodextrin–dye 'threaded' complex, (b) polyethylene glycol, (c) polypropylene glycol and (d) polyisobutylene*

With β-cyclodextrin the situation is reversed, with no complex being formed with polyethylene glycol, although a complex was formed with polypropylene glycol.[56] Both the α- and the β-cyclodextrins seem to incorporate two polymer repeat units within each cyclodextrin unit. Complexes are also formed between polypropylene glycol and γ-cyclodextrin.[56] Since β-cyclodextrin does not form complexes with polyethylene glycol, we would not initially expect complexation to occur with the even larger γ-cyclodextrin, but studies showed that in fact a complex was formed, though it was quite different to that formed by α-cyclodextrin in that two polymer chains passed through the macrocycle.[57]

Other polymers that have been utilised are poly(oligoimino methylenes)[58] and the hydrophobic, water-insoluble polyisobutylene (Figure 6.8d).[59] This latter polymer formed complexes with both β- and γ-cyclodextrin but not the α-cyclodextrin molecule. Interestingly, the γ-cyclodextrin gave better yields with higher-molecular-weight polymers whereas the β-cyclodextrin gave better yields with lower-molecular-weight polymers. The γ-cyclodextrin was studied further and it was shown that three polymer repeat units were included in each macrocycle. The complexes were more stable than many of the earlier complexes, and although no de-threading occurred even upon boiling, the addition of the strong hydrogen-bonding compound urea caused the complex to disassociate, indicating that hydrogen-bonding between the cyclodextrins plays a major role in complex stabilisation. Polymethylvinylether also formed complexes with γ-cyclodextrin.[60] A variety of other polymers have also been included, such as oligoethylene, which forms complexes with α-cyclodextrin,[61] and polypropylene,[61] which forms complexes with β- and γ-cyclodextrin. Polyesters such as polycaprolactone (Figure 6.9a) and others have also been shown to form inclusion complexes with all three major cyclodextrins.[62,63] These polymers also gave the highest yields of complex for polymers with molecular weights of about 1000. Again the γ-cyclodextrin was shown to be capable of binding two polymer chains within the macrocycle, whereas the smaller cyclodextrins were found to bind just one; a schematic of the structure of the

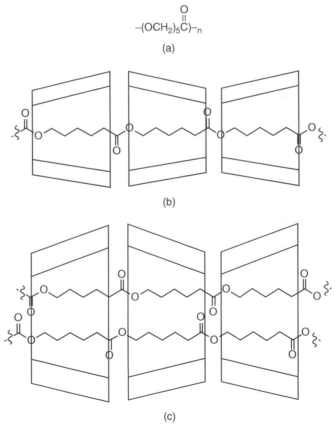

$$-(OCH_2)_5\overset{\displaystyle O}{\overset{\|}{C}})-_n$$

(a)

(b)

(c)

Figure 6.9   *(a) Polycaprolactone, and the structure of the polymer complexed by (b) α- and (c) γ-cyclodextrin*

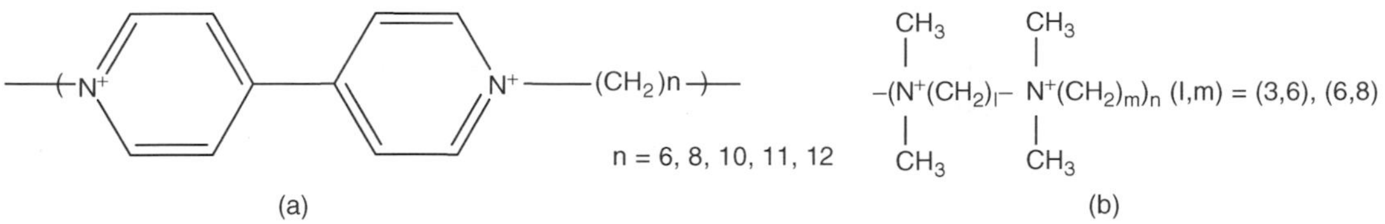

$-(N^+(CH_2)_n)-$ n = 6, 8, 10, 11, 12

(a)

$-(N^+(CH_2)_l- N^+(CH_2)_m)_n$ (l,m) = (3,6), (6,8)

with CH_3 substituents

(b)

Figure 6.10 *(a) Viologen containing polymers and (b) polyionenes*

polycaprolactone–cyclodextrin complexes is shown in Figure 6.9b. β-cyclodextrins were found to give much poorer yields of complex than either of the other two cyclodextrins, and the same was observed for other polyesters such as polyethylene adipate.

Ionic polymers based on viologens have also been studied;[64] the polymer with an octamethylene (but not hexamethylene) spacer (Figure 6.10a) formed complexes with α-cyclodextrin and the corresponding decamethylene-spaced polymer with β-cyclodextrin. Binding was found to be stronger to β-cyclodextrin. Another series of ionic polymers, the ionenes (Figure 6.10b), were also found to be bound by γ-cyclodextrin.[64] For the interested reader, these types of polyrotoxanes and their properties have been reviewed by Harada *et al.*[61]

The vast majority of compounds that have been complexed with cyclodextrins have been organic, but some inorganic compounds have been shown to form complexes with cyclodextrins. It has long been known that a simple test for starch is to add a solution of iodine and potassium iodide to the material in question; the presence of starch is confirmed by a deep blue-black colour. Amylose forms a helical structure and it is thought that linear triiodides and higher polyiodides can form inside the helices. It is the presence of these species that is responsible for the colouration. Since cyclodextrins are cyclic versions of amylose, various groups have studied their interactions with iodine.

X-ray crystallographic studies[65] have been made of crystals formed by an interfacial reaction between aqueous α-cyclodextrin and iodine in ether. The red-brown complex indicates that the cyclodextrins adopt a cage structure, with the iodide species in the centre of the ring, and it should be noted that there is also some water of hydration. From the bond distances observed and the packing structure, it is obvious that the iodine molecules are enclosed by the cyclodextrin and that there is no iodine–iodine interaction. However, when crystals are formed from aqueous solution of the cyclodextrin combined with iodine and iodide, a very different blue-black complex is formed.[66] Several different structures might be obtained depending on which particular iodide salt was used, although all display a channel-type structure, in which the cyclodextrin moieties are arranged together to in effect form a tubular structure containing polyiodide ions of various sizes. It is these polyiodides that give the complex its colour. The electrical conductivity of this complex has been shown to be much higher in the direction parallel to the channels than perpendicular to them. Both the β- and γ-cyclodextrins also give similar complexes, as does the much larger 26-membered macrocycle mentioned earlier in this chapter. When this macrocycle is complexed with iodine–iodide, it is seen to accommodate triiodide ions in its helices.[67] Very few of the larger cyclodextrins have been shown to form complexes, but the cyclic nonamer does form complexes with several cyclic ketones.[68] X-ray crystal structures have been obtained for the complex with cycloundecanone and show the cyclodextrin forms a head-to-head dimer enclosing four cycloundecanone molecules.

Other inorganic species have been complexed with cyclodextrins. When α-cyclodextrin is crystallised from water in which krypton is dissolved at high pressure, krypton molecules can be included in the

cavity.[69] Potassium acetate[70] can be co-crystallised with α-cyclodextrin to give a channel-type structure, although the potassium ions are located between cyclodextrin molecules not included in the cavity.
Organometallic species have also been complexed, with for example ferrocene forming a complex in which
it is encapsulated between two α-cyclodextrin molecules.[71]

Cyclodextrins have been used to solubilise C_{60} and the other fullerenes in water. Boiling solid C_{60}
in aqueous solutions of γ-cyclodextrin led to formation of mixtures of fullerene complexes with either
one or two cyclodextrin molecules.[72] Prolonged heating also led these species to form aggregates in
solution, which could be dispersed back into the original complexes by addition of more cyclodextrin.

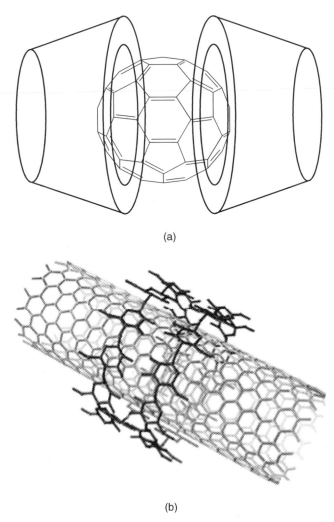

(a)

(b)

Figure 6.11   *(a) 2 : 1 C60–γ-cyclodextrin complex and (b) carbon nanotube/12-membered cyclodextrin.
(Reproduced by permission of the Royal Society of Chemistry [81])*

Under these conditions neither α- nor β-cyclodextrin nor C_{70} gave complex formation. Simple ball-milling of γ-cyclodextrin and C_{60} powders[73] also gave formation of a 2:1 complex (Figure 6.11a) with high solubility and better preservation of the fullerene structure. Dissolving β-cyclodextrin and C_{60} into an organic medium allowed formation of a 2:1 cyclodextrin–fullerene complex, which could be re-dispersed in water and was stable for a period of months, although less stable than the complex with γ-cyclodextrin.[74] There has also been a recent report of the synthesis of a 2:1 γ-cyclodextrin–C_{70} complex.[75]

6.3   Cyclodextrins of Other Sizes

Although the relative ease of isolation and purification means that the majority of cyclodextrin work has been performed on the hexamer, heptamer and octamer, other ring sizes have been synthesised, as already mentioned. The cyclic hexamer is the smallest cyclodextrin that can be synthesised using the standard enzymatic procedures, although chemical synthesis of the cyclic pentamer,[76] which was thought by many workers to be too strained to exist, was finally achieved by a step-wise procedure.

Large-ring cyclodextrins have also been investigated and we have already mentioned some of them. These are synthesised by enzymatic means as before, although the utilisation of different enzymes and conditions can lead to a mixture of larger ring sizes being produced. One major problem is that a mixture is generally formed, requiring expensive and time-consuming chromatographic methods to purify, and this has limited the study of these larger ring systems. The largest to be isolated that we have found in the literature is a cyclodextrin with 39 units, deduced by mass spectrometry and NMR.[77] No X-ray structure has been obtained for this material however, and the largest cyclodextrin for which a crystal structure has been obtained is the 26-unit cyclodextrin mentioned earlier.[12,67] Much larger macrocycles have been detected in mixtures. Potato D-enzyme, for example, catalyses the conversation of amylose to cyclodextrins.[78] Chromatographic and mass spectral investigations of these mixtures have shown the presence of cyclodextrins with rings sizes ranging from 17 up to a few hundred glucose units, but it is likely that separations of the individual higher-molecular-weight compounds from this mixture would be well nigh impossible.

Like their smaller brethren, the large cyclodextrins can accommodate a variety of guests. Their binding however is usually not as effective since the larger cyclodextrins are more flexible and do not contain such well-defined cavities. There are exceptions to this and as discussed before, δ-cyclodextrin forms crystalline complexes with cycloundecanone.[68] Also, workers have made studies of fullerene complexation with γ- and δ-cyclodextrins,[79] formed by the mechanochemical method mentioned earlier.[73] Whereas C_{60} formed stable complexes with both cyclodextrins,[79] the complex of δ-cyclodextrin with C_{70} was much more stable than the complex with C_{60}. C_{70} could also be solvated to a much greater extent in water by δ-cyclodextrin than by γ-cyclodextrin.[80] Besides fullerenes, other carbon-based systems have been investigated, such as carbon nanotubes. These systems can be quite intractable, being insoluble in most solvents and requiring often aggressive chemical modification. However, a large cyclodextrin (12 glucose units) allowed the solubilisation of 1.2 nm-diameter carbon nanotubes in water.[81] It is possible that the cyclodextrins are just adsorbed onto the nanotube surface, but the use of NMR and modelling studies indicate that instead two cyclodextrin units are threaded onto the nanotubes as shown in Figure 6.11b. The 26-member cyclodextrin has also been shown by X-ray crystallography to act as a host for iodine, as mentioned before, and also to form complexes with long-chain compounds[82] such as undecanoic acid and dodecanol. The cyclodextrin has a structure containing two helices, each of which hosts a guest.

6.4   Modification Reactions of Cyclodextrins

Chemical modification of cyclodextrins (usually of the hydroxyl groups) has been used to create a wide library of compounds. There are a variety of reasons for modifying the cyclodextrins, such as changing the binding properties, changing the solubility of the cyclodextrins or allowing them to be coupled together to make *bis* or multiple cyclodextrin structures. Much of the early work up to 1982 on chemical modification of cyclodextrins has been extensively reviewed[83] and a few examples will be given below.

The simplest type of reaction is to modify all the hydroxyl groups simultaneously. For example, combining the parent cyclodextrin, methyl iodide and a solid base such as KOH in dimethyl sulfoxide[84] leads to almost quantitative yields of permethylated cyclodextrins. A simplistic view would be that substituting the hydroxyl groups would reduce water solubility, although in fact the opposite is true; what is also unusual is that solubility in water is lower at elevated temperatures. Crystals can be grown by heating an aqueous solution of permethylated cyclodextrin, and when examined by X-ray crystallography[85] the structures of all three major cyclodextrins are shown to have a much reduced cavity volume. They can be thought of as bowl-shaped rather than doughnut-shaped. This method works well for a number of alkyl halides and has led to the production of a variety of totally or partially substituted cyclodextrins as commercial products, such as perhydroxypropyl cyclodextrin (Kleptose HPB) and cyclodextrins substituted with 4-sulfonatobutyl ethers (Captisol), both of which are used in drug-delivery applications. One versatile method for fully substituting cyclodextrins is to firstly synthesise the fully substituted allyl ether, which can be obtained in good yield using sodium hydride and allyl bromide, and then oxidising the allyl groups with osmium tetroxide to give the 2,3-propanediol-substituted compound.[86] This can easily be converted to the hydroxyethyl-substituted cyclodextrin, which can be oxidised to give cyclodextrins substituted with carboxymethyl groups.

Esterification reactions have also been used to modify the cyclodextrins, and in fact reactions such as acetylation with acetic anhydride/pyridine[87] were used as early as 1935 to help make the mixtures of cyclodextrins easier to separate and characterise. Other ester functions are utilised, with for example all the hydroxyl groups of cyclodextrin capable of being reacted to give benzoate esters,[88] which allows the selective removal of these ester groups from the primary hydroxyls to give cyclodextrins substituted at just the 2- and 3-positions (Figure 6.12). This of course then opens up the possibility of further modification at the 6-position to synthesise further asymmetrically substituted cyclodextrins. A wide range of other ester groups have also been used, as previously reviewed.[83] X-ray-structure analyses of the fully acetylated and propyl and butyl esters analogues of β-cyclodextrin[89] show that the macrocycle does not adopt the conical form but instead is elliptically distorted to form a nonplanar boat-shaped structure.

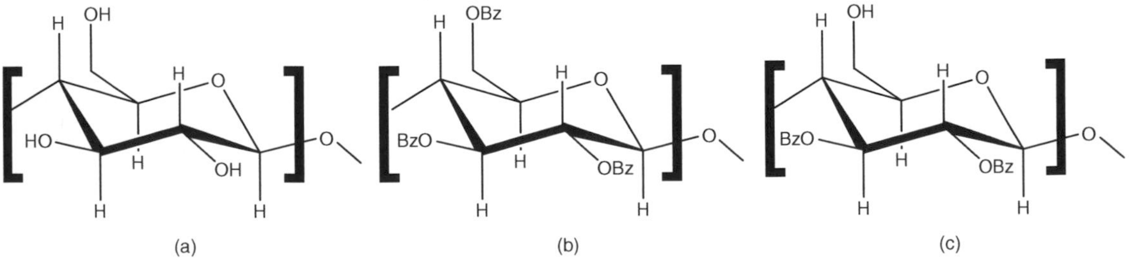

Figure 6.12   (a) Cyclodextrin, (b) cyclodextrin fully benzoated and (c) cyclodextrin benzoated at the 2- and 3-positions

A variety of attempts have been made to nitrate the hydroxyl groups of the cyclodextrins,[83] but as yet the presence of either a single partially nitrated compound or fully nitrated products appears not to have been proved.

Although many of the simple peralkylation and acylation reactions are of interest, a much more complex problem is that of selective reactions of the —OH groups. Simply reacting a cyclodextrin with, for example, six equivalents of a reagent could very easily lead to a product that is a complex mixture of cyclodextrins with differing degrees of substitution and different substitution patterns. Even the smallest common cyclodextrin has 18 hydroxyl groups, leading to a huge number of potential partial reaction products. For such applications as supramolecular chemistry, the products of any reaction must be pure and well characterised. This problem would appear at first to be insurmountable but there are several factors that make it somewhat easier. First, the three types of hydroxyl group are not identical in their behaviour. The hydroxyl groups at the 6-position are both the most basic and the most nucleophilic of these groups, with those at the 2-position actually being the most acidic and those at the 3-position being the hardest to access. Because of this, a number of reaction schemes have been devised, which allow us to substitute either all of one particular hydroxyl position or to synthesise cyclodextrins with just one or several positions substituted. This approach allows the substitution of different types of hydroxyl with differing functional groups.

6.4.1 Modification of the 6-hydroxyls

One reaction that is of great use is the formation of mesylates or tosylates of cyclodextrins. Since these are excellent leaving groups, they open up the possibility of a great number of potential substitutions. The primary hydroxyl groups of cyclodextrins appear to undergo preferential reaction with toluene sulfonyl chloride[90] and after chromatography the hexatosylated α-cyclodextrin can be obtained (Figure 6.13a). Simple displacement with azide gives the hexaazido compound (Figure 6.13b), which can then be catalytically hydrogenated to give the amino compound (Figure 6.13c), which can be converted to the acetamide (Figure 6.13d). A similar strategy is used by selectively activating the primary hydroxyl groups with a bulky triphenyl phosphonium salt and then carrying out nucleophilic substitution with azide or ammonia.[88] Reactions have also been reported with other amines to give for example the N-methyl and N-ethyl substituted derivatives.[83]

The primary hydroxyl group at the 6-position is also amenable to silylation. Reaction of α-cyclodextrin with t-butyl dimethylsilyl chloride[91] gives the compound shown in Figure 6.13e. The remaining secondary hydroxyls can then be converted to their methyl ethers or acetate esters. These derivatives can have the silyl-protecting group removed and the primary hydroxyls mesylated or tosylated, and the resulting compounds converted to the hexaazide, hexachloride, hexabromide or hexaiodide. The hexaglucopyranosyl derivative can also be synthesised from the tosylate.[91] Further work using the silyl-protecting agent has led to the synthesis of many partially esterified and etherified α-cyclodextrins along with similar products derived from the β- and γ-analogues.[92]

The periodo compounds (Figure 6.13f) can easily be synthesised by reaction of cyclodextrins with iodine and triphenyl phosphine.[93] The presence of the easily replaceable iodine atom makes the synthesis of a wide variety of derivatives possible, such as cyclodextrins that have been substituted with amino acid groups.[93] Other workers have used this method to synthesise cyclodextrins substituted in the 6-position with propargyl[94] groups (Figure 6.13g), which can then be used to link sugar groups to the cyclodextrin, alkyl thioether[95] groups and sugar units attached by a sulfide linker.[96] The attachment of further sugar units such as glucose, mannose and rhamnose via a thioether link has been used to synthesise what can be thought of as first-generation dendrimers based on cyclodextrin.[96]

For many applications, however, it may be necessary to react just one of the hydroxyl groups of a cyclodextrin. In this context, if the synthesis of a linear cyclodextrin polymer by a free-radical method

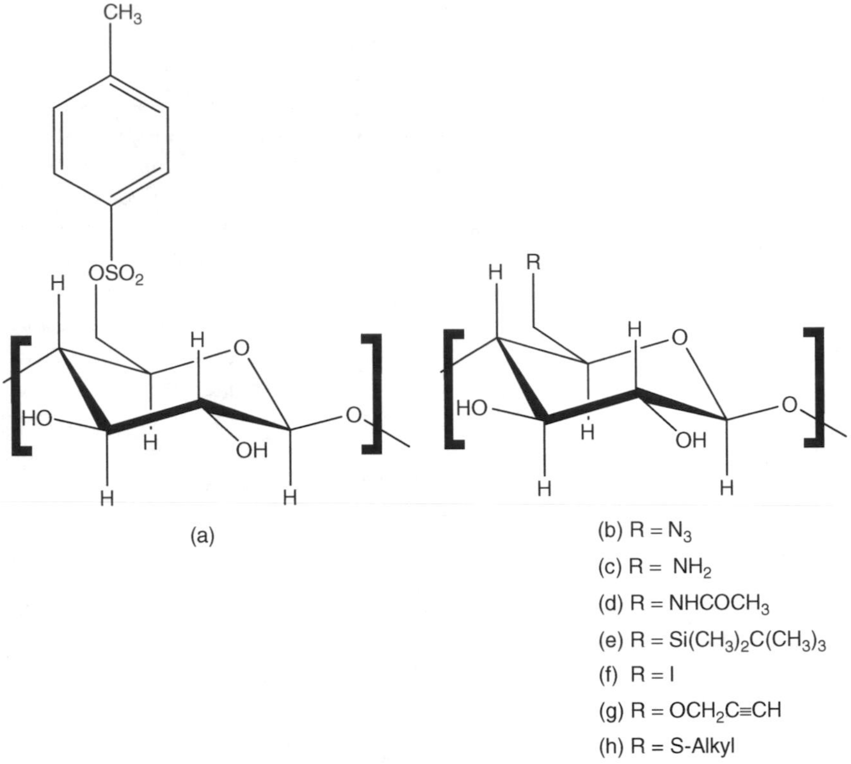

Figure 6.13   *Cyclodextrins substituted at the 6-position*

is required, there must only be one polymerisable group per cyclodextrin, or else crosslinking will occur. These types of reaction do exist, with for example the reaction of either α- or β-cyclodextrin with tosyl chloride under alkaline aqueous conditions[97] giving the mono-tosylated derivative (tosylation takes place at one of the primary hydroxyls). This of course immediately opens up a wide range of possible derivatives, as discussed earlier for the multi-tosylated derivatives. Most mono-substituted cyclodextrins originate from the tosyl derivatives by direct substitution or conversion of the tosyl to other reactive moieties such as amino groups.

Another possible reaction for modification of the 6-position is to oxidise the primary alcohol to an aldehyde. A mild oxidation[98] of all three of the major cyclodextrins with Dess–Martin periodinane led to the formation of the monoaldehydes in good yields and these compounds could then be further coupled to a variety of species. It proved possible to form di- and trialdehydes, but these reactions were much more difficult. One possibility is that the relatively bulky reagent forms a complex with the cyclodextrin and hinders further attack.

Making *bis*-substituted cyclodextrins is even more complex, since even if the reaction is selective for the 6-position, there are three different positional isomers, as shown in Figure 6.14 for β-cyclodextrin (α-cyclodextrin also has three isomers and γ-cyclodextrin four). This means either a selective synthetic scheme or effective purification is required. However, success has been achieved; β-cyclodextrin for example can be modified with t-butyl dimethylsilyl chloride to give a disubstituted product as the A,D positional isomer. This can then be peracetylated at all the other hydroxyls, the silyl groups removed and

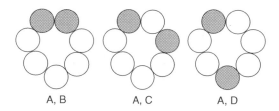

A, B   A, C   A, D

Figure 6.14   *Possible positional isomers of* bis-*substituted β-cyclodextrin*

two sugar units coupled to the cyclodextrin.[99] Finally, the acetyl esters can be hydrolysed back to the hydroxyls. Cyclodextrins can also be substituted by tosyl chloride and then separated by chromatographic methods to give all three disubstituted β-cyclodextrin isomers and all four γ-cyclodextrin isomers.[100] The tosyl groups can then be replaced by anthranilic acid groups and used as fluorescent sensors for various guests. It has also proved possible to replace just one of the tosyl groups with a dansyl group to obtain a *bis*-substituted cyclodextrin in which the two substituents differ.[101]

If a disulfonyl chloride such as mesitylene disulfonyl chloride is used, it is capable of reacting with two primary A,B hydroxyl groups of the same cyclodextrin[102] to give a capped cyclodextrin (Figure 6.15a). This compound can then be reacted with imidazole to break one of the capping bonds, giving a cyclodextrin substituted with two different groups in an A,B arrangement. In the case of a short mesityl spacer, it is only capable of bridging between adjacent sugar residues, but longer rigid spacers should be capable of bridging between A,C or A,D sugar units. This was shown to be possible by utilising other aromatic disulfonyl chlorides. Two very effective bridging agents are benzophenone disulfonyl chloride (Figure 6.15b), which was shown to form A,C bridges with β-cyclodextrin, and stilbene disulfonyl chloride (Figure 6.15c), which formed A,D bridges.[103] What is also of interest is that it proved possible to synthesise the doubly capped cyclodextrin using benzophenone sulfonyl chloride (Figure 6.16). Other workers have also used the similar azobenzene disulfonyl chlorides to form A,D-bridged β-cyclodextrin and A,E-bridged γ-cyclodextrin.[100] When a monotosylated β-cyclodextrin was reacted with a rigid disulfonyl, a number of trisubstituted species were obtained[104] (the structure of one isomer being shown in Figure 6.16b) and these could be converted into other trisubstituted cyclodextrins.

Although highly interesting species in themselves, these capped cyclodextrins also serve a dual purpose since the capping groups are still sulfonate esters and are therefore easily displaced to give other cyclodextrins with controlled substitution of the primary alcohol rim. The use of these agents and their conversion to a wide range of functional groups is a major reason that cyclodextrins can be incorporated within many of the intricate structures that will be detailed within this and other chapters.

6.4.2   Modification of the 2- or 3-positions

There appears to be much less literature on direct selective modification of these hydroxyls, probably because they are less nucleophilic secondary alcohols and more sterically hindered. Many of the cyclodextrins substituted in these positions have in fact been synthesised by multi-step procedures which involve protection of the C6-hydroxyls first, followed by reactions at the 2- and/or 3-positions and then deprotection of the C6 groups. There are however some protocols for direct modification at these positions.

Reaction of γ-cyclodextrin with aromatic sulfonyl imidazoles,[105] catalysed by molecular sieves, gives a product in which just one of the 2-hydroxyls has been converted to the sulfonyl ester. Similar results have been obtained using a triazole sulfonyl and sodium hydride[106] and the use of a base in the reaction is probably the major factor in determining selectivity since the 2-hydroxyl is the most acidic. A similar

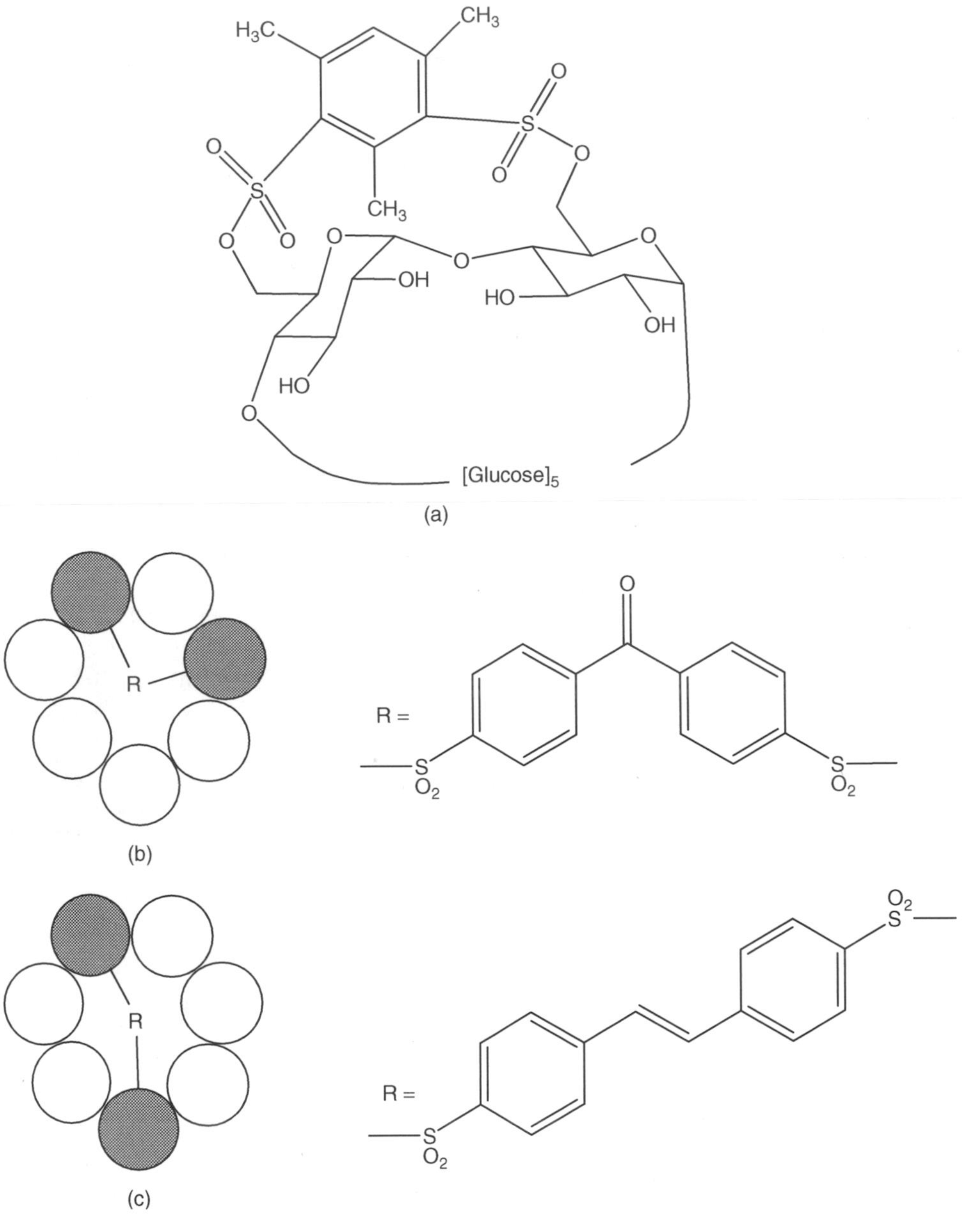

(a)

(b)

(c)

Figure 6.15 β-*cyclodextrins capped with (a) mesitylene, (b) benzophenone and (c) stilbene groups*

process was utilised with β-cyclodextrin, where NaOH and benzyl bromide in DMSO were used to synthesis the monobenzyl ether of the cyclodextrin.[107] This compound proved difficult to isolate so was converted into the fully methylated derivative with MeI/NaOH, then the benzyl ether was removed by hydrogenation to give the cyclodextrin with one 2-hydroxyl group. This group was then converted to a variety of functional groups. Reactions at the 3-position have also proved possible and for example when naphthalene sulfonyl

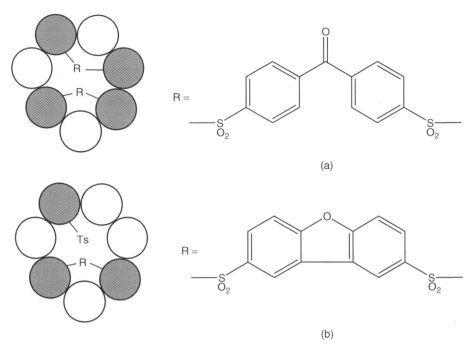

Figure 6.16   (a) β-cyclodextrins doubly capped with benzophenone and (b) triply substituted β-cyclodextrin

chloride is reacted with β-cyclodextrin in acetonitrile,[108] a mixture of the mono-substituted derivative along with the A,C and A,D disubstituted derivatives is obtained. The production of these derivatives rather than the expected 6-hydroxyl-substituted products has been explained as being due to the formation of inclusion complexes prior to reaction, which places the 3-hydroxy and sulfonyl chloride moieties in close proximity. Steric hindrance between bulky naphthalene units is thought to account for the minimal formation of the A,B isomer. Later work on this reaction showed it could also be used to selectively synthesise a trisubstituted β-cyclodextrin,[109] with the three substituents in the A,C,E positions. A *p*-nitrophenyl tosylate reagent has also been found to react with β-cyclodextrin to give the mono-tosylate in low yield (10%), where again the 3-hydroxyl is tosylated.[110]

A different approach involves converting the diol unit formed by the 2- and 3-hydroxyls into an epoxide. This reactive species can then be derivatised with a variety of reagents. We have already described the mono-, di- and tri-substitution of the secondary rim of cyclodextrins with various sulfonyl groups. These can be cyclised to the epoxides (Figure 6.17a) under alkaline conditions to give cyclodextrins containing one or more epoxy units.[111] The ring-opening reaction of a number of cyclodextrin epoxides to give cyclodextrins substituted at mainly the 2-position, but with minor substitution at 3-positions, have been described.[111] Groups attached to these rings include imidazole, azido, amino, thioether or iodo moieties. Either one, two or three substituents can usually be achieved, depending on the parent epoxide. The same group took γ-cyclodextrin and protected the 6-hydroxyls by silylation, converted it fully into the epoxide form (Figure 6.17b) and then reduced this with $LiAlH_4$ to give the 3-deoxy-cyclo-mannin compound.[112]

Substitution of the cyclodextrins affects their conformation and binding properties. Monosubstitution does not greatly affect the shape of the cyclodextrin macrocycle, but cyclodextrins substituted with suitable groups can display self-inclusion; that is, the substituent can bend around and be included in the cyclodextrin

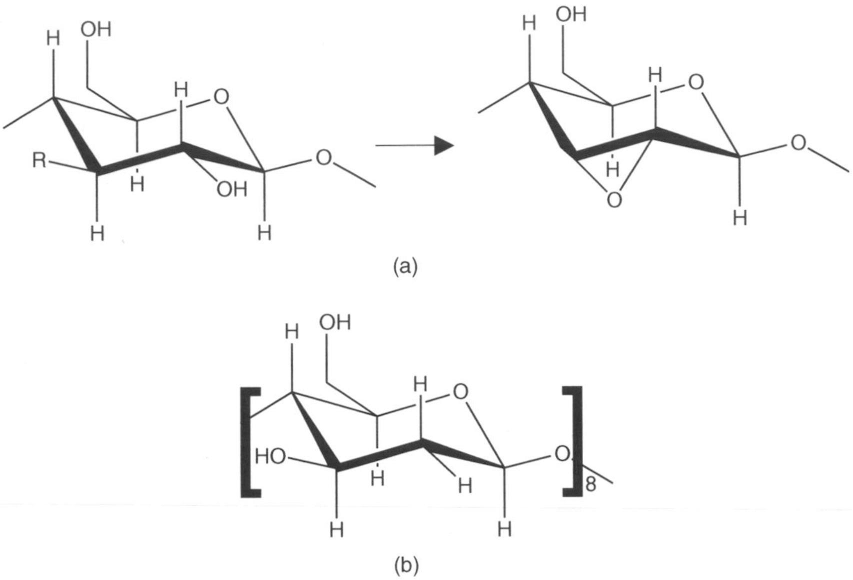

(a)

(b)

Figure 6.17   *(a) Conversion of cyclodextrin units to epoxides and (b) structure of per(3-deoxy)-γ-cyclomannin*

cavity or can form a complex with an adjacent cyclodextrin. For example, a β-cyclodextrin substituted at the primary hydroxyl with a phenylethylamino group[113] shows intermolecular inclusion where the aromatic group is incorporated into an adjacent cyclodextrin (Figure 6.18a). The choice of linker and terminal group, although not having much effect on the cyclodextrin structure itself, can have large effects on the way they pack together and variation of the substituent group can cause the cyclodextrins to pack in channel-type, cage-type or helical structures.[113] When however β-cyclodextrin is substituted at the primary hydroxyl with a imidazole with a t-butoxy-terminated side chain,[114] the substituent 'turns around' and the t-butyl group is included in the cavity of the cyclodextrin (Figure 6.18b) in what has been termed a 'sleeping swan' arrangement. This type of behaviour is of interest because it raises the possibility of attaching reporter groups to the cyclodextrin ring that self-include but can be displaced by suitable guests. If the reporter group is fluorescent for example, binding of guests will cause a concurrent change in spectra. Doubly substituted cyclodextrins also show some interesting behaviour; a β-cyclodextrin with two pyridylethylamino substituents[115] attached to vicinal 6-positions shows a structure in which one of the pyridyl units self-includes in the primary hydroxyl end while the other is included into the cavity of an adjacent cyclodextrin via the secondary end. These leads to the formation in the crystal of long polymeric-like structures stabilised by these host–guest interactions. Again, many of these types of compound have been reviewed in detail elsewhere.[83]

Persubstituted cyclodextrins have also been shown to form solid-state complexes similar to the cyclodextrins. There can be noticeable changes due to the differences both in steric hindrance and also hydrogen-bonding. For example, the fully permethylated ethers of the cyclodextrins have a much narrower primary end and cannot act as hydrogen-bond donors. The crystal structure of the 1 : 1 *p*-nitrophenol complex with permethylated α-cyclodextrin has the phenolic guest orientated in opposite direction to the complex with the parent cyclodextrin; that is, with the hydroxyl group pointing to the primary end and the nitro group protruding from the secondary end of the macrocycle.[116] Although other molecules, such as *p*-iodoaniline, adopt the same configuration in the permethylated cyclodextrin, they cannot insert themselves as deeply

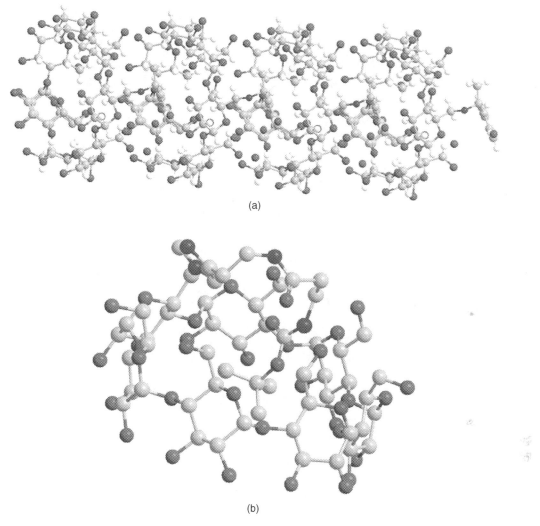

(a)

(b)

Figure 6.18   *(a) Intermolecular and (b) intramolecular formation of cyclodextrin self-inclusion complexes*

there as they can into the parent cyclodextrin.[117] However, long linear molecules such as ethyl laurate can also be included into permethylated β-cyclodextrin and notwithstanding the narrower primary rim, can still protrude through the primary opening and interact with the adjacent cyclodextrin,[118] although there is some distortion of the macrocyclic structure.

Cyclodextrins which have been 'capped' by bridging again usually retain their conical shape. A 6A,6D-diaminosubstituted methoxylated α-cyclodextrin[119] for example can be reacted with a 2,2′-bipyridyl diacid chloride under high dilution conditions to give the bisamide, where the bipyridyl unit bridges across the primary end of the cyclodextrin. The cyclodextrin retains its conical shape in the crystal (Figure 6.19a). Again, when a shorter 1,2-phenylene diphospite bridge is used, the cyclodextrin still retains its conical shape.[120] However, when the two amines are complexed with a single platinum atom[121] the short bridge causes the cyclodextrin to become distorted into a rectangular shape (Figure 6.19b).

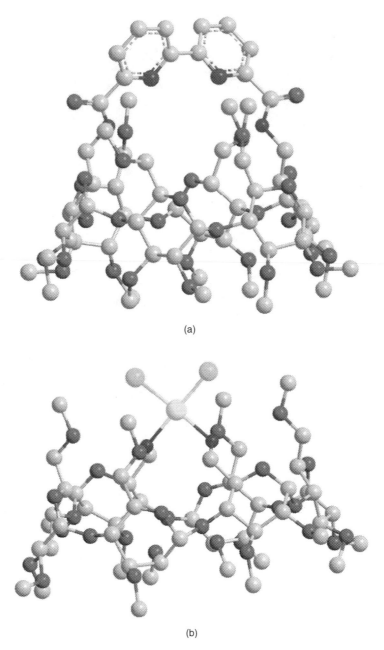

(a)

(b)

Figure 6.19   (a) Crystal structures of a bipyridyl-capped cyclodextrin and (b) a Pt complex-capped cyclodextrin

6.5   Selectivity of Cyclodextrins

A huge number of cyclodextrins have been synthesised, not only by varying the ring sizes but by chemically modifying the parent compounds. Between them these macrocycles have been shown to bind a wide variety of targets. However, binding is often not enough and for a great many applications selective binding is required. This section will discuss how the cyclodextrins can display selectivity, either in the solid state, for example by the preferential crystallisation of cyclodextrins with specific guests, in solution, or in heterogeneous interactions between solid or supported cyclodextrins and the liquid or vapour phase, as demonstrated by the use of cyclodextrins in chromatographic separations.

There are many examples of geometric selectivity; for example, α-cyclodextrin forms a complex with *p*-nitrophenol[122] in which the aromatic guest is deeply embedded in the cavity, while *m*-nitrophenol[123] is only partially embedded and *o*-nitrophenol forms no crystalline inclusion complex at all with. In an early chromatographic separation experiment, both α- and β-cyclodextrin and their permethylated derivatives were adsorbed onto Chromosorb W and utilised as packing material for a gas chromatography column.[124] A wide variety of volatile species can be separated by the cyclodextrin columns, with both the size and polarity of the analyte affecting its retention, indicating that both inclusion into and hydrogen-bonding with the cyclodextrins contribute to guest-binding. With the methylated cyclodextrins the hydrogen-bonding is reduced and the geometry of the analyte has a much greater effect, with branched and tertiary alcohols being complexed much less efficiently than their linear analogues. Similar effects are found with the xylenes, with the α-cyclodextrin binding most strongly to the *p*-isomer, followed by the *m*-isomer and finally the *o*-isomer. The differences are less with β-cyclodextrin and in fact the order also changes, with *o*-xylene being retained the most and *m*-xylene the least. Similarly, 1,2,4-trimethylbenzene was retained more strongly than the 1,2,3 or 1,3,5, isomers. A much wider range of chemicals have been separated using cyclodextrin-modified columns, many of which will be found in the Bibliography.

One of the most intensely studied areas of cyclodextrin selectivity comes from the fact that all cyclodextrins are inherently chiral, due to the presence of the chiral D-glucose units. This means that complexation of a cyclodextrin with a chiral guest will lead essentially to the formation of diastereoisomers with potentially different physical properties. Again, selectivity can be seen in the solid state and in separation techniques.

Attempts to utilise co-crystallisation of racemic compounds with cyclodextrins to achieve enantiomeric separation have met with varying degrees of success. Some compounds such as racemic 1-phenylethanol (Figure 6.20a) simply crystallise with α-cyclodextrin with both isomers in a 50:50 ratio; that is, without any selectivity. X-ray analysis of crystals made with pure enantiomers shows that both give a very similar crystal structure, with some disorder of the guest position.[125] However, when the drug compound fenoprofen (Figure 6.20b) is co-crystallised with β-cyclodextrin, a 3:1 excess of the S-isomer crystallises out.[126] X-ray studies[127] show that the inclusion compounds of the cyclodextrin with the two isomers have quite different crystal structures. The cyclodextrins form a dimer linked at the secondary faces and each dimer contains two guests *of the same isomer*; that is, either R,R or S,S, but not R,S. For the S isomer the two guests adopt a head-to-tail configuration, with the carboxylic acid groups of the guest forming hydrogen bonds to the cyclodextrin, whereas the R isomers adopt a head-to-head arrangement and hydrogen-bond with included water molecules. Methylated cyclodextrins can show similar behaviour,[128] permethylated α-cyclodextrin forming complexes with both enantiomers of mandelic acid (Figure 6.20c), although the R isomer is much more deeply embedded than the S isomer and forms a channel-type structure, whereas the S enantiomeric forms a layered structure. Repeated crystallisations of 1-(p-bromophenyl) ethanol (Figure 6.20d) as a host–guest complex with permethylated β-cyclodextrin[129] also allows enantiomeric separation. X-ray studies have shown the R isomer to be more deeply included in the cyclodextrin cavity than the S isomer. The replacement of the bromo substituent with other halides greatly affects the separation behaviour.[130]

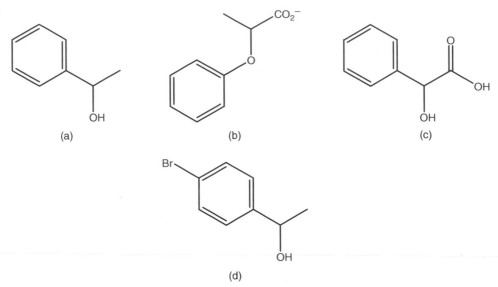

Figure 6.20   *Structures of (a) 1-phenylethanol, (b) fenoprofen, (c) mandelic acid and (d) 1-(4-bromophenyl)-ethanol*

Besides these solid-state studies, there much work has been done on the separation of enantiomers by utilisation of cyclodextrins in chromatographic techniques. A wide variety of chiral GC and HPLC columns are commercially available which utilise both the major cyclodextrins and various derivatives such as methylated cyclodextrins as a chiral stationary phase. A vast amount of work has already been performed on this subject, such as for example the commercial database Chirbase GC, accessed in mid-2009, which claimed experimental details for over 24 000 protocols and over 8000 compounds. We will only provide a very brief overview of this field here.

Gas chromatography is a method where mixtures of gases or other volatile compounds are passed with an inert carrier gas such as helium (known as the mobile phase) down a long column which contains a solid phase such as silica, polymer beads or some other solid material. Alternatively, the column can be a very thin capillary, where its walls are the stationary phase. Gas or vapour molecules absorb and desorb repeatedly from the stationary phase. This means the time or retention on the column is dependent on the relative strengths of interaction between the stationary phase and the components of the mobile phase. If two enantiomeric compounds are passed down a column they will undergo identical interactions and elute from the column simultaneously unless the column materials themselves have been modified in some way to be chiral. Materials for a chiral column must be stable, nonvolatile and easily available in a pure enantiomeric form. The relative ease of synthesis and stability of the cyclodextrins along with their inherent chirality makes them very suitable for this purpose. Schurig reviews much of the history and current applications of chiral GC.[131]

One of the earliest cyclodextrin-containing GC columns utilised permethylated β-cyclodextrin dissolved in polysiloxane and absorbed onto the walls of a glass capillary column to chirally resolve volatile compounds such as 2,5-dimethyl tetrahydrofuran or α-pinene.[132] What is interesting is that the order of elution of isomers does not just depend on the chiral match between eluent and cyclodextrin. Cyclodextrins, being natural products, are only available in a single enantioform, but the order of elution of the R and S enantiomers is dependent on the size of the cyclodextrin and using a β-cyclodextrin column can lead to elution of the enantiomers in the opposite order to a α-cyclodextrin column.[133] Similarly, the same size of cyclodextrin modified with hydrophobic pentyl chains can lead to opposite elution when a cyclodextrin with permethylated 2-hydroxyethyl substituents is used.[133] It appears that a combination of factors affect their interaction, chiral recognitions and separation behaviour, such as the polarity, steric bulk, orientation and degree of substitution of the cyclodextrins and the eluents.

High-performance liquid chromatography is a similar method to GC except that the mobile phase is a solvent or solvent mixture and the stationary phase a packed powder column. In Chapter 3 we discussed how chiral crown ethers adsorbed onto solid phases could be used to separate different enantiomers of amino acid derivatives. Cyclodextrins have also been utilised in both HPLC and thin-layer chromatography application.[134] This is a versatile technique and can cope with a number of sample formats. Racemic drugs for example can be enantiomerically separated and analysed even from serum samples[135] using a β-cyclodextrin chemically bound to silica column, which is commercially available.[136] Normal phase applications in which the analytes are in organic solution can also be performed on cyclodextrin columns. Modification of β-cyclodextrin with a variety of substituents, including the addition of chiral 1-naphthylethyl groups to generate additional stereocentres, can be attained and these materials when immobilised on silica provid suitable solid phases for HPLC.[137] It is interesting that the modified cyclodextrins show better resolution for some analytes under normal phase conditions than native cyclodextrins, indicating that inclusion and aromatic interactions can both contribute to separation. In the case of aromatic analytes, the elution order is determined by the configuration of the naphthyl group.

6.6 Multiple Cyclodextrin Systems

6.6.1 *Bis* and multi-cyclodextrins

Once specific functionalisation of cyclodextrins was developed, it then became possible to link together cyclodextrins in a controlled manner. Since many cyclodextrins form 2 : 1 complexes with guests, a number of groups began to investigate cyclodextrin dimers, reasoning that if the two host macrocycles are covalently linked, there should be a cooperative effect, leading to much stronger binding and to the structure being better defined, since the covalent linker will decrease the freedom of movement of the cyclodextrins and mean the guest is grasped at both ends, not just one. The appearance of reproducible synthetic schemes to produce mono- and di-substituted cyclodextrins allowed this idea to become reality.

There are several different arrangements in which cyclodextrins can be attached to form dimeric systems. They can be linked at the primary or the secondary rim and by one or more linkers. Linking chains can be rigid or flexible, long or short – and where there are two or more linking chains, they can be the same or different. The manner of these linkages and the way the cyclodextrins can be held at various distances apart and at different angles has major effects on the binding properties of the systems. Much of the early work on cyclodextrin dimers and their application as specific binding agents can be found in the Bibliography and in the review by Breslow *et al.*[138]

Some of the earliest work published involved using amines to bridge between two cyclodextrins. In this reaction,[139] β-cyclodextrin was capped with a *bis*-sulfonyl chloride (Figure 6.21a) in a similar manner to the capped cyclodextrins previously discussed. This was then reacted with excess ethylene diamine to give the tetra-amine shown in Figure 6.21b, which was reacted with a further equivalent of the *bis*-sulfonate to give the cyclodextrin dimer in Figure 6.21c. This compound was shown to complex methyl orange with a binding constant six times that of the cyclodextrin tetra-amine. An alternative synthesis of *bis*-cyclodextrins involved the synthesis of both α- and β-cyclodextrins bearing a single mercapto (—SH) group, which could be oxidised to give the dimer linked by a single disulfide (S—S) bond.[140] The *bis*-β-cyclodextrin

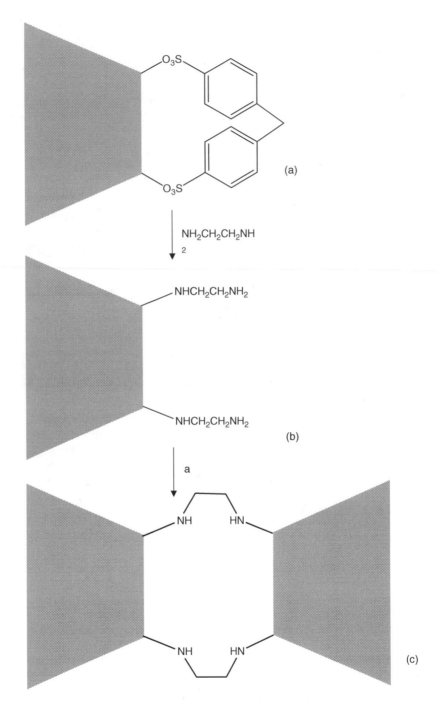

Figure 6.21   *Synthesis of a* bis-*cyclodextrin*

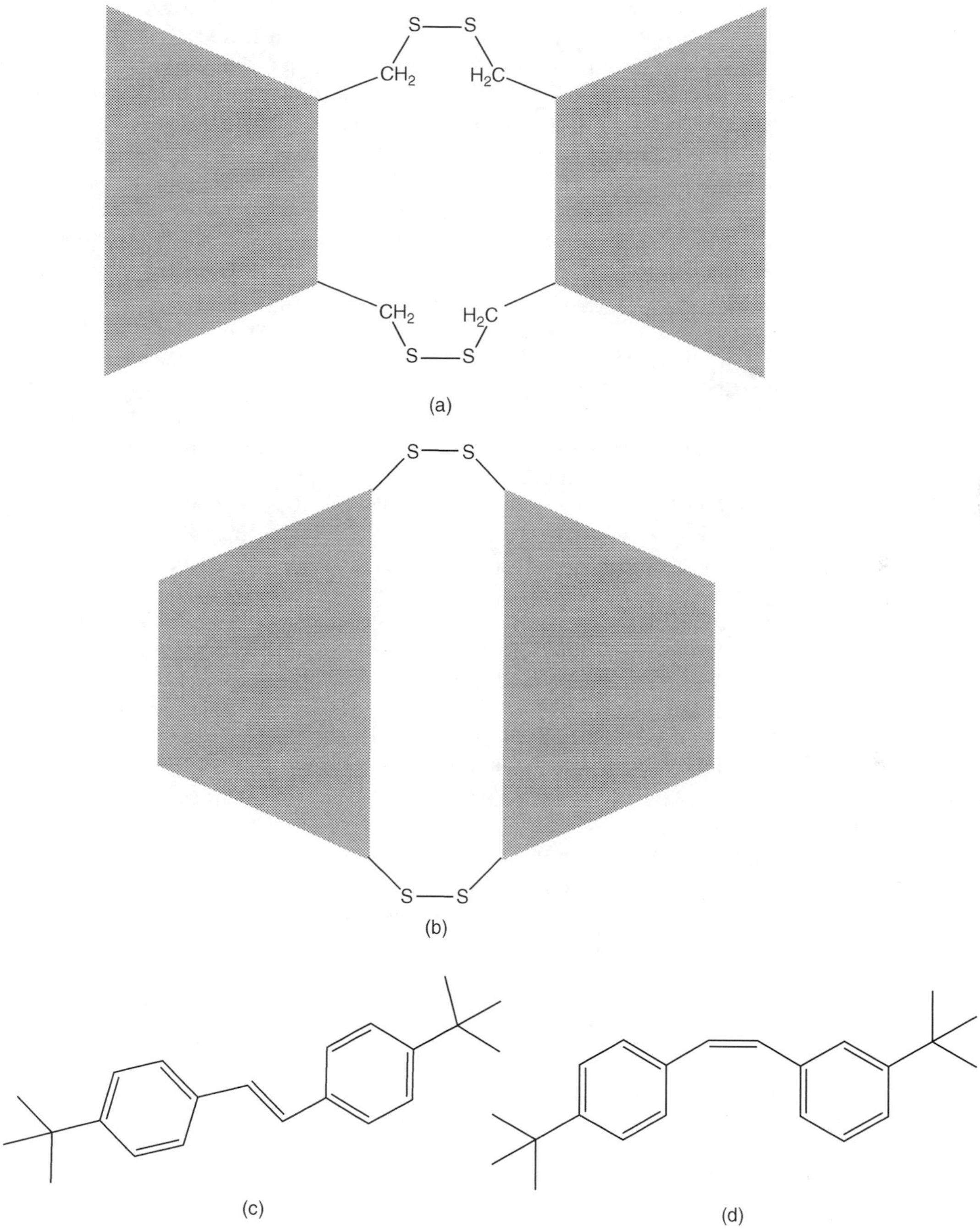

Figure 6.22 *Structures of bis-cyclodextrins and their guests*

(Figure 6.22a) was shown to bind ethyl orange 224 times more strongly than the parent cyclodextrin. This demonstrates the considerable effect of collaborative binding by two linked cyclodextrin units. A double γ-cyclodextrin also displayed extremely high affinity for methyl and ethyl orange,[141] forming 2 : 1 complexes with association constants several hundred times higher than that of the parent cyclodextrin.

So far the dimers discussed have been linked via the primary rim, but a rigid terephthaloyl spacer has been used to couple two cyclodextrins via the secondary rim,[142] as has a disulfide link (Figure 6.22b). Within the same work *bis*-cyclodextrins[142] bridged between the primary rims by thioanthone links were studied. A variety of aromatic guests capped with t-butyl groups were examined and in many cases extremely high binding constants were obtained. The size, shape and flexibility of the guests affected their binding, with longer guests tending to bind more strongly to the cyclodextrin with the longer ester link and the smaller guests binding more strongly to the system shown in Figure 6.22a. Similarly, the *trans* isomer of 4,4'-t-butylstilbene (Figure 6.22c) bound much more strongly to the compound in Figure 6.22a than did the *cis* isomer (Figure 6.22d). However, the cyclodextrin linked at the secondary rim by a short disulfide link showed little enhancement and it is possible that the short linkage renders the binding sterically unfavourable.

One problem about a host with single links is that the cyclodextrin complexes may not be held in an optimum configuration for binding and in fact could easily adopt configurations in which it is impossible for a guest to simultaneously bind to two rings. Multiple links would preorganise the host more and possibly enhance binding, in the same way that cryptands usually bind more strongly to their guests than crown ethers. We have already described how functional groups can be introduced onto neighbouring glucose units of the same cyclodextrin via A,B substitution. Through these groups it should prove possible to synthesise doubly linked cyclodextrins. One of the first compounds of this type was synthesised by taking β-cyclodextrin with iodo groups in the A,B positions, substituting these for imidazole groups and then reacting the *bis*-imidazole units with another di-iodo cyclodextrin[142] to give the structure shown in Figure 6.23a. Two isomers were obtained from this reaction, which were called the occlusive (Figure 6.23b, where the A position of one ring is bound to the B of the other) and the aversive (Figure 6.23c, where A binds to A and B to B). The occlusive was found to bind to double-ended substrates up to a hundred times more strongly than the aversive, although the binding constants were not as strong as for the singly bridged *bis*-cyclodextrins.[142]

This *bis*-cyclodextrins mentioned above all have either just one linker or two equivalent linkers. However, asymmetric compounds have also been synthesised. The di-iodo β-cyclodextrins mentioned above for example can be reacted first with naphthalene-2,6-dithiol to give a singly linked *bis*-cyclodextrin;[143] the remaining iodo groups can be converted to thiols and then oxidised to form a second disulfide bridge (Figure 6.24a). As before, occlusive and aversive isomers can be obtained and separated by chromatography. As shown in Figure 6.24b,c, the two isomers are not face to face. The aversive isomer has the two cyclodextrin rings in a conformation in which they cannot bind cooperatively to the same substrate and displays binding constants typical of monomeric cyclodextrins. However, the occlusive isomer, because of the two different linker lengths, has a form which the authors referred to as a 'clam shell'. This allows it to bind suitable substrates in which complexation takes place with both cyclodextrin rings, such as the flexible *bis*-(t-butylphenyl)ester shown in Figure 6.25a, with a binding constant of 10^{10} M^{-1} (comparable to a strong antibody–antigen interaction). In the case of rigid substrates, not only is the size of the substrate important but so is its shape and for example the compounds shown in Figure 6.25b,c are similar in size but the bent compound (Figure 6.25b) displays a binding constant at least 4000 times higher than its linear analogue.

Some *bis*-cyclodextrins do not necessarily display strong host–guest interactions. This can often be because of interactions between the two macrocycles or with the linker groups. Cyclodextrins modified with a single amino group on the secondary ring were linked together by reaction with an activated

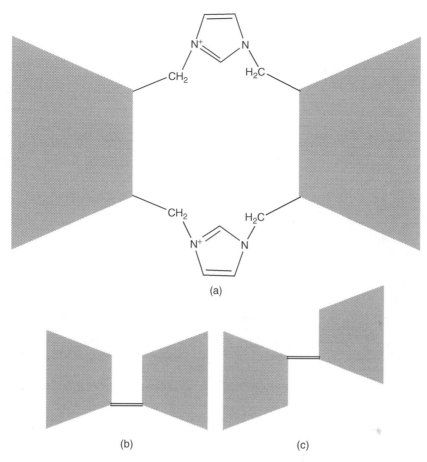

Figure 6.23   *(a) The structure of doubly attached* bis-*cyclodextrins, and the (b) occlusive and (c) aversive isomers*

dicarboxylic acid to give the structures shown in Figure 6.26. NMR studies of the β-cyclodextrin dimers in solution showed that when an eight-carbon spacer was used, the spacer group was actually partially incorporated into one of thee cyclodextrin cavities.[144] The nature of this synthesis meant that it was possible to make an unsymmetrical α-β-cyclodextrin dimer. In the case of this species, the linker was included in the β-cavity, whereas from solution studies on monomeric cyclodextrins, inclusion of alkyl chains into the α-cavity would be expected. In other work by this group, rigid bipyridyl spacers were also utilised (Figure 6.26b).[145] All these dimeric species bound a variety of aromatic guests much more strongly than single cyclodextrins. In the case of the aliphatic spacers, the short two-carbon-chain spacer led to stronger complexation than the eight-carbon analogue, indicating that closeness of the cyclodextrins and/or lack of self-inclusion aids complexation.[145] The symmetric β-β dimer also showed stronger binding than the α-β cyclodextrin dimer. High affinity was noted for the bipyridine-bridged dimer for porphyrins and no self-inclusion was seen to occur. Other workers have noted self-inclusion of an azobenzene spacer into permethylated *bis*-cyclodextrins.[146]

Formation of *bis*-cyclodextrins by linking together at the secondary rims via disulfide binding at either the 2- or the 3-position of cyclodextrins has been reported.[147] Other workers have used a variety of nitrogen- and sulfur-containing chains to link together β-cyclodextrins and showed that they bound aromatic guests

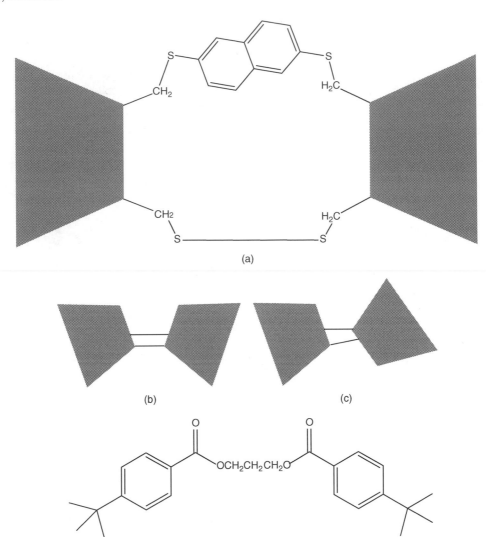

Figure 6.24   *(a) The structure of asymmetric* bis-*cyclodextrins, and the (b) occlusive and (c) aversive isomers*

with association constants up to 20 times those of β-cyclodextrin.[148] Selenium-containing chains[149] have also been used, with -Se-(CH$_2$)$_3$-Se- and 1,2-diselenobenzene units being used to bridge between two β-cyclodextrin units via the primary or in one case the secondary rims. In recognition studies with aromatic dyes, the dimers displayed higher binding constants than the native monomer. Interestingly, the binding properties were enhanced by complexation of the selenium atoms with Pt(IV). In one case this allowed the synthesis of a species bearing four cyclodextrin moieties. Later work by the same group utilised linkers containing amino and seleno groups of different lengths and formed *mono-* and *bis-*Pt complexes of these species.[150] The binding constants for various dyes were evaluated and found to follow some trends, with for example the binding constants for 1-anilino-naphthalene-6-sulfonate decreasing as the length of the linking group increased. A system utilising Te-Te bridges has also been synthesised and shown to behave as a mimic for the enzyme glutathione peroxidase.[151]

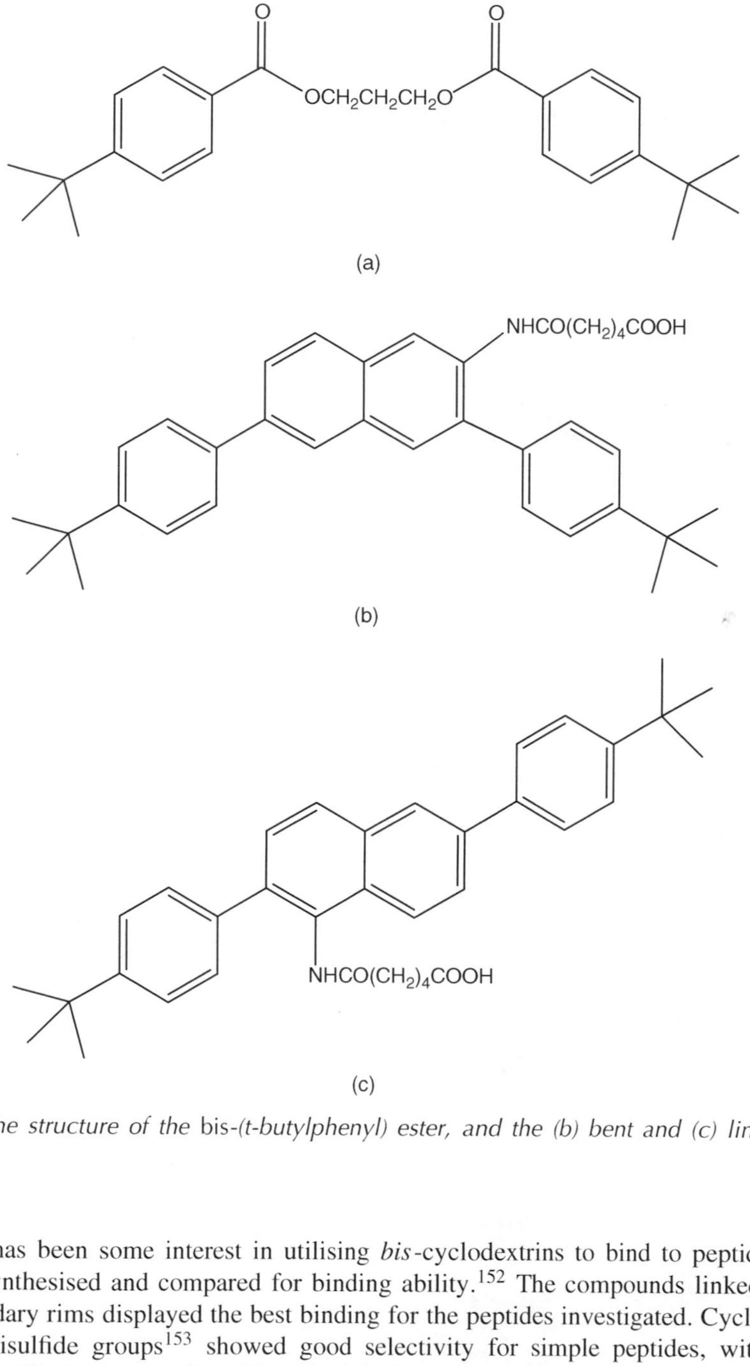

Figure 6.25   *(a) The structure of the bis-(t-butylphenyl) ester, and the (b) bent and (c) linear isomers of an aromatic guest*

Recently there has been some interest in utilising *bis*-cyclodextrins to bind to peptides, with a series of dimers being synthesised and compared for binding ability.[152] The compounds linked by ester groups between the secondary rims displayed the best binding for the peptides investigated. Cyclodextrins bridged with amino and disulfide groups[153] showed good selectivity for simple peptides, with for example a selectivity of 5.0 for Gly-Leu versus Leu-Gly, which is thought to be due to interactions between protonated amine groups in the tether and the peptides. The binding and transport of saccharides has also been of interest and for example two β-cyclodextrins were linked via a flexible ether linkage[154] and shown to transport various monosaccharides, but not disaccharides, across a liquid membrane.

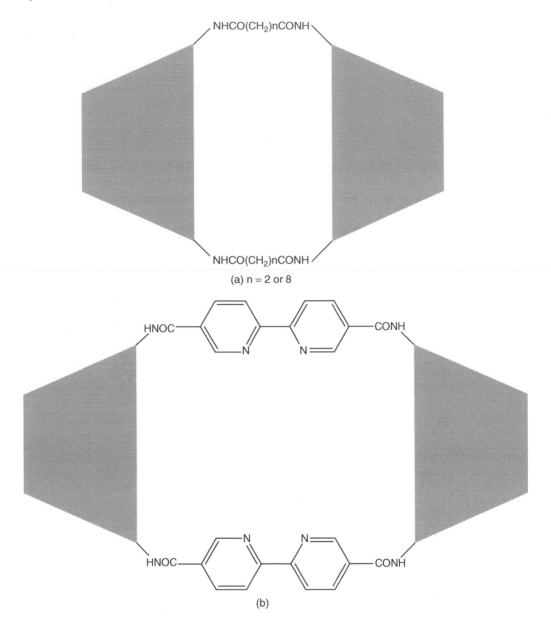

Figure 6.26   *The structure of the* bis-*cyclodextrins with (a) flexible and (b) bipyridyl linkers*

An A,D-diamino permethylated cyclodextrin[155] was bridged using rigid biphenyl units and found to form singly bridged and doubly bridged dimers along with a cyclic trimer (Figure 6.27). When complexation was undertaken with an anthracene derivative, the highest binding constant was observed for the trimer and the lowest for the doubly bridged compound, indicating that the higher preorganisation and entropic advantage of this compound is countered by the greater difficulty of opening up the duplex to allow access by the dye. The doubly bridged system also showed excimer formation upon irradiation in water, which indicates

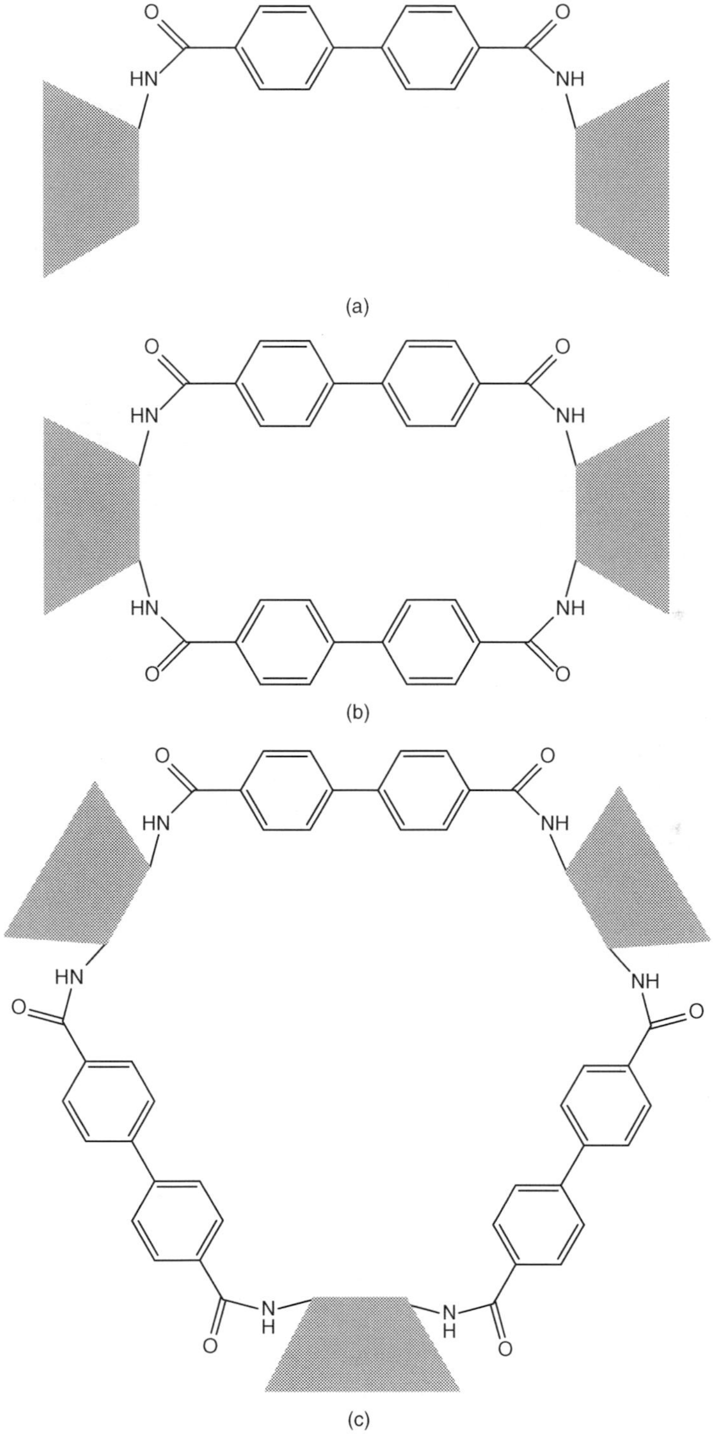

Figure 6.27 *Structures of the (a) singly and (b) doubly bridged* bis-cyclodextrins *and (c)* tris-cyclodextrin

that the biphenyl units are held close together in a rigid structure; this does not occur in organic solutions or for the other cyclodextrins in this study. Other workers have also synthesised cyclodextrin trimers and shown them to bind to trimeric and dimeric amino acid amides, with the trimeric species binding more strongly.[156] Cyclodextrin trimers containing two α-cyclodextrins linked via one β-cyclodextrin[157] have also been used to catalyse an esterification reaction. An ester compound is bound between the two α-cyclodextrins in an arrangement similar to a 2 : 1 inclusion complex and a catalytic imidazole compound is bound by the β-cyclodextrin in an organised manner so that the imidazole is held close to the ester (Figure 6.28). When a long diphenyl-substituted ester is used there is a noticeable enhancement (30%) in the initial rate of hydrolysis but there is no enhancement when a short ester with a single phenyl substituent is used, indicating that preorganisation of the two reagents enhances their relative reactivity.

A switchable cyclodextrin dimer has been made by linking together two β-cyclodextrin units, each containing a single amino group, directly via a photochromic dithienylethene linker.[158] This material can be switched photochemically between a relatively flexible (open) and a rigid (closed) form. Binding of this compound to tetrakis-sulfonatophenyl porphyrin is 35 times stronger in the open form than in the closed. However, if flexible trimethylene spacers are incorporated between the cyclodextrins and the dithienylethene unit, irradiation has minimal effect on binding since both open and closed forms are flexible enough to bind to the guest. This opens up the possibility of systems which can bind and release guests on application of an external stimulus.

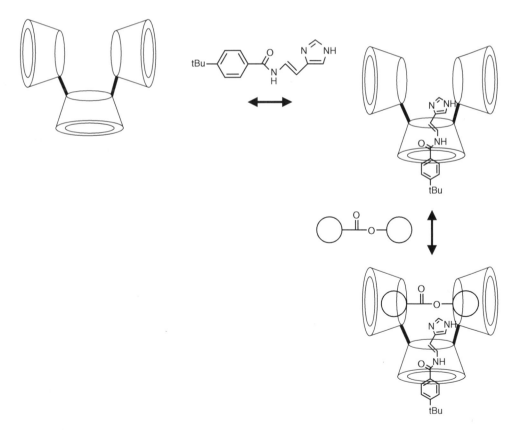

Figure 6.28   *Structure of a tris-cyclodextrin preorganising substrates for reaction*

6.7 Polymeric Cyclodextrins

There are several methods for making polymeric cyclodextrins. One, which we discussed earlier in this chapter, simply involves taking a long polymer molecule and threading the annular cyclodextrin molecules along it, thus forming a molecular necklace. Reaction of the end of the polymer with bulky stopper groups can prevent dethreading. Other methods include covalent linking of cyclodextrins and assembly using noncovalent forces to form supramolecular poly-cyclodextrins.

6.7.1 Covalently linked cyclodextrin polymers

One of the simplest ways to make a polymeric cyclodextrin is to utilise the reaction hydroxyl groups directly, for instance by reaction with epichlorohydrin. Early work in this field was hampered by the fact that due to the large number of hydroxyl groups on a cyclodextrin, the resultant polymers were crosslinked and insoluble.[159] However, by carefully controlling the reaction conditions,[160] β-cyclodextrin could be condensed with epichlorohydrin to give soluble polymers with molecular weights in the range $10^4–10^6$.

Once the synthesis of mono-functionalised cyclodextrins became possible, they could be grafted onto preformed polymers, with for example the mono-tosylate of β-cyclodextrin being capable of being grafted onto polyvinylamine.[161] This process allowed a variety of polymers with different levels of grafting to be synthesised. Although generally only a few of the amine groups were substituted, the final polymer had a dense, folded conformation due to the multiple hydrogen bonds between the polymer and the cyclodextrin units. Addition of a naphthalene-based dye caused formation of 2 : 1 cyclodextrin–guest complexes, confirming that the cyclodextrin units are held close together. These polymers have been used to catalyse the hydrolysis of *p*-nitrophenyl acetate, with the results indicating that the cyclodextrin and amine units cooperate in this process.[162] Polyallylamine was also used as a substrate and grafted with γ-cyclodextrins that had been substituted on either the primary or the secondary rims.[163] Polyamines are susceptible to protonation and hydrogen-bonding, meaning that the product polymers are highly folded in a similar manner to the polyvinylamines, although they are not highly soluble. To address this problem, free amine groups were acylated, giving uncharged and highly water-soluble polymers. Other workers also utilised polyethylenimine as the base polymer, grafted on β-cyclodextrin and then studied the effects of the resultant material on the hydrolysis of various aromatic esters.[164] Reaction rates were greatly enhanced because the polymer both acts as a complexing agent via the cyclodextrin and provides nucleophilic amines for the hydrolysis.

As an alternative to polyamine-type materials, β-cyclodextrin can be converted to an alkoxide with LiH and then reacted with the anhydride units of a methyl vinyl ether–maleic anhydride copolymer.[165] The resulting material contains β-cyclodextrin grafted via its 2-hydroxyl group, is water-soluble and can be obtained in a variety of polymer : cyclodextrin ratios from 6% to 84% substituted. Similar cyclodextrins can be grafted onto polyvinyl pyrrolidinone–maleic anhydride copolymers, although binding to adamantane-type derivatives is less favourable than for the parent cyclodextrins.[166]

Instead of grafting onto a preformed polymer, cyclodextrins can instead be synthesised containing a polymerisable group, which can then be used to synthesise the polymer. One example[167] took both α- and β-cyclodextrins and treated them with a variety of *m*-nitrophenyl esters under conditions which caused mono-substitution of the cyclodextrin at the 2-hydroxyl. Polymerisable groups grafted onto the cyclodextrins included acryloyl [$CH_2= CHCO_2$-] and N-acrylyl-6-aminocaproyl [$CH_2= CHCONH(CH_2)_5CO_2$-]. These materials could then be polymerised using a free-radical reaction, either by themselves or with other water-soluble monomers to give copolymers. The same group[168] also showed that the hydrolysis of *p*-nitrophenyl acetate and benzoate was catalysed by the resultant polymers, with an enhanced efficiency in comparison to the monomers. However, as the degree of substitution decreased (and thereby the average distance along the polymer chain between cyclodextrin units increased), the enhancement decreased. This indicates the presence of 2 : 1 host–guest complexes enhances the catalysis. Furthermore the polymer was found to be less effective than the monomer at binding the acetate ester but more effective for the benzoate ester, again indicating cooperative effects between two cyclodextrin rings in the case of the larger guest. Again, studies with the fluorescent dye 2-*p*-toluidinylnaphthalene-6-sulfonate showed that monomeric cyclodextrins had 1 : 1 host–guest complexation at low concentrations and 2 : 1 complexation at high concentrations, whereas the acryloyl polymer showed exclusively 2 : 1 complexation, again confirming that the cooperative effect is caused by the macrocycles being located on a polymer chain.[169] Fluorescence studies using naphthalene-based probes have also been performed on epichlorohydrin–cyclodextrin copolymers. They show that complexes with polymeric systems are much stronger than those with monomeric cyclodextrins.[170]

Reaction of cyclodextrins with maleic anhydride gave a material with up to five polymerisable groups per macrocycle. This could be copolymerised with N-isopropylacrylamide to give a pH- and temperature-sensitive hydrogel.[171] This hydrogel was capable of binding methyl orange, as was a second polymer made by combining cyclodextrin, epichlorohydrin and maleic anhydride.[172]

Conjugated polymers have also had cyclodextrin units attached, and for example a β-cyclodextrin with a single amino group was condensed with (4-carboxyphenyl)acetylene and then polymerised to give the polyphenylacetylene[173] shown in Figure 6.29. This polymer had a right-handed helical structure and in DMSO solution gave an intense red colour. However, heating the solution to 80 °C changed the colour to yellow (cooling reversed the colour change) and caused a switching of the helix from right- to left-handed. Addition of water or alcohols to the DMSO solution lowered the temperature required for this switch.

Dendrimers have been used as an alternative to polymers. For example, β-cyclodextrin can be attached to polyethylenimine dendrimers to give species that display novel behaviour.[174] Protonation of the composite occurs much less readily than for the parent dendrimer and it has been proposed that the complex has a very compact structure, with the dendrimer tightly wrapped around the cyclodextrin. The combination of complexing agent and amine nucleophiles means that these composites readily catalyse the hydrolysis of several esters. Other workers grafted carboxy-modified β-cyclodextrin onto the commercial PAMAM polyamine dendrimers and showed the resultant conjugates to have higher complexing abilities towards certain dyes than either the cyclodextrin or the dendrimer alone.[175]

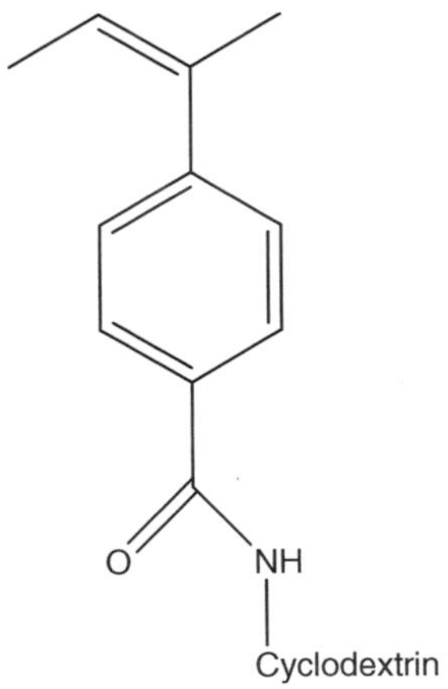

Figure 6.29   *Structure of a* β-*cyclodextrin-substituted polyphenylacetylene*

Crosslinked polymers have also been investigated for their ability to serve as molecularly imprinted polymers. Toluene-2,4-diisocyanate was used to crosslink β-cyclodextrin in the presence of various steroids.[176] The resultant polymers had the steroids washed out and were then shown to display a preference for binding those steroids over similar compounds. This is thought to be due to the cyclodextrins being mutually orientated so that they cooperatively bind those target guests that are too large to be included in the cavity of one cyclodextrin molecule. Similarly, acryloyl-substituted cyclodextrins were copolymerised with bisacrylamides and templated with various antibiotics or peptides, again to give molecularly imprinted polymers.[177] Epichlorohydrin-crosslinked β-cyclodextrins could also be used as molecularly imprinted polymers for creatinine.[178]

6.7.2 Noncovalently linked cyclodextrin polymers

Other interactions have been used to link together cyclodextrins in various assemblies. In fact, in crystalline cyclodextrins the individual macrocycles, often with included guests, can assemble in channel-type structures; these can be thought of as 'polymeric' cyclodextrins held together by hydrogen-bonding and other interactions. Again as previously described, long guests such as polyethylene glycols can hold together numerous cyclodextrins, threaded like beads on a necklace.

More complex systems have been formulated to assemble cyclodextrins in a variety of architectures. One of the simplest is formed by attaching a single aromatic tether to the primary ring of β-cyclodextrin. These types of system in solution can form head-to-tail or even tail-to-tail dimeric structures, where the tether of one macrocycle is actually included within the cavity of another.[179] When crystallised, these types of material can form helical or channel-type structures. When β-cyclodextrins substituted with single naphthalene sulfonates were used, head-to-head dimers were formed in solution,[180] where both naphthalenes were encapsulated within the opposing macrocycle, entering via the secondary rim.

Trimers have also been shown to form in solution. When a cinnamoyl substituent was used (Figure 6.30a) the β-cyclodextrin was found to be relatively insoluble in water but could be solubilised as a host–guest complex with compounds such as *p*-iodophenol, indicating that in the absence of guests, the cyclodextrins form a polymer.[181] However, when α-cyclodextrin was substituted, the resultant material was soluble, with the predominant species in solution appearing to be a trimer. Similar behaviour was observed for the system shown in Figure 6.30b, which could be reacted with bulky 'stopper' groups to give the material in Figure 6.30c. This precipitated as a solid product[181] and was shown to be mainly a trimer with a structure the authors termed a 'daisy chain' (Figure 6.30d). Polymeric versions with molecular weights of up to 18 000 could also be obtained. Other workers used tethered grouos to give 'polymeric' cyclodextrins.[113,115] Use of azobenzene tethers and naphthalene 'stoppers' allowed formation of a much larger superstructure[182] containing five cyclodextrin rings, again in a daisy-chain formation.

As an interesting variation on these themes, it has proved possible to incorporate both α- and β-cyclodextrins as alternating units in the same polymer.[183] When an α-cyclodextrin substituted with an adamantyl group is combined with a β-cyclodextrin substituted with a *t*-butyl phenyl group, the adamantyl group preferentially binds within the β-cyclodextrin cavity. The remaining *t*-butyl phenyl tether is then incorporated within the α-cyclodextrin. This is shown schematically in Figure 6.31, with the resultant polymers having molecular weights as high as 15 000.

6.8   Cyclodextrins Combined with Other Macrocyclic Systems

Not only have cyclodextrins been combined together, they have also been combined with a number of other macrocyclic systems. Just a few examples will be given here, but these show the possibilities that

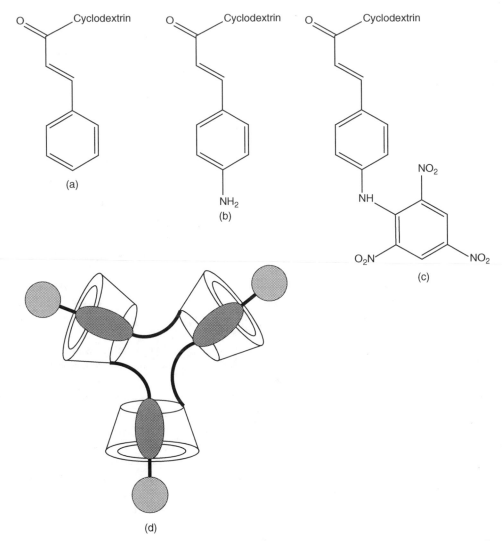

Figure 6.30 *(a,b,c) Structures of cyclodextrins substituted with cinnamoyl derivatives and (d) a 'daisy chain' trimer*

can be explored. Crown ethers have been attached to cyclodextrins, with one of the earliest examples being the compound shown in Figure 6.32a, in which a diaza-18-crown-6 moiety has been attached at the primary ring.[184] This compound shows a 7–10 times higher affinity for aromatic ammonium ions than the parent cyclodextrin, which is thought to be due to a cooperative effect in which the cyclodextrin binds the aromatic moiety and the crown the ammonium ion. Multiply substituted cyclodextrins have also been synthesised (Figure 6.32b), where seven crown-ether units are attached to a single β-cyclodextrin unit.[185] Again these have been shown to bind to ammonium cations, but in the case of monovalent ammonium ions the binding is poorer than that of the unsubstituted crown ethers, due to steric crowding and unfavourable ion–ion interactions. In the case of divalent dialkylammonium ligands, an increase in binding constant of two orders of magnitude is seen for certain host–guest pairs, indicating the possibility of chelate effects.

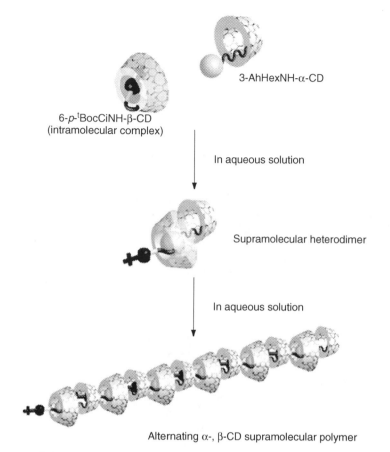

Figure 6.31 *Structures of alternating α-, β-cyclodextrin polymers. (Reprinted with permission from.[183] Copyright 2004 American Chemical Society)*

Other crown ether-based systems have also been synthesised, such as that shown in Figure 6.32c, where the cyclodextrin has been attached by the primary rim. In the same work[186] *bis*-cyclodextrins were synthesised in which the cyclodextrins were linked by crown ethers via either the primary (Figure 6.32d) or the secondary rim (Figure 6.32e). Complexation studies with fluorescent guests showed that the primary substituted cyclodextrins displayed increased binding to anionic guests, whereas the secondary-rim-linked *bis*-cyclodextrin displayed enhanced binding to linear dyes, probably due to cooperative effects between the two cyclodextrin units. Other workers[187] have also synthesised cyclodextrins substituted with either benzo-15-crown-5 or benzo-18-crown-6 at the primary or secondary rims, and showed that the cyclodextrins modified at the secondary rim displayed superior complexation properties towards tryptophan and that use of the larger crown ether further enhanced binding.

Calixarene–cyclodextrin conjugates[188] have also been synthesised, such as those shown in Figure 6.33a,b. It was thought that they would complement each other well, since calixarene binding is often electrostatic in nature and cyclodextrin binding is often caused by hydrophobic effects. Complex formation was investigated with various fluorescent dyes and increases of binding coefficient towards such dyes as sodium 2-(p-toluidino)-6-naphthalenesulfonate and acridine red were observed. Synthesis

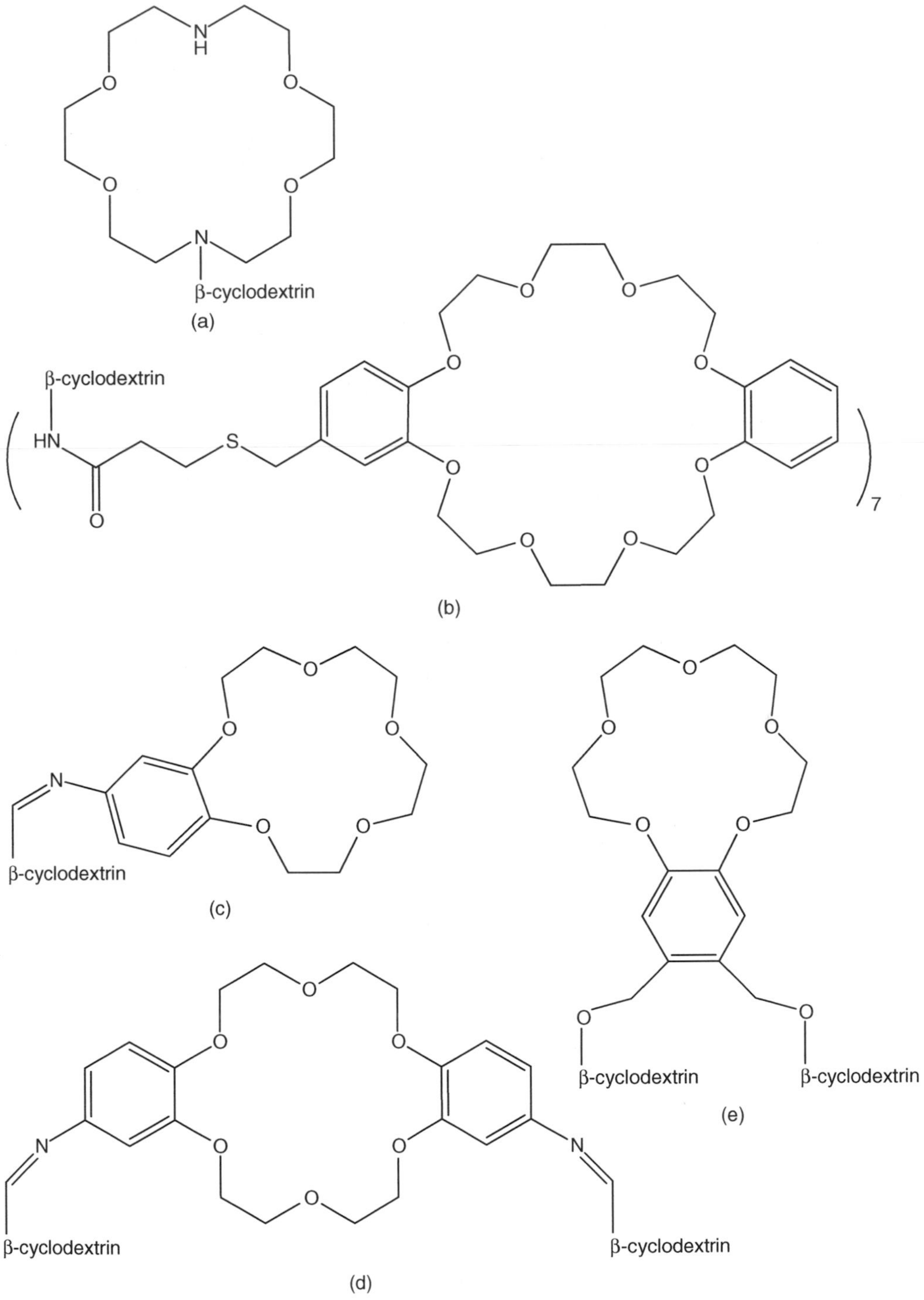

Figure 6.32   *Structures of crown-substituted β-cyclodextrins*

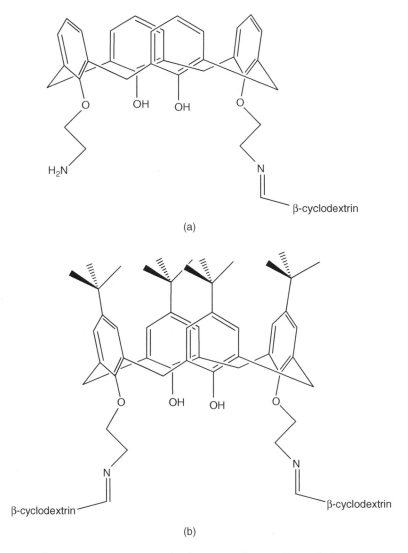

(a)

(b)

Figure 6.33 *Structures of calixarene-substituted β-cyclodextrins*

of a *bis*-benzaldehyde-substituted γ-cyclodextrin and its reaction with pyrrole[189] allowed the one-pot construction of a cage molecule containing four cyclodextrin and two porphyrin units. Fluorescence measurements appeared to confirm the cofacial arrangement of the porphyrin moieties.

6.9 Therapeutic Uses of Cyclodextrins

The discovery of the therapeutic activity of an organic compound is often only the first step in the development of a successful pharmaceutical product; there are then a wide number of formulation problems to be solved. First is the intrinsic solubility of the drug compound, which is simple if it is

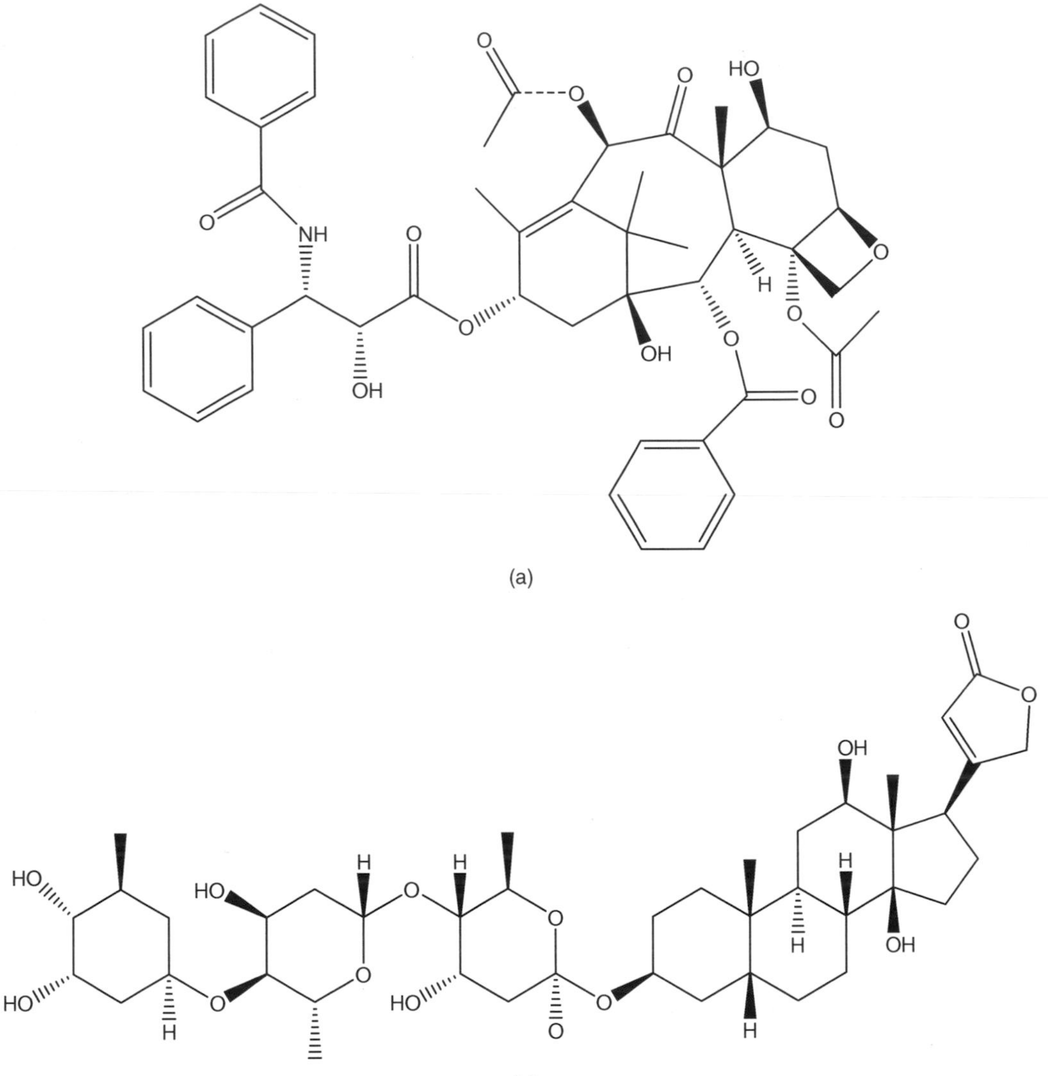

(a)

(b)

Figure 6.34   *Structures of (a) paclitaxel and (b) digoxin*

readily soluble in plain water, but many drugs are not. The use of other solvents is unfortunately not really an option for pharmaceutical applications since the majority of solvents are toxic. A typical example is paclitaxel (Figure 6.34a), usually marketed under the name Taxol, which is a potent anticancer drug. The pharmaceutical system has severe side effects, some of which may not actually be due to the drug itself but to the solvating agents. Insolubility can be an even worse problem if the drug is intended for intravenous application since whilst the stomach has the capability to deal with a wide range of materials, the cardiovascular system is far less tolerant.

Even if the compound is soluble in water, this may still not be enough since often the desired form of application is a tablet, which must dissolve after being swallowed. Many factors, such as the particle size, salt form and level of hydration, can affect the rate of solution of an intrinsically soluble compound. Also, the crystal form can vary depending on how the compound has been isolated and this too can have drastic effects on solubility.

Besides solubility, other factors can also render a drug unsuitable. Stability of the compound is often vital, both on-shelf (because products with a short shelf-life can easily end up being thrown away or used after they have become ineffective) and *in vivo*, if the drug compound is to be taken orally it must not be inactivated by exposure to digestive juices or acidic environments. If the drug is taken orally it must be capable of being adsorbed by the digestive system and into the bloodstream. Once in the bloodstream it must usually be capable of crossing cell barriers to be of any use.

In many instances a simple one-off administration of a drug will suffice, for example taking an analgesic in response to a headache. Often the medical condition may require continuous medication over a long period of time. Also, many therapeutic compounds have an optimum concentration within the body; too low and no benefit is derived, too high and unwanted side effects or toxicity can occur, endangering the health of the patient. An example of this is digoxin (Figure 6.34b), a cardiac drug widely used in the treatment of various heart conditions such as atrial fibrillation and atrial flutter,[190] with a narrow therapeutic range of $0.8-2.0$ ng ml^{-1}. When a drug is ingested or injected, the level within the patient's blood tends to rise towards a plateau and then after some time to fall again to zero. A more efficient application method would give a stable level of treatment. At the time of writing, approximately 15% of the current world pharmaceutical market consists of products which utilise a carrier system,[191] and a wide variety of drug-carrier and -delivery systems are under investigation, with several reviews published on the subject.[192-195] We will discuss here the principles of drug delivery and give examples of a few current commercial and research applications of these materials; more details can be found in the References.[192-195]

Ingestion is the most popular form of drug delivery, because of its simplicity and convenience. There are a number of problems with this technique, however, as discussed above; irritation of the bowel is a common side effect, for example. A drug can also be dispersed, usually via a nebuliser, and directly inhaled. Especially suitable for treating respiratory diseases, this technique is widely used for the treatment of asthma. However, delivery via this method can still be adversely affected by the barrier between air and blood within the lung. A drug can be incorporated within a transdermal patch, similar to the nicotine patches used to relieve the 'cravings' of people attempting to stop smoking. Patches often need to be applied only once every several days, are noninvasive and painless, and can be safely applied to unconscious patients. The skin barrier does lead to slow penetration rates however and therefore in many instances only relatively low dosage levels can be attained. A drug can also be administered by injection, but this technique has the disadvantage of being invasive as well as being painful. Drug-carrier systems can be utilised with this method, to prevent degradation of the drug within the bloodstream for example. Alternatively, a drug–carrier composite can be surgically implanted close to an affected site.

Cyclodextrins are of interest in such situations because of their complexing ability: many water-insoluble compounds can be dissolved in water by encapsulating them inside the cavity of a cyclodextrin. Being encapsulated within a cyclodextrin can protect the guest from being degraded by the environment, as well as enhancing its shelf and operational stability. Similarly, many compounds undergo polymorphic transitions from one crystalline form to another upon storage, which can modify their efficacy, and encapsulation can prevent this. Hydrophobic-modified cyclodextrins can be used to encapsulate water-soluble drugs and act as sustained-release agents. Other advantages of cyclodextrins include their well-defined structure, low toxicity and relatively simple chemical modification procedures, allowing production of tailored cyclodextrins. The reader is referred to a series of reviews of cyclodextrins as drug-delivery systems,[196-198] and to a 1999 special issue of the journal *Advanced Drug Delivery Reviews* on the subject (see Bibliography).

Initially cyclodextrins were actually thought to be toxic; in early work it was reported that rats fed on cyclodextrins would only eat them in small quantities and soon died.[199] Later work did not reproduce this and it is now thought possible that the cyclodextrins used had been isolated as inclusion complexes containing toxic organic solvents that were responsible for the effects noted.[3] Once it had been finally established that cyclodextrins were not as toxic as first thought, their use in therapeutic applications was made possible. However, cyclodextrins are not completely benign – some of their effects are briefly reviewed by Uekama[200] and include the removal of cholesterol from cell membranes by methylated cyclodextrins, which leads to cell lysis.

There are numerous ways in which cyclodextrins can be utilised, a few of which will be listed here. Hydrophilic cyclodextrins have been used to solubilise a range of water-insoluble drugs, such as Captisol®, which is a β-cyclodextrin substituted with a number of sulfobutyl ether groups ionised at physiological pHs and rendering this compound highly water-soluble. Commercial applications for this material include the formulation Vfend® (marketed by Pfizer), which is a mixture of Captisol® and the antifungal active ingredient voriconazole, which can be dissolved in water and applied intravenously. A similar formulation is Zeldox, which contains the active ingredient ziprasidone, an antipsychotic agent. Captisol has also been used within the delivery of budenoside via a nasal spray to relieve allergic symptoms and asthma.[201] Another β-cyclodextrin derivative that is being investigated for similar applications is the hydroxypropyl ether (Kleptose HPB®). Digoxin and other digitalis glycosides have been shown to form 1 : 4 complexes with γ-cyclodextrin and to lead, after oral administration, to 5.4-fold increases in plasma drug levels in dogs.[202] A series of other drug–cyclodextrin complexes have been approved for use in various countries, as listed by Loftsson and Duchêne[198] and by French.[199] These include Nitropen® (a tablet form of nitroglycerin with β-cyclodextrin), Meiact (a tablet containing the antibiotic cephalosporin with α-cyclodextrin) and Aerodiol® (a nasal spray containing estradiol and methylated β-cyclodextrin).

One potential application is the delayed-release agent, where a drug is released at a desired time or in a desired place. For example, if one wanted to deliver a drug to the colon, simply swallowing it might not suffice since it would have to run the gauntlet of the stomach and intestinal tract before it reached its desired target and might not survive the journey. A stable host–drug complex might be engineered to pass through this system and then degrades in the colon to release the drug. For example, cyclodextrins covalently substituted with an anti-inflammatory agent, 4-biphenylylacetic acid, were found to release 95% of the drug within one to two hours of exposure to rat colonic contents, and were unaffected by stomach or intestine contents.[203]

Gene delivery using cyclodextrins is also being widely investigated. A variety of cationic, neutral and anionic cyclodextrins have been studied and in several cases shown to enhance adenoviral-mediated gene transfer to intestinal cells,[204] with positively charged cyclodextrins being the most efficient. Combinations of cyclodextrins and commercial PAMAM dendrimers were shown to be up to 200 times more effective at DNA transfer than the dendrimer alone.[205] Enhancements were also found for conjugates of dendrimers covalently bound to cyclodextrins (1 : 1 ratio), with α-cyclodextrin being the most effective.[206] Transfection activity was 100 times greater for the conjugate than for either the dendrimer alone or a physical mix of dendrimer and α-cyclodextrin.

These applications have only employed simple cyclodextrin complexes, but far more complex formulations are described in the literature and have been reviewed by Trichard *et al.*[207] These include incorporating cyclodextrins into emulsions, microcapsules, polymeric nanospheres, nanocapsules and liposomes. Again the reader is advised to study the Bibliography for further information.

6.10   Other Uses of Cyclodextrins

Besides the pharmaceutical applications, there are a vast range of other potential uses for cyclodextrins, many of which are now commercial products. A 2009 search of the European patent database reveals almost 9500 patents for cyclodextrins, with an almost identical number in the US patent database, about 3000 in the Japanese and 1000 in the world patent database. The Bibliography and other reviews mentioned in this chapter describe many of these potential applications in detail and a brief overview will be given here. The cyclodextrins are the only group of supramolecular entities to so far have significant widespread commercial applications. The main reason for this is that they can easily be manufactured on a large scale, meaning they have relatively low cost. Combined with their stability and low toxicity, this has formed the basis of their success.

One obvious potential application for cyclodextrins is as adsorbents. Among the simplest is their use as deodorants, because of their ability to complex molecules and essentially remove them from the environment. Products such as the Febreze range of household odour-removal products, marketed by Proctor and Gamble, contain cyclodextrins as the active ingredient. Other applications include the removal of unwanted compounds from food and drink. For example, early-season Florida grapefruit have a high bitterness due to high levels of compounds such as nomilin and limonin (Figure 6.35); passing grapefruit juice through β-cyclodextrin polymers, usually as fluidised beds, removes these compounds and can improve its taste.[208] Similarly, green tea contains catechins, which have health-promoting effects but unfortunately also impart a very bitter taste. Addition of cyclodextrins reduces the bitter flavour whilst also helping to stabilise the catechins.

Cyclodextrins often find applications as stabilising agents. Often an unstable compound can become entrapped inside a cyclodextrin and in effect become shielded. This effect has been commercialised and for example Wacker Fine chemicals have produced a range of cyclodextrin complexes for use in the personal-care industry.[209] They currently market cyclodextrin-encapsulated products such as retinol (Figure 6.36a), which is used within the skin-care industry as an anti-aging agent. Retinol is highly sensitive to UV light and oxygen and retains <10% activity after six months of storage, but when complexed with two γ-cyclodextrin molecules is stabilised to such an extent that it retains >90% of its activity after two years. When the cosmetic is applied to the skin the retinol is released onto its surface. Other skin-care products of a similar nature include vitamin E (tocophenol), which is supplied as a 1:2 complex[210] with β- or γ-cyclodextrin (Figure 6.36b), linoleic acid (Figure 6.36c), as a 1:4 complex with α-cyclodextrin, and tea-tree oil, complexed with β-cyclodextrin.

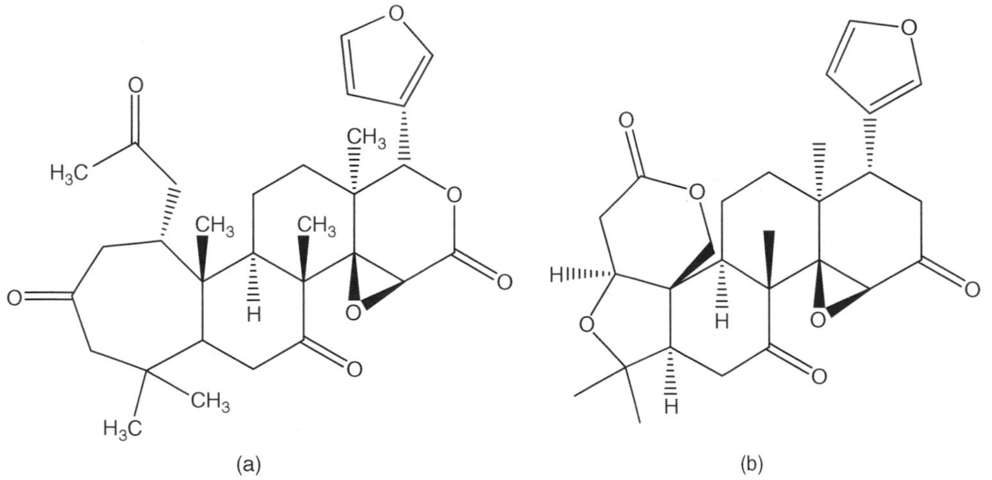

(a)   (b)

Figure 6.35 *Structures of (a) nomilin and (b) limonin*

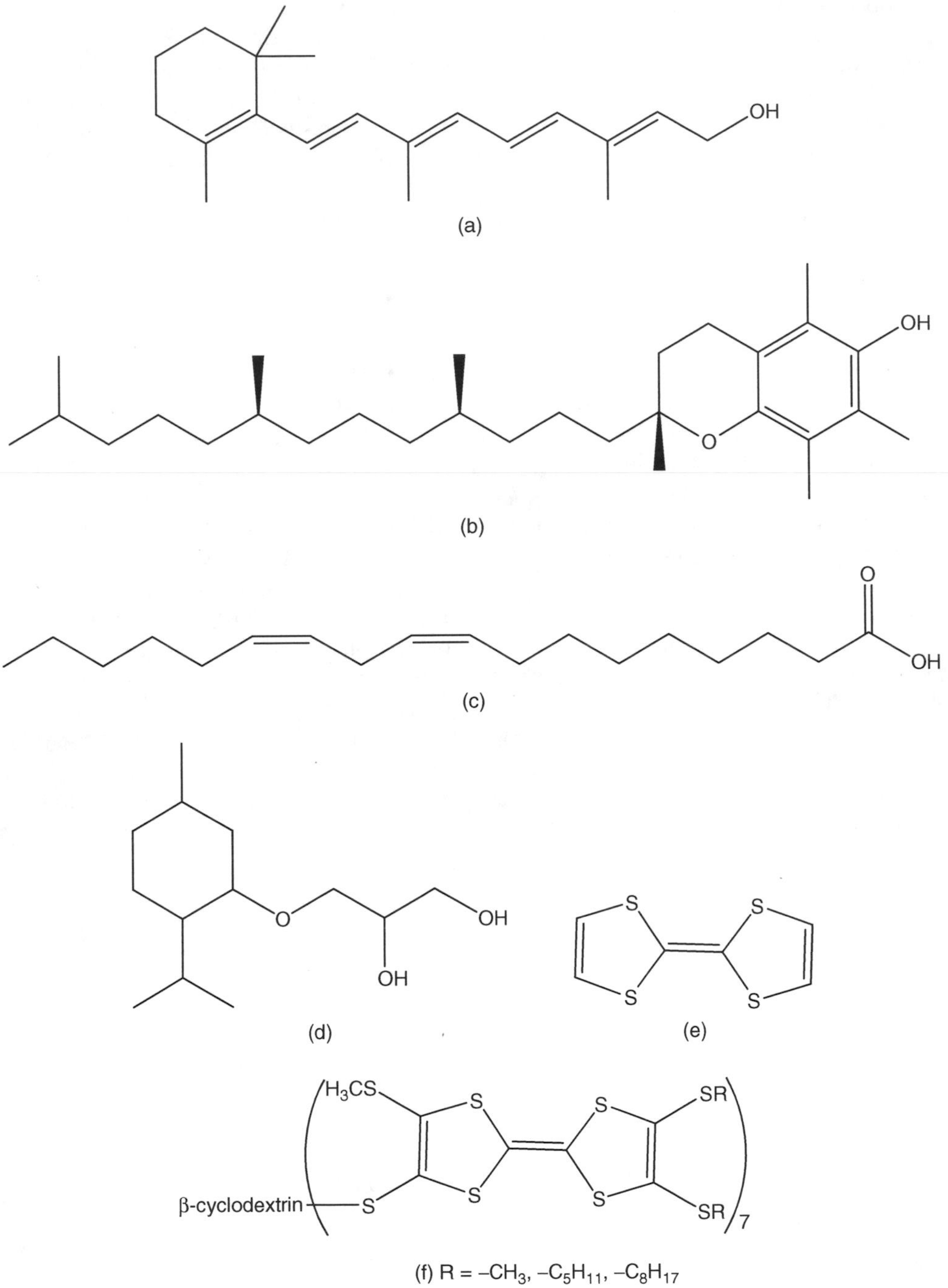

(a)

(b)

(c)

(d)   (e)

(f) R = –CH$_3$, –C$_5$H$_{11}$, –C$_8$H$_{17}$

Figure 6.36 Structures of (a) retinol, (b) tocopherol, (c) linoleic acid, (d) 3-1-menthoxypropane-1,2-diol, (e) tetrathiafulvalene and (f) tetrathiafulvalene-substituted cyclodextrin

Another product that utilises cyclodextrins as entrapment agents is Bounce, which is marketed by Proctor and Gamble. This is a fabric sheet added to washing during a dryer cycle and contains various perfumes which are stabilised by encapsulation inside cyclodextrins. During the dryer cycle these complexed perfumes are transferred to the clothes, and some are then released by a combination of heat and water vapour from drying clothes.[211] When the fabric is remoistened by perspiration or other moisture, fragrance is again released from the complex to give an impression of freshness. Often the lifetime of this process is extended by using a mixture of perfumes of differing volatilities.

Slower release of flavour compounds can also be obtained using cyclodextrins. For example, a chewing gum containing the mint flavouring 3-1-menthoxypropane-1,2-diol (Figure 6.36d) was formulated with and without β-cyclodextrin;[212] the gum that contained the cyclodextrin demonstrated longer-lasting flavour and 'coolness' and gave a much firmer texture (one disadvantage of 3-1-menthoxypropane-1,2-diol as a flavour agent is that it can soften the gum). Similar flavour components characteristic of spices such as garlic, cinnamon and ginger can be complexed with cyclodextrins for incorporation into baking products. Not only does this enhance the stability, but as less of these flavour components are lost during cooking due to reduced volatility, a smaller amount needs to be incorporated initially.

Cyclodextrins can also be used to mask scents and flavours. Beneficial dietary supplements such as fish oils and garlic oils often have unpleasant odours, but complexation with cyclodextrins removes these and allows the product to be formulated as a tablet or powder. Typical commercial products include OmegaDry (manufactured by Wacker Chemie AG), which is a fish oil–γ-cyclodextrin conjugate, and Cardiomax garlic tablets (Seven Seas Health Care), in which pure essential oil of garlic is formulated with cyclodextrins.

One major commercial application of cyclodextrins is in cholesterol removal. Steroids form complexes with cyclodextrins,[176] and cholesterol is no exception, forming complexes with β-cyclodextrin and a variety of *bis*- and polycyclodextrins.[213] This process is the basis of many low-cholesterol foods currently available on the market, as reviewed by Szente and Szejtli.[214] For example, molten butter can be mixed with β-cyclodextrin, which forms a complex with the cholesterol present in the butter and can easily be removed. Cholesterol can be removed in a similar way from cheese.[215] Typical commercial low-cholesterol products include Balade butter and Simply Eggs, a liquid-egg product made from whole eggs.

We have discussed at some length how cyclodextrins can aid in drugs and other formulations to improve properties such as solubility and stability. This approach is not just limited to these applications and one large area undergoing investigation is the use of cyclodextrins to modify agrochemicals. Modern agriculture requires a wide range of chemicals, with fertilisers and pesticides being two of the most obvious. Many of the problems associated with drug compounds, such as solubility, difficulty of controlled release and stability, are also applicable to agrochemicals. Because of the large-scale availability of cyclodextrins and their affinity for a wide range of compounds – especially the pesticides, with their associated problems of toxicity and controlled application – work has been undertaken on agrochemical–cyclodextrin formulations. So far though this represents only a small fraction of the total cyclodextrin literature and no large-scale commercial applications yet exist. Much of this work has been reviewed by Morillo.[216]

Several novel products based on cyclodextrins are currently being marketed in Japan, as reviewed by Hashimoto.[217] These include textiles modified with cyclodextrin complexed with squalene or linoleic acid to help treat conditions such as dermatitis. A wide range of other compounds such as perfumes, insecticides and so on complexed with cyclodextrins have been incorporated into textiles. Some workers have incorporated cyclodextrins complexed with antifungal agents (*o*-methoxcinnamaldehyde) into socks to successfully treat athlete's foot. The use of cyclodextrin-based agents to modify textiles has been reviewed by Szejtli.[218] Wrapping materials which include cyclodextrin products have also been made. The spice wasabi has been shown to have antibacterial properties, and stabilised by cyclodextrins it has been incorporated into food wrapping materials to suppress mould and bacterial growth.[217]

The stability and chirality of cyclodextrins has led to their incorporation as the recognition element in chiral chromatography columns. The literature contains many papers describing the separation of a vast number of compounds into their enantiomers using chiral GC or LC methods with cyclodextrin-based columns. A number of commercial columns are available, such as the Cyclobond and Chiraldex series marketed by Advanced Separation Technologies (Astec). Chiraldex columns are capillary GC columns with a variety of stationary phases based on all three of the main cyclodextrins, which have then been modified to give various etherified or trifluoroacetylated products. These can be used for separation of aromatic and nonaromatic enantiomers. Cyclobond is a series of HPLC columns utilising a packed column containing silica onto which cyclodextrins have been chemically grafted via an ether linkage. Again various cyclodextrin sizes and substitution patterns are available. Similarly, Agilent supplies a number of chiral columns such as the CycloSil-B GC capillary column, in which the stationary phase contains heptakis (2,3-di-O-methyl-6-O-t-butyl dimethylsilyl)-ß-cyclodextrin, the similar Cyclodex column containing the parent ß-cyclodextrin and the HP-Chiral ß GC column containing permethylated ß-cyclodextrin. Phenomenex produces the chiral HPLC Chiradex columns based on silica solid phases modified with ß-cyclodextrin. There are a number of other cyclodextrin-based columns available on the market.

We mentioned earlier in this chapter how certain reactions can be catalysed by the presence of cyclodextrins. Although the study of cyclodextrins as catalysts has increased greatly in recent years,[219] and the effects of cyclodextrins on the rates and sterospecificities of many reactions are being intensively investigated, as yet no major commercial applications have been developed. The catalysis can take several forms, via direct bond formation with the cyclodextrin, by acid base catalysis of the hydroxyl groups or by the cyclodextrins simply accommodating one or more substrates within their cavity and acting as a miniature reaction vessel. Cyclodextrins have also shown an ability to indirectly aid electrochemical reactions; tetrathiafulvalene for example (Figure 6.36e) is a versatile mediation for electrochemical reactions but is insoluble in water. It readily complexes with the three major cyclodextrins[220] and as the 1 : 1 complex with ß-cyclodextrin has been shown to effectively mediate the transfer of electrons between the reaction centre of immobilised glucose oxidase and the electrode, with no mediator present the enzyme is essentially insulated, with no electron transfer being observed.

Because of their sometimes selective complexation behaviour in solution, attempts have been made to immobilise cyclodextrins onto various surfaces for use as sensors. For example, 3-methyl thiophene can be electrodeposited to form a conducting polymer containing embedded γ-cyclodextrin[221] and the resultant modified electrode has been shown to be capable of detecting chlorpromazine and some neurotransmitters at levels as low as 10^{-7} M. Similarly, both α- or β-cyclodextrins can be incorporated into polypyrrole films and the resultant polymer membranes have been shown to display high permeability to metal ions, with the β-cyclodextrin film being the most permeable except in the case of copper.[222]

Attaching thiol or disulfide groups to the cyclodextrins allows their attachment directly to a gold surface. Early work utilised ß-cyclodextrins functionalised with seven sulfur-containing groups, but the resultant films were very permeable and gave behaviour indicating poor packing of the cyclodextrin moieties. When cyclodextrins with lesser degrees of substitution (usually 2–4 thiol groups on the primary rim) or monofunctionalised cyclodextrins with both short and long spacers were utilised,[223] the structure of the film and the orientation of the cyclodextrin macrocycle were found to be highly dependent on the substitution. Cyclic voltammetry showed that cyclodextrins with longer spacer chains gave less permeable films, indicating better packing and that monofunctionalised cyclodextrins packed more efficiently than those which were multifunctionalised. Other characterisation techniques indicated that the monofunctionalised cyclodextrins with long spacer chains packed with the cyclodextrin orientated with the macrocycle almost perpendicular to the plane of the gold surface, whereas for the other systems the cyclodextrins were parallel to the substrate.

Other cyclodextrin mono-thiols were immobilised at gold electrodes and interrogated by AC impedance.[224] Near-perfect packing efficiencies of the macrocycles on the surface were observed and the film was shown to respond to the presence of anilinonaphthalene sulfonate in solution. Other work[225] also showed the response of these types of monolayer to a variety of substrates and further demonstrated that mixed monolayers of the cyclodextrins with simple alkyl thiols were of higher quality than films of cyclodextrin alone. A cyclodextrin monofunctionalised at the primary rim with a thiol group and at the secondary rim with an anthraquinone unit also formed monolayers on gold, with fast electron transfer being observed between the anthraquinone unit and the substrate even though the cyclodextrin ring was between them.[226] This process could be affected by the presence of a naphthalene guest in the cyclodextrin cavity. It is not necessary to covalently bind a mediator to the cyclodextrin and simple inclusion is sufficient; for instance, electron transfer has been demonstrated for thiolated cyclodextrin layers containing a mediator – methylene blue – bound within the cavity.[227] Similar work on various quinine mediators in solution has shown that electron transfer only occurs when the mediator is of a size that allows it to enter the cyclodextrin cavity.[228]

Other workers studied thiolated derivatives of the three main cyclodextrins[229] and showed that α-cyclodextrins blocked the access of ferrocene carboxylic acid to the electrode while the larger macrocycles allowed access, although access could then be blocked by complexation with electroinactive guests, demonstrating molecular recognition. Positional isomerisation was also found to be important. The four positional isomers of a dithio derivative of γ-cyclodextrin[230] for example were synthesised and assembled on gold surfaces, and when the two anchor groups were well separated (i.e. the 6A,6E isomer), near perfect coverage of the gold electrode was seen. Lesser coverage was however observed for the 6A,6D (89%), 6A,6C (86%) and 6A, 6B (60%) isomers.

As an alternative to direct immobilisation of a cyclodextrin, a gold surface was modified with thiooctic acid and α-, β- or γ-cyclodextrins modified with protonated amino groups electrostatically adsorbed onto this surface.[231] The resultant system was shown to bind ferrocene carboxylate. The Langmuir–Blodgett technique also offers an alternative method for assembling cyclodextrins in layers. For example, a β-cyclodextrin substituted with seven tetrathiafulvalene moieties (Figure 6.36f) and the corresponding γ-cyclodextrin can be deposited as an LB film,[232] with the tetrathiafulvalene units showing the potential to act as mediators.

Electrochemical methods have also been used to assemble cyclodextrins onto surfaces, and a potential controlled adsorption of β-cyclodextrin onto gold led to the formation of cyclodextrin 'nanotubes' on the surface,[233] with the cyclodextrin cavities facing sideways rather than upwards in the tubes.

Cyclodextrins have also been utilised in the construction of a great many supramolecular systems, such as rotoxanes and catenanes, and have also been applied to the development of a number of molecular machines and devices. This will be discussed in Chapter 9.

6.11   Conclusions

Since nature has been synthesising organic chemicals for far longer than mankind, it is fitting that one of the oldest and most commonly used groups of chemicals described within this book is of natural origin. Without the enzymatic processes available to synthesise these systems, they would remain a little-known academic curiosity, if indeed they had even been discovered at all. Instead cyclodextrins are so easily available that we use them in a wide variety of commercial applications.

The ease of obtaining cyclodextrins at inexpensive prices and the wide variety of reactions that can be used to selectively modify these compounds indicates that besides their current applications, a wide range of future uses beckon. It is almost inevitable that the large numbers of intriguing papers published on these macrocycles every year will lead to many new advances in fields as diverse as pharmaceuticals, household applications, food modification and health, along with more esoteric applications such as sensors and molecular devices. The use of cyclodextrins in drug formulations could well provide a whole new range of orally applicable drugs combined with prolonged therapeutic effects. They could also be instrumental in the development of new specific site- or time-controlled delivery drugs for, for example colon-specific delivery, and in advancing fields such as gene delivery. Other potential future applications could include complex separation and fractionation techniques, incorporation into various matrices to give smart materials and the mitigation or better still prevention of environmental pollution.

Within this chapter we have discussed the physical, chemical and functional properties of a selection of what is a wide-ranging array of compounds. It is to be hoped that the advances made in this field will continue and that we may begin to approach molecules with selectivities and specificities similar to the wide array of proteins, enzymes, antibodies and nucleic acids that can be found in the natural world.

Bibliography

Dodziuk H. Cyclodextrins and Their Complexes. Wiley-VCH; 2006.
Cragg P. A Practical Guide to Supramolecular Chemistry. Wiley-Blackwell; 2005.
Steed JW, Atwood JL. Supramolecular Chemistry. John Wiley and Sons; 2000, 2009.
Uekema K. *Advanced Drug Delivery Reviews*. 1999; **36**: 1–141, and papers within.
D'Souza VT, Lipkowitz KB. *Chemical Reviews*. **98**: 1741–2076, and papers within.
Szejtli J, Osa T. Comprehensive Supramolecular Chemistry. Volume 3. Cyclodextrins. Elsevier; 1996.
Szeitli J. Cyclodextrin Technology. Kluwer Academic; 1996.

References

1. Villiers A. Sur la transformation de la fécule en dextrine par le ferment butyrique [On the transformation of starch to dextrin by the butyric ferment]. *Compt Rend Fr Acad Sci.* 1891; 435–438.
2. Schardinger F. Bildung kristallisierter Polysaccharide (Dextrine) aus Stärke kleister durch Microbien [Production of crystalline polysaccharide (Dextrin) from starch paste by microorganisms]. *Zentralbl Bakteriol Parasitenkd Infektionskr Hyg II.* 1911; **29**: 188.
3. Szejtli J. Introduction and general overview of cyclodextrin chemistry. *Chem Rev.* 1998; **98**: 1743–1753.
4. Biwer A, Antranikian G, Heinzle E. Enzymatic production of cyclodextrins. *Appl Microbiol Biotechnol.* 2002; **59**: 609–617.
5. Manor PC, Saenger W. Topography of cyclodextrin inclusion complexes. III. Crystal and molecular structure of cyclohexaamylose hexahydrate, the water dimer inclusion complex. *J Am Chem Soc.* 1974; **96**: 3630–3639.
6. Chacko KK, Saenger W. Topography of cyclodextrin inclusion complexes. 15. Crystal and molecular structure of the cyclohexaamylose-7.57 water complex, form III. Four- and six-membered circular hydrogen bonds. *J Am Chem Soc.* 1981; **103**: 1708–1715.
7. Maclennan JM, Stezowski JJ. The crystal structure of uncomplexed-hydrated cyclooctaamylose. *Biochem Biophys Res Commun.* 1980; **92**: 926–932.
8. Fujiwara T, Tanaka N, Kobayashi S. Structure of δ-cyclodextrin 13.75H2O. *Chem Lett.* 1990; **19**: 739–742.
9. Endo T, Nagase H, Ueda H, Kobayashi S, Shiro M. Crystal structure of cyclomaltodecaose (epsilon-cyclodextrin) at 203 K. *Anal Sci.* 1999; **15**: 613–614.
10. Jacob J, Gebler K, Hoffmann D, Sanbe H, Koizumi K, Smith SM, Takaha T, Saenger W. Band-flip and kink as novel structural motifs in a-(1'4)-D-glucose oligosaccharides. Crystal structures of cyclodeca- and cyclotetradecaamylose. *Carbohyd Res.* 1999; **322**: 228–246.
11. Harata K, Endo T, Ueda H, Nagai T. X-ray structure of i-cyclodextrin. *Supramol Chem.* 1998; **9**: 143–150.

12. Gessler K, Usón I, Takaha T, Krauss N, Smith SM, Okada S, Sheldrick GM, Saenger W. V-amylose at atomic resolution: X-ray structure of a cycloamylose with 26 glucose residues (cyclomaltohexaicosaose). *PNAS.* 1999; **96**: 4246–4251.

13. Connors KA. The stability of cyclodextrin complexes in solution. *Chem Rev.* 1997; **97**: 1325–1357.

14. Connors KA. Population characteristics of cyclodextrin complex stabilities in aqueous solution. *J Pharm Sci.* 1998; **84**: 843–848.

15. Gidley MJ, Stephen MB. ^{13}C-C.p.-m.a.s. n.m.r. studies of frozen solutions of $(1 \to 4)$-α–glucans as a probe of the range of conformations of glycosidic linkages: the conformations of cyclomaltohexaose and amylopectin in aqueous solution. *Carbohyd Res.* 1988; **183**: 126–130.

16. Koehler JEH, Saenger W, van Gunsteren WF. Conformational differences between α-cyclodextrin in aqueous solution and in crystalline form: a molecular dynamics study. *J Mol Biol.* 1988; **203**: 241–250.

17. Lipkowitz KB. Symmetry breaking in cyclodextrins: a molecular mechanics investigation. *J Org Chem.* 1991; **56**: 6357–6367.

18. Matsui Y, Mochida K. Binding forces contributing to the association of cyclodextrin with alcohol in an aqueous solution. *Bull Chem Soc Jpn.* 1979; **10**: 2808–2814.

19. Satake I, Ikenoue T, Takeshita T, Hayakawa K, Maeda T. Conductometric and potentiometric studies of the association of α-cyclodextrin with ionic surfactants and their homologs. *Bull Chem Soc Jpn.* 1985; **58**: 2746–2750

20. Satake I, Yoshida S, Hayakawa K, Maeda T, Kusumoto Y. Conductometric determination of the association constants of β-cyclodextrin with amphiphilic ions. *Bull Chem Soc Jpn.* 1986; **59**: 3991–3993.

21. Sanemasa I, Osajima T, Deguchi T. Association of C5–C9 normal alkanes with cyclodextrins in aqueous medium. *Bull Chem Soc Jpn.* 1990; **63**: 3814–3819.

22. Tee OS, Gadosy TA, Giorgi JB. The binding of alkyl chains to -cyclodextrin and hydroxypropyl–cyclodextrin. *J Chem Soc. Perkin Trans.* 1993; **2**: 1705–1706.

23. Bastos M, Briggner L-E, Shehattaa I, Wadsö I. The binding of alkane-α,ω-diols to α-cyclodextrin: a microcalorimetric study. *J Chem Thermodyn.* 1990; **22**: 1181–1190.

24. Hingerty B, Saenger W. Topography of cyclodextrin inclusion complexes. 8. Crystal and molecular structure of the alpha-cyclodextrin-methanol-pentahydrate complex. Disorder in a hydrophobic cage. *J Am Chem Soc.* 1976; **98**: 3357–3365.

25. Saenger W, McMullan RK, Fayos J, Mootz D. Topography of cyclodextrin inclusion complexes. IV. Crystal and molecular structure of the cyclohexaamylose-1-propanol-4.8 hydrate complex. *Acta Cryst B.* 1974; **30**: 2019–2028.

26. Harata K. The structure of the cyclodextrin complex. IV. The crystal structure of α-cyclodextrin–sodium 1-propanesulfonate nonahydrate. *Bull Chem Soc Jpn.* 1977; **50**: 1259–1266.

27. Hamilton JA. Structure of inclusion complexes of cyclomaltoheptaose (cycloheptaamylose): crystal structure of the 1-adamantanemethanol adduct. *Carbohyd Research.* 1985; **142**: 21–47.

28. Hamilton JA, Sabesan MN. Structure of a complex of cycloheptaamylose with 1-adamantanecarboxylic acid. *Acta Cryst B.* 1982; **38**: 3063–3069.

29. Steiner T, Koellner G, Saenger W. A vibrating flexible chain in a molecular cage: crystal structure of the complex cyclomaltoheptaose (β-cyclodextrin)-1,4-butanediol·6.25H2O*1. *Carbohyd Research.* 1992; **228**: 321–332.

30. Geßler K, Steiner T, Koellner G, Saenger W. Crystal structures of cyclomaltoheptaose (β-cyclodextrin) complexed with ethylene glycol·8.0H2O and glycerol·7.2H2O*1. *Carbohyd Research.* 1993; **249**: 327–344.

31. Bergeron RJ, Channing MA, Gibeily GJ, Pillor DM. Disposition requirements for binding in aqueous solution of polar substrates in the cyclohexaamylose cavity. *J Am Chem Soc.* 1977; **99**: 5146–5151.

32. Buvari A, Barcza L. Complex formation of phenol, aniline, and their nitro derivatives with p-cyclodextrin. *J Chem Soc. Perkin Trans.* 1988; **II**(11): 543–545.

33. Connors KA, Lipari JM. Effect of cycloamyloses on apparent dissociation constants of carboxylic acids and phenols: equilibrium analytical selectivity induced by complex formation. *J Pharm Sci.* 1976; **65**: 379–383.

34. Connors KA, Lin S-F, Wong AB. Potentiometric study of molecular complexes of weak acids and bases applied to complexes of -cyclodextrin with para-substituted benzoic acids. *J Pharm Sci.* 1982; **71**: 217–222.

35. Bonora GM, Fornasier R, Scrimin P, Tonellato U. Hydrolytic cleavage of p-nitrophenyl alkanoates in aqueous solutions of cyclodextrins. *J Chem Soc. Perkin Trans.* 1985; **II**: 367–369.

36. VanEtten RL, Sebastian JF, Clowes GA, Bender ML. Acceleration of phenyl ester cleavage by cycloamyloses: a model for enzymic specificity. *J Am Chem Soc.* 1967; **89**: 3242–3253.

37. Harata K. The structure of the cyclodextrin complex. II. The crystal structure of α-cyclodextrin-methyl orange (2 : 1) complex. *Bull Chem Soc Jpn.* 1976; **49**: 1493–1501.

38. Kamitori S, Muraoka S, Kondo S, Okuyama K. Crystal structures of two forms of a 2 : 1 cyclomaltohexaose (α-cyclodextrin)/4,4′-biphenyldicarboxylic acid inclusion complex. *Carbohyd Research.* 1998; **312**: 177–181.

39. Fujiki M, Deguchi T, Sanemasa I. Association of naphthalene and its methyl derivatives with cyclodextrins in aqueous medium. *Bull Chem Soc Jpn.* 1988; **61**: 1163–1167.

40. Fujita K, Ejima S, Imoto T. α/β-selectivity in hydrolyses of α- or β- naphthyl acetates in the presence of cycloamyloses. *Tet Lett.* 1984; **25**: 3587–3590.

41. Añibarro M, Geßler K, Usón I, Sheldrick GM, Saenger W. X-ray structure of β-cyclodextrin-2,7-dihydroxy-naphthalene·4.6 H2O: an unusually distorted macrocycle. *Carbohyd Research.* 2001; **333**: 251–256.

42. Schneider H-J, Blatter T, Simova S. NMR and fluorescence studies of cyclodextrin complexes with guest molecules containing both phenyl and naphthyl units. *J Am Chem Soc.* 1991; **113**: 1996–2000.

43. Eftink MR, Andy ML, Bystrom K, Perlmutter HD, Kristol DS. Cyclodextrin inclusion complexes: studies of the variation in the size of alicyclic guests. *J Am Chem Soc.* 1989; **111**: 6765–6772.

44. Tamaki T, Kokubu T. Acceleration of the photodimerization of water-soluble anthracenes included by β- and γ-cyclodextrins. *J Inclus Phenom Macro Chem.* 1984; **2**: 815–822.

45. de la Pena AM, Ndou T, Zung JB, Warner IM. Stoichiometry and formation constants of pyrene inclusion complexes with beta- and gamma-cyclodextrin. *J Phys Chem.* 1991; **95**: 3330–3334.

46. Takahashi N, Gombojav B, Yoshinari T, Takahashi Y, Nagasaka S, Yamamoto A, Goto T, Kasuya A. Luminescences of pyrene single crystal and pyrene molecules inserted in a molecular vessel of cyclodextrin. *J Phys Soc Jpn.* 2007; **76**: 703.

47. López DL, Barroso SRO, Díez LMP. Selective determination of benzo(ghi)perylene in β-cyclodextrin medium. *Fres J Anal Chem.* 1990; **337**: 366–368.

48. Pistolis G, Malliaris A. Evidence for highly selective supramolecular formation between perylene/γ-CD and pyrene/γ-CD complexes in water. *J Phys Chem B.* 2004; **108**: 2846–2850.

49. Hamai S. Inclusion complexes of γ-cyclodextrin with coronene in aqueous methanol. *J Inclus Phenom Macro Chem.* 1991; **11**: 55–61.

50. Cloudy P, Létoffé JM, Germain P, Bastide JP, Bayol A, Blasquez S, Rao RC, Gonzalez B. Physicochemical characterization of cholesterol-beta cyclodextrin inclusion complexes. *J Therm Anal Calor.* 1991; **37**: 2497–2506.

51. Kamitori S, Hirotsu K, Higuchi T. Crystal and molecular structures of double macrocyclic inclusion complexes composed of cyclodextrins, crown ethers, and cations. *J Am Chem Soc.* 1987; **10**: 2409–2414.

52. Vögtle F, Müller WM. Complexes of γ-cyclodextrin with crown ethers, cryptands, coronates, and cryptates. *Angew Chem Int Ed.* 1987; **18**: 623–624.

53. Yonemura H, Saito H, Matsushima S, Nakamura H, Matsuo T. Anomalously stable cyclodextrin complexes of phenothiazine-viologen linked compounds with a long spacer chain. *Tet Lett.* 1989; **30**: 3143–3146.

54. Harada A, Kamachi M. Complex formation between poly(ethylene glycol) and α-cyclodextrin. *Macromolecules.* 1990; **23**: 2821–2823.

55. Harada A, Li J, Kamachi M. The molecular necklace: a rotaxane containing many threaded α-cyclodextrins. *Nature.* 1992; **356**: 325–327.

56. Harada A, Kamachi M. Complex formation between cyclodextrin and poly(propylene glycol). *J Chem Soc Chem Commun.* 1990; 1322–1323.

57. Harada A, Li J, Kamachi M. Double-stranded inclusion complexes of cyclodextrin threaded on poly(ethylene glycol). *Nature.* 1994; **370**: 126–128.

58. Wenz G, Keller B. Threading cyclodextrin rings on polymer chains. *Angew Chem Int Ed.* 1992; **31**: 197–199.

59. Harada A, Li J, Suzuki S, Kamachi M. Complex-formation between polyisobutylene and cyclodextrins: inversion of chain-length selectivity between beta-cyclodextrin and gamma-cyclodextrin. *Macromolecules.* 1995; **26**: 5267–5268.

60. Harada A, Li J, Kamachi M. Complex-formation between poly(methyl vinyl ether) and gamma-cyclodextrin. *Chem Lett.* 1993; **2**: 237–240.

61. Harada A, Okada M, Kawaguchi Y, Kamachi M. Macromolecular recognition: new cyclodextrin polyrotaxanes and molecular tubes. *Polym Adv Technol.* 1999; **10**: 3–12.

62. Kawaguchi Y, Nishiyama T, Okada M, Kamachi M, Harada A. Complex formation of poly(epsilon-caprolactone) with cyclodextrins. *Macromolecules.* 2000; **33**: 4472–4477

63. Harada A, Nishiyama T, Kawaguchi Y, Okada M, Kamachi M. Preparation and characterization of inclusion complexes of aliphatic polyesters with cyclodextrins. *Macromolecules.* 2000; **33**: 7115–7118.

64. Harada A, Adachi H, Kawaguchi Y, Okada M, Kamachi M. Complex formation of cyclodextrins with cationic polymers. *Poly J.* 1996; **28**: 159–163.

65. McMullan RK, Saenger W, Fayos J, Mootz D. Topography of cyclodextrin inclusion complexes *1. Part II. The iodine-cyclohexa-amylose tetrahydrate complex; its molecular geometry and cage-type crystal structure. *Carbohyd Res.* 1973; **31**: 211–227.

66. Noltemeyer M, Saenger W. Topography of cyclodextrin inclusion complexes. 12. Structural chemistry of linear.alpha.-cyclodextrin-polyiodide complexes. X-ray crystal structures of (.alpha.-cyclodextrin)2. LiI3.I2.8H2O and (.alpha.-cyclodextrin)2.Cd0.5).I5.27H2O. Models for the blue amylose-iodine complex. *J Am Chem Soc.* 1980; **102**: 2710–2722.

67. Nimz O, Geßler K, Uson I, Laettig S, Welfle H, Sheldrick GM, Saenger W. X-ray structure of the cyclomalto-hexaicosaose triiodide inclusion complex provides a model for amylose–iodine at atomic resolution. *Carbohyd Res.* 2003; **338**: 977–986.

68. Harata K, Akasaka H, Endo T, Nagase H, Ueda H. X-ray structure of the δ-cyclodextrin complex with cycloun-decanone. *J Chem Soc Chem Commun.* 2002; 1968–1969.

69. Saenger W, Noltemeyer M. Topographie der Cyclodextrin-Einschlußverbindungen, VII. Röntgenstrukturanalyse des -Cyclodextrin. Krypton-Pentahydrats: Zum Einschlußmechanismus des Modell-Enzyms. *Chem Ber.* 1976; **109**: 503–517.

70. Hybl A, Rundle RE, Williams DE. The crystal and molecular structure of the cyclohexaamylose-potassium acetate complex. *J Am Chem Soc.* 1985; **87**: 2779–2788.

71. Odagaki Y, Hirotsu K, Higuchi T, Harada A, Takahashi S. X-ray structure of the α-cyclodextrin–ferrocene (2 : 1) inclusion compound. *J Chem Soc. Perkin Trans.* 1990; **1**: 1230–1231.

72. Andersson T, Westman G, Wennerström O, Sundahl M. NMR and UV–VIS investigation of water-soluble fullerene-60–γ-cyclodextrin complex. *J Chem Soc. Perkin Trans.* 1994; **2**: 1097–1101.

73. Braun T, Buvári-Barcza Á, Barcza L, Konkoly-Thege I, Fodor M, Migali B. Mechanochemistry: a novel approach to the synthesis of fullerene compounds. Water soluble buckminsterfullerene–γ-cyclodextrin inclusion complexes via a solid-solid reaction. *Solid State Ionics.* 1994; **74**: 47–51.

74. Murthy CN, Geckeler KE. Stability studies on the water-soluble beta-cyclodextrin [60]fullerene inclusion complex. *Fullerenes Nanotubes and Carbon Nanostructures.* 2002; **10**: 91–98.

75. Ikeda A, Matsumoto M, Akiyama M, Kikuchi J, Ogawa T, Takeya T. Direct and short-time uptake of [70]fullerene into the cell membrane using an exchange reaction from a [70]fullerene–γ-cyclodextrin complex and the resulting photodynamic activity. *J Chem Soc Chem Commun.* 2009; 1547–1549.

76. Nakagawa T, Ueno K, Kashiwa M, Watanabe J. The stereoselective synthesis of cyclomaltopentaose: a novel cyclodextrin homologue with DP five. *Tet Lett.* 1994; **35**: 1921–1924.

77. Taira H, Nagase H, Endo T, Ueda H. Isolation, purification and characterization of large-ring cyclodextrins (CD36~CD39). *J Inclus Phenom Macro Chem.* 2006; **56**: 23–28.

78. Takaha T, Yanase M, Takata H, Okada S, Smith SM. Potato D-enzyme catalyzes the cyclization of amylose to produce cycloamylose, a novel cyclic glucan. *J Biol Chem.* 1996; **271**: 2902–2908.

79. Süvegh K, Fujiwara K, Komatsu K, Marek T, Ueda T, Vértes A, Braun T. Positron lifetime in supramolecular gamma- and delta-cyclodextrin-C60 and -C70 compounds. *Chem Phys Lett.* 2001; **344**: 263–269.

80. Furuishi T, Endo T, Nagase H, Ueda H, Nagai T. Solubilization of C-70 into water by complexation with delta-cyclodextrin. *Chem Pharm Bull.* 1998; **46**: 1658–1659.

81. Dodziuk H, Ejchart A, Anczewski W, Ueda H, Krinichnaya E, Dolgonos G, Kutner W. Water solubilization, determination of the number of different types of single-wall carbon nanotubes and their partial separation with respect to diameters by complexation with -cyclodextrin. *J Chem Soc Chem Commun.* 2003; 986–987.

82. Nimz O, Gessler K, Usón I, Sheldrick GM, Saenger W. Inclusion complexes of V-amylose with undecanoic acid and dodecanol at atomic resolution: X-ray structures with cycloamylose containing 26 d-glucoses (cyclo-hexaicosaose) as host. *Carbohyd Res.* 2004; **399**: 1427–1437.

83. Croft AP, Bartsch R.A. Synthesis of chemically modified cyclodextrins. *Tetrahedron.* 1983; **39**: 1417–1474.

84. Ciucanu I, Kerek F. A simple and rapid method for the permethylation of carbohydrates. *Carbohyd Res.* 1984; **131**: 209–217.

85. Steiner T, Saenger W. Closure of the cavity in permethylated cyclodextrins through glucose inversion, flipping, and kinking. *Angew Chem Int Ed.* 1998; **37**: 3404–3407.

86. Kraus T, Buděšínský M, Závada J. General approach to the synthesis of persubstituted hydrophilic and amphiphilic β-cyclodextrin derivatives. *J Org Chem.* 2001; **66**: 4595–4600.

87. Freudenberg K, Jacobi R. Über Schardingers Dextrine aus Stärke. *Justus Liebigs Ann Chem.* 1935; **518**: 102–108.

88. Boger J, Corcoran RJ, Lehn J-M. Cyclodextrin chemistry: selective modification of all primary hydroxyl groups of α- and β-cyclodextrins. *Helv Chim Acta.* 1978; **61**: 2190–2218.

89. Añibarro M, Gessler K, Usón I, Sheldrick GM, Harata K, Uekama K, Hirayama F, Abe Y, Saenger W. Effect of peracylation of β-cyclodextrin on the molecular structure and on the formation of inclusion complexes: an X-ray study. *J Am Chem Soc.* 2001; **123**: 11854–11862.

90. Umezawa S, Tatsuta K. Studies of aminosugars. XVIII. Syntheses of amino derivatives of schardinger α-dextrin and raffinose. *Bull Chem Soc Jpn.* 1968; **41**: 464–468.

91. Takeo K, Uemura K, Mitoh H. Derivatives of α-cyclodextrin and the synthesis of 6-O–D-glucopyranosyl–cyclodextrin. *J Carbohy Chem.* 1988; **7**: 293–308.

92. Takeo K, Mitoh H, Uemura K. Selective chemical modification of cyclomalto-oligosaccharides via tert-butyldimethylsilylation. *Carbohyd Res.* 1989; **187**: 203–221.

93. Ashton PR, Königer R, Stoddart JF, Alker D, Harding VD. Amino acid derivatives of β-cyclodextrin. *J Org Chem.* 2001; **61**: 903–908.

94. Calvo-Flores FG, Isac-García J, Hernández-Mateo F, Pérez-Balderas F, Calvo-Asín JA, Sanchéz-Vaquero E, Santoyo-González G. 1,3-dipolar cycloadditions as a tool for the preparation of multivalent structures. *Org Lett.* 2000; **2**: 2499–2502.

95. Kobayashi K, Kajikawa1 K, Sasabea H, Knoll W. Monomolecular layer formation of amphiphilic cyclodextrin derivatives at the air/water interface. *Thin Solid Films.* 1999; **349**: 244–249.

96. Ortega-Caballero F, Giménez-Martínez JJ, García-Fuentes L, Ortiz-Salmerón E, Santoyo-González F, Vargas-Berengue A. Binding affinity properties of dendritic glycosides based on a β-cyclodextrin core toward guest molecules and concanavalin A. *J Org Chem.* 2001; **66**: 7786–7795.

97. Takahashia K, Hattoria K, Toda F. Monotosylated α- and β-cyclodextrins prepared in an alkaline aqueous solution. *Tet Lett.* 1984; **25**: 3331–3334.

98. Cornwell MJ, Huff JB, Bieniarz C. A one-step synthesis of cyclodextrin monoaldehydes. *Tet Lett.* 1995; **36**: 8371–8374.

99. Ikuta A, Mizuta N, Kitahata S, Murata T, Usui T, Koizumi K, Tanimoto T. Preparation and characterization of novel branched β-cyclodextrins having β-D-galactose residues on the non-reducing terminal of the side chains and their specific interactions with peanut (Arachis hypogaea) agglutinin. *Chem Pharm Bull.* 2004; **52**: 51–56.

100. Narita M, Hamada F, Suzuki I, Osa T. Variations of fluorescent molecular sensing for organic guests by regioselective anthranilate modified β- and γ-cyclodextrins. *J Chem Soc. Perkin Trans.* 1998; **2**: 2751–2758.

101. Narita M, Hamada F. The synthesis of a fluorescent chemo-sensor system based on regioselectively dansyl-tosyl-modified β- and γ-cyclodextrins. *J Chem Soc. Perkin Trans.* 2000; **2**: 823–832.

102. Yuan D-Q, Yamada T, Fujita K. Amplification of the reactivity difference between two methylene groups of cyclodextrins via a cap. *Chem Commun.* 2001; 2706–2707.

103. Tabushi I, Kuroda Y, Yokota K, Yuan LC. Regiospecific A,C- and A,D-disulfonate capping of.beta.-cyclodextrin. *J Am Chem Soc.* 1981; **103**: 711–712.

104. Atsumi M, Izumida M, Yuan D-Q, Fujita K. Selective synthesis and structure determination of 6A,6C,6E-tri(O-sulfonyl)-β-cyclodextrins. *Tet Lett.* 2000; **41**: 8117–8120.

105. Teranishi K, Tanabe S, Hiramatsu M, Yamada T. Convenient regioselective mono-2-O-sulfonation of cyclomaltooctaose. *Biosc Biotech Biochem.* 1998; **62**: 1249–1252.

106. Law H, Baussanne I, Fernández JMG, Defaye J. Regioselective sulfonylation at O-2 of cyclomaltoheptaose with 1-(p-tolylsulfonyl)-(1H)-1,2,4-triazole. *Carbohyd Res.* 2003; **338**: 251–253.

107. Suzuki M, Nozoe Y. Facile preparation of mono-2-O-modified eicosa-O-methylcyclomaltoheptaoses (β-cyclodextrins). *Carbohyd Res.* 2002; **337**: 2393–2397.

108. Fujita K, Tahara T, Imoto T, Koga T. Regiospecific sulfonation onto C-3 hydroxyls of beta-cyclodextrin: preparation and enzyme-based structural assignment of 3A,3C and 3A3D disulfonates. *J Am Chem Soc.* 1986; **108**: 2030–2034.

109. Fujita K, Tahara T, Yamamura H, Imoto T, Koga T, Fujioka T, Mihashi K. Specific preparation and structure determination of 3A,3C,3E-tri-O-sulfonyl-.beta.-cyclodextrin. *J Org Chem.* 1990; **55**: 877–880.

110. Ueno A, Breslow R. Selective sulfonation of a secondary hydroxyl group of β-cyclodextrin. *Tet Lett.* 1982; **34**: 3451–3454.

111. Yuan D-Q, Tahara T, Chen W-H, Okabe Y, Yang C, Yagi Y, Nogami Y, Fukudome M, Fujita K. Functionalization of cyclodextrins via reactions of 2,3-anhydrocyclodextrins. *J Org Chem.* 2003; **68**: 9456–9466.

112. Yang C, Yuan D-Q, Nogamib Y, Fujita K. Per(3-deoxy)-γ-cyclomannin: a non-glucose cyclooligosaccharide featuring inclusion properties. *Tet Lett.* 2003; **44**: 4641–4644.

113. Harata K, Takenaka Y, Yoshida N. Crystal structures of 6-deoxy-6-monosubstituted – cyclodextrins: substituent-regulated one-dimensional arrays of macrocycles. *J Chem Soc. Perkin Trans.* 2001; **2**: 1667–1673.

114. Impellizzeri G, Pappalardo G, D'Alessandro F, Rizzarelli E, Saviano M, Iacovino R, Benedetti E, Pedone C. Solid state and solution conformation of 6-[4-[N-tert-butoxycarbonylN-(N-ethyl)propanamide]imidazolyl]-6-deoxycyclo maltoheptaose: evidence of self-inclusion of the boc group within the β-cyclodextrin cavity. *Eur J Org Chem.* 2000; 1065–1076.

115. Saviano M, Benedetti E, di Blasio B, Gavuzzo E, Fierro O, Pedone C, Iacovino R, Rizzarelli E, Vecchio G. Difunctionalized -cyclodextrins: synthesis and X-ray diffraction structure of 6I,6II-dideoxy-6I,6II-bis [2-(2-pyridyl)ethylamino]–cyclomaltoheptaose. *J Chem Soc. Perkin Trans.* 2001; **2**: 946–952.

116. Harata K, Uekama K, Otagiri M, Hirayama F. The structure of the cyclodextrin complex. XV. Crystal structure of hexakis(2,3,6-tri-O-methyl)-α-cyclodextrin–p-nitrophenol (1 : 1) complex monohydrate. *Bull Chem Soc Jpn.* 1982; **55**: 3905–3910.

117. Harata K, Uekama K, Otagiri M, Hirayama F. The structure of the cyclodextrin complex. XI. Crystal structure of hexakis(2,3,6-tri-O-methyl)-α-cyclodextrin–p-iodoaniline monohydrate. *Bull Chem Soc Jpn.* 1982; **55**: 407–410.

118. Mentzafos D, Mavridis IM, Schenk H. Crystal structure of the 1 : 1 complex of heptakis(2,3,6-tri-O-methyl)cyclomaltoheptaose (permethylated β-cyclodextrin) with ethyl laurate. *Carbohyd Res.* 1994; **253**: 39–50.

119. Armspach D, Matt D, Kyritsakas N. Anchoring a helical handle across a cavity: the first 2,2'-bipyridyl-capped α-cyclodextrin capable of encapsulating transition metals. *Polyhedron.* 2001; **20**: 663–668.

120. Engeldinger É, Armspach D, Matt D, Toupet L, Wesolek M. Synthesis of large chelate rings with diphosphites built on a cyclodextrin scaffold: unexpected formation of 1,2-phenylene-capped α-cyclodextrins. *Comp Rend Chim.* 2002; **5**: 359–372.

121. Armspach D, Matt D. Metal-capped α-cyclodextrins: squaring the circle. *Inorg Chem.* 2001; **40**: 3505–3509.

122. Harata K. The structure of the cyclodextrin complex. V. Crystal structures of α-cyclodextrin complexes with p-nitrophenol and p-hydroxybenzoic acid. *Bull Chem Soc Jpn.* 1977; **50**: 1416–1424.

123. Harata K, Uedaira H, Tanaka J. The structure of the cyclodextrin complex. VI. The crystal structure of α-cyciodextrin–m-nitrophenol (1 : 2) complex. *Bull Chem Soc Jpn.* 1978; **51**: 1627–1634.

124. Mráz J, Feltl L, Smolková-Keulemansová E. Cyclodextrins and methylated cyclodextrins as stationary phases in gas-solid chromatography. *J Chromatogr.* 1984; **286**: 17–22.

125. Harata K. The structure of the cyclodextrin complex. XII. Crystal structure of α-cyclodextrin-1-phenylethanol (1 : 1) tetrahydrate. *Bull Chem Soc Jpn.* 1982; **55**: 1367–1371.

126. Hamilton JA, Chen L. Crystal structure of an inclusion complex of beta-cyclodextrin with racemic fenoprofen: direct evidence for chiral recognition. *J Am Chem Soc.* 1988; **110**: 5833–5841.

127. Hamilton JA, Chen L. Crystal structures of inclusion complexes of beta-cyclodextrin with S-(+)- and (R)-(−)-fenoprofen. *J Am Chem Soc.* 1988; **110**: 4379–4391.

128. Harata K, Uekama K, Otagiri M, Hirayama F. The structure of the cyclodextrin complex. XIX. Crystal structures of hexakis(2,3,6-tri-O-methyl)-α-cyclodextrin complexes with (S)- and (R)-mandelic acid. Chiral recognition through the induced-fit conformational change of the macrocyclic ring. *Bull Chem Soc Jpn.* 1987; **60**: 497–502.

129. Grandeury A, Petit S, Gouhier G, Agasse V, Coquerel G. Enantioseparation of 1-(p-bromophenyl)ethanol by crystallization of host–guest complexes with permethylated β-cyclodextrin: crystal structures and mechanisms of chiral recognition. *Tet Asymm.* 2003; **14**: 2143–2152.

130. Grandeury A, Tisse S, Gouhier G, Agasse V, Petit S, Coquerel G. Crystallization of supramolecular complexes as an alternative route for the separation of racemic p-X-phenylethanol. *Chem Eng Tech.* 2003; **26**: 354–358.

131. Schurig V. Separation of enantiomers by gas chromatography. *J Chroma A.* 2001; **906**: 275–299.

132. Schurig V, Nowotny H-P. Separation of enantiomers on diluted permethylated β-cyclodextrin by high-resolution gas chromatography. *J Chroma A.* 1988; **441**: 155–163.

133. Armstrong DW, Li W, Pitha J. Reversing enantioselectivity in capillary gas chromatography with polar and nonpolar cyclodextrin derivative phases. *Anal Chem.* 1990; **62**: 214–217.

134. Ward TJ, Armstrong DW. Improved cyclodextrin chiral phases: a comparison and review. *J Liq Chroma.* 1986; **9**: 407–423.

135. Haginaka J, Wakai J. Beta-cyclodextrin-bonded silica for direct injection analysis of drug enantiomers in serum by liquid chromatography. *Anal Chem.* 1990; **62**: 997–1000.

136. Stalcup AM, Williams KL. Determination of enantiomers in human serum by direct injection onto a β-cyclodextrin hplc bonded phase. *J Liq Chroma.* 1992; **15**: 29–37.

137. Armstrong DW, Stalcup AM, Hilton ML, Duncan JD, Faulkner JR Jr, Chang SC. Derivatized cyclodextrins for normal-phase liquid chromatographic separation of enantiomers. *Anal Chem.* 1990; **62**: 1610–1615.

138. Breslow R, Halfon S, Zhang B. Molecular recognition by cyclodextrin dimers. *Tetrahedron.* 1995; **51**: 377–388.

139. Tabushi I, Kuroda Y, Shimokawa K. Duplex cyclodextrin. *J Am Chem Soc.* 1979; **101**: 1614–1615.

140. Fujita K, Ejima S, Imoto T. Fully collaborative guest binding by a double cyclodextrin host. *J Chem Soc Chem Commun.* 1984; 1277–1278.

141. Fujita K, Ejima S, Imoto T. 1 : 2 host–guest binding by double γ-cyclodextrin. *Chem Lett.* 1985; **14**: 11–12.

142. Breslow R, Greenspoon N, Guo T, Zarzycki R. Very strong binding of appropriate substrates by cyclodextrin dimers. *J Am Chem Soc.* 1989; **111**: 8296–8297.

143. Breslow R, Chung S. Strong binding of ditopic substrates by a doubly linked occlusive C1 clamshell as distinguished from an aversive C2 loveseat cyclodextrin dimer. *J Am Chem Soc.* 1990; **112**: 9659–9660.

144. Birlirakis N, Henry B, Berthault P, Venema F, Nolte RJM. High resolution 1H-NMR study on self-complexation phenomena in cyclodextrin dimers. *Tetrahedron.* 1998; **54**: 3513–3522.

145. Venema F, Nelissen HFM, Berthault P, Birlirakis N, Rowan AE, Feiters MC, Nolte RJM. Synthesis, conformation, and binding properties of cyclodextrin homo and heterodimers connected through their secondary sides. *Chem Eur J.* 1998; **4**: 2237–2250.

146. Yamada T, Fukuhara G, Kaneda T. Molecular magic: formation of a self-inclusion complex from a dumbbell-shaped permethylated β-cyclodextrin derivative. *Chem Lett.* 2003; **32**: 534–535.

147. Fukudome M, Okabe YI, Yuan D-Q, Fujita K. Cyclodextrin-accelerated cleavage of phenyl esters: is it the 2-hydroxy or the 3-hydroxy that promotes the acyl transfer? *J Chem Soc Chem Commun.* 1999; 1045–1046.

148. Yamamura H, Yamada S, Kohno K, Okuda N, Araki S, Kobayashi K, Katakai R, Kano K, Kawai M. Preparation and guest binding of novel -cyclodextrin dimers linked with various sulfur-containing linker moieties. *J Chem Soc Perkin Trans.* 1999; **1**: 2943–2948.

149. Liu Y, You C-C, Chen Y, Wada T, Inoue Y. Molecular recognition studies on supramolecular systems. 25. Inclusion complexation by organoselenium-bridged bis(β-cyclodextrin)s and their platinum(IV) complexes. *J Org Chem.* 1999; **64**: 7781–7787.

150. Liu Y, Li L, Zhang H-Y, Song Y. Synthesis of novel bis(β-cyclodextrin)s and metallobridged bis (β-cyclodextrin)s with 2,2′-diselenobis(benzoyl) tethers and their molecular multiple recognition with model substrates. *J Org Chem.* 2003; **68**: 527–536.

151. Ren X, Xue Y, Liu J, Zhang K, Zheng J, Luo G, Guo C, Mu Y, Shen J. A novel cyclodextrin-derived tellurium compound with glutathione peroxidase activity. *Chem Bio Chem.* 2002; **3**: 356–363.

152. Breslow R, Yang Z, Ching R, Trojandt G, Odobel F. Sequence selective binding of peptides by artificial receptors in aqueous solution. *J Am Chem Soc.* 1998; **120**: 3536–3537.

153. Liu Y, Yang Y-W, Song Y, Zhang H-Y, Ding F, Wada TO, Inoue Y. Residue- and sequence-selective binding of nonaromatic dipeptides by bis(cyclodextrin) with a functional tether. *Chem Bio Chem.* 2004; **5**: 868–874.

154. Ikeda H, Matsuhisa A, Ueno A. Efficient transport of saccharides through a liquid membrane mediated by a cyclodextrin dimer. *Chem Eur J.* 2003; **9**: 4907–4910.

155. Sasaki K, Nagasaka M, Kuroda Y. New cyclodextrin dimer and trimer: formation of biphenyl excimer and their molecular recognition. *J Chem Soc Chem Commun.* 2001; 2630–2631.

156. Leung DK, Atkin JH, Breslow R. Synthesis and binding properties of cyclodextrin trimers. *Tet Lett.* 2001; **42**: 6255–6258.

157. Nakajima H, Sakabe Y, Ikeda H, Ueno A. *Bioorg Chem Med Lett.* 2004; **14**: 1783–1786.

158. Mulder A, Jukovi A, van Leeuwen FWB, Kooijman H, Spek AL, Huskens J, Reinhoudt D.N. Photocontrolled release and uptake of a porphyrin guest by dithienylethene-tethered β-cyclodextrin host dimers. *Chem Eur J.* 2004; **10**: 1114–1123.

159. Wiedenhof N, Lammers JNJJ, van Panthaleon Van Eck CL. Properties of cyclodextrins. Part III. Cyclodextrin-epichlorhydrin resins: preparation and analysis. *Starch.* 1969; **21**: 119–123.

160. Renard E, Deratani A, Volet G, Sebille B. Preparation and characterization of water soluble high molecular weight β-cyclodextrin-epichlorohydrin polymers. *Eur Polym J.* 1997; **33**: 49–57.

161. Martel B, Leckchiri Y, Pollet A, Morcellet M. Cyclodextrin-poly(vinylamine) systems. I. Synthesis, characterization and conformational properties. *Eur Polym J.* 1995; **31**: 1083–1088.

162. Martel B, Morcellet M. Cyclodextrin-poly(vinylamine) systems. II. Catalytic hydrolysis of p-nitrophenyl acetate. *Eur Polym J.* 1995; **31**: 1083–1088.

163. Ruebner A, Statton GL, James MR. Synthesis of a linear polymer with pendent γ-cyclodextrins. *Macromol Chem Phys.* 2000; **201**: 1185–1188.

164. Suh J, Lee SH, Zoh KD. A novel host containing both binding site and nucleophile prepared by attachment of beta-cyclodextrin to poly(ethylenimine). *J Am Chem Soc.* 1992; **114**: 7916–7917.

165. Renard E, le Volet G, Amiel C. Synthesis of a novel linear water-soluble β-cyclodextrin polymer. *Polym Int.* 2005; **54**: 594–599.

166. Weickenmeier M, Wenz G. Cyclodextrin sidechain polyesters: synthesis and inclusion of adamantane derivatives. *Macromol Rapid Commun.* 1996; **17**: 731–736.

167. Harada A, Furue M, Nozakura S. Cyclodextrin-containing polymers. 1. Preparation of polymers. *Macromolecules.* 1976; **9**: 701–704.

168. Harada A, Furue M, Nozakura S. Cyclodextrin-containing polymers. 2. Cooperative effects in catalysis and binding. *Macromolecules.* 1976; **9**: 705–710.

169. Harada A, Furue M, Nozakura S. Interaction of cyclodextrin-containing polymers with fluorescent compounds. *Macromolecules.* 1977; **10**: 676–681.

170. Werner TC, Warner IM. The use of naphthalene fluorescence probes to study the binding sites on cyclodextrin polymers formed from reaction of cyclodextrin monomers with epichlorohydrin. *J Inclus Phenom Macro Chem.* 1994; **18**: 385–396.

171. Liu Y-Y, Fan X-D. Synthesis and characterization of pH- and temperature-sensitive hydrogel of N-isopropylacrylamide/cyclodextrin based copolymer. *Polymer.* 2002; **43**: 4997–5003.

172. Liu Y-Y, Fan X-D. Preparation and characterization of a novel responsive hydrogel with a β-cyclodextrin-based macromonomer. *J Appl Poly Sci.* 2002; **89**: 361–367.

173. Yashima E, Maeda K, Sato O. Switching of a macromolecular helicity for visual distinction of molecular recognition events. *J Am Chem Soc.* 2001; **123**: 8159–8160.

174. Suh J, Haha SS, Lee SH. Dendrimer poly(ethylenimine)s linked to β-cyclodextrin. *Bioorg Chem.* 1997; **25**: 63–75.

175. Maiko M, Hiroshi I, Munetaka N. Synthesis of BETA-cyclodextrin modified dendrimer and study of its ability in inclusion-complex formation by fluorescence and absorption spectroscopies. *J Spectroscopical Soc Jpn.* 2003; **52**: 19–23.

176. Hishiya T, Shibata M, Kakazu M, Asanuma H, Komiyama M. Molecularly imprinted cyclodextrins as selective receptors for steroids. *Macromolecules.* 1999; **32**: 2265–2269.

177. Asanuma H, Akiyama T, Kajiya K, Hishiya T, Komiyama M. Molecular imprinting of cyclodextrin in water for the recognition of nanometer-scaled guests. *Anal Chim Acta.* 2001; **435**: 25–33.

178. Tsai H-A, Syu M-J. Synthesis of creatinine-imprinted poly(β-cyclodextrin) for the specific binding of creatinine. *Biomaterials.* 2005; **26**: 2759–2766.

179. Liu Y, Fan Z, Zhang H-Y, Yang Y-W, Ding F, Liu S-X, Wu X, Wada T, Inoue Y. Supramolecular self-assemblies of β-cyclodextrins with aromatic tethers: factors governing the helical columnar versus linear channel superstructures. *J Org Chem.* 2003; **68**: 8345–8352.

180. Park JW, Song HE, Lee SY. Facile dimerization and circular dichroism characteristics of 6-O-(2-sulfonato-6-naphthyl)-β-cyclodextrin. *J Phys Chem B.* 2002; **106**: 5177–5183.

181. Hoshino T, Miyauchi M, Kawaguchi Y, Yamaguchi H, Harada A. Daisy chain necklace: tri[2]rotaxane containing cyclodextrins. *J Am Chem Soc.* 2000; **122**: 9876–9877.

182. Kaneda T, Yamada T, Fujimoto T, Sakata Y. The first [5]supercyclodextrin whose cyclopentameric array is held only by a mechanical bond. *Chem Lett.* 2001; **30**: 1264–1265.

183. Miyauchi M, Harada A. Construction of supramolecular polymers with alternating α-, β-cyclodextrin units using conformational change induced by competitive guests. *J Am Chem Soc.* 2004; **126**: 11418–11419.

184. Park JW, Lee SY, Park KK. Molecular recognition of organic ammonium ions by diaza-crown ether-modified β-cyclodextrin in aqueous media. *Chem Lett.* 2000; **29**: 594–595.

185. Fulton DA, Cantrill SJ, Stoddart JF. Probing polyvalency in artificial systems exhibiting molecular recognition. *J Org Chem.* 2002; **67**: 7968–7981.

186. Liu Y, Yang Y-W, Li L, Chen Y. Cooperative molecular recognition of dyes by dyad and triad cyclodextrin–crown ether conjugates. *Org Biomol Chem.* 2004; **2**: 1542–1548.

187. Suzuki I, Obata K, Anzai J-I, Ikeda H, Ueno A. Crown ether-tethered cyclodextrins: superiority of the secondary-hydroxy side modification in binding tryptophan. *J Chem Soc. Perkin Trans.* 2000; **2**: 1705–1710.

188. Liu Y, Chen Y, Li L, Huang G, You C-C, Zhang H-Y, Wada T, Inoue Y. Cooperative multiple recognition by novel calix[4]arene-tethered β-cyclodextrin and calix[4]arene-bridged bis(β-cyclodextrin). *J Org Chem.* 2001; **66**: 7209–7215.

189. Chen W-H, Yan J-M, Tagashira Y, Yamaguchi M, Fujita K. Cage molecules with multiple recognition cavities: quadruply cyclodextrin-linked cofacial porphyrins. *Tet Lett.* 1999; **40**: 891–894.

190. Terra SG, Washam JB, Dunham GD, Gattis WA. Therapeutic range of Digoxin's efficacy in heart failure: what is the evidence? *Pharmacotherapy.* 1999; **19**: 1123–1126.

191. Roco MC, Bainbridge WS. Converging Technologies for Improving Human Performance. National Science Foundation Report. Dordrecht: Kluwer Academic; 2002.

192. Langer R. Biomaterials and biomedical engineering. *Chem Eng Sci.* 1995; **50**: 4109–4121.

193. Langer R. Drug delivery and targeting. *Nature.* 1998; **392**(suppl): 5–10.

194. Kaparissides C, Alexandridou S, Kotti K, Chaitidou S. Recent advances in novel drug delivery systems. *J Nanotech.* 2006. Available from: 10.2240/azojono0111.

195. Davis F, Higson SPJ. Carrier systems and biosensors for biomedical applications. In: Tissue Engineering using Ceramics and Polymers. Woodhead; 2007.

196. Uekama K, Otagiri M. Cyclodextrins in drug carrier systems. *CRC Crit Rev Ther Drug Carr Syst.* 1987; **3**: 1–40.

197. Uekama K. Pharmaceutical applications of cyclodextrins and their derivatives. In: Dodziuk H, editor. Cyclodextrins and Their Complexes. Wiley-VCH; 2006.

198. Loftsson T, Duchêne D. Cyclodextrins and their pharmaceutical applications. *Intl J Pharmaceut.* 2007; **329**: 1–11.

199. French D. The Schardinger dextrins. *Adv Carbohydr Chem.* 1957; **12**: 189–260.

200. Uekama K. Design and evaluation of cyclodextrin-based drug formulation. *Chem Pharma Bull.* 2004; **52**: 900–915.

201. Salapatek A, Mccue S, Benz M, Eisfeld BS, Mancini D, Patel P, Pipkin JD, Zimmerer R. Captisol-enabled (R) Budesonide nasal spray significantly reduces total nasal (TNSS) and non-nasal symptoms (TNNSS) compared to Rhinocort Aqua (R) and placebo in ragweed allergic patients in an environmental exposure chamber (EEC) model. *Annals Allergy Asthma Immunol.* 2008; **100**: A93.

202. Uekama K, Fujinaga T, Hirayama F, Otagiri M, Yamasaki M, Seo H, Hashimoto T, Tsuruoka M. Improvement of the oral bioavailability of digitalis glycosides by cyclodextrin complexation. *J Pharmaceut Sci.* 1983; **72**: 1338–1341.

203. Uekama K, Minami K, Hirayama F. 6A-O-[(4-biphenylyl)acetyl]-α-, -β-, and -γ-cyclodextrins and 6A-deoxy-6A-[[(4-biphenylyl)acetyl]amino]-α-, -β-, and -γ-cyclodextrins: potential prodrugs for colon-specific delivery. *J Med Chem.* 1997; **40**: 2755–2761.

204. Croyle MA, Roessler BJ, Hsu C-P, Sun R, Amidon GL. Beta cyclodextrins enhance adenoviral-mediated gene delivery to the intestine. *Pharm Res.* 1998; **15**: 1348–1355.

205. Roessler BJ, Bielinska AU, Janczak K, Lee I, Baker JR Jr. Substituted β-cyclodextrins interact with PAMAM dendrimer–DNA complexes and modify transfection efficiency. *Biochem Biophyl Res Commun.* 2001; **283**: 124–129.

206. Arima H, Kihara F, Hirayama F, Uekama K. Enhancement of gene expression by polyamidoamine dendrimer conjugates with α-, β-, and γ-cyclodextrins. *Bioconjug Chem.* 2001; **12**: 476–484.

207. Trichard L, Duchene D, Bochot A. Cyclodextrins in dispersed systems. In: Dodziuk H, editor. Cyclodextrins and Their Complexes. Wiley-VCH; 2006.

208. Shaw PE, Buslig BS. Selective removal of bitter compounds from grapefruit juice and from aqueous solution with cyclodextrin polymers and with Amberlite XAD-4. *J Agric Food Chem.* 1986; **34**: 837–840.

209. http://www.wacker.com/cms/media/publications/downloads/6266_EN.pdf [accessed 2009 Aug 12].

210. Patent application DE 10200657 A.

211. Wang L. Dryer sheets. *Chem Eng News.* 2008; **15**: 47.

212. Patel MH, Hvizdos SA. Cooling agent/cyclodextrin complex for improved flavor release. US Patent 5,165,943. 1991.

213. Breslow R, Zhang B. Cholesterol recognition and binding by cyclodextrin dimers. *J Am Chem Soc.* 1996; **118**: 8495–8496.

214. Szente L, Szejtli J. Cyclodextrins as food ingredients. *Trends Food Sci Techn.* 2004; **1**: 137–142.

215. Kwak HS, Jung CS, Shim SY, Ahn J. Removal of cholesterol from cheddar cheese by β-cyclodextrin. *J Agric Food Chem.* 2002; **50**: 7293–7298.

216. Morillo E. Applications of cyclodextrins in agrochemistry. In: Dodziuk H, editor. Cyclodextrins and Their Complexes. Wiley-VCH; 2006.

217. Hashimoto H. Present status of industrial application of cyclodextrins in Japan. *J Incl Phenom Macro Chem.* 2002; **44**: 57–62.

218. Szejtli J. Cyclodextrins in the textile industry. *Starch/Stärke.* 2003; **55**: 191–196.

219. Komiyama M, Monflier F. Cyclodextrin catalysis. In: Dodziuk H, editor. Cyclodextrins and Their Complexes. Wiley-VCH; 2006.

220. Schmidt PM, Brown RS, Luong JHT. Inclusion complexation of tetrathiafulvalene in cyclodextrins and bio-electroanalysis of the glucose-glucose oxidase reaction. *Chem Eng Sci.* 1995; **50**: 1867–1876.

221. Bouchta D, Izaoumen N, Zejli H, El Kaoutit M, Temsamani KR. A novel electrochemical synthesis of poly-3-methylthiophene-γ-cyclodextrin film: application for the analysis of chlorpromazine and some neurotransmitters. *Biosens Bioelec.* 2005; **20**: 2228–2235.

222. Reece DA, Ralph SF, Wallace GG. Metal transport studies on inherently conducting polymer membranes containing cyclodextrin dopants. *J Membr Sci.* 2005; **249**: 9–20.

223. Nelles G, Weisser M, Back R, Wohlfart P, Wenz G, Mittler-Neher S. Controlled orientation of cyclodextrin derivatives immobilized on gold surfaces. *J Am Chem Soc.* 1996; **118**: 5039–5046.

224. Henke C, Steinem C, Janshoff A, Steffan G, Luftmann H, Sieber M, Galla H-J. Self-assembled monolayers of monofunctionalized cyclodextrins onto gold: a mass spectrometric characterization and impedance analysis of host–guest interaction. *Anal Chem.* 1996; **68**: 3158–3165.

225. Weisser M, Nelles G, Wenz G, Mittler-Neher S. Guest–host interactions with immobilized cyclodextrins. *Sens Actuat B.* 1997; **38–39**: 58–67.

226. Stine KJ, Andrauskas DM, Khan AR, Forgo P, D'Souza VT. Electrochemical study of self-assembled monolayers of a β-cyclodextrinmethyl sulfide covalently linked to anthraquinone. *J Electro Anal Chem.* 1999; **465**: 209–218.

227. Chmurski K, Koralewska A, Temeriusz A, Bilewicz R. Catalytic Au electrodes based on SAMs of per (6-deoxy-6-thio-2,3-di-O-methyl)-β-cyclodextrin. *Electroanalysis.* 2004; **16**: 1407–1412.

228. Lee J-Y, Park S-M. Electrochemistry of guest molecules in thiolated cyclodextrin self-assembled monolayers: an implication for size-selective sensors. *J Phys Chem B.* 1998; **102**: 9940–9945.

229. Suzuki I, Murakami K, Anzai J-I, Osa T, He P, Fang Y. Preparation and molecular recognition of cyclodextrin monolayers with different cavity size on a gold wire electrode. *Mat Sci Eng.* 1998; **C6**: 19–25.

230. Suzuki I, Murakami K, Anzai J-I. Cyclic voltammetric studies with gold wire electrodes covered with lipoyl γ-cyclodextrins: positional isomerism effect of the modified cyclodextrins on monolayer quality and electrochemical responses on soluble probes. *Mat Sci Eng.* 2001; **C17**: 143–148.

231. Wang Y, Kaifer AE. Interfacial molecular recognition: binding of ferrocenecarboxylate to β-aminocyclodextrin hosts electrostatically immobilized on a thioctic acid monolayer. *J Phys Chem B.* **102**: 9922–9927.

232. Le Bras Y, Sallé M, Leriche P, Mingotaud C, Richomme P, Møller J. Functionalization of the cyclodextrin platform with tetrathiafulvalene units: an efficient access towards redox active Langmuir–Blodgett films. *J Mater Chem.* 1997; **7**: 2393–2396.

233. Ohira A, Ishizaki T, Sakata M, Taniguchi I, Hirayama C, Kunitake M. Formation of the 'nanotube' structure of β-cyclodextrin on Au(III) surfaces induced by potential controlled adsorption. *Coll Surf A.* 2000; **169**: 29–33.

7

Cyclotriveratrylenes and Cryptophanes

7.1 Introduction

We have already discussed the calixarenes and the related resorcinarenes within Chapter 4 of this work. Calixarenes have phenol units as their basic building blocks, while resorcinarenes are based on 1,3-dihydroxy benzene (resorcinol). Another similar range of compounds is that of the cyclotriveratrylenes, which have a structure containing derivatives of 1,2-dihydroxy benzene or catechol, meaning they might also be thought of as calix[3]catecholarenes. Alternatively, they can be classified as members of the [1.1.1]orthocyclophane family.

These compounds have been shown to assume rigid crown-type structures and to form inclusion complexes with a variety of guests. One advantage of these compounds over many other macrocycles is that it is relatively simple to synthesise derivatives which are inherently chiral and then to separate the enantiomers. Many other macrocycles are much more difficult to substitute to give inherently chiral structures and in the case of cyclodextrins only one enantiomer is usually available.

Another feature of these systems is that it is relatively simple to link two units together to form cryptophanes, which are hosts for a variety of organic compounds. These hosts display a high degree of preorganisation, which leads to high binding coefficients and good selectivity.

7.2 Synthesis of Cyclotriveratrylenes

Probably the first synthesis of these types of compound came from initial work by Arthur Ewins,[1] who reacted 3,4-methylenedioxybenzyl chloride (Figure 7.1a) under acidic conditions to give what he thought was a dihydroanthracene structure (Figure 7.1b). Similar work by Gertrude Robinson[2] in 1915 was reported where homoveratryl alcohol (Figure 7.1c) was condensed by a sulfuric–glacial acetic acid mixture to give what was thought to be the dimeric structure shown in Figure 7.1d. An identical product could be obtained by the acid-catalysed condensation of veratrole (1,2-dimethoxy benzene) with formaldehyde, a synthesis analogous to those reported for calixarenes. The supposed structure was supported by chemical reactions in which the condensate was decomposed with, for example, nitric acid to give 6-nitroveratic acid and 4,5-dinitroveratrole.

Macrocycles: Construction, Chemistry and Nanotechnology Applications, First Edition.   Frank Davis and Séamus Higson.
© 2011 John Wiley & Sons, Ltd. Published 2011 by John Wiley & Sons, Ltd.

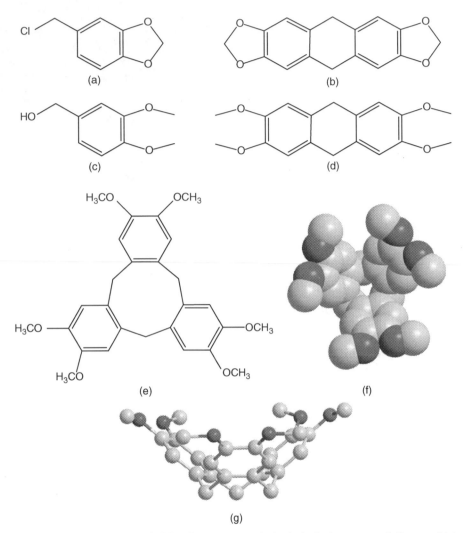

Figure 7.1 *Structures of (a) 'piperonyl chloride', (c) veratryl alcohol, (b,d) suspected dimers of (a) and (c), and (e,f) X-ray crystal structure of cyclotriveratrylene*

Later work disputed the dihydroanthracene nature of the product and other structures were proposed such as cyclic hexamers. However, it was finally established that a cyclic trimeric structure (Figure 7.1e) was the correct one by determination of its molecular weight and chemical reaction products.[3] In this paper Lindsey also first coined the name 'cyclotriveratrylene' (CTV) and with a combination of molecular modelling and NMR measurements deduced that the molecule existed in a 'crown' conformation. Other workers obtained mass spectra and NMR which confirmed this result.[4]

Final proof of the structure[5] came from an X-ray crystal study which clearly showed a trimeric crown-shaped structure (Figure 7.1f,g). Both benzene and water co-crystallised as guests within this structure (not shown), the benzene being incorporated into the cavity and the water hydrogen-bonding to the methoxy

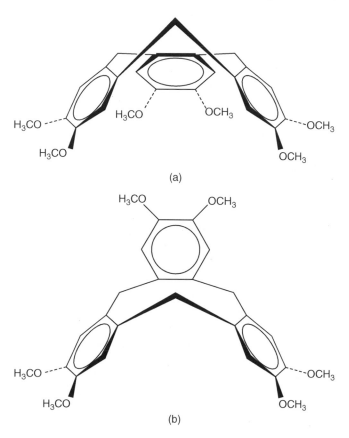

Figure 7.2   *Structures of (a) 'crown' and (b) 'saddle' forms of cyclotriveratrylene*

groups. The NMR studies showed that the methylene linker groups were asymmetric, that is the two methylene protons were not equivalent,[4] leading to splitting of the signal into a quartet. This means that the crown structure is rigid, at least on the NMR timescale, as confirmed by a three-dimensional crystal structure (Figure 7.1f), which shows that the narrow end of the cavity is practically closed and that the ring hydrogens practically touch. Work on a cyclotriveratrylene that had been isotopically substituted to render it chiral[6] showed that inversion of the macrocycle can occur, with the calculated lifetimes for racemisation being of the order of 960 days, 36 days and 3 minutes at 0 °C, 20 °C and 100 °C, respectively.

Besides the crown conformer, a possible 'saddle' conformer of CTV has been proposed; racemisation of the crown conformer would progress via this structure. However, it was not isolated until 2004, when it was found that rapid quenching with ice of hot solutions or melts of CTV compounds led to the formation of mixtures of crown and saddle conformers that could be separated chomatographically,[7] with up to 105 mg (from a 700 mg quenched sample) of the saddle conformers being obtained with >99% purity.[7] The two structures are compared in Figure 7.2. NMR studies showed that the saddle form is much more flexible than the crown. In chloroform solution a crown–saddle equilibrium can be observed, with up to 10% of the CTV being in the saddle form and the half-life of the interconversion being about a day. More polar solvents favour the crown form even more. Under the correct conditions, solid samples of the saddle form are stable for months at room temperature.

Although the trimer is the major product, evidence of a tetrameric product was also observed in the mass spectra.[4] Later work succeeded in isolating the tetramer and NMR studies showed it to be much more conformationally mobile in solution than the trimer, probably because of the larger, more flexible central ring.[8] Much of the early work on CTVs has been extensively reviewed elsewhere.[9] From the fact that direct reaction of veratryl alcohol and the condensation reaction between veratrole and formaldehyde give identical products, it would appear that the active component in this reaction is the veratryl cation, which is generated under acidic conditions and then condenses with itself. Another starting material for this synthesis is N-tosyl veratrylamine and its derivatives, which have been shown to form the cyclic trimer upon reaction with perchloric acid;[10] it was also shown that use of organic solvents increased the yield of the cyclic tetramer.

The condensation of veratrole-type compounds with formaldehyde is one method of making CTVs, but a preferred technique is the earlier method using substituted benzylic alcohols. Formaldehyde has been condensed with a variety of aromatic compounds as described in this and previous chapters. For the formation of CTVs, the formaldehyde condensation seems to be restricted to aromatic rings bearing two electron-donating groups. A single pure product can moreover only be obtained if the aromatic compound is symmetrical, such as veratrole,[2] 1,2-diethoxybenzene or benzocrown ethers;[11] see for example Figure 7.3a,b. These condense with formaldehyde to form crown ether-substituted CTVs, which form crystalline 1 : 3 complexes with NaSCN and KSCN. When a nonsymmetrical compound such as that shown in Figure 7.4a is used, however, condensation with formaldehyde gives two structures, one with C1 symmetry (Figure 7.4a) and one with C3 symmetry (Figure 7.4b), in a statistical 3 : 1 ratio.[9] Both these structures are chiral and are formed as the racemate. Other aldehydes such as acetaldehyde tend to form products with dihydroanthracene-type structures rather than trimers.[3]

Using benzylic alcohol-based starting materials, however, it is possible to both vary the structure of the starting aromatic species more and synthesise the C3 symmetric substituted analogues of CTV without any C1 isomer. A number of disubstituted benzyl alcohols have been studied and reviewed by Collet.[9] The structure of the benzyl compound has major effects on the efficiency of the reaction. Once the benzyl cation is formed, it must attack a second benzylic compound *ortho* to the benzyl group to give a dimer as shown in Figure 7.5a. This dimer can either be deprotonated and react with a third unit or else it may be attacked by another benzyl cation to give a trimer. This then undergoes an internal reaction to attain ring closure. For this reaction to occur, the position *ortho* to the benzyl group must be susceptible to electrophilic attack, which requires the presence of a strongly electron-donating group *para* to this position (X in Figure 7.5). As an example, 4-bromo-3-methoxy benzyl alcohol readily forms a cyclic trimer, whereas the corresponding

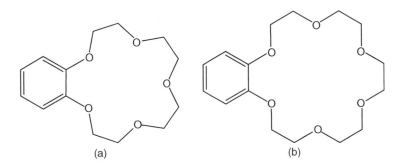

(a)   (b)

Figure 7.3   *Structures of cyclotriveratrylene-forming crown ethers*

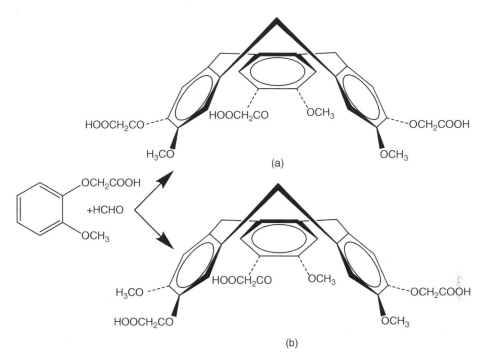

Figure 7.4   *Structures of substituted cyclotriveratrylene with (a) C1 and (b) C3 symmetry*

3-bromo-4-methoxy benzyl alcohol with a bromine atom in the X position does not.[12] From this we would expect vanillyl alcohol (Figure 7.6a) to form a CTV-like structure – instead intractable tars are formed. Possibly the free hydroxyl group increases reactivity at the 5-position, causing crosslinking in a similar manner to that in Bakelite formation (Chapter 4). The allyl ether of vanillyl alcohol (Figure 7.6b) however does form a cyclic trimer.[13] Similarly, if the 4-position is unprotected, as in 3-methoxybenzyl alcohol, only a small amount of trimer can be isolated, again indicating perhaps that this position is attacked, leading to polymeric products.[9]

The facile synthesis and isolation of the pure C3 isomeric form also means that the substitution reaction is, on the timescale of this reaction, essentially irreversible. If the methylene linkage were broken and reformed, as for example happens during many calixarene and resorcinarenes synthetic processes (Chapter 4), then the C1 form would become a sizeable contributor to the isolated products and perhaps even the major form. Again, with this reaction the use of organic solvents increases the production of higher analogues of CTVs and for example when the monomer shown in Figure 7.6b is reacted in 65% perchloric acid, almost pure trimer is isolated (45% yield), whereas in acetic acid–perchloric acid 3 : 4 mixture, the trimer, tetramer and pentamer are obtained in 28%, 20% and 7% yield, respectively.[9]

Several trisubstituted benzyl alcohols have also been successfully trimerised. For example, 2,3,4-trimethoxy benzyl alcohol can be reacted under acidic conditions[9] to give the CTV structure shown in Figure 7.6d. Later work isolated the trimer from this in both the 'boat' and the 'saddle' forms.[14] These two conformers are both inherently chiral, so the crown form can be further separated into its enantiomers using chiral HPLC (the rapid interconversion of the two saddle enantiomers prevents their separation). Another interesting compound is the hydroxyl-substituted benzyl alcohol,[15] which, unlike vanillyl alcohol, can be trimerised in aqueous HCl, since all positions *ortho* and *para* to the hydroxyl

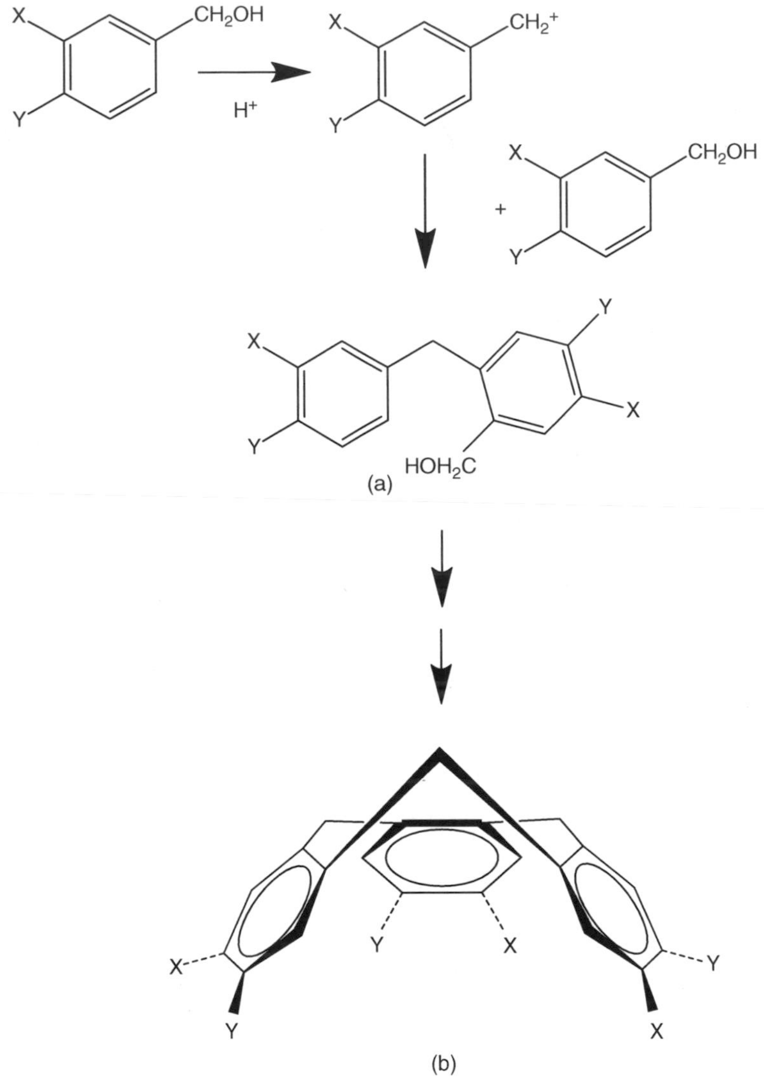

Figure 7.5   *Synthesis of CTV via benzylic alcohol trimerisation*

group are substituted. The resulting structures (Figure 7.6e) are quite similar to calix[3]arenes, except they have phenolic groups 'outside' the macrocycle. However, NMR studies of these compounds indicated they did not have a rigid crown conformer, and no inclusion compounds were isolated.

CTVs can also be synthesised by a step-by-step procedure in which two aromatic rings are combined to form a biphenyl methane unit, which can then be reacted with a third ring to form a cyclic trimer. This is of interest since it allows the synthesis of CTVs in which one ring bears different substituents to another, or even of CTVs containing three different aromatic rings. One of the earliest examples[3] is shown in Figure 7.7a, where a bischloromethyl-substituted diphenylmethane was condensed with a variety of disubstituted benzenes, including 2-methyl anisole and 1-benzyloxy-2-methoxy benzene, to give CTVs in

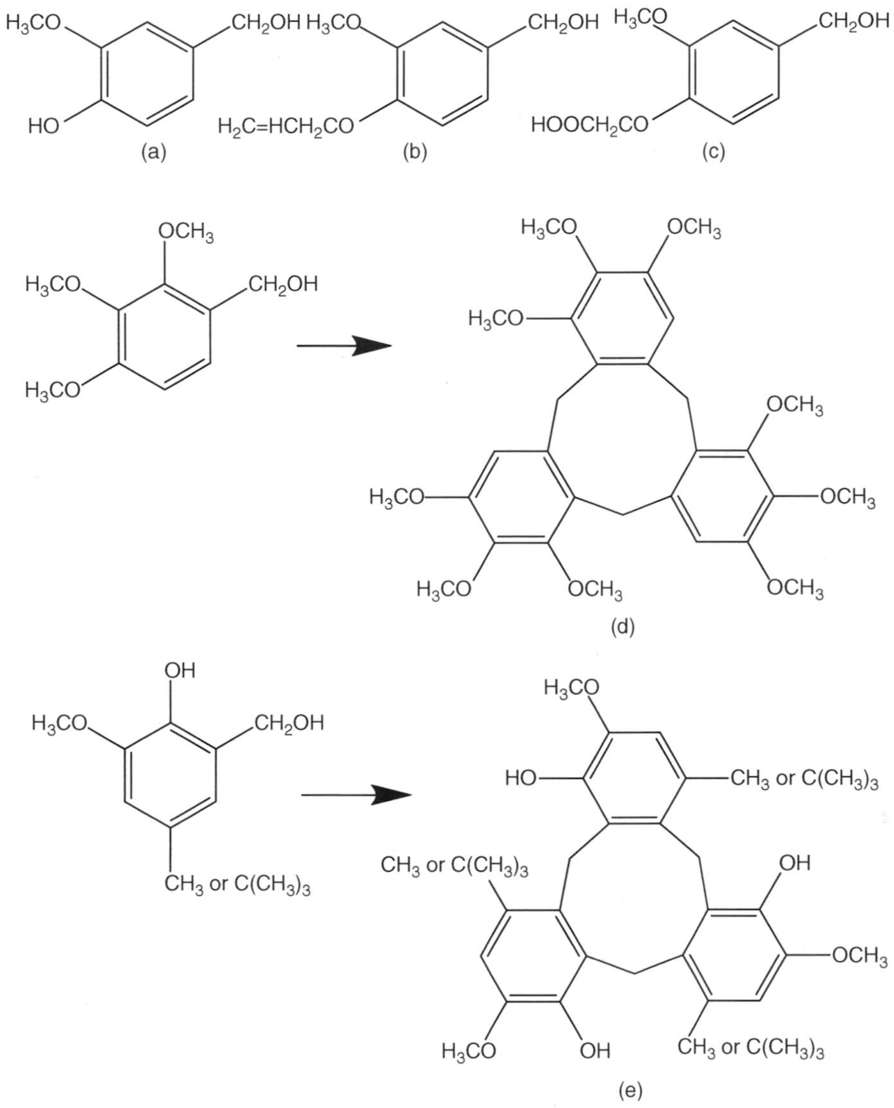

Figure 7.6   *Benzylic alcohols utilised in synthesis of CTVs*

which one or two of the methoxy groups are replaced by other substituents. The monosubstituted CTVs are actually chiral and the compound containing a single benzyloxy substituent (Figure 7.7e) was successfully resolved on a chiral cellulose acetate column into its two enantiomers.[16]

This procedure can be used in a wider context to generate CTV-type structures. For example, the *bis*-hydroxymethyl diphenylmethane compound shown in Figure 7.7f can be reacted with benzene under acidic conditions to give a CTV-type structure, but without the peripheral substituents.[17] This structure was shown by NMR studies to adopt a rigid crown structure. Analogues synthesised from diphenyl units containing different bridging groups can also be made when carbonyl or ether bridging groups are used

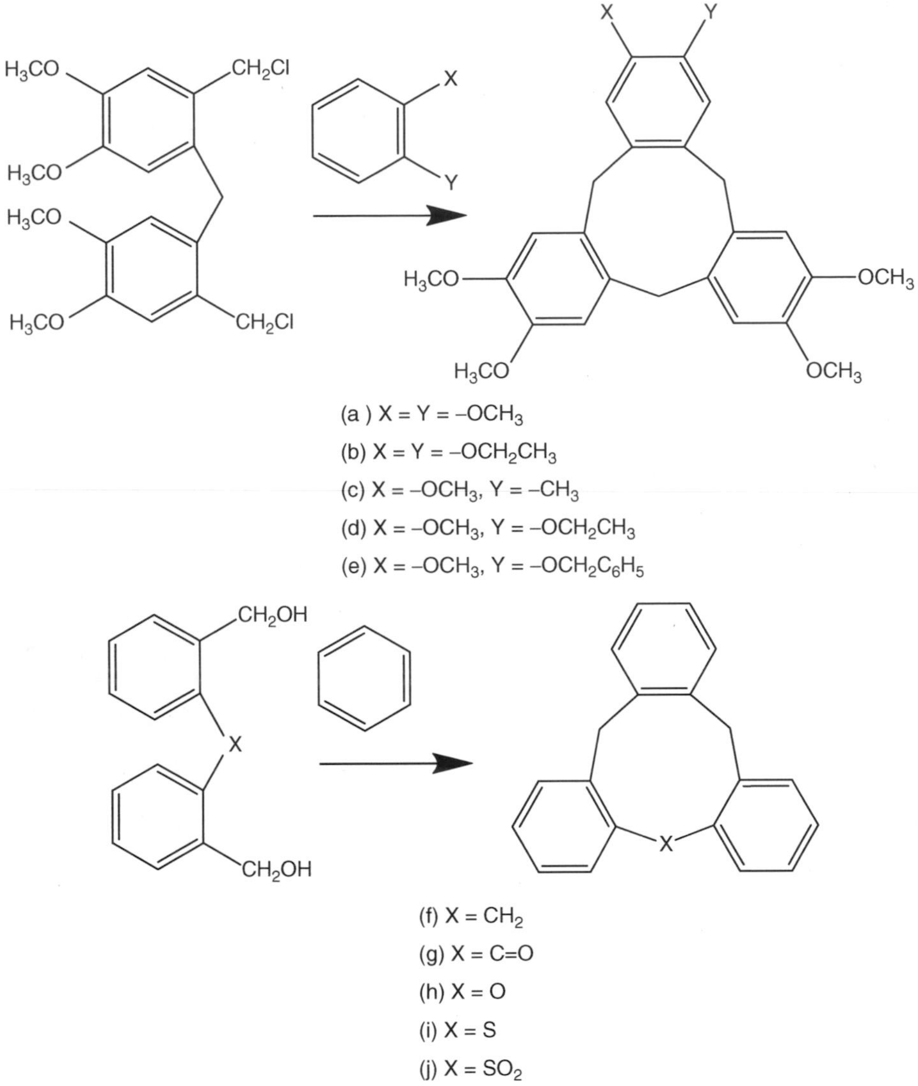
H3CO
CH2Cl
H3CO
H3CO
H3CO
CH2Cl
X
Y
X
Y
H3CO
OCH3
H3CO
OCH3
(a) X = Y = –OCH3
(b) X = Y = –OCH2CH3
(c) X = –OCH3, Y = –CH3
(d) X = –OCH3, Y = –OCH2CH3
(e) X = –OCH3, Y = –OCH2C6H5
CH2OH
X
CH2OH
X
(f) X = CH2
(g) X = C=O
(h) X = O
(i) X = S
(j) X = SO2

Figure 7.7 *Diphenyl units in the synthesis of CTVs*

(Figure 7.7g,h); a macrocycle that is flexible in solution results. Macrocycles containing thia and sulfonyl bridging units (Figure 7.7i,j), however, show the same rigidity as the hydrocarbon analogue.

An intensive study of the effects of the reaction conditions on the ratio of trimer : tetramer formation was performed using a variety of synthetic methods.[18] CTV was determined to be the kinetically favoured cyclisation product and to be formed almost exclusively when 3,4-*bis*(methyloxy)-benzyl chloride is reacted stoichiometrically with $AgBF_4$, in dichloromethane. Cyclooligomerisation of the corresponding benzyl alcohol also leads to high yields of CTV, especially when the reaction is performed in a nonsolvent and catalysed with trifluorosulfonic or perchloric acid. Maximum yields of the tetramer (55%) are obtained by using dilute solutions and a large excess of trifluoracetic acid as catalyst. In many cases higher cyclics

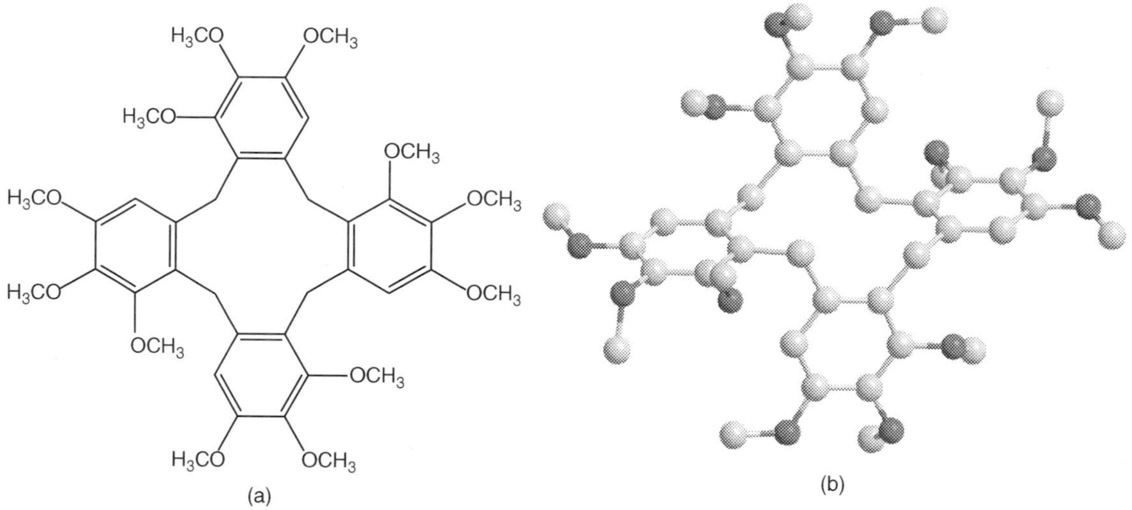

Figure 7.8   *Tetramers based on 3,4,5-trimethoxy benzyl alcohol*

are also formed. The tetramer formed from 2,3,4-trimethoxy benzyl alcohol (Figure 7.8a) can also be synthesised in reasonable yields[19] and has been shown by NMR to exist in solution in a mixture of two forms, crown and saddle – whereas X-ray studies show the saddle conformation is preferred in the solid crystal (Figure 7.8b). The tetramer analogue of CTV has previously been shown to adopt just the crown conformer[8] in solution.

As an interesting alternative to many of the earlier syntheses, two new methods were reported which unlike many of those described above do not require volatile inorganic solvents or inorganic acid catalysts. One utilised an ionic liquid based on a tributylhexylammonium/*bis*(trifluoromethanesulfonyl)amide salt as a 'green solvent' for the condensation of 3,4-dimethoxy benzyl alcohol, with *p*-toluenesulfonic acid as the catalyst.[20] Heating these compounds with the ionic liquid (80 °C, 4 hours) gave a viscous paste, which upon addition of methanol gave 89% of crystalline CTV and, upon removal of methanol under vacuum, allowed the ionic liquid to be recycled. In the same work, the authors also demonstrated that simply grinding solid monomer and catalyst intermittently over a period of two days at room temperature gave up to 54% of pure CTV product. Similar results were also obtained using the monomer shown in Figure 7.6b, giving a *tris*-allyl CTV.

7.3   Modification of Cyclotriveratrylenes

There are several methods for the modification of CTVs. One of the most popular involves modification of the peripheral substituents. Treatment of CTV with boron tribromide[3] allows for complete removal of the methoxy groups to give what can be termed a cyclotricatecylene (Figure 7.9a), which can then be reacted with acetic anhydride to give the hexa-acetate (Figure 7.9b). Later work showed that the phenolic compound forms host–guest complexes with a number of guests in a variety of stoichiometries. Water, DMF and DMSO form crystalline complexes in which each host complexes three guest molecules and can also form 1 : 2 host–guest complexes with acetone and isopropyl alcohol.[21] Guests can only be removed by prolonged heating under vacuum. No stable complex was observed for ethanol or methyl ethyl ketone.

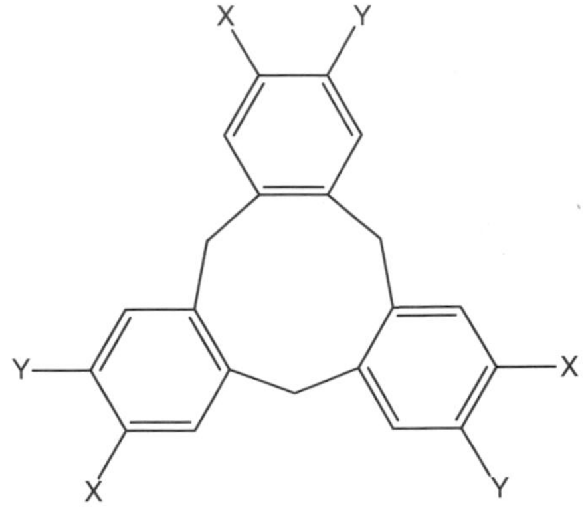

(a) X = Y = –OH
(b) X = Y = –OCOCH$_3$
(c) X = Y = –OCO–p–C$_6$H$_4$O(CH$_2$)$_{11}$CH$_3$
(d) X = –OCH$_3$, Y = –OCH$_2$CH=CH$_2$
(e) X = –OCH$_3$, Y = –OH
(f) X = –OCH$_3$, Y = –H
(g) X = –OH, Y = –H
(h) X = –H, Y = –H
(i) X = –D, Y = –H
(j) X = –OCH$_3$, Y = –Br
(k) X = –OCH$_3$, Y = –Li
(l) X = –OCH$_3$, Y = –CO$_2$CH$_2$CH$_3$
(m) X = –OCH$_3$, Y = –CH$_2$OH
(n) X = –OCH$_3$, Y = –CH$_2$O.CO.NH

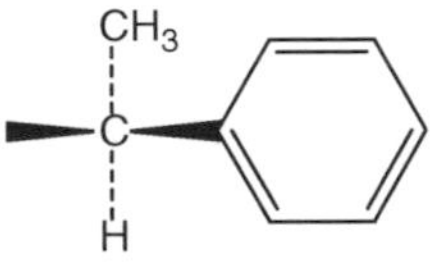

Figure 7.9   *Substituted cyclotriveratrylenes*

X-ray studies of the complex with isopropyl alcohol (Figure 7.10a) showed the macrocycle to adopt an approximate crown conformation, with the solvent molecules sandwiched between rows. Exact C3 symmetry was not found, possibly due to hydrogen-bonding effects. The authors claim that the phenolic compound forms more and better-defined complexes than are found for the parent CTV. For example, no complex was observed for CTV with DMF or acetone, but complexes could be observed with ethyl acetate, ethanol and methyl ethyl ketone, though usually with lower than 1 : 3 stoichiometry. The difference in behaviour of these two hosts is probably a combination of macrocycle rigidity and the fact the phenolic compound acts as a hydrogen-bond donor, unlike the methoxy compound.

Crystalline complexes have been found between cyclotriveratrylene and a variety of other organic species, although in many cases there was not an exact ratio between host and guest, indicating perhaps that a mixture of complexes was formed.[22] In early work, two types of unit cell structure were found for complexes of CTV, depending on whether guests were linear (such as butyric acid) or bulkier (such as chloroform or benzene).[22] Water and a range of organic solvents have been shown to be incorporated in a similar manner, with for example CTV having been shown by X-ray crystallography to form complexes with water, toluene, bromobenzene, chloroform, acetone and dimethoxyethane.[23] Water and chloroform

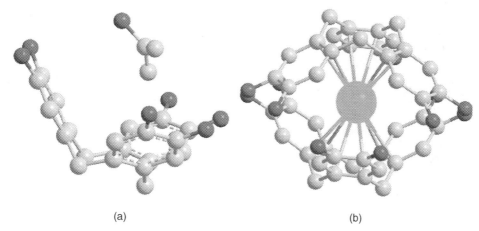

(a)   (b)

Figure 7.10   *Crystal structure of cyclotricatechylene with (a) isopropyl alcohol and (b) clam-shell structure with rubidium*

were found to act as hydrogen-bond donors towards the methoxy groups and there were some distortions from C3 symmetry in many of the complexes. In the same work cyclotricatechylene was shown to form an inclusion complex with DMF and water which displayed an alternating hydrophobic/hydrophilic bilayer structure. Other workers[24] also formed solid-state complexes of CTV with a variety of chlorinated species; in the main, these again showed inclusion of the guests between the host CTV units and in the case of both 1,1,1-trichloroethane and 1,1,2-trichloroethane, structures were formed in which one solvent molecule occupied the cavity.

Cyclotricatechylene has been shown to be capable of complexing cations, as demonstrated in this recent work.[25] A novel clam shell-like structure was observed when cyclotricatechylene was mixed with guandinium and rubidium chloride. The phenolic compounds lost several protons to become anionic in nature and formed a 2 : 1 host–guest complex containing an entrapped rubidium cation.[25] X-ray crystal structures have conclusively demonstrated that a 'bare', unsolvated Rb^+ ion is located at the centre of the complex, making contact with six aromatic rings that are face-on to it (Figure 7.10b). Similar structures could be formed with caesium or tetramethylammonium guests but no 'clam shell' could be obtained with potassium; this cation may be too small to fit snugly inside such a complex. A larger tetraethylammonium guest led to formation of an open clam structure. Electrospray mass spectrometry indicated that these structures are stable in aqueous solution, suggesting that the metal ions 'prefer' the interior of the clam shell to being solvated in water. Cyclotetracatechylene has been shown to form complexes[26] and in this context the macrocycle for example was shown to form a 1 : 6 complex with DMF and a mixed 1 : 2 : 2 complex with pyridine and methanol. X-ray crystal structures determined a chair conformation for the cyclic tetramer.

The ability to convert the CTV into the hexa-phenol compound allows the synthesis of a wide range of derivatives, in manners similar to those described earlier for calixarenes and other compounds of this nature. Oxidation of the phenolic compound to give a quinone-type structure was attempted but gave dark polymeric products.[3] Alkylation and acylation of the phenolic groups gave hexa-alkyl and hexa-acyl CTV derivatives. These are of interest since many of them (for example the compound shown in Figure 7.9c) display liquid-crystal properties.[27] The combination of a rigid core with flexible side chains enabled these compounds to stack, with the conical CTV units being embedded in each other, 0.48 nm apart[28] – the same distance as in crystals of CTV itself. A wide variety of alkyl and acyl side chains were utilised to give liquid crystals which adopt a columnar mesophase over wide temperature domains. What is also of interest is that within the column all the CTV units are aligned in the same direction, which leads to the axis of the column being polar and to the potential for aligning the crystals with an electric field. Similarly, a chiral trimer such as that shown in Figure 7.6d, but with the methoxy units replaced by octanoyloxy chains, could be synthesised and separated into its enantiomers by HPLC.[29] Both the racemic mixture and the pure enantiomers were shown to form columnar hexagonal liquid-crystal phases. The chiral mesophases were found to be less well-ordered and more tightly packed than the racemate, as well as showing a lesser tendency to crystallise.

The cyclic tetramers may also be liquid-crystalline materials and for example a number of octa-ether and octa-ester derivatives of the tetrameric version of CTV demonstrate thermotropic columnar mesophases, despite their relatively flexible structures compared to CTV.[30] Twelve octa-acyl derivatives with chains from 5 to 16 carbons in length were all found to show mesophases that were biaxial in nature, whereas the octa-alkyl derivatives were uniaxial.[31]

The trimer obtained from 1-allyl-2-methoxy benzene[13] (Figure 7.9d) is often used as a starting material for a wide range of CTV-type compounds. It has the chiral C3 nature of these types of material. The allyl ether could easily be removed by catalytic hydrogenation[13] to give the triphenol (Figure 7.9e), which could then be separated into its optical isomers. The triphenol could be reacted with 1-chloro-2-phenyltetrazole under basic conditions to give the tetrazole ether, which could then be removed by hydrogenolysis to give the compound shown in Figure 7.9f. Further reaction with BBr_3 gave the triphenol shown in Figure 7.9g and this could be reacted with 1-chloro-2-phenyltetrazole and further reacted with either hydrogen or deuterium to give the tribenzylene compounds shown in Figure 7.9h,i. The cyclotribenzylene with the deuterium substituent is very interesting because it is isotopically chiral and if prepared as a single enantiomer displays an optical rotation of $[\alpha]_{365} = 8.9°$, which is much higher than noted for many other isotopically substituted compounds. This has been proposed to be due to an exciton mechanism where dipoles within the benzene chromophores give rise to the optical rotation.[9,12]

The tribromo compound (Figure 7.9j), which can be synthesised directly,[12] has also been used as the starting material for many compounds. The methoxy groups can be removed to give the triphenol.[12] The bromide groups can be reacted to give the lithium derivatives (Figure 7.9k), which can be converted to a range of other compounds,[32] for example reaction with ethyl chloroformate gives the triester (Figure 7.9l), which can be hydrolysed to the acid (Y = —COOH) or reduced to the alcohol (Y = —CH$_2$OH; Figure 7.9m), which can then be converted to the chloride (Y = —CH$_2$Cl) and then the thiol (Y = —CH$_2$SH). Later work utilised a triphenol grafted with bulky aromatic side chains[33] to give the structure shown in Figure 7.11a, which showed mesophasic behaviour from room temperature up to 150 °C. Both racemic and optically active forms were synthesised and it is thought they could well orientate under a strong electric field.

Other long-chain compounds that have been attached include polyethylene oxide chains. Attempts to react the cyclotricatechylene directly with ethylene oxide only resulted in intractable material, but the macrocycle could be substituted with ethylene oxide tosylates to give CTVs bearing six polyethylene oxide side chains.[34] These 'octopus' molecules (Figure 7.11b) proved effective at solubilising alkali metal and ammonium salts in organic solvents, approximately as well as 18-crown-6. The preorganisation caused by the CTV skeleton has been shown to be important for good complexation and similar compounds based on a flexible calix[4]resorcinarene framework did not solubilise the salts.[29] A similar compound (Figure 7.11c) named 'hexapus' was synthesised[35] and shown to form micelles in aqueous solution, with about nine macrocycles per micelle. Molecular models indicate it has a variable hydrophobic cavity and it was shown both to form complexes with the dye phenol blue and to solubilise naphthalene in water. It was also capable of binding p-nitrophenyl butyrate and inhibiting its base-catalysed hydrolysis. Similarly, the CTV shown in Figure 7.9e could be derivatised to give CTVs bearing three carboxylic acid groups. These could be converted to acid chlorides and then reacted with 1,7,13-triaza-18-crown-6 under high

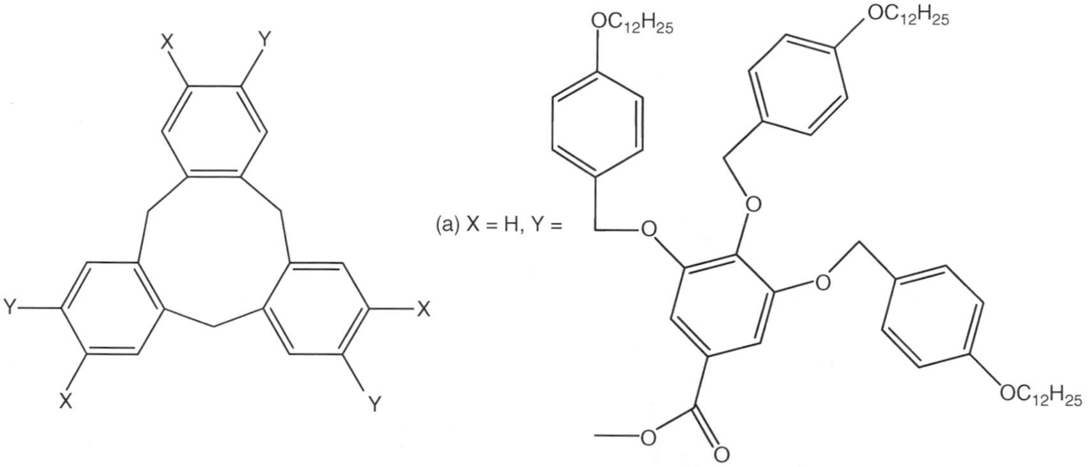

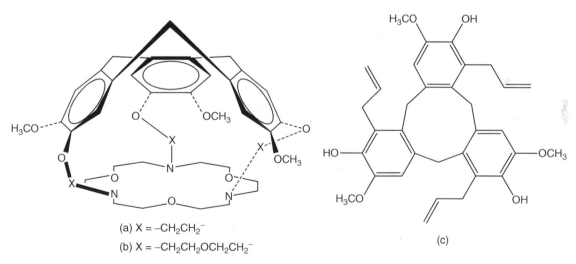

(a) X = H, Y =

(b) X = Y = $-O(CH_2CH_2O)_nCH_2CH_3$ n = 2, 3 or 4

(c) X = Y = $-O(CH_2)_{10}CO_2^-$

Figure 7.11   *Substituted derivatives of cyclotricatechylene*

(a) X = $-CH_2CH_2^-$

(b) X = $-CH_2CH_2OCH_2CH_2^-$

(c)

Figure 7.12   *(a,b) Cyclotriveratrylene/azacrown compound and (c) Claisen rearrangement of a cyclotriveratry-lene*

dilution conditions[36] as described earlier to give triamide compounds, which could then be reduced to give the CTV-crown speleands shown in Figure 7.12a,b. Methylammonium cation was found to associate strongly with the compound in Figure 7.12a, probably via a combination of hydrogen bonding between the azacrown and the ammonium and inclusion of the methyl group in the CTV cavity.

The triallylether ether derivative (Figure 7.9d) when refluxed in dimethylaniline undergoes a facile Claisen rearrangement, generating the compound shown in Figure 7.12c. NMR studies show a single sharp peak for the methylene protons, indicating that the compound is conformationally mobile and probably adopts the saddle conformation.[37] The presence of both phenol groups and ally groups makes this material especially amenable to further modification.

7.4   Synthesis of Optically Active Cyclotriveratrylenes

We have already discussed the synthesis of C3 triveratrylenes, macrocycles which display an inherent chirality, but most syntheses produce racemic mixtures of the two enantiomers. There has been a great deal of research into obtaining optically pure enantiomers of these compounds. One method is simply to separate the enantiomers using chiral chromatographic techniques;[14,16] other approaches involve using other chiral materials to either influence the reaction to form an excess of one enantiomer or aid in their separation.

Modification of the C3 CTVs has been used to effect enantiomeric separation by the formation of diasterioisomers. When a benzyl alcohol containing a chiral substituent is trimerised[38,39], two diasterioisomers are formed, as shown in Figure 7.13. Because these are not simple optical isomers, they are usually formed in different amounts and can be separated by chromatographic methods. The presence of the reactive ester group has also allowed their conversion into other derivatives. As an alternative to chromatographic separation, the triphenol (Figure 7.9e) can be synthesised as a racemic mixture and then esterified with a

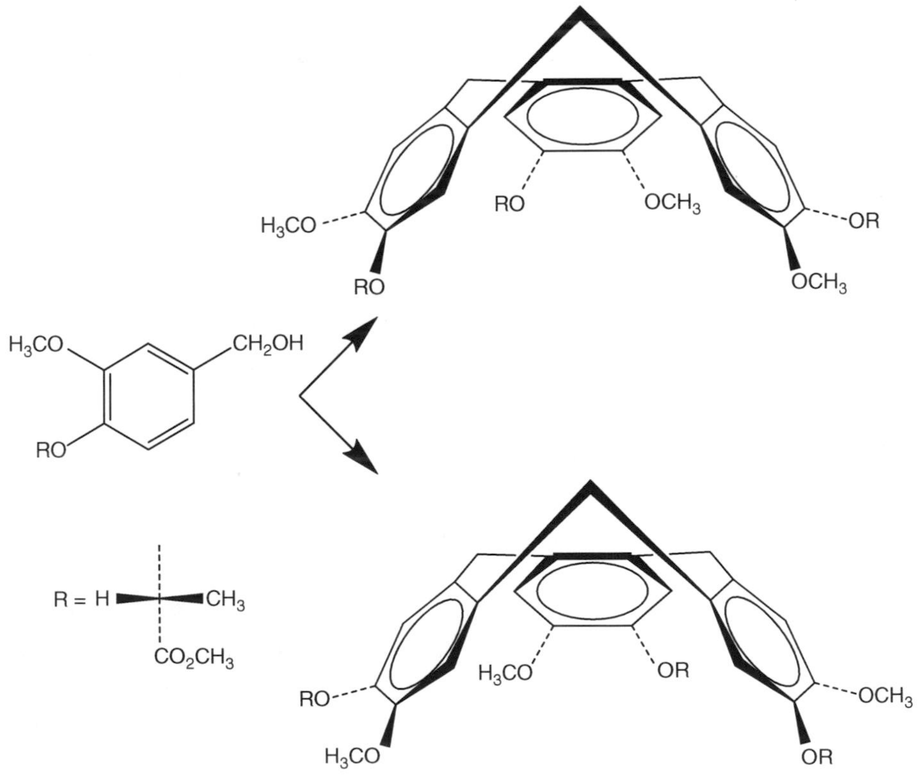

Figure 7.13   *Chirally substituted triveratrylene diasteroisomers*

chirally pure compound such as the acid chloride of campanic acid.[40] This results in a 50:50 mixture of the two diastereoisomers, which can be separated by normal (not chiral) chromatography on silica and then reductively cleaved with lithium aluminium hydride back to the original triphenol enantiomers, but in an optically pure state. As an alternative, R-(+)-2-phenoxypropionic acid can be used as a resolving agent in a similar manner.[40] A number of other triphenols can be obtained by this method and can of course be derivatised by many of the reactions mentioned above.

Differential crystallisation of diastereoisomers has been used for many years as a method of separating enantiomers. One example of this within CTV chemistry is the reaction of racemic CTV triol (Figure 7.9n) with a chiral isocyanate[9] to give the diastereoisomeric compound shown in Figure 7.9o. This compound when heated for 10 days gives 80% of a single crystalline diastereoisomer.[9] Obviously, under heating in solution, the two diastereoisomers interconvert via inversion of the CTV ring and the observed product crystallises out preferentially, thereby driving the equilibrium towards conversion to this isomer. Alternatively, both diastereoisomers can be obtained via chromatographic separation.

7.5   Modification of the Bridging Groups

Since the early days of CTV compounds, even before their structures were fully elucidated, it has been known that they undergo oxidation reactions of the bridging methylene groups. In fact, formation of the monoketone (Figure 7.14a) was one of the factors that enabled deduction of the trimeric structure of CTV.[3] For example, oxidation of CTV with chromic acid gave the monoketone, which was shown by a combination of IR and NMR spectral studies to have a flexible saddle structure that allows conjugation of the carbonyl group with the aromatic rings.[41] These derivatives could be further reacted and for example reduction of the ketone with hydrides gave the secondary alcohol (Figure 7.14e), which is not conjugated, appears to be synthesised in both crown and saddle forms, and upon heating adopts purely the crown conformation with the —OH equatorial to the central ring. Reaction with methyl lithium gave a flexible tertiary alcohol (Figure 7.14f), which appeared to exist in the saddle form. Heating this compound caused loss of water to form 5-methylenecyclotriveratrylene (Figure 7.14b), which in solution is an interconverting

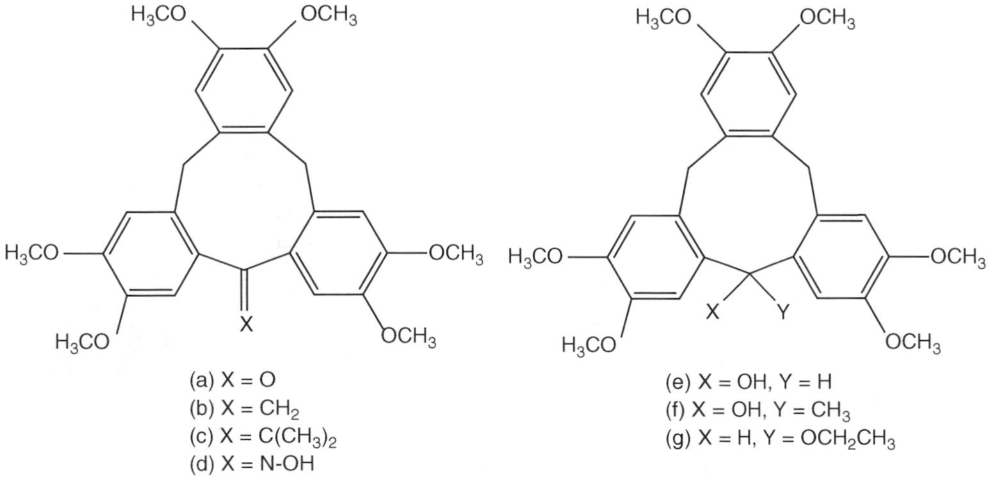

(a) X = O
(b) X = CH$_2$
(c) X = C(CH$_3$)$_2$
(d) X = N-OH

(e) X = OH, Y = H
(f) X = OH, Y = CH$_3$
(g) X = H, Y = OCH$_2$CH$_3$

Figure 7.14   *Structures of cyclotriveratrylenes modified at the bridging positions*

mixture of saddle and crown forms in a 1 : 7 ratio.[41] Similar behaviour occured when the ketone was reacted with isopropyl magnesium bromide to give the compound shown in Figure 7.14c (the intermediate tertiary alcohol is in this case too unstable to be isolated). Other reactions include the acid-catalysed etherification of the saddle form of Figure 7.14e with alcohols to give compounds like that shown in Figure 7.14g for the reaction with ethanol. These compounds again were synthesised in an unstable flexible form which upon heating converted to a stable, rigid crown form with the ether groups in an equatorial position.[41] The same reactions occurred with the crown conformations of Figure 7.14e but at a much slower rate.

The monoketone could also be reacted with hydroxylamine to give the oxime shown in Figure 7.14d. This can exist in both crown and saddle forms,[42] with the crown form being favoured in polar solvents (DMSO, acetone, acetonitrile) and slightly favoured in dioxane and the saddle form favoured in chloroform. The triketone of CTV was also reported but later work showed that it was not stable and underwent an internal rearrangement to give a lactone structure with loss of the CTV macrocycle.[43] The triketone of cyclotribenzylene has been reported, as discussed above.[17]

Cyclotriveratrylene has been shown to be a relatively poor host in solution. This is thought to be due to the relative shallowness of the cavity compared to many other families of host compounds. This means that the CTV molecule does not provide much shielding, for instance of a hydrophobic guest from a polar solvent. The aromatic rings also are only capable of relatively weak $CH-\pi$ and $\pi-\pi$ interactions. This can also be seen in the solid state, where the bowl of the CTV is usually occupied by the base of another CTV macrocycle, with any guests tending to be located in channels between the molecules rather than in the cavity, as was shown for isopropyl alcohol.[20]

There are exceptions to this rule, with for example the C_{60} structure having a size and shape that complements the CTV macrocycle very well. Mixing toluene solutions of CTV and C_{60} led to formation of a black crystalline material that contained CTV and the fullerene in a 2 : 3 ratio.[44] Although pure C_{70} could not be induced to form a complex, using mixtures of C_{60} and C_{70} led to formation of complex materials containing both fullerenes. An X-ray crystal structure of C_{60}/CTV was somewhat difficult to obtain due to the solvent and excess C_{60} being disordered and acting as space fillers. However, the crystal structure obtained clearly demonstrated formation of a ball-and-socket-type complex between host and guest, with the distance between the two aromatic species being at the Van der Waals limit, similar to the plane-stacking distance in graphite. This interaction has also been utilised to immobilise fullerenes on gold surfaces. CTVs with thiooctic acid substituents (Figure 7.15a) will form complexes with C_{60} in toluene.[45] These CTV derivatives spontaneously form monolayers on gold and the resulting layers have been shown to bind the fullerene. A CTV derivative with thioether side chains (Figure 7.15b) did not form complexes in solution but still bound the fullerene in the monolayer.

CTV trimers substituted with dendrimer branches (Figure 7.15c shows the simplest of this family of compounds but higher generations were also used) have also been shown to form complexes with fullerenes. When the dendrimer branches alone were used, no complexation was observed, indicating the CTV was the recognition site.[46] Similar dendrimers containing polyethylene glycol chains (Figure 7.15d) were capable of solvating C_{60} into water,[47] thereby opening up the possibility of using these complexes in biological systems. Recent studies on polybenzyl ether dendrimers similar to those above with a CTV core have also shown 1 : 1 binding of C_{60} and spectral studies indicate the fullerene is in close proximity to the CTV unit.[48] It has also proved possible, by using a suitable CTV derivative containing long alkyl chains (Figure 7.15e), to generate 2 : 1 CTV–fullerene complexes that display liquid-crystal properties when cast from benzene solution.[49]

Other workers have used CTVs as the basis for the design of capsules to envelop fullerenes. For instance, a CTV substituted with three long chains containing pyrimidinone units (Figure 7.15f) formed a dimeric hydrogen-bonded capsule in certain organic solvents.[50] This material had high affinities for fullerenes and showed high selectivity for C_{70} over C_{60}. The capsule structure could be disrupted by polar solvents such as

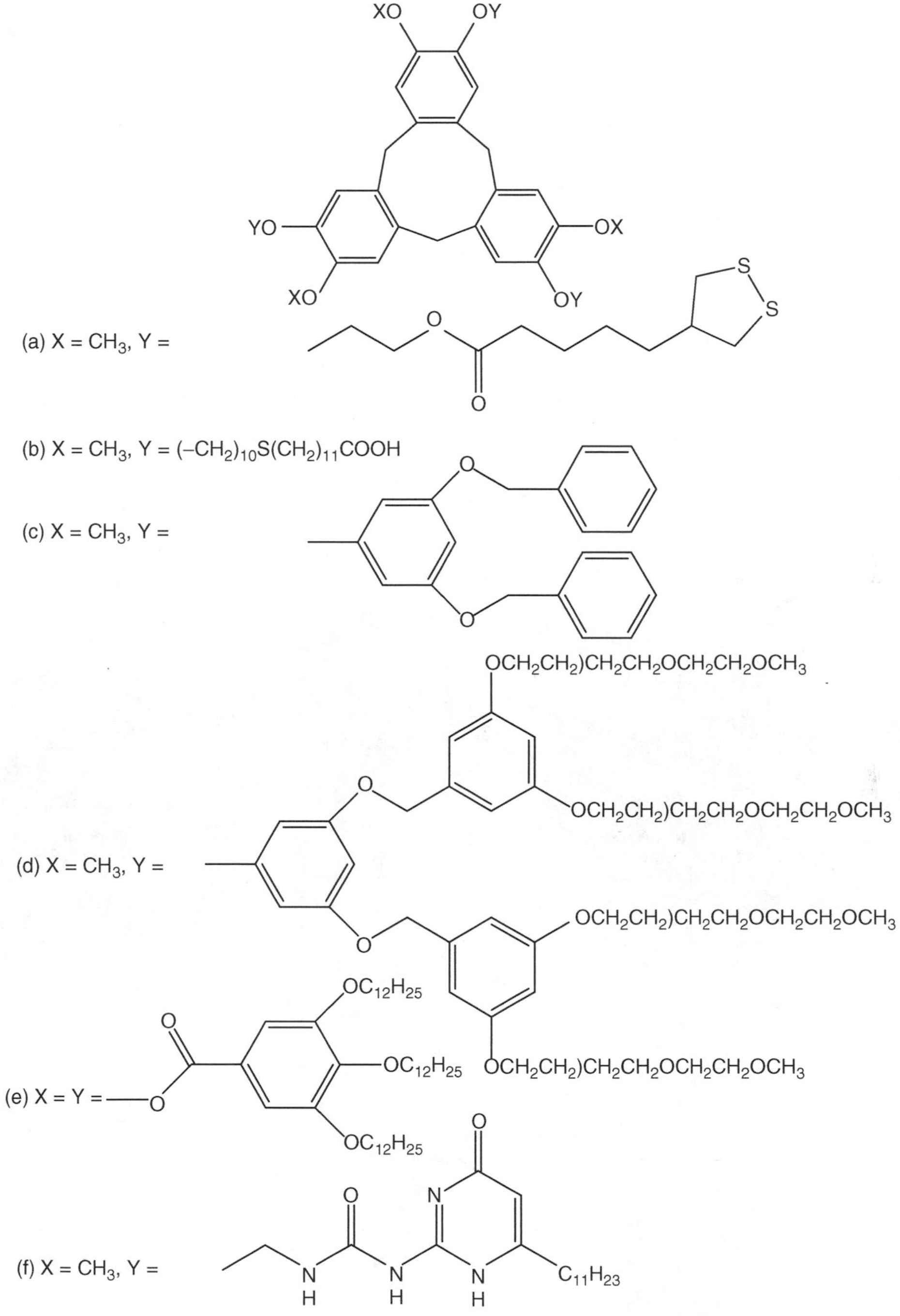

(a) X = CH₃, Y =

(b) X = CH₃, Y = (–CH₂)₁₀S(CH₂)₁₁COOH

(c) X = CH₃, Y =

(d) X = CH₃, Y =

(e) X = Y = —O

(f) X = CH₃, Y =

Figure 7.15   *Structures of (a,b) monolayer-forming CTVs, (c,d) polybenzyl ether dendrimers, (e) a liquid-crystal CTV and (f) a pyrimidinone-substituted capsule-forming CTV*

THF, thereby releasing the bound guest. Utilising such simple solid–liquid extractions, it proved possible to obtain C_{70} with a purity of 97% from fullerite after only two repeats of the extraction process. Later work also showed that C_{84} could be also preferentially extracted, having an affinity constant approximately ten times higher than that of the lower fullerene analogues.[51] Purities as high as 85% could be obtained using solid–liquid extractions as above, allowing a relatively inexpensive and rapid method of obtaining this material.

Another complimentary-shaped compound that forms inclusion complexes with CTV is 1,2-dicarbadodecaborane(12), which has been shown to form 1 : 1 complexes with CTV in both solution and the solid state.[52] These two compounds have been shown to form crystalline 2 : 1 complexes in which one carborane is incorporated into a CTV bowl, forming CH–π bonds to the aromatic rings, while the other resides between three CTV bowls, held by hydrogen bonds to the methoxy groups (Figure 7.16a,b). These complexes can exist as infinite arrays with two-dimensional hexagonal-grid or helical-chain topologies. The resultant materials are capable of binding other species as well, including fullerenes and various solvents. Figure 7.16c shows the structure of the quaternary system $(C_{70})(o$-carborane)(CTV)(1,2-dichlorobenzene). Later work by the same groups showed the CTV/carborane

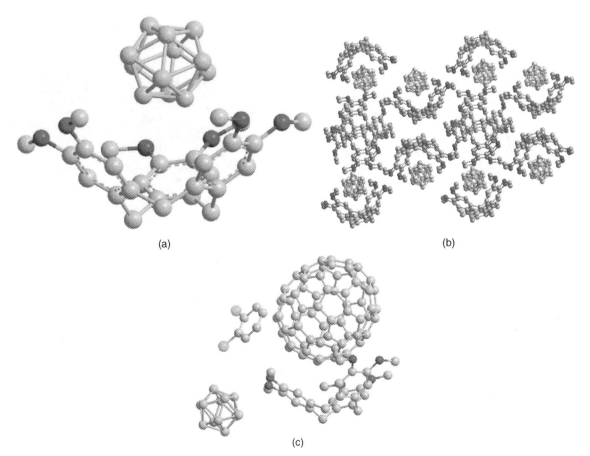

(a)

(b)

(c)

Figure 7.16 *(a) Crystal structure of carborane/CTV complex, (b) crystal packing and (c) C_{70}/carborane/CTV/1,2-dichlorobenzene complex*

system also formed supramolecular assemblies with alkali metal ions and DMF, which showed chelation of the metal ions by methoxy groups, and also inclusion of DMF into CTV cavities.[53] A nonspherical anionic cobalt-*bis*-carbolide could also be crystallised in 1 : 1 ratio with CTV and was shown to form a two-dimensional Na-CTV coordination polymer with a channel structure containing the anion.[54] Some of the CTV cavities were occupied by solvent (trifluoroethanol) molecules. Mass spectral studies showed that clusters of CTV with various guests could be obtained in the gas phase. Other cationic species could also be incorporated into these *bis*-carbolide structures and for example silver ions in this context could be incorporated into a CTV/cobalt *bis*-carbolide complex, with strong interaction between the Ag(I) ion and the CTV ring system.[55] Sodium[2,2,2]cryptate could also be incorporated and again X-ray crystal structures showed strong interactions between the cryptate and CTV.[56]

7.6   Modification of the Aromatic Rings with Organometallic Groups

We have already shown for macrocyclic aromatics such as calixarenes that they can be modified with organometallic moieties to give species capable of binding anions. One of the earliest compounds of this type was synthesised by reaction of a ruthenium cymene organometallic compound with CTV. Both the monosubstituted and trisubstituted compounds, such as that shown in Figure 7.17a, could be isolated.[57] X-ray studies (Figure 7.17b) of the trisubstituted CTV show that it retains its bowl-shaped structure and that a tetrafluoroborate ion is encapsulated within the cavity, albeit with a high degree of disorder.

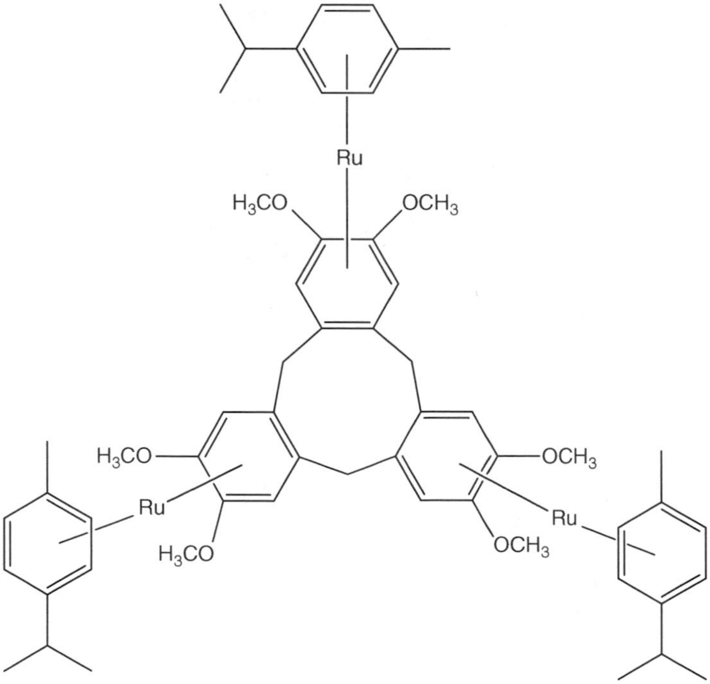

(a) 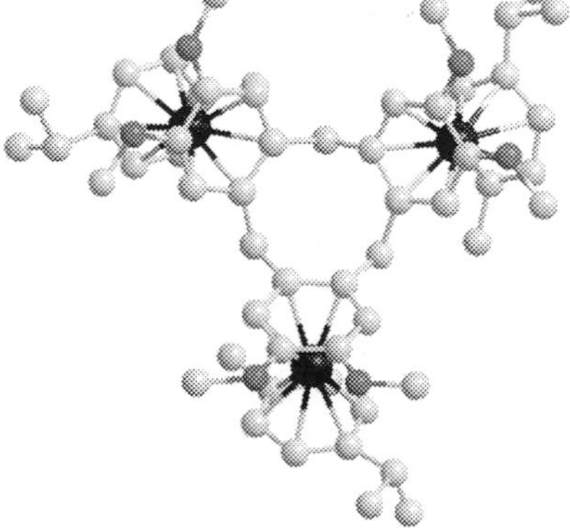   (b)

Figure 7.17   *(a) Structure of triruthenium/CTV complex and (b) X-ray crystal structure*

A wider range of compounds were reported in a later paper[58] in which CTV was substituted with both the ruthenium species above and the pentamethylcyclopentadienyl iridium compound Ir(C₅Me₅). Mono-, di- and trisubstituted species were obtained. The presence of the organometallic substituents had major effects, for example in the monosubstituted compounds, where the presence of the substituent disrupted the classic columnar packing of these materials. This meant the cavity of the CTV was more accessible in the crystal and it was found to be capable of binding a variety of guests, such as nitromethane and various anions. With di- and trimetallated species, anions were found to bind strongly in the CTV cavity, stabilised no doubt by the partial positive charges on the CTV rings. Very strong binding was found between the tri-iridium species and tetrafluoroborate. Complex anions such as ReO_4^- and TcO_4^- were found to bind very strongly to the diruthenium-substituted CTV[59] with inclusion of the anion in the cavity (Figure 7.18a). Other species that were synthesised included a *bis*-CTV complex in which a ruthenium atom was sandwiched between two CTV units (Figure 7.18b).

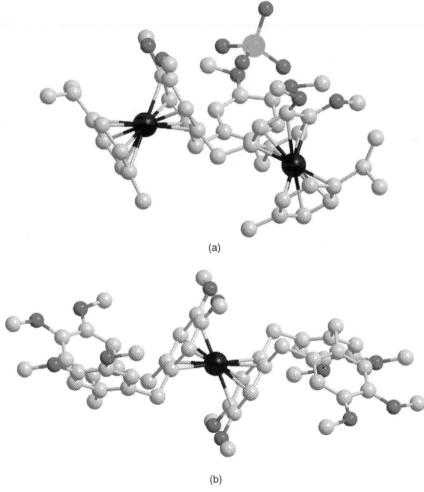

(a)

(b)

Figure 7.18 *X-ray crystal structures of (a) bis-ruthenium/CTV complex with ReO_4^- and (b) a bis-CTV Ru sandwich compound*

As an alternative method of developing anion sensors, the attachment of cationic metallocene groups as peripheral groups was studied.[60] Racemic CTV triphenol (Figure 7.9e) was reacted with a sandwich compound of the ferricinium type, where iron is sandwiched between a benzene and a cyclopentadiene ring. The resultant tricationic CTV (Figure 7.19a) was shown to display high affinities for the PF_6^- ion. X-ray studies (Figure 7.19b, R = Cl) show the material to contain a deep cavity, with the anion bound deep within it. More anions and diethyl ether molecules were incorporated between the CTV units. This complex was also found to bind halide ions in solution and there appeared to be some preference for iodide, although this was not conclusively established. Using these materials, there is also the possibility of separating out the optical isomers and achieving chiral recognition of anions. Later work with these types of complex again showed them to be capable of extracting pertechnetate and perrhenate anions from

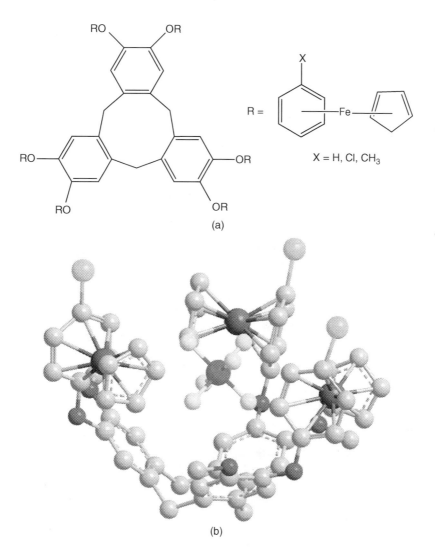

(a)

(b)

Figure 7.19 *(a) Structure and (b) X-ray crystal structures of a ferricinium-substituted CTV*

0.9% saline solution into nitromethane.[61] This high efficiency in selectively extracting pertechnetate is of great interest as pertechnetate is a major concern in the clean-up of nuclear sites, especially since it easily migrates to the surrounding environment.

7.7   Selective Binding Applications of Cyclotriveratrylenes

As described in other chapters, suitably substituted macrocyclic compounds display selective binding properties, and CTVs are no exception. Although CTVs themselves are relatively nonselective, a wide range of substituted CTVs have been synthesised and their selective binding to many substrates studied. One of the earliest group of compounds of this nature utilised a basic CTV triphenol,[62] the hydroxyl groups of which were substituted with sidearms terminated with bidentate groups such as catecholate, pyridylamino or bipyridine, examples of which are shown in Figure 7.20a–c. The presence of three of these groups, each with two binding sites, allows each macrocycle to bind a metal ion with octahedral coordination. Iron is bound especially strongly to these species.

Recently CTVs have been used to develop sensors for various species. A sensor for Hg(II) ions[63] has been developed using a CTV substituted with three chemochromic azobenzene groups (Figure 7.20d). In solution this gave a clear visual colour change (yellow to red/orange), with a visual detection limit of 5×10^{-6} mol L^{-1} and a linear range of adsorption to Hg concentration up to 2×10^{-4} mol L^{-1}. There was minimal interference from other metal ions and the CTV could be immobilised onto a PET film to give a strip test. A similar CTV with quinoxaline substituents (Figure 7.20e) has been shown to act as a selective fluorescent chemosensor for copper.[64] Detection limits of 10^{-5} mol L^{-1} could be observed for this macrocycle when immobilised in a polymeric membrane. Anion sensing has also been developed, as seen for example when a CTV tri-amide (Figure 7.20f) substituted with disulfide groups was immobilised on gold electrodes.[65] Using AC impedance techniques, reversible binding of acetate could be detected with some interference from dihydrogen phosphate and minimal effect from bromide, chloride, sulfate and nitrate. A CTV substituted with three phosphonate groups (Figure 7.20g) has also been developed and in water and physiological media it was shown to selectively bind acetylcholine[66] in a range of 0–0.125 mol L^{-1} as determined by the enhancement of CTV fluorescence in the presence of the analyte.

The use of CTVs as scaffolds for assembling peptide chains has also been investigated. A library of 40 different compounds based on a CTV triphenol substituted with a variety of amino acids and di- and higher peptides was synthesised using combinatorial methods.[67] One of these macrocycles containing two leucine units per side chain was shown to selectively bind an Ac-Ala-Ala tripeptide. Larger libraries have been developed; a CTV triphenol (Figure 7.20a) for example could be reacted with benzyl bromoacetate to give the monosubstituted CTV (Figure 7.21a) along with smaller amounts of the di- and trisubstituted product.[68] This could then be further reacted to give the Boc- or CBz-protected amino compound (Figure 7.21b). This compound could then have the benzyl group removed to give the carboxylic acid, which could be bound to a solid substrate. Combinatorial methods were used to substitute the two free side chains, generating a 2197 strong library of CTV-based compounds containing peptide side chains. Similar methods have been used to substitute CTV triphenols with PEG chains, to which amino acids glycoconjugates could be bound.[69]

The synthesis of selectively substituted CTVs has been studied, via both the synthesis of selected enantiomers and the selective substitution of the methoxy units. One elegant example took the simple hexamethoxy CTV and demethylated it using a variety of procedures.[70] Although many procedures such as demethylation with HI gave complex mixtures, more controlled reactions could give pure products and reaction with a large excess of $TiCl_4$ under the correct conditions (room temperature, 24 hours) led to formation of a CTV in which a single methoxy group was converted to a phenol. The compound shown in

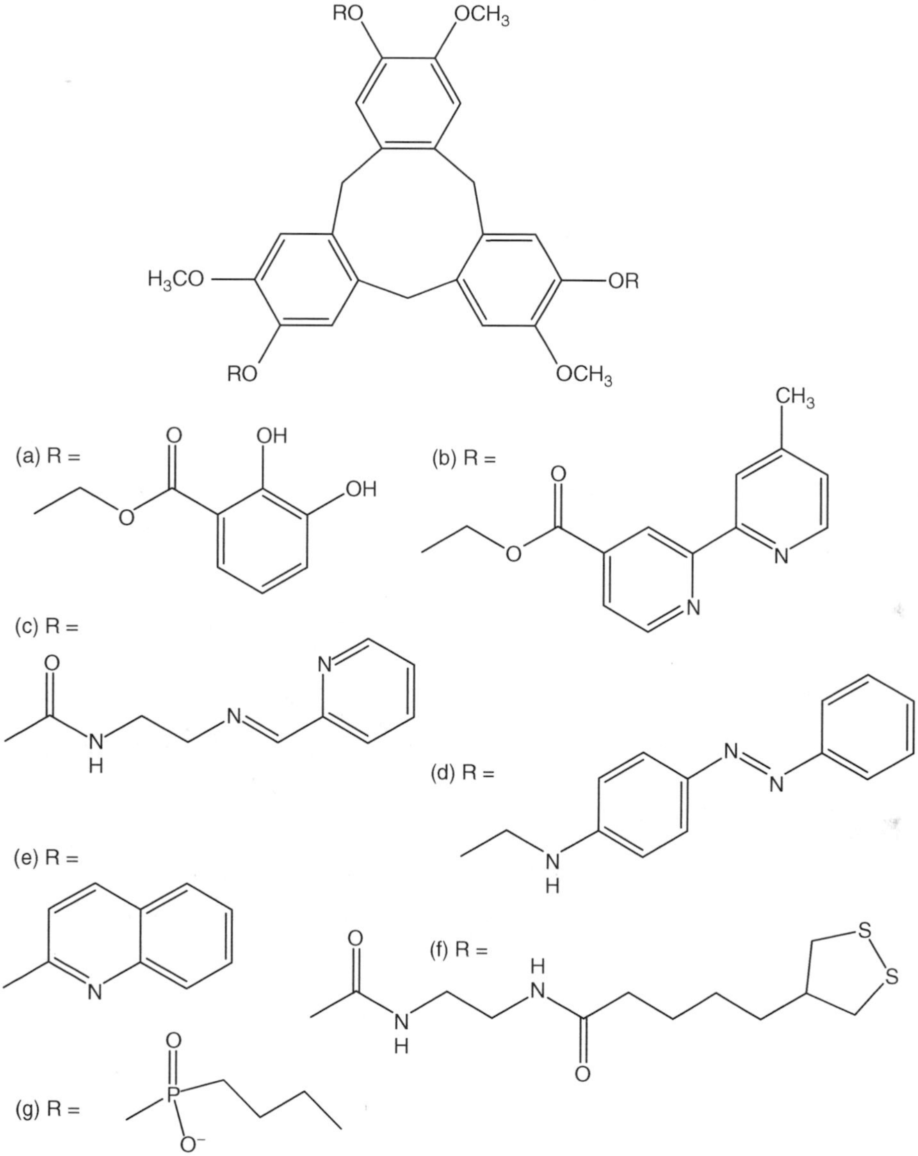

Figure 7.20 *Structures of substituted CTVs*

Figure 7.9e was obtained by reaction with AlCl₃, and with BBr₃ a mixture could be obtained containing the compound shown in Figure 7.22a. The phenol groups present in these products can be converted into a wide range of derivatives. Another synthesis involved the use of the racemic compound shown in Figure 7.22b (synthesised from thiovanillin), which could be co-crystallised with the chiral amine cinchonidine and separated into its enantiomers.[71] These could then be cyclised to form CTVs (a typical one is shown in Figure 7.22c) which were diasterisomeric in nature and therefore separable by chromatographic methods. The sulfur substituents could then be converted into a variety of other functional groups such as —OH.

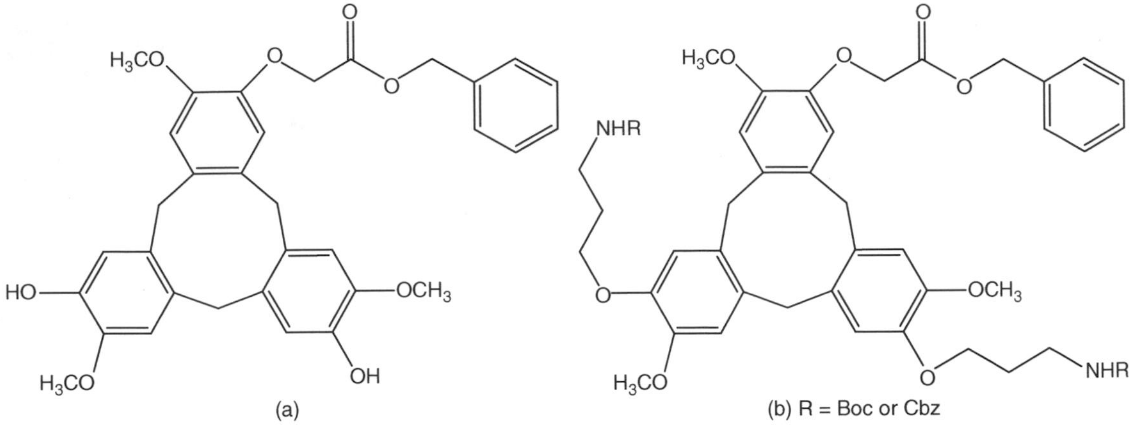

Figure 7.21 *Structures of (a) mono-substituted CTV and (b) its protected diamino derivative*

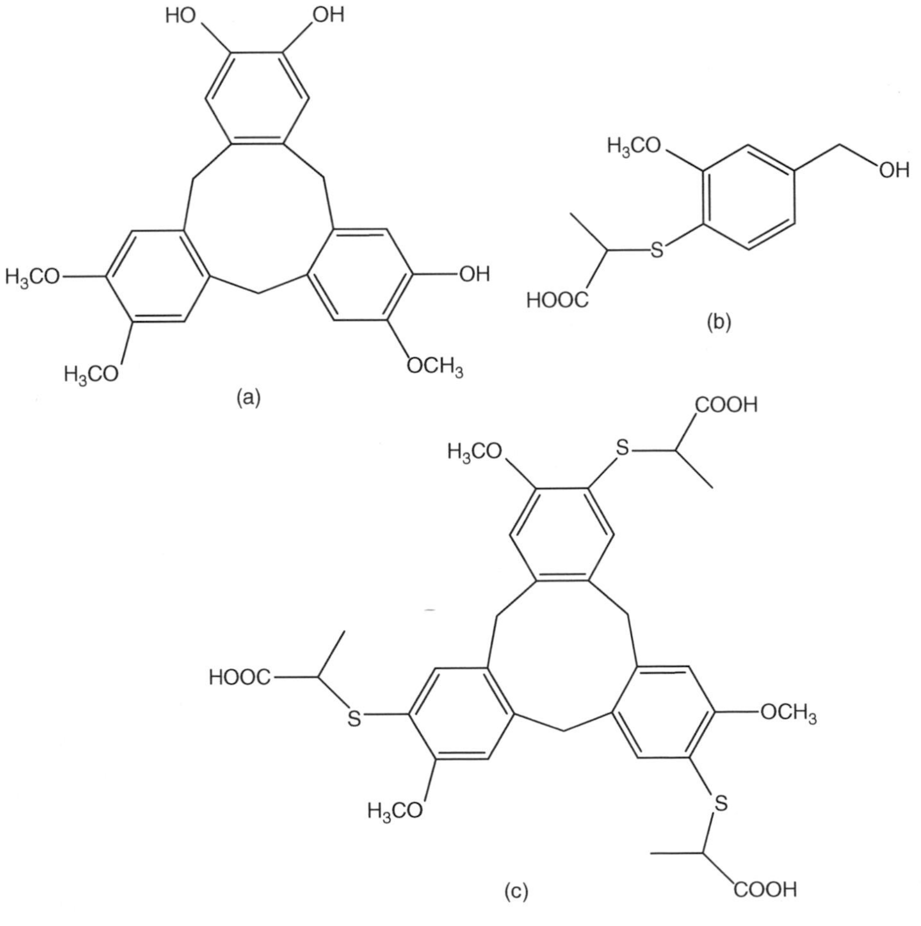

Figure 7.22 *Structures of (a) partially substituted CTV, (b) thiovanillin derivative and (c) CTV from 7.22b*

CTV structures containing metal atoms bound to chelating groups attached to the CTV skeleton have been widely studied. One of the simplest is cyclotricatechylene, which can be reacted with a platinum (II) chloride complex with either diphenylphosphinobenzene or diphenylphosphinoferrocene[72] to give the materials shown in Figure 7.23a,b. X-ray structural studies of the diphenylphosphinobenzene compound have shown that the CTV retains its convex shape and that within the extended cavity formed by the CTV and the ligands, four guest dimethylacetamide units are located. Another family of compounds is based on the triamino CTV shown in Figure 7.24a. This can be synthesised by the trimerisation of 3-methoxy-4-acetamidobenzyl alcohol,[73] which occurs in excellent yield, followed by hydrolysis to the salt of the triamino compound.[74] These then can be reacted with a variety of aldehydes such as salicylaldehyde to give Schiff bases as shown in Figure 7.24b. These bases tend not to form complexes with metal centres

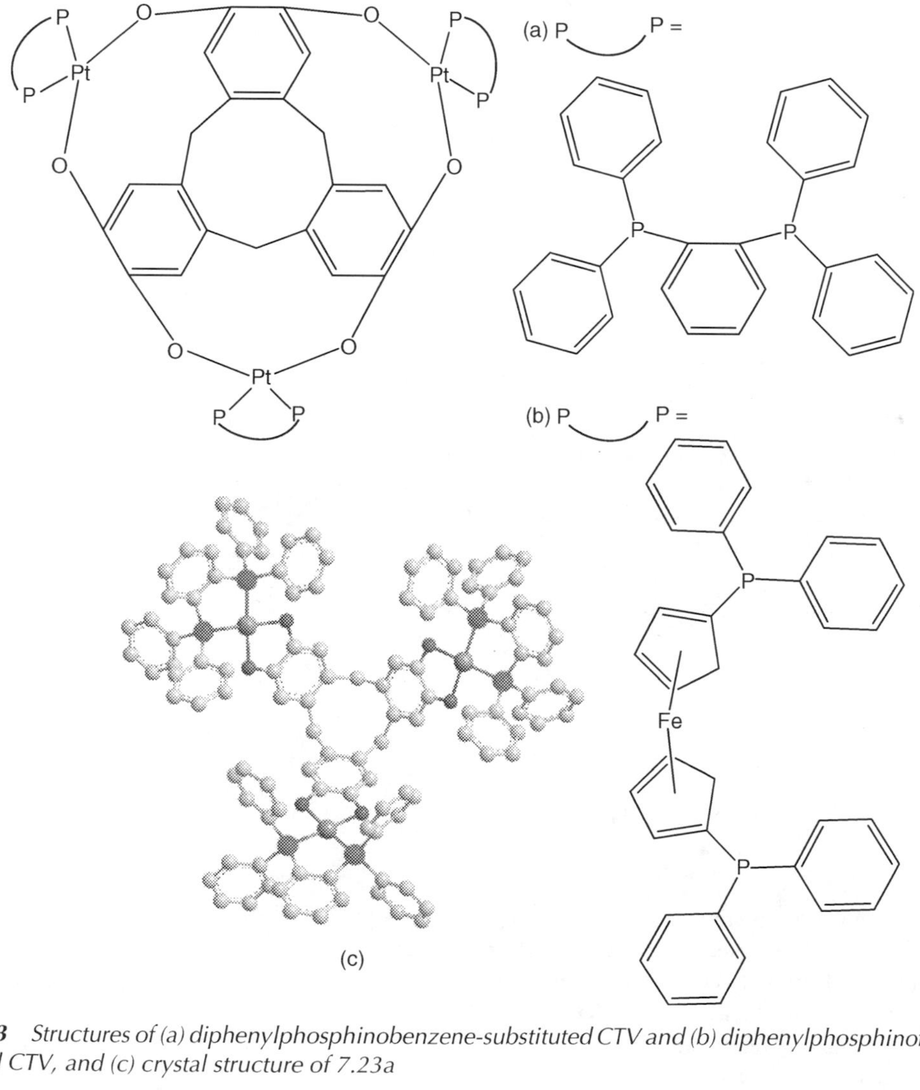

Figure 7.23   *Structures of (a) diphenylphosphinobenzene-substituted CTV and (b) diphenylphosphinoferrocene--substituted CTV, and (c) crystal structure of 7.23a*

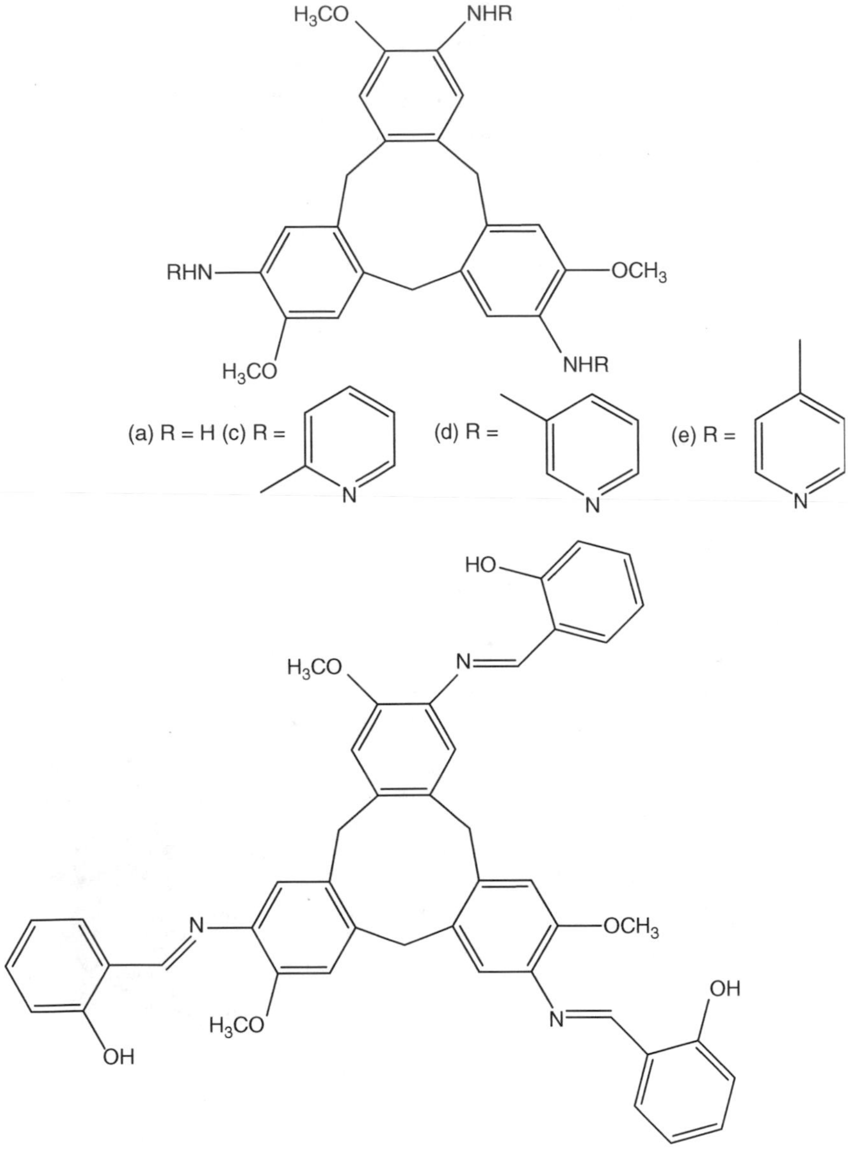

(a) R = H (c) R =   (d) R =   (e) R =

(b)

Figure 7.24   *Structures of triamino CTV derivatives*

very well, perhaps due to steric effects of the methoxy group; this can be removed by boron tribromide and complexes with metal such as nickel can then be obtained.

The triamino-substituted CTV has been used as a feedstock for many other compounds. For example, various pyridine carboxaldehydes have been reacted with the CTV to form Schiff bases.[75] These tend to hydrolyse quite easily after synthesis, so they can be reduced to give the amines shown in Figure 7.24c–e. The compounds can be crystallised themselves or can form complexes with a variety of metallic species.

With silver, a variety of topologies can be observed, for example the 3-pyridyl derivative (Figure 7.24d) forms a dimeric 2:2 complex with a capsular shape, as shown by X-ray crystallography of the trifluoromethyl sulfonate salt (Figure 7.25a). However, when crystallised at the hexafluoroantimonate instead of a capsule, a one-dimensional polymeric structure is observed (Figure 7.25b). When the 4-pyridyl compound (Figure 7.24e) is complexed with silver ions, tetrahedral species containing four CTV units can be observed, both in solution and when crystallised from the solid state;[75] Figure 7.25c shows the structure

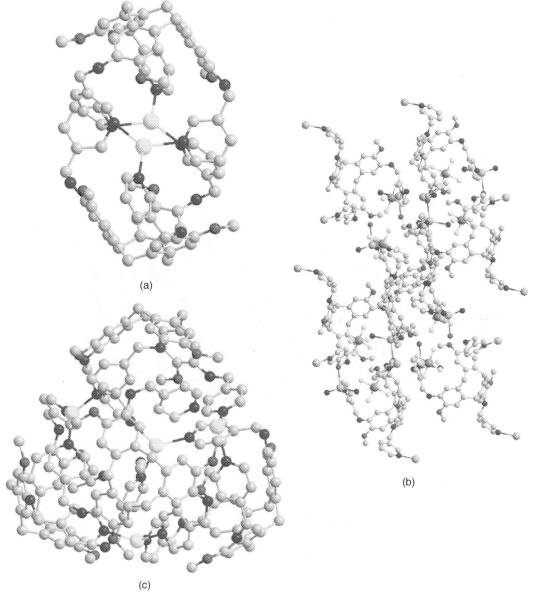

(a)

(b)

(c)

Figure 7.25   *Structures of tripyridyl CTV derivatives complexed with silver ions*

of the hexafluorophosphate, clearly demonstrating the tetrahedral arrangement of the CTV units. When the 4-pyridyl material is crystallised with aliphatic dinitrile compounds such as glutaronitrile, coordination polymers are formed with a structure which consists of an extended network incorporating roughly rectangular channels. These dinitriles are too bulky to be incorporated into the cavity of the tetrahedron and this could be why a polymeric structure is preferred.

Using different metals can lead to even more complex structures. Palladium can be complexed with an isonicotinyl derivative (Figure 7.26a) to give a Pd_6CTV_8-type structure.[76] This has been described as 'a stella octangular structure, a stellation of an octahedron that occurs when the edges of the octahedron are

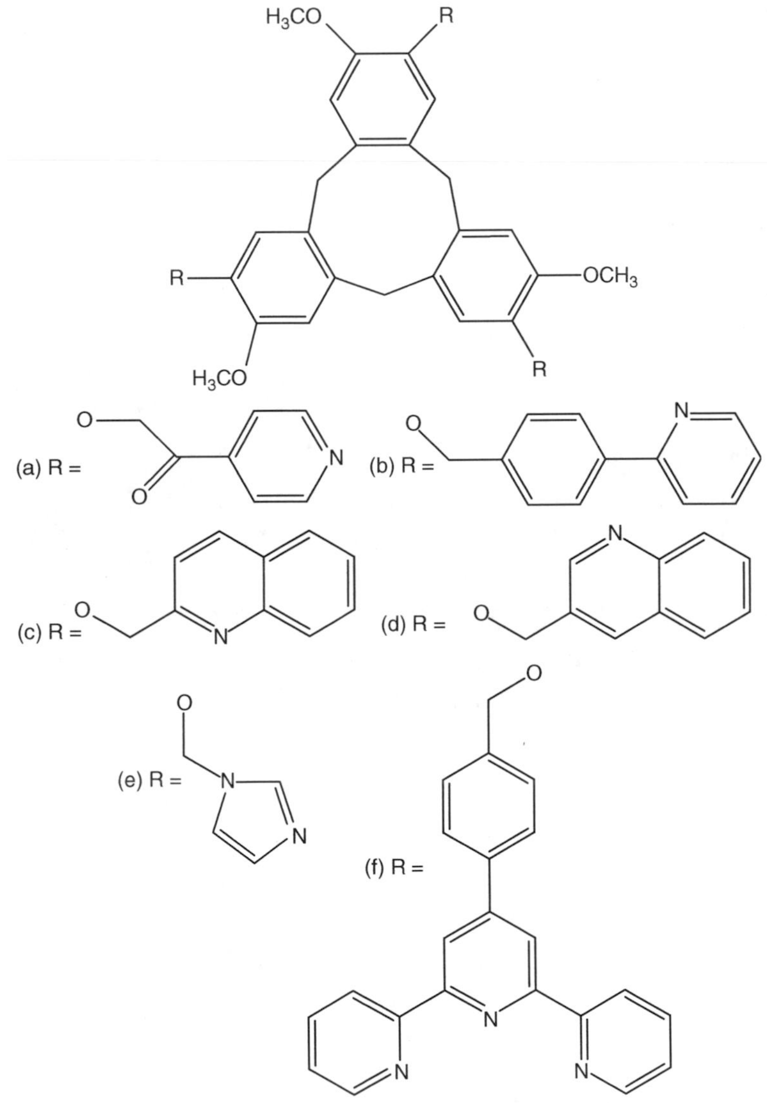

Figure 7.26   *Structures of trisubstituted CTV derivatives*

extended until they intersect at points to produce a starlike prism'.[76] Larger side chains such as pyridyl-benzyl, quinolyl, imidazole and terpyridyl[77] have been attached to CTV structures (Figure 7.26b–e); several of these species undergo what the authors have termed 'handshaking', where the side chain of one CTV is included as a guest in the cavity of another. Compounds that undergo this interaction include the 2-pyridylbenzyl compound (Figure 7.26b), the 2-quinolyl compound (Figure 7.26c) and the imidazole (Figure 7.26e), all as the silver complexes. However, they are quite different in nature, with X-ray studies showing that the 2-pyridylbenzyl and the imidazole-substituted CTVs form one-dimensional polymers, whereas the 2-quinolyl-substituted CTV exists as a tetrahedral structure. The terpyridyl struc-ture forms a Cu(II)$_3$CTV complex with CuCl$_2$ and its crystal structure (Figure 7.27) clearly shows an example of the handshake, with a dimeric structure in which the benzyl-terpyridyl-CuCl$_2$ arm of one complex within the dimer occupies the molecular cavity of the other and vice versa, giving a racemic dimer.

Earlier we mentioned how the presence of bulky dinitrile guests had an effect on the structures of these substituted CTVs[75] and a similar effect has also been found for bulky carborane anions. When a 4-pyridyl-substituted CTV (Figure 7.24e) is complexed with cadmium acetate in the presence of *o*-carborane, the globular guest occupies the CTV cavity and the resultant structure is that of a two-dimensional coordination polymer.[78] However, in the absence of the bulky guest, only simple Cd$_3$CTV complexes are formed.

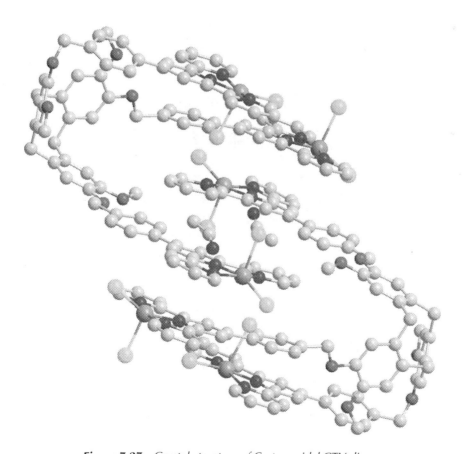

Figure 7.27 *Crystal structure of Cu-terpyridyl CTV dimer*

The tendency of many of these CTV species to aggregate has led to their investigation as possible gelating agents. Substituted CTVs have been shown to act as gelling agents for a variety of organic solvents. For example, CTVs containing primary amine or nicotamic pendent substituents (Figure 7.28a,b) have been synthesised;[79] whereas the compound in Figure 7.28a only acted as a gelling agent for acetonitrile and 2-propanol, the compound in Figure 7.28b was shown to form opaque gels with 20 common solvents and water. Robust, opaque and stable gels could be formed by as little as 14 mg ml^{-1} of the CTV in a few seconds. Analysis of these gels showed the formation of ribbons made up of assemblies of fibres 4–5 nm in diameter. FTIR studies showed that hydrogen bonding between the amide groups was responsible for the formation and stability of these systems. CTV–metal complexes also form gels, such

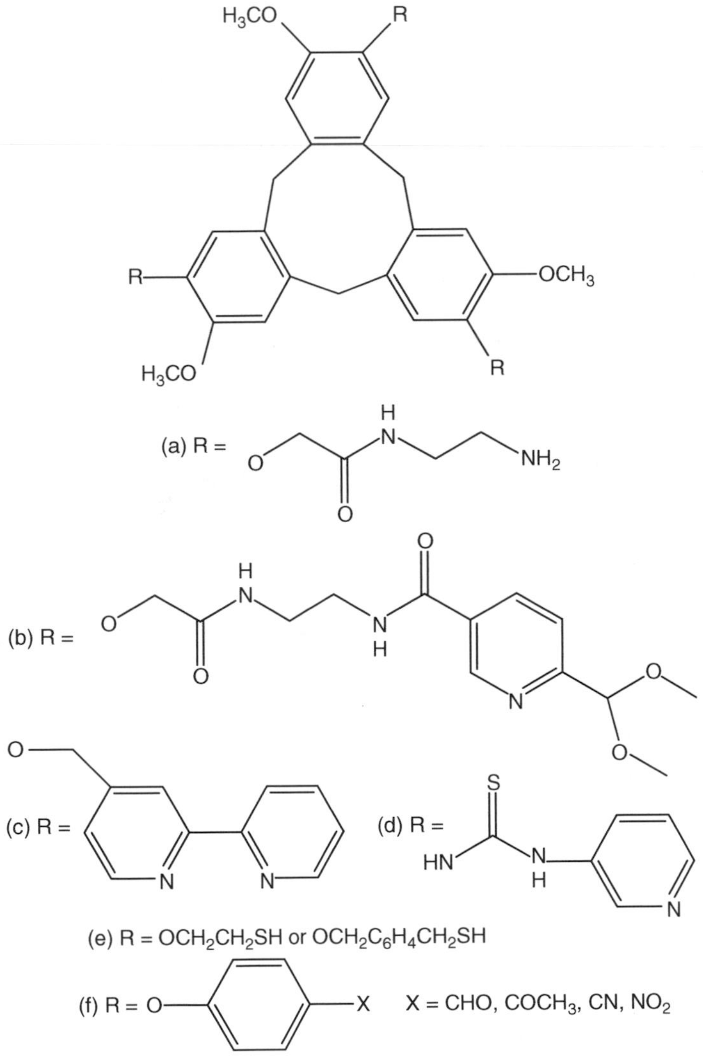
H3CO
R
R
OCH3
H3CO
R
(a) R =
H
N
O
O
NH2
(b) R =
O
H
N
O
O
N
H
N
O
O
(c) R =
O
N
N
(d) R =
S
HN
N
H
N
(e) R = OCH2CH2SH or OCH2C6H4CH2SH
(f) R = O
X
X = CHO, COCH3, CN, NO2

Figure 7.28 *Structures of trisubstituted CTV derivatives that act as gelling or selective binding agents*

as the nicotinyl-containing CTV (Figure 7.26a), which when complexed with CuBr$_2$ forms a dark-green gel with acetonitrile.[80] Similarly, the compound in Figure 7.28c forms gels with copper(II) and DMF, or DMSO if complexed with AgSbF$_6$. Again, microscopy has shown these gels to be fibrous in nature. The thiourea derivative (Figure 7.28d) forms gels in trifluoroethanol and does not complex transition metals.

CTVs substituted with mercaptoethyl groups or thiolaromatic groups (Figure 7.28e) have also been synthesised and shown to bind cubic iron–sulfur cluster compounds,[81] which can act as synthetic models for similar clusters found inside certain proteins. When cluster compounds with small substituents are bound, they are incorporated into the cavity, but with larger substituents although binding still occurs there is no incorporation. A variety of other substituted CTVs have also been made; for example CTVs with either three or all six of the methoxy groups replaced by ethyl, propyl, ally or propargyl groups have all been synthesised and their structures characterised;[82] many of these demonstrate either intra-cavity host–guest binding or self-stacking structures. A library of CTVs substituted with various dendrimer chains was synthesised and shown to adopt crown-like conformations.[83] These materials spontaneously self-assembled to give a variety of structures including helical pyramidal columns and supramolecular spheres, which could be racemic or chiral.

Other workers have looked at extending the cavity by reacting the CTV triphenol with fluorobenzenes bearing electron-withdrawing groups.[84] Several derivatives could be synthesised (Figure 7.28f) and further chemistry could be performed, such as reduction of the nitro or cyano groups to the corresponding amines and conversion into acetamides, phthalimides and so on. Sugar units have also been attached to CTVs (Figure 7.29a), thereby rendering them water-soluble.[85] A CTV substituted with three acid groups (Figure 7.29b) has also been synthesised and shown to induce the folding of collagen peptides into a triple helix. Recent work described a CTV substituted with multiple pyridine groups (Figure 7.29c) that was synthesised and shown to bind copper ions;[87] it was shown to form adducts with carbon monoxide, sulfur and oxygen. These materials could possibly serve as models for oxygen-processing copper proteins.

As an alternative to condensation with formaldehyde, trifluoromethyl acetaldehyde could be condensed with veratrole to give a mixture of cyclic and linear products.[88] When BF$_3$ was used as the catalyst, it proved possible to isolate CTVs with trifluoromethyl groups attached to the methylene bridges (Figure 7.29d).

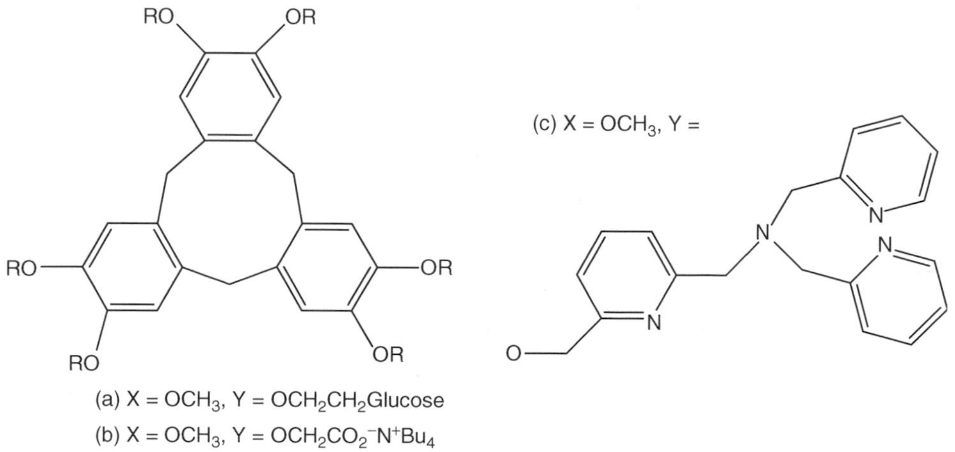

(a) X = OCH$_3$, Y = OCH$_2$CH$_2$Glucose
(b) X = OCH$_3$, Y = OCH$_2$CO$_2^-$N$^+$Bu$_4$
(c) X = OCH$_3$, Y =

Figure 7.29   *(a–d) Structures of trisubstituted CTV derivatives and (e) a CTV with trifluoromethyl bridging groups*

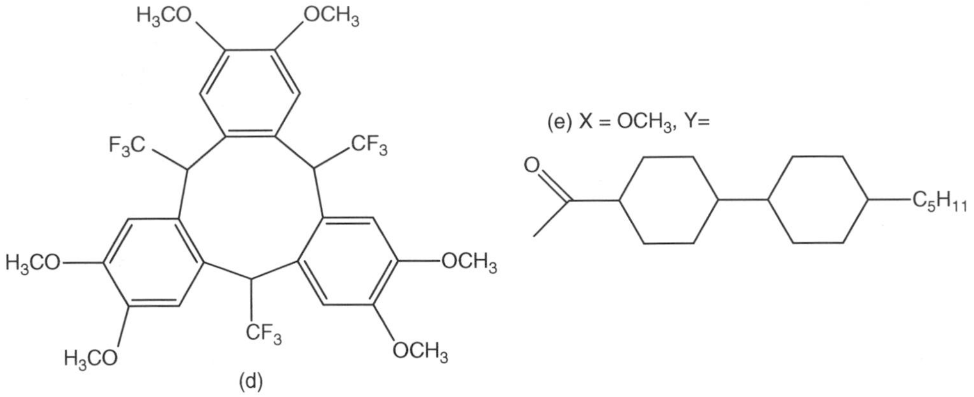

(e) X = OCH₃, Y=

(d)

Figure 7.29   *(continued)*

Liquid-crystal groups have been attached to CTVs, a typical structure being depicted in Figure 7.29e, which was shown to form columnar smectic liquid-crystal phases.[89] A cyclotetraveratrylene derivative was also found to form a columnar phase. Other workers have incorporated cyclotetraveratrylene derivatives into branched polyethers and shown formation of columnar mesophases.[90] A tetrameric analogue of the CTV shown in Figure 7.24d has also been developed and was shown by NMR to adopt a 'sofa' conformation in solution and to be flexible at room temperature.[91] The tetramer was found to form complexes with Cu, Ag and Pd; the structure of the Ag complex is shown in Figure 7.30. This shows the cyclic tetramer to exist in a distorted cup formation and to possess a binding cavity.

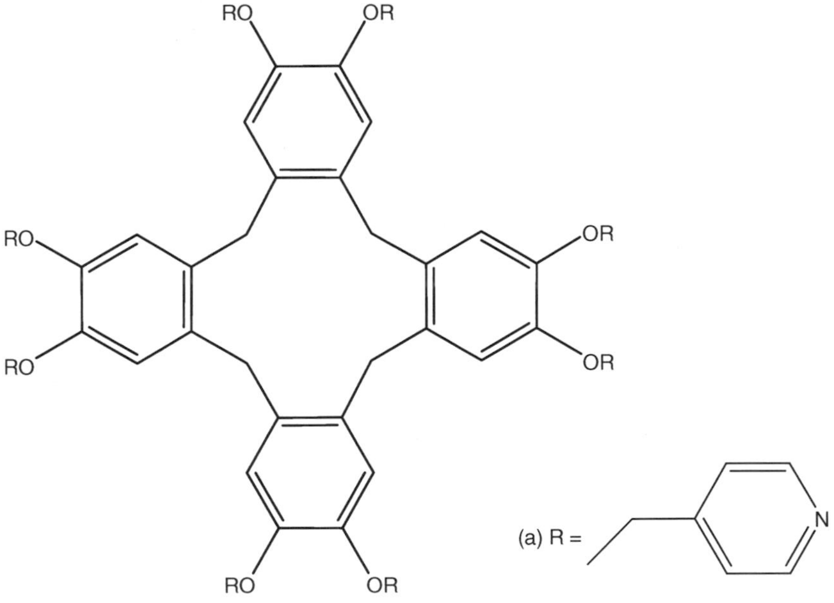

(a) R =

Figure 7.30   *(a) Structure and (b) crystal structure of a cyclic tetramer–Ag complex*

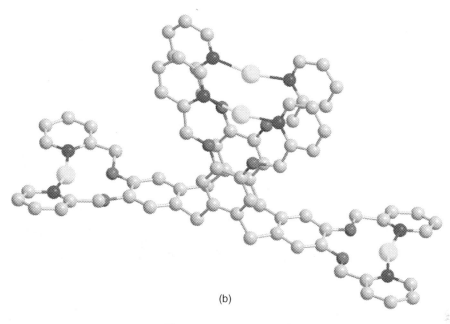

(b)

Figure 7.30   *(continued)*

7.8   Analogues of CTV

A rigid analogue of CTV in which the methylene groups are linked together via a carbon bridge (Figure 7.31a) has been synthesised by a multistep approach.[92] This compound was shown to adopt a somewhat different conformation than CTV, with significant out-of-plane distortion of the structure. The bridging groups proved amenable to substitution, the compound being capable of being converted into the tribromo, trimethyl or trihydroxy derivatives.

Heterocyclic analogues of CTVs have been synthesised and cyclic trimers, for example, can be obtained from various pyrrole derivatives, which react with formaldehyde to give the products shown in Figure 7.31b–d in reasonable yields.[93] A more recent paper[94] details the iodine-catalysed trimerisation of the monomer shown in Figure 7.31g to give the cyclic trimer in Figure 7.31e. NMR studies of this compound show it to exist in the crown conformation and it has substituents that are easily modified chemically. Other workers synthesised the cyclic tripyrrole (Figure 7.31f), which was shown by NMR studies to adopt the crown conformation in chloroform but to be conformationally mobile in pyridine solution.[95] When crystallised from chlorobenzene, the trimer was shown to adopt a saddle conformation (Figure 7.31h).

We have already mentioned the calix[3]indoles in Chapter 5. An N-methyl indole methanol compound cyclises under acid catalysis to give the trimer shown in Figure 7.32a. The crystal structure (Figure 7.32b) shows this material to adopt the saddle conformation.[96] Other N-substituted trimers could also be obtained using this process, as could trimers derived from indoles with a variety of substituents on the benzene ring.

Thiophene analogues of CTVs have also been synthesised, with 2,5-dimethyl thiophene condensing with formaldehyde to give the cyclic trimer[97] shown in Figure 7.32c and 2,5-dimethyl furan and pyrrole just giving complex tarry mixtures that cannot be purified. NMR studies show that the trimer is conformationally mobile and probably exists in a saddle-like form. A small amount of tetramer is obtained as a by-product.

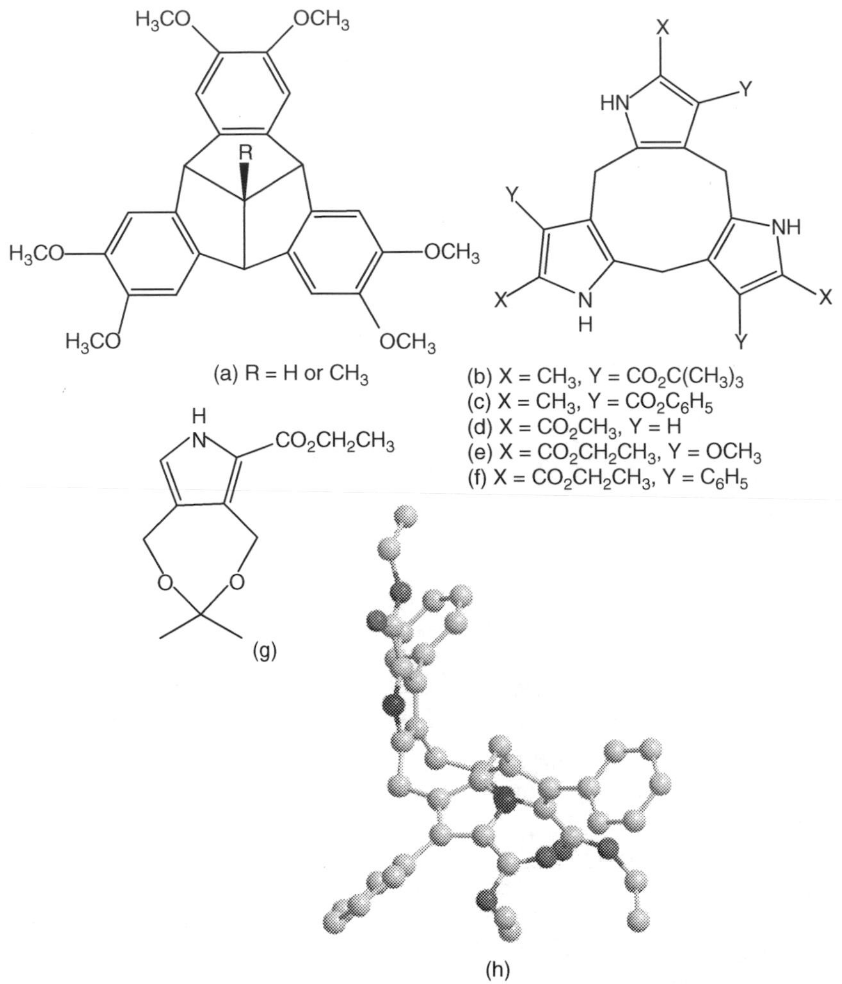

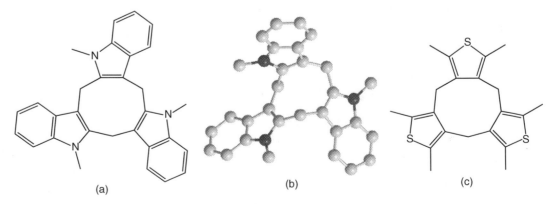

H3CO
OCH3
R
H3CO
OCH3
H3CO
OCH3
(a) R = H or CH3
X
HN
Y
Y
NH
X
N
H
X
Y
(b) X = CH3, Y = CO2C(CH3)3
(c) X = CH3, Y = CO2C6H5
(d) X = CO2CH3, Y = H
(e) X = CO2CH2CH3, Y = OCH3
(f) X = CO2CH2CH3, Y = C6H5
H
N
CO2CH2CH3
O
O
(g)
(h)

Figure 7.31 *(a) Rigid analogue of CTV, (b–f,h) heterocyclic analogues of CTVs and (g) pyrrole monomer*

N
N
N
S
S
S
(a)
(b)
(c)

Figure 7.32 *(a) Structure and (b) crystal structure of an indole trimer; (c) structure of a thiophene trimer*

7.9   Synthesis and Structure of Cryptophanes

We have already commented on the relatively poor solution binding of CTV compared to many other macrocyclic systems. Enhancements in binding have been made by chemical substitution, for example in 'hexapus' and the metallated CTVs. Another obvious method of improving complex formation and stability is to deepen the cavity. Within this chapter we have already described how complexation with various metal species with the CTV rim has enabled formation of deeper cavities, as has substitution of the rim. Another approach is to increase the depth, rigidity and degree of preorganisation of the cavity by covalent bridging. We have already demonstrated this effect for calixarenes and other macrocycles; Donald Cram was again one of the earliest to investigate this approach.[98] CTV was demethylated to give the hexaphenol, which was then bridged using 1,3-*bis*(chloromethyl)arenes to give structures such as that shown in Figure 7.33. A variety of materials were synthesised and one was successfully crystallised (Figure 7.33b). NMR studies

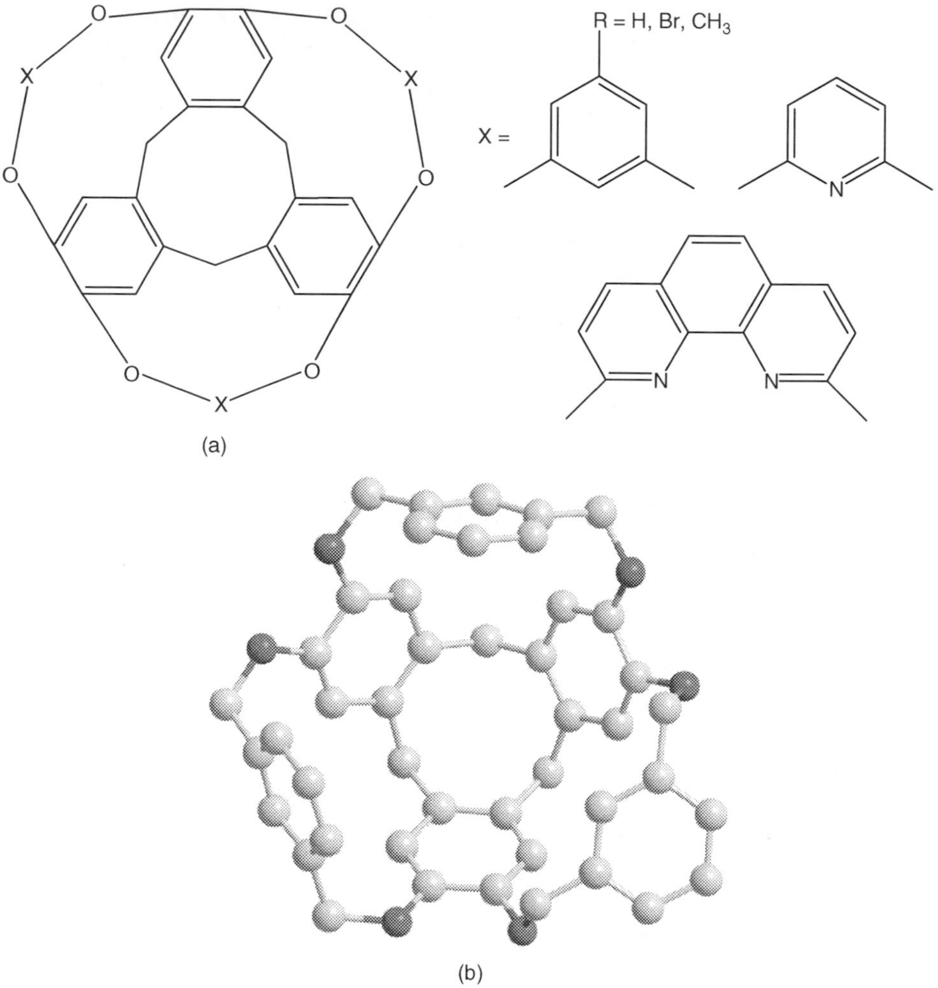

(a)

(b)

Figure 7.33   *(a) Structure of bridged CTV and (b) X-ray crystal structure of 7.32a (R = H)*

showed that the CTV retained its rigid structure, although the bridging aromatic units were much more mobile. In the X-ray structure two of the xylylene bridges were orientated approximately perpendicular to the plane of the macrocycle, and one approximately parallel. Calculations also indicate that the deeper cavity CTVs will form complexes with higher affinities than native CTV.

We have described in previous chapters how calixarenes and other macrocyclic systems can be linked together to form 'capsules' that can temporarily or permanently entrap suitable guests. Several approaches have also been utilised to link together CTVs to construct a series of compounds for which the name 'cryptophane' has been coined and about which a number of reviews have been written.[9,99,100] CTVs are linked via the outer-rim substituents to form dimeric capsules. Much of this the work on cryptophanes has been carried out by the group of André Collét of the Collège de France and the Université Claud Bernard-Lyon. Collét worked for quite some time in the laboratory of another worker featured prominently within this book, Jean-Marie Lehn, but he will be remembered in his own right as a pioneer in the field of cryptophanes and their parents, the cyclotriveratrylenes.

One of the earliest examples of cryptophane synthesis utilised the CTV triphenol (Figure 7.9h), which could be substituted to give the structure shown in Figure 7.34a. The three benzyl alcohol units could then be cyclised[101] to give a product in 25% yield (when R = H) along with 5% of a second product. When R = OMe, the yields increased to >80% for a single product. This good yield of product, rather than formation of a crosslinked polymeric material, indicates that prearrangement of the reactive benzyl alcohol units onto a preformed CTV skeleton aids cyclisation within the same molecule rather than intermolecular reactions. This can be thought of as another example of the template effect.

There were two possible structures (known as *syn* and *anti*) for this compound, as shown generically for cryptophanes in Figure 7.35. In the case of the material with R = H, two products were obtained. The cavitand nature of the major product was proved by the fact that a molecule of dichloromethane was encapsulated within the cavity, as were the similar molecules bromochloromethane and dibromomethane.

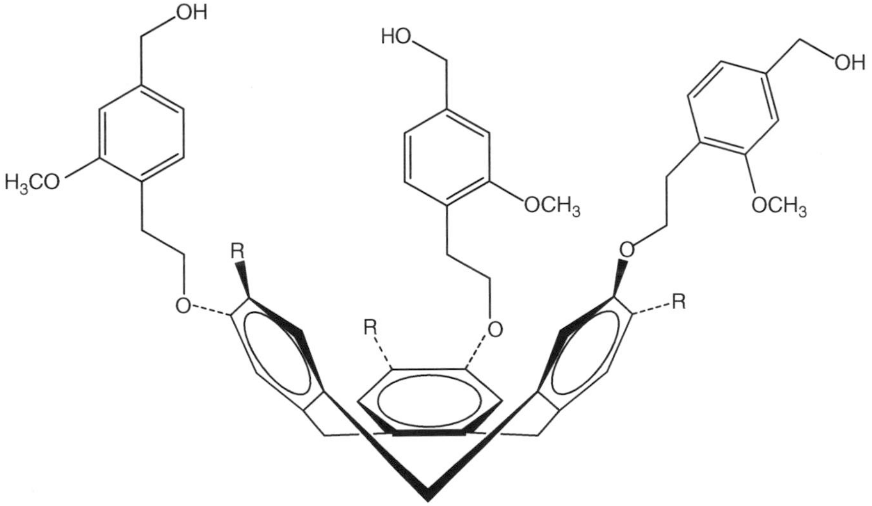

(a)

Figure 7.34 *Structures of (a) substituted CTV and (b) cryptophane (R = H or CH₃)*

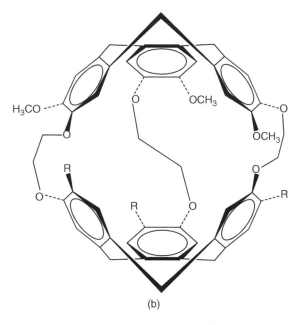

(b)

Figure 7.34   (continued)

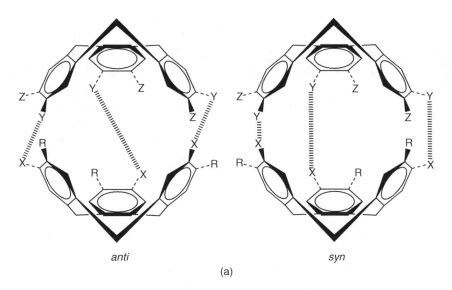

anti

syn

(a)

Figure 7.35   (a) Structures of syn and anti cryptophane and (b) crystal structure of CH_2Cl_2 clathrates

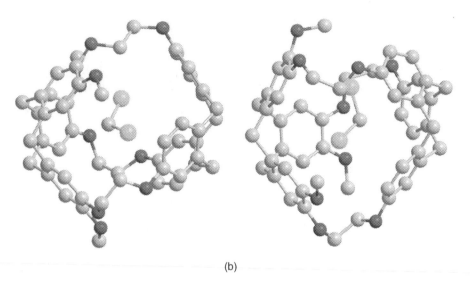

(b)

Figure 7.35   *(continued)*

NMR studies showed the binding coefficient for the dichloro compound was 12 times that of the dibromo compound. An X-ray structure was obtained for this material and demonstrated that the cryptophane existed in the *anti* conformation, as shown in Figure 7.35b, with the minor product being the *syn* compound. When R = OMe, yields were greatly increased and only the *anti* product was obtained. The *syn* form also forms complexes with dichloromethane (although not as strongly as the *anti*) and discriminates between it and dibromomethane.[102] Crystal structures of this clathrate were obtained and showed that there are some differences between the *anti* and *syn* forms in the solid phase, one being that the CTV units in the *syn* isomer are nearly completely eclipsed whereas in the *anti* isomer they are staggered by about 52°.

Besides the *syn* and *anti* isomers, there are a number of possible conformers since CTVs can be conformationally flexible. The structure determined above gives the so-called *out-out* conformer.[101] Other conformations[100] are shown schematically in Figure 7.36: *in-in, in-out* and *out-saddle*.

Because of the difficulty of naming these cryptophanes and due to the complexity of their IUPAC names, many of these species were named chronologically. Therefore, the earliest synthesised cryptophane (Figure 7.37 R = OMe) was named cryptophane A (in the *anti* conformation) and the corresponding *syn* isomer cryptophane B, even though this has not been isolated. The cryptophane shown in Figure 7.35 R = H, is cryptophane C in the *anti* form and D in the *syn* form. Cryptophanes E and F with —O(CH$_2$)$_3$O— bridging groups were next to be discovered; others such as cryptophanes O and P with —O(CH$_2$)$_5$O— bridging groups followed. As the number of cryptophanes increased, this convention soon became untenable, but even in the most recent papers many of the earlier compounds are referred to by their familiar names. Another system that has been used for the most common form of cryptophanes, those with simple alkyl bridging chains, is to simply give the number of carbon atoms in each chain; in this nomenclature cryptophane A becomes cryptophane 222 and cryptophane E becomes cryptophane 333. Mixed systems also exist, such as cryptophane 223 with two dimethylene and one trimethylene spacer.

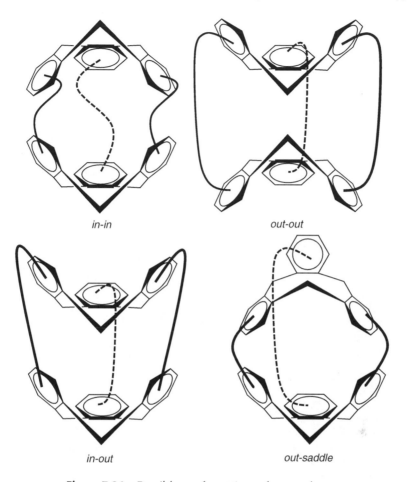

in-in

out-out

in-out

out-saddle

Figure 7.36 *Possible conformations of cryptophanes*

A direct method[9] of synthesising cryptophanes involves two benzyl alcohol units being joined together by a spacer group and then both macrocyclic rings forming simultaneously, as shown in Figure 7.38. This synthetic method is more simple than the 'template' synthesis and does not require high-dilution conditions. Unfortunately, it has the disadvantages that yields are often very low (2–5% for cryptophane A) and it is not possible to synthesise cryptophanes with two different macrocyclic rings, since using an unsymmetrical dibenzyl alcohol would lead to intractable mixtures of multiple cryptophanes. However, other bridging groups have led to more amenable syntheses, with some cryptophanes such as cryptophane E being obtained in 10–20% yields.[9]

Cryptophanes generally adopt the *out-out* conformation and therefore tend to exist as quasi-spherical units with both CTV units adopting cone conformations. However, they are not completely rigid and CTV units have been shown to interconvert or adopt a saddle conformation.[7] Numerous X-ray structures have been obtained for cryptophanes, as shown earlier in Figure 7.35. Similarly, cryptophane E has also

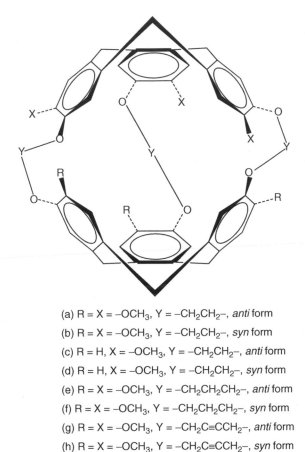

(a) R = X = –OCH$_3$, Y = –CH$_2$CH$_2$–, *anti* form

(b) R = X = –OCH$_3$, Y = –CH$_2$CH$_2$–, *syn* form

(c) R = H, X = –OCH$_3$, Y = –CH$_2$CH$_2$–, *anti* form

(d) R = H, X = –OCH$_3$, Y = –CH$_2$CH$_2$–, *syn* form

(e) R = X = –OCH$_3$, Y = –CH$_2$CH$_2$CH$_2$–, *anti* form

(f) R = X = –OCH$_3$, Y = –CH$_2$CH$_2$CH$_2$–, *syn* form

(g) R = X = –OCH$_3$, Y = –CH$_2$C≡CCH$_2$–, *anti* form

(h) R = X = –OCH$_3$, Y = –CH$_2$C≡CCH$_2$–, *syn* form

(o) R = X = –OCH$_3$, Y = –(CH$_2$)$_5$–, *anti* form

(p) R = X = –OCH$_3$, Y = –(CH$_2$)$_5$–, *syn* form

Figure 7.37   *Naming and structures of the earliest cryptophanes (*anti *form shown)*

been obtained as its dichloromethane clathrate[103] (Figure 7.39a) and a similar structure has been shown to occur for the chloroform clathrate.[104] Interestingly, the dynamics of the two inclusion complexes have been shown to be quite different in NMR studies; chloroform gave a broad peak, indicating slow reorientation within the cavity, whereas dichloromethane gave a sharp peak, indicating fast motion of the guest molecule.[103] Cryptophane A has also recently been successfully crystallised as the chloroform clathrate and was shown to display complex behaviour, with three different crystal structures being obtained.[105] These all contained a chloroform molecule bound within the cavity; Figure 7.39b shows the simple 1 : 1 complex. Two others exist, with either 2 : 1 or 3.5 : 1 guest–host ratios; however, the additional chloroform molecules are not within the cavity but between the cryptophane molecules. Cryptophane O also displays an *out-out* structure (Figure 7.39c) and was crystallised as part of a study on cryptophanes linked with bridges of 4–10 methylene units;[106] at 250 °C this cryptophane was shown to be capable of inversion to give an *in-out* structure.[100] NMR studies have indicated that in the larger members of this system, there is also equilibration between *in-out* and *out-out* structures.[106]

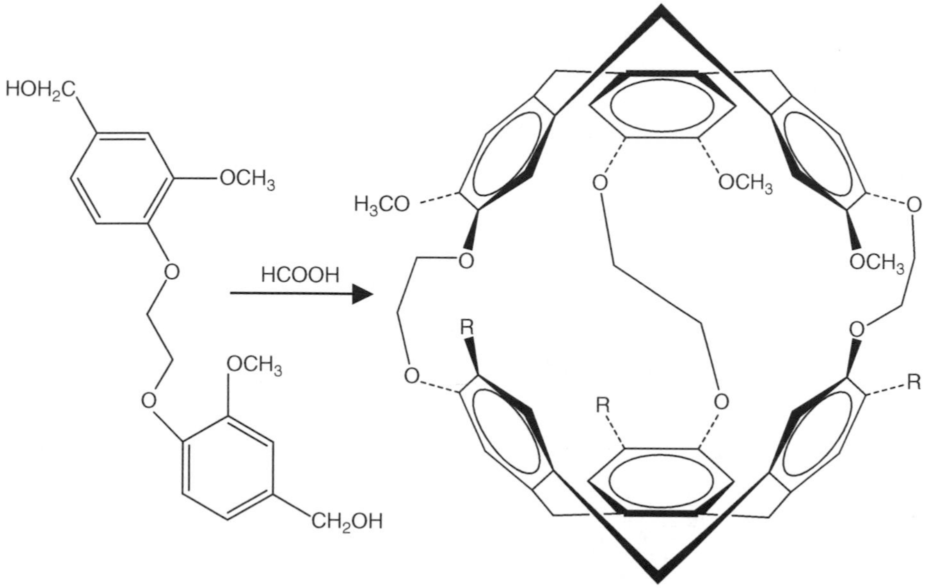

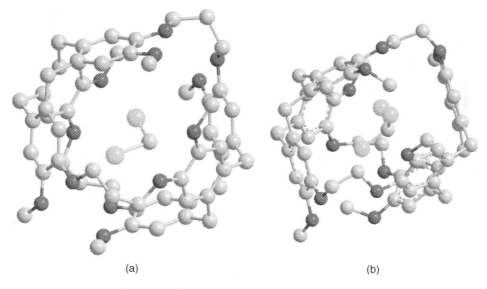

Figure 7.38 *Direct synthesis of cryptophane A*

Figure 7.39 *X-ray structures of (a) cryptophane E + CH₂Cl₂, (b) cryptophane A + CHCl₃ and (c) cryptophane O*

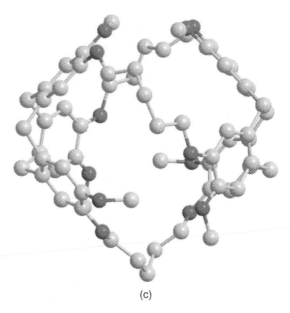

(c)

Figure 7.39   *(continued)*

Other workers have synthesised dibenzyl alcohol with a variety of bridges, such as the *m*-xylyl-bridged compounds shown in Figure 7.40. After condensation in formic acid, the compound in Figure 7.39a could be obtained in 21% yield in its chiral form, whilst the achiral *syn* form was obtained from the same reaction in 14% yield.[107] X-ray structures of the *syn* form showed the ester functions orientated towards the centre of the cryptophane (Figure 7.41a). The ester groups could then be hydrolysed to give the free carboxylic acids. This triacid derivative was shown in chloroform solution to complex methanol (with stability constants for the 1 : 1 complex of 7500 L mol^{-1}) and ethanol (41 L mol^{-1}) but not isopropanol, acetonitrile, nitromethane, acetaldehyde or acetone. The triacid also acted as an efficient extractor for a variety of metal ions into chloroform, probably due to the acid groups being directed towards the centre of the cavity, and showed a preference for Ca^{2+} and Sr^{2+} ions. High affinities were also noted for the lanthanides ytterbium and europium. A variation of this synthesis has been used to generate a cryptophane with three *exo* ester groups (Figure 7.40c) in 10–12% yield exclusively in the racemic *anti* conformation.[108] This system was found to form a strong 1 : 3 complex with THF, which could be shown by X-ray crystallography to contain an encapsulated THF molecule, with more THF located between the cryptophanes (Figure 7.41b). Prolonged heating caused loss of first the interstitial and then the encapsulated THF to give an 'empty' cryptophane. Interestingly, this was seen to exist as a mixture of 'empty' host and a second material, which could be separated by chromatography. This second material was shown to be cryptophane that had 'imploded'. NMR and X-ray studies showed the two CTV macrocycles to be non-equivalent, where one CTV is in the cup conformation and the other is in the saddle conformation (Figure 7.41c), the so-called *out-saddle* conformation.

Similar CTVs were synthesised first by the direct method but later more effectively by the template method,[109] *with o*-xylyl, *p*-xylyl or diethylenoxy linking groups as shown in Figure 7.40d–g. A recent paper has also been published detailing the use of scandium triflate[110] in the synthesis of CTVs and cryptophanes; in the case of cryptophane A this was synthesised via the template method, with yields being increased to as high as 51%. This method also reduced the formation of side products and allowed

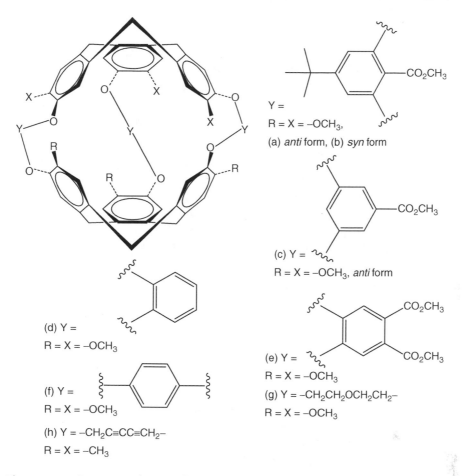

Y =

R = X = –OCH$_3$,

(a) *anti* form, (b) *syn* form

(c) Y =

R = X = –OCH$_3$, *anti* form

(d) Y =

R = X = –OCH$_3$

(e) Y =

R = X = –OCH$_3$

(f) Y =

R = X = –OCH$_3$

(g) Y = –CH$_2$CH$_2$OCH$_2$CH$_2$–

R = X = –OCH$_3$

(h) Y = –CH$_2$C≡CC≡CH$_2$–

R = X = –CH$_3$

Figure 7.40 *Structures of cryptophanes (*anti *form shown) with various bridging groups*

the synthesis of cryptophanes that could not be obtained via the classical strong acid conditions, such as benzyl ether substituted compounds.

Another method used to synthesise cryptophanes is to couple together substituted CTVs. For example, CTV triphenols can be substituted with propargyl bromide to give a CTV bearing three acetylenic groups.[111] Two CTVs can then be coupled together using copper-catalysed oxidation of these groups to give the structure shown in Figure 7.40h in both the *anti* and *syn* forms, albeit after chromatographic separation, with yields of 4% and 2% respectively. X-ray crystal structures show that the CTV units are staggered at 120° in the *anti* isomer and 60° in the *syn* isomer. The *anti* isomer was shown to strongly complex dichloromethane in hexachloroacetone solution; other guests that readily complexed included toluene, chloroform, propylene oxide and cubane. Within limits, the cavity size appears to be adaptable by bridging-unit folding to accommodate the steric requirements of the guest. Noncovalent interactions have also been utilised to link together CTVs into cryptophane-type molecules; pyridine-substituted CTVs for example have been linked by silver atoms to give dimers and tetramers.[75] As an alternative, palladium species have also been utilised to link together CTVs.[112] A tribromo-substituted CTV can be coupled with diethyl(4-pyridyl)borane to give the CTV shown in Figure 7.42a; this can then be reacted with a palladium complex

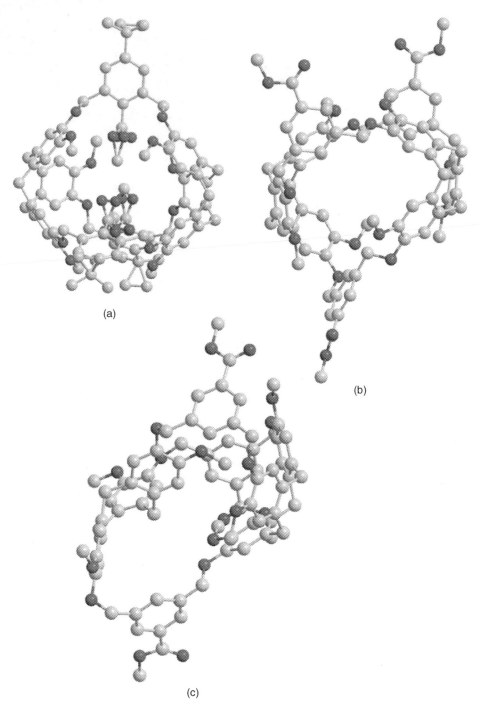

Figure 7.41 *X-ray structures of cryptophanes, showing (a) the syn trimester, (b) THF clathrate (solvent not shown) and (c) empty 'imploded'*

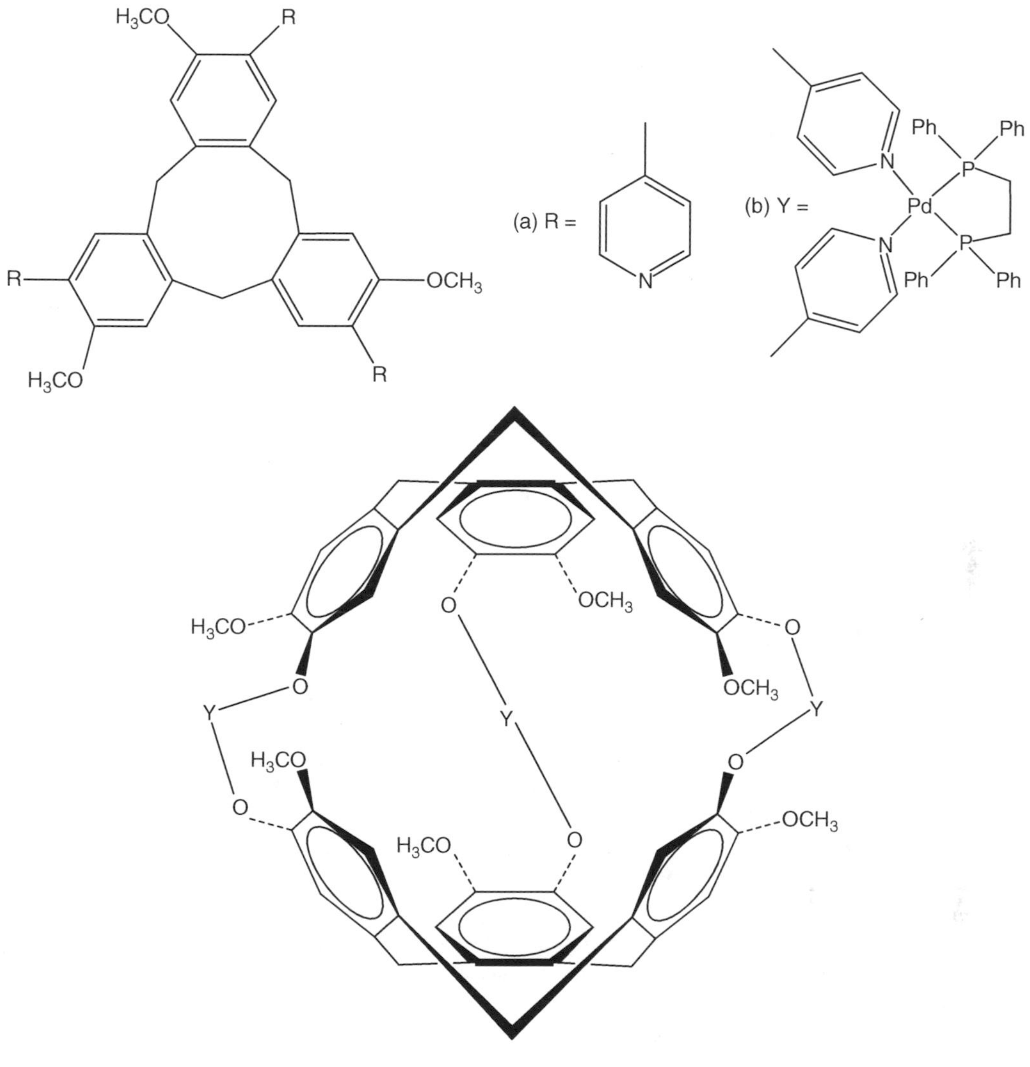

Figure 7.42   *Structures of palladium-linked cryptophanes*

to give cryptophanes like that shown in Figure 7.42b. If a racemic mixture of CTVs is used as the feedstock, a mixture of both racemates and the *meso* cryptophane is produced; using a chirally resolved CTV leads to formation of chiral cryptophanes.

7.10   Modification of Cryptophanes

Cryptophanes can be chemically modified in the same manner as CTVs. We have already noted how CTVs can be demethylated with a variety of reagents to give the phenolic compounds, and the same reaction can be performed on cryptophanes. However, there is a complication in that too vigorous conditions result in the similar reaction occurring for the ether-linking groups and destroying the cryptophane structure. One of

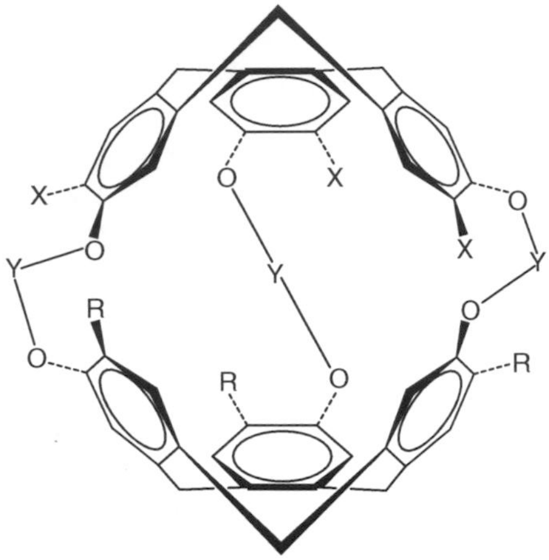

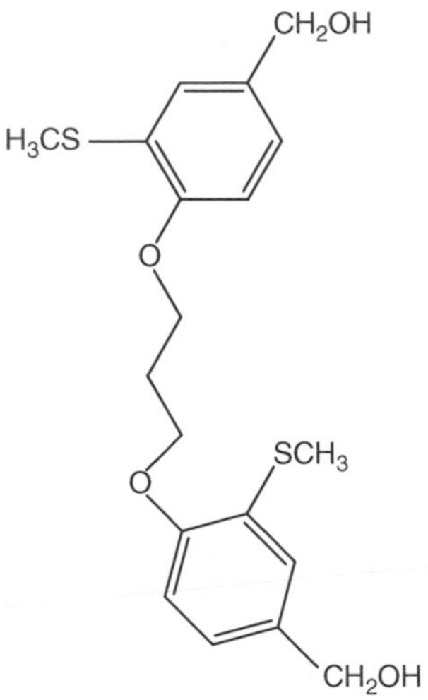

(a) R = X = –OCH$_2$COOH, Y = –CH$_2$CH$_2$–
(b) R = X = –SCH$_3$, Y = –CH$_2$CH$_2$CH$_2$–
(c) R = X = –OCH$_3$, Y = –CH$_2$–
(d) R = –SCH$_2$CONH(CH$_2$)$_6$OH,X = –OCH$_3$, Y = –(CH$_2$)$_4$–
(e) R = –SCH$_2$CONHCH$_{16}$H$_{33}$,X = –OCH$_3$, Y = –(CH$_2$)$_4$–

Figure 7.43   *Structures of substituted cryptophanes*

the earliest successful examples[113] involved taking cryptophane A, demethylating the methoxy groups with lithium diphenylphosphide, substituting the phenol groups with methyl bromoacetate and hydrolysing to give the hexa-acid shown in Figure 7.43a. This compound proved to be soluble in water as the sodium salt and reversibly complexed dichloromethane and chloroform in these conditions, with chloroform binding more strongly.

A direct synthesis using a *bis*-thiovanillin derivative (Figure 7.43f) has allowed the synthesis of a cryptophane E analogue,[114] with —SMe replacing OMe substituents (Figure 7.43b). The presence of the —SMe groups reduces the negative charge density in the aromatic rings of the cyclophane and obstructs the entrances to the cavity. This leads to (in comparison to cyclophane E) much slower guest-exchange rates and lower association constants with guests such as chloroform, trimethylammonium and tetramethyammonium salts.

A triphenol CTV (Figure 7.9e) could be reacted with a benzyl alcohol substituted with an allyl ether to give a CTV in which just one of the phenolic groups is etherified.[115] This could then be substituted with two more veratryl units and cyclised to give the cyclophane shown in Figure 7.44a. The allyl ether could be deprotected by palladium-catalysed hydrogenation to give a cyclophane bearing a single phenolic group (Figure 7.44b). Reacting this material (named a 'cryptophanol') with 1,10-diiododecane gives the *bis*-cryptophane shown in Figure 7.44c. An unsymmetrical *bis*-cryptophane in which one of the units is partially deuterated has also been reported. The cryptophanol could also be reacted with the chiral (−)-camphanic acid chloride, which since the cryptophanol is also racemic gives rise to two diastereoisomeric materials.[116] These could be separated by crystallisation and hydrolysed to give both enantiomers with an enantiomeric excess >98%. The cryptophanols could then be reacted with methyl iodide to give optically pure cryptophane A.[117] Vibrational circular dichroism measurements have given the absolute configuration and shown that both 'empty' (in tetrachloroethylene solution) and filled (chloroform or dichloromethane

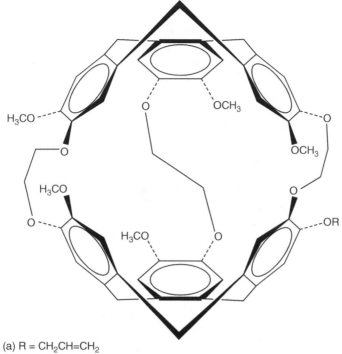

(a) R = CH₂CH=CH₂
(b) R = H

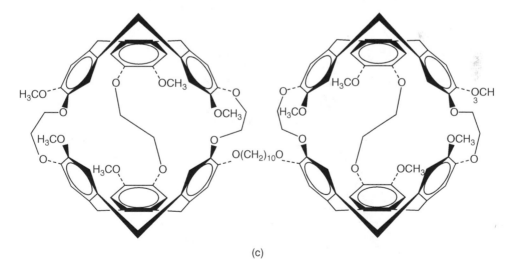

(c)

Figure 7.44 *Cryptophanols and* bis-*cryptophanes*

solution) cryptophane have similar configurations. A series of derivatives of optically pure cryptophanol have also been synthesised, where the free phenol was substituted with a variety of groups or used to link two cryptophanes together via a decyl chain.[118] Again vibrational circular dichroism measurements were used to determine the absolute configuration.

Larger assemblies of CTVs and cryptophanes are also known. Cryptophanes can be linked to form one-dimensional coordination polymers, with for example the cryptophane with three ester groups on the xylyl linker[108] (Figure 7.40c) capable of being hydrolysed to give the tricarboxylic acid.[119] When combined with copper ions, this material forms a linear coordination polymer.

An optically pure CTV with three aldehyde substituents (Figure 7.45a) was synthesised and could then be reacted with a number of diamines.[120] When 1,2-diaminocyclohexane was used, a chiral cryptophane was formed (Figure 7.45b). However, when linear amines such as 1,4-diaminobenzene (12 equivalents) were combined with eight equivalents of the CTV, a chiral nanocube (Figure 7.45c) was formed, with the eight CTV units forming the corners of the cube and the amines the edges. The structure of the cube is deduced from NMR and MALDI-TOF mass spectra.

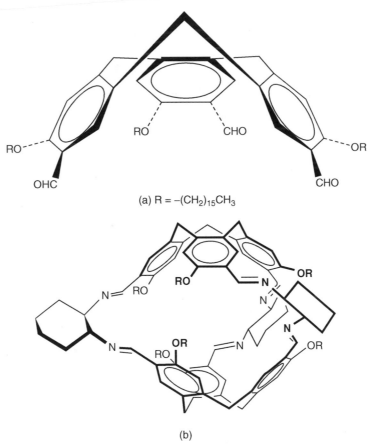

(a) R = $-(CH_2)_{15}CH_3$

(b)

Figure 7.45 *(a) CTV, (b) diimine-linked cryptophane and (c) nanocube (Reprinted with permission from[120]. Copyright 2008 American Chemical Society)*

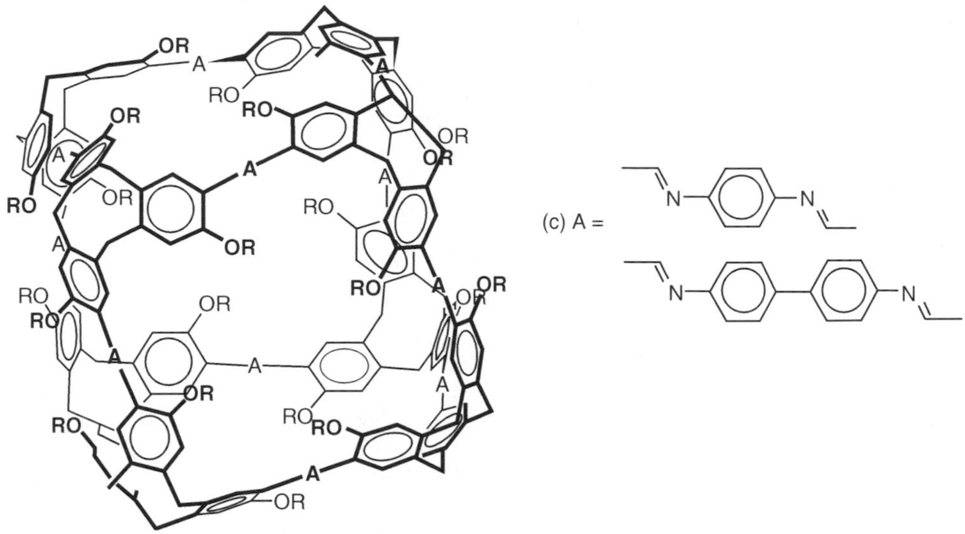

(c) A =

Figure 7.45   *(continued)*

We have already discussed the substitution of CTVs with metalloorganic groups and some similar work has been performed on the cryptophanes. Racemic cryptophane E can be reacted with cyclopentadienyl ruthenium to give the hexaruthenylated compound shown in Figure 7.46a as triflate or antimony hexafluoride salts, which are highly water/air-stable compounds.[121] The presence of the organometallic moieties renders the cavity relatively electron-poor and facilitates the binding of anions. Triflate is encapsulated within the cavity, as shown by the X-ray crystal structure (Figure 7.46b), as is the hexafluroantiminonate (Figure 7.46c). Other anions that can be successfully incorporated are PF_6^- and trifluoroacetate.

7.11   Complexes with Cryptophanes

As detailed above, a number of solvents and ions have already been shown to associate with cryptophanes. The three-dimensional structure of these macrocyclic compounds makes them much more suited as hosts than the parent CTVs and a wide variety of host–guest complexes have been synthesised, as detailed elsewhere.[9,99,100] Examples of some will be given here.

Since cryptophanes are often inherently chiral, attempts have been made to form complexes with chiral guests. Cryptophane C was used as a host for bromochlorofluoromethane[122] and NMR studies showed that when a single enantiomer host was used, the NMR spectra of the complexed guests gave two sets of peaks, indicating formation of diasteroisomeric complexes. Theoretical calculations were made on this system and confirmed the experimental observations that the (−) enantiomer of bromochlorofluoromethane[123] was more strongly bound to the (−) enantiomer of cryptophane C.

Cryptophane E, with slightly longer linking chains that cryptophane A, forms complexes with a variety of halomethanes, such as dichloromethane, chloroform and bromoform.[124] Chloroform showed the strongest binding, followed by dichloromethane and then bromoform, with only weak complexation of carbon tetrachloride and acetone. Binding constants for these guests were higher than for cryptophane A. Other

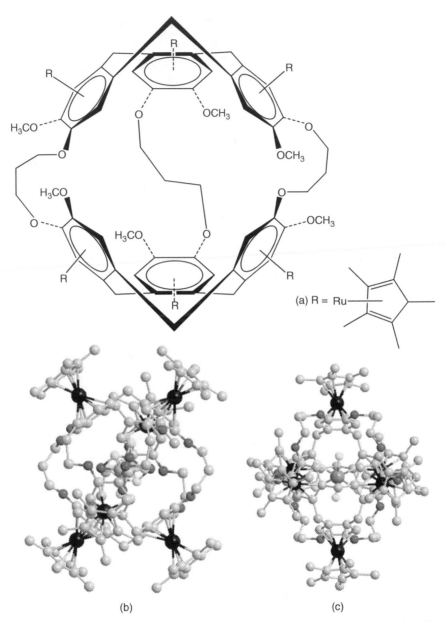

(a) R = Ru—

Figure 7.46 *(a) Hexaruthenium cryptophane, and the X-ray structure of (b) triflate and (c) hexafluroantimi-nonate salts*

work showed that cryptophane A forms an inclusion complex with methane, even managing to hold on to this volatiles guest at room temperature.[125] Cryptophane E also binds a molecule of isobutane from 1,1,2,2-tetrachloroethane solution (this solvent is too big to enter the cavity). When substituted with acid groups, the resultant water-soluble cryptophane E compound binds isobutane from water over 500 times more strongly than in organic solvent and forms weaker complexes with butane and isobutene.[99,126] A detailed description of the overall binding of cryptophanes C and E has recently been published.[100] The even larger cryptophane-O was demethylated and converted into its hexaacid derivative; this was shown to complex tetrahedral molecules such as tetramethyl and tetraethyl derivatives of silicon, germanium, tin and lead.[127] A variety of hydrocarbons can also be bound by cryptophanes with diethylenoxy or aromatic linking groups (Figure 7.40d,e,g). Depending on the structures of the cryptophanes and whether they are *anti* or *syn*, a number of hydrocarbons such as 3-methylpentane or 2,2,3-trimethylbutane can be encapsulated.[109] What is noticeable in many of these host–guest interactions is that many of the cryptophanes seem to prefer tetrahedral guests.

Charged tetrahedral guests have also been shown to form 1 : 1 complexes with cryptophanes. The aromatic bridged cryptophanes in Figure 7.40d,e were shown to bind a number of cationic ammonium species, one of the strongest-binding being methyltriethyl ammonium.[128] Cryptophanes with three —OMe groups replaced by —SMe groups have been shown to bind trimethylammonium and tetramethyammonium salts,[114] again with preference for the tetramethylammonium compound (binding constant 475 000 M^{-1}). Molecular dynamics calculations suggest that incorporation of a tetramethyl ammonium cation within a cryptophane E cavity causes the cryptophane to stretch and the cage region to become more rigid, thereby slowing the fluctuations needed for release of the guest, stabilising the complex and extending its lifetime.[129] A great deal of experimental and theoretical work has been conducted to determine the driving forces for molecular recognition by these species. For example, similar theoretical studies on cryptophane E encapsulating tetramethylammonium or neopentane species indicated better ordering of the cationic species within the cavity and that the neopentane tumbles relative to the cryptophane much faster than the tetramethylammonium moiety.[130]

The effects of cryptophane size and substitution on the binding of cationic substrates have been investigated for acetylcholine and other ammonium compounds.[131] Cryptophane E in chloroform bound tetramethylammonium ions most strongly with the replacement of one of the methyl groups with a longer hydrocarbon chain (Et, Pr, Bu and so on), causing large decreases in binding. Similarly, its hexaacid derivative bound ammonium salts from water, albeit with lower binding constants, and showed similar patterns. The larger cryptophane O, as its hexaacid derivative, bound these species as well, but with much less sizable differences in binding and some preference for larger ions such as tetraethylammonium and acetylcholine. The same cryptophane was also shown to bind a number of piperidine-based free-radical species and electron-spin resonance studies showed the radicals were located within the hydrophobic cavity of the host.[132] Some studies were also carried out on diethylenetrioxy-bridged cryptophanes (Figure 7.40g), showing their ability to extract metals from water with a preference for caesium.

The complex between cryptophane E and chloroform has also been investigated by NMR studies and shown to essentially behave as a single molecule on the NMR timescale; that is, there is no motion of the guest relative to the host within the cavity.[133] The spectra also show evidence of strongly anisotropic (directional) interaction between the cyclotriveratrylene unit of the host and the hydrogen of the guest. These types of structure were confirmed by Raman studies on crystalline samples of chloroform complexes of cryptophane A and E.[105] However, NMR studies of the dichloromethane complex indicated fast rotation of the guest relative to the host, possibly because of the smaller size of the guest.[103] Spectral analysis indicated that the C—H bond of the guest pointed towards the centre of one of the CTV units and an increased red shift for the cryptophane E complex confirmed stronger binding of the guest than for cryptophane A. A solid-state carbon NMR study of the crystalline cryptophane E–chloroform complex demonstrated that the mobilities of the guest were similar in both solid-state and solution complexes.[134] There have also been studies on chloroform and dichloromethane complexes with cryptophane E,[135] and the thermodynamics and kinetics of this binding process have been evaluated. The same group studied the binding of these halomethanes to cryptophane A and the intermediate cryptophane (often known as cryptophane 223) and determined the rotational freedom of the guests within the cavities.[136] Other workers using thiomethylated

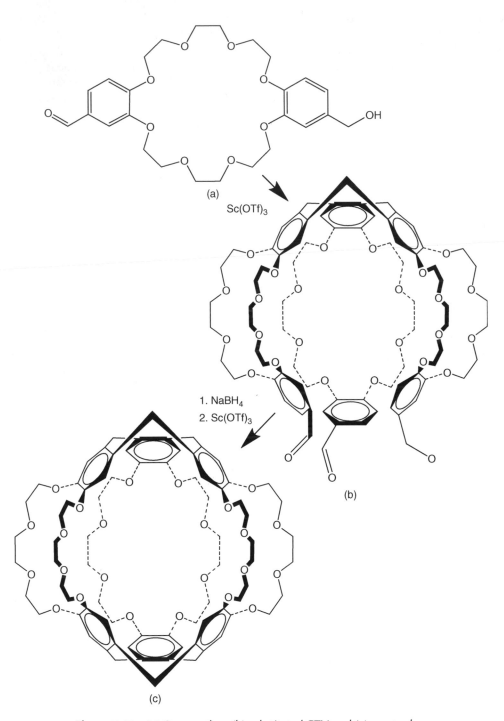

Figure 7.47   *(a) Crown ether, (b) substituted CTV and (c) cryptophane*

cryptophane E (Figure 7.43b) bound either a CHFClBr or a CDFClBr guest and showed by NMR that the guest motion was slowed upon complexation.[137]

Fluorescence measurements have been made on cryptophane A, which has been shown to give an emission peak at 320–330 nm.[138] Strong complex formation with dichloromethane was demonstrated and chloroform and carbon tetrachloride were also studied, but they appeared to simply quench fluorescence.

We have seen how the molecular size of the cryptophanes affects their binding; recently some work has been carried out on expanding the variety of cryptophanes and the study of their binding. A large cryptophane containing six linking chains was constructed, as shown in Figure 7.47. This structure can be considered to be a combination of a cryptophane and a crown ether.[139] A dibenzo crown ether (Figure 7.47a) substituted with a benzyl alcohol-type group can be cyclised using scandium triflate to give the CTV substituted with three crown-ether units (Figure 7.47b). The aldehyde substituents can be reduced to give a second set of benzyl alcohols, which can undergo a second cyclisation to give the cryptophane in Figure 7.47c. Two cationic aromatic guests were studied: dimethyldiazapyrenium and 4,4′-biphenylbisdiazonium ions (Figure 7.48a,b). NMR studies showed formation of 1 : 1 complexes and this was confirmed by the crystal

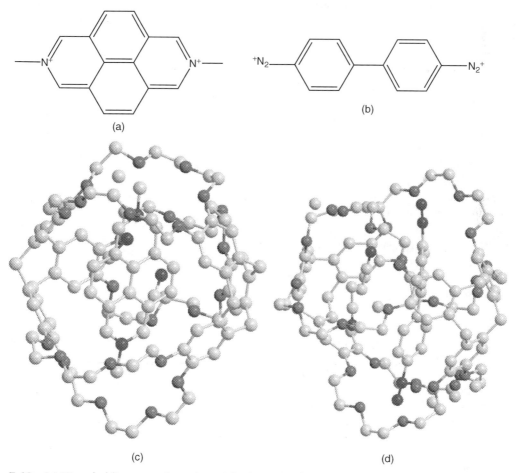

Figure 7.48 (a) Dimethyldiazapyrenium, (b) 4,4′-biphenyl bisdiazonium and (c,d) complexes with the cryptophane in Figure 7.46c

structures (Figure 7.48c,d). For the dimethyldiazapyrenium guests, the pyridinium rings were threaded through two of the crown-ether rings.

At the other extreme, the smallest of this family, cryptophane 111 (Figure 7.43c), can be synthesised by coupling two CTV triphenol units together with bromochloromethane.[140] This smallest cryptophane bound xenon very strongly in organic solvents, 10 000 M^{-1}; in the next section we will describe xenon–cryptophane complexes. Later studies investigated the formation of complexes of cryptophane 111 with other small molecules.[141] Both methane and hydrogen were shown to bind to the cryptophane, whereas ethane and ethylene were much more weakly complexed. These results indicate the potential for cryptophane 111 to act as a selective host for small gases and perhaps to lead to a range of size- and shape-selective gas-sensing cryptophanes.

7.12   Cryptophane–Xenon Complexes

We have already described the extremely high affinity of cryptophane 111 for xenon, but many other xenon–cryptophane complexes are known. What makes these complexes more than just an academic curiosity is that they can be utilised in MRI and biosensing. Nuclei of the stable xenon isotope ^{129}Xe have a spin of 1/2, making them suitable for NMR detection. When nuclei of this type are placed in a strong magnetic field, they can align with or against it. Usually a small excess of nuclei align with the field – this is referred to as polarisation (typically 0.001% of the maximum value at room temperature, even in the strongest magnets). However, in the case of xenon this can be increased to over 50% of the maximum possible value by mixing with nitrogen and alkali metal (usually rubidium) vapour and irradiating with a circularly polarised laser. The laser polarises the valence electrons of the metal atoms and this polarisation can be transferred to the xenon nuclei in a process known as hyperpolarisation. In the case of ^{129}Xe, even after the laser irradiation ceases and the xenon is separated from the metal vapour, hyperpolarisation can be maintained for long periods of time, ranging from several seconds for xenon atoms dissolved in blood[142] to several hours in the gas phase[143] and several days in deeply frozen solid xenon.[144] This enhanced hyperpolarisation greatly increases the detectability of xenon via MRI by factors of up to 10 000. This has led to its use in studies of the flow of gases within the lungs,[145] for example, and trials are being carried out on its use in imaging other tissues. Laser polarisers for xenon are available commercially. There are several problems, however, such as the fact that dissolved xenon gas loses its hyperpolarisation rapidly in blood and there are no covalent compounds of xenon suitable for use. While compounds such as xenon fluorides and oxides exist, they are not suitable for clinical use.

One possible approach is to encapsulate the hyperpolarised xenon within an organic compound such as a cryptophane. The cryptophane can encapsulate the xenon, in effect rendering it into an easily handled form, thereby increasing its solubility compared to just dissolving the gas in a solvent due to high binding constants, whilst also protecting it from environmental influences. Water-soluble cryptophanes could be used to increase biocompatibility.

One of the first reports of xenon encapsulation in cryptophanes described the reversible 1 : 1 complexation of xenon in cryptophane A.[146] NMR studies (^{1}H and ^{129}Xe) clearly showed the binding of the xenon atoms with a large upfield shift of the xenon nuclei and a binding constant of about 3900 M^{-1}, four times that observed for chloroform and 20 times that for methane. Binding was also stronger than for the corresponding complexes between xenon and α-cyclodextrin. The strong binding was thought to be due to a good size match between the cryptophane and guest, high London forces between the polarisable guest and the electron-rich aromatic rings, and the minimal loss of entropy of the guest. Utilising cyclophane A in which either or both of the methyl groups and the linking chains had been deuterated allowed further refinement of the binding process.[147] The chemical shift of the xenon nuclei was a direct function of the level of deuteration of the cryptophane and it was also shown that all complexation took place between cryptophane and 'free' xenon and that there was minimal movement of xenon directly from one cryptophane to another in a collision-type mechanism.

The possibility of encapsulating xenon led to a series of works on synthesising new cryptophanes in an attempt to provide a 'tailor-made' cavity; an example is the deuterated cryptophanes above, in which either the methoxy groups were replaced by —OCD_3 groups or the linking chains were perdeuterated.[148] The chemical shifts of bound xenon were found to depend on the ratio of hydrogen to deuterium. Besides these, a more complex series was synthesised in which two types of linking chain were used in the same cryptophane, as shown when a CTV triphenol (Figure 7.9e) was used as the starting material. First this was monosubstituted with a benzyl alcohol unit by reacting the CTV with an alkyl iodide in 1 : 1 ratio (Figure 7.49a, yield 30%); further reaction with a protected deuterated benzyl alcohol substituted the remaining phenol units to give the structure shown in Figure 7.49b, which could then be cyclised using typical conditions to give the cryptophane with one protonated and two deuterated linking chains. The same method can be used to synthesise a wide variety of asymmetrical cryptophanes with different linking groups between CTV units. For example, cryptophanes 223, 233 and 224 linked together by —$O(CH_2)_nO$ chains (where the numbers refer to the number of carbons (n) in each individual linking group) were synthesised,[149] albeit with lower yields than cryptophane A, possibly due to the asymmetry increasing the number of undesirable side-products. Cryptophanes 223 and 233 bound xenon as a 1 : 1 complex with slow exchange and binding constants of 2810 M^{-1} and 810 M^{-1} respectively, lower than for cryptophane A. Xenon appears to bind to cryptophane 224, but with rapid exchange between solution and bound states; the relatively poor binding of this cryptophane may be due to the increased flexibility of the host.

Bis-cryptophanes have been synthesised, similar to those in Figure 7.40c, but where one or both of the cryptophanes is deuterated.[150] The *bis*-cryptophanes bind two Xe atoms with similar complexation behaviour to cryptophane A. In the case of the asymmetric *bis*-cryptophane in which one of the units is deuterated, at low temperatures it proved possible to distinguish the bound xenon atoms by NMR, depending on the cavity in which they were located (a difference in chemical shift of 1.16 ppm). Grafting a (−)-camphanic acid unit onto a racemic cryptophanol gives rise to diasterisomeric cryptophanes. When these materials are complexed with laser-polarised xenon, NMR studies prove capable of distinguishing between the two diastereoisomers, the chemical shifts showing a difference of 7 ppm.[151] Modelling of the two diastereoisomers indicates that the more flexible of the two can distort its cavity more effectively to accommodate the xenon guest.

Water-soluble cryptophane hexaacids are soluble in water at biological pH. A series of these compounds (based on cryptophanes A, E, 223 and 233) were synthesised and their complexation behaviours with xenon studied.[152] Binding constants as high as 6900 M^{-1} for the cryptophane A hexaacid in water were observed, making these compounds potential candidates for biosensing.

Once the feasibility of utilising cryptophanes as hosts for xenon was demonstrated, a number of applications were tested. For example, a cryptophanol substituted with a single carboxylic acid group was first coupled with a peptide chain and then capped with a substituent bearing a biotin unit[153] to give the structure shown in Figure 7.50a. Biotin is well known to undergo extremely strong and selective binding to the protein avidin. When laser-polarised xenon was incorporated into the cryptophane A unit on this molecule and the resultant 'functionalised' xenon moiety was combined with avidin, a change in chemical shift of the ^{129}Xe NMR signal combined with broadening was observed. This did not occur when avidin that had had its biotin binding site blocked with native biotin was used. Binding to avidin appeared to have minimal effect on the xenon–cryptophane binding. What we therefore have developed is a 'biosensing' molecule which can selectively detect avidin. Combined with laser polarisation of the xenon, with its resultant enhancement in sensitivity, this could potentially form the basis of an NMR 'biosensor'. The possible advantages of these types of biosensor include their ability to simultaneously detect multiple analytes,

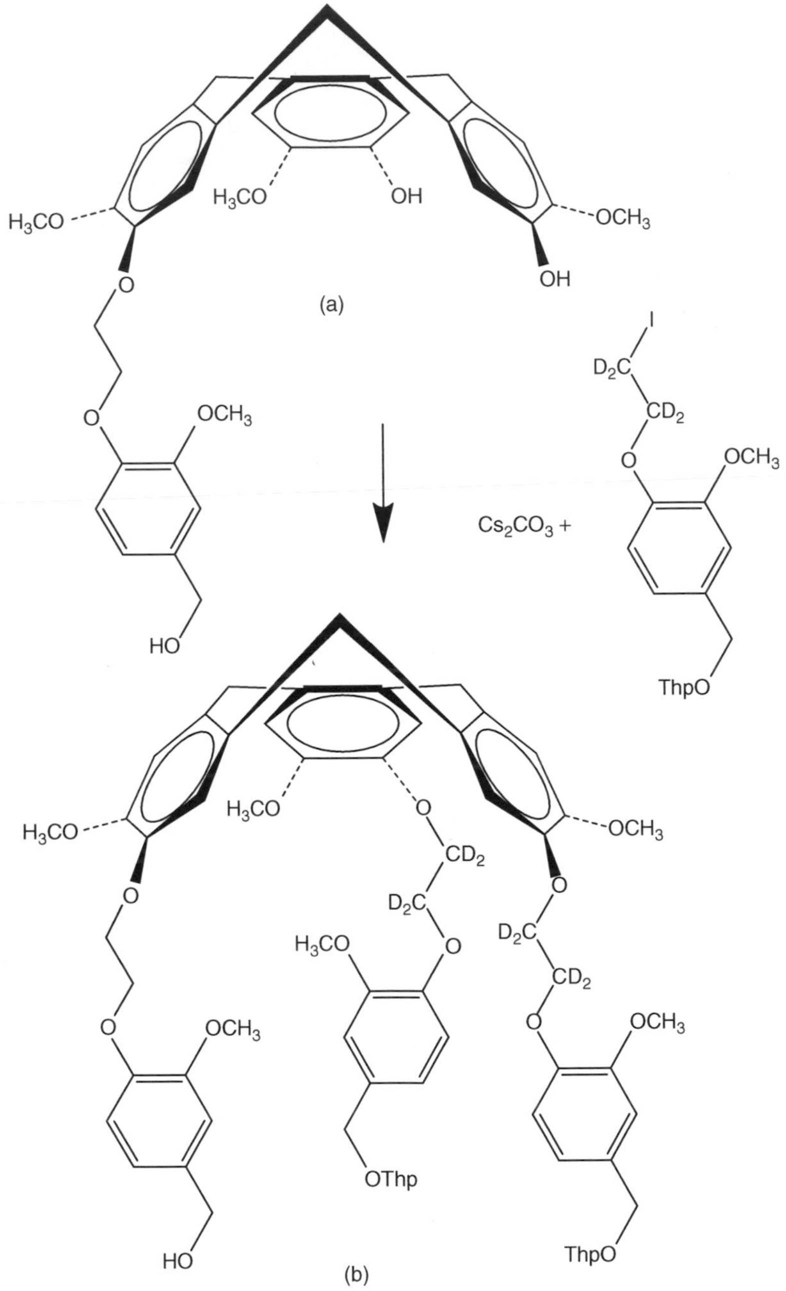

Figure 7.49   *(a) Monosubstituted and (b) asymmetric CTVs*

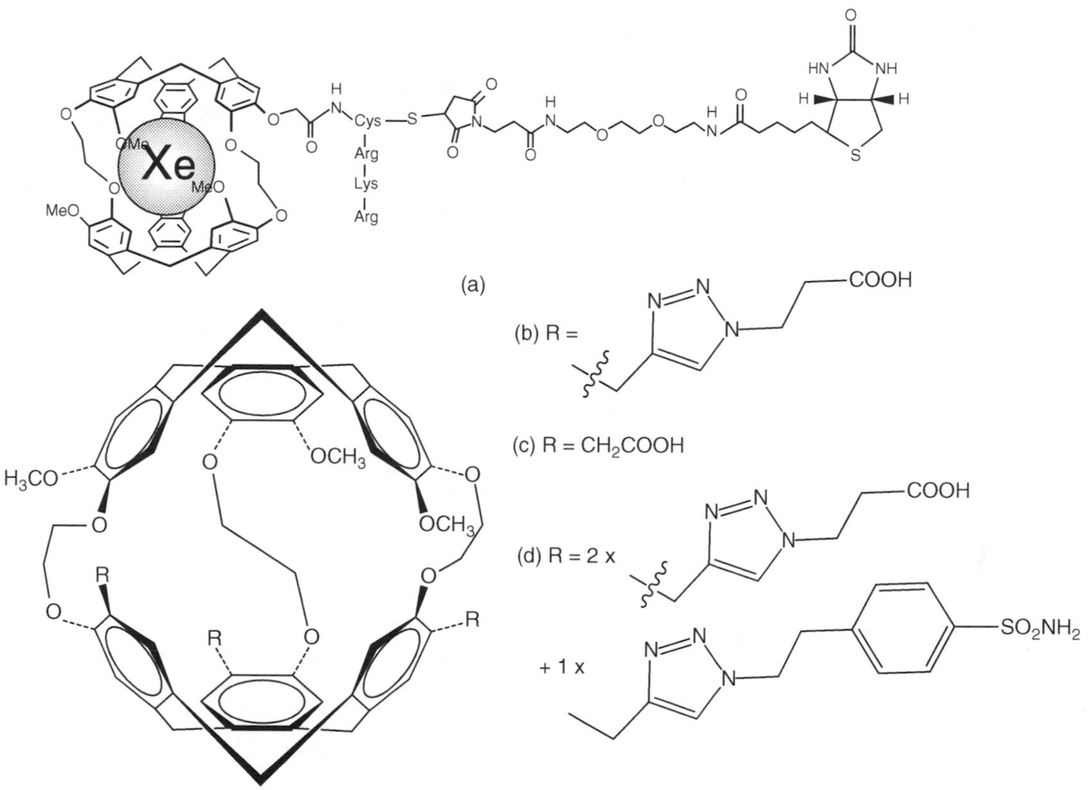

Figure 7.50 (a) Biotinylated cryptophane (Reprinted with permission from[153]. Copyright 2004 American Chemical Society) and (b,c,d) triazole-, triacid- and sulfonamidate-substituted cryptophanes

their applicability to *in vivo* spectroscopy and imaging, and the possibility of 'remote' amplified detection. Different cryptophane/peptide diastereoisomers affected the xenon chemical shift, allowing 'fine-tuning' of the shift and so raising the possibility of the simultaneous use of multiple 'biosensing' molecules, each specific to a different target.

Xenon biosensors have been utilised for investigation into various biological interactions. The width of the xenon signal can act to limit the sensitivity of this biosensor and therefore attempts have been made to optimise the signal. Various biotinylated xenon@cryptophane conjugates ('xenon@cryptophane' being used to describe xenon encapsulated within a cryptophane molecule) with different peptide and linking groups were synthesised.[154] Both the line width and sensitivity of chemical shift to avidin binding of the xenon biosensor were found to be inversely proportional to linker length. Coupling of a xenon@cryptophane complex to a peptide chain led to the development of a biosensor for the detection of metalloproteinase.[155] Enzymatic cleavage of the peptide (which was a substrate for the proteinase) led to a measurable change in the ^{129}Xe spectra (change in chemical shift of 0.5 ppm). Similarly, an optically pure cryptophanol could be substituted with a spacer and then a single-stranded 20-mer oligonucleotide chain.[156] Binding of the counterpart to the oligonucleotide caused a shift in ^{129}Xe spectra. Use of laser hyperpolarisation meant that low concentrations of probe ($2\,\mu$M) could be utilised.

Further design of the cryptophane core to maximise the effects of binding on the xenon NMR has been undertaken. A water-soluble cryptophane (Figure 7.50b) substituted with triazole units has been synthesised[157] and its complexation with xenon in water and human plasma has been studied by fluorescence. A binding coefficient of $30\,000$ M^{-1} in water was obtained. Further work utilised a triacid cryptophane (Figure 7.50c), which was shown by NMR to exist in both crown-crown and crown-saddle (<5%) conformations and displayed an even higher xenon binding coefficient of $33\,000$ M^{-1} in water.[158] The authors hypothesise that the presence of ionisable groups near the cryptophane core improves xenon binding. A range of 12 further cryptophanes with varying substituents, bridging groups and *anti* and *syn* conformations were synthesised, and their xenon-binding behaviour determined.[159] Many of these gave different xenon NMR chemical shifts, allowing for the possibility of multiplexed sensors.

Cryptophanes have also been conjugated with the enzyme human carbonic anhydrase II.[160] A cryptophane was synthesised (Figure 7.50d) that conjugates via a sulfonamidate anion to the active Zn^{2+} site of the enzyme. An X-ray crystal structure of the enzyme–cryptophane conjugate has been obtained, which verified the structure and encapsulation of the xenon. Later work[161] showed that the binding to the enzyme affected the NMR spectra of the xenon nuclei. Biosensor dissociation coefficients could be obtained for the conjugates and the chemical shifts of the xenon allowed differentiation between the I and II isozymes. Xenon biosensors may potentially determine human diseases characterised by protein biomarkers.

7.13   Other Uses of Cryptophanes

We have already detailed the use of xenon-based cryptophanes as sensors, but other cryptophanes also have sensing abilities for a number of species. Carboxylic acid-substituted cryptophanes have been deposited onto quartz crystal microbalance (QCM) crystals[162] and shown to respond to ammonia in moist air with minimal cross-sensitivity to H_2S, CO_2, N_2O or CH_4. Cryptophane A can be deposited as a solvent-cast film or spin-coated along with polyvinyl chloride onto silicon substrates.[163] Alternatively it can be deposited as a Langmuir–Blodgett film along with a phospholipid or a porphyrin. The potential differences of the films can then be measured, and in the case of the LB films, measurable differences are observed when samples are exposed to a methane atmosphere, indicating their potential as a methane sensor.

Cryptophanes A and E can be dispersed in a polysiloxane matrix and then dip-coated onto optical fibres.[164] Optical measurements can then be made on the fibres during exposure to various levels of methane. An evanescent wave is generated in this format which interacts with the fibre coating and allows measurement of its refractive index. This refractive index is found to increase upon exposure to methane of the cryptophane A-loaded system, with a detection limit of 2% v/v. High levels of other alkanes (>15% v/v) do interfere, probably due to solubilisation in the polymer matrix. Cryptophane E composites can also be used to detect methane but are not as sensitive (detection limit of 6% v/v). Recently a similar system using optical fibres with cryptophane A incorporated into silicone cladding was used to detect methane.[165] Decreases of mode-filtered light intensity were shown to correlate with methane concentrations in the range 0–16% v/v with a detection limit of 0.15% v/v and a response time of five minutes. Minimal interferences were observed for oxygen, hydrogen and carbon dioxide, although halomethanes caused interference.

Cryptophane A can also be deposited from THF onto QCM crystals by an electrospray method.[166] This sensor has proved to be selective and highly sensitive to methane (detection limit 0.05% v/v), with response and recovery times <30 seconds, but becomes saturated at levels above 0.2% v/v. Cryptophane A can also be deposited as a composite with polysiloxane onto an SPR chip and has proved suitable for the detection of dissolved methane in an aqueous environment;[167] it is selective and reversible and is suitable for detection at low concentrations (1–300 nM), typical of open ocean environments, with detection limits lower than 0.2 nM.

Cryptophane A is capable of being deposited as monolayers when mixed with phospholipids, as previously described.[163] Cryptophanes C and E, along with two larger analogues with —O(CH$_2$)$_9$O— or

—O(CH$_2$)$_{10}$O— bridging units, can also be spread as monolayers at the air–water interface.[168] High surface pressures cannot be achieved with these species, probably because of their lack of a classical amphiphile structure, with a hydrophilic head group and a long hydrophobic alkyl chain. Cryptophane C does not appear to form a stable monolayer but the other materials form monomolecular layers when the surface pressure is between 0 and 10 mN m^{-1}. It is thought the weakness of the ether oxygen–water interaction and the readiness of these materials to crystallise lowers monolayer stability. However, when longer side chains are incorporated, much more stable monolayers are formed. Cryptophanes substituted with three hexanols or hexadecyls (Figure 7.43d,e) can be synthesised and have been studied as Langmuir monolayers.[169] The highest-quality monolayers are obtained for cryptophanes substituted with three hexadecyl groups, these being stable up to 27 mN m^{-1}. Higher pressure leads to the irreversible formation of aggregates.

It has also proved possible to incorporate cryptophanes into liquid crystals. Cryptophane A can be mixed with and solvated into commercial liquid crystals and the resultant systems have been studied by NMR,[170] showing higher degrees of spatial orientation and decreased mobility compared to bulk solutions. Later work on this cryptophane and the *bis*-cryptophane again shows enhanced ordering for both hosts.[171] The *bis*-cryptophane gives very similar results to the parent moiety despite differences in size and mobility. By analysis of chloroform molecules complexed within the cryptophane, it has been deduced in both cases that the C3 axis of the cryptophane lies perpendicular to the liquid-crystal director.

Charge-transfer products have also been formed by cryptophanes. Normally charge-transfer complexes are formed between two flat aromatic molecules such as tetrathiafulvalene and tetracyanoquinodimethane. However, nonplanar macrocyclic systems can be incorporated into these systems. For example, cryptophane E can be electrochemically oxidised to give a radical cation and forms a stable 1:1 salt with PF$_6$$^-$, which can even be crystallised and binds a chloroform molecule within the cavity.[172] A recent report demonstrates that charge-transfer complexes can also be formed between C$_{60}$ and either cryptophane A or E. NMR, adsorption spectroscopy, fluorescence and electrochemical studies confirmed complex formation.[173]

7.14 Hemicryptophanes

Hemicryptophanes are species similar to cryptophanes except that one of the CTV rings is replaced by a different macrocyclic system. Several examples will be given here, such as the CTV–azacrown compound mentioned earlier[36] and the compound shown in Figure 7.51a, in which a CTV has been synthesised with three 'tentacles' that are each 'capped' with a thiophosphoryl trihydrazide unit.[174] This compound has a nonpolar CTV unit and a more polar hydrazide moiety, making it a ditopic receptor with a potential metal binding site. This material forms a crystalline inclusion compound with a toluene atom located in the cavity. Similarly, a hemicryptophane synthesised from a chiral CTV and a chiral trialkanolamine unit can be synthesised (Figure 7.51b), with all the stereoisomers characterised and absolute configurations determined by CD spectra.[175] The compound can also be reacted to give the vanadium oxide complex (Figure 7.51c).

Recently it was found that both the *syn* and *anti* forms of this vanadyl complex could be obtained, and both were found to be effective oxidation catalysts, catalysing for example the oxidation of sulfides to sulfoxide by cumyl hydroperoxide.[176] In these complexes the vanatrane unit has a propeller-like shape and is chiral, leading to the formation of diasteroisomers. Interestingly, it was found that the solvent controls the preferential clockwise or anticlockwise orientation of the atrane moiety, with for example benzene giving the opposite orientation to DMSO.[177] An X-ray crystal structure of this hemicryptophane has been obtained (Figure 7.51d). Unusually a clathrate is not formed; instead the molecule is compact in shape, with one of the linking units occupying the cavity and pushing the atrane unit outwards.

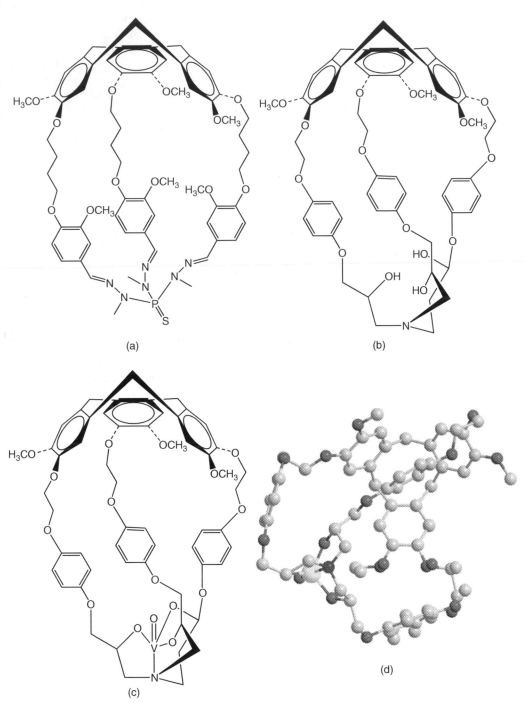

Figure 7.51   Hemicryptophanes capped with (a) thiophosporyltrihydrazide, (b) trialkanol amine and (c) vanadyl complex, and (d) the X-ray crystal structure of a hemicryptophane

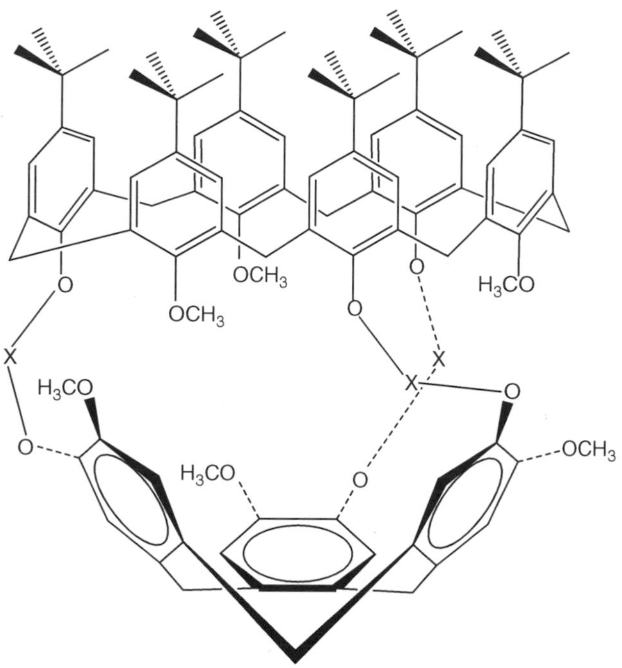

$X = -(CH_2)_2-, -(CH_2)_3-, -(CH_2)_4-, -(CH_2)_5-, -(CH_2)_6-,$
$-CH_2CH_2OCH_2CH_2-, -(CH_2CH_2OCH_2CH_2)_2-$

Figure 7.52   *Structure of a cryptacalix[6]arene*

Mixed macrocyclic systems can also be synthesised; for example 1,3,5-trimethoxy-t-butyl calix[6]arene can be substituted to give a calix with three benzyl alcohol substituents on the lower rim. These can then be cyclised to give a 'cryptacalix[6]arene' (Figure 7.52) with a calix[6]arene joined by three alkyl bridges to a cyclotriveratrylene.[178] NMR studies indicate that both macrocycles are in a cone or crown form and that there is a large cavity in the cavitand.

7.15   Conclusions

We have described the evolution of the cyclotriveratrylenes and cryptophanes. The cyclotriveratrylenes provide a wide range of building blocks, often with inherent chirality, that can be used to assemble many different cryptophanes. These have a variety of potential applications, including as sensors for species such as methane. They also provide an opportunity for a study of the effects of confinement on a variety of small molecules. Importantly, they can be used to effectively functionalise xenon, thereby opening up the possibility for its use in clinical settings or as a biosensor. The combination of cryptophane chemistry and xenon hyperpolarisation offers the potential for the production of a wide range of materials for multiplexed biosensing and/or magnetic imaging.

Much of the chemistry of cryptophanes has been concerned with modifying the outside of the cavities to optimise properties such as solubility. However, there is also potential to introduce functional groups within the cavity, in an attempt to generate a selective or reactive site within the inner space to form a nanoreactor.

Some success has also been attained in generating intermediate structures, where a cyclotriveratrylene has been combined with another macrocyclic system to form a hemicryptophane.

Up to now the compounds described have mainly been utilised within the academic community. However, the improvements in synthesis and purification, along with the development of methods for incorporating these species in devices or *in vivo* applications, promises that these materials will soon join many of the other families of macrocyclic compounds described within this work in being used within a variety of scientific, commercial and medical settings.

Bibliography

Cragg P. A Practical Guide to Supramolecular Chemistry. Wiley-Blackwell; 2005.

Steed JW, Atwood JL. Supramolecular Chemistry. John Wiley and Sons; 2000, 2009.

Lehn J-M. Supramolecular Chemistry. New York: Wiley-VCH; 1995.

Brotin T, Dutasta J-P. Cryptophanes and their complexes: present and future. *Chem Rev.* 2009; **109**: 85–130

Collet A. Dutasta J-P. Lozach B. Canceill J. Cyclotriveratrylenes and cryptophanes: their synthesis and applications to host–guest chemistry and to the design of new materials. *Top Curr Chem.* 1993; **165**: 104–129.

Collet A. Cyclotriveratrylenes and cryptophanes. *Tetrahedron.* 1987; **43**: 5725–5759.

References

1. Ewins AJ. CLXVIII. The action of phosphorus pentachloride on the methylene ethers of catechol derivatives. Part V. Derivatives of protocatechuyl alcohol and protocatechuonitrile. *J Chem Soc. Trans.* 1909; **95**: 1482–1488.

2. Robinson GM. XXX. A reaction of homopiperonyl and of homoveratryl alcohols. *J Chem Soc. Trans.* 1915; **107**: 267–276.

3. Lindsey AS. 316. The structure of cyclotriveratrylene (10,15-dihydro-2,3,7,8,12,13-hexamethoxy-5H-tribenzo[a,d,g]cyclononene) and related compounds. *J Chem Soc.* 1965; 1685–1692.

4. Erdtman H, Haglid F, Ryhage R. Macrocyclic condensation products of veratrole and resorcinol. *Acta Chem Scand.* 1964; **18**: 1249–1254.

5. Cerrini S, Giglio E, Mazza F, Pavel NV. The crystal structure of the inclusion compound between cycloveratril, benzene and water. *Acta Cryst.* 1979; **B35**: 2605–2609.

6. Collet A, Gabard J. Optically active (C3)-cyclotriveratrylene-d9. Energy barrier for the 'crown to crown' conformational interconversion of its nine-membered ring system. *J Org Chem.* 1980; **45**: 5400–5401.

7. Zimmermann H, Tolstoy P, Limbach H-H, Poupko R, Luz Z. The saddle form of cyclotriveratrylene. *J Phys Chem B.* 2004; **108**: 18772–18778.

8. White JD, Gesner BD. Cyclotetraveratrylene: characterization and conformational properties. *Tetrahedron.* 1974; **30**: 2273–2277.

9. Collet A. Cyclotriveratrylenes and cryptophanes. *Tetrahedron.* 1987; **43**: 5725–5759.

10. Umezawa B, Hoshino O, Hara H, Ohyama K, Mitsubayashi S, Sakakibara J. Chemistry of cyclotriveratrylene. I. Formation of cyclotriveratrylene from veratrylamine N-tosylates. *Chem Pharm Bull.* 1969; **17**: 2240–2244.

11. Frensch K, Vögtle F. Ligandstruktur und Komplexierung. LII. Notiz über Kronenether mit Triveratrylen-Gerüst. *Liebigs Ann Chim.* 1979; 2121–2123.

12. Canceill J, Collet A. Synthesis and optical-activity of p-(−)-2,7,12-tribromo-3,8,13-trihydroxycyclotribenzylene and related-compounds: evidence for the nonorthogonality of the b2u and b1u transitions in polysubstituted benzenes and its relevance to the exciton chirality method. *Nouv J Chim.* 1986; **10**: 17–23.

13. Canceill J, Collet A, Gottarelli G. Optical activity due to isotopic substitution. Synthesis, stereochemistry, and circular dichroism of (+)- and (−)-[2,7,12-2H3]cyclotribenzylene. *J Am Chem Soc.* 1984; **106**: 5997–6003.

14. Luz Z, Poupko R, Wachtel EJ, Zheng H, Friedman N, Cao X, Freedman TB, Nafie LA, Zimmermann H. Structural and optical isomers of nonamethoxy cyclotriveratrylene: separation and physical characterization. *J Phys Chem A.* 2007; **111**: 10507–10516.

15. Manville JF, Troughton GE. Synthesis, structure, and conformation of 10,15-dihydro-1,6,11-trihydroxy-2,7,12-trimethoxy-4,9,14-trimethyl-5H-tribenzo[a,d,g]cyclononene and its tripropyl analog. *J Org Chem.* 1973; **38**: 4278–4281.

16. Lüttringhaus A, Peters KC. Conformational enantiomerism of a derivative of cyclotriveratrylene. *Angew Chem Int Ed.* 1966; **5**: 593–594.

17. Sato T, Uno K. Medium-sized cyclophanes. Part XV. 10,15-dihydro-5H-tribenzo-[a,d,g]cyclononene and analogues. *J Chem Soc. Perkin Trans.* 1973; **1**: 895–900.

18. Percec V, Cho CG, Pugh C. Cyclotrimerization versus cyclotetramerization in the electrophilic oligomerization of 3,4-bis(methyloxy)benzyl derivatives. *Macromolecules,* 1991; **24**: 3227–3234.

19. Maliniak A, Luz Z, Poupko R, Krieger C, Zimmermann H. Dodecamethoxyorthocyclophane: conformational and dynamic properties studied by proton 2D exchange NMR. *J Am Chem Soc.* 1990; **112**: 4277–4283.

20. Scott JL, MacFarlane DR, Raston CL, Teoh CM. Clean, efficient syntheses of cyclotriveratrylene (CTV) and tris-(O-allyl)CTV in an ionic liquid. *Green Chem.* 2000; **2**: 123–126.

21. Hyatt JA, Duesler EN, Curtin DY, Paul IC. Stoichiometric inclusion compounds of cyclotriveratrylene and cyclotricatechylene with small neutral molecules: X-ray crystal tructure of cyclotricatechylene Di-2-propanolate. *J Org Chem.* 1980; **45**: 5074–5079.

22. Caglioti V, Liquori AM, Gallo N, Giglio E, Scrocco M. Clathrate compounds of cycloveratril. *J Inorg Nucl Chem.* 1958; **8**: 572–576.

23. Steed JW, Zhang H, Atwood JL. Inclusion chemistry of cyclotriveratrylene and cyclotricatechylene. *Supramol Chem.* 1996; **7**: 37–45.

24. Caira MR, Jacobs A.. Nassimbeni LR. Inclusion compounds of cyclotriveratrylene (2,3,7,8,12,13-hexamethoxy-5,10-dihydro-15Htribenzo[a,d,g] cyclononene) with chlorinated guests. *Supramol Chem.* 2004; **16**: 337–342.

25. Abrahams BF, FitzGerald NJ, Hudson TA, Robson R, Waters T. Closed and open clamlike structures formed by hydrogen-bonded pairs of cyclotricatechylene anions that contain cationic meat. *Angew Chem Int Ed.* 2009; **48**: 3129–3132.

26. Barbour LJ, Steed JW, Atwood JL. Inclusion chemistry of cyclotetracatechylene. *J Chem Soc. Perkin Trans.* 1995; **II**: 857–860.

27. Malthete J, Collet A. Liquid-crystals with a cone-shaped cyclotriveratrylene core. *Nouv J Chim.* 1985; **9**: 151–153.

28. Levelut AM, Malthete J, Collet A. X-ray structural study of the mesophases of some cone-shaped molecules. *J Physique.* 1986; **47**: 351–357.

29. Luz Z, Poupko R, Wachtela EJ, Zimmermann H. Mesomorphic properties of the neat enantiomers of a chiral pyramidic liquid crystal. *Phys Chem Chem Phys.* 2009; **11**: 9562–9568.

30. Zimmermann H, Poupko R, Luz Z, Billard J. Tetrabenzocyclododecatetraene: A new core for mesogens exhibiting columnar mesophases. *Liq Cryst.* 1988; **6–7**: 759–770.

31. Spielberg N, Sarkar M, Luz Z, Poupko R, Billard J, Zimmermann H. The discotic mesophases of octaalkyloxy- and octaalkanoyloxyorthocyclophanes. *Liq Cryst.* 1988; **15**: 311–330.

32. Cram DJ. Cavitands: organic hosts with enforced cavities. *Science.* 1983; **219**: 1177–1183.

33. Malthete J, Collet A. Inversion of the cyclotribenzylene cone in a columnar mesophase: a potential way to ferroelectric materials. *J Am Chem Soc.* 1987; **109**: 7544–7545.

34. Hyatt JA. Octopus molecules in the cyclotriveratrylene series. *J Org Chem.* 1978; **43**: 1808–1811.

35. Menger FM, Takeshita M, Chow JF. Hexapus, a new complexing agent for organic molecules. *J Am Chem Soc.* 1981; **103**: 5938–5939.

36. Canceill J, Collet A, Gabard J, Kotzyba-Hibert F, Lehn J-M. Speleands: macropolycyclic receptor cages based on binding and shaping sub-units. Synthesis and properties of macrocycle cyclotriveratrylene combinations. Preliminary communication. *Helv Chim Acta.* 1982; **65**: 1894–1897.

37. Canceill J, Gabard J, Collet A. (C3)-tris-(O-allyl)-cyclotriguaiacylene, a key intermediate in cyclotriveratrylene chemistry: short and efficient synthesis of cyclotriguaiacylene. *Chem Soc Chem Commun.* 1983; 122–123.

38. Collet A, Jacques J. Chromophores chiraux possedant la symetrie C3: synthese de derives optiquement actifs du cyclotriveratrylene. *Tet Lett.* 1978; **15**: 1265–1268.
39. Collet A, Gabard J, Jacques J, Cesario M, Guilhem J, Pascard C. Synthesis and absolute configuration of chiral (C) cyclotriveratrylene derivatives. Crystal structure of (M)–()-2,7,12-triethoxy-3,8,13-tris-[(R)-1–methoxycarbonyl ethoxy]-10,15-dihydro-5H-tribenzo[a,d,g]-cyclononene. *J Chem Soc. Perkin Trans.* 1980; **I**: 1630–1638.
40. Canceill J, Collet A, Gabard J, Gottarelli G, Spada GP. Exciton approach to the optical activity of C3-cyclotriveratrylene derivatives. *J Am Chem Soc.* 1985; **107**: 1299–1308.
41. Cookson RC, Halton B, Stevens IDR. Conformation in the cyclotriveratrylene series. *J Chem Soc B.* 1968; 767–774.
42. French DC, Lutz MR, Lu C, Zeller M, Becker DP. A thermodynamic and kinetic characterization of the solvent dependence of the saddle-crown equilibrium of cyclotriveratrylene oxime. *J Phys Chem A.* 2009; **113**: 8258–8267.
43. Baldwin JE, Kelly DP. The structure of the triketone from cyclotriveratrylene: an unusual transannular rearrangement. *J Chem Soc Chem Commun.* 1968; 1664–1665.
44. Steed JW, Junk PC, Atwood JL, Barnes MJ, Raston CL, Burkhalter RS. Ball and socket nanostructures: new supramolecular chemistry based on cyclotriveratrylene. *J Am Chem Soc.* 1994; **116**: 10346–10347.
45. Zhang S, Palkar A, Fragoso A, Prados P, de Mendoza J, Echegoyen L. Noncovalent immobilization of C_{60} on gold surfaces by SAMs of cyclotriveratrylene derivatives. *Chem Mater.* 2005; **17**: 2063–2068.
46. Nierengarten J-F, Oswald L, Eckert J-F, Nicoud J-F, Armaroli N. Complexation of fullerenes with dendritic cyclotriveratrylene derivatives. *Tet Lett.* 1999; **40**: 5681–5684.
47. Rio Y, Nierengarten J-F. Water soluble supramolecular cyclotriveratrylene–[60]fullerene complexes with potential for biological applications. *Tet Lett.* 2002; **43**: 4321–4324.
48. Lijanova IV, Maturano JF, Chavez JGD, Montes KES, Ortega SH, Klimova T, Martinez-Garcia M. Synthesis of cyclotriveratrylene dendrimers and their supramolecular complexes with fullerene C_{60}. *Supramol Chem.* 2009; **21**: 21–34.
49. Nierengarten J-F. Supramolecular encapsulation of [60]fullerene with dendritic cyclotriveratrylene derivatives. *Full Nano Carb Nanostr.* 2005; **13**: 229–242.
50. Huerta E, Metselaar GA, Fragoso A, Santos E, Bo C, de Mendoza J. Selective binding and easy separation of C70 by nanoscale self-assembled capsules. *Angew Chem Int Ed.* 2007; **46**: 202–205.
51. Huerta E, Cequier E, de Mendoza J. Preferential separation of fullerene[84] from fullerene mixtures by encapsulation. *J Chem Soc Chem Commun.* 2007; 5016–5018.
52. Blanch RJ, Williams M, Fallon GD, Gardiner MG, Kaddour R, Raston CL. Supramolecular complexation of 1,2-dicarbadodecaborane(12). *Angew Chem Int Ed.* 1997; **36**: 504–506.
53. Hardie MJ, Godfrey PD, Raston CL. Self-assembly of grid and helical hydrogen-bonded arrays incorporating bowl-shaped receptor sites that bind globular molecules. *Chem Eur J.* 1999; **5**: 1828–1833.
54. Hardie MJ, Raston CL, Wells B. Altering the inclusion properties of CTV via crystal engineering: CTV, carborane and DMF supramolecular assemblies. *Chem Eur J.* 2000; **5**: 1828–1833.
55. Hardie MJ, Raston CL. Supramolecular chemistry of anionic cobalt(III) bis(dicarbollide) and cyclotriveratrylene in the solid phase and gas phase. *Angew Chem Int Ed.* 2000; **39**: 3835–3839.
56. Ahmad R, Hardie MJ. Variable Ag(I) coordination modes in silver cobalt(III) bis(dicarbollide) supramolecular assemblies with cyclotriveratrylene host molecules. *Cryst Growth Des.* 2003; **3**: 493–499.
57. Hardie MJ, Raston CL. Solid state supramolecular assemblies of charged supermolecules (Na[2.2.2]cryptate)$^+$ and anionic carboranes with host cyclotriveratrylene. *J Chem Soc Chem Commun.* 2001; 905–906.
58. Holman KT, Halihan MM, Jurisson SS, Atwood JL, Burkhalter RS, Mitchell AR, Steed JW. Inclusion of neutral and anionic guests within the cavity of π-metalated cyclotriveratrylenes. *J Am Chem Soc.* 1996; **118**: 9567–9576.
59. Holman KT, Halihan MM, Steed JW, Jurisson SS, Atwood JL. Hosting a radioactive guest: binding of $^{99}TcO_4^-$ by a metalated cyclotriveratrylene. *J Am Chem Soc.* 1995; **117**: 7848–7849.
60. Holman KT, Orr GW, Atwood JL, Steed JW. Deep cavity [CpFe(arene)]$^+$ derivatized cyclotriveratrylenes as anion hosts. *Chem Commun.* 1998; 2109–2110.

61. Gawenis JA, Holman KT, Atwood JL, Jurisson SS. Extraction of pertechnetate and perrhenate from water with deep-cavity [CpFe(arene)]+-derivatized cyclotriveratrylenes. *Inorg Chem.* 2002; **41**: 6028–6031.

62. Veriot G, Dutasta J-P, Matouzenko G, Collet A. Synthesis of C3–cyclotriveratrylene ligands for iron(II) and iron(III) coordination. *Tetrahedron.* 1995; **51**: 389–400.

63. Nuriman, Kuswandia B, Verboom W. Selective chemosensor for Hg(II) ions based on tris[2-(4-phenyldiazenyl) phenylaminoethoxy] cyclotriveratrylene in aqueous samples. *Anal Chim Acta.* 2009; **655**: 75–79.

64. Moriuchi-Kawakami T, Sato J, Shibutani Y. C3-functionalized cyclotriveratrylene derivative bearing quinolinyl group as a fluorescent probe for Cu^{2+}. *Anal Sci.* 2009; **25**: 449–452.

65. Zhang S, Echegoyen L. Selective Anion Sensing by a tris-amide CTV derivative: 1H NMR titration, self-assembled monolayers, and impedance spectroscopy. *J Am Chem Soc.* 2005; **127**: 2006–2011.

66. Dumartin M-L, Givelet C, Meyrand P, Bibal B, Gosse I. A fluorescent cyclotriveratrylene: synthesis, emission properties and acetylcholine recognition in water. *Org Biomol Chem.* 2009; **7**: 2725–2728.

67. van Wageningen AMA, Liskamp RMJ. Solution phase combinatorial chemistry using cyclotriveratrylene based tripodal scaffolds. *Tet Lett.* 1999; **40**: 9347–9351.

68. Chamorro C, Liskamp RMJ. Approaches to the solid phase of a cyclotriveratrylene scaffold-based tripodal library as potential artificial receptors. *J Combin Chem.* 2003; **6**: 794–801.

69. van Ameijde J, Liskamp RMJ. Synthesis of novel trivalent amino acid glycoconjugates based on the cyclotriveratrylene ('CTV') scaffold. *Org Biomol Chem.* 2003; **1**: 2661–2669.

70. Chakrabarti A, Chawla HM, Hundalb G, Pant N. Convenient synthesis of selectively substituted tribenzo[a,d,g]cyclononatrienes. *Tetrahedron.* 2005; **61**: 12323–12329.

71. Costante J, Garcia C, Collet A. Key intermediates in cyclotriveratrylene chemistry: a new route to the enantiomers of C3–cyclotriphenolene and cryptophane–C chirality. 1997;9: 446–453.

72. Bohle DS, Stasko D. Extended rim redox active tris-metallodioxolene derivatives of cyclotricatechylene. *J Chem Soc Chem Commun.* 1998; 567–568.

73. Garcia C, Malthete J, Collet A. Key Intermediates in cyclotriveratrylene chemistry: synthesis of new C3–cyclotriveratrylene with nitrogen substituents. *Bull Soc Chim Fr.* 1993; **130**: 93–95.

74. Bohle DS, Stasko D. Salicylaldiminato derivatives of cyclotriveratrylene: flexible strategy for new rim-metalated CTV complexes. *Inorg Chem.* 2000; **39**: 5768–5770.

75. Sumby CJ, Fisher J, Prior TJ, Hardie MJ. Tris(pyridylmethylamino)cyclotriguaiacylene cavitands: an investigation of the solution and solid-state behaviour of metallo-supramolecular cages and cavitand-based coordination polymers. *Chem Eur J.* 2006; **12**: 2945–2959.

76. Ronson TK, Fisher J, Harding LP, Hardie MJ. Star-burst prisms with cyclotriveratrylene-type ligands: a [Pd6L8]12+ stella octangular structure. *Angew Chem Int Ed.* 2007; **46**: 9086–9088.

77. Carruthers C, Ronson TK, Sumby CJ, Westcott A, Harding LP, Prior TJ, Rizkallah P, Hardie MJ. The dimeric 'hand-shake' motif in complexes and metallo-supramolecular assemblies of cyclotriveratrylene-based ligands. *Chem Eur J.* 2008; **14**: 10286–10296.

78. Carruthers C, Fisher J, Harding LP, Hardie MJ. Host–guest influence on metallo-supramolecular assemblies with a cyclotriveratrylene-type ligand. *J Chem Soc. Dalton Trans.* 2010; **39**: 355–357.

79. Bardelang D, Camerel F, Ziessel R, Schmutzc M, Hannon MJ. New organogelators based on cyclotriveratrylene platforms bearing 2-dimethylacetal-5-carbonylpyridine fragments. *J Mater Chem.* 2008; **18**: 489–494.

80. Westcott A, Sumby CJ, Walshaw RD, Hardie MJ. Metallo-gels and organo-gels with tripodal cyclotriveratrylene-type and 1,3,5-substituted benzene-type ligands. *New J Chem.* 2009; **33**: 902–912.

81. van Strijdonck GPF, van Haare JAEH, van der Linden JGM, Steggerda JJ, Nolte RJM. Novel subsite-differentiated [4Fe-4S] clusters based on cyclotriveratrylene. *Inorg Chem.* 1994; **33**: 999–1000.

82. Ahmad R, Hardie MJ. Synthesis and structural studies of cyclotriveratrylene derivatives. *Supramol Chem.* 2005; **18**: 29–38.

83. Percec V, Imam MR, Peterca M, Wilson DA, Heiney PA. Self-assembly of dendritic crowns into chiral supramolecular spheres. *J Am Chem Soc.* 2009; **131**: 1284–1304.

84. Arduini A, Calzavacca F, Demuru D, Pochini A, Secchi A. Synthesis of cavity extended cyclotriveratrylenes. *J Org Chem.* 2004; **69**: 1386–1389.

85. Thomas RM, Iyengar DS. Synthesis of a novel sugar cyclotriveratrylene by introduction of Oglycosyl groups. *Synth Comm.* 1999; **29**: 2507–2513.

86. Rump ET, Rijkers DTS, Hilbers HW, de Groot PG, Liskamp RMJ. Cyclotriveratrylene (CTV) as a new chiral triacid scaffold capable of inducing triple helix formation of collagen peptides containing either a native sequence or Pro-Hyp-Gly repeats. *Chem Eur J.* 2002; **20**: 4613–4621.

87. Maiti D, Woertink JS, Ghiladi RA, Solomon EI, Karlin KD. Molecular oxygen and sulfur reactivity of a cyclotriveratrylene derived trinuclear copper(I) complex. *Inorg Chem.* 2009; **48**: 8342–8356.

88. Guy A, Doussot J, Falguierea A, Prieur B, Bachet B. New functionalised derivatives of cyclotriveratrylene: synthesis of tris-trifluoromethyl cyclotriveratrylenes. *Bull Soc Chim Fr.* 1996; **133**: 1005–1010.

89. Lunkwitz R, Tschierske C, Diele S. Formation of smectic and columnar liquid crystalline phases by cyclotriveratrylene (CTV) and cyclotetraveratrylene (CTTV) derivatives incorporating calamitic structural units. *J Mater Chem.* 1997; **7**: 2001–2011.

90. Percec V, Cho CG, Pugh C, Tomazos D. Synthesis and characterization of branched liquid-crystalline polyethers containing cyclotetraveratrylene-based disklike mesogens. *Macromolecules.* 1992; **25**: 1164–1176.

91. Sumby CJ, Gordon KC, Walsh TJ, Hardie MJ. Synthesis and complexation of multiarmed cycloveratrylene-type ligands observation of the 'boat' and 'distorted-cup' conformations of a cyclotetraveratrylene derivative. *Chem Eur J.* 2008; **14**: 4415–4425.

92. Harig M, Neumann B, Stammler H-G, Kuck D. 2,3,6,7,10,11-hexamethoxytribenzotriquinacene: synthesis, solid-state structure, and functionalization of a rigid analogue of cyclotriveratrylene. *Eur J Org Chem.* 2004; 2381–2397.

93. Treibs A, Kreuzer F-H, Haberle N. Tripyrryl-trismethanes (hexahydro-cyclononatripyrroles). *Liebigs Ann Chem.* 1970; **733**: 37–43.

94. Steüpien M, Sessler JL. A facile synthesis of cyclononatripyrroles. *Org Lett.* 2007; **9**: 4785–4787.

95. Uno H, Fumoto Y, Inoue K, Ono N. NMR and X-ray analyses of triethyl 3,7,11-triphenylcyclonona[1,2-b;4,5-b0;7,8-b00]tripyrrole-2,6,10-tricarboxylate:reinvestigation of crown vs saddle conformation of cyclononatripyrroles. *Tetrahedron.* 2003; **59**: 601–605.

96. Santoso M, Somphol K, Kumar N, Black DSC. Synthesis of indolocyclotriveratrylenes. *Tetrahedron.* 2009; **65**: 5977–5983.

97. Meth-Cohn O. Novel macrocycles from 2,5-dimethylthiophene and related systems. *Tet Lett.* 1973; **14**: 91–94.

98. Cram DJ, Weiss J, Helgeson RC, Knobler CB, Dorigo AE, Houk KN. Design, synthesis, and comparison of crystal, solution, and calculated structures within a new family of cavitands. *J Chem Soc Chem Commun.* 1988; 407–409.

99. Collet A, Dutasta J-P, Lozach B, Canceill J. Cyclotriveratrylenes and cryptophanes: their synthesis and applications to host-guest chemistry and to the design of new materials. *Top Curr Chem.* 1993; **165**: 104–129.

100. Brotin T, Dutasta J-P. Cryptophanes and their complexes: present and future. *Chem Rev.* 2009; **109**: 85–130.

101. Canceill J, Cesario M, Collet A, Guilhem J, Pascard C. A new bis-cyclotribenzyl cavitand capable of selective inclusion of neutral molecules in solution: crystal structure of its CH_2Cl_2 cavitate. *J Chem Soc Chem Commun.* 1985; 361–363.

102. Canceill J, Cesario M, Collet A, Guilhem J, Riche C, Pascard C. Selective recognition of neutral molecules: 1H NMR study of the complexation of CH_2Cl_2 and CH_2Br_2 by cryptophane-D in solution and crystal structure of its CH_2Cl_2 cavitate. *J Chem Soc Chem Commun.* 1986; **3**: 39–341.

103. Petrov O, Tosner Z, Csolregh I, Kowalewski J, Sandstr D. Dynamics of chloromethanes in cryptophane-E inclusion complexes: a 2H solid-state NMR and X-ray diffraction study. *J Phys Chem A.* 2005; **109**: 4442–4451.

104. Canceill J, Cesario M, Collet A, Guilhem J, Lacombe L, Lozach B, Pascard C. Structure and properties of the cryptophane E-$CHCl_3$ complex, a stable Van der Waals molecule. *Angew Chem Int Ed.* 1989; **28**: 1246–1248.

105. Cavagnat D, Brotin T, Bruneel J-L, Dutasta J-P, Thozet A, Perrin M, Guillaume F. Raman microspectrometry as a new approach to the investigation of molecular recognition in solids: chloroform–cryptophane complexes. *J Phys Chem B.* 2004; **108**: 5572–5581.

106. Garcia C, Aubry A, Collet A. Stereoselectivity in the template-directed synthesis of D-3 (chiral) and C-3h (achiral) cryptophanes with long O(CH2)(n)O spacer bridges. *Bull Soc Chim Fr.* 1996; **133**: 853–867.

107. Roesky CEO, Weber E, Rambusch T, Stephan H, Gloe K, Czugler M. A new cryptophane receptor featuring three endo-carboxylic acid groups: synthesis, host behavior and structural study. *Chem Eur J.* 2003; **9**: 1104–1112.

108. Mough ST, Goeltz JC, Holman KT. Isolation and structure of an 'imploded' cryptophane. *Angew Chem Int Ed.* 2004; **43**: 5631–5635.

109. Akabori S, Takeda M, Miura M. The complexing abilities of diethyleneoxy- and xylene-bridged cryptophanes with alkanes. *Supramol Chem.* 1999; **10**: 253–262.

110. Brotin T, Roy V, Dutasta J-P. Improved synthesis of functional CTVs and cryptophanes using Sc(OTf)$_3$ as catalyst. *J Org Chem.* 2005; **70**: 6187–6195.

111. Cram DJ, Tanner ME, Keipert SJ, Knobler CB. Two chiral [1.1.1]orthocyclophane units bridged by three biacetylene units as a host which binds medium-sized organic guests. *J Am Chem Soc.* 1991; **113**: 8909–8916.

112. Zhong Z, Ikeda A, Shinkai S, Sakamoto S, Yamaguchi K. Creation of novel chiral cryptophanes by a self-assembling method utilizing a pyridyl-Pd(II) interaction. *Org Lett.* 2001; **3**: 1085–1087.

113. Canceill J, Lacombe L, Collet A. Water-soluble cryptophane binding lipophilic guests in aqueous solution. *J Chem Soc Chem Commun.* 1987; 219–221.

114. Garcia C, Humilière D, Riva N, Collet A, Dutasta J-P. Kinetic and thermodynamic consequences of the substitution of SMe for OMe substituents of cryptophane hosts on the binding of neutral and cationic guests. *Org Biomol Chem.* 2003; **1**: 2207–2216.

115. Darzac M, Brotin T, Bouchub D, Dutasta J-P. Cryptophanols, new versatile compounds for the synthesis of functionalized cryptophanes and polycryptophanes. *J Chem Soc Chem Commun.* 2002; 48–49.

116. Brotin T, Barbe R, Darzac M, Dutasta J-P. Novel synthetic approach for optical resolution of cryptophanol-A: a direct access to chiral cryptophanes and their chiroptical properties. *Chem Eur J.* 2003; **9**: 5784–5792.

117. Brotin T, Cavagnat D, Dutasta J-P, Buffeteau T. Vibrational circular dichroism study of optically pure cryptophane-A. *J Am Chem Soc.* 2006; **128**: 5533–5540.

118. Cavagnat D, Buffeteau T, Brotin T. Synthesis and chiroptical properties of cryptophanes having C1-symmetry. *J Org Chem.* 2008; **73**: 66–75.

119. Mough ST, Holman KT. A soft coordination polymer derived from container molecule ligands. *J Chem Soc Chem Commun.* 2008; 1407–1409.

120. Xu D, Warmuth R. Edge-directed dynamic covalent synthesis of a chiral nanocube. *J Am Chem Soc.* 2008; **130**: 7520–7521.

121. Fairchild RM, Holman KT. Selective anion encapsulation by a metalated cryptophane with a π-acidic interior. *J Am Chem Soc.* 2005; **127**: 16364–16365.

122. Canceill J, Lacombe L, Collet A. Analytical optical resolution of bromochlorofluoromethane by enantioselective inclusion into a tailor-made 'cryptophane' and determination of its maximum rotation. *J Am Chem Soc.* 1985; **107**: 6993–6996.

123. Costante-Crassous J, Marrone TJ, Briggs JM, McCammon JA, Collet A. Absolute configuration of bromochlorofluoromethane from molecular dynamics simulation of its enantioselective complexation by cryptophane-C. *J Am Chem Soc.* 1997; **119**: 3818–3823.

124. Canceill J, Lacombe L, Collet A. New cryptophane forming unusually stable inclusion complexes with neutral guests in a lipophilic solvent. *J Am Chem Soc.* 1986; **108**: 4230–4232.

125. Garel L, Dutasta J-P, Collet A. Complexation of methane and chlorofluorocarbons by cryptophane-A in organic solution. *Angew Chem Int Ed.* 1993; **32**: 1169–1171.

126. Canceill J, Lacombe L, Collet A. Molecular recognition: formation of a stable inclusion complex between isobutane and cryptophane-E in a lipophilic medium. *Comp Rend Acad Sci II.* 1987; **304**: 815–818.

127. Garel L, Dutasta J-P, Collet A. Complexation of tetraalkylated derivativesd of silicium, germanium, tin and lead by a water soluble cryptophane. *New J Chem.* 1993; **32**: 1169–1171.

128. Miura M, Yuzawa S, Takeda M, Takeda M, Habata Y, Tanese T, Akabori S. Syntheses of aromatic bridged cryptophanes and their complexing abilities with alkyl ammonium cations. *Supramol Chem.* 1996; **8**: 53–66.

129. Kirchhoff PD, Dutasta J-P, Collet A, McCammon JA. Structural fluctuations of a cryptophane-tetramethylammonium host–guest system: a molecular dynamics simulation. *J Am Chem Soc.* 1997; **119**: 8015–8022.

130. Kirchhoff PD, Dutasta J-P, Collet A, McCammon JA. Dynamic and rotational analysis of cryptophane host–guest systems: challenges of describing molecular recognition. *J Am Chem Soc.* 1999; **121**: 381–390.

131. Garel L, Lozach B, Dutasta J-P, Collet A. Remarkable effect of receptor size in the binding of acetylcholine and related ammonium ions to water-soluble cryptophanes. *J Am Chem Soc.* 1993; **115**: 11652–11653.

132. Garel L, Vezin H, Dutasta J-P, Collet A. Piperidine aminoxyl radicals as EPR probes for exploring the cavity of a water-soluble cryptophane. *J Chem Soc Chem Commun.* 1996; 719–720.

133. Lang J, Dechter JJ, Effemey M, Kowalewski J. Dynamics of an inclusion complex of chloroform and cryptophane-E: evidence for a strongly anisotropic Van der Waals bond. *J Am Chem Soc.* 2001; **123**: 7852–7858.

134. Tosner Z, Petrov O, Dvinskikh SV, Kowalewski J, Sandstrom D. A 13C solid-state NMR study of cryptophane-E: chloromethane inclusion complexes. *Chem Phys Lett.* 2004; **388**: 208–211.

135. Aski SN, Takacs Z, Kowalewski J. Inclusion complexes of cryptophane-E with dichloromethane and chloroform: a thermodynamic and kinetic study using the 1D-EXSY NMR method. *Magn Reson Chem.* 2008; **46**: 1135–1140.

136. Aski SN, Lo AYH, Brotin T, Dutasta J-P, Eden M, Kowalewski J. Studies of inclusion complexes of dichloromethane in cryptophanes by exchange kinetics and 13C NMR in solution and the solid state. *J Phys Chem C.* 2008; **112**: 13873–13881.

137. Crassous J, Hediger S. Dynamics of CHFClBr and CDFClBr inside a thiomethylated cryptophane, studied by 19F-1H CSA-DD cross-correlated relaxation and 2H quadrupolar relaxation measurements. *J Phys Chem A.* 2003; **107**: 10233–10240.

138. Zhang C, Shen W, Wen G, Chaoa J, Qin L, Shuang S, Donga C, Choi MMF. Spectral study on the interaction of cryptophane-A and neutral molecules CH_nCl_{4-n} (n = 0, 1, 2). *Talanta.* 2008; **76**: 235–240.

139. Li M-J, Lai C-C, Liu Y-H, Peng S-M, Chiu S-H. Two guest complexation modes in a cyclotriveratrylene-based molecular container. *J Chem Soc Chem Commun.* 2009; 5814–5816.

140. Fogarty HA, Berthault P, Brotin T, Huber G, Desvaux H, Dutasta J-P. A cryptophane core optimized for xenon encapsulation. *J Am Chem Soc.* 2007; **129**: 10332–10333.

141. Chaffee KE, Fogarty HA, Brotin T, Goodson BM, Dutasta J-P. Encapsulation of small gas molecules by cryptophane-111 in organic solution. 1. Size and shape-selective complexation of simple hydrocarbons. *J Phys Chem A.* 2009; **113**: 13675–13684.

142. Wolber J, Cherubini A, Leach MO, Bifone A. On the oxygenation-dependent 129Xe T1 in blood. *NMR Biomed.* 2000; **13**: 234–237.

143. Chann B, Nelson IA, Anderson LW, Driehuys B, Walker TG. 129Xe–Xe molecular spin relaxation. *Phys Rev Lett.* 2002; **88**: 113–201.

144. von Schulthess GK, Smith H-J, Pettersson H, Allison DJ. Untitled. In: The Encyclopaedia of Medical Imaging. Taylor & Francis; 1998.

145. Albert MS, Balamore D. Development of hyperpolarized noble gas MRI. *Nucl Inst Meth Phys Res A.* 1998; **402**: 441–453.

146. Bartik K, Luhmer M, Dutasta J-P, Collet A, Reisse J. ^{129}Xe and ^{1}H NMR study of the reversible trapping of xenon by cryptophane-A in organic solution. *J Am Chem Soc.* 1998; **120**: 784–791.

147. Brotin T, Lesage A, Emsley L, Collet A. ^{129}Xe NMR spectroscopy of deuterium-labeled cryptophane-A xenon complexes: investigation of host–guest complexation dynamics. *J Am Chem Soc.* 2000; **121**: 1171–1174.

148. Brotin T, Devic T, Lesage A, Emsley L, Collet A. Synthesis of deuterium-labeled cryptophane-A and investigation of Xe@cryptophane complexation dynamics by 1D-EXSY NMR experiments. *Chem Eur J.* 2001; **7**: 1561–1573.

149. Brotin T, Dutasta JP. Xe@cryptophane complexes with C2 symmetry: synthesis and investigations by 129Xe NMR of the consequences of the size of the host cavity for xenon encapsulation. *Eur J Org Chem.* 2003; 973–984.

150. Darzac M, Brotin T, Rousset-Arzel L, Bouchub D, Dutasta J-P. Synthesis and application of cryptophanol hosts: ^{129}Xe NMR spectroscopy of a deuterium-labeled (Xe)$_2$@bis-cryptophane complex. *New J Chem.* 2004; **28**: 502–512.

151. Huber JG, Dubois L, Desvaux H, Dutasta J-P, Brotin T, Berthault P. NMR study of optically active mono-substituted cryptophanes and their interaction with xenon. *J Phys Chem A*. 2004; **108**: 9608–9615.

152. Huber G, Brotin T, Dubois L, Desvaux H, Dutasta J-P, Berthault P. Water soluble cryptophanes showing unprecedented affinity for xenon: candidates as NMR-based biosensors. *J Am Chem Soc*. 2006; **128**: 6239–6246.

153. Spence MM, Ruiz EJ, Rubin SM, Lowery TJ, Winssinger N, Schultz PRG, Wemmer DE, Pines A. Development of a functionalized xenon biosensor. *J Am Chem Soc*. 2004; **126**: 15287–15294.

154. Lowery TJ, Garcia S, Chavez L, Ruiz EJ, Wu T, Brotin T, Dutasta J-P, King DS, Schultz PG, Pines A, Wemmer DE. Optimization of xenon biosensors for detection of protein interactions. *Chembiochem*. 2006; **7**: 65–73.

155. Wei Q, Seward GK, Hill PA, Patton B, Dimitrov IE, Kuzma NN, Dmochowski IJ. Designing ^{129}Xe NMR biosensors for matrix metalloproteinase detection. *J Am Chem Soc*. 2006; **128**: 13274–13283.

156. Roy V, Brotin T, Dutasta J-P, Charles M-H, Delair T, Mallet F, Huber G, Desvaux H, Boulard Y, Berthault P. A cryptophane biosensor for the detection of specific nucleotide targets through xenon NMR spectroscopy. *ChemPhysChem*. 2007; **8**: 2082–2085.

157. Hill PA, Wei Q, Eckenhoff RG, Dmochowski IJ. Thermodynamics of xenon binding to cryptophane in water and human plasma. *J Am Chem Soc*. 2007; **129**: 9262–9263.

158. Hill PA, Wei Q, Troxler T, Dmochowski IJ. Substituent effects on xenon binding affinity and solution behavior of water-soluble cryptophanes. *J Am Chem Soc*. 2009; **131**: 3069–3077.

159. Huber G, Beguin L, Desvaux H, Brotin T, Fogarty HA, Dutasta J-P, Berthault P. Cryptophane–xenon complexes in organic solvents observed through NMR spectroscopy. *J Phys Chem A*. 2008; **112**: 11363–11372.

160. Aaron JA, Chambers JM, Jude KM, Di Costanzo L, Dmochowski IJ, Christianson DW. Structure of a 129Xe–cryptophane biosensor complexed with human carbonic anhydrase II. *J Am Chem Soc*. 2008; **130**: 6942–6943.

161. Chambers JM, Hill PA, Aaron JA, Han Z, Christianson DW, Kuzma NN, Dmochowski IJ. Cryptophane xenon-129 nuclear magnetic resonance biosensors targeting human carbonic anhydrase. *J Am Chem Soc*. 2009; **131**: 563–569.

162. Schramm U, Roesky CEO, Winter S, Rechenbach T, Boeker P, Schulze Lammers P, Weber E, Bargon J. Temperature dependence of an ammonia sensor in humid air based on a cryptophane-coated quartz microbalance. *Sens Actuat B*. 1999; **57**: 233–237.

163. Souteyrand E, Nicolas D, Martin JR, Chauvet JP, Perez H. Behaviour of cryptophane molecules in gas media. *Sens Actuat B*. 1996; **33**: 182–187.

164. Benounis M, Jaffrezic-Renault N, Dutasta J-P, Cherif K, Abdelghani A. Study of a new evanescent wave optical fibre sensor for methane detection based on cryptophane molecules. *Sens Actuat B*. 2005; **107**: 32–39.

165. Wu S, Zhang Y, Li Z, Shuang S, Dong C, Choi MMF. Mode-filtered light methane gas sensor based on cryptophane A. *Anal Chim Acta*. 2009; **633**: 238–243.

166. Sun P, Jiang Y, Xie G, Dua X, Hu J. A room temperature supramolecular-based quartz crystal microbalance (QCM) methane gas sensor. *Sens Actuat B*. 2009; **141**: 104–108.

167. Boulart C, Mowlem MC, Connelly DP, Dutasta J-P, German CR. A novel, low-cost, high performance dissolved methane sensor for aqueous environments. *Opt Expr*. 2008; **16**: 12607–12617.

168. Gambut L, Chauvet J-P, Garcia C, Berge B, Renault A, Rivieire S, Meunier J, Collet A. Ellipsometry, Brewster angle microscopy and thermodynamic studies of monomolecular films of cryptophanes at the air–water interface. *Langmuir*. 1996; **12**: 5407–5412.

169. Gosse I, Chauvet J-P, Dutasta J-P. Synthesis and interfacial properties of amphiphatic cryptophanes. *New J Chem*. 2005; **29**: 1549–1554.

170. Marjanska M, Goodson BM, Castiglione F, Pines A. 2. Inclusion complexes oriented in thermotropic liquid-crystalline solvents studied with carbon-13 NMR. *J Phys Chem B*. 2003; **107**: 12558–12561.

171. Chaffee KE, Marjanska M, Goodson BM. NMR studies of chloroform@cryptophane-A and chloroform@bis-cryptophane inclusion complexes oriented in thermotropic liquid crystals. *Sol St Nuc Mag Res*. 2006; **29**: 104–122.

172. Renault A, Talham D, Canceill D, Batail P, Collet A, Lajzerowicz J. Cryptophane radical cations as components of 3-dimensional charge transfer salts. *Angew Chem Int Ed.* 1989; **28**: 1249–1250.

173. Zhang C, Shen W, Fan R, Zhang G, Shangguan L, Chao J, Shuang S, Dong C, Choi MMF. Study of the contact charge transfer behavior between cryptophanes (A and E) and fullerene by absorption, fluorescence and 1H NMR spectroscopy. *Anal Chim Acta.* 2009; **650**: 118–123.

174. Gosse I, Dutasta J-P, Perrin M, Thozet A. A thiophosphorylated hemicryptophane: structure of the toluene inclusion complex. *New J Chem.* 1999; **23**: 545–548.

175. Gautier A, Mulatier J-C, Crassous J, Dutasta JP. Chiral trialkanolamine-based hemicryptophanes: synthesis and oxovanadium complex. *Org Lett.* 2005; **7**: 1207–1210.

176. Martinez A, Dutasta J-P. Hemicryptophane–oxidovanadium(V) complexes: lead of a new class of efficient supramolecular catalysts. *J Catalysis.* 2009; **267**: 188–192.

177. Martinez A, Robert V, Gornitzka H, Dutasta J-P. Controlling helical chirality in atrane structures: solvent-dependent chirality sense in hemicryptophane-oxidovanadium(V) complexes. *Chem Eur J.* 2010; **16**: 520–527.

178. Janssen RG, Verboom W, van Duynhoven JPM, van Velzen EJJ, Reinhoudt DN. Cryptocalix[6]arenes: molecules with a large cavity. *Tet Lett.* 1994; **35**: 6555–6558.

8

Cucurbiturils

8.1 Introduction

One recurrent theme throughout this book appears to be the serendipity of the discovery of many of these macrocyclic compounds. Cyclodextrins are natural products, whereas crown ethers and calixarenes were discovered by workers who were attempting to synthesise entirely different materials. The family of macrocycles discussed within this chapter was also discovered almost by accident.

In 1905 workers in the group of Robert Behrend studied the acidic condensation of glycoluril (itself a condensation product between urea and glyoxal) and formaldehyde, which gave rise to an almost insoluble solid material which is now known as Behrend's polymer.[1] From this type of reaction we would expect an intractable cross-linked polymer to form. The product from this particular reaction however could be dissolved in and recrystallised from hot concentrated sulfuric acid to give a crystalline product in yields of up to 70%, which was shown to be a condensation product – with one glycoluril reacting with two equivalents of formaldehyde. This product was shown to display high chemical stability and to form co-crystals with a variety of salts, including silver nitrate and potassium permanganate, as well as with organic species such as methylene blue.

After the initial studies, very little work was done and the actual structure and properties of this product remained a mystery until the 1980s. It was then that the group of Mock repeated this synthesis and isolation, demonstrating by FTIR that the carbonyl remained intact and providing a simple NMR spectrum which indicated high symmetry.[2] Final determination of the structure was obtained by X-ray crystallography of the calcium sulfate complex of the compound. The condensation reaction was shown to give a cyclic hexamer (Figure 8.1a); it is surprising that a single product is obtained in such good yields from this reaction in which 24 bonds are generated. This is thought to be a consequence of favourable strain coupled with an abundance of hydrogen-bonding interactions. It is probable that the initial reaction generates a wide range of cyclic and polymeric products and that under the drastic conditions needed to dissolve these products, rearrangement occurs to give the thermodynamically favoured hexamer.

The hexamer bears a resemblance in shape to a pumpkin, which has led to the commonly used name of 'cucurbituril' (from the Latin *cucurbitaceae*) first used in an early paper by Freeman *et al.*[2] Six glycoluril

Macrocycles: Construction, Chemistry and Nanotechnology Applications, First Edition.   Frank Davis and Séamus Higson.
© 2011 John Wiley & Sons, Ltd. Published 2011 by John Wiley & Sons, Ltd.

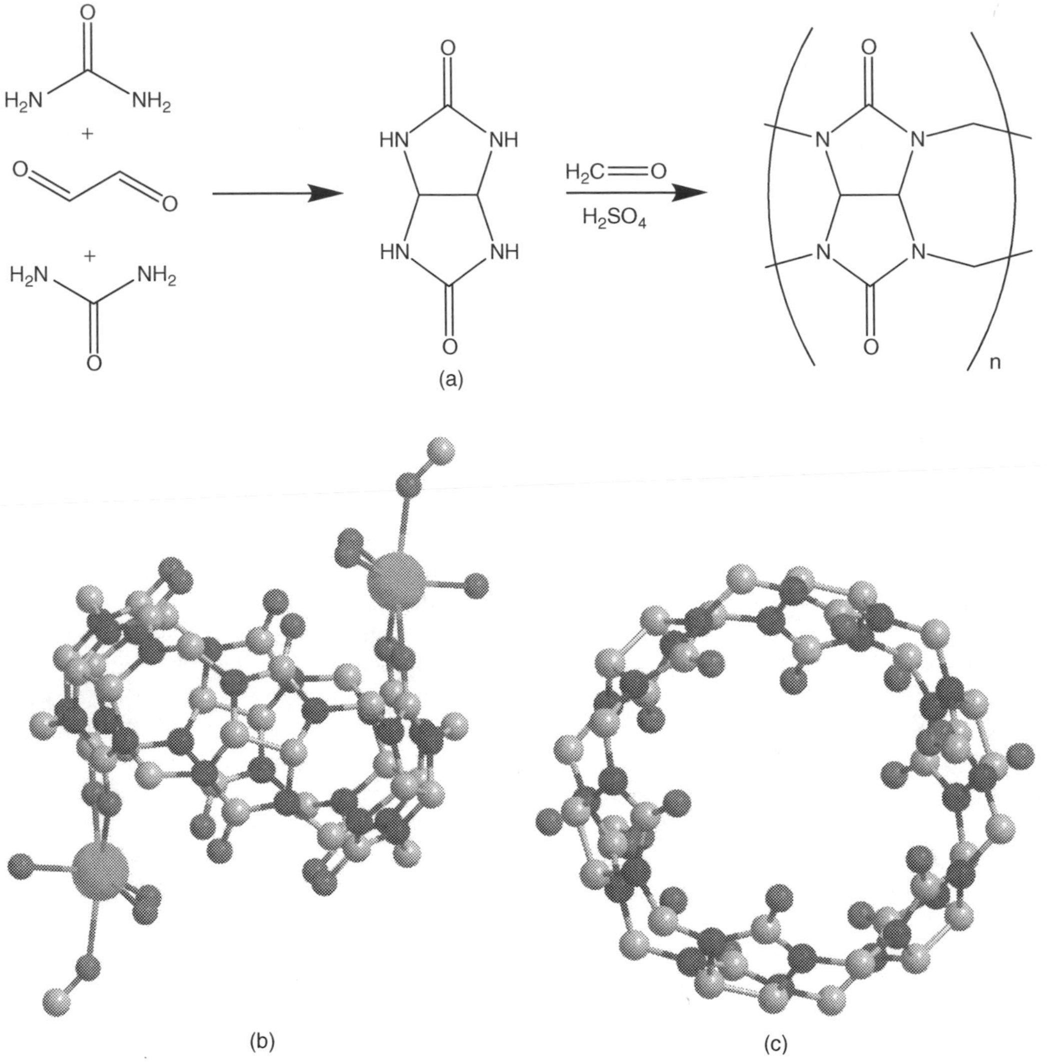

(a)

(b)

(c)

Figure 8.1   (a) Synthesis and (b) crystal structure of cucurbituril complexed with calcium, and (c) 'top' view of cavity (sulfate and water of crystallisation removed for clarity)

units are linked together by paired methylene bridges (unlike other macrocycles, which are linked together by single bridges); the methylene groups are orientated so that the C—H bonds point outward from the macrocyclic structure. Unlike many of the other macrocyclic compounds described within this book, there is no distinct upper and lower rim and the cyclic hexamer is symmetrical. The structure displays an internal cavity some 0.55 nm in diameter, capped with 0.4 nm portals formed by the carbonyl groups (Figure 8.1b). In the crystal structure the calcium ions coordinate to the carbonyl oxygens. There was also some early description of binding within the work of Freeman *et al.*, with for example cyclopentanemethyl amine being shown to form a complex with the hexamer in formic acid, while the larger cyclohexanemethyl amine did not.

Much of the early work on cucurbiturils has been extensively reviewed, as shown within the Bibliography and two recent and exhaustive works by Lagona *et al*. and Isaacs.[3,4] Initial work with cucurbiturils utilised just the hexamer, which can be abbreviated as CB[6]. This is the thermodynamically most stable isomer and the easiest to synthesise in a pure form; for many years it was the only member of this family available. Later work found that if lower temperatures and otherwise milder conditions were utilised, a variety of compounds could be synthesised, usually as kinetically controlled mixtures. For example, glycoluril could be condensed with formaldehyde in 9 M sulfuric acid at 75 °C for 24 hours and then at 100 °C for 12 hours.[5] Electrospray ionisation mass spectrometry demonstrated the presence of a mixture of CB[n] (n = 5–11). A typical batch of the mixture contains 60% of CB[6], 10% of CB[5], 20% of CB[7] and 10% of other higher CB homologues. It is thought this is due to formation of linear condensation products, which then cyclise to form the family of macrocycles. Careful fractionation and crystallisation has allowed the isolation of CB[5], CB[7] and CB[8] and determination of their X-ray crystal structures, which were similar to that of CB[6].

Other workers also examined the synthesis of cucurbiturils under a wide variety of acidic condensation conditions[6] and demonstrated formation of CB[5]–CB[10]. Several general rules for the production of CB[n] compounds were deduced. First, strong acids such as sulfuric or hydrochloric acid are required; trifluoroacetic acid, for example, is not strong enough to catalyse formation of cucurbiturils. Lower concentrations of starting materials (glycoluril) lead to relatively higher yields of CB[5] and CB[6] compared to the higher homologues. It was also found that whilst CB[5], CB[6] and CB[7] are stable under hot, acidic conditions, CB[8] reacts to give a mixture of CB[5]–CB[8]. The effects of metal ions on CB[n] formation were also studied to see whether a template effect existed.[7] Some differences were noted, namely that potassium ions lead to higher yields of CB[5] whereas lithium ions catalyse the increased production of the higher members of this family compared to when the reaction is carried out just in concentrated HCl.

The same group also found evidence of what from the molecular weight was a CB[15] compound, but the NMR indicated presence of a CB[10] moiety plus another species which could not be separated by multiple recrystallisation.[8] X-ray crystallography finally solved this conundrum (Figure 8.2) and demonstrated that the compound consisted of a CB[10] macrocycle with a CB[5] macrocycle encapsulated within it as a guest. The smaller CB[5] was shown by NMR studies to freely revolve within the larger macrocycle. Multigram syntheses of this complex proved possible. Isolation of pure CB[10] from this complex by simple crystallisation or other methods was impossible, but addition of another strongly binding guest molecule, melamine diamine, displaced the CB[5]. The second guest could then be reacted with acetic anhydride, thereby converting the amino groups to amides and causing loss of guest[9] to give pure CB[10].

The cucurbiturils all display similar structures, with a rigid symmetric cyclic structure approximately 0.9 nm in height containing a central cavity. Table 8.1 lists the characteristics of the different cucurbiturils and shows how the central cavity increases dramatically in size with the incorporation of more glycoluril units. The range of the cavity sizes of these compound spans and exceeds that of the cyclodextrins. It should also be noted that the portal diameters are in all cases smaller than the cavity, adding steric hindrance to the binding and unbinding of some guests. The cucurbiturils all display high stability, only decomposing at high temperatures (>350–400 °C) and being stable in strongly acidic solutions.

The discovery of the synthesis of other members of the cucurbituril family besides CB[6] greatly increased the interest in these compounds, since they now offered a wide variety of potential host groups with wide-ranging cavity sizes and shapes. There are some differences in physical properties between members of this family, most noticeably that the odd-numbered CB[5] and CB[7] both have reasonable (20–30 mM L^{-1}) solubilities in water,[10] comparable to β-cyclodextrin, whereas CB[6] and CB[8] are much less soluble (<0.01 mM L^{-1}). The aqueous solubilities can be greatly enhanced by using either strong acid conditions or metal salts. For example, CB[6] has a solubility of 61 mM L^{-1} in 50 : 50 water–formic acid, with CB[5] having similar solubility[3] and CB[7] a solubility of 700 mM L^{-1}. Addition of salts can

Table 8.1   *Structural and physical parameters of cucurbiturils*

Cucurbituril	Cavity Width (nm)	Portal Width (nm)	Cavity Volume (nm^3)
CB[5]	0.44	0.24	0.082
CB[6]	0.58	0.39	0.164
CB[7]	0.73	0.54	0.279
CB[8]	0.88	0.69	0.479
CB[10]	1.13–1.24	0.95–1.06	0.870

also greatly increase solubility; alkali, alkaline and ammonium ions for example were all shown to form 2:1 complexes with CB[6] and to greatly enhance its solubility in water.[10] The binding constants obtained between the CB[6] compounds and the cations were found to be higher than for similar metal–crown ether complexes. Cucurbiturils are also usually insoluble in organic solvents and so there has been some work on the synthesis of derivatives of CB[n] to enhance solubility.[11]

8.2   Complexation Behaviour of Simple Cucurbiturils

8.2.1   Cucurbit[5]uril

The smallest of the cucurbituril family, CB[5] can bind a variety of small guests within the cavity or externally via the carbonyl units. For example, alkali metals, ammonium ions and alkaline earth metals all bind to CB[5] to form complexes.[12] Smaller cations were found to have somewhat higher binding constants, for example sodium bound more strongly than caesium. Ammonium and hexylamine chloride also bound with very similar strengths, indicating that binding was between the carbonyl and ammonium group, with no inclusion of the alkyl chain in the cavity. Similar work describes the complex formation of CB[5] with a variety of transition metal ions,[13] showing that the complexes are generally less stable than those with CB[6]. The adduct of CB[5] with Cu(II) was crystallised and its structure determined by X-ray crystallography.[14] Crystallisation of CB[5] with a molybdenum oxo-species[15] gave rise to a structure in which the inorganic species capped the two ends of the cucurbituril (Figure 8.2b). Interestingly, two sodium ions and one chloride ion were included within the CB[5] cavity as a Na-Cl-Na unit. Lanthanide ions, along with cadmium, barium and potassium, were also shown to cap CB[5] units and form molecular capsules in which a chloride ion could be trapped;[16,17] the structure of a dysprosium-capped[16] CB[5] is shown in Figure 8.3a. While these species tend to crystallise as isolated molecules, the potassium complex actually forms a one-dimensional polymeric chain in the solid state.[17]

CB[5] and its lanthanum-capped complex were examined for their ability to complex various anions. Fluorescence studies showed the capped complex had a marked preference for chloride ion over nitrate, while the uncapped parent molecule demonstrated a preference for the nitrate ion.[18] The structures of these materials were confirmed crystallographically (Figure 8.3b shows the CB[5]/La/NO$_3^-$ complex). Asymmetric capped complexes in which CB[5] is capped at one end by a potassium ion and at the other by a lanthanum species have also been isolated and shown again to incorporate a chloride ion.[19] Complexes between CB[5] and uranyl ions have also been reported.[20]

Organic compounds have also been shown to form complexes with CB[5]. NMR studies showed that hexamethylene tetramine forms capped complexes with CB[5] (and CB[7]) but not CB[6] or CB[8]. With CB[5], both 1:2 complexes in which the hexamethylene tetramine caps both ends of the CB[5] and

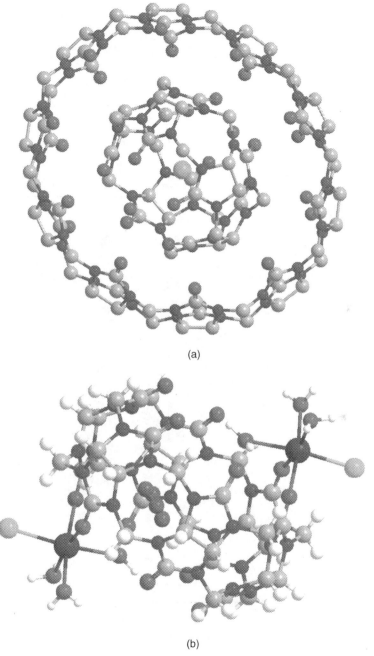

(a)

(b)

Figure 8.2 *(a) Crystal structure of CB[5]@CB[10] (some ions and water removed for clarity) and (b) crystal structure of CB[5]/Mo complex*

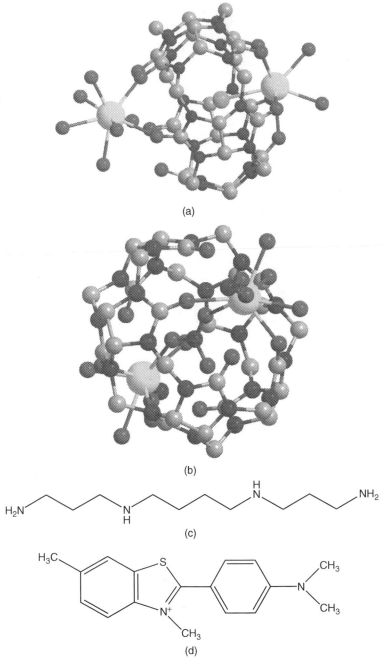

Figure 8.3 *(a) Crystal structure of CB[5]/Dy/Cl complex, (b) crystal structure of CB[5]/La/NO$_3^-$ complex, (c) spermine and (d) thioflavin T*

1 : 1 complexes (probably one-dimensional polymers in nature, with the hexamethylene tetramine bridging between CB[5] units) can be formed.[21] CB[5] also forms 2 : 1 portal complexes with both 2-aminomethyl pyridine and 2(2-aminoethyl) pyridine.[22] Other materials that also form complexes include lysine, which forms a 2 : 1 complex with CB[5], and the peptide pentalysine, which by mass spectral studies forms a 1 : 1 complex, though this cannot be threaded through the macrocycle but instead appears to bridge around it,[23] in contrast to CB[6] in which the pentapeptide is threaded through the macrocycle. However, when CB[5] is combined with the linear tetraamine spermine (Figure 8.3c), it appears that the amine does thread through the macrocycle.[24] This has been confirmed by reacting the complex with benzoyl chloride to give the rotaxane.

The aromatic compound thioflavin T (Figure 8.3d) has also recently been shown to form 1 : 1 and 2 : 1 host–guest complexes with CB[5], again via portal binding.[25] The binding of the first macrocycle to the guest has been shown to occur much more readily than binding of the second host, probably because the initial binding occurs via a strong ion–dipole interaction between the host and guest molecules, whereas the 2 : 1 complex formation is mainly driven by weaker forces like hydrophobic interaction. CB[5] also forms weakly bound 1 : 1 complexes with cyclodextrins in water, with the interaction between CB[5] and α-cyclodextrin being the strongest.[26] CB[5] and CB[6] give very similar results, indicating that ring size has only minimal effects, probably because the cucurbiturils are too large to be accommodated within the cyclodextrin cavity.

8.2.2   Cucurbit[6]uril

The cyclic hexamer was the first member of this family of macrocycles to be isolated and is the material on which most complexation studies have been carried out. The presence of a larger cavity than CB[5] and wider portals means that in addition to capped CB[6], a wide variety of species have been obtained in which guests are complexed within the central cavity. CB[6] has a pKa of 3.02, meaning that it can be thought of as a weak base. This means that in many of the media in which it is soluble, such as 50 : 50 formic acid–water, the macrocycle is in fact protonated and any guest must compete with H_3O^+ in binding. This means that comparison of binding constants in different media can be misleading.

Numerous metal ions have been shown to bind to CB[6]. Studies performed on CB[6] in formic acid–water mixture showed a maximum in solubility at 60% formic acid.[27] Complexation studies showed high affinities for alkali and alkaline earth metals, higher than observed for 18-crown-6 for most metals. The barium complex proved soluble enough to be studied under a range of acid concentrations and it was found that the $\log_{10}$ of the binding constants were much higher in water (5.23) than in 50% formic acid (2.83).

Other work studied the dissolution of CB[6] in salt solutions. CB[6] is appreciably soluble in sodium sulfate solution[28] and can be crystallised to give a tetra-sodium complex. This complex has been shown to be capable of binding THF within the cavity, the structure of which is shown in Figure 8.4a. Two sodium atoms are above and two below the CB[6] units, which along with water of crystallisation form 'lids' to the barrels, and the THF unit is included within the cavity (only one 'lid' is shown for clarity). NMR studies indicate rapid exchange of bound sodium ions with those in free solution. Addition of trifluoroacetic acid leads to loss of the bound THF. Other bound guest molecules include benzene, cyclopentanone and furan. Later work by the same group[29] also showed that xenon could form a stable 1 : 1 complex with CB[6]. A variation on this approach used caesium ions to help solvate CB[6] and showed by X-ray crystallography that a single caesium ion resides at each portal.[30] This species was also found to complex THF with a higher binding constant. This stronger binding was explained by the crystal structure, which demonstrated Cs-O binding between the ion and the oxygen of the THF (Figure 8.4b). Other workers demonstrated by thermochemical measurements that CB[6] formed complexes with a variety of neutral molecules, such as

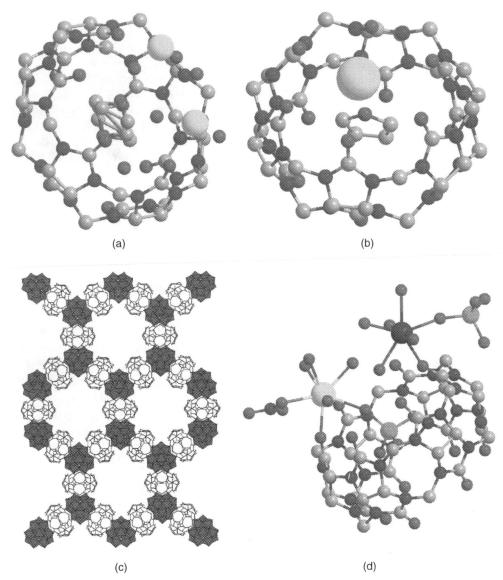

(a)

(b)

(c)

(d)

Figure 8.4 *(a) Crystal structure of CB[6]/Na/THF complex, (b) crystal structure of CB[6]/Cs/THF complex, (c) hexagonal structure of Al complex with CB[6] and (d) crystal structure of CB[6]/Sm/U/ReO$_4$ complex (Reprinted from[38] with kind permission from Springer Science + Business Media)*

aliphatic acids, alcohols and nitriles.[31] Although the formation of inclusion complexes was shown, in these cases the guests were all shown to bind relatively weakly to the CB[6].

In other work this group studied the binding of CB[6] and CB[5] to alkali metal, alkaline earth and ammonium ions[10,12] and showed that CB[6] generally forms much stronger complexes than the smaller analogue (binding constants tend to be 10–100 times higher) and displays a particularly high affinity for barium. Similarly, CB[6] binds transition metal ions[13] as or more strongly than CB[5] and forms weak complexes with cyclodextrins.[26] Lanthanide ions have also been shown to bind to CB[6] to form 1 : 1 complexes, with neodymium having a slightly higher binding constant and the rest of the lanthanides all giving very similar binding.[32]

A wide variety of CB[6]/metal compounds have been demonstrated, both in solution and in the solid state. CB[6] (and CB[8]) have been crystallised from strontium nitrate solution to give a quasi-polymeric structure where the CB[6] unit is capped by two strontium metals and the water and nitrate groups then act as bridging ligands to other CB[6]/Sr complexes.[33] The same group showed that both 1 : 1 and 2 : 1 complexes of Sc(III), Eu(III) and Gd(III) with CB[6] can be crystallised.[34] The lanthanides have been shown in the crystal structure to coordinate to the carbonyl oxygens; in the case of the 2 : 1 Gd/CB[6] complex a pyridine molecule is complexed within the CB[6] cavity. The same group also observed similar behaviour for Pr(III) and Nd(III), with both 1 : 1 and 2 : 1 complexes being successfully crystallised,[35] and successfully obtained crystal structures of 1 : 1 complexes of CB[6] with La(III)[36] and Gd(III).[37]

Polynuclear aluminium aquo complexes have also been shown to form complexes with CB[6]. When $Al_{13}O_{56}$ 'Keggin'-type ions were utilised,[38] 1 : 1 coordination compounds with CB[6] could be obtained in which the crystal structure was actually formed from a series of hexagons, each hexagon containing six $Al_{13}O_{56}$ and six CB[6] units (Figure 8.4c). The $Al_{13}O_{56}$ units are at the vertices of the hexagons, which are distorted analogously to cyclohexane in a chair conformation due to the tetrahedral shape of $Al_{13}O_{56}$. In the crystal structure, these layers are packed in stacks in such a way that the $Al_{13}O_{56}$ polycation of one layer lies virtually above the centre of the hexagonal ring of another layer. Polynuclear lanthanide aquo complexes have also been successfully crystallised as complexes with CB[6], with for example CB[6] and isonicotinic acid having been co-crystallised with $Ln_4(OH)_8$ complexes[39] (where Ln = La, Pr, Dy, Ho, Er or Yb). These complexes have been shown to form a sandwich-type structure and in the case of Eu are luminescent. Addition of silver ions led to the formation of heterometallic coordination polymers. In other work the luminescence and magnetic properties of the Eu(III) and Tb(III) complexes were studied.[40]

Other workers have recently reported the crystallisation of CB[6] with lanthanide, uranyl and perrhenate ions to give a series of compounds with lanthanide and uranyl ions bound at the cucurbituril rim – and a perrhenate ion actually encapsulated within the macrocycle.[41] A typical structure is shown for the samarium complex in Figure 8.4d; similar structures have been obtained for Lu, Gd and Eu.

Much stronger binding was observed when cationic guests were used. A variety of amino compounds were combined with CB[6] in 50% formic acid and the results of many of these experiments are summarised by Lagona *et al.*[3] Initial work by the group of Mock studied a series of aliphatic and aromatic amines and diamino compounds and their complexes with CB[6] in 50% formic acid.[42,43] For example, the binding constant of primary amines has been shown to increase as chain length increases up to butylamine (binding constant of 10^5 M^{-1}), after which it begins to fall again.[43] Addition of a branching methyl group reduces binding, with for example the binding constant of 4-methylbutylamine being about a third of that of butylamine, and branching methyl groups nearer the amino unit lead to even greater reductions in binding. In the case of α-methyl amines, NMR studies have shown that the substituent and α-methyl group are both outside the cavity.

Relatively strong binding was observed for some amines capped with cycloalkylgroups; C-cyclopentyl methylamine for example has a binding constant of 330 000 M^{-1}. Molecular modelling appears to show a good fit of the cyclopentyl moiety to the cucurbituril cavity. Similar results were obtained for the cyclobutyl analogues, but the cyclopropyl and cyclohexyl versions of this material had much lower binding constants. Aromatic compounds could also be bound, with 4-methyl benzylamine showing weak binding (binding constant of 320 M^{-1}) and the 2-methyl or 3-methyl benzyl analogues showing no binding, indicating that the aromatic ring is complexed within the macrocycle and that the presence of *ortho* or *meta* methyl groups increases steric hindrance to a level at which binding is no longer possible.

The strongest binding constants were observed for diamino compounds and it was deduced that the diamino compound is often 'threaded' through the cucurbituril macrocycle, thereby allowing interaction of both ammonium groups with the two sets of carbonyl groups.[42,43] Again this factor is length-dependent, with binding constants peaking for 1,6-diaminohexane, which has a binding constant of $2\,800\,000$ M^{-1}, and 1,5-diaminopentane, which has a binding constant of $2\,400\,000$ M^{-1}. However, hexylamine binds to CB[6] with a binding constant of 2300 M^{-1}, and 6-aminohexanol has an even weaker binding constant of 1200 M^{-1}. This shows the high preference of CB[6] for cationic groups. NMR studies show that the diamino compounds (except for 1,3-diaminopropane) are threaded through the macrocycle, with the centre four methylene groups undergoing chemical shifts due to shielding. The presence of heteroatoms in the alkyl chain has also been studied; the replacement of the central methylene unit of 1,5-diaminopentane with a sulfur atom for example leads to a six-fold fall in binding constant, but when an oxygen atom is incorporated instead, the binding constant drops by a factor of 450. This is thought to be due to solvation effects on the uncomplexed guest and the more hydrophilic ether compound interacting more strongly with the solvent mixture than the thioether or methylene analogues. Polyamines such as spermine and spermidine have also been shown to strongly associate with the macrocycles.[42]

NMR studies have been made on many of these complexes and show that the speed of complexation is related to the molecular volume of the guest, not the thermodynamic stability of the final complex.[44] Relaxation studies also demonstrate that the guests appear to move freely within their hosts. Kinetic studies on the complexation of C-cyclohexyl methylamine with CB[6] under a variety of pH conditions show that the binding constant is highest between pH 3 and 10, the region in which the CB[6] is unprotonated and the amine is protonated.[45] At high pH the rates of ingress and egress of the guest dramatically increase by a factor of 300; this is thought to be due to participation of the free amine in the binding process.

Much of the pioneering work on cucurbit[6]uril complexes came from the group of Buschmann. For example, a series of ω-amino acids and ω-amino alcohols were shown to form complexes with CB[6]; thermochemical measurements showed that the number of methylene groups in the amino acids had only minimal effects on complexation unless the alkyl chain was eight or more methylene groups in length, whereupon the binding constant decreased.[46] The amino alcohols showed different behaviour, with 3-aminopropanol having the strongest association, and the binding constant falling for longer alkyl-chain compounds. Further work also studied the interactions of various dipeptides with CB[6] and showed binding constants that varied with the peptide structure.[47] Later work proved that inclusion complexes were not formed, possibly because of unfavourable interactions between the polar peptide bond and the relatively hydrophobic cavity of CB[6]; instead one ammonium group is bound by the portal carbonyl and the rest of the peptide extends outwards.[48]

The same group also studied α,ω-dicarboxylic acids and diols – these also formed complexes with CB[6] but with binding constants in the range 170–570 M^{-1}, depending on the guest structure[49] – as well as the diamides formed by reacting α,ω-diamines with carboxylic acid chlorides,[50] which were shown to form threaded complexes with CB[6], although binding constants were much lower than the parent diamino compounds. What is of interest is that the number of methylene groups in the parent diamine had only minimal effect on binding strength, which was instead dependent on the length of acid chloride substituent, with longer substituents such as butanoyl increasing the binding constant. The use of cycloalkyl or aromatic acid chloride substituents led to blocking of the binding of the resultant diamides. The effects of various metal salts on complexation of cyclohexylmethyl amine and hexylamine hydrochlorides were also studied and it was shown that the binding of amines decreased as salt concentrations increased, although no specificity was observed.[51]

Studies have also been made on using guests to noncovalently link cucurbiturils with other macrocyclic species. For example, complexes of CB[6] with spermine and spermidine have been formulated and combined with various crown ethers. NMR and other measurements have determined that the macrocycle can encapsulate the polyamines with in the case of spermidine one of the terminal amino groups being located outside the cavity and in the case of spermine, the central two amino groups coordinating to the CB[6] carbonyls to form a threaded structure with two external amino groups.[52] These free amino groups can then form complexes with crown ethers. The CB[6]–spermidine composite forms 1 : 1 complexes with 12-crown-4, 15-crown-5 and 18-crown-6 with binding constants of about 130–170 M^{-1}. In the case of spermine, the binding constants are somewhat higher (200–1000 M^{-1}) and a second crown will bind, giving 1 : 1 : 2 CB[6]–spermine–crown complexes. Similarly, a 1 : 1 complex can be formulated between CB[6] and dihexylammonium where one hexyl chain is incorporated into the CB[6] cavity and the other is external.[53] This second chain can then be incorporated into a cyclodextrin cavity, thereby giving a 1 : 1 : 1 CB[6]–dihexylammonium–cyclodextrin complex. All three common cyclodextrins have been used and form complexes, with binding being strongest for β-cyclodextrin, where the complex is stable enough to be detected by mass spectrometry. NMR studies indicate much stronger hydrogen-bonding between the two macrocycles than in the complexes formed with α- or γ-cyclodextrin. Further work showed that a series of dialkyl and diarylamines could be used to generate these complexes,[54] with both homogenous and heterogeneous complexes being possible.

A series of complexes of cucurbit[6]uril and α- and β-cyclodextrins with nonionic surfactants and polyethylene glycols were also studied; CB[6] was shown to bind to all of these and to bind more strongly to polyethylene glycols with molecular weight >400 than to a lower-molecular-weight oligomer.[55] Binding to CB[6] was generally weaker than to cyclodextrins; again this is thought to be due to competition from binding of protons at the portal rim. Mixed systems in which both CB[6] and α-cyclodextrin are simultaneously threaded onto PEG 2000 chains have also been synthesised, although more cyclodextrin molecules than CB[6] molecules are found in the final product.[56] Complexes have been obtained between CB[6] and either diazacrown ethers or cryptands, but binding constants are much lower (550–560 M^{-1}) than for classical linear diamines and it is thought that no inclusion complex is formed – instead one of the amine groups is simply bound at the portal.[57]

A variety of organic molecules such as toluene, naphthalene, aniline and 1,2-dichlorobenzene have been shown to be bound by CB[6] from the vapour phase.[58] Desorption experiments indicate that in the case of toluene, complex formation takes place, rather than simple adsorption. Attempts have also been made to form complexes with *bis*-aromatic systems based on bipyridine. Although simple 4,4-bipyridine and alkylated bipyridinium salts did not form complexes with CB[6], when two bipyridine units were linked by an alkyl chain of four or more methylene groups, formation of a threaded 1 : 1 complex was observed.[59] Other work described the complexation of a series of azo dyes with CB[6]. It was found that a number of dyes could form complexes with CB[6] in formic acid solution and that as the acid concentration decreased, the binding constant increased,[60] thereby demonstrating again the effects of competitive proton binding at the cucurbituril rims. For example, the binding constant of Direct Orange 40 (Figure 8.5a) increased from approximately 60 M^{-1} in 40% formic acid to 1500 M^{-1} in 10% formic acid. In other work, two azo compounds (4,4′-diaminoazobenzene and 4-amino-4′-nitroazobenzene) were studied. Both were shown to bind to CB[6], with the diamino compound displaying the stronger binding.[61]

Another aromatic compound that has been shown to form complexes with CB[6] is the fluorescent compound 2-anilinonaphthalene-6-sulfonate (Figure 8.5b), which has a binding constant of approximately 50 M^{-1} in 0.2 M Na_2SO_4. The fluorescence of the dye is increased by a factor of five[62] and the authors proposed that the mode of inclusion involves the phenyl group of the dye, because the relatively small size of the cucurbituril cavity precludes incorporation of the naphthalene moiety. Solid fluorescent 2 : 1 guest–host complexes between another compound of this family, 1-anilinonaphthalene-8-sulfonate (Figure 8.5c), and CB[6] could also be obtained, but in this case X-ray crystal structures (Figure 8.5d) demonstrated that the guest molecules were between the CB[6] molecules rather than part of an inclusion complex.[63]

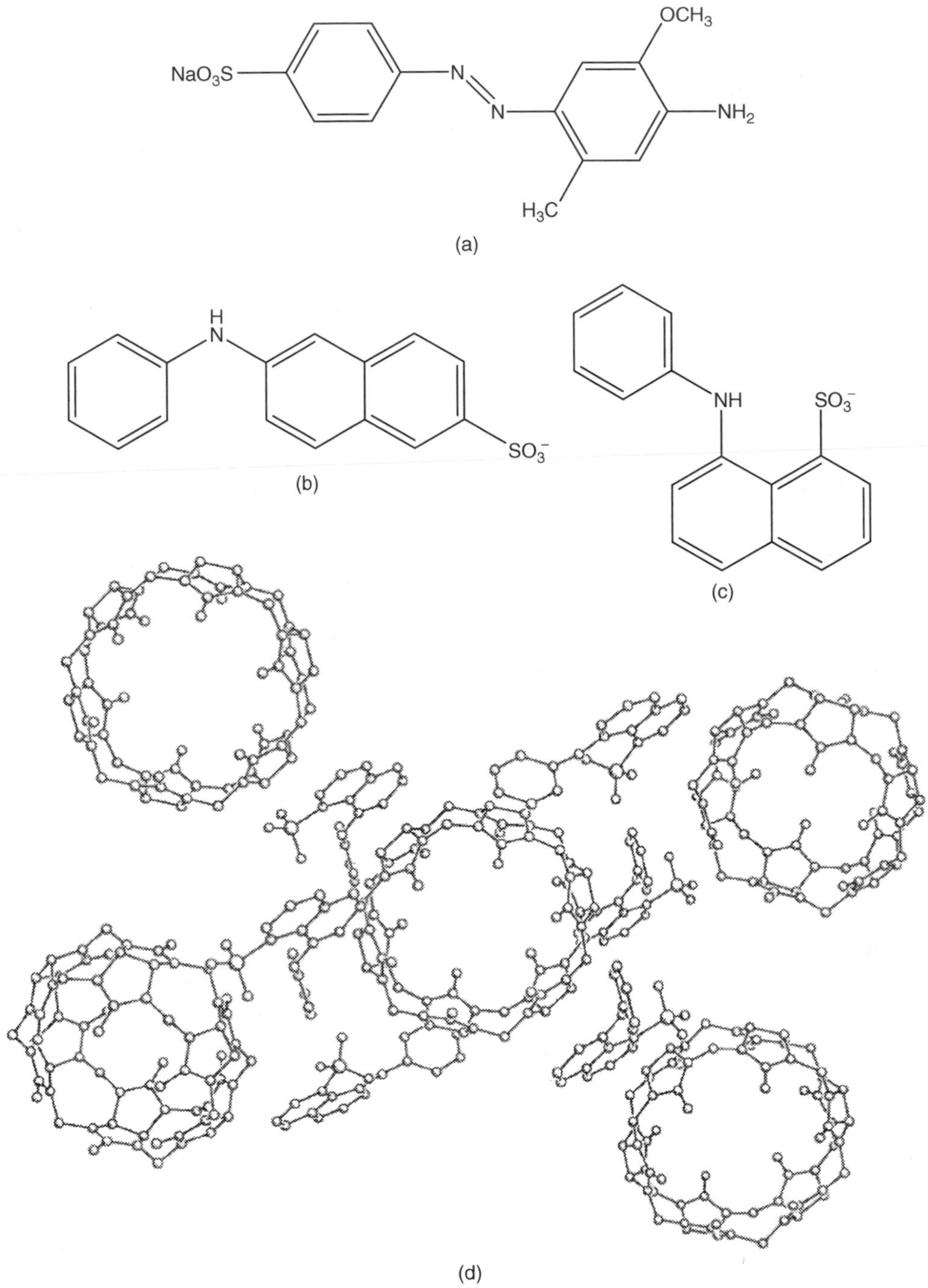

(a)

(b)

(c)

(d)

Figure 8.5   *Structures of (a) Direct Orange 40, (b) 2-anilino naphthalene-6-sulfonate and (c) 1-anilinonaph-thalene-8-sulfonate, and (d) crystal structure of 2 : 1 complex of 8.5c with CB[6] (Reprinted with permission from[63]. Copyright 1999 American Chemical Society)*

Cucurbit[6]uril has been shown to bind diaminohexane very strongly, so aromatic variations on this molecule have been synthesised. For example, a 1,6-hexylenedipyridinium guest dication forms a 1:1 complex with CB[6] and this can be crystallised to clearly show threading of the guest through the macrocycle.[64] The hexyl chain accommodates the hydrophobic CB[6] cavity, with the two pyridyl groups protruding from the end to give a pseudorotaxane (Figure 8.6a). A number of similar compounds in which

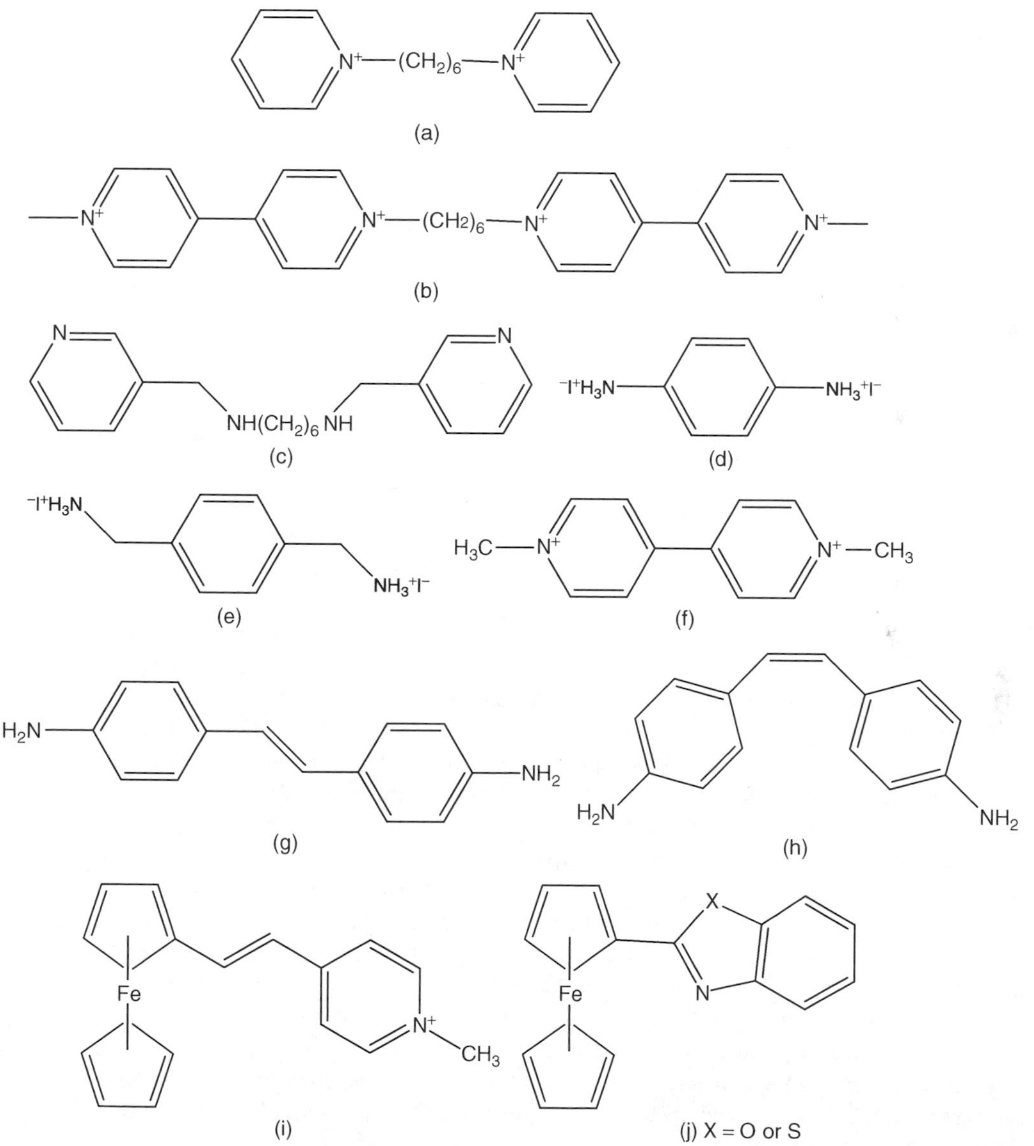

Figure 8.6 *Structures of (a) CB[6]–hexylenedipyridinium, (b) bis-bipyridine compound, (c) bis-pyridine compound, (d) p-phenylene diammonium diiodide, (e) p-xylylene diammonium diiodide, (f) methyl viologen, (g) trans and (h) cis-4,4′-diaminostilbene, (i) (E)-1-ferrocenyl-2-(1-methyl-4-pyridinium)ethylene and (j) 2-ferrocenyl benzoxazole/benzthiazole*

the pyridinium rings were substituted with cyano groups have also been complexed with CB[6] and shown to form pseudorotaxane structures.[65] CB[6] has also been shown to form a 1 : 1 threaded complex with the *bis*-bipyridine compound shown in Figure 8.6b, where the hexyl chain is incorporated within the cavity.[66]

Other workers studied α,ω-diaminoalkyl compounds with different lengths of alkyl chain and where the amino groups had been reacted to give amides bearing 2-, 3- or 4-pyridyl groups.[67] A variety of structures, including pseudorotaxanes, were seen by NMR and X-ray crystallography but some of the complexes also displayed an external binding of guests. Pyridine-substituted amines such as that shown in Figure 8.6c also formed pseudorotaxanes with CB[6] and when complexed with various metals such as copper, cobalt or silver formed one-dimensional coordination polymers as shown by NMR and X-ray studies.[68]

Smaller guests have also been shown to include in the cavity of CB[6], for example a 1 : 1 complex of p-phenylene diammonium diiodide (Figure 8.6d) could be crystallised and the aromatic guest was shown to be incorporated into the cavity.[69] The similar p-xylylene diammonium diiodide (Figure 8.6e) also formed an inclusion compound with CB[6], and X-ray crystal studies showed a large ellipsoidal deformation of the cucurbit[6]uril skeleton.[70] When the three isomers (*o*, *m*, *p*) of phenylene diamine were all complexed with CB[6], the *para* isomer was shown to form an inclusion complex, whereas the other two formed external complexes in which the guests were bound at the portals.[71] The *para* isomers could not be displaced from the complex by *tert*-butylamine, whereas the others exchanged with the amine.

Imidazole units have been shown to form complexes with CB[6]; the ionic liquids 1-ethyl-3-methyl imidazolium bromide and 1-ethyl-3-methyl imidazolium bromide for example could be complexed with the macrocycle and greatly increased its solubility in water.[72] There were differences between the binding of the two guests: the ethyl compound was initially thought to form a 2 : 1 guest–host complex with rapid exchange kinetics, whereas the butyl derivative formed a 1 : 1 complex with slow exchange between bound and free guests. Further work[73] demonstrated that only 1 : 1 complexes were formed for both imidazolium compounds and that in the case of the ethyl derivative the imidazolium unit was encapsulated within the cavity, whereas the longer-chain analogue had its imidazolium located at the portal with the butyl chain encapsulated.

Xenon has been shown to form complexes with CB[6], albeit with much lower binding constants (~210 M^{-1}) than are observed for the cryptophanes described in the previous chapter.[29] Detailed NMR studies reveal that exchange between free and bound xenon is slow on the ^{129}Xe NMR timescale but fast on the ^{1}H NMR timescale.[74] Mass spectral studies have been carried out on CB[6] and show that inclusion complexes with amines can exist in the gas phase.[75] CB[6] has been shown to form a 1 : 1 complex with 1,4-diaminobutane in which the diamine is threaded through the molecule; however, the smaller CB[5] forms 2 : 1 guest–host complexes, probably where a guest is bound at each rim via one of its amino groups.

Other gases have been shown to bind to CB[6]. Sulfur hexafluoride binds to CB[6] at 298 K with a binding constant of 31 000 M^{-1} for SF$_6$ dissolved in water.[76] This is one of the largest constants seen for a neutral guest. ^{19}F NMR also demonstrated very slow exchange kinetics. Density functional calculations have also been performed, showing that completely encapsulated SF$_6$ is the most energetically favoured.[77] Another guest that has been successfully bound to CB[6] is acetylene.[78] The method used to synthesise CB[6] was found to be crucial: CB[6] synthesised using HCl catalysis was found to be highly porous and to contain one-dimensional channels within a honeycomb-like structure. These channels and cavities were estimated to form approximately 24% of the volume. Acetylene could be adsorbed into this porous structure, and even under room temperature and pressure, CB[6] could adsorb 6.1% of its own weight of acetylene.

Butylamine and methacrylic acid could be combined to give the salt butylammonium methacrylate, which formed a 1 : 1 complex with CB[6]. This complex could then be copolymerised with acrylamide to give a number of copolymers with enhanced rigidity and higher glass transition temperatures.[79] Thiols and disulfides such as aminoethane thiol or cysteamine also formed complexes with CB[6], where the thiol/disulfide was encapsulated within the cavity.[80] This encapsulation led to modification of the reactive properties of these sulfur-containing groups, with thiols showing drastically decreased reactivity when treated with several oxidants, and the CB[6]-bound disulfide also exhibited hindered reactivity with reducing agents.

8.2.3 Cucurbit[7]uril

Since CB[7] has a very similar chemistry and shape to CB[6], they display a great similarity in much of their complexation behaviour, for example in the binding of metal ions to the rims. Instead of simply repeating much of the work for CB[6] with a slightly larger macrocycle, we will instead concentrate on how CB[7] and larger macrocycles differ from CB[6]. The main differences of course are the larger cavity size and larger portal size. These enable CB[7] to bind a range of guests that are simply too large to 'fit' inside the CB[6] cavity and this 'enhanced' binding will be the subject of this section.

A much wider range of aromatic structures are capable of being complexed with a CB[7] cavity, such as viologen compounds. For example, methyl viologen (Figure 8.6f) forms a 1 : 1 complex with CB[7], as shown by spectroscopic and electrochemical methods.[81] Binding of the cationic forms of the viologen was shown to be favoured over the reduced, uncharged species. Other workers also showed good binding of the intermediate mono-cationic material.[82] The same group studied the effects of salts on this equilibrium and showed that metal ions, especially calcium, reduce the binding of the viologen, probably due to competitive effects.[83] Viologens with longer alkyl substituents instead of methyl were also studied and shown to bind in two different modes: short-chain viologens were bound via interactions between CB[7] and the aromatic unit, while longer side chains such as butyl displaced the viologen from the cucurbituril cavity.[84] Use of amino-substituted side chains followed by reaction of the terminal amines with bulky 'stopper' groups allowed the formation of rotaxane structures. Viologens substituted with dendrimers have also been shown to form complexes with CB[7], as shown by NMR and mass spectroscopy.[85] *Bis*(pyridinium)-1,4-xylylene species and benzyl-substituted viologens could also be incorporated within these complexes, with both internal and external binding modes being observed.[86] The effect of the larger cavity is demonstrated for the compound shown in Figure 8.6b, which, when complexed with one equivalent of CB[6] or CB[7], forms a 1 : 1 complex in which the hexyl chain is inside the cavity, but when excess CB[7] is used a 1 : 2 complex is formed, with each of the two viologen moieties being included in the cavity of a separate CB[7] unit.[66] No 1 : 2 complex however can be obtained for CB[6].

Two fluorescent dyes which formed complexes with CB[6], 2-anilinonaphthalene-6-sulfonate and 1-anilinonaphthalene-8-sulfonate, also formed complexes with CB[7]. In the case of the 2,6 isomer the phenyl group was shown to be included in the CB[7] cavity, whereas in the case of the other isomer, a 2 : 1 host–guest structure with the dye sandwiched between two CB[7] units was proposed.[87] The *trans* isomer of 4,4'-diaminostilbene (Figure 8.6g) could also be encapsulated in CB[7]; what is of interest is that when the complex is irradiated the stilbene can switch to the *cis* isomer[88] (Figure 8.6h). However, whereas the *cis* isomer in solution isomerises back to the more stable *trans* form, in the complex the *cis* isomer is quite stable at room temperature over a period of 30 days, although at higher temperatures isomerisation takes place. This stabilisation is thought to be due to strong host–guest interactions such as hydrogen-bonding between the two protonated amine termini of the *cis* guest and the portal carbonyl groups of the host. Another example of CB[7] affecting *cis–trans* isomerisation occurs when the macrocycle is complexed with 4,4'-diaminoazobenzene[89] or the analogue in which one benzene ring is replaced by a naphthalene. Whereas the azobenzenes are yellow in colour, the complexes are purple, and the spectra indicate the azobenzenes are in the thermodynamically unfavoured *cis* conformation. There is thermal isomerisation between *cis* and *trans* azobenzene and it appears that enhanced binding to the CB[7] compensates for the energy cost of the *cis* conformation. The azobenzenes can be displaced by certain amines, leading to a colour change back to yellow and the development of a potential colourimetric test for these compounds.

Other aromatic species that have been incorporated include the organometallic compounds ferrocene and cobalticene.[90] Strong 1 : 1 binding was observed between CB[7] and the oxidised cationic forms of these compounds. In the case of cobalticinium, a binding constant of at least 10^6 M^{-1} was observed. Ferrocene moieties substituted with carboxylic acid-terminated dendrimers were also shown to form complexes, via binding of the ferrocene unit, with CB[7]. This binding was shown in the case of the first-generation dendrimer to be pH-dependent.[91] At pH 2, clear evidence for binding was observed, but at pH 7 the terminal carboxylate groups are deprotonated and repelled by the CB[7] rim, meaning no complexation occurs. Higher-generation dendrimers in which the acid groups are remote from the ferrocene unit bind at both pHs. Cobalticenium-substituted dendrimers have also been studied and the second-generation dendrimer shown to undergo optimal binding.[92]

A very stable complex (binding constant of 10^{12} M^{-1} in water) was observed between CB[7] and (E)-1-ferrocenyl-2-(1-methyl-4-pyridinium)ethylene cation (Figure 8.6i). Photochemical studies showed that isomerisation to the Z isomer was completely inhibited by complexation.[93] Other cationic ferrocenes such as trimethylammonio)methylferrocene[94] (Figure 8.6j) have also been shown to form extremely stable complexes with CB[7]. A recent paper studied complexation of CB[7] with several cationic and neutral ferrocene derivatives and demonstrated the formation of highly stable inclusion complexes (K > 10^7 M^{-1}) in all cases.[95] Neutral ferrocene species such as heterocyclic 2-ferrocenyl benzothiazole and 2-ferrocenyl benzoxazole (Figure 8.6j) also form complexes with CB[7], with the resultant NMR and UV/Vis studies indicating that the ferrocene is included in the cavity, not the heterocycle.[96]

The cationic guests *bis*(diethylsulfonium)-p-xylylene (Figure 8.7a) and the related *bis*(tetrahydrothiophenium)-p-xylylene have also been shown by mass spectrometric, NMR and UV/Vis spectroscopic studies[97] to form complexes with CB[7]. Although these monomers can be polymerised to form poly(phenylene vinylene), the complexes did not undergo this reaction. However, they could be pre-polymerised to give a polyelectrolyte which formed complexes with CB[7] and then converted to the polymer (Figure 8.7b) at much lower temperatures than the untreated polyelectrolyte. CB[6] did not show this behaviour. Other cationic guests capable of being bound by CB[7] include the cationic species R_4N^+, R_4P+ and R_3S^+ (R = methyl, ethyl, propyl and so on). Whereas simple ammonium ions tend to bind at the portal, these more charge-diffuse ions bind primarily within the cavity of CB[7] with stability constants that are dependent on the size and coordination number of the central atom, as well as the size and hydrophobicity of the alkyl group.[98] Tetraethyl ammonium ion is bound especially strongly to CB[7] and there is good selectivity of the macrocycle for different ions.

Larger ring systems have been shown to be capable of binding within the cavity, and for example 2,7-dimethyldiazapyrenium (Figure 8.7c) has been shown to form a highly stable 1 : 1 complex with CB[7], where NMR and mass spectrometry measurements show formation of an inclusion complex with some distortion of the host into an elliptical shape to accommodate the guest.[99] Multiple-ring systems such as the 1,4-phenylene *bis*-imidazolium (Figure 8.7d) and diphenyl-*bis*-imidazolium (Figure 8.7e) have also been shown to form complexes with CB[7] with high association constants.[100] The presence of a variety of aromatic, imidazolium and alkyl groups allows formation with CB[7] of 2 : 1, 1 : 1 and 1 : 2 host–guest complexes with varying binding configurations. Later work also demonstrated that the types of compound shown in Figure 8.7e form mixed heterocyclic systems in which the presence of two types of binding site allows these guests to form heterogeneous complexes where each guest simultaneously binds to CB[7] and a cyclodextrin.[101] In these complexes the *bis*-imidazolium unit is bound by the CB[7] and one of the aromatic units by a cyclodextrin, and the NMR spectra clearly show that there is cooperative supramolecular interaction, probably due to hydrogen-bonding between the two macrocycles. Multiple cationic species have been used as guests in CB[7], such as compounds containing up to four imidazolium units, which have

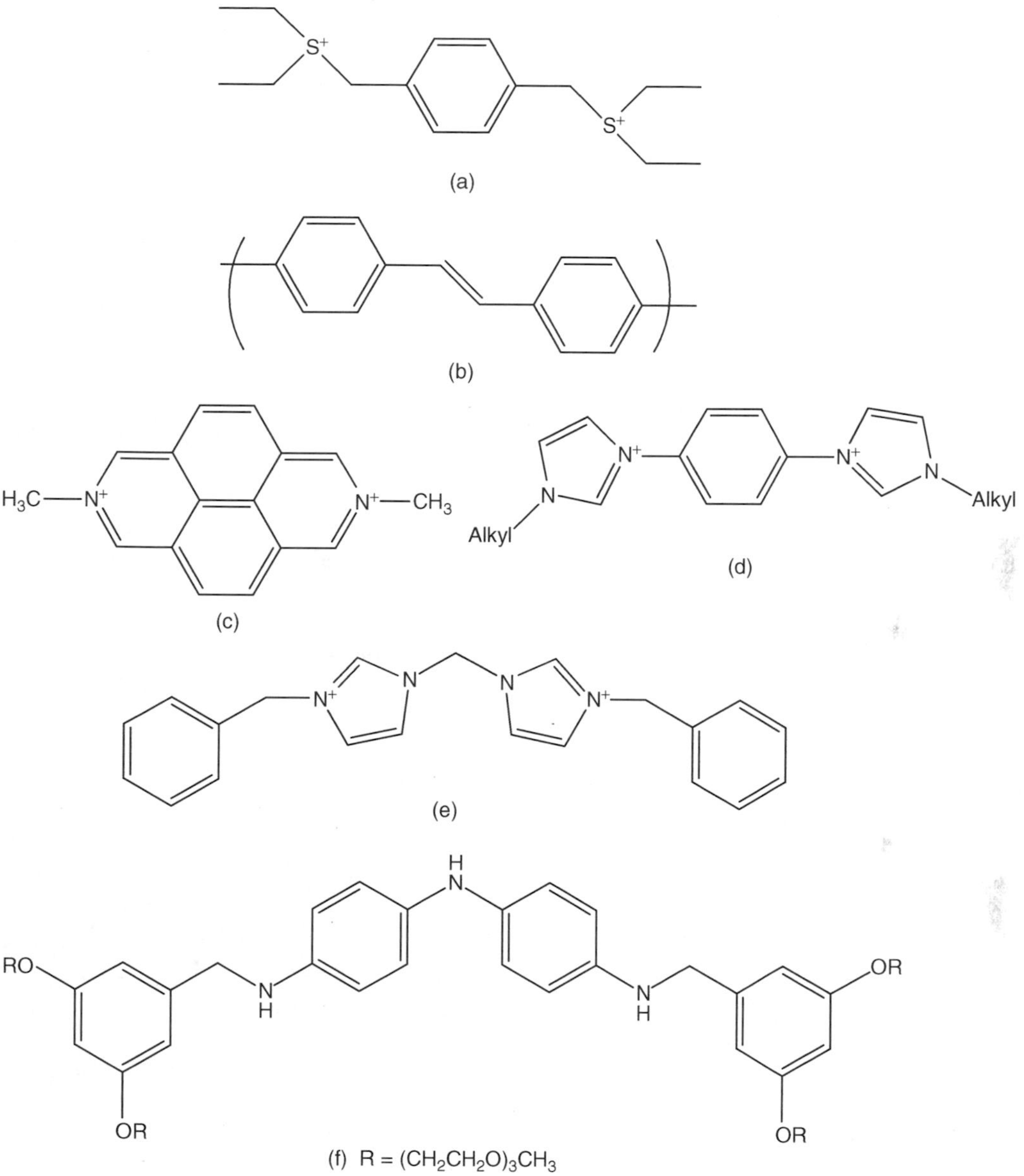

Figure 8.7   *(a) Bis(diethyl sulphonium)-p-xylylene, (b) poly(phenylenevinylene), (c) 2,7 dimethyldiazapyrenium, (d) a typical 1,4-phenylene bis-imidazolium, (e) diphenyl-bis-imidazolium compound and (f) a typical oligoaniline*

been shown to form complexes with one, two or three CB[7] units threaded onto a single guest.[102] NMR studies show that the 1,4-xylylene unit is complexed within the CB[7] cavity.

Other heterocyclic compounds that have been studied as complexes with CB[7] include N-substituted 4-benzoylpyridinium monocations, which were shown to undergo electron-transfer reactions whilst bound in the CB[7] cavity.[103] The fluorescent material 4′,6-Diamidino-2-phenylindole also formed a stable 1 : 1 complex with CB[7] and the exchange of the guest with various ionic liquids based on

1-alkyl-3-methylimidazolium was studied, with the hexyl derivative having the strongest affinity for the macrocycle.[104] Nicotine was also shown to form complexes with CB[7] by competitive-binding experiments with methylene blue, and this could be used as an assay for nicotine in cigarettes.[105] Trimethyl ammonium-substituted anthraquinones bound to CB[7], although with binding constants of only about 1000 M^{-1}, indicating a less than ideal fit between the anthraquinone residue and the CB[7] cavity,[106] though the electrochemical behaviour of the anthraquinone was strongly affected by binding.

Dicationic species have already been shown to form threaded complexes with cucurbiturils. A variety of materials which are basically alkyl chains substituted with trialkylammonium or trialkyl phosphonium complexes have been shown to form complexes with CB[7]. At lower guest concentrations, 1 : 1 complexes are formed in which the alkyl chain is threaded through the macrocycle.[107] However, at higher CB[7] concentrations 2 : 1 host–guest complexes are formed where each CB[7] unit encapsulates a $-^+NR_3$ or $-^+PR_3$ headgroup. Similar behaviour is seen for chains with terminal pyridinium groups where 1 : 1 and 2 : 1 complexes are formed.[108] The 2 : 1 complexes are less stable due to steric and electronic repulsions between the two CB[7] hosts. The aromatic amine di(4-aminophenyl)amine forms complexes with CB[7] which can then be reacted with aromatic aldehydes to give threaded complexes of oligoanilines like that shown in Figure 8.7f. If bulky groups are used, stable rotaxanes can be obtained and the oligoaniline can undergo oxidation reactions both chemically and electrochemically,[109] with encapsulation within the macrocycle stabilising the radical cation form of the guest. Later work showed that oligoaniline salts of different length could be threaded through the cavities of CB[7] macrocycles.[110] Pure complexes could be obtained with up to four CB[7] units threaded onto an oligoaniline unit.

In this and other chapters, we have mentioned complexation of many chromophores and fluorophores with cucurbiturils, and a wide variety of dye molecules have been studied for inclusion within these systems since incorporation of a dye within a macrocycle can modify its properties. For instance, by insulating the chromophores from the environment, a cucurbituril can improve fluorescence lifetimes for many dyes or can disrupt the formation of dimers or other aggregates. A range of dye complexes with CB[7] and the effects on their properties are reviewed by Koner and Nau.[111] Rhodamine dyes are one of the most important dye groups and complexation of rhodamine 6G (Figure 8.8a), which formed the basis of the first dye lasers, with CB[7] leads to many beneficial effects, such as greatly enhanced fluorescence lifetime, prevention of aggregation or surface adsorption and a reduction in photobleaching.[112] Similar effects are seen for hydrophobic derivatives of rhodamine,[113] where the dye is substituted with alkyl side chains. Complexation with CB[7] reduces aggregation and enhances fluorescence. Cyanine dyes have also been complexed with CB[6] and CB[7], with the larger macrocycle having higher binding constants, forming a threaded structure with the dyes and completely disrupting the formation of aggregates.[114] Competing adamantine guests can be used to control the degree of dye aggregation, as can addition of anionic polyelectrolytes.[115]

The fluorescent porphyrin in Figure 8.8b has been shown to strongly bind four CB[7] units in aqueous solution[116] with effects on both adsorption and fluorescence spectra. Neutral Red (Figure 8.8c) was bound into the cavity of CB[7]; this binding could be affected by metal ions, which could be used to control the transfer of the dye from CB[7] into the binding pocket of the protein bovine serum albumin.[117] Acridine orange also bound to CB[7], with a binding constant of 17 000 M^{-1}, and increased fluorescence activity.[118] NMR and mass spectral studies confirmed incorporation of the dye into the cavity and disruption of dye aggregation.[119] Other dyes that have recently been reported to form complexes with CB[7] include lumichrome,[120] where shifts in the fluorescence spectra demonstrate that complexation stabilises a thermodynamically unfavoured isoalloxazine-type structure, and 3,3′-diethylthia carbocyanine iodide dye,[121] where again complexation shifts the fluorescence. Methylene blue has also been shown to form complexes with CB[7] and laser photolysis allows generation of its excited triplet state with almost a doubling of the lifetime compared to the aqueous dye.[122] Complexation with the macrocycle also protects the triplet state

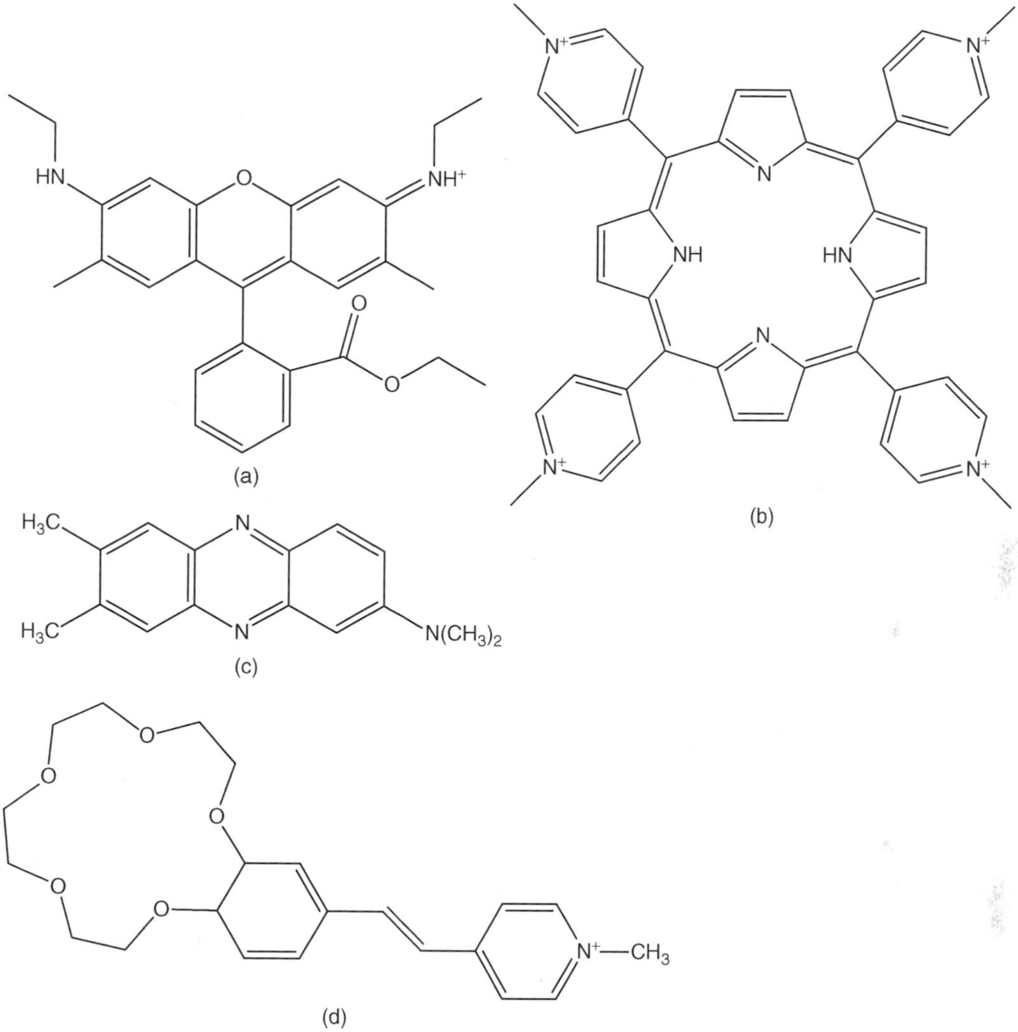

Figure 8.8 *(a) Rhodamine 6G, (b) 5,10,15,20-Tetrakis(4-N-methylpyridyl)porphyrin, (c) Neutral Red and (d) a typical crown-substituted styryl dye*

from oxygen. Complexation of seven members of this family of tricyclic dyes and the determination of association constants and effects on excited states have recently been reported.[123]

Styryl and *bis*-styryl dyes substituted with crown ethers, as shown in Figure 8.8d, have also been studied.[124] The mono-styryl dyes have been shown to form 1:1 complexes with CB[7], whereas the *bis*-styryl compounds can bind either one or two equivalents of CB[7], with large effects on the spectra and excited-state lifetimes upon complexation. Metal ions have been shown to form complexes with the dye–CB[7] composites, though binding is to the CB[7] portals rather than the crown-ether units. Dendrimers which have been modified with multiple naphthalene units at their periphery have been synthesised and combined with CB[7] in solution.[125] These light-harvesting materials have been shown to have 70–100% higher energy-transfer efficiencies compared to the uncomplexed dendrimers. Thioflavin

T (Figure 8.3d) has also been shown to form 1 : 1 and 1 : 2 complexes with CB[7]. What is of interest is the effect of metal ions on these systems. Addition of metal ions to the 1 : 1 complex causes a drop in fluorescence, due to the metal ions competing with the dye for CB[7] binding, but with the 1 : 2 complex the opposite effect is observed.[126] Sodium ions give a 160-fold increase and calcium ions a 270-fold increase in fluorescence intensity; this is thought to be due to the CB[7] units assembling into a supramolecular capsule and protecting the dye, with the metal ions forming a 'seal' at the portals of the cavity.

A series of works included 2,3-diazabicyclo[2,2,2]oct-2-ene in CB[7] and demonstrated the close fit of the guest within the host, and from the fluorescence behaviour of the guest showed that the polarisability of the CB[7] cavity is extremely low.[127,128] The interaction of fluorescent probes with DNA and nucleotides has been the subject of recent research and the CB[7] complexes described above have been shown, compared to other fluorescent probes, to respond very selectively to different nucleotides.[129]

Besides these large macrocyclic systems, small molecules have been bound within CB[7] units, and uranyl species have been shown to form solid-state complexes with CB[7] and acidic species such as perrhenic, phosphoric and polycarboxylic acids.[130] Molecular complexes and one-dimensional polymeric species have been characterised. Incorporation of nitrate ions also allows the formation of two-dimensional polymeric species.[131] Other workers studied the complexation of a wide variety of small, nonionic guests within CB[7] and demonstrated that hydrophobic and dipole–quadrupole interactions are responsible for the binding of the guests.[132] Inorganic species can also be complexed within the CB[7] and for example Figure 8.9a shows the X-ray crystal structure of the 1 : 1 complex of CB[7] with cis-$SnCl_4(OH_2)_2$, where the tin unit is completely encapsulated within the macrocyclic cavity.[133] CB[7] can also be incorporated into capillary electrophoresis experiments and used to help separate positional isomers or aromatic compounds such as nitrotoluenes.[134]

Multinuclear platinum complexes such as that shown in Figure 8.9b have been used in the treatment of many human cancers. One drawback of these materials is their high toxicity. Encapsulation of the complex in CB[7] has been shown to mitigate this somewhat by reducing the reactivity of the platinum centres without compromising their cytotoxicity.[135] A crystal structure of this complex has been obtained (Figure 8.9c) and shows encapsulation of the Pt species within the cavity[136] and enhanced stability, with the complex being stable up to 290 °C. Another species that can be bound into CB[7] is oxaliplatin (Figure 8.9d), with encapsulation of the drug resulting in a large stability enhancement and decreases in reactivity towards amino acids, which the authors suggest may reduce unwanted side effects caused by protein-binding of the platinum drug.[137]

Many more compounds of therapeutic interest have been studied as guests for CB[7]. For example, riboflavin and CB[7] have been shown to form 1 : 1 complexes both in solution and in the solid state.[138] Vitamin B12 can also bind to CB[7] via an axial 5,6-dimethylbenzimidazole nucleotide base to form highly stable complexes and can under the correct conditions more weakly bind a second CB[7] unit.[139] Other nucleotides also interact with CB[7], such as adenine and its derivatives,[140] which bind to the macrocycle with binding constants in the order of 1800–6000 M^{-1}.

The histamine H2-receptor antagonist ranitidine (Figure 8.9e) has been shown to bind to CB[7] with binding constants as high as 10^8 M^{-1} for the diprotonated species, falling to 10^3 M^{-1} for the neutral molecule.[141] NMR studies show the central chain is complexed by the macrocycle. The aquated forms of molybdocene dichloride and titanocene dichloride, which have both been shown to have antitumour properties, can be encapsulated by both CB[7] and CB[8] to form 1 : 1 host–guest complexes in which both the cyclopentadienyl ligands and the metal centre are positioned deep within the cucurbituril cavity.[142] An encapsulated molybdocene complex was shown by *in vitro* cell studies to be more active than the corresponding free metallocene; CB[7] (but not CB[8]) slowed protonolysis of the metallocenes. The clinically important natural alkaloid berberine also forms 1 : 1 complexes with CB[7], leading to a 500-fold increase in fluorescence,[143] as do the alkaloids palmatine and dehydrocorydaline, again with dramatic

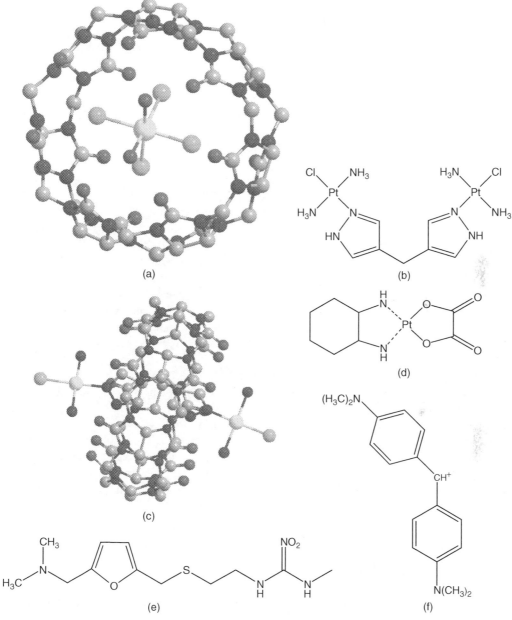

Figure 8.9 *(a) Crystal structure of CB[7]/SnCl₄(OH)₂, (b) platinum complex with anticancer properties, (c) crystal structure of 8.9b with CB[7], (d) oxaliplatin, (e) ranitidine and (f) 4,4′-bis(dimethylamino) diphenylmethane carbocation*

increases in fluorescence.[144] Palmatine has been shown to bind more strongly than dehydrocorydaline, with deeper encapsulation in the cavity. The heterocyclic compound norharmane[145] also binds strongly to CB[7].

A wide variety of the biologically important cholines and phosponiumcholines[146] have been studied for their binding to CB[7]. Drug compounds have also been studied to see if encapsulation with CB[7] in the solid state leads to enhanced stabilisation. For example, atenolol, glibenclamide, memantine and paracetamol can all be stabilised in the solid state by complexation.[147] Similarly, the local anaesthetics procaine, tetracaine, procainamide, dibucaine and prilocaine all form complexes with CB[7], with a concurrent rise in their pKa values.[148]

Binding of various molecules within CB[7] has been shown to affect their chemical reactivity. When a *bis*-imidazolium compound was included as a 1 : 1 complex in CB[7], the exchange of the protons in the 2-positions of the imidazolium unit with deuterium was shown to be inhibited by the macrocycle.[149] The stability of the 4,4′-*bis*(dimethylamino) diphenylmethane carbocation (Figure 8.9f) in water can be greatly enhanced by complexation with CB[7], with up to 90% of the guest existing in the intensely blue carbocation form.[150] Nitroxide derivatives of imidazoline and pyrrolidine[151] can also be stabilised by CB[7]. However, when ester derivatives of cadaverine are included in CB[7] the rate of their acid hydrolysis is seen to increase by up to 300-fold,[152] possibly as a result of the macrocycle stabilising the positively charged reaction intermediate.

CB[7] has been shown to affect the activity of a variety of protease enzymes by forming complexes with their substrates.[153] The macrocycle has also been shown to display interesting templating properties; when gold complexes are reduced in the presence of CB[7], gold nanoparticles <1 nm in diameter are formed, which are encapsulated within the cavity.[154] No evidence for this process is observed for CB[5] or CB[6]. Gold nanoparticles stabilised with CB[7] have been shown to display much higher selectivity and lower chemical reactivity than uncomplexed materials.

Other compounds that have been studied include *o*-carborane, which has been shown to form a 1 : 1 complex with CB[7] that the authors describe as resembling a molecular ball bearing.[155] Other large molecules have also been shown to interact with cucurbiturils; milling together CB[7] and C_{60} led to formation of a 1 : 2 host–guest complex in which the fullerenes interact weakly with each portal of the CB[7] macrocycle.[156] Other workers[157] also reported formation of this complex by stirring solid C_{60} and an aqueous solution of CB[7]. Carbon nanotubes[158] can also be solubilised by CB[7] (but not CB[5]). Experiments showed that CB[7] is adsorbed on the surface of the nanotubes in place of the formation of threaded structures.

8.2.4   Cucurbit[8]uril

In some ways CB[8] simply behaves like a larger version of the previously described macrocycles. However, it does also display some unique behaviour of its own. The larger cavity means that unlike its smaller brethren, CB[8] is capable of simultaneously accommodating two guests, for example two different aromatic rings. This has led to studies of several types of association within this host. For example, if two different aromatic guests are encapsulated, charge-transfer interactions can occur. Alternatively, two molecules of the same guest can be accommodated. One of the earliest observations of this was the formation of complexes of the *bis*-imidazolium naphthalene compound (Figure 8.10a), which was shown to form 1 : 1 complexes with CB[6] and CB[7], but with CB[8] a 1 : 2 host–guest complex was preferred.[159] The binding of a second guest molecule was strongly favoured by π–π stacking interactions, so much so that when the guest and CB[8] were mixed in a 1 : 1 ratio, a mixture of empty and 1 : 2 complexes was obtained, with no 1 : 1 complex.

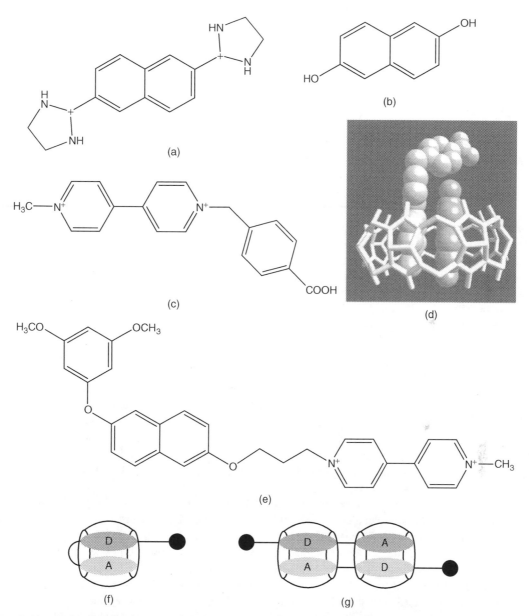

Figure 8.10 *(a) Bis-imidazolium naphthalene, (b) 2,6-dihydroxy naphthalene, (c) substituted viologen and (d) crystal structure of CB[8] complex with 8.10b + 8.10c (Copyright Wiley-VCH Verlag GmbH & Co. KGaA. Reproduced with permission[160]). (e) Viologen–dihydroxynaphthalene compound and (f,g) 1:1 and 2:2 complexes of CB[8]/8.10e (Reproduced by permission of the Royal Society of Chemistry[162])*

When a relatively electron-poor aromatic species such as a viologen is mixed with CB[8] and an electron-rich aromatic system such as 2,6-dihydroxy naphthalene (Figure 8.10b), formation of a ternary complex in which the host binds one molecule of each guest is observed.[160] When a substituted viologen is used (Figure 8.10c), the crystal structure of the ternary complex can be obtained, which clearly shows inclusion of both aromatic species within the cavity (Figure 8.10d). When viologens with long hydrocarbon substituents are used, formation of the ternary complex is accompanied by the formation of large, highly stable vesicles.[161] Further to this work, a novel compound has been synthesised (Figure 8.10e) which contains both viologen and naphthalene units.[162] This has a number of possible host–guest complexes with CB[8], such as a 2 : 2 complex where the viologen of one guest binds together with the naphthalene unit of another within a cavity, as shown in Figure 8.10f. However, the true structure has been shown by NMR studies to be a 1 : 1 complex where the guest adopts a folded conformation with both its aromatic units within the same macrocycle (Figure 8.10g). In later work, guests containing naphthalene and dipyridyliumylethylene could be induced to form 1 : 1 and 2 : 2 complexes with CB[8], with the stoichiometry being dependent on the linkers used to connect the two aromatic units.[163] When a ruthenium-containing unit is attached to one of this group of compounds, irradiation of the resultant folded complex with CB[8] leads to photoinduced intramolecular electron transfer and generation of a viologen radical cation.[164]

The tetrathiafulvalene radical cation can also be incorporated as a dimer within CB[8]. The neutral molecule is not encapsulated by CB[8] and if a preformed 1 : 2 complex is treated with ascorbic acid to reduce the radical, the complex is destroyed.[165] Viologens only form 1 : 1 complexes with CB[8], since the electrostatic repulsion between two viologen units would destabilise any 1 : 2 complex. However, when the viologen units are reduced electrochemically to the radical cation form, formation of 1 : 2 complexes can be observed,[166] along with production of empty CB[8]. Formation of radical cation dimers has also been observed for viologens with dendrimers attached.[167] When CB[8] is complexed with a compound containing viologen and dihydroxybenzene units, a folded 1 : 1 complex is obtained. However, when one equivalent of diethyl viologen is added and the viologen units are reduced to the radical cation form, a dynamic system evolves consisting of an equilibrium mixture of 1 : 2 complexes with two viologen units encapsulated by the macrocycle.[168] A ruthenium bipyridyl complex in which one of the bipyridyl moieties is substituted with a phenol unit can also form a ternary complex with CB[8] and methyl viologen.[169] Irradiation of this species leads to intermolecular electron transfer and formation of a 1 : 2 : 1 ruthenium–viologen–CB[8] complex in which the viologen exists as the radical cation.

The complexation of a series of azastilbene molecules with CB[8] followed by UV irradiation leads to the dimerisation of these molecules.[170] In simple aqueous solution, the azastilbenes yield a mixture of products of geometric isomerisation, cyclisation and hydration. However, when complexed with CB[8], the predominant product is that of dimerisation, with predominantly a single dimer being obtained for each compound. Similarly, a series of naphthalene esters form complexes with CB[8], and when irradiated, dimerisation of the naphthyl units occurs with much higher efficiency than in simple solution.[171] Again this is thought to be due to formation of 1 : 2 complexes, bringing the naphthyl units close together and facilitating dimerisation. A compound containing two naphthyl units (Figure 8.10g) has been shown to form a 'folded' 1 : 1 complex with CB[8]; when irradiated again it undergoes an intramolecular dimerisation reaction with a conversion of 98%, which is not observed in solution.[172]

Coumarins (Figure 8.11a) also form complexes with cucurbiturils. In the case of CB[7] a 1 : 1 complex is formed, but in the case of CB[8] a 1 : 2 complex can be isolated, and the macrocycle has been shown by X-ray crystallography (Figure 8.11b) to encapsulate the coumarins, although there is disorder in the complex due to there being various possible locations for the guests inside the host.[173] Irradiation of CB[8]–coumarin complexes leads to dimerisation reactions, with the CB[8] both facilitating the reaction and leading to the preferential formation of *syn* isomers.[174]

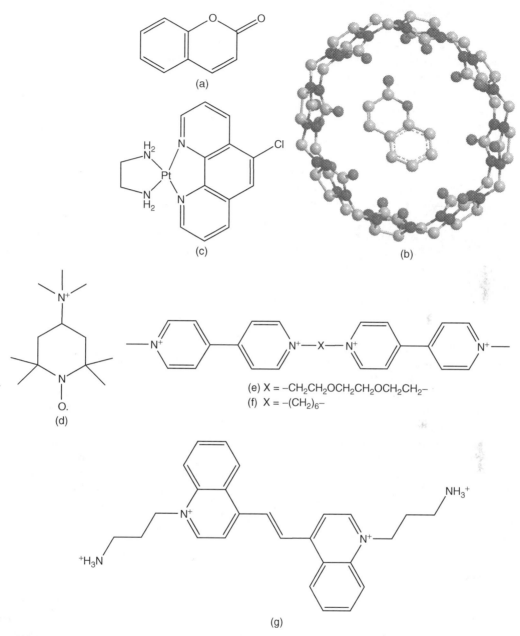

Figure 8.11   *(a) Coumarin, (b) crystal structure of CB[8] complex with coumarin, (c) platinum-based DNA intercalator, (d) nitroxide radical moiety, (e,f) bis-viologen compounds and (g) bis-quinoline compound*

A number of other guests have been shown to form complexes with CB[8]. Host–guest complexes with 1 : 2 stoichiometry have been formed with imidazole derivatives substituted with benzyl groups, whereas the smaller CB[6] and CB[7] only form 1 : 1 complexes.[175] A series of phenanthrolines were studied and shown to form complexes with CB[8], again with two guests bound within the cavity.[176] Different isomers can be formed where the guests have different relative orientations and there are π–π stacking interactions between the guests. A platinum-based DNA intercalating compound (Figure 8.11c) has been shown to form complexes with CB[6], CB[7] and CB[8]. In the case of CB[6], binding has been shown to increase the cytotoxicity of the metal complexes and the complex shows fast exchange processes.[177] CB[7] has been shown to bind the ancillary ligand of the complex, whereas CB[8] binds the phenanthroline moiety; in both cases cytotoxicity is reduced.

The fluorescent dye acridizinium forms complexes with both CB[7] and CB[8]. In the case of CB[8], a 1 : 2 complex is formed, where the two guest molecules form a nonfluorescent, noncovalent dimer complex. Adding CB[7] competitively extracts the dye and restores the fluorescence.[178] The aromatic amino acids tryptophan, phenylalanine and tyrosine all form complexes with CB[8] and its complex with methyl viologen, whereas none of the other of the twenty genetically encoded amino acids bind,[179] showing some selectivity of the CB[8] unit. A recent interesting application of these interactions utilised proteins that had a genetically incorporated N-terminal phenylalanine-glycine-glycine peptide. Mixing of this protein with CB[8] led to the macrocycle binding two of these endgroups, thereby forming a noncovalently bound protein dimer.[180] This could be monitored by utilising fluorescent proteins and detecting energy transfer between the two units of the dimer. A series of cationic diaryl ureas containing phenyl or biphenyl units were also studied and shown to form either 1 : 2 or 2 : 2 host–guest complexes with CB[8], with the conformation of the urea group being dependent on the binding.[181]

A number of paramagnetic nitroxide probes (such as that shown in Figure 8.11d) have been shown to form 1 : 1 complexes with CB[7] and CB[8]. Interestingly, in the case of CB[8], the complexes have been shown to interact with each other in solution and the EPR spectra are consistent with the formation of trimers in aqueous solution. The formation of these trimeric species is dependent on the concentration of the complexes and affected by the addition of NaCl.[182] CB[8] has also been shown to catalyse the oxidation of benzyl alcohols to aldehydes by *p*-iodoxy benzoic acid.[183] Although the *p*-iodoxy benzoic acid is a mild oxidant suitable for the generation of aldehydes, its use has been hindered by its poor solubility. This is mitigated however by the use of CB[8] as a 'nano-reactor', helping to bring together the reagents and facilitate the reaction. Presence of 0.1 equivalents of CB[8] almost doubles the reaction yield.

Larger, more complex assemblies containing CB[8] have been successfully obtained. Dendrimers which have been substituted with either a single viologen or a *para*-dialkoxybenzene unit can be synthesised and shown to form complexes with CB[8], where one dendrimer of each type is linked via interactions of its aromatic groups with CB[8] and the other dendrimers.[184] However, electrochemical reduction of the viologens to the radical cation leads to formation of complexes where one CB[8] binds two viologen dendrimers. Similarly, linear polymers with one end substituted with a viologen or naphthyloxy unit can be self-assembled into AB block copolymers by binding of their end groups in CB[8], meaning that the blocks are held together by supramolecular interactions rather than covalent bonds.[185] A selection of *mono*- or *bis*-viologen compounds such as that shown in Figure 8.11e can also be synthesised and assembled with CB[8] and naphthyloxy compounds to give a number of complexes of different, controlled stoiciometries.[186] These are stable enough to be detected by mass spectrometry. A *bis*-viologen compound (Figure 8.11f) has been shown to form a threaded compound with CB[8], with its hexamethylene linking group included in the cavity.[187] However, electrochemical reduction of the viologens to the radical cations leads to formation of a looped structure, with both viologen units incorporated within the cavity. Materials of this type have been proposed as potential molecular actuators since the complex undergoes a large change in molecular shape and size in response to an electrochemical stimulus.

Quinolines have been shown to form complexes with cucurbiturils. A variety of quinolines can be incorporated into the cavities of CB[7] and CB[8] and display room-temperature phosphorescence.[188] This effect is strongest when the ratio of guest to host is 1:1. Unsaturated *bis*-quinolines substituted with aminoalkyl chains (Figure 8.11g) can form 2:1, 1:1 or 1:2 complexes with CB[8]. When a 1:1 mixture of the quinoline compound and CB[8] are crystallised under irradiation, the less-stable Z-isomer of the quinoline is observed by X-ray crystallography.[189] Other species bound include Proflavin,[190] an acridine-type molecule that forms strong complexes with CB[7] and CB[8] but not CB[6]. Thioflavin T can also be hosted by CB[8] and forms a 2:1 guest–host complex, but incorporation of calcium ions instead leads to formation of the 2:2 complex, which displays intense excimer emission at 570 nm. This is thought to be due to closer association between the dye units when incorporated into a 2:2 complex and opens up the possibility of using calcium as a 'switch' to turn the excimer fluorescence on and off.[191]

Numerous complexes of CB[8] with various inorganic ions have been isolated. These display similar properties and structures to their smaller analogues so they will not be described here. Two compounds that have recently been characterised will be described however, due to their novelty. CB[8] has been shown to form complexes with bismuth; the X-ray structure shown in Figure 8.12a is the first complex obtained for cucurbiturils with bismuth and shows portal-type binding.[192] The complex ion Co(ethylene diamine)$_2$Cl$_2$ has been shown to form a 1:1 complex with CB[8] and in this case the ion is completely encapsulated (Figure 8.12b), also existing in the *trans* form, whereas in solution the *cis* form tends to dominate.[193] Prolonged heating in an evacuated tube is required to convert the *trans* to the *cis* form. Inclusion within the macrocycle increases the thermal stability of the metal complex and hinders isomerisation into the *cis* form.

The relatively large cavity of the CB[8] moiety enables it to incorporate smaller macrocycles. The amine macrocycles described in Chapter 3, cyclen (Figure 3.7b) and cyclam (Figure 3.12a), can both be successfully incorporated into CB[8]. Crystal structures of both free and metal complexes of these two macrocycles can be obtained.[194] In both cases the inner macrocycle is titled with respect to the cucurbituril, as shown by the X-ray structures of the CB[8]–cyclen complex (Figure 8.12c). Metal complexes of these composites can be obtained – another example of a 'Russian doll' complex, with a guest inside a guest inside a host, as shown for CB[8]–cyclam–Cu (Figure 8.12d). Later work obtained nickel complexes of CB[8]–cyclam[195] and showed that they could be synthesised directly from the metal complex plus CB[8]. Complexes with CB[8] with copper and nickel cyclams in a variety of conformations can be separated and characterised.[196] Copper complexes have been shown to exist in both the *trans*-I and *trans*-II forms, whereas the nickel complexes prefer the more energetically favourable *trans*-III form.

8.2.5 Cucurbit[10]uril

This is the largest of the family to be isolated and much less work has been done on this system. The earliest example of guest complexation actually occurred during the synthesis of this compound,[8] when it was found to form as a complex with CB[5]. It took complexation with a synthetic melamine diamino compound (Figure 8.13a), followed by reaction with acetic anhydride,[9] to enable the isolation of pure CB[10]. Once this compound had been isolated, it was shown to form complexes with various guests.

The X-ray crystal structure of the 1:2 host–guest complex with the synthetic melamine diamino compound is shown in Figure 8.13b; as can clearly be seen, two U-shaped guests occupy the central cavity.[9] The chiral binaphthyl compounds shown in Figure 8.13c also form a 1:2 complex with CB[10]. What is interesting is that if a racemic mixture of the guest is used, the homochiral complexes appear to be preferred over complexes in which one of each enantiomer is bound within the macrocycle. A series of larger macrocycles were tested to see if they can be bound by CB[10], but the only success in this work

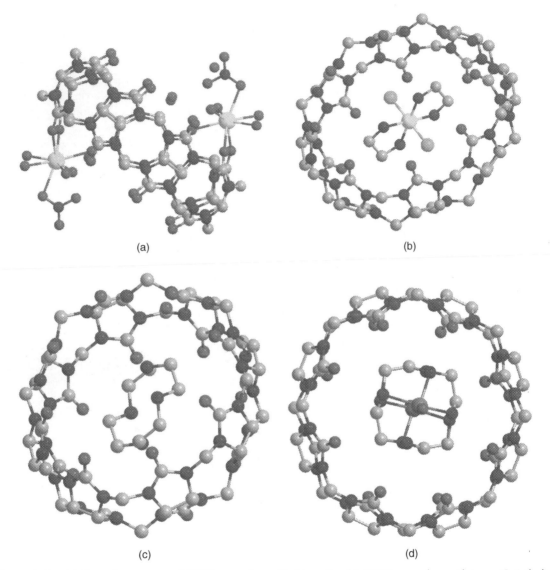

Figure 8.12   *(a) Crystal structure of CB[8] complex with bismuth, (b) CB[8] complex with* trans-Co(ethylene diamine)₂Cl₂, *(c) complex with cyclen and (d) complex with Cu(cyclam)*

occurred with the tetraamine-substituted calix[4]arene (Figure 8.13d). This was shown to be bound as a mixture of cone and alternate isomers. When a second guest was added, in this case a substituted adamantine, formation of a ternary complex was seen and the calixarene adopted a cone conformation. Notably, the calixarene alone does not bind the adamantine guests.

Later work studied the binding of the tetracationic porphyrin shown in Figure 8.8b, along with its metallated derivatives, to CB[10]. Mass spectral studies supported the formation of 1 : 1 complexes.[197]

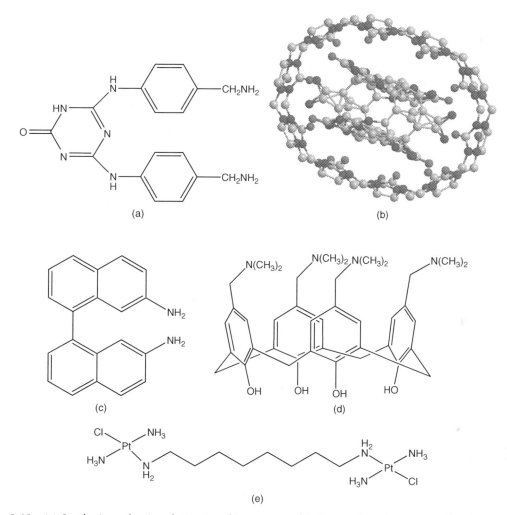

Figure 8.13   *(a) Synthetic melamine derivative, (b) structure of 1:2 complex of CB[10] with 8.13a, (c) chiral binaphthyl guest, (d) calix[4]arene guest and (e) bis-platinum complex*

NMR studies indicated that as a result of binding there is considerable flattening of the CB[10] macrocycle. Smaller guests such as pyridines and quinoline can also be incorporated to give ternary complexes which benefit from $\pi-\pi$ stacking between the aromatic systems. Especially strong binding is observed for 4,4'-bipyridine. A pair of binuclear platinum and ruthenium complexes were also combined with CB[10] and shown to undergo complexation.[198] Both guests formed threaded complexes with CB[10], the complex being shown to be unsuitable as a slow-release agent for the platinum complex (Figure 8.13e) but suitable for controlled slow release of the ruthenium complex. It was also noted within this paper that 1,12-diamino-dodecane could be used to displace CB[5] from the CB[5]@CB[10] complex and that the 1,12-diaminododecane could then be removed with a sodium hydroxide–methanol solution.

8.3   Modification of Cucurbiturils

The cucurbituril macrocycle is somewhat less amenable to modification than many of the other families described within this group, mainly due to its lack of readily modified groups such as hydroxyl. However, derivatives of CB[n] have been synthesised, either by using modified glycoluril units in the synthesis of the macrocycle or by using chemical modification of a preformed CB[n] unit. Many of these homologues of cucurbituril have been reviewed by Lee *et al.*[11]

One of the earliest 'substituted' cucurbiturils was decamethylcucurbit[5]uril (Me10CB[5]). This compound (Figure 8.14a) was first synthesised and fully characterised in 1992 by condensing formaldehyde with dimethylglycoluril and took a year to crystallise.[199] The crystal structure of this compound showed it to be very similar to the parent CB[5], with almost identical cavity and portal sizes. Binding to metal ions has been studied and formation of 1 : 1 complexes via portal binding was demonstrated, with exceptionally high affinity for Pb^{2+} ($>10^9$ M^{-1}); this is thought to be due to a good size match between the ion and the portal.[200] Binding of small solvent molecules with encapsulation in the cavity has also been shown by mass spectrometry, as has binding of gases such as nitrogen and oxygen.[201] In these systems, the portals of the macrocycle are capped with ammonium ions, preventing escape of the guests. It was demonstrated that when a competing host for the ammonium ions, 18-crown-6, is introduced, this capping is lost and the gases can escape. Later work also studied further adsorption of gases to this macrocycle.[202] Whereas gases that were much smaller (helium, neon, hydrogen) or larger (methane, krypton, xenon) than the cavity could not be adsorbed, reversible adsorption was observed for carbon monoxide and dioxide, nitrous and nitric oxide and argon. Repeated cycles of adsorption and desorption of these gases could be observed, indicating their potential use as gas storage agents. Threaded structures of decamethyl CB[5] have also been observed with the amine spermine.[24]

Partially methylated versions of these compounds have also been obtained. When monomethyl glycoluril (Figure 8.14b) is reacted with paraformaldehyde and HCl, two major products are obtained.[203] Although a number of isomers are possible for each macrocycle, separation and analysis of the products showed them each to be single isomers, one of CB[5] and one of CB[6]. The crystal structures of these show that the most thermodynamically stable forms are obtained. The partially methylated cucurbiturils are much more soluble in water and organic solvents than their parent compounds. By reacting unsubstituted monomethyl and dimethyl glycolurils in various proportions and under various conditions, a variety of partially methylated cucurbit[5]urils can be obtained. For example, tetra-, hexa- and nonamethyl CB[5] compounds have been synthesised and isolated.[204] X-ray structures show that the methyl groups all point outwards; therefore they only have minimal effects on the cavity properties. Supramolecular assemblies of these macrocycles with hexafluoroplatinate ions were also constructed and their structures deduced. When complexed with strontium ions, the metals were bound at the portals, thereby capping the macrocycles and leading to formation of supramolecular capsules which encapsulated anions.[205]

One major problem with substituted cucurbiturils is that they can often only be obtained as complex mixtures, which require separation. One method is to use chromatography, using diaminoalkyl-substituted polystyrene beads as the solid phase.[206] This enables the isolation of the fully methylated macrocycle, dodecamethyl CB[6], from a reaction mixture in which the major product is decamethyl CB[5], albeit in only 0.2% yield. Later work enabled the isolation of this derivative in a much better yield, and its crystal structure with 1,4-dihydroxy benzene was obtained, showing inclusion of the guest within the cavity.[207] Another substituted cucurbituril can be obtained when diphenyl glycoluril (Figure 8.14c) is reacted with unsubstituted glycoluril to give mainly the cyclic hexamer where the macrocycle bears two phenyl substituents,[208] with small amounts of the tetraphenyl CB[6] being detected by mass spectrometry. No higher substituted derivatives were obtained. A rotaxane derivative with spermine was also synthesised. These fully substituted CB[5] and CB[6] compounds and the partially substituted materials described below

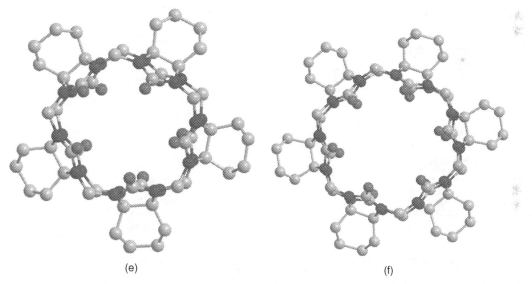

Figure 8.14 *(a) Decamethyl CB[5], (b) partially methylated CB[n], (c) diphenyl glycoluril, (d) structure of cyclohexano CB[n], and crystal structures of (e) pentamer and (f) hexamer*

showed much improved solubility in water and other solvents over their unsubstituted analogues. This means it is now possible to investigate the binding of these hosts in neutral water without requiring high formic acid or salt concentrations.

Reaction of a dimethylglycoluril derivative with a glycoluril dimer with no methyl groups gave tetramethyl–CB[6], in which two opposite dimethylglycoluril groups are incorporated into the hexamer.[209] Interestingly, this material has an ellipsoidal cavity, as shown in its complex with 2,2′-bipyridine, which binds with its long axis parallel to the cavity. This compound also formed 1:1 complexes with phenyl and benzyl pyridines and it was observed that the phenyl moiety was encapsulated within the cavity, with the pyridyl unit interacting with the portal carbonyl.[210] Other guests which formed a 1:1 inclusion complex with this macrocycle were 3-amino-5-phenyl pyrazole,[211] again with the phenyl group encapsulated, and α-ω-alkyl dipyridines, with one pyridyl group encapsulated by the CB[6] unit.[212] The amino acid glutamate was also shown to form a threaded 1:1 inclusion complex with this macrocycle,[213] whereas in the case of *p*-nitrophenol a solid-state exclusion complex was formed, with the guest molecules being incorporated between cucurbituril units rather than complexed within the cavity.[214]

Another useful monomer for synthesising substituted CBs is cyclohexanoglycoluril,[215] which can be condensed with formaldehyde and then under acidic conditions converted into a mixture of substituted CB[5] and CB[6], as shown in Figure 8.14d. The pentamer is the major product compared to the hexamer and these can be separated by fractional crystallisation in final yields of 16% and 2% respectively. Alternatively, they can also be separated by the chromatographic procedure described above.[206] Crystal structures have been obtained for both isomers and clearly show formation of a macrocyclic structure, decorated on the outside with cyclohexyl rings (Figure 8.14e,f).

The cyclohexyl–CB[5] macrocycle binds potassium ions in a 1:2 ratio, not 1:1 as found for CB[5] in formic acid solution, thus providing evidence for the binding of protons to cucurbituril portals in acidic solution. Ion-selective electrodes have been constructed using these species, with cyclohexyl–CB[5] showing good selectivity for lead ions, whereas cyclohexyl–CB[6] allows construction of specific electrodes for acetylcholine with good selectivity over choline.[216]

Symmetric compounds of this type can be attained by condensing the diether of cyclohexanoglycoluril with the dimer of glycoluril to give a water-soluble CB[6] bearing two cyclohexyls located on glycoluril units opposite one another across the macrocycle.[217] X-ray crystal structures such as that of the cobalt complex show that the CB[6] cavity is elliptical in shape, similar to the tetramethyl–CB[6] compounds described above (Figure 8.15a). Inclusion of a bipyridine unit with its long axis parallel to the cavity has also been demonstrated. Other symmetrical cucurbiturils can be synthesised, such as a tricyclohexyl–CB[6] in which the substituents are located on alternate glycoluril units.[218] The crystal structures of this compound with a number of guests have been obtained; Figure 8.15b shows the complexation of a bipyridine compound. Cyclopentyl rings can also be used to substitute CB[6] units; three different compounds are obtained with two, three and four cyclopentyl units.[219] Symmetrical structures, such as a 1,3,5-trisubstituted structure or a 1,2,4,5-tetrasubstituted structure (the crystal structure shown in Figure 8.15c), can be synthesised. Alternatively, a less symmetric 1,3-disubstituted structure can be obtained; the crystal structure of the 1:1 complex with dihexyl bipyridinium is shown in Figure 8.15d, where one of the hexyl chains is complexed within the cavity.

So far the cucurbiturils have been synthesised by simple cyclisation reactions. With many of the macrocycles described in other chapters, stepwise build-up of the ring system has been used as an alternative method to synthesise less symmetrical macrocycles. This approach has also been applied to the synthesis of CB derivatives. For example, a series of linear dimers, trimers and tetramers based on substituted glycoluril compounds have been synthesised.[220] These can be cyclised with phthalhydrazide (Figure 8.16a) to give a series of macrocycles such as those shown in Figure 8.16b. The ester groups can easily be replaced by other groups such as carboxylic acid or amides. The crystal structure of the butanamide derivative is given in Figure 8.16c and clearly shows formation of an elongated cavity compared to that of simple cucurbiturils.

These compounds were further studied because of the presence of such a suitable binding cavity, the potential for π–π stacking interactions, which increased the binding possibilities, and their inherent fluorescence, which should aid detection of any binding.[221] The presence of benzene enhances host fluorescence and shows a binding constant of about 7000 M^{-1}, whereas when a fluorescent guest, the much larger Nile Red molecule (Figure 8.16d), is used, suppression of guest fluorescence is observed and a binding constant of about 8×10^6 M$^-$ is found. These binding constants are superior to those observed with any of the cyclodextrins.

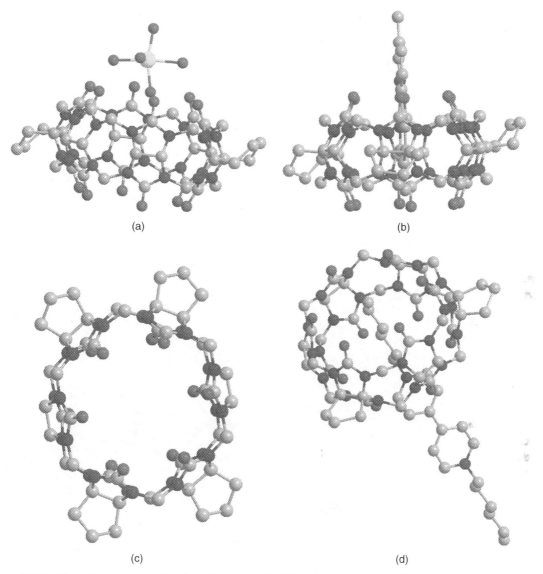

Figure 8.15 *Crystal structures of (a) 1,4-dicyclohexyl–CB[6] cobalt complex, (b) 1,3,5-tricyclohexyl–CB[6], bipyridine complex, (c) 1,2,4,5-tetracyclopentyl–CB[6] and (d) 1,3-dicyclopentyl–CB[6] dihexyl bipyridinium complex*

There has been a great deal of work performed on building up these glycoluril linear oligomers and great insights into the chemistry of cucurbituril formation have been made. Crystal structures have been obtained showing how these oligomers often adopt curved conformations, such as C shapes. This gives an insight into why cyclisation occurs, and often with such specificity. As the reaction proceeds, the oligomers build up in length and adopt curved shapes, which increase the likelihood of cyclisation. A detailed discussion is beyond the scope of this chapter, so we will recommend the Bibliography and the extensive reviews by Lagona *et al.* and Isaacs.[3,4]

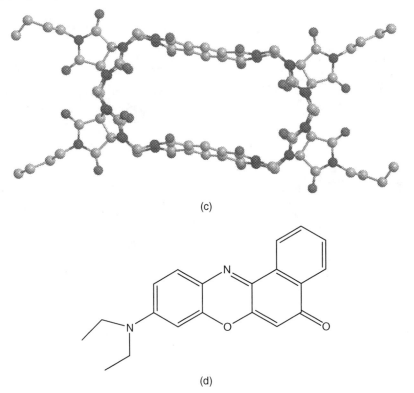

R = –CO₂CH₂CH₃, –COOH or
–CON(CH₂)₃CH₃

Figure 8.16 (a) Phthalhydrazide, (b) aromatic ring incorporation into cucurbiturils, (c) structure of 8.15c and (d) Nile Red

As an alternative to the direct synthesis of substituted cucurbiturils from specially synthesised monomer units, with all of the attendant problems of control of ring size and formation of complex mixtures, it might be possible to simply take whichever CB[n] macrocycle you wish and chemically modify it to put on any functional groups. Unfortunately, the CB[n] unit is much more resistant to modification than many of the other families of macrocycles described within this book. However, some reactions have been formulated which can be used to modify these ring systems. One of the first was to simply oxidise CB[6] with hot potassium persulfate solution.[222] This gives perhydroxycucurbit[6]uril (Figure 8.17a) in 45% yield, which is soluble in DMF and DMSO and has a crystal structure typical of CB[6], with the hydroxyl groups pointing outwards. Besides improving the solubility of the compound, the presence of the hydroxyl groups allows further modifications, such as for example conversion into allyl ethers, which can then be converted into thioethers. Other alkyl ethers and esters have also been synthesised. Perhydroxy CB[5] can be synthesised by the same procedure, though the larger CB[7] and CB[8] give very poor (<5%) yields.

Various applications have been demonstrated for these compounds. When silylated glass slides substituted with thiol groups were placed in solutions of the allyl derivative of CB[6] and irradiated, the macrocycles were chemically grafted to the surfaces of the slides. Imaging of these was possible by complex formation with a fluorescent spermine derivative.[222] Alternatively, long-chain substituted CB[6] can be emulsified to form nanospheres. Similarly, these long-chain derivatives can be incorporated into lipid membranes and shown to transport metal ions, with transport activities for the CB[6] derivative being in the order

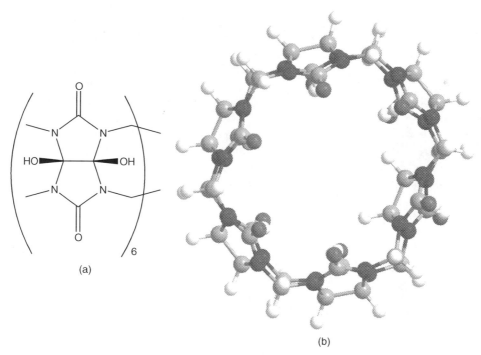

(a)

(b)

Figure 8.17 *(a) Perhydroxy CB[6] and crystal structures of (b) inverted CB[6], (c) nor-seco-CB[6] and (d) nor-sec-o-CB[10] (1,4-diaminobenzene complex)*

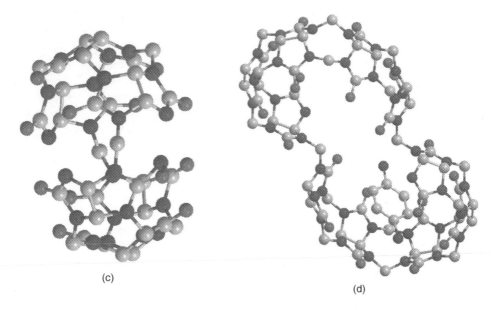

(c)

(d)

Figure 8.17   *(continued)*

$Li^+ > Cs^+ \sim Rb^+ > K^+ > Na^+$, whereas for the CB[5] derivative lithium was transported more readily than sodium but none of the other alkali metals, possibly because they were too big to pass through the portals of the CB[5] unit.[223]

There have also been variations on the basic cucurbituril framework isolated from reaction mixture. Inverted cucurbiturils, where one of the glycoluril units is turned inward and directs its methane hydrogens into the cavity, have been synthesised. Inverted CB[6] and CB[7] have been successfully isolated and their structures (shown in Figure 8.17b for inverted CB[6]) determined by X-ray crystallography.[224] The inwards-pointing hydrogens reduce the size of the cavity, and there is also flattening of the macrocycle. Inverted cucurbiturils tend to bind less effectively than normal cucurbiturils. Under acidic conditions, inverted cucurbiturils are shown to rearrange to their normal counterparts.[225]

When the amount of formaldehyde in the reaction was restricted, an interesting analogue of CB[6] was isolated.[226] The nor-seco cucurbit[6]uril was shown to have a structure in which essentially one pair of methylene binding units is absent. This system forms a twisted structure (Figure 8.17c), which is inherently chiral. It binds amino acids and amino alcohols with some enantiomeric selectivity. A larger version of this compound, nor-seco cucurbit[10]uril, has also been successfully synthesised[227] in up to 15% yield. X-ray studies of the phenylene diamine complex (Figure 8.17d) show the structure to contain two binding activities, each roughly equivalent in size to a CB[6] or CB[7]. This system binds guests such as amines in 1:2 ratio and larger guests in 1:1 ratio, where the guest occupies both cavities. The macrocycle can expand or contract to fit the guests. When several guests are presented to the macrocycle, it will bind two of the same guest in preference to a mixed system; this indicates that binding of the first guest preorganises the second cavity to bind a second guest. This macrocycle was studied for its potential to form supramolecular polymers, since it was thought that a host with two cavities and guests which contained two adamantylammonium ions separated by aromatic spacing groups could form polymeric systems held together by noncovalent interactions.[228] However, these systems appeared to prefer to form 1:1, 2:2 or oligomeric complexes.

8.4   Uses of Cucurbiturils

The high binding affinities and potential selectivities of this family of macrocycles has attracted much attention, as has their ability to be included in supramolecular systems with the potential for use as molecular machines. Here we will attempt to give an overview of some of these uses.

One potential use of these materials is as selective binding agents, since cucurbiturils often display very high binding coefficients for certain guests. There have been several elegant examples of self-sorting, where a series of guests and hosts are combined in the same solution and selective binding occurs, with each host complexing a specific guest. For example, a wide variety of guests were tested for their affinities to CB[6], CB[7] and CB[8]. Most (but not all) of the guests were amino compounds with a variety of substituents.[229] Wide variations in binding coefficients were noted for the various host–guest pairs; adamantane derivatives for example displayed strong binding to CB[7]. This enables the hosts to pick out various guests from mixed solutions. Mixed-host systems also display similar behaviour; for example, a mixed 12-component system of cucurbiturils, cyclodextrins and cryptands with various guests was mixed and shown to resolve itself into six individual host–guest pairings, with each host complexing its preferred guest from the mixture.[230] This self-sorting behaviour approaches the specificity seen in biological systems.

There have been other recent examples of self-sorting systems. For example, the compound shown in Figure 8.18a forms 1 : 2 guest–host complexes with CB[6], where each CB[6] is bound to a diaminomethyl triazole unit.[231] However, with CB[7] and CB[8] a 1 : 1 complex is formed with encapsulation of the central dodecyl chain and interactions of the portals with the amine units. When a mixture of CB[6] and CB[8] is mixed with the guest, a 2 : 1 : 1 complex is formed with a central CB[8] unit complexing the alkyl chain of the guest, with the resulting arrangement then binding two further CB[6] units at each end of the guest. However, if the 2 : 1 complex of CB[6]–guest is combined with excess CB[8], no binding of CB[8] is noted, demonstrating that there is good binding to the guest endgroups by the bulky CB[6] units, which act as stoppers and prevent the CB[8] from complexing the alkyl chains. Another example of this occurred for the spermine derivative shown in Figure 8.18b, which could bind CB[7] around the central diamino butane unit of the guest and a CB[6] unit at each end.[232] Therefore, if first CB[7] and then excess CB[6] is sequentially added to the guest, the complex pseudorotaxane shown in Figure 8.18c will form, since the larger CB[7] unit slips more easily over the endgroups of the guest. However, on heating to 90 °C, free exchange occurs and the CB[7] unit is replaced by a CB[6] since this displays a higher binding constant. This is an example of kinetic versus thermodynamic self-sorting.

Cucurbiturils have also been utilised as building blocks in larger supramolecular assemblies such as rotaxanes and catenanes (see Chapter 9). Dioctyl viologen (Figure 8.19a) forms a 1 : 1 complex with CB[7] where one of the octyl chains is complexed with the cavity. However, addition of excess α-cyclodextrin leads to formation of a 2 : 1 : 1 complex where the CB[7] unit has been 'driven' along to encapsulate the viologen unit, with each of the octyl side chains forming a complex with a cyclodextrin unit.[233] The ruthenium-substituted *bis*-viologen compound (Figure 8.19b) binds a CB[8] unit on the viologen closest to the metal complex.[234] Addition of more CB[8] leads to binding of a second CB[8] on the other viologen. Irradiation leads to light-induced electron transfer, which converts the terminal viologen into its radical cation, meaning that two of these units can be incorporated into a single cavity. The resultant supramolecular assembly consists of two ruthenium units linked together by three CB[8] units, as shown in Figure 8.19b. Polymeric assemblies are also possible, with for example aniline being capable of being complexed with CB[7] and then chemically oxidised to give a polyaniline–CB[8] composite material of approximate ratio of two aniline units per cucurbituril unit, and with the CB[7] units being threaded on the polyaniline chain.[235] The composite displays greater solubility in water and the radical cation form of polyaniline has been shown to be much more stable in the composite than in the pure polymer.

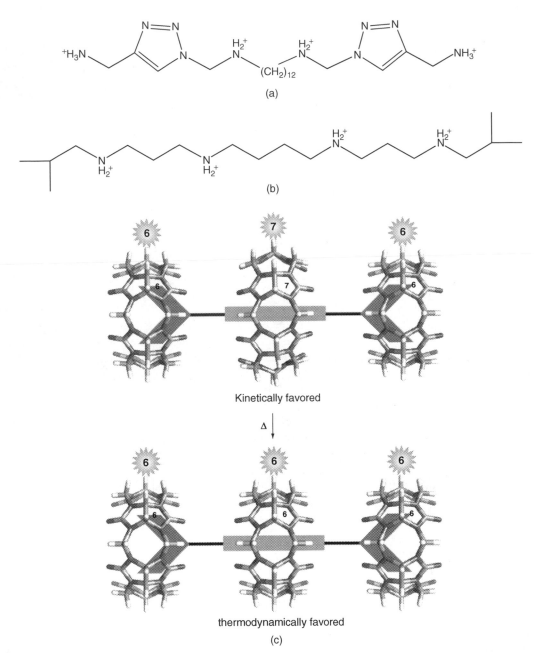

Figure 8.18 (a) Bis-triazole guest, (b) spermine derivative and (c) kinetic versus thermodynamic self-sorting on a spermine guest (Reprinted with permission from[232]. Copyright 2009 American Chemical Society)

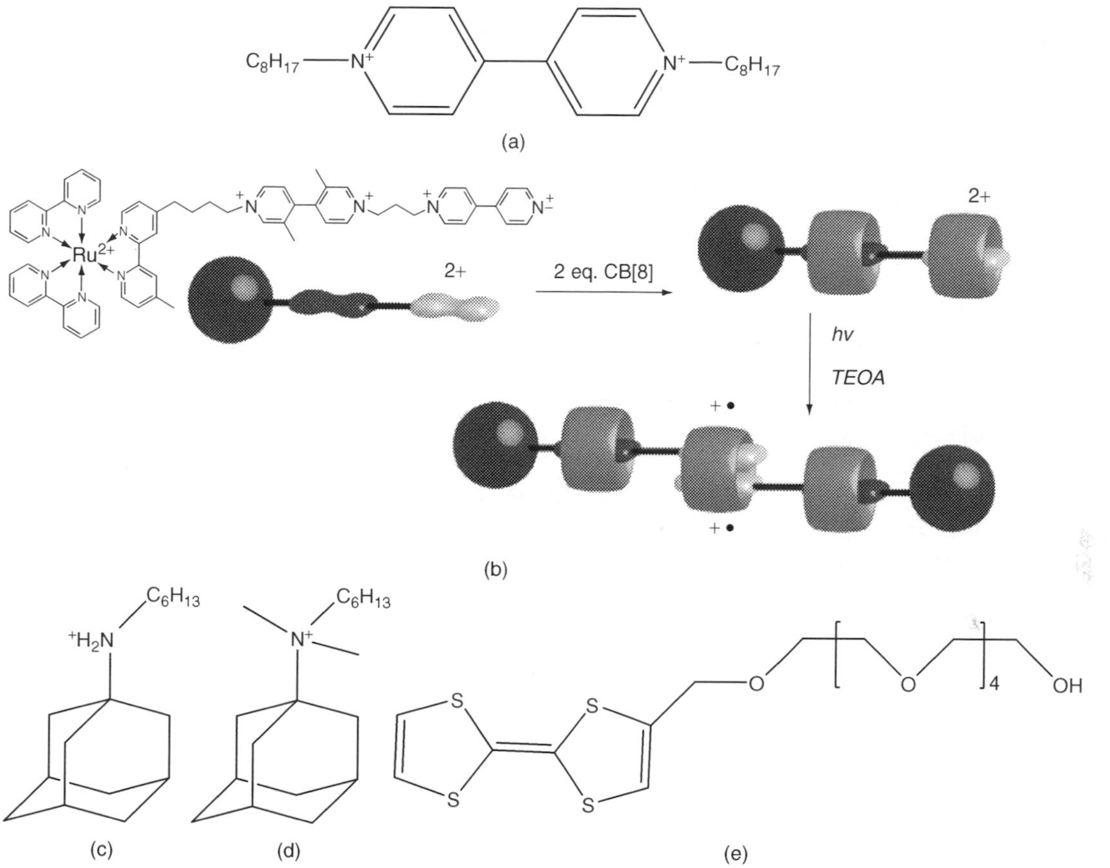

Figure 8.19 *(a) Dioctyl viologen, (b) Ru(bipy)₃ bis-viologen guest and binding with CB[8] (reproduced with permission of the Royal Society of Chemistry[234]), (c,d) amino adamantane derivatives and (e) tetrathiafulvalene guest*

A pH-controlled molecular switch system has been developed in solution.[236] Hexylamino adamantane (Figure 8.19c) and hexyldimethylamino adamantane (Figure 8.19d) were mixed in solution with β-cyclodextrin and CB[6]. This mixture was self-sorting and at pH < 7 the dimethyl derivative preferred to bind to the cyclodextrin and the less-crowded dihydrogen compound to CB[6]. At high pH, however, this amino compound became deprotonated and the two guests then switched places, with the neutral compound binding to β-cyclodextrin and the cationic dimethyl compound to the CB[6]. Electrochemical switching systems have also been demonstrated, and in this context the complex formation between CB[8], dimethyl viologen and the tetrathiafulvalene (Figure 8.19e) has been studied.[237] When the cucurbituril is mixed with one equivalent each of the two compounds, a heterocomplex is preferred where one molecule each of viologen and tetrathiafulvalene are encapsulated in the cavity. Addition of a reducing agent (sodium dithionite, $Na_2S_2O_4$) causes conversion of the viologen into its radical cation and a new 2 : 1 viologen–CB[8] complex is formed, plus 'empty' CB[8]. When ferric perchlorate is used the tetrathiafulvalene is oxidised and this molecule forms a 2 : 1 complex with CB[8].

Larger assemblies of cucurbiturils have also been formed. We have already described how CB[6] substituted with allyl groups can be synthesised;[222] later work showed that when these compounds were irradiated in methanol solution along with a variety of dithiols, addition reactions occurred between the allyl and thiol groups to give thioethers.[238] Surprisingly, the product was not a crosslinked intractable material; instead, polymeric nanocapsules were observed to form, with yields of up to 87%. After isolation by dialysis, these were seen to have diameters of about 110 ± 30 nm. Variation of solvent and dithiol allowed capsules between 50 nm and 600 nm in diameter to be obtained. Amine-substituted dyes were shown to bind to the CB[6] units from solution and fluorescence results indicated the shells of the capsules were essentially one molecule thick. Endocytosis into human oral cancer KB cells of substituted nanocapsules was also observed. Later work utilised these capsules as hosts for quantum dots. If the reaction between CB[6] and dithiols is performed in a solution of ZnS/CdSe quantum dots, the resultant nanocapsules are shown to be fluorescent due to entrapped dots.[239]

Alternatively, cucurbiturils can be attached to the surface of solid materials by other means. Figure 8.20a–c shows the work done on the development of molecular 'nanovalves'. Mesoporous silica can be loaded with dyes such as Rhodamine B and then derivatised at its surface with alkyne groups. These can be converted to triazole units by a CB[6]-catalysed reaction, which leads to a surface that is covered in tightly packed CB[6] groups, which prevent the dyes from escaping.[240] Raising the pH removes the CB[6]–guest interaction and allows controlled release of the Rhodamine B into solution. The authors suggest that tailoring the linking groups 'will enable the development of CB[6]-based nanovalves for *in vivo* applications using the natural variations in pH that exist within healthy and diseased cells in living systems'. Other workers chemically grafted allyl-substituted CB[7] onto surfaces coated with vinyl groups.[241] These surfaces were then used to immobilise biological moieties (such as glucose oxidase) which had been substituted with ferrocenemethylammonium groups. These substituents displayed high affinity for CB[6] (binding constants of 10^{12} M^{-1}) and it was proposed that the system could serve as an alternative to the classical avidin–biotin immobilisation method. The immobilised enzyme was shown electrochemically to retain its activity, allowing its use within a glucose sensor. A thiolated dipyridiniumethylene compound (Figure 8.21a) could be reacted with a gold surface, binding via its thiol group. When this surface was exposed to CB[8] along with a difunctional guest with electron-donor and acceptor groups separated by a rigid spacer (Figure 8.21b), a multilayer system was built up where a CB[8] unit formed a complex at the surface with the dipyridiniumethylene unit and the naphthalenoxy group of the other guest. This then left free viologen units at the surface to complex more CB[8], which of course could then bind a second guest via the naphthalenoxy group.[242] The resultant charge-transfer interactions between the two guests stabilised these systems and meant that after a day, a layer approximately four to five CB[8] units thick had grown on the surface. Longer immersion times did not further increase thickness, as shown by FTIR and SPR studies. Other self-assembly systems include complexes between CB[6] and butylamine toluene sulfonate salts, with these forming hydrogels that spontaneously form a fibrous structure and are thermally responsive.[243]

Some studies have been made of potential medical applications of cucurbiturils. We have already discussed the potential for cryptophanes as hosts for xenon and its use in medical imaging. CB[6] has a cavity suitable in size for the binding of xenon, and the water-soluble cyclohexyl derivatives (Figure 8.14d) have been studied as potential hosts.[244] Xenon forms a 1:1 complex with CB[6] and has a binding constant of 3400 M^{-1} in water and a much higher relaxation time than many water-soluble cryptophanes. Complexes of CB[6] with 1,6-*bis*(imidazol-1-yl)hexane have also been shown to cleave DNA in physiological environments.[245] It appears that the complex cleaves the phosphate esters, and it has been shown to catalyse the hydrolysis of *bis*(2,4-dinitrophenyl) phosphate. In other work, the grafting of thiolglycosides was used

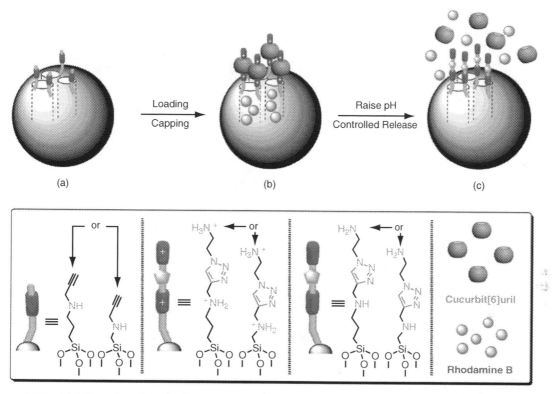

Figure 8.20 *(a) Alkyne-functionalised mesoporous silica nanoparticles MCM-41 are loaded (b) with rhodamine B (RhB) molecules and capped with CB[6] during the CB[6]-catalysed alkyne–azide 1,3-dipolar cycloadditions. RhB molecules are released (c) by switching off the ion–dipole interactions between the CB[6] rings and the bisammonium stalks upon raising the pH value (Copyright Wiley-VCH Verlag GmbH & Co. KGaA. Reproduced with permission[240])*

to attach sugar units to allyl-substituted cucurbiturils.[246] These polysugars were shown to bind strongly to concanavalin A. The mannose-substituted CB[6] was shown to potentially act as a gene delivery agent for hepaocytes.[247]

A number of reactions have been shown to be catalysed by cucurbiturils, one being the formation of triazoles from alkynes and aliphatic azides. One of the earliest examples[248] of this reaction is that shown in Figure 8.21c, which is accelerated by a factor of 55 000 by CB[6]. For example, diazides and dialkynes can be combined with CB[6] to give oligomeric triazoles, whereas in the absence of the macrocycle, no triazole is formed.[249] The oligomer is threaded through the macrocycles, but it can be dethreaded by treating with base. By selecting the correct alkynes and azides it is possible using CB[6]-catalysed reactions to synthesise a wide range of *bis-* and *tris*-triazoles in a regiospecific manner.[250]

Another potential application for cucurbiturils, due to their ability to bind a wide range of substances, is in the clean-up of waste streams. Complex ions such as chromates, dyes and many other pollutants can be bound from water by cucurbiturils. A detailed description is outside the scope of this chapter but much of the work has been reviewed.[3]

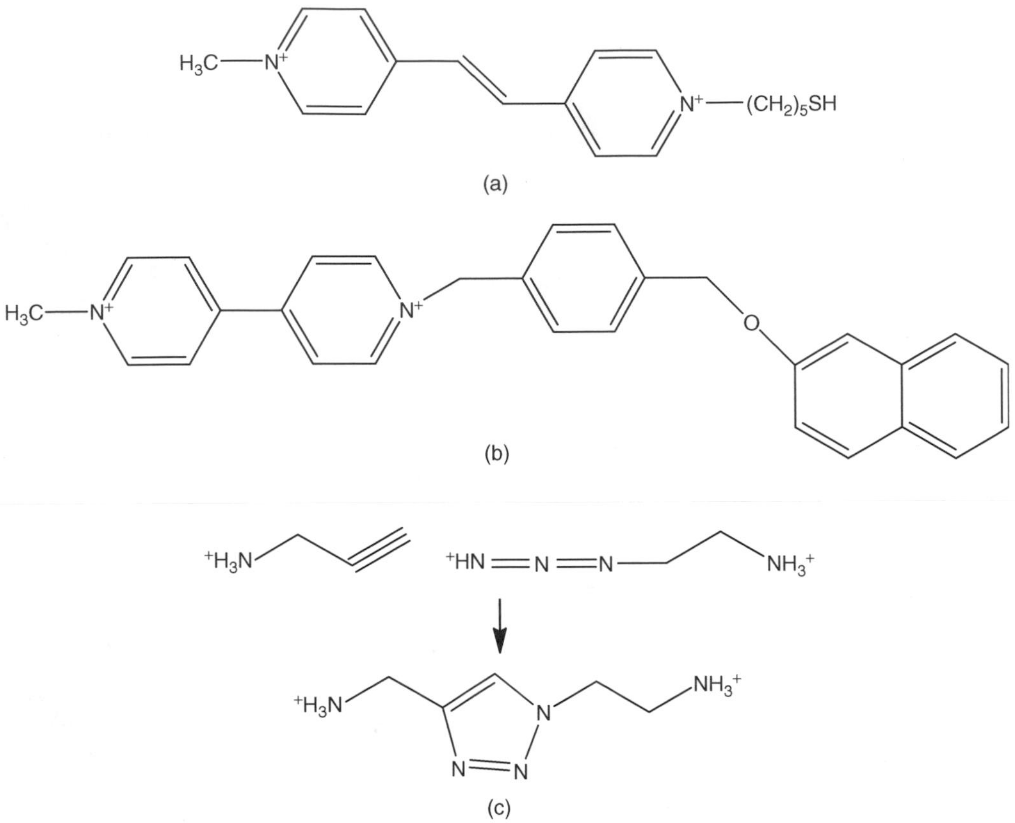

(a)

(b)

(c)

Figure 8.21   *(a) Thiol dipyridiniumethylene anchor, (b) difunctional guest and (c) reaction of alkynes and azides to form triazoles*

8.5   Hemicucurbiturils

As an interesting adjunct to cucurbiturils, a condensation product can also be obtained by reacting ethylene urea with formaldehyde (Figure 8.22a). Mixing equimolar amounts of the two reagents in 4N HCl at room temperature gives the cyclic hexamer in 94% yield.[251] The X-ray structure (Figure 8.22b) clearly demonstrates the cyclic structure, an alternate arrangement of the ethylene urea units and the presence of a chloride ion in the cavity. The macrocycle can also accommodate small species such as thiocyanate ion, formamide and propargyl alcohol. A larger macrocycle, the dodecamer, can be obtained in excellent yield by heating an equimolar mixture of ethyleneurea and 37% formalin in 1N HCl at 55 °C for three hours. This compound can be recrystallised from chloroform and has been shown to be the cyclic dodecamer, again in an alternate configuration (Figure 8.22c), and to include several chloroform molecules in its crystal lattice. Later work with the cyclic hexamers showed them to be quite specific metal-binding agents, forming complexes with Co^{2+}, Ni^{2+} and UO_2^{2+} but not with alkali or alkaline earth metals, silver or ammonium ions.[252]

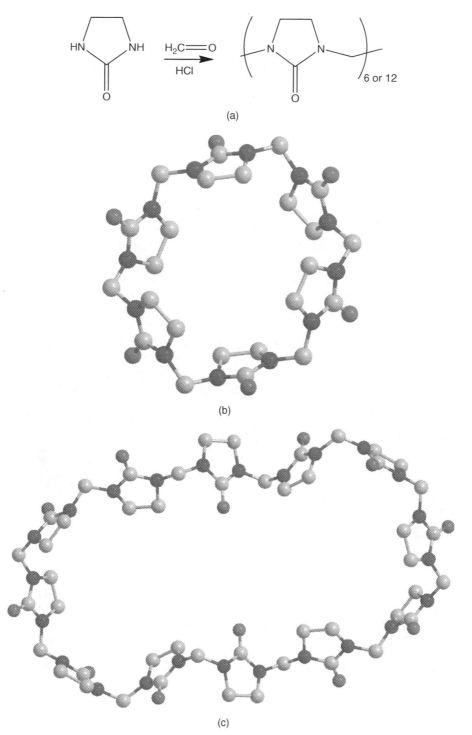

(a)

(b)

(c)

Figure 8.22 *(a) Synthesis of hemicucurbiturils and crystal structures of (b) hemicucurbit[6]uril and (c) hemicu-curbit[12]uril*

8.6   Conclusions

The supramolecular chemistry of cucurbituril means that it has been one of the most intensely investigated family of cavitands recently, due to its multiple modes of molecular recognition, relative ease of synthesis and availability in a range of ring sizes. Cucurbiturils not only bind strongly to alkali and alkaline earth metal cations and ammonium ions but are capable of encapsulating a wide range of aliphatic and aromatic molecules. They display high binding affinities and high selectivities, allowing self-sorting of hosts and guests in complex mixed systems.

The potential for the use of these species in molecular machines is obvious. So far we have seen examples of molecular shuttles utilising these macrocycles and controlled-release applications. The ability of the larger cucurbiturils to bind two guests at once enables the formation of a range of novel charge-transfer species, which in turn enables the binding together of two or more macrocycles. A range of intricate supramolecular structures including cucurbiturils have been devised, some of which will be described in Chapter 9.

Many of the early problems with the CB[n] series, such as poor solubility, have been addressed, either by use of salts or by chemical substitution of the macrocycle. Since the beginning of the century, the number of papers on this family of molecules has risen greatly, aided by the commercial availability of many of its members. New synthetic procedures have been developed for the substitution of these materials, which can only widen their field of use. Cucurbiturils have already been investigated for their application in such fields as waste remediation, catalysis, gas purification and gene transfection. They also show great promise in the field of nanotechnology, with their potential for use in self-assembly, molecular machines and self-assembled monolayers. Systems that are responsive to external stimuli such as pH, along with chemical, electrochemical and photochemical stimuli, have been developed.

Although for years very little work was done on this family of molecules, a recent Scopus search simply for 'cucurbituril' (March 2010) gave 1453 hits, of which 858 were since 2006. This demonstrates how the field has expanded over the last few years and shows that cucurbiturils will compete with the other families of macrocycles for many of the applications proposed for these types of compound.

Bibliography

Kim K, Ho HY, Selvapalam N. Cucurbiturils: Chemistry, Supramolecular Chemistry and Applications. Imperial College Press; 2010.

References

1. Behrend R, Meyer E, Rusche F. I. Ueber Condensationsproducte aus Glycoluril und Formaldehyd. *Justus Liebigs Ann Chem.* 1905; **339**: 1–37.
2. Freeman WA, Mock WL, Shih N-Y. Cucurbituril. *J Am Chem Soc.* 1981; **103**: 7367–7368.
3. Lagona J, Mukhopadhyay P, Chakrabarti S, Isaacs L. The cucurbit[n]uril family. *Angew Chem Int Ed.* 2005; **44**: 4844–4870.
4. Isaacs L. Cucurbit[n]urils: from mechanism to structure and function. *J Chem Soc Chem Commun.* 2009; 619–629.
5. Kim J, Jung I-S, Kim S-Y, Lee E, Kang J-K, Sakamoto S, Yamaguchi K, Kim K. New cucurbituril homologues: syntheses, isolation, characterization, and X-ray crystal structures of cucurbit[n]uril (n = 5, 7, and 8). *J Am Chem Soc.* 2000; **122**: 540–541.
6. Day A, Arnold AP, Blanch RJ, Snushall B. Controlling factors in the synthesis of cucurbituril and its homologues. *J Org Chem.* 2001; **66**: 8094–8100.

7. Day AI, Blanch RJ, Coe A, Arnold AP. The effects of alkali metal cations on product distributions in cucurbit[n]uril synthesis. *J Inclus Phenom Macrocyc Chem.* 2002; **43**: 247–250.

8. Day AI, Blanch RJ, Arnold AP, Lorenzo S, Lewis GR, Dance I. A cucurbituril-based gyroscane: a new supramolecular form. *Angew Chem Int Ed.* 2002; **41**: 275–277.

9. Liu S, Zavalij PY, Isaacs L. Cucurbit[10]uril. *J Am Chem Soc.* 2005; **127**: 16798–16799.

10. Buschmann H-J, Cleve E, Schollmeyer E. Cucurbituril as a ligand for the complexation of cations in aqueous solutions. *Inorg Chim Acta.* 1992; **193**: 93–97.

11. Lee JW, Samal S, Selvapalam N, Kim H-J, Kim K. Cucurbituril homologues and derivatives: new opportunities in supramolecular chemistry. *Acc Chem Res.* 2003; **36**: 621–630.

12. Buschmann H-J, Cleve E, Jansen K, Wego A, Schollmeyer E. Complex formation between cucurbit[n]urils and alkali, alkaline earth and ammonium ions in aqueous solution. *J Inclus Phenom Macrocyc Chem.* 2001; **40**: 117–120.

13. Buschmann H-J, Cleve E, Jansen K, Schollmeyer E. Determination of complex stabilities with nearly insoluble host molecules: cucurbit[5]uril, decamethylcucurbit[5]uril and cucurbit[6]uril as ligands for the complexation of some multicharged cations in aqueous solution. *Anal Chim Acta.* 2001; **437**: 157–163.

14. Liu SM, Huang ZX, Wu XJ, Liang F, Wu CT. Synthesis and crystal structure of macrocyclic cavitand Cucurbit[5]uril and its supramolecular adduct with Cu(II). *Chin J Chem.* 2004; **22**: 1208–1210.

15. Samsonenko DG, Gerasko OA, Virovets AV, Fedin VP. Synthesis and crystal structure of a supramolecular adduct of trinuclear molybdenum oxocluster with macrocyclic cavitand cucurbit[5]uril containing the included ionic associate Na$^+$..Cl$^-$..Na$^+$. *Russ Chem Bull Int Ed.* 2005; **54**: 1557–1562.

16. Zhang Y-Q, Zeng JP, Zhu Q-J, Xue S-F, Tao Z. Molecular capsules formed by three different cucurbit[5]urils and some lanthanide ions. *J Mol Struct.* 2009; **929**: 167–173.

17. Liu JX, Long LS, Huang RB, Zheng LS. Molecular capsules based on cucurbit[5]uril encapsulating 'naked' anion chlorine. *Cryst Growth Des.* 2006; **6**: 2611–2614.

18. Liu JX, Long LS, Huang RB, Zheng LS. Interesting anion-inclusion behavior of cucurbit[5]uril and its lanthanide-capped molecular capsule. *Inorg Chem.* 2007; **46**: 10168–10173.

19. Liu J, Gu Y, Lin R, Yao W, Liu X, Zhu J. Anion encapsulation by Ln(III)/K(I) heterobismetal-capped cucurbit[5]uril. *Supramol Chem.* 2010; **22**: 130–134.

20. Thuery P. Uranyl ion complexes with cucurbit[5]uril: from molecular capsules to uranyl-organic frameworks. *Cryst Growth Des.* 2009; **9**: 1208–1215.

21. Shen Y, Xue S, Zhao Y, Tao QZZ. NMR study on self-assembled cage complex of hexamethylenetetramine and cucurbit[n]urils. *Chinese Sci Bull.* 2003; **48**: 2694–2697.

22. Fu H-Y, Xue S-F, Zhu Q-J, Tao Z, Zhang J-X, Day AI. Investigation of host–guest compounds of cucurbit[n=5–8]uril with some ortho aminopyridines and bispyridine. *J Inclus Phenom Macrocy Chem.* 2005; **52**: 101–107.

23. Zhang H, Grabenauer M, Bowers MT, Dearden DV. Supramolecular modification of ion chemistry: modulation of peptide charge state and dissociation behavior through complexation with cucurbit[n]uril (n = 5, 6) or α-cyclodextrin. *J Phys Chem A.* 2009; **113**: 1508–1517.

24. Wego A, Jansen K, Buschmann H-J, Schollmeyer E, Dopp D. Synthesis of cucurbit[5]uril-spermine-[2]rotaxanes. *J Inclus Phenom Macrocy Chem.* 2002; **43**: 201–205.

25. Choudhury SD, Mohanty J, Upadhyaya HP, Bhasikuttan AC, Pal H. Photophysical studies on the noncovalent interaction of thioflavin T with cucurbit[n]uril macrocycles. *J Phys Chem B.* 2009; **113**: 1891–1898.

26. Buschmann H-J, Cleve E, Jansen K, Wego A, Schollmeyer E. The determination of complex stabilities between different cyclodextrins and dibenzo-18-crown-6, cucurbit[6]uril, decamethylcucurbit[5]uril, cucurbit[5]uril, p-tert-butylcalix[4]arene and p-tert-butylcalix[6]arene in aqueous solutions using a spectrophotometric method. *Mat Sci Eng.* 2001; **14**: 35–39.

27. Buschmann H-J, Jansen K, Meschke C, Schollmeyer E. Thermodynamic data for complex formation between cucurbituril and alkali and alkaline earth cations in aqueous formic acid solution. *J Solution Chem.* 1998; **27**: 135–140.

28. Jeon Y-M, Kim J, Whang D, Kim K. Molecular container assembly capable of controlling binding and release of its guest molecules: reversible encapsulation of organic molecules in sodium ion complexed cucurbituril. *J Am Chem Soc.* 1996; **118**: 9790–9791.

29. Haouaj ME, Ko YH, Luhmer M, Kim K, Bartik K. NMR investigation of the complexation of neutral guests by cucurbituril. *J Chem Soc. Perkin Trans.* 2001; **2**: 2104–2107.

30. Whang D, Heo J, Park JH, Kim K. A molecular bowl with metal ion as bottom: reversible inclusion of organic molecules in cesium ion complexed cucurbituril. *Angew Chem Int Ed.* 1998; **37**: 78–80.

31. Buschmann H-J, Jansen K, Schollmeyer E. Cucurbituril as host molecule for the complexation of aliphatic alcohols, acids and nitriles in aqueous solution. *Thermochim Acta.* 2000; **346**: 33–36.

32. Buschmann H-J, Jansen K, Schollmeyer E. Cucurbit[6]uril as ligand for the complexation of lanthanide cations in aqueous solution. *Inorg Chem Comm.* 2003; **6**: 531–534.

33. Gerasko OA, Virovets AV, Samsonenko DG, Tripolskaya AA, Fedin VP, Fenske D. Synthesis and crystal structures of sumpramolecular compounds of cucurbit[n]urils (n = 6, 8) with polynuclear strontium aqua complexes. *Russ Chem Bull Int Ed.* 2003; **52**: 585–593.

34. Tripolskaya AA, Mainicheva EA, Mitkina TV, Gerasko OA, Yu D, Naumov, Fedin VP. Sc(III), Eu(III), and Gd(III) complexes with macrocyclic cavitand cucurbit[6]uril: synthesis and crystal structures. *Russ J Coord Chem.* 2005; **31**: 768–774.

35. Mainicheva EA, Tripolskaya AA, Gerasko OA, Yu D, Naumov, Fedin VP. Synthesis and crystal structures of PrIII and NdIII complexes with the macrocyclic cavitand cucurbit[6]uril. *Russ Chem Bull Int Ed.* 2006; **55**: 1566–1573.

36. Triposkaya AA, Gerasko OA, Yu D, Naumov, Fedin VP. Crystal structure of a La(III) complex with macrocyclic cavitand cucurbit[6]uril. *J Struct Chem.* 2007; **48**: 949–953.

37. Triposkaya AA, Mainicheva EA, Gerasko OA, Yu D, Naumov, Fedin VP. Synthesis and crystal structure of a supramolecular adduct of the aqua nitrate complex of gadolinium $[Gd(NO_3)(H_2O)_7]^{2+}$ with macrocyclic cavitand cucurbit[6]uril. *J Struct Chem.* 2007; **48**: 547–551.

38. Mainicheva EA, Gerasko OA, Sheludyakova LA, Yu D, Naumov, Naumova MI, Fedin VP. Synthesis and crystal structures of supramolecular compounds of polynuclear aluminum(III) aqua hydroxo complexes with cucurbit[6]uril. *Russ Chem Bull Int Ed.* 2006; **55**: 267–275.

39. Gerasko OA, Mainicheva EA, Naumova MI, Neumaier M, Kappes MM, Lebedkin S, Fenske D, Fedin VP. Sandwich-type tetranuclear lanthanide complexes with cucurbit[6]uril: from molecular compounds to coordination polymers. *Inorg Chem.* 2008; **47**: 8869–8880.

40. Gerasko OA, Mainicheva EA, Naumova MI, Yurjeva OP, Alberola A, Vicent C, Llusar R, Fedin VP. Tetranuclear lanthanide aqua hydroxo complexes with macrocyclic ligand cucurbit[6]uril. *Eur J Inorg Chem.* 2008; 416–424.

41. Theury P. Uranyl–lanthanide heterometallic complexes with cucurbit[6]uril and perrhenate ligands. *Inorg Chem.* 2009; **48**: 825–827.

42. Mock WL, Shih N-Y. Structure and selectivity in host–guest complexes of cucurbituril. *J Org Chem.* 1986; **51**: 4440–4446.

43. Mock WL, Shih N-Y. Organic ligand–receptor interactions between cucurbituril and alkylammonium ions. *J Am Chem Soc.* 1988; **110**: 4706–4710.

44. Mock WL, Shih N-Y. Dynamics of molecular recognition involving cucurbituril. *J Am Chem Soc.* 1989; **111**: 2697–2699.

45. Marquez C, Nau WM. Two mechanisms of slow host-guest complexation between cucurbit[6]uril and cyclohexylmethylamine: pH-responsive supramolecular kinetics. *Angew Chem Int Ed.* 2001; **40**: 3155–3160.

46. Buschmann H-J, Jansen K, Schollmeyer E. The formation of cucurbituril complexes with amino acids and amino alcohols in aqueous formic acid studied by calorimetric titrations. *Thermochim Acta.* 1998; **317**: 95–98.

47. Buschmann H-J, Schollmeyer E, Mutihac L. The formation of amino acid and dipeptide complexes with α-cyclodextrin and cucurbit[6]uril in aqueous solutions studied by titration calorimetry. *Thermochim Acta.* 2003; **399**: 203–208.

48. Buschmann H-J, Mutihac L, Mutihac R-C, Schollmeyer E. Complexation behavior of cucurbit[6]uril with short polypeptides. *Thermochim Acta.* 2005; **430**: 79–292.

49. Buschmann H-J, Jansen K, Schollmeyer E. The complex formation of α,ω-dicarboxylic acids and α,ω-diols with cucurbituril and a-cyclodextrin: the first step to the formation of rotaxanes and polyrotaxenes of the polyester type. *Acta Chim Slov.* 1999; **46**: 405–411.

50. Jansen K, Buschmann H-J, Zliobaite E, Schollmeyer E. Steric factors influencing the complex formation with cucurbit[6]uril. *Thermochim Acta.* 2002; **385**: 177–184.

51. Buschmann H-J, Mutihac L, Schollmeyer E. Complex formation of cucurbit[6]uril with amines in the presence of different salts. *J Incl Phenom Macrocyc Chem.* 2008; **61**: 343–346.

52. Buschmann H-J, Mutihac L, Schollmeyer E. Complex formation of crown ethers with the cucurbit[6]uril–spermidine and cucurbit[6]uril–spermine complex in aqueous solution. *J Incl Phenom Macrocyc Chem.* 2005; **53**: 85–88.

53. Rekharsky MV, Yamamura H, Kawai M, Osaka I, Arakawa R, Sato A, Ko YH, Selvapalam N, Kim K, Inoue Y. Sequential formation of a ternary complex among dihexylammonium, cucurbit[6]uril, and cyclodextrin with positive cooperativity. *Org Lett.* 2006; **8**: 815–818.

54. Buschmann H-J, Mutihac L, Schollmeyer E. The formation of homogeneous and heterogeneous 2:1 complexes between dialkyl- and diarylammonium ions and α-cyclodextrin and cucurbit[6]uril in aqueous formic acid. *Thermochim Acta.* 2009; **495**: 28–32.

55. Buschmann H-J, Jansen K, Schollmeyer E. Cucurbituril and α- and β-cyclodextrins as ligands for the complexation of nonionic surfactants and polyethyleneglycols in aqueous solutions. *J Incl Phenom Macrocyc Chem.* 2000; **37**: 231–236.

56. Buschmann H-J, Wego A, Jansen K, Schollmeyer E, Dopp D. Polyethylene glycol as string for the simultaneous complexation of a-cyclodextrin and cucurbit[6]uril. *J Incl Phenom Macrocyc Chem.* 2005; **53**: 183–189.

57. Buschmann H-J, Mutihac L, Jansen K, Schollmeyer E. Cucurbit[6]uril as ligand for the complexation of diamines, diazacrown ethers and cryptands in aqueous formic acid. *J Incl Phenom Macrocyc Chem.* 2005; **53**: 281–284.

58. Dantz DA, Meschke C, Buschmann H-J, Schollmeyer E. Complexation of volatile organic molecules from the gas phase with cucurbituril and β-cyclodextrin. *Supramol Chem.* 1998; **9**: 79–83.

59. Buschmann H-J, Meschke C, Schollmeyer E. Formation of pseudorotaxanes between cucurbituril and some 4,4′-bipyridine derivatives. *An Quim Int Ed.* 1998; **94**: 241–243.

60. Buschmann H-J, Schollmeyer E. Cucurbituril and α-cyclodextrin as hosts for the complexation of organic dyes. *J Inclus Phenom Mol Recog Chem.* 1997; **29**: 167–174.

61. Neugebauer R, Knoche W. Host–guest complexes of cucurbituril with 4-amino-4′-nitroazobenzene and 4,4′-diaminoazobenzene in acidic aqueous solutions. *J Chem Soc. Perkin Trans.* 1998; **II**: 529–534.

62. Wagner BD, Fitzpatrick SJ, Gill MA, MacRae AI, Stojanovic N. A fluorescent host–guest complex of cucurbituril in solution: a molecular Jack O'Lantern. *Can J Chem.* 2001; **79**: 1101–1104.

63. Wagner BD, MacRae AI. The lattice inclusion compound of 1,8-ANS and cucurbituril: a unique fluorescent solid. *J Phys Chem B.* 1999; **103**: 10114–10119.

64. Xiao X, Zhang Y-Q, Tao Z, Xue S-F, Zhu Q-J. Bis(cucurbit[6]uril) bis(hexane-1,6-diyldipyridinium) tetrabromide tridecahydrate. *Acta Cryst.* 2007; **E63**: o389–o391.

65. Gao Y, He X-Y, Wang Z-B, Zhang F, Li Y-Z, Chen H-L. A new type of pseudorotaxanes based on cucurbit[6]uril and biscyanopyridyl alkane compounds. *Supramol Chem.* 2009; **21**: 699–706.

66. Yuan L, Wang R, Macartney DH. Binding modes of cucurbit[6]uril and cucurbit[7]uril with a tetracationic bis(viologen) guest. *J Org Chem.* 2007; **72**: 4539–4542.

67. Lu H, Mei L, Zhang G, Zhou X. Interaction between cucurbit[6]uril and bispyridinecarboxamide. *J Inclus Phenom Macro Chem.* 2007; **59**: 81–90.

68. Wang Z-B, Zhao M, Li Y-Z, Chen H-L. Metal ion-assisted assembly of one-dimensional polyrotaxanes incorporating cucurbit[6]uril. *Supramol Chem.* 2008; **20**: 689–696.

69. Huang W-H, Zavalij PY, Isaacs L. Cucurbit[6]uril p-phenylenediammonium diiodide decahydrate inclusion complex. *Acta Cryst.* 2007; **E63**: o1060–o1062.

70. Huang W-H, Zavalij PY, Isaacs L. Cucurbit[6]uril p-xylylenediammonium diiodide decahydrate inclusion complex. *Acta Cryst.* 2007; **E64**: o1321–o1322.

71. Dearden DV, Ferrell TA, Asplund MC, Zilch LW, Julian RR, Jarrold MF. One ring to bind them all: shape-selective complexation of phenylenediamine isomers with cucurbit[6]uril in the gas phase. *J Phys Chem A.* 2009; **113**: 989–997.

72. Liu L, Zhao N, Scherman OA. Ionic liquids as novel guests for cucurbit[6]uril in neutral water. *J Chem Soc Chem Commun.* 2008; 1070–1072.

73. Kolman V, Marek R, Strelcova Z, Kulhanek P, Necas M, Svec J, Sindelar V. Electron density shift in imidazolium derivatives upon complexation with cucurbit[6]uril. *Chem Eur J.* 2009; **15**: 6926–6931.

74. El Haouaj M, Luhmer M, Ko YH, Kim K, Bartik K. NMR study of the reversible complexation of xenon by cucurbituril. *J Chem Soc. Perkin Trans.* 2001; **II**: 804–807.

75. Zhang H, Paulsen ES, Walker KA, Krakowiak KE, Dearden DV. Cucurbit[6]uril pseudorotaxanes: distinctive gas-phase dissociation and reactivity. *J Am Chem Soc.* 2003; **125**: 9284–9285.

76. Fusaro L, Locci E, Lai A, Luhmer M. On the binding of SF6 to cucurbit[6]uril host: density functional investigations. *J Phys Chem B.* 2008; **112**: 15014–15020.

77. Pinjari RV, Gejji SP. NMR study of the reversible trapping of SF6 by cucurbit[6]uril in aqueous solution. *J Phys Chem A.* 2010; **114**: 2338–2343.

78. Lim S, Kim H, Selvapalam N, Kim K-J, Cho SJ, Seo G, Kim K. Cucurbit[6]uril: organic molecular porous material with permanent porosity, exceptional stability, and acetylene sorption properties. *Angew Chem Int Ed.* 2008; **47**: 3352–3355.

79. Huang X, Tan Y, Wang Y, Yang H, Cao J, Che Y. Synthesis, characterization, and properties of copolymer of acrylamide and complex pseudorotaxane monomer consisting of cucurbit[6]uril with butyl ammonium methacrylate. *J Polym Sci A: Polym Chem.* 2008; **46**: 5999–6008.

80. Berbeci LS, Wang W, Kaifer AE. Drastically decreased reactivity of thiols and disulfides complexed by cucurbit[6]uril. *Org Lett.* 2008; **10**: 3721–3724.

81. Kim H-J, Jeon W-S, Ko YH, Kim K. Inclusion of methylviologen in cucurbit[7]uril. *Proc Nat Acad Sci.* 2002; **99**: 5007–5011.

82. Ong W, Gomez-Kaifer M, Kaifer AE. Cucurbit[7]uril: a very effective host for viologens and their cation radicals. *Org Lett.* 2002; **4**: 1791–1794.

83. Ong W, Kaifer AE. Salt effects on the apparent stability of the cucurbit[7]uril-methyl viologen inclusion complex. *J Org Chem.* 2004; **69**: 1383–1385.

84. Moon K, Kaifer AE. Modes of binding interaction between viologen guests and the cucurbit[7]uril host. *Org Lett.* 2004; **6**: 185–188.

85. Ong W, Kaifer AE. Molecular encapsulation by cucurbit[7]uril of the apical 4,4′-bipyridinium residue in Newkome-type dendrimers. *Angew Chem Int Ed.* 2003; **42**: 2164–2167.

86. Sindelar V, Moon K, Kaifer AE. Binding selectivity of cucurbit[7]uril: bis(pyridinium)-1,4-xylylene versus 4,4′-bipyridinium guest sites. *Org Lett.* 2004; **6**: 2665–2668.

87. Wagner BD, Stojanovic N, Day AI, Blanch RJ. Host properties of cucurbit[7]uril: fluorescence enhancement of anilinonaphthalene sulfonates. *J Phys Chem B.* 2003; **107**: 10741–10746.

88. Choi S, Park SH, Ziganshina AY, Ko YH, Lee JW, Kim K. A stable cis-stilbene derivative encapsulated in cucurbit[7]uril. *J Chem Soc Chem Commun.* 2003; 2176–2177.

89. Wu J, Isaacs L. Cucurbit[7]uril complexation drives thermal trans–cis-azobenzene isomerization and enables colorimetric amine detection. *Chem Eur J.* 2009; **15**: 11675–11680.

90. Sobransingh D, Kaifer AE. Binding interactions between the host cucurbit[7]uril and dendrimer guests containing a single ferrocenyl residue. *J Chem Soc Chem Commun.* 2005; 5071–5073.

91. Sobransingh D, Kaifer AE. New dendrimers containing a single cobaltocenium unit covalently attached to the apical position of newkome dendrons: electrochemistry and guest binding interactions with cucurbit[7]uril. *Langmuir.* 2006; **22**: 10540–10544.

92. Wang R, Yuan L, Macartney DH. Stabilization of the (E)-1-ferrocenyl-2-(1-methyl-4-pyridinium)ethylene cation by inclusion in cucurbit[7]uril. *Organometallics.* 2006; **25**: 1820–1823.

93. Yuan L, Macartney DH. Kinetics of the electron self-exchange and electron-transfer reactions of the (trimethylammonio)methylferrocene host–guest complex with cucurbit[7]uril in aqueous solution. *J Phys Chem B.* 2007; **111**: 6949–6954.

94. Cui L, Gadde S, Li W, Kaifer AE. Electrochemistry of the inclusion complexes formed between the cucurbit[7]uril host and several cationic and neutral ferrocene derivatives. *Langmuir.* 2009; **25**: 13763–13769.

95. Feng K, Wu L-Z, Zhang L-P, Tung C-H. Cucurbit[7]uril-included neutral intramolecular charge-transfer ferrocene derivatives. *J Chem Soc. Dalton Trans.* 2007; 3991–3994.

96. Ong W, Kaifer AE. Unusual electrochemical properties of the inclusion complexes of ferrocenium and cobaltocenium with cucurbit[7]uril. *Organometallics.* 2003; **22**: 4181–4183.

97. Ling Y, Kaifer AE. Complexation of poly(phenylenevinylene) precursors and monomers by cucurbituril hosts. *Chem Mater.* 2006; **18**: 5944–5949.

98. St-Jacques AD, Wyman IW, Macartney DH. Encapsulation of charge-diffuse peralkylated onium cations in the cavity of cucurbit[7]uril. *J Chem Soc Chem Commun.* 2008; 4936–4938.

99. Sindelar V, Cejas MA, Raymo FM, Kaifer AE. Tight inclusion complexation of 2,7-dimethyldiazapyrenium in cucurbit[7]uril. *New J Chem.* 2005; **29**: 280–282.

100. Noujeim N, Jouvelet B, Schmitzer AR. Formation of inclusion complexes between 1,1′-dialkyl-3,3′-(1,4-phenylene)bisimidazolium dibromide salts and cucurbit[7]uril. *J Phys Chem B.* 2009; **113**: 16159–16168.

101. Leclercq L, Noujeim N, Sanon SH, Schmitzer AR. Study of the supramolecular cooperativity in the multirecognition mechanism of cyclodextrins/cucurbituril/disubstituted diimidazolium bromides. *J Phys Chem B.* 2008; **112**: 14176–14184.

102. Samsam S, Leclercq L, Schmitzer AR. Recognition of 1,4-xylylene binding sites in polyimidazolium cations by cucurbit[7]uril: toward pseudorotaxane assembly. *J Phys Chem B.* 2009; **113**: 9493–9498.

103. Thangavel A, Rawashdeh AMM, Sotiriou-Leventis C, Leventis N. Simultaneous electron transfer from free and intercalated 4-benzoylpyridinium cations in cucurbit[7]uril. *Org Lett.* 2009; **11**: 1595–1598.

104. Miskolczy Z, Biczok L, Megyesi M, Jablonkai I. Inclusion complex formation of ionic liquids and other cationic organic compounds with cucurbit[7]uril studied by 4′,6-diamidino-2-phenylindole fluorescent probe. *J Phys Chem B.* 2009; **113**: 1645–1651.

105. Zhou Y, Yu H, Zhang L, Xu H, Wu L, Sun J, Wang L. A new spectrofluorometric method for the determination of nicotine base on the inclusion interaction of methylene blue and cucurbit[7]uril. *Microchim Acta.* 2009; **164**: 63–68.

106. Sindelar V, Parker SE, Kaifer AE. Inclusion of anthraquinone derivatives by the cucurbit[7]uril host. *New J Chem.* 2007; **31**: 725–728.

107. Wyman IW, Macartney DH. Host–guest complexes and pseudorotaxanes of cucurbit[7]uril with acetylcholinesterase inhibitors. *J Org Chem.* 2009; **74**: 8031–8038.

108. Wyman IW, Macartney DH. Cucurbit[7]uril host–guest and pseudorotaxane complexes with a,x-bis(pyridinium)alkane dications. *Org Biomol Chem.* 2009; **7**: 4045–4051.

109. Eelkema R, Maeda K, Odell B, Anderson HL. Radical cation stabilization in a cucurbituril oligoaniline rotaxane. *J Am Chem Soc.* 2007; **129**: 12384–12385.

110. Yin J, Chi C, Wu J. Efficient preparation of separable pseudo[n]rotaxanes by selective threading of oligoalkylammonium salts with cucurbit[7]uril. *Chem Eur J.* 2009; **15**: 6050–6057.

111. Koner AL, Nau WM. Cucurbituril encapsulation of fluorescent dyes. *Supramol Chem.* 2006; **19**: 55–66.

112. Mohanty J, Nau WM. Ultrastable rhodamine with cucurbituril. *Angew Chem Int Ed.* 2005; **44**: 3750–3754.

113. Halterman RL, Moore JL, Mannel LM. Disrupting aggregation of tethered rhodamine B dyads through inclusion in cucurbit[7]uril. *J Org Chem.* 2008; **73**: 3266–3269.

114. Gadde S, Batchelor EK, Weiss JP, Ling Y, Kaifer AE. Control of H- and J-aggregate formation via host–guest complexation using cucurbituril hosts. *J Am Chem Soc.* 2008; **130**: 17114–17119.

115. Gadde S, Batchelor EK, Kaifer AE. Controlling the formation of cyanine dye H- and J-aggregates with cucurbituril hosts in the presence of anionic polyelectrolytes. *Chem Eur J.* 2009; **15**: 6025–6031.

116. Mohanty J, Bhasikuttan AC, Choudhury SD, Pal H. Noncovalent interaction of 5,10,15,20-tetrakis(4-N-methylpyridyl)porphyrin with cucurbit[7]uril: a supramolecular architecture. *J Phys Chem B.* 2008; **112**: 10782–10785.

117. Shaikh M, Mohanty J, Bhasikuttan AC, Uzunova VD, Nau WM, Pal H. Salt-induced guest relocation from a macrocyclic cavity into a biomolecular pocket: interplay between cucurbit[7]uril and albumin. *J Chem Soc Chem Commun.* 2008; 3681–3683.

118. Zhou Y, Yu H, Zhang L, Sun J, Wu L, Lu Q, Wang L. Host properties of cucurbit[7]uril: fluorescence enhancement of acridine orange. *J Incl Phenom Macrocyc Chem.* 2008; **61**: 259–264.

119. Liu J, Jiang N, Ma J, Du X. Insight into unusual downfield NMR shifts in the inclusion complex of acridine orange with cucurbit[7]uril. *Eur J Org Chem.* 2009; 4931–4938.

120. Miskolczy Z, Biczók L, Görner H. Tautomerization of lumichrome promoted by supramolecular complex formation with cucurbit[7]uril. *J Photochem Photobiol A: Chem.* 2009; **207**: 47–51.

121. Petrov N, Ivanov DA, Golubkov DV, Gromov SP, Alfimov MV. The effect of cucurbit[7]uril on photophysical properties of aqueous solution of 3,3′-diethylthiacarbocyanine iodide dye. *Chem Phys Lett.* 2009; **480**: 96–99.

122. Gonzalez-Bejar M, Montes-Navajas P, García H, Scaiano JC. Methylene blue encapsulation in cucurbit[7]uril: laser flash photolysis and near-IR luminescence studies of the interaction with oxygen. *Langmuir.* 2009; **25**: 10490–10494.

123. Montes-Navajas P, García H. Complexes of basic tricyclic dyes in their acid and basic forms with cucurbit[7]uril: determination of pKa and association constants in the ground and singlet excited state. *J Photochem Photobiol A: Chem.* 2009; **204**: 97–101.

124. Fedorova OA, Chernikova EY, Fedorov YV, Gulakova EN, Peregudov AS, Lyssenko KA, Jonusauskas G, Isaacs L. Cucurbit[7]uril complexes of crown-ether derived styryl and (bis)styryl dyes. *J Phys Chem B.* 2009; **113**: 10149–10158.

125. Zeng Y, Li Y, Li M, Yang G, Li Y. Enhancement of energy utilization in light-harvesting dendrimers by the pseudorotaxane formation at periphery. *J Am Chem Soc.* 2009; **131**: 9100–9106.

126. Choudhury SD, Mohanty J, Pal H, Bhasikuttan AC. Cooperative metal ion binding to a cucurbit[7]uril–thioflavin T complex: demonstration of a stimulus-responsive fluorescent supramolecular capsule. *J Am Chem Soc.* 2010; **132**: 1395–1401.

127. Marquez C, Nau WM. Polarizabilities inside molecular containers. *Angew Chem Int Ed.* 2001; **40**: 4387–4390.

128. Mohanty J, Nau WM. Refractive index effects on the oscillator strength and radiative decay rate of 2,3-diazabicyclo[2.2.2]oct-2-ene. *Photochem Photobio Sci.* 2004; **3**: 1026–1031.

129. Marquez C, Pischel U, Nau WM. Selective fluorescence quenching of 2,3-diazabicyclo[2.2.2]oct-2-ene by nucleotides. *Org Lett.* 2003; **5**: 3911–3914.

130. Thuery P, Masci B. Uranyl ion complexation by cucurbiturils in the presence of perrhenic, phosphoric, or polycarboxylic acids: novel mixed-ligand uranyl-organic frameworks. *Cryst Growth Des.* 2010; **10**: 716–725.

131. Thuery P. Uranyl ion complexes of cucurbit[7]uril with zero-, one- and two-dimensionality. *Cryst Eng Comm.* 2009; **11**: 1150–1156.

132. Wyman IW, Macartney DH. Cucurbit[7]uril host–guest complexes with small polar organic guests in aqueous solution. *Org Biomol Chem.* 2008; **6**: 1796–1801.

133. Lorenzo S, Day A, Craig D, Blanch R, Arnold A, Dance I. The first endoannular metal halide–cucurbituril: cis-SnCl$_4$(OH$_2$)$_2$@cucurbit[7]uril. *Cryst Eng Comm.* 2001; **49**: 1–7.

134. Xu L, Liu S-M, Wu C-T, Feng Y-Q. Separation of positional isomers by cucurbit[7]uril-mediated capillary electrophoresis. *Electrophoresis.* 2004; **25**: 3300–3306.

135. Wheate NJ, Day AI, Blanch RJ, Arnold AP, Cullinane C, Collins JG. Multi-nuclear platinum complexes encapsulated in cucurbit[n]uril as an approach to reduce toxicity in cancer treatment. *J Chem Soc Chem Commun.* 2004; 1424–1425.

136. Kennedy AR, Florence AJ, McInnes FJ, Wheate NJ. A chemical preformulation study of a host–guest complex of cucurbit[7]uril and a multinuclear platinum agent for enhanced anticancer drug delivery. *J Chem Soc. Dalton Trans.* 2009; 7695–7700.

137. Jeon YJ, Kim S-Y, Ko YH, Sakamoto S, Yamaguchi K, Kim K. Novel molecular drug carrier: encapsulation of oxaliplatin in cucurbit[7]uril and its effects on stability and reactivity of the drug. *Org Biomol Chem.* 2005; **3**: 2122–2125.

138. Zhou Y, Sun J, Yu H, Wu L, Wang L. Inclusion complex of riboflavin with cucurbit[7]uril: study in solution and solid state. *Supramol Chem.* 2009; **495**: 501–509.

139. Wang R, MacGillivray BC, Macartney DH. Stabilization of the base-off forms of vitamin B12 and coenzyme B12 by encapsulation of the a-axial 5,6-dimethylbenzimidazole ligand with cucurbit[7]uril. *J Chem Soc. Dalton Trans.* 2009; 3584–3589.

140. Huang Y, Xue S-F, Zhu Q-J, Zhu T. Inclusion interactions of cucurbit[7]uril with adenine and its derivatives. *Supramol Chem.* 2008; **20**: 279–287.

141. Wang R, Macartney DH. Cucurbit[7]uril host–guest complexes of the histamine H2-receptor antagonist ranitidine. *Org Biomol Chem.* 2008; **6**: 1955–1960.

142. Buck DP, Abeysinghe PM, Cullinane C, Day AI, Collins JG, Harding MM. Inclusion complexes of the antitumour metallocenes Cp2MCl2 (M = Mo, Ti) with cucurbit[n]urils. *J Chem Soc. Dalton Trans.* 2008; 2328–2334.

143. Megyesi M, Biczok L, Jablonkai I. Highly sensitive fluorescence response to inclusion complex formation of berberine alkaloid with cucurbit[7]uril. *J Phys Chem C.* 2008; **112**: 3410–3416.

144. Li C, Li J, Jia X. Selective binding and highly sensitive fluorescent sensor of palmatine and dehydrocorydaline alkaloids by cucurbit[7]uril. *Org Biomol Chem.* 2009; **7**: 2699–2703.

145. Wang R, Wyman IW, Wang S, Macartney DH. Encapsulation of a B-carboline in cucurbit[7]uril. *J Incl Phenom Macrocyc Chem.* 2009; **64**: 233–237.

146. Wyman IW, Macartney DH. Cucurbit[7]uril host–guest complexes of cholines and phosphonium cholines in aqueous solution. *Org Biomol Chem.* 2010; **8**: 253–260.

147. McInnes FJ, Anthony NG, Kennedy AR, Wheate NJ. Solid state stabilisation of the orally delivered drugs atenolol, glibenclamide, memantine and paracetamol through their complexation with cucurbit[7]uril. *Org Biomol Chem.* 2010; **8**: 765–773.

148. Wyman IW, Macartney DH. Host–guest complexations of local anaesthetics by cucurbit[7]uril in aqueous solution. *Org Biomol Chem.* 2010; **8**: 247–252.

149. Wang R, Yuan L, Macartney DH. Inhibition of C(2)-H/D exchange of a bis(imidazolium) dication upon complexation with cucubit[7]uril. *J Chem Soc Chem Commun.* 2006; 2908–2910.

150. Wang R, Macartney DH. Cucurbit[7]uril stabilization of a diarylmethane carbocation in aqueous solution. *Tet Lett.* 2008; **49**: 311–314.

151. Kirilyuk I, Polovyanenko D, Semenov S, Grigorev I, Gerasko O, Fedin V, Bagryanskaya E. Inclusion complexes of nitroxides of pyrrolidine and imidazoline series with cucurbit[7]uril. *J Phys Chem B.* 2010; **114**: 1719–1728.

152. Klock C, Dsouza RN, Nau WM. Cucurbituril-mediated supramolecular acid catalysis. *Org Lett.* 2009; **11**: 2595–2598.

153. Hennig A, Ghale G, Nau WM. Effects of cucurbit[7]uril on enzymatic activity. *J Chem Soc Chem Commun.* 2007; 1614–1616.

154. Corma A, Garcia H, Montes-Navajas P, Primo A, Calvino JJ, Trasobares S. Gold nanoparticles in organic capsules: a supramolecular assembly of gold nanoparticles and cucurbituril. *Chem Eur J.* 2007; **13**: 6359–6364.

155. Blanch RJ, Sleeman AJ, White TJ, Arnold AP, Day AI. Cucurbit[7]uril and o-carborane self-assemble to form a molecular ball bearing. *Nano Lett.* 2002; **2**: 147–149.

156. Constabel F, Geckeler KE. Solvent-free self-assembly of C60 and cucurbit[7]uril using high-speed vibration milling. *Tet Lett.* 2004; **45**: 2071–2073.

157. Constabel F, Geckeler KE. Nanoencapsulation of [60]fullerene with the cavitand cucurbit[7]uril. *Fullerenes, Nanotubes and Carbon Nanostructures.* 2004; **12**: 811–818.

158. Ogoshi T, Inagaki A, Yamagishi T, Nakamoto Y. Defection-selective solubilization and chemically-responsive solubility switching of single-walled carbon nanotubes with cucurbit[7]uril. *J Chem Soc Chem Commun.* 2008; 2245–2247.

159. Mu TW, Liu L, Zhang KC, Guo QX. A theoretical study on the stereoisomerism in the complex of cucurbit[8]uril with 2, 6-bis(4, 5-dihydro-1himidazol-2-yl)naphthalene. *Chin Chem Lett.* 2001; **12**: 783–786.

160. Kim H-J, Heo J, Jeon WS, Lee E, Kim J, Sakamoto S, Yamaguchi K, Kim K. Selective inclusion of a hetero-guest pair in a molecular host: formation of stable charge-transfer complexes in cucurbit[8]uril. *Angew Chem Int Ed.* 2001; **40**: 1526–1529.

161. Jeon YJ, Bharadwaj PK, Choi SW, Lee JW, Kim K. Supramolecular amphiphiles: spontaneous formation of vesicles triggered by formation of a charge-transfer complex in a host. *Angew Chem Int Ed.* 2002; **41**: 4474–4476.

162. Lee JW, Kim K, Choi SW, Ko YH, Sakamoto S, Yamaguchi K, Kim K. Unprecedented host-induced intramolecular charge-transfer complex formation. *J Chem Soc Chem Commun.* 2002; 2692–2693.

163. Ko YH, Kim K, Kim E, Kim K. Exclusive formation of 1:1 and 2:2 complexes between cucurbit[8]uril and electron donor–acceptor molecules induced by host-stabilized chargetransfer interactions. *Supramol Chem.* 2007; **19**: 287–293.

164. Zou D, Andersson S, Zhang R, Sun S, Åkermark B, Sun L. A host-induced intramolecular charge-transfer complex and light-driven radical cation formation of a molecular triad with cucurbit[8]uril. *J Org Chem.* 2008; **73**: 3775–3783.

165. Ziganshina AY, Ko YH, Jeon WS, Kim K. Stable π-dimer of a tetrathiafulvalene cation radical encapsulated in the cavity of cucurbit[8]uril. *J Chem Soc Chem Commun.* 2004; 806–807.

166. Jeon WS, Kim H-J, Lee C, Kim K. Control of the stoichiometry in host–guest complexation by redox chemistry of guests: Inclusion of methylviologen in cucurbit[8]uril. *J Chem Soc Chem Commun.* 2002; 1828–1829.

167. Moon K, Grindstaff J, Sobransingh D, Kaifer AE. Cucurbit[8]uril-mediated redox-controlled self-assembly of viologen-containing dendrimers. *Angew Chem Int Ed.* 2004; **43**: 5496–5499.

168. Zhang T, Sun S, Liu F, Fan J, Pang Y, Sun L, Peng X. Redox-induced partner radical formation and its dynamic balance with radical dimer in cucurbit[8]uril. *Phys Chem Chem Phys.* 2009; **11**: 11134–11139.

169. Sun S, Andersson S, Zhang R, Sun L. Unusual partner radical trimer formation in a host complex of cucurbit[8]uril, ruthenium(II) tris-bipyridine linked phenol and methyl viologen. *Chem Commun.* 2010; **46**: 463–465.

170. Maddipatla MVSN, Kaanumalle LS, Natarajan A, Pattabiraman M, Ramamurthy V. Preorientation of olefins toward a single photodimer: cucurbituril-mediated photodimerization of protonated azastilbenes in water. *Langmuir.* 2007; **23**: 7545–7554.

171. Lei L, Luo L, Wu X-L, Liao G-H, Wu L-Z, Tung C-H. Cucurbit[8]uril-mediated photodimerization of alkyl 2-naphthoate in aqueous solution. *Tet Lett.* 2008; **49**: 1502–1505.

172. Wu X-L, Luo L, Lei L, Liao G-H, Wu L-Z, Tung C-H. Highly efficient cucurbit[8]uril-templated intramolecular photocycloaddition of 2-naphthalene-labeled poly(ethylene glycol) in aqueous solution. *J Org Chem.* 2008; **73**: 491–494.

173. Wang R, Bardelang D, Waite M, Udachin KA, Leek DM, Yu K, Ratcliffe CI, Ripmeester JA. Inclusion complexes of coumarin in cucurbiturils. *Org Biomol Chem.* 2009; **7**: 2435–2439.

174. Pemberton BC, Barooah N, Srivatsava DK, Sivaguru J. Supramolecular photocatalysis by confinement-photodimerization of coumarins within cucurbit[8]urils. *J Chem Soc Chem Commun.* 2010; **46**: 225–227.

175. Feng Y, Xue S-F, Fan Z-F, Zhang Y-Q, Zhu Q-J, Tao Z. Host–guest complexes of some cucurbit[n]urils with the hydrochloride salts of some imidazole derivatives. *J Incl Phenom Macrocyc Chem.* 2009; **64**: 121–131.

176. Fu H, Xue S, Mu L, Du Y, Zhu Q, Tao Z, Zhang J, Day A. Host–guest complexes of cucubit[8]uril with phenanthrolines and some methyl derivatives. *Sci China Ser B Chem.* 2005; **48**: 305–314.

177. Kemp S, Wheate NJ, Wang S, Collins JG, Ralph SF, Day AI, Higgins VJ, Aldrich-Wright JR. Encapsulation of platinum(II)-based DNA intercalators within cucurbit[6,7,8]urils. *J Biol Inorg Chem.* 2007; **12**: 969–979.

178. Wang R, Yuan L, Ihmels H, Macartney DH. Cucurbit[8]uril/cucurbit[7]uril controlled off/on fluorescence of the acridizinium and 9-aminoacridizinium cations in aqueous solution. *Chem Eur J.* 2007; **13**: 6468–6473.

179. Rajgariah P, Urbach AR. Scope of amino acid recognition by cucurbit[8]uril. *J Incl Phenom Macrocyc Chem.* 2008; **62**: 251–254.

180. Nguyen HD, Dang DT, van Dongen JLJ, Brunsveld L. Protein dimerization induced by supramolecular interactions with cucurbit[8]uril. *Angew Chem Int Ed.* 2010; **49**: 895–898.

181. Chakrabarti S, Isaacs L. Cucurbit[8]uril controls the folding of cationic diaryl ureas in water. *Supramol Chem.* 2008; **20**: 191–199.

182. Jayaraj N, Porel M, Ottaviani MF, Maddipatla MVSN, Modelli A, Da Silva JP, Bhogala BR, Captain B, Jockusch S, Turro NJ, Ramamurthy V. Self aggregation of supramolecules of nitroxides@cucurbit[8]uril revealed by EPR spectra. *Langmuir.* 2009; **25**: 13820–13832.

183. Cong H, Zhao F-F, Zhang J-X, Zeng X, Tao Z, Xue S-F, Zhu Q-J. Rapid transformation of benzylic alcohols to aldehyde in the presence of cucurbit[8]uril. *Catal Comm.* 2009; **11**: 167–170.

184. Wang W, Kaifer AE. Electrochemical switching and size selection in cucurbit[8]uril-mediated dendrimer self-assembly. *Angew Chem Int Ed.* 2006; **45**: 7042–7046.

185. Rauwald U, Scherman OA. Supramolecular block copolymers with cucurbit[8]uril in water. *Angew Chem Int Ed.* 2008; **47**: 3950–3953.

186. Deroo S, Rauwald U, Robinson CV, Scherman OA. Discrete, multi-component complexes with cucurbit[8]uril in the gas-phase. *J Chem Soc Chem Commun.* 2009; 644–646.

187. Jeon WS, Ziganshina AY, Lee JW, Ko YH, Kang J-K, Lee C, Kim K. A [2]pseudorotaxane-based molecular machine: reversible formation of a molecular loop driven by electrochemical and photochemical stimuli. *Angew Chem Int Ed.* 2003; **42**: 4097–4100.

188. Mu L, Yang X-B, Xue S-F, Zhua Q-J, Tao Z, Zeng X. Cucurbit[n]urils-induced room temperature phosphorescence of quinoline derivatives. *Anal Chim Acta.* 2007; **597**: 90–96.

189. Kuzmina LG, Vedernikov AI, Lobova NA, Howard JAK, Strelenko YA, Fedin VP, Alfimov MV, Gromov SP. Photoinduced and dark complexation of unsaturated viologen analogues containing two ammonium tails with cucurbit[8]uril. *New J Chem.* 2006; **30**: 458–466.

190. Kemp S, Wheate NJ, Stootman FH, Aldrich-Wright JR. The host–guest chemistry of proflavine with cucurbit[6,7,8]urils. *Supramol Chem.* 2007; **19**: 475–484.

191. Mohanty J, Choudhury SD, Upadhyaya HP, Bhasikuttan AC, Pal H. Control of the supramolecular excimer formation of thioflavin T within a cucurbit[8]uril host: a fluorescence on/off mechanism. *Chem Eur J.* 2009; **15**: 5215–5219.

192. Sokolov MN, Mitkina TV, Gerasko OA, Fedin VP, Virovets AV, Llusar R. Coordination of bimuth(III) to cucurbit[8]uril: preparation and X-ray structure of [{Bi(NO$_3$)(H$_2$O)$_5$}$_2$(Q8)][Bi(NO$_3$)$_3$(H$_2$O)$_4$]$_2$[Bi(NO$_3$)$_5$]$_2$·Q8·19H$_2$O. *Z Anorg Allg Chem.* 2003; **629**: 2440–2442.

193. Mitkina TV, Sokolov MN, Naumov DY, Kuratieva NV, Gerasko OA, Fedin VP. Jørgensen complex within a molecular container: selective encapsulation of trans-[Co(en)2Cl2]+ into cucurbit[8]uril and influence of inclusion on guest's properties. *Inorg Chem.* 2006; **45**: 6950–6955.

194. Kim S-Y, Jung I-S, Lee E, Kim J, Sakamoto S, Yamaguchi K, Kim K. Macrocycles within macrocycles: cyclen, cyclam, and their transition metal complexes encapsulated in cucurbit[8]uril. *Angew Chem Int Ed.* 2001; **40**: 2119–2121.

195. Mitkina TV, Yu D, Naumov, Gerasko OA, Dolgushin FM, Vicent C, Llusar R, Sokolov MN, Fedin VP. Inclusion of nickel(II) and copper(II) complexes with aliphatic polyamines in cucurbit[8]uril. *Russ Chem Bull Int Ed.* 2004; **53**: 2519–2524.

196. Hart SL, Haines RI, Decken A, Wagner BD. Isolation of the trans-I and trans-II isomers of CuII(cyclam) via complexation with the macrocyclic host cucurbit[8]uril. *Inorg Chim Acta.* 2009; **362**: 4145–4151.

197. Liu S, Shukla AD, Gadde S, Wagner BD, Kaifer AE, Isaacs L. Ternary complexes comprising cucurbit[10]uril, porphyrins, and guests. *Angew Chem Int Ed.* 2008; **47**: 2657–2660.

198. Pisani MJ, Zhao Y, Wallace L, Woodward CE, Keene FR, Day AI, Collins JG. Cucurbit[10]uril binding of dinuclear platinum(II) and ruthenium(II)complexes: association/dissociation rates from seconds to hours. *J Chem Soc. Dalton Trans.* 2010; **39**: 2078–2086.

199. Flinn A, Hough GC, Stoddart JF, Williams DJ. Decamethylcucurbit[5]uril. *Angew Chem Int Ed.* 1992; **31**: 1475–1477.

200. Zhang XX, Krakowiak KE, Xue G, Bradshaw JS, Izatt RM. A highly selective compound for lead: complexation studies of decamethylcucurbit[5]uril with metal ions. *Ind Eng Chem Res.* 2000; **39**: 3516–3520.

201. Kellersberger KA, Anderson JD, Ward SM, Krakowiak KE, Dearden DV. Encapsulation of N$_2$, O$_2$, methanol, or acetonitrile by decamethylcucurbit[5]uril(NH$_4^+$)$_2$ complexes in the gas phase: influence of the guest on 'lid' tightness. *J Am Chem Soc.* 2001; **123**: 11316–11317.

202. Miyahara Y, Abe K, Inazu T. Molecular sieves: lid-free decamethylcucurbit[5]uril absorbs and desorbs gases selectively. *Angew Chem Int Ed.* 2002; **41**: 3020–3023.

203. Lin J, Zhang Y, Zhang J, Xue S, Zhu Q, Tao Z. Synthesis of partially methyl substituted cucurbit[n]urils with 3a-methyl-glycoluril. *J Mol Struct.* 2008; **875**: 442–446.

204. Lu L-B, Yu D-H, Zhang Y-Q, Zhu Q-J, Xue S-F, Tao Z. Supramolecular assemblies based on some new methyl-substituted cucurbit[5]urils through hydrogen bonding. *J Mol Struct.* 2008; **885**: 70–75.

205. Zhou F-G, Wua L-H, Lu X-J, Zhang Y-Q, Zhu Q-J, Xue S-F, Tao Z. Molecular capsules based on methyl-substituted cucurbit[5]urils and strontium-capped. *J Mol Struct.* 2009; **927**: 14–20.

206. Sasmal S, Sinha MK, Keinan E. Facile purification of rare cucurbiturils by affinity chromatography. *Org Lett.* 2004; **6**: 1225–1228.

207. Lu L-B, Zhang Y-Q, Zhu Q-J, Xue S-F, Tao Z. Synthesis and X-ray structure of the inclusion complex of dodecamethylcucurbit[6]uril with 1,4-dihydroxybenzene. *Molecules.* 2007; **12**: 716–722.

208. Isobe H, Sato S, Nakamura E. Synthesis of disubstituted cucurbit[6]uril and its rotaxane derivative. *Org Lett.* 2002; **4**: 1287–1289.

209. Zhou Y, Xue S-F, Zhu Q-J, Tao Z, Zhang J-X, Wei Z, Long L, Hu M, Ziao H-P, Day A. Synthesis of a symmetrical tetrasubstituted cucurbit[6]uril and its host–guest inclusion complex with 2,2′-bipyridine. *Chin Sci Bull.* 2004; **49**: 1111–1116.

210. Cong H, Tao L-L, Yu Y-H, Tao Z, Yang F, Zhao Y-J, Xue S-F, Lawrance GA, Wei G. Interaction between tetramethylcucurbit[6]uril and some pyridine derivates. *J Phys Chem A.* 2007; **111**: 2715–2721.

211. Feng Y, Xiao X, Xue S-F, Zhang Y-Q, Zhu Q-J, Tao Z, Lawrance GA, Wei G. Host–guest complex of a water-soluble cucurbit[6]uril derivative with the hydrochloride salt of 3-amino-5-phenylpyrazole. *Supramol Chem.* 2008; **20**: 517–525.

212. Xiao X, Zhang Y-Q, Zhu Q-J, Xue S-F, Tao Z. Host–guest complexes of a water soluble cucurbit[6]urilderivative with some dications of 1,ω-alkyldipyridines: ¹H NMR and X-ray structures. *Sci China Ser B: Chem.* 2009; **52**: 475–482.

213. Yi J-M, Zhang Y-Q, Cong H, Xue S-F, Tao Z. Crystal structures of four host–guest inclusion complexes of a,a0,d,d0-tetramethylcucurbit[6]uril and cucurbit[8]uril with some L-amino acids. *J Mol Struct.* 2009; **933**: 112–117.

214. Zheng L-M, Zhang Y-Q, Zeng J-P, Qiu Y, Yu D-H, Xue S-F, Zhu Q-J, Tao Z. Structure of supramolecular assemblies formed by α,δ-tetramethylcucurbit[6]uril and 4-nitrophenol. *Molecules.* 2008; **13**: 2814–2822.

215. Zhao J, Kim H-J, Oh J, Kim S-Y, Lee JW, Sakamoto S, Yamaguchi K, Kim K. Cucurbit[n]uril derivatives soluble in water and organic solvents. *Angew Chem Int Ed.* 2001; **40**: 4233–4235.

216. Lagona J, Fettinger JC, Isaacs L. Cucurbit[n]uril analogues. *Org Lett.* 2003; **5**: 3745–3747.

217. Zheng L, Zhu J, Zhang Y-Q, Zhu Q-J, Xue S-F, Tao Z, Zhang J-X, Zhou X, Wei Z, Long L, Day AI. Opposing substitution in cucurbit[6]urils forms ellipsoid cavities: the symmetrical dicyclohexanocucurbit[6]uril is no exception highlighted by inclusion and exclusion complexes. *Supramol Chem.* 2008; **20**: 709–716.

218. Ni X-L, Zhang Y-Q, Zhu Q-J, Xue S-F, Tao Z. Crystal structures of host–guest complexes of meta-tricyclohexyl cucurbit[6]uril with small organic molecules. *J Mol Struct.* 2008; **876**: 322–327.

219. Wu L-H, Ni X-L, Wu F, Zhang Y-Q, Zhu Q-J, Xue S-F, Tao Z. Crystal structures of three partially cyclopentano-substituted cucurbit[6]urils. *J Mol Struct.* 2009; **920**: 183–188.

220. Lagona J, Fettinger JC, Isaacs L. Cucurbit[n]uril analogues. *Org Lett.* 2003; **5**: 3745–3747.

221. Wagner BD, Boland PG, Lagona J, Isaacs L. A cucurbit[6]uril analogue: host properties monitored by fluorescence spectroscopy. *J Phys Chem B.* 2005; **109**: 7686–7691.

222. Jon SY, Selvapalam N, Oh DH, Kang J-K, Kim S-Y, Jeon YJ, Lee JW, Kim K. Facile synthesis of cucurbit[n]uril derivatives via direct functionalization: expanding utilization of cucurbit[n]uril. *J Am Chem Soc.* 2003; **125**: 10186–10187.

223. Jeon YJ, Kim H, Jon S, Selvapalam N, Oh DH, Seo I, Park C-S, Jung SR, Koh D-S, Kim K. Artificial ion channel formed by cucurbit[n]uril derivatives with a carbonyl group fringed portal reminiscent of the selectivity filter of K⁺ channels. *J Am Chem Soc.* 2004; **126**: 15944–15945.

224. Isaacs L, Park S-K, Liu S, Ko YH, Selvapalam N, Kim Y, Kim H, Zavalij PY, Kim G-H, Lee H-S, Kim K. The inverted cucurbit[n]uril family. *J Am Chem Soc.* 2005; **127**: 18000–18001.

225. Liu S, Kim K, Isaacs L. Mechanism of the conversion of inverted CB[6] to CB[6]. *J Org Chem.* 2007; **72**: 6840–6847.

226. Huang W-H, Zavalij PY, Isaacs L. Chiral recognition inside a chiral cucurbituril. *Angew Chem Int Ed.* 2007; **46**: 7425–7427.

227. Huang W-H, Liu S, Zavalij PY, Isaacs L. Nor-seco-cucurbit[10]uril exhibits homotropic allosterism. *J Am Chem Soc.* 2006; **128**: 14744–14745.

228. Nally R, Isaacs L. Toward supramolecular polymers incorporating double cavity cucurbituril hosts. *Tetrahedron.* 2009; **65**: 7249–7258.

229. Liu S, Ruspic C, Mukhopadhyay P, Chakrabarti S, Zavalij PY, Isaacs L. The Cucurbit[n]uril family: prime components for self-sorting systems. *J Am Chem Soc.* 2005; **127**: 15959–15967.

230. Mukhopadhyay P, Wu A, Isaacs L. Social self-sorting in aqueous solution. *J Org Chem.* 2004; **69**: 6157–6164.

231. Celtek G, Artar M, Scherman OA, Tuncel D. Sequence-specific self-sorting of the binding sites of a ditopic guest by cucurbituril homologues and subsequent formation of a hetero[4]pseudorotaxane. *Chem Eur J.* 2009; **15**: 10360–10363.

232. Masson E, Lu X, Ling X, Patchell DL. Kinetic vs thermodynamic self-sorting of cucurbit[6]uril, cucurbit[7]uril, and a spermine derivative. *Org Lett.* 2009; **11**: 3798–3801.

233. Liu Y, Li X-Y, Zhang H-Y, Li C-J, Ding F. Cyclodextrin-driven movement of cucurbit[7]uril. *J Org Chem.* 2007; **72**: 3640–3645.

234. Andersson S, Zou D, Zhang R, Sun S, Sun L. Light driven formation of a supramolecular system with three CB[8]s locked between redox-active Ru(bpy)3 complexes. *Org Biomol Chem.* 2009; **7**: 3605–3609.

235. Liu Y, Shi J, Chen Y, Ke C-F. A polymeric pseudorotaxane constructed from cucurbituril and aniline, and stabilization of its radical cation. *Angew Chem Int Ed.* 2008; **47**: 7293–7296.

236. Chakrabarti S, Mukhopadhyay P, Lin S, Isaacs L. Reconfigurable four-component molecular switch based on pH-controlled guest swapping. *Org Lett.* 2007; **9**: 2349–2352.

237. Hwang I, Ziganshina AY, Ko YH, Yun G, Kim K. A new three-way supramolecular switch based on redox-controlled interconversion of hetero- and homo-guest-pair inclusion inside a host molecule. *J Chem Soc Chem Commun.* 2009; 416–418.

238. Kim D, Kim E, Kim J, Park KM, Baek K, Jung M, Ko YH, Sung W, Kim HS, Suh JH, Park CG, Na OS, Lee D-K, Lee KE, Han SS, Kim K. Direct synthesis of polymer nanocapsules with a noncovalently tailorable surface. *Angew Chem Int Ed.* 2007; **46**: 3471–3474.

239. Li M, Zaman MB, Bardelang D, Wu X, Wang D, Margeson JC, Leek DM, Ripmeester JA, Ratcliffe CI, Quan, Yang LB, Yu K. Photoluminescent quantum dot–cucurbituril nanocomposites. *J Chem Soc Chem Commun.* 2009; 6807–6809.

240. Angelos S, Yang Y-W, Patel K, Stoddart JF, Zink JI. pH-responsive supramolecular nanovalves based on cucurbit[6]uril pseudorotaxanes. *Angew Chem Int Ed.* 2008; **47**: 2222–2226.

241. Hwang I, Baek K, Jung M, Kim Y, Park KM, Lee D-W, Selvapalam N, Kim K. Noncovalent immobilization of proteins on a solid surface by cucurbit[7]uril-ferrocenemethylammonium pair, a potential replacement of biotin–avidin pair. *J Am Chem Soc.* 2007; **129**: 4170–4171.

242. Kim K, Kim D, Lee JW, Ko YH, Kim K. Growth of poly(pseudorotaxane) on gold using host-stabilized charge-transfer interaction. *J Chem Soc Chem Commun.* 2004; 848–849.

243. Yang H, Tan Y, Wang Y. Fabrication and properties of cucurbit[6]uril induced thermo-responsive supramolecular hydrogels. *Soft Matter.* 2009; **5**: 3511–3516.

244. Kim BS, Ko YH, Kim Y, Lee HJ, Selvapalam N, Lee HC, Kim K. Water soluble cucurbit[6]uril derivative as a potential Xe carrier for ^{129}Xe NMR-based biosensors. *J Chem Soc Chem Commun.* 2008; 2756–2758.

245. Huo F-J, Yin C-X, Yang P. The crystal structure, self-assembly, DNA-binding and cleavage studies of the [2]pseudorotaxane composed of cucurbit[6]uril. *Bioorg Med Chem Lett.* 2007; **17**: 932–936.

246. Kim J, Ahn Y, Park KM, Kim Y, Ko YH, Oh DH, Kim K. Carbohydrate wheels: cucurbituril-based carbohydrate clusters. *Angew Chem Int Ed.* 2007; **46**: 7393–7395.

247. Kim SK, Park KM, Singha K, Kim J, Ahn Y, Kim K, Kim WJ. Galactosylated cucurbituril-inclusion polyplex for hepatocyte-targeted gene delivery. *J Chem Soc Chem Commun.* 2010; **46**: 692–694.

248. Mock WL, Irra TA, Wepsiec JP, Manimaran TL. Cycloaddition induced by cucurbituril: a case of Pauling principle catalysis. *J Org Chem.* 1983; **48**: 3619–3620.

249. Krasia TC, Steinke JHG. Formation of oligotriazoles catalysed by cucurbituril. *J Chem Soc Chem Commun.* 2002; 22–23.

250. Tuncel D, Steinke JHG. The synthesis of [2], [3], and [4]rotaxanes and semirotaxanes. *J Chem Soc Chem Commun.* 2002; 496–497.

251. Miyahara Y, Goto K, Oka M, Inazu T. Remarkably facile ring-size control in macrocyclization: synthesis of hemicucurbit[6]uril and hemicucurbit[12]uril. *Angew Chem Int Ed.* 2004; **43**: 5019–5022.

252. Buschmann H-J, Zielesny A, Schollmeyer E. Hemicucurbit[6]uril a macrocyclic ligand with unusual complexing properties. *J Inclus Phenom Macrocyc Chem.* 2006; **54**: 181–185.

9

Rotaxanes and Catenanes

9.1 Introduction

For many years chemists speculated on the possibility of mechanically interlocked structures in which individual molecules are held together in a supramolecular arrangement by mechanical interactions rather than by chemical bonds (sometimes termed a 'mechanical bond'). Two examples of this type of structure are rotaxanes and catenanes. There is an aesthetic pleasure in designing and constructing systems of this type that is shared by many scientists, with a quote in one of the major reviews of this subject asking 'How can anyone not find these beautiful molecules fascinating?!'.[1] As this chapter will describe, besides their artistic appeal, there are also a wide range of potential applications for these systems, such as 'molecular machines' and other nanotechnology applications.

A rotaxane is a structure in which a linear molecule is threaded through a cyclic molecule, with the name 'rotaxane' being derived from the Latin for 'wheel' (*rota*) and 'axle' (*axis*). Should the linear molecule be free to slip in and out of the macrocycle, the resultant structure is called a 'pseudorotaxane', as shown schematically in Figure 9.1a. Capping the linear molecule with two bulky end groups prevents this slippage, since the capping groups cannot slip through the macrocycle, giving rise to a rotaxane structure (Figure 9.1b). Rotaxanes are usually named by the total number of units, with the structure shown in Figure 9.1b for example being a [2]rotaxane while that in Figure 9.1c is a [3]rotaxane.

Another mechanically interlocked system is the catenane, in which two or more macrocyclic rings are threaded through each other as shown in Figure 9.1d,e. These systems take their name from the Latin *catena* meaning 'chain' and are named by the number of rings, as shown by the structures in Figure 9.1d,e, which are a [2]catenane and a [3]catenane respectively.

There has been a great deal of work done on these types of interlinked systems and a detailed review is far beyond the scope of this chapter. Instead we will attempt to give an overview of these systems and pick out examples of various types, rather than detailing every structure that has been constructed by a number of groups across the world. Similarly, much attention has been given to the topologies of these systems as described mathematically, which again we will not discuss in detail. For further investigation into these topics, we would recommend that the reader consults the Bibliography. Several books on these molecules have been written and there have been numerous reviews,[1,2] including a themed issue of *Chemical Society*

Macrocycles: Construction, Chemistry and Nanotechnology Applications, First Edition. Frank Davis and Séamus Higson.
© 2011 John Wiley & Sons, Ltd. Published 2011 by John Wiley & Sons, Ltd.

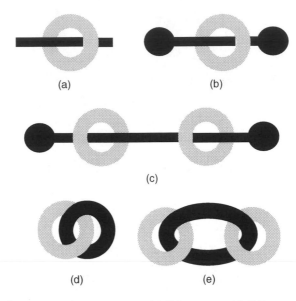

Figure 9.1   *(a) Pseudorotaxane, (b) [2]rotaxane, (c) [3]rotaxane, (d) [2]catenane and (e) [3]catenane*

Reviews published in 2009 dedicated to the work of one of the leaders of this field, Professor Jean-Pierre Sauvage. Sauvage was a PhD student of another figure mentioned frequently within this book, Professor Jean-Marie Lehn, and published a thesis on cryptands and cryptates.[3] However, his interest turned to the making of molecules with novel topologies and he was the first to report synthesis of catenanes in good yield, using a metal templating process which will be described later. This led to a long and distinguished career which included the synthesis of many catenanes and other novel topologies such as molecular knots.

9.2   Rotaxanes

9.2.1   Synthesis using statistical methods

Within this book we have already described numerous pseudorotaxane compounds, in which macrocycles encapsulate a variety of linear guests within their cavities, with the ends of the guests sticking out. Especially suitable for this are the cyclodextrins and the cucurbiturils, with which a wide variety of linear guests have been complexed. The larger systems can even accommodate more than one linear chain within their cavities. Polyrotaxanes have also been described earlier within this work, such as for example the complexes between polyethylene glycol and cyclodextrins that are described in Chapter 6. Within this section we will describe a number of other rotaxanes, attempting to provide a glimpse of the wide variety of structures available.

The first rotaxanes were synthesised using classical chemical techniques. If a macrocyclic compound and a linear-chain compound are in solution together, then if the macrocycle is large enough, statistically a fraction of the linear chains will be threaded through the macrocycle (albeit usually quite a small fraction in the absence of host–guest binding interactions). Should the ends of this pseudorotaxane then be chemically modified by bulky groups, the chain will in effect be trapped and unable to de-thread from the macrocycle. This method was first adopted by the Harrisons, who reasoned that although the amount of rotaxane synthesised by a single reaction was small, more would be obtained if the reaction could be repeated multiple times. They therefore attached a C_{30} macrocyclic ring to a solid support, namely a Merrifield-type resin.[4] This was treated with decane-1,10-diol to form the pseudorotaxane and capped with triphenylmethyl chloride. The process was repeated 70 times and then the macrocycle was cleaved from the resin. The resultant rotaxane was obtained as an oil in 6% yield and its rotaxane structure was elucidated by chromatography and spectroscopy. Chemical cleavage of the macrocycle or removal of the stopper groups could be used to separate the two components.

Later work studied simple hydrocarbon rings with a variety of ring sizes. A mixture of C_{14}—C_{40} macrocycles was heated to 120 °C with the triphenylmethyl substituted decane-1,10-diol and it was found that the ring could slip over the end groups at this elevated temperature[5] to form small amounts of rotaxanes, which on cooling could be separated by chromatographic methods. However, only the rotaxane with a C_{29} ring could be isolated, indicating that smaller rings could not slip over the end groups and that larger rings probably did form rotaxanes but could easily slip back off, meaning these systems were not stable at room temperature. Another method within the same paper used a small amount of acid, which reversibly removes and reattaches the stopper groups, allowing threading; this allowed rotaxanes to form with rings in the range C_{25}—C_{29} which were stable at room temperature and below C_{29} were stable at 120 °C. By using larger stopper groups, similar rotaxanes with ring sizes up to C_{34} could be obtained.[6]

The major drawback to these methods was the extremely low yields obtained because of the low probability of a macrocycle being threaded at any one time. One answer to this was to utilise other interactions to prearrange a pseudorotaxane in high yield and then cap the axle to give a good yield of final product. Higher yields of rotaxanes can be obtained when there is a degree of recognition between the chains; a macrocyclic crown compound, dibenzo-58.2-crown-19.4, for example, was mixed with polyethylene glycol (molecular weight 400) at 120 °C for 30 minutes and then capped with trityl units, each substituted with a bromomethyl group. Up to 15% of the [2]rotaxane could be obtained and although the crown ethers could slip over the capping groups, the rotaxane structure was relatively stable, indicating interactions between ring and threading chain.[7] Coupling of the bromomethyl groups allowed isolating in 23% yield, which was shown to be a [4]rotaxane (three rings and one chain). Other workers[8] utilised the method of reversibly removing and attaching stopper groups on an alkyl chain in the presence of cycloalkanes and by optimisation of ring and chain sizes could obtain rotaxanes in yields of up to 11.3%.

Because of low yields and purification problems, workers in the field turned their attention to 'directed' syntheses. The group of Schill pioneered the use of directed covalent synthesis of rotaxanes and other interlinked structures. Although many of these syntheses were highly complex and required numerous steps and repeated purification, they are an elegant method of making these systems. We will show here part of one of these syntheses (Figure 9.2); many more are included in Schill's book *Catenanes, Rotaxanes, Knots* (see Bibliography). The *ansa*-compound shown in Figure 9.2a can be synthesised by a multi-step procedure; as can be seen it contains a macrocyclic ring where two chemical groups on opposing sides of the ring are bridged by a substituted aromatic group.[9] The aromatic group contains two long alkyl substituents, which are capped by large groups. The bridging aromatic group is then cleaved from the macrocycle but cannot escape from inside the ring because of the presence of the bulky stopper groups, thereby forming a rotaxane, as confirmed by mass spectrometry.

9.2.2 Rotaxanes containing cyclodextrins, cucurbiturils and calixarenes

One of the earliest attempts to form rotaxanes via noncovalent binding interactions involved the use of cyclodextrins. We have already noted in Chapter 6 how many compounds form pseudorotaxanes with cyclodextrins, where a linear-chain compound is bound by the macrocycle with both end groups protruding from the cavity. Obviously, if these molecules can be successfully capped, a rotaxane will be formed. One

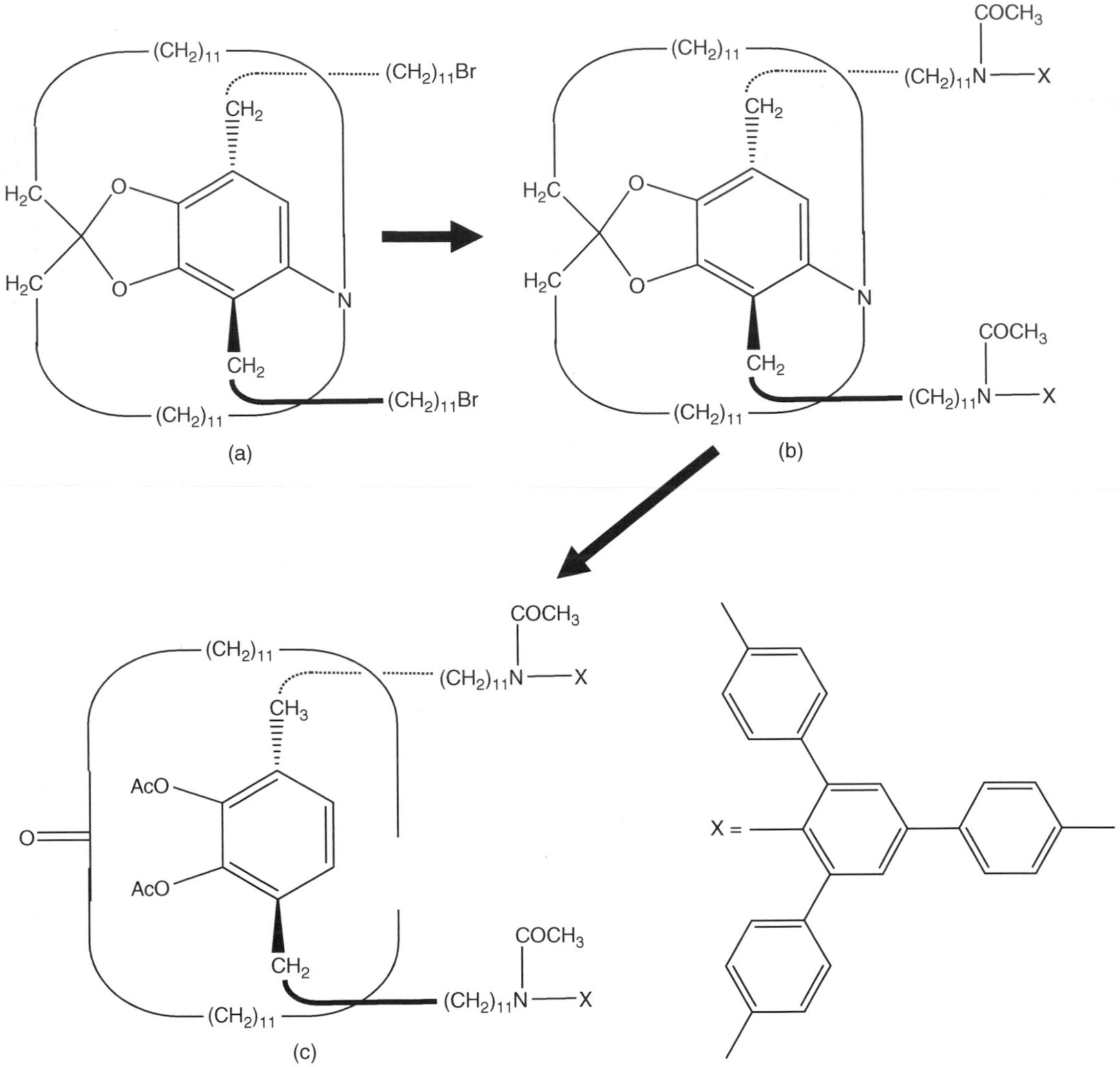

Figure 9.2   *Latter stages of rotaxane-directed synthesis: (a) ansa-compound, (b) attachment of capping groups and (c) formation of rotaxane*

of the first demonstrations of this involved complexing either α- or β-cyclodextrin with 1,10-diaminodecane or 1,12-diaminododecane, followed by reaction with *cis*-[CoCl$_2$(en)$_2$]Cl to give a rotaxane capped with the inorganic cobalt species.[10] Other workers used dibromoalkanes and thiol-containing cobalt complexes.[11] Similarly, a biphenyl unit substituted with two alkyl ammonium chloride units (Figure 9.3a) was included in the cavity of heptakis-(2,6-di-O-methyl)-β-cyclodextrin to give an aqueous solution which precipitated when a bulky anion (tetraphenyl boron) was added to give a structure in 71% yield, which although not strictly a rotaxane, was stable in acetone but disassociated when amines were added.[12] Later work utilised the same method along with a porphyrin with four aryloxy side chains (Figure 9.3b) as the axle, which could form an analogue of a [3]rotaxane containing two cyclodextrins rings attached to an aryloxy

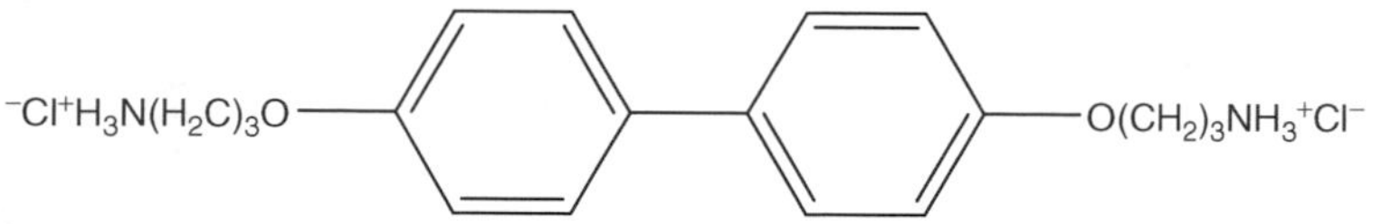

$^-Cl^+H_3N(H_2C)_3O$ ─── ... ─── $O(CH_2)_3NH_3^+Cl^-$

(a)

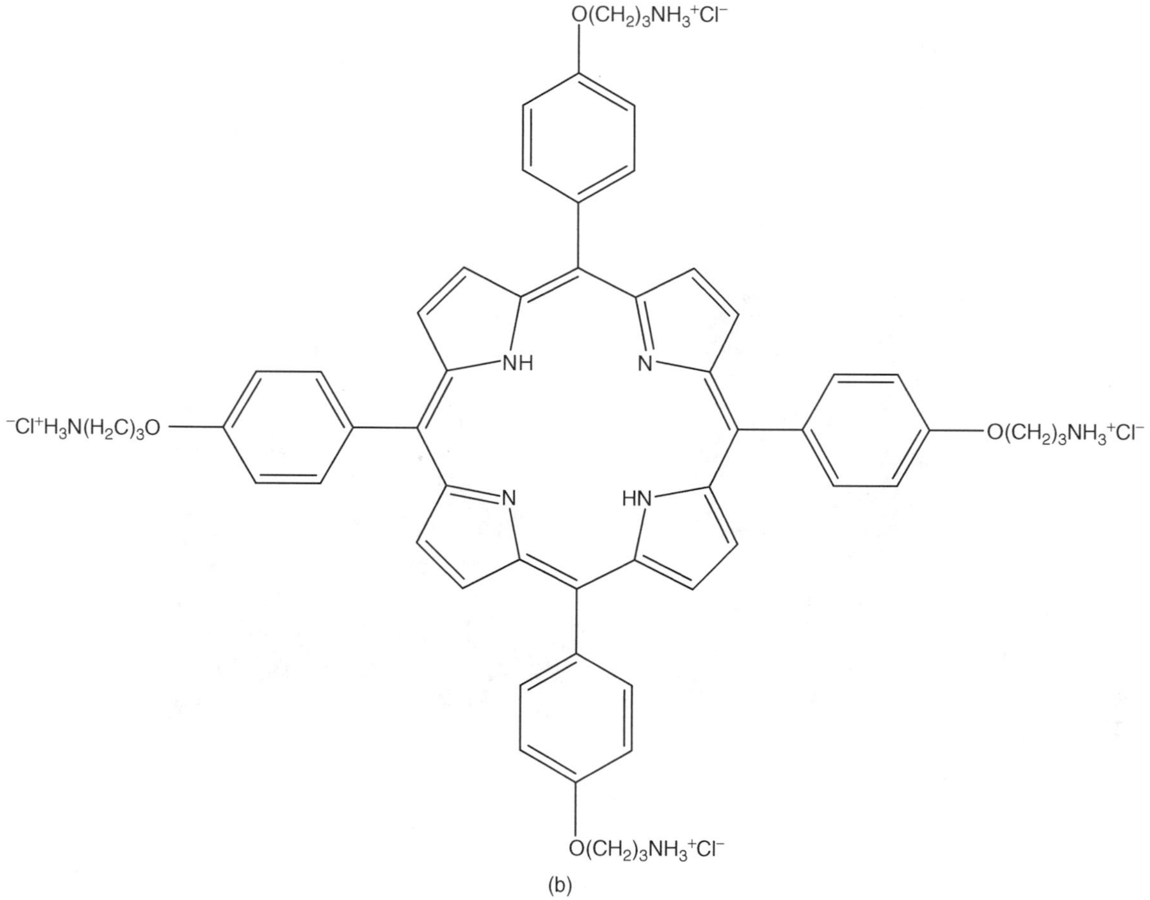

$O(CH_2)_3NH_3^+Cl^-$

$^-Cl^+H_3N(H_2C)_3O$ ───

─── $O(CH_2)_3NH_3^+Cl^-$

─NH   N─

N   HN─

$O(CH_2)_3NH_3^+Cl^-$

(b)

Figure 9.3 *(a) Axle for cyclodextrin-based rotaxane and (b) porphyrin axle*

group on opposite sides of the porphyrin unit.[13] Potentially this compound might act as a binder for four cyclodextrins, but it appears that steric hindrance prevents more than two macrocycles binding.

Cucurbiturils have also been utilised in the construction of pseudorotaxanes and rotaxanes. We have already described in Chapter 8 how many linear and aromatic molecules are complexed by cucurbiturils of various sizes, with the guest being threaded through the cavity. This of course is a pseudorotaxane and requires only capping to form a rotaxane. Much of the work in this field has been carried out by the group of Kimoon Kim.[14] For instance, diamines can be threaded though cucurbiturils and then metal ions used to link together the terminal groups of the pseudorotaxanes to form polyrotaxanes. Various polymeric architectures can be constructed by varying the axle and the linking ions, with for example linear, zigzag and helical polymers being capable of being formed.[14] More complex architectures such

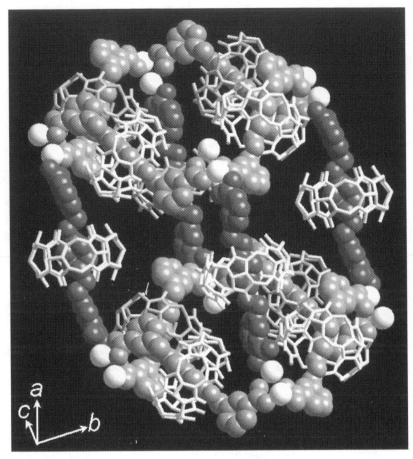

Figure 9.4 *A three-dimensional polyrotaxane obtained using cucurbituril units (Reproduced by permission of The Royal Society of Chemistry[14])*

as two-dimensional hexagonal or square-grid polyrotaxanes can also be synthesised, and when lanthanide ions are used as linkers, three-dimensional polyrotaxanes such as that shown in Figure 9.4 can be obtained.

Many of the early rotaxanes were quite symmetrical, but later work developed rotaxanes with two different binding sites on the same axle. What happens then is that the macrocycle can 'choose' which site to bind to. External stimuli can affect these systems, giving us the molecular equivalent of a switch, where the structure of the molecule can be reversibly changed by stimuli such as a change in pH. An example is the compound shown in Figure 9.5a, which forms a [2]rotaxane with cucurbit[6]uril.[15] At low pH, all three amine groups are protonated and since the binding constant of a diaminohexyl group for CB[6] is higher than that of a diaminobutyl, the CB[6] binds as shown and the resultant complex is fluorescent. Increasing the pH causes deprotonation of the aryl amino group and the CB[6] unit will then bind more strongly to the diprotonated diaminobutyl unit and 'shuttle' along the axle to give a nonfluorescent complex. Another system of interest uses the axle shown in Figure 9.5b, which forms a pseudorotaxane with CB[6] in which the macrocycle resides on the central diaminobutyl unit.[16] Deprotonation of these amine units leads to the rapid relocation of the CB[6] onto one of the viologen units. However, addition of acid does not immediately reverse this process; the movement back of the CB[6] unit is exceedingly slow and requires

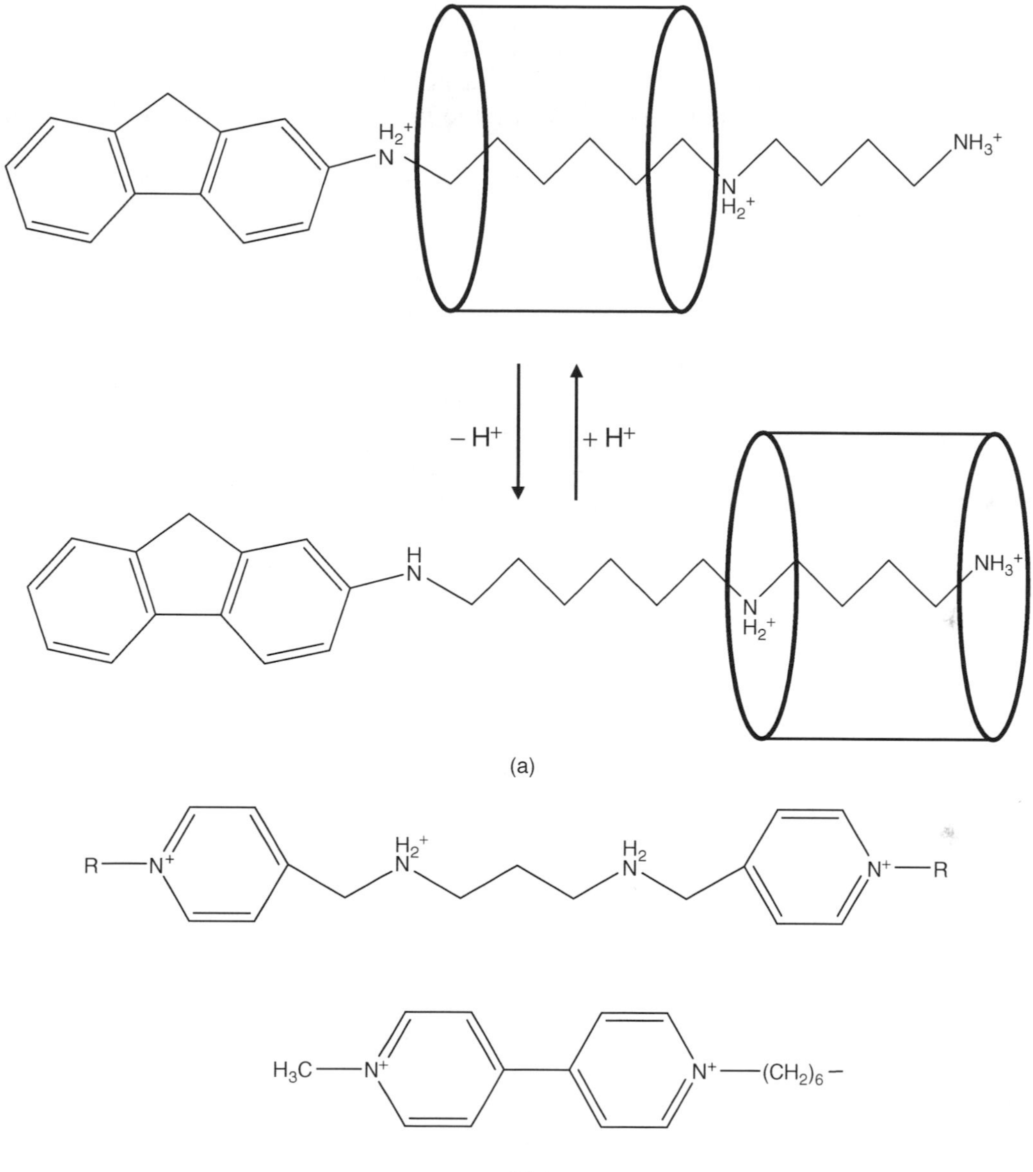

(a)

(b)

Figure 9.5 *(a) Switchable pseudorotaxanes containing cucurbituril units with fluorescence responses and (b) axle of a kinetically controlled switchable pseudorotaxane*

heating to 80 °C before it occurs at a reasonable rate. A [3]rotaxane with two CB[6] units complexed with a long *bis*-triazole compound (Figure 8.18b) can be synthesised.[17] This compound has a CB[6] unit complexed to each triazole, but addition of base causes deprotonation and one of the macrocycles migrates to the central chain. Addition of further base causes the second chain to also switch to forming a complex with the central alkyl chain, whereas addition of acid and heat reverses this process and regenerates the original structure, giving a molecular three-way switch. A variety of compounds of this nature have been studied and switching has been shown to depend on the nature of the aliphatic spacer.[18] A tetraphenyl porphyrin unit substituted with triazole groups has been used as the central unit for a [5]rotaxane and 5-pseudorotaxanes containing four CB[6] units.[19] These have been demonstrated to undergo pH-dependent switching, in which the macrocycles bind to the triazole units at low pH and the phenyl units at higher pH.

Ionic interactions have also been utilised to assemble rotaxanes. CB[6] and CB[7] can both be threaded with 1,6-diaminohexane and the resulting complex capped with tetraphenyl boron ions to give rotaxanes in 80% yield.[20] CB[5] cannot be threaded and therefore forms a structure in which the diamino compounds bridge between the macrocycles. The synthesis of polyrotaxanes containing cucurbituril units threaded on linear and polyaromatic polymer chains such as polyaniline has already been described in Chapter 8. Other more complex architectures have also been synthesised; commercial dendrimers for example can be substituted so that they contain terminal diaminobutyl groups. These are then combined with CB[6] to construct dendrimer pseudopolyrotaxanes tipped with rigid-shell CB[6] units.[21] As an alternative method, two dendrimers can be synthesised, one bearing a single viologen unit at the dendrimer apex, the other a 1,5-dioxynaphthalene unit. Combination of these dendrimers with CB[8] in solution leads to a hetero-pseudorotaxane in which the viologen and 1,5-dioxynaphthalene moieties are both bound by the CB[8] and one of each dendrimer is the stopper group.[22]

Calixarenes have also been shown to form threaded structures, with for example a calix[6]arene having been shown to form pseudorotaxanes with viologen guests.[23] One interesting example is the use of tetra-urea-substituted calix[4]arenes such as those shown in Figure 9.6. These can be synthesised in looped or open-chain forms; when mixed together in aprotic solvents, heterodimers form exclusively where the linear substituents are threaded through the looped ones.[24] Further addition of bulky stopper groups leads to the formation of a fourfold [2]rotaxane as shown in Figure 9.6. Calixarenes have also been used as the stopper groups in rotaxane synthesis.[25]

9.2.3   Rotaxanes based on π–electron interactions

It has been long known that there are strong interactions between electron-rich and electron-poor aromatic systems. We have already noted in Chapter 3 how aromatic crown ethers can form stable complexes with electron-poor systems such as paraquat. One particularly strong interaction is between *bis-p*-phenylene-34-crown-10 and paraquat as the hexafluorophosphate salt (Figure 9.7a,b). As can be seen from the crystal structure,[26] the paraquat unit is located in the cavity of the cyclic crown ether and stabilised by both π–π interactions and C—H—O hydrogen bonds between the crown-ether oxygen atoms and the relatively acid aryl protons. Similar interactions can be noted for the complex in which a cyclic diparaquat unit is complexed with a dialkoxy benzene unit (Figures 9.7d,e). Use of a suitable capping agent allows capping of this pseudorotaxane with trisopropylsilyl groups to give the true rotaxane, the crystal structure of which is shown in Figure 9.8a. In an extensive paper by Anelli *et al.*[26] the syntheses of a number of rotaxanes and a [2]catenane were reported. These noncovalent charge-transfer interactions often lead to a characteristic orange colour for these complex materials and have been utilised to construct a large range of some of the most intricate interlinked structures developed to date – in particular the catenanes, which will be discussed in Section 9.3. Much of this was pioneered by such workers as Angel Kaifer and especially Sir J. Fraser Stoddart.

We have already described the threading approach to making these compounds, but the π–π inter-actions can also preorganise the components in such as way to allow clipping of the outer ring onto the central axle. When the linear compound shown in Figure 9.7d was 'pre-capped' with triisopropy-lsilyl groups, it could be combined with the compound shown in Figure 9.8b and xylylene dibromide to synthesise the rotaxane in 14% yield.[26] This indicates that the diviologen compound and the alkoxy

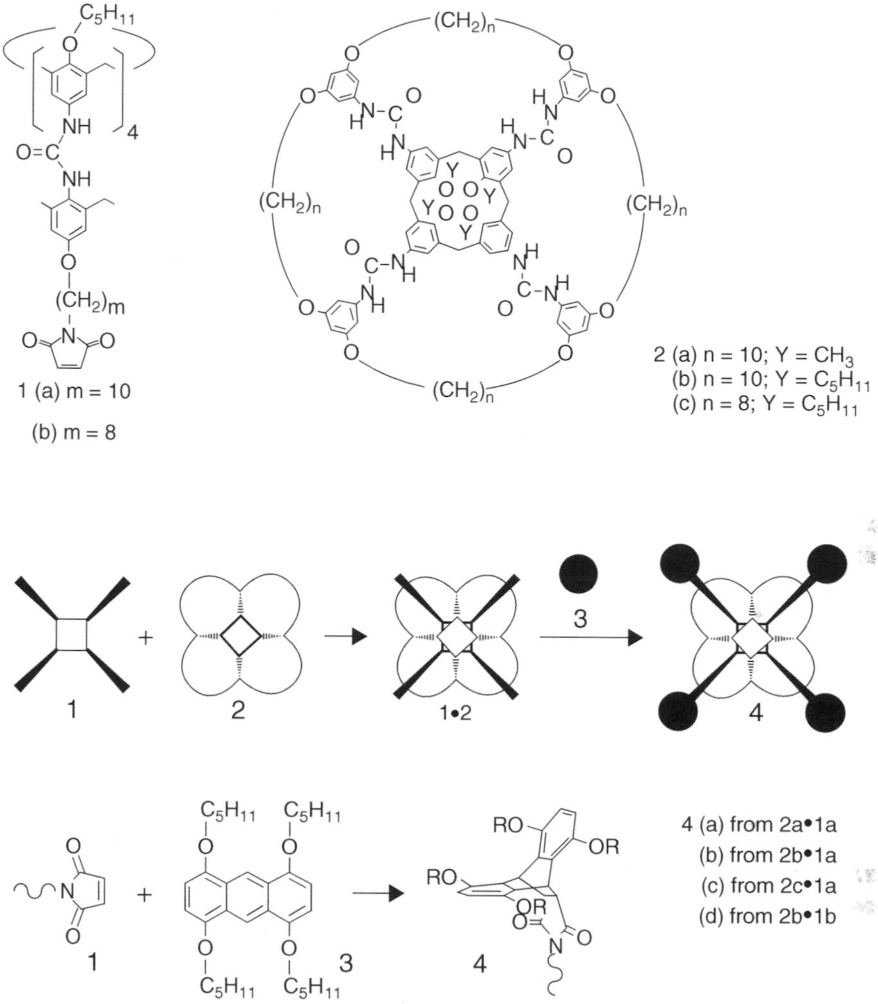

Figure 9.6 *Tetra [2]rotaxanes obtained using calixarene units (Reprinted with permission from[24]. Copyright 2005 American Chemical Society)*

benzene axle preorganise to some extent before cyclisation occurs. This process can be enhanced by increasing the number of electron-rich units on the axle; an axle containing two dialkoxy benzene units (Figure 9.8c), for example, was used to form a rotaxane with the paraquat macrocycle via the same clipping process.[27] Not only were yields increased to 32% but this system behaved as a molecular shuttle in that the cationic macrocycle could reside on either of the two electron-rich units. At room temperature the NMR spectra of the dialkoxybenzenes are equivalent, indicating fast shuttling of the wheel along the axle. Cooling to $-50\,^{\circ}$C slows down this movement and the complexed and uncomplexed rings can be distinguished.

Longer versions of these rotaxanes, containing dialkoxybenzene units separated by polyether chains and stoppered with adamantyl groups, were synthesised and studied using the clipping procedure shown in Figure 9.8b. Interestingly, yields increased from 1% for the rotaxane with two dialkoxybenzene units up

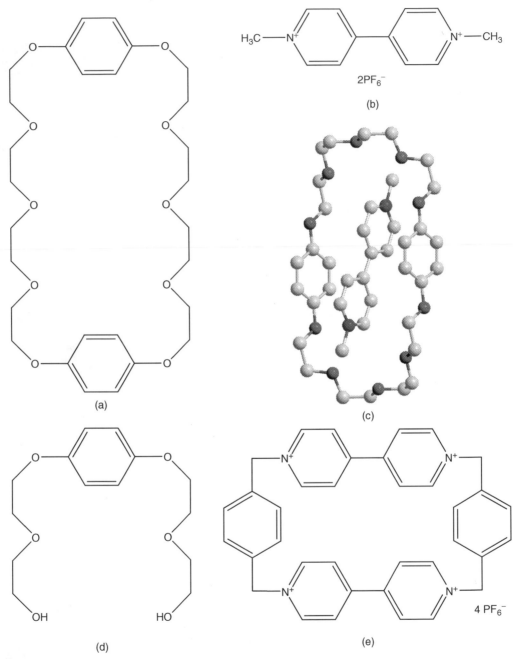

Figure 9.7 *(a)* Bis-p-phenylene-34-crown-10, *(b)* paraquat hexafluorophosphate, *(c)* crystal structure of the 1 : 1 complex, *(d)* a linear dialkoxy benzene and *(e)* a cyclic diparaquat compound

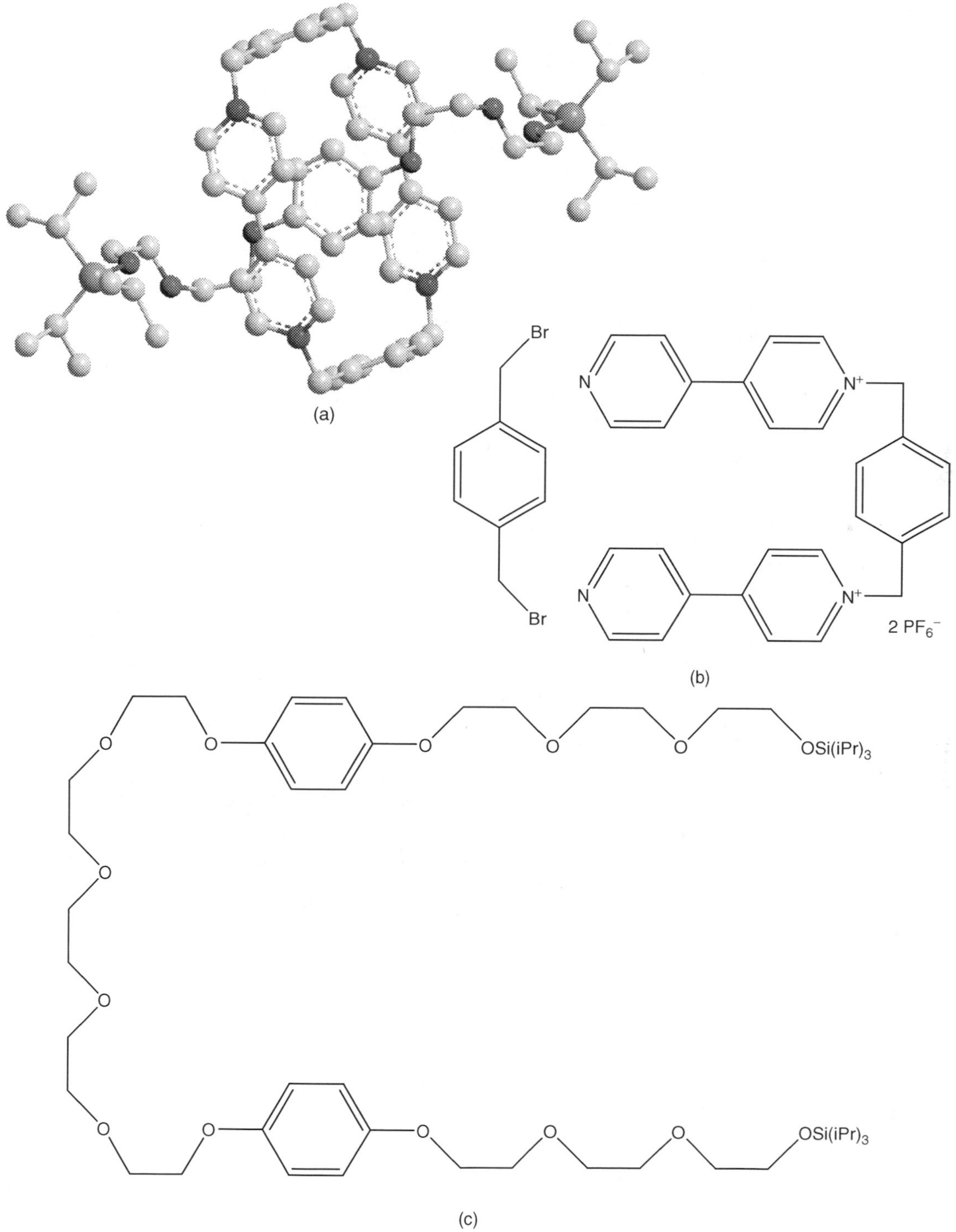

(a)

(b)

(c)

Figure 9.8   *(a) Crystal structure of a cyclic diparaquat rotaxane with a dialkoxybenzene axle, (b) the clipping reaction and (c) an axle with two electron-rich moieties*

to 40% for one containing five.[28] Analysis of the pentamer showed that the in the [2]rotaxane, the cyclic paraquat units reside equally on the three central dialkoxybenzene units, whereas at higher temperatures shuttling occurs. Interestingly, despite the presence of multiple electron-rich sites, no [3]rotaxanes or higher species were observed.

Once rotaxanes with multiple sites had been synthesised, an obvious advance was to make asymmetric systems with two or more potential different binding sites in order to study which sites are preferred and whether outside stimuli can be used to switch the binding. One of the earliest successful switchable systems involved the cyclic paraquat (Figure 9.7d) as the wheel, with the axle being a compound containing two dialkoxybenzene units along with a central tetrathiafulvalene unit (Figure 9.9a). This could be switched by solvent effects; in acetone the macrocycle enveloped one of the dialkoxybenzene units, whereas in DMSO the tetrathiafulvalene acted as the guest.[29] Similarly, a rotaxane axle containing benzidine and bisphenol units was synthesised (Figure 9.9b) and could be utilised to synthesise a [2]rotaxane in which the paraquat macrocycle preferentially occupied the more electron-rich benzidine site. However, this system could be reversibly switched so that the macrocycle enveloped the bisphenol moiety. This process could be attained both chemically by protonation with trifluoroacetic acid (and reversed by pyridine) and electrochemically by oxidising the benzidine to the radical cation.[30]

The rotaxanes described above contain electron-rich axles and electron-poor wheels. However, the reverse arrangement is also possible, such as for example the structure shown in Figure 9.10a, where viologen units are used to make the axles.[31] This can be assembled along with the cyclic crown shown in Figure 9.7a to give a rotaxane in 23% yield. Shuttling of the macrocycle between the viologen groups has been demonstrated to occur about 300 000 times per second at room temperature, much faster than the systems above, and it is necessary to cool this system to −80 °C to slow it to the point where the complexed and uncomplexed viologens can be visualised by NMR. A similar rotaxane (where one of the t-butyl units has been removed from the stopper) can be assembled by slippage; heating the axle and the excess of the crown ether at 55 °C for 10 days leads to formation of rotaxanes with one or two crown-ether units complexed with the axle.[32] The proportions of the [2] and [3]rotaxane can be controlled by the ratios of axle and crown, and the [3]rotaxane can be obtained in yields of 55%. The same procedure was applied to another similar rotaxane (where one of the t-butyl units had been replaced by an isopropyl group) and allowed formation of the [2] and [3]rotaxane with the same crown ether as above.[33] When the [2]rotaxane was treated with another somewhat larger crown ether (Figure 9.10b), a [3]rotaxane could be obtained with one of each crown ether complexed with the axle. A three-way axle (Figure 9.11) could also be constructed and then used to assemble rotaxanes using the slippage method.[34] Heating the axle at 50 °C in acetonitrile for 10 days in the presence of two equivalents of Figure 9.7a led to isolation of the [2], [3] and [4]rotaxanes after chromatography in yields of 46, 26 and 6%, respectively. Yields of the higher rotaxanes could be increased by using excesses of the crown ether.

9.2.4   Other synthetic methods for rotaxane synthesis

Other noncovalent interactions have also been used in the assembly of rotaxanes. For example, when the anthracene amine hydrochloride (Figure 9.12a) is reacted with an acylating agent (Figure 9.12b) an amide is formed. It is known that crown ethers can form stable complexes with ammonium salts; if such a reaction is carried out in the presence of dibenzo-24-crown-8, under optimum conditions up to 22% of a [2]rotaxane can be obtained, where the resultant ammonium salt compound (Figure 9.12c) is threaded through the crown.[35] This rotaxane is formed best in a mixed water/chloroform media and it is thought that the crown first complexes the amine salt and the acylating agent then approaches from the opposite side of the macrocycle, reacts and acts as a stopper. The rotaxane is stable enough to be crystallised and X-ray studies prove the threaded structure (Figure 9.12d). When a anthracene thiol (Figure 9.13a) is mixed

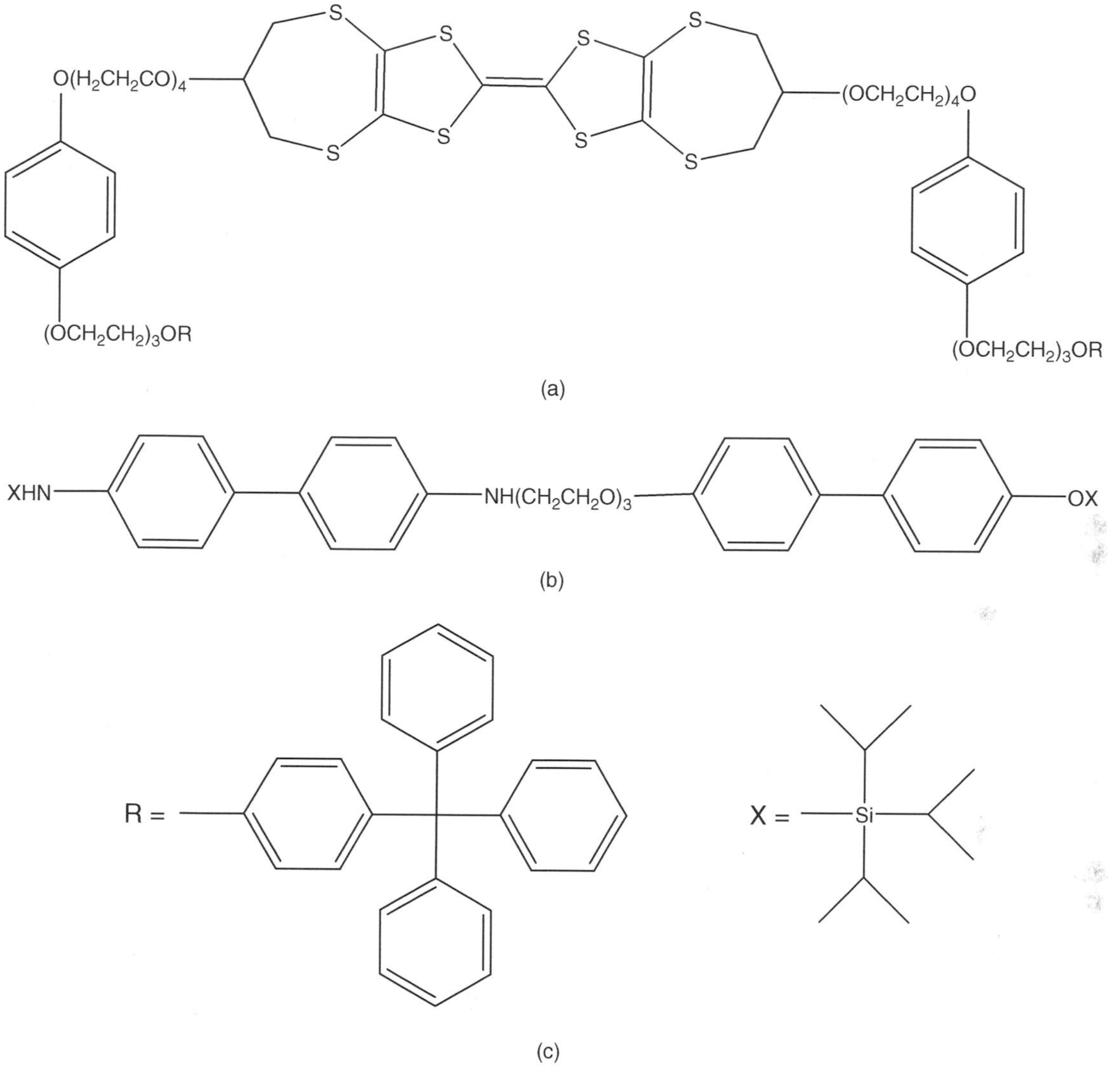

(a)

(b)

(c)

Figure 9.9 *(a) Structure of a tetrathiafulvalene rotaxane axle, (b) structure of the benzidene/bisphenol rotaxane axle and (c) an axle with two electron-poor viologen moieties*

with dibenzo-24-crown-8 and then oxidised with iodine to convert the thiols to disulfides, this resulting in a [3]rotaxane with the two anthracene stoppers joined by a disulfide bond (Figure 9.13b), trapping between them two crown-ether units.[36] What is interesting is the very high yield of this reaction, 84%. X-ray crystal structures again confirm the rotaxane structure (Figure 9.13c).

These crown ether–ammonium interactions have been pursued as a route to rotaxanes and pseudoro-taxanes. For example, dibenzo-24-crown-8 forms a 1:1 pseudorotaxane[37]-like complex with dibenzyl ammonium hexafluorophosphate (Figure 9.14a) and when this is mixed with the diammonium compound shown in Figure 9.14b, a 3-pseudorotaxane is formed in which each ammonium unit is enveloped by a

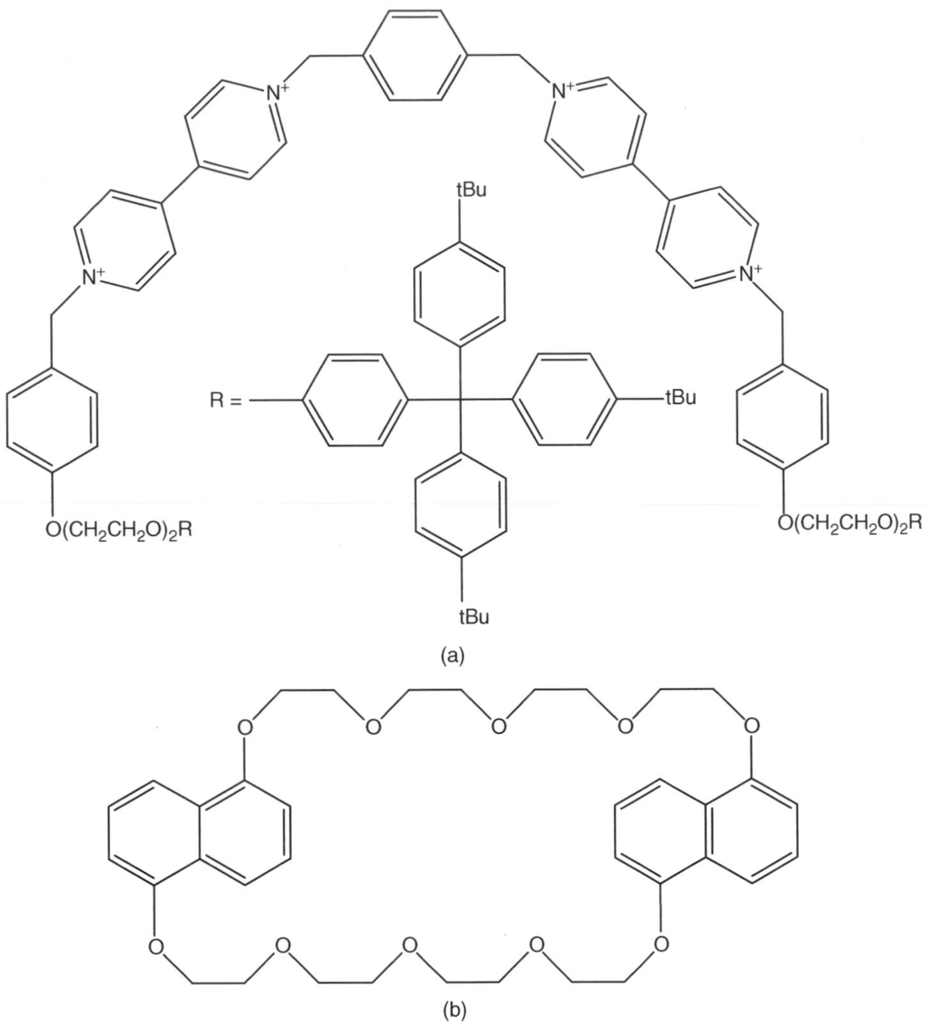

(a)

(b)

Figure 9.10 *(a) Structure of an axle with two electron-poor viologen moieties and (b) a naphthalene-containing crown ether*

crown ether.[38] When larger crowns such as dibenzo-34-crown-10 are combined with the dibenzyl ammonium cation, they have been shown to also form [3]rotaxanes, but containing one ring and two axles. With the diammonium compound an even more complex 4-pseudorotaxane is observed in the solid state (Figure 9.14c), containing two rings connected by two axles.[38] Further expanding the size of the crown ethers[39] allows construction of [4]rotaxanes (one macrocycle and three ammonium compounds) and the 5-pseudorotaxane with four ammonium groups (Figure 9.14d). This compound contains a central hexafluorophosphate ion, which is thought to assist in the formation of this supramolecular system and opens up the possibility of using it as an anion receptor.

Using ammonium axles with reactive groups such as azides allows construction of pseudorotaxanes which can be further reacted to give true rotaxanes.[40,41] In an elegant process an axle containing an ammonium and a viologen unit is synthesised in the presence of dibenzo-24-crown-8 to give the rotaxane

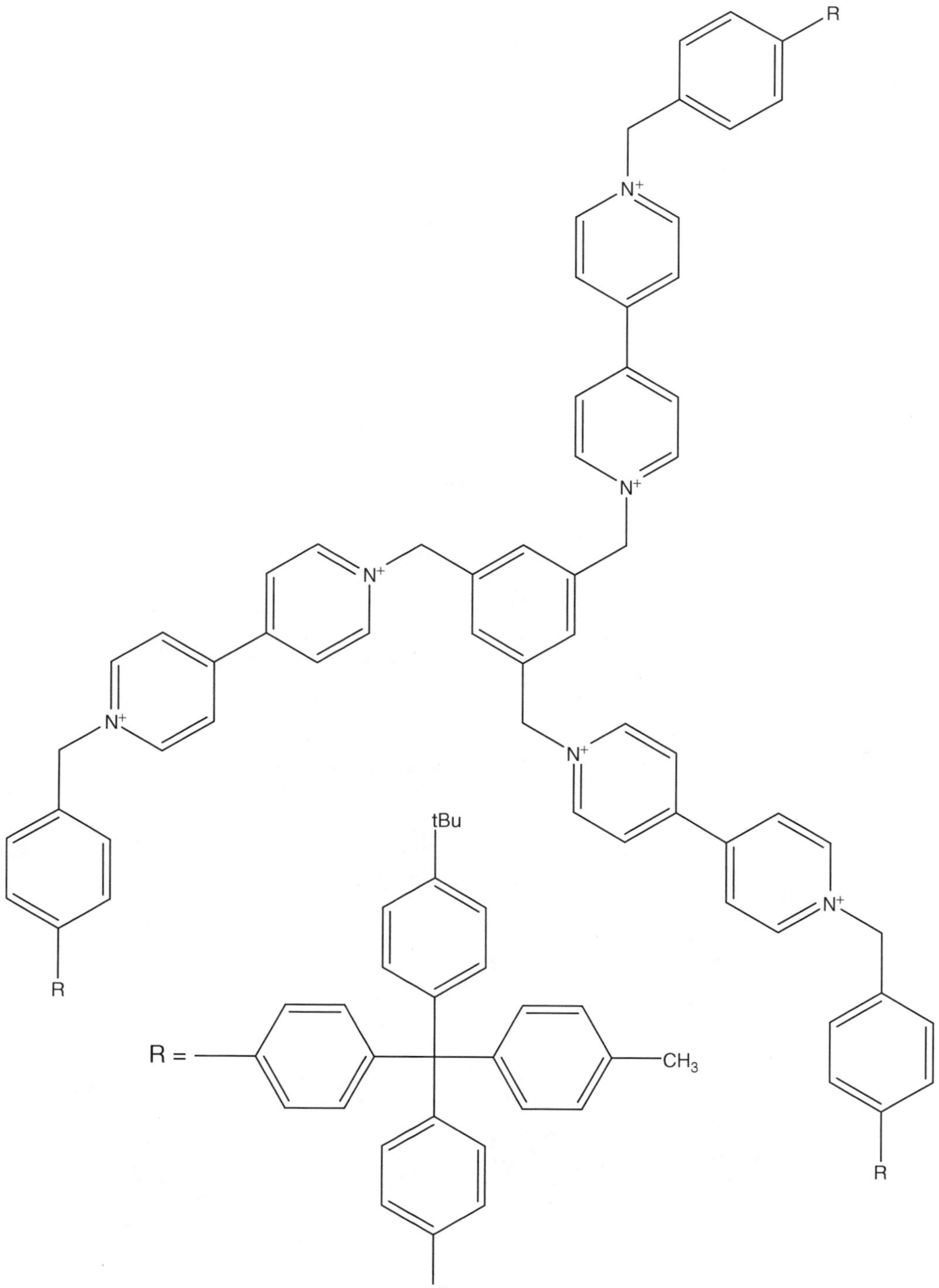

Figure 9.11 *Structure of a three-way axle*

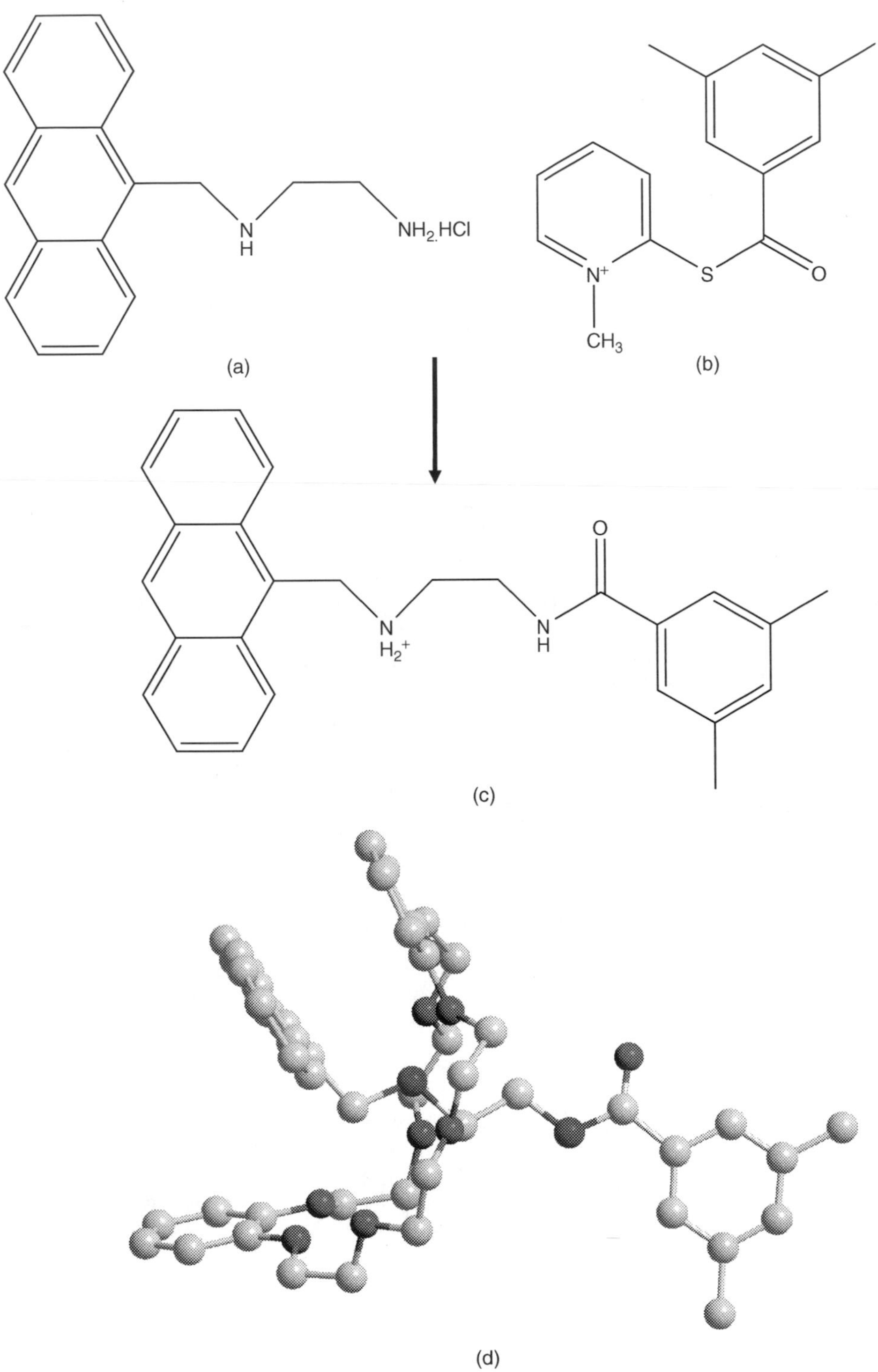

(a)

(b)

(c)

(d)

Figure 9.12   *(a–c) Acylations of an anthracene amine and (d) X-ray crystal structure of the resultant rotaxane*

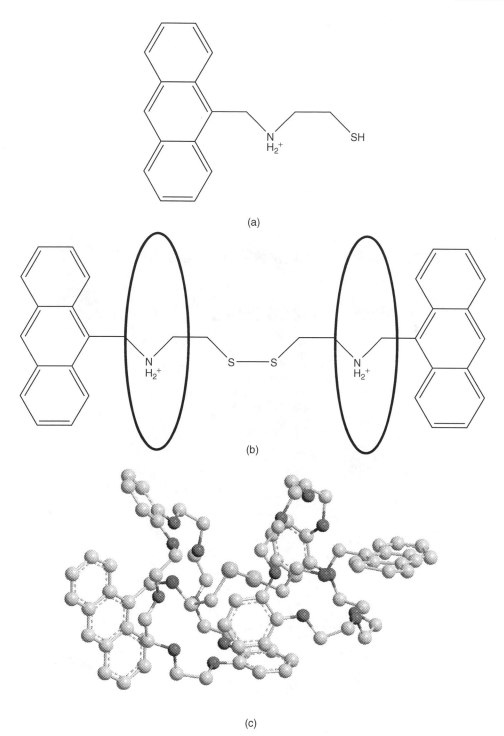

(a)

(b)

(c)

Figure 9.13 (a) Anthracene amine thiol, (b) product of the oxidation reaction in the presence of diben-zo-18-crown-6 and (c) X-ray crystal structure of the resultant [3]rotaxane

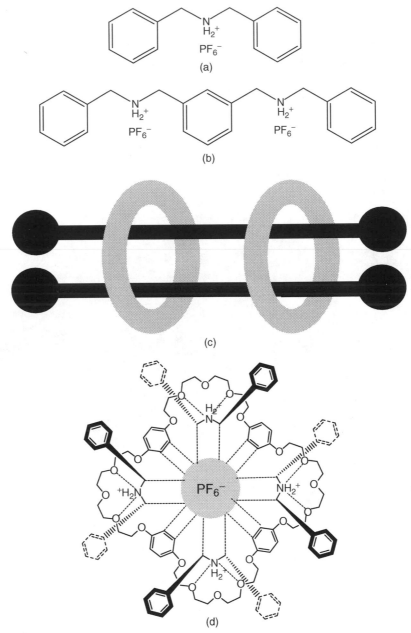

Figure 9.14 (a) Dibenzyl ammonium hexafluorophosphate, (b) diammonium axle, (c) schematic of the larger [4]rotaxane (two rings, two axles) and (d) the [5]rotaxane (one ring, four axles) (Reprinted with permission from[42]. Copyright 1998 American Chemical Society)

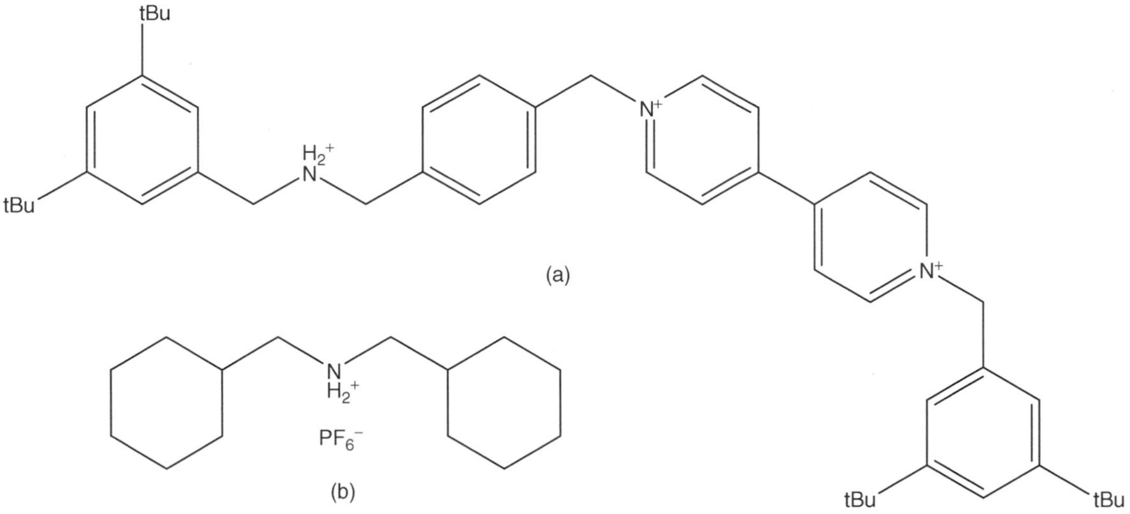

Figure 9.15   *(a) Axle containing ammonium and viologen binding sites and (b) dicyclohexyl ammonium hexafluorophosphate*

shown in Figure 9.15a. The crown ether has been shown to envelop the cationic ammonium group, but when an amine compound is added, the ammonium group is deprotonated and the crown ether then shuttles along the axle to occupy the viologen site.[41] This transition is accompanied by the appearance of a red colour due to charge-transfer interactions and can be reversed by adding trifluoroacetic acid. The slipping procedure can also be used to assemble rotaxanes of this type and for example the axle shown in Figure 9.15b can be heated with dibenzo-24-crown-8 in 90% yield.[42]

Some workers have utilised amide-containing macrocycles to form rotaxanes by the slipping method. For example, the macrocycle shown in Figure 9.16a can be melted with the axle (Figure 9.16d) to give a [2]rotaxane.[43] A similar material can be obtained with sulfonyl linkers in the macrocycle. A very interesting reaction occurs when the axle shown in Figure 9.17a is combined in high-dilution conditions with isophthaloyl dichloride and *p*-xylylene diamine.[44] Initial expectations were that this would synthesise a simple polyamide, but the peptide axle appears to template the reaction and leads to formation of a [2]rotaxane in yields of up to 30%. The presence of multiple cooperative hydrogen bonds is proposed to be an explanation for this process. The macrocycle structure (Figure 9.17b) has been confirmed by NMR, as well as threading of the axle through the macrocycle. NMR has also demonstrated that the macrocycle shuttles between the two identical binding sites of the axle. Other workers have also shown that similar reactions can be performed in the presence of a diamide compound containing two large blocking groups (Figure 9.17c). The presence of this compound preorganises the monomers and affords cyclisation to form the [2]rotaxane,[45] with hydrolysis of the end groups allowing the release of the macrocycle (Figure 9.17b). Without the presence of the axle a mixture of polymeric, cyclic and even catenated products forms.

Threading has also been shown to be a suitable method for rotaxane formation, again indicating that preorganisation of the components is occurring. When the macrocycle shown in Figure 9.16a is combined with isophthaloyl chloride and a bulky (4-aminophenyl)triphenylmethane stopper is added, a [2]rotaxane is formed in good yield. It is thought the monoamide is formed first – this forms a complex with the macrocycle due to good hydrogen-bonding interactions – and then addition of a second stopper leads to

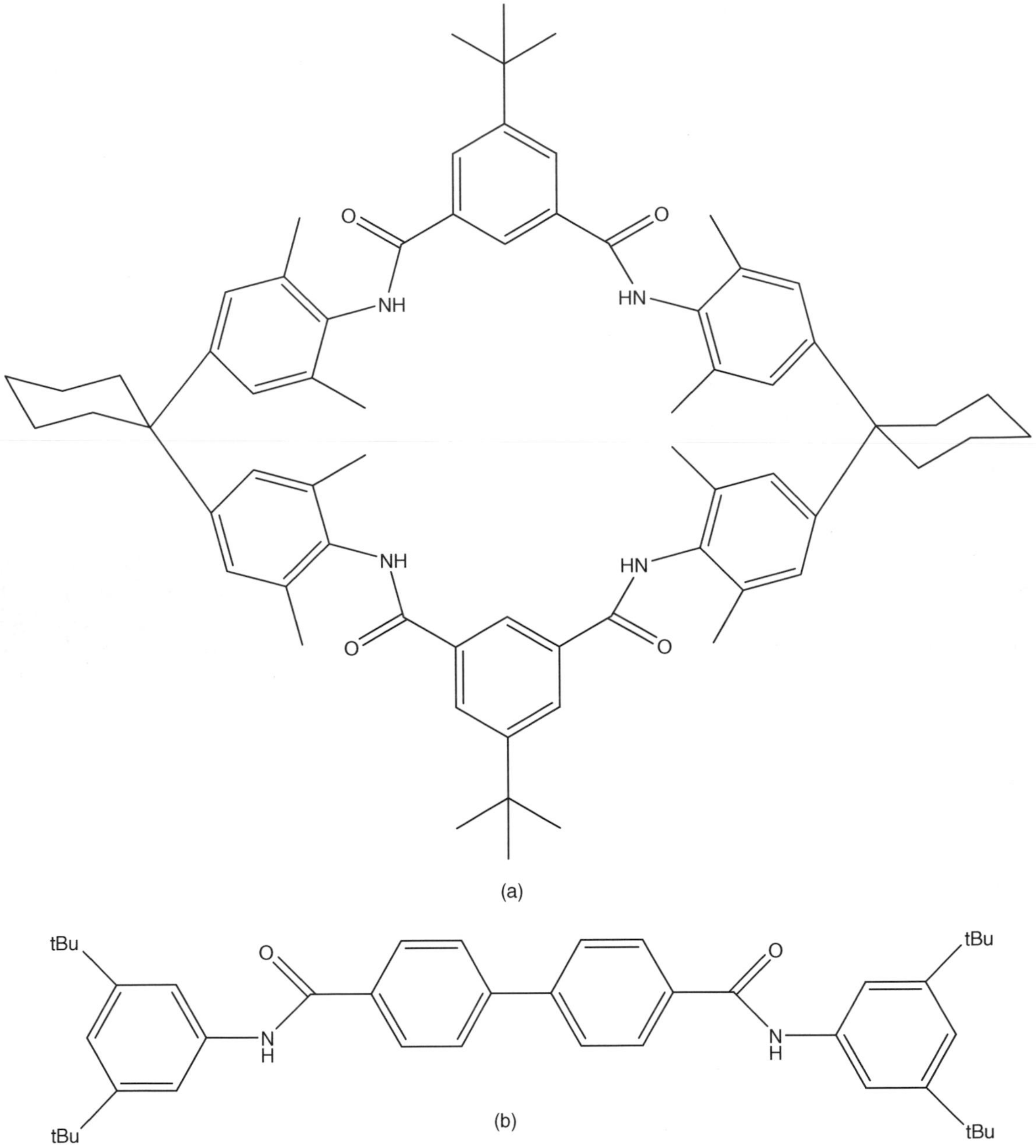

(a)

(b)

Figure 9.16   *(a) Amide macrocycle and (b) corresponding axle*

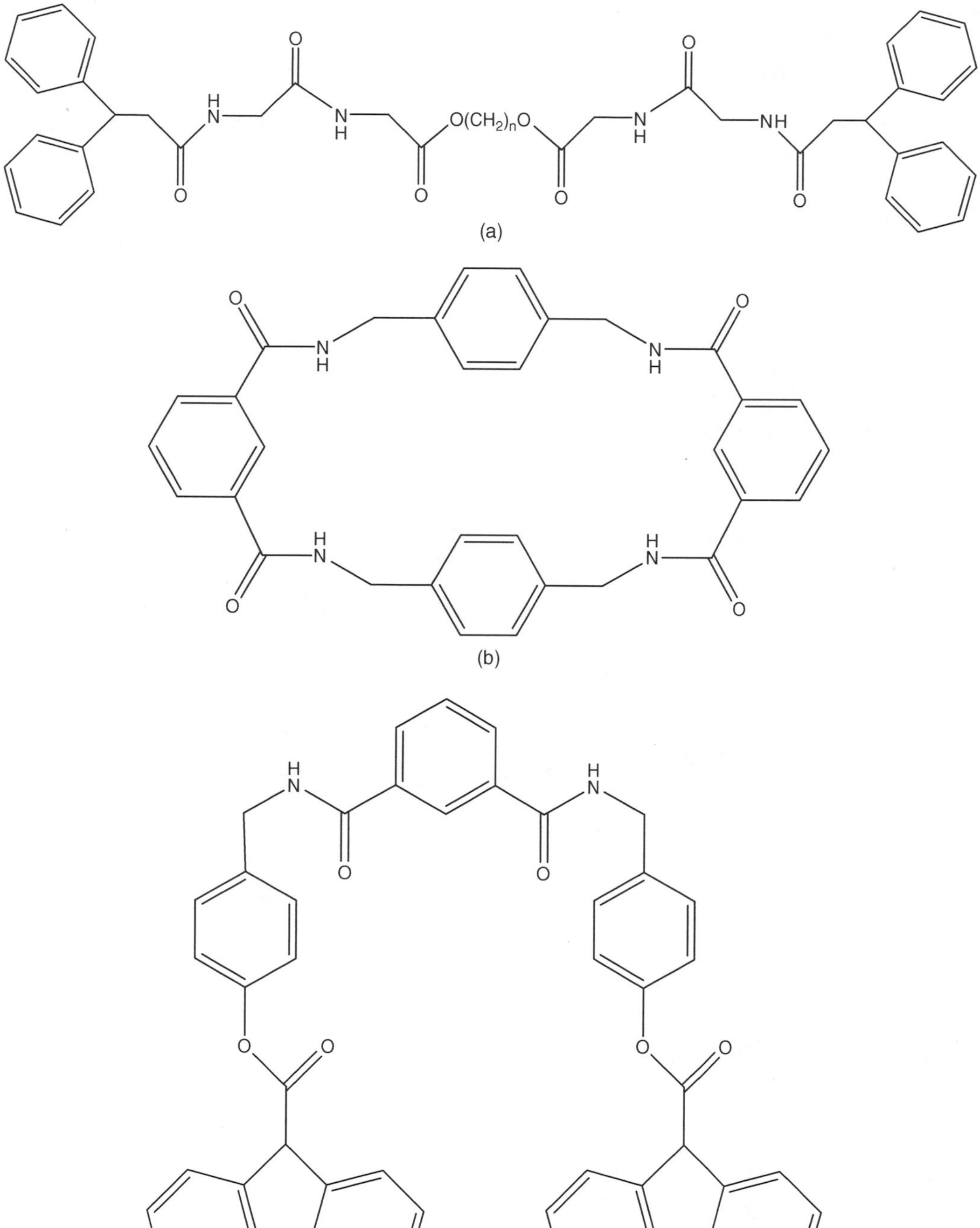

Figure 9.17 (a) Peptide axle (n = 11), (b) structure of the corresponding macrocycle formed by the templated reaction and (c) amide axle

rotaxane formation.[46] This method has proven useful for the synthesis of a variety of rotaxanes. For example, the isophthalamide unit can be replaced by heteroaromatic units,[47] substituted benzene rings, terephthaloyl,[47] biphenyls, azobenzenes, benzophenone,[48] benzene sulfonamides[47] and nonaromatic systems such as fumaroyl.

These amide-based rotaxanes have proven amenable to being assembled into larger supermolecules; for example an axle containing two isophthaloyl units can be combined with excess macrocycle and then stoppered to give the [3]rotaxane with two macrocycles on the axle.[49] The analogous rotaxane in which one of the macrocycles' amide groups is replaced by a sulfonamide group has been synthesised[50] and since sulfonamides are relatively acidic compared to amides, it has proved possible to couple two macrocycles together to give a *bis*-[2]rotaxane (Figure 9.18a). Similarly, a [2]rotaxane in which one of the axle amide groups is replaced by a sulfonamide group has been synthesised,[50] thereby allowing coupling of two axles together to give the *bis*-[2]rotaxane shown in Figure 9.18b. Variations on this theme using [2]rotaxanes in which the axle has two potential binding sites also exist, thereby allowing shuttling along the axle. Coupling of these via the macrocycle gives a *bis*-[2]rotaxane that can potentially exist in any of three states (Figure 9.18c). Asymmetric *bis*-[2]rotaxane systems with one long axle containing two binding sites and one shorter single-site axle can also be obtained.[50]

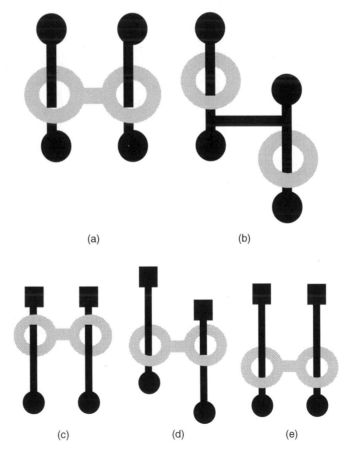

Figure 9.18   *Schematics of (a)* bis-[2]rotaxane via coupling of macrocycles, (b) bis-[2]rotaxane via coupling of axles and (c–e) three potential states of bis-[2]rotaxane with axles that have two binding sites*

Many different functional groups have been incorporated into rotaxanes, either as substituents covalently bound to the rotaxane or by simple physical incorporation as a wheel or axle, leading to potential modifications of the behaviour of the components. For example, it is possible to synthesise oligoalkynes like that shown in Figure 9.19a, but these compounds are relatively unstable and tend to crosslink. When synthesised in the presence of α-cyclodextrin or its silylated derivative, rotaxanes have been isolated containing one or two cyclodextrin units, which stabilise the resultant systems, prevent crosslinking and improve their solubility.[51] Other workers have again used coupling reactions to form rotaxanes by first forming pseudorotaxanes of dibenzo-24-crown-8 with an allyl-substituted ferrocenyl compound (Figure 9.19b) and then coupling it using a Ru-catalysed cross-metathesis reaction with an acrylate ester containing a bulky ferrocene or dimethyl benzene substituent to give [2]rotaxanes.[52] A ferrocene substituted with two acrylate groups can also be utilised to give a [3]rotaxane with the axle structure shown in Figure 9.19c and containing two crown-ether macrocycles. The same capping reaction has also been used along with derivatives of dibenzo-24-crown-8 in which one of the benzo units is replaced by a ferrocene unit to give a series of [2]rotaxanes with ammonium axles capped with groups such as anthryl or ferrocenyl.[53] Fluorescence experiments have shown that energy-transfer reactions between ferrocenyl and anthryl groups depend on the relative positions of these two groups and can be affected by shuttling of the ferrocene macrocycle along the axle.

Crown ethers can also be used to synthesise one of the smallest rotaxanes,[54] where the pseudorotaxane of dipropargylammonium tetrafluoroborate and 21-crown-7 is simply ball-milled with 1,2,4,5-tetrazine to give the rotaxane with the axle shown in Figure 9.20a in a 81% yield. X-ray crystallography demonstrates the rotaxane structure (Figure 9.20b). Other work has also demonstrated solvent-free syntheses for rotaxanes by ball-milling crown ethers with *bis*(4-formylbenzyl) ammonium compounds as well as 1,8-diaminonaphthalene.[55] The crown ether and ammonium compound forms a pseudorotaxane, which then undergoes a condensation reaction with 1,8-diaminonaphthalene to give a [2]rotaxane with the axle structure shown in Figure 9.20c. Using a *tris*-ammonium compound based around a triphenylbenzene core also allows efficient synthesis of the [4]rotaxane. Other workers have formed a [2]rotaxane using dibenzo-24-crown-8 and a protonated tertiary amine (Figure 9.20d) and showed that in its ionic form the amine is complexed by the macrocycle and that deprotonation to the neutral compound causes a shuttling of the macrocycle away from the amino group onto the central benzene ring, while still retaining the rotaxane structure.[56] Later work also alkylated the amino group of the rotaxane with a variety of functional groups, and these compounds displayed pH-controlled shuttling, the crown binding the protonated ammonium group but shuttling onto a benzene ring when deprotonated.[57]

Other workers have developed 'one-pot' syntheses for rotaxanes. Later in this chapter we will describe how copper complexes of phenanthroline have been used to assemble catenanes. Similar strategies have been used for rotaxanes;[58] for example, a crown ether containing phenanthroline and fullerene units can be complexed with a second phenanthroline compound and then capped with a variety of bulky end groups such as porphyrins or ferrocenes; Figure 9.21a shows the reaction scheme. The resultant systems exhibit high flexibility and display novel electron-transfer mechanisms. Rotaxanes with porphyrin-containing wheels and fullerene axles have also been synthesised (Figure 9.21b) and shown to undergo electron-transfer processes.[59] In these systems a relatively long-lived radical-ion pair can be generated by irradiation, with rotaxanes with shorter axles undergoing this reaction more efficiently. Other workers have synthesised [3]rotaxanes[60] with two truxene-containing crown-ether wheels as electron donors and an axle containing a oligo(paraphenylenevinylene) unit as the acceptor (Figure 9.21b). Efficient energy transfer from the wheels to the axis is observed even in dilute solution and the solid state. It is thought not only that the [3]rotaxane topology promotes energy transfer, but that it prevents intermolecular aggregation in the solid state. Molecules of this type have been proposed as light-harvesting molecular systems. As an alternative to these systems, an axle containing fullerene groups has been synthesised and used as the

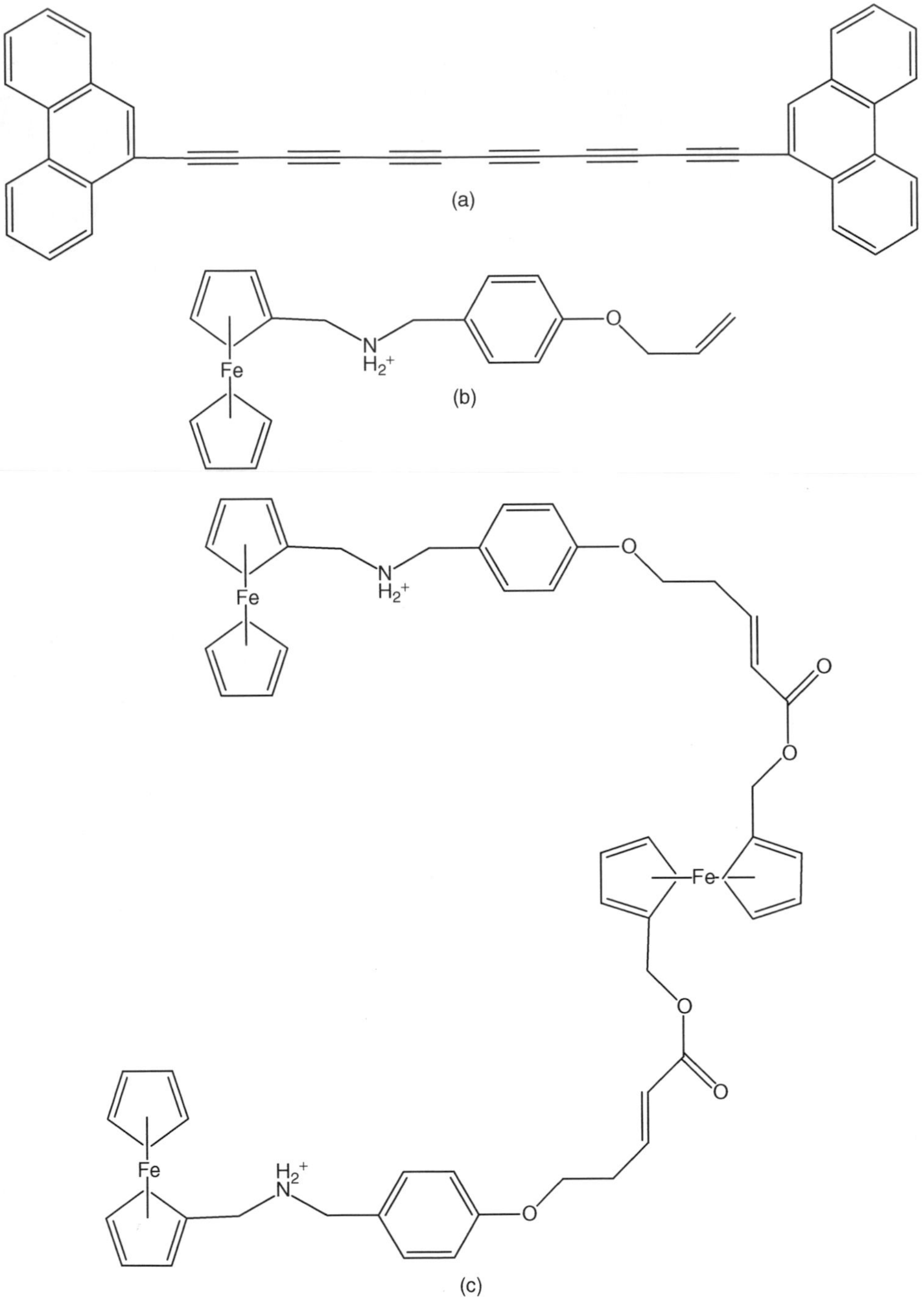

Figure 9.19   *(a) Structure of an oligoalkyne axle, (b) a ferrocene-containing axle and (c) an axle containing three ferrocene units*

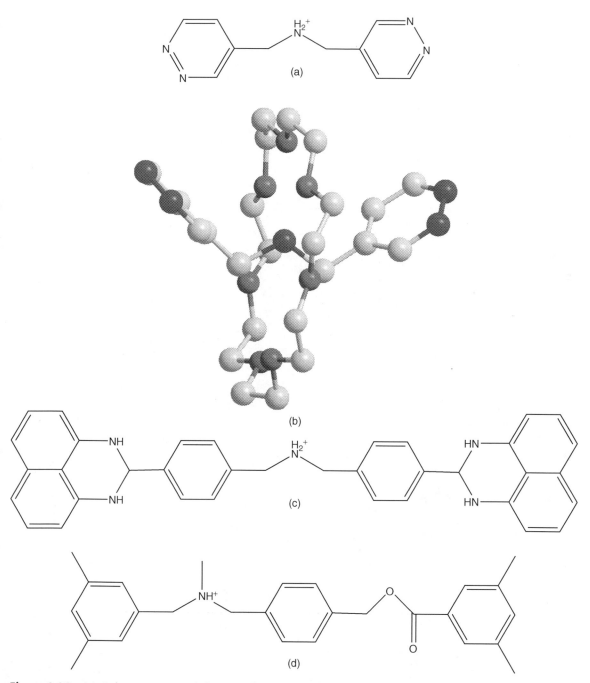

Figure 9.20 *(a) Axle structure and (b) crystal structure of the resultant rotaxane with 21-crown-7, (c) axle structure formed by dialdehyde/1,8-diaminonaphtalene condensation and (d) an axle containing a tertiary ammonium group*

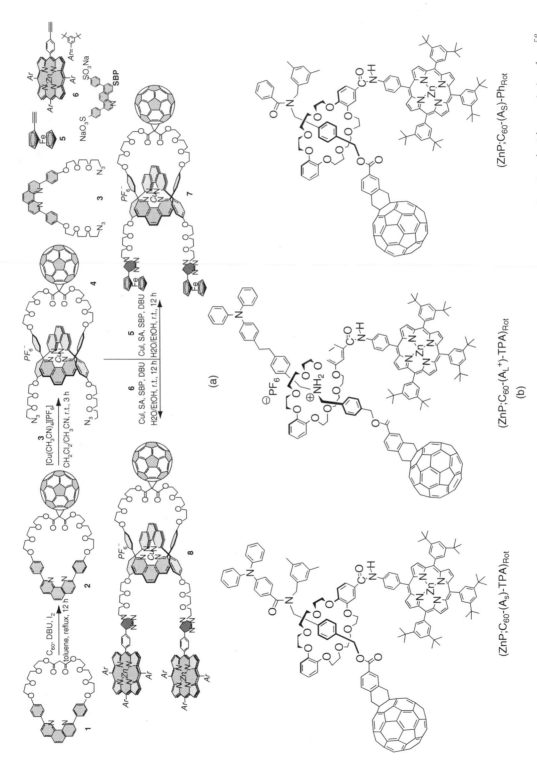

Figure 9.21 (a) One-pot synthetic strategy for the synthesis of a variety of fullerene-containing rotaxanes (Reprinted with permission from[58]. Copyright 2009 American Chemical Society), (b) rotaxanes containing porphyrin and fullerene groups (Reprinted with permission from[59]. Copyright 2009 American Chemical Society) and (c) structure of a light-harvesting rotaxane (Copyright Wiley-VCH Verlag GmbH & Co. KGaA. Reproduced with permission[60])

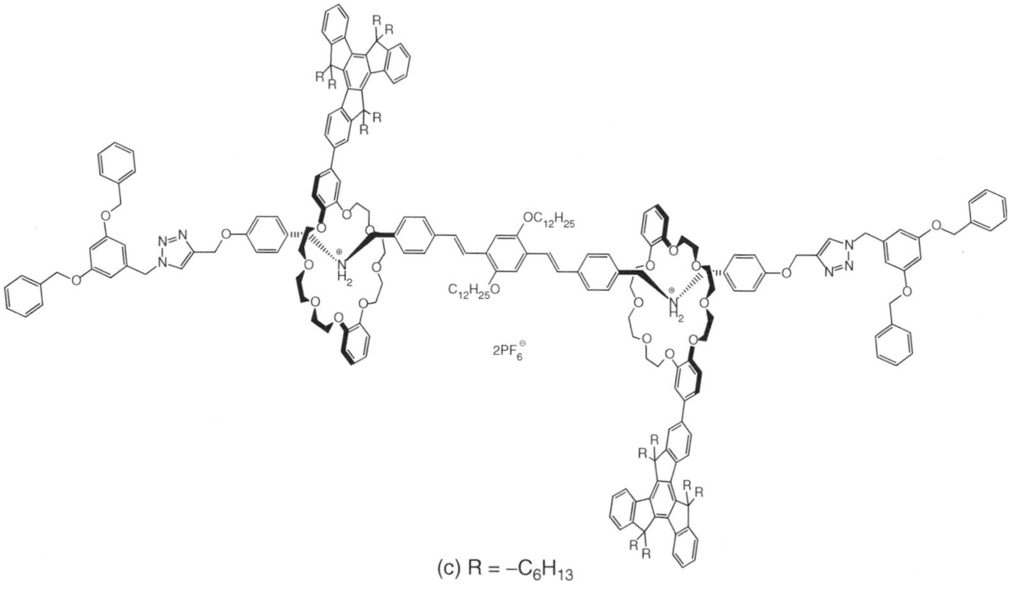

(c) R = –C$_6$H$_{13}$

Figure 9.21   *(continued)*

template for the reaction of a ferrocene-substituted isophthaloyl chloride with *p*-xylylenediamine to give the [2]rotaxane.[61] These rotaxanes have also been modified to contain porphyrin units.

More complex systems can also be synthesised using many of the procedures described above. Clipping reactions have been used to assemble a number of rotaxanes with crown-ether wheels and ammonium axles. Use of an axle containing four ammonium groups allowed formation of a [5]rotaxane containing four crown-ether units.[62] In the same paper, axles containing two or four ammonium units were used to template the clipping reactions of *bis*-crown ethers, thereby forming rectangular [4]rotaxanes, which are shown schematically in Figure 9.22.

Figure 9.22   *Schematic of a rectangular [4]rotaxane*

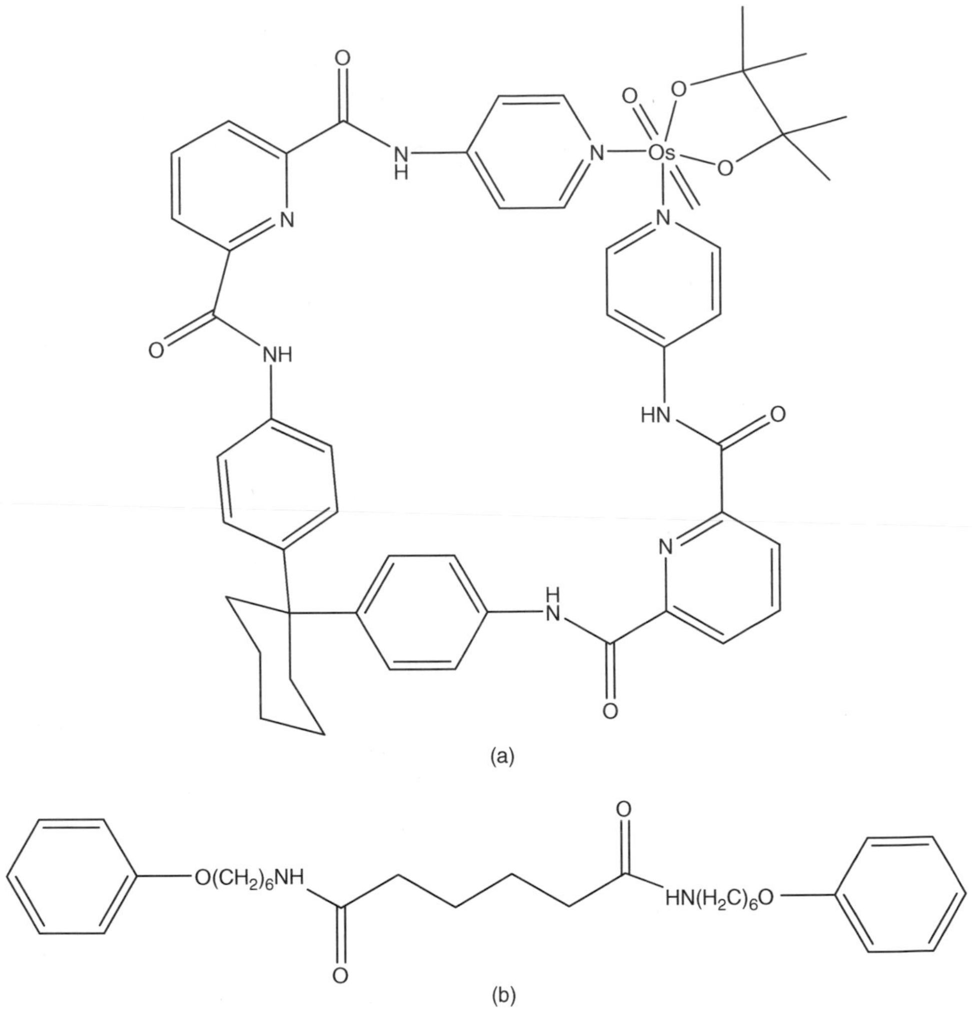

(a)

(b)

Figure 9.23 *(a) An osmium-based macrocycle and (b) a diamide axle used in rotaxane formation*

Metal atoms have been incorporated into many rotaxane systems, as reviewed by Suzaki *et al.*[63] Usually this is via the axle, for example by utilising diaminoalkanes as the axle and capping with cobalt complexes.[10,11] Other workers have utilised organometallic species, for example by attaching ferrocene groups to the axle or macrocycle.[52,61,63] Organometallic or inorganic macrocycles have also been utilised; for example, the macrocycle shown in Figure 9.23a can be assembled in solution with an axle containing two amide groups (Figure 9.23b) to give a pseudorotaxane.[64] Axles with larger trityl end groups can also be utilised and show much slower association and dissociation kinetics. It is thought that in the case of the smaller stopper group the axle simple threads through the macrocycle, whereas in the case of the trityl end group this cannot occur and rotaxane formation proceeds via reversible dissociation and reassociation of the organometallic coordination bonds. In recent work an inorganic wheel consisting of seven chromium atoms and one cobalt atom, held together by coordination with pivalic acid and fluoride anions, was synthesised and shown to form a [2]rotaxane with an ammonium axle.[65] X-ray crystal structures clearly show templating of the macrocycle formation around the central ammonium group of the axle (Figure 9.24a). Using an axle containing two ammonium groups led to formation of a [2]rotaxane in which the macrocycle shuttles between the two binding sites, along with a [3]rotaxane containing two inorganic rings. Replacement of the cobalt species with copper led to formation of larger inorganic rings through which two axles could thread, enabling construction of a [4]rotaxane (Figure 9.24b).

Larger macrocycles have been used to synthesise [3]rotaxanes with one wheel and two axles. As discussed in Chapter 6, γ-cyclodextrin, the largest of the common members of this family, is capable of simultaneously threading two guests. This has made it suitable for the construction of a hetero[3]rotaxane by inclusion of stilbene diboronic acid into the γ-cyclodextrin cavity and capping with terphenyl groups, followed by inclusion of a cyanine dye diboronic guest and further capping.[66] This is the first example of a [3]rotaxane containing two different axles. Other workers reported the high-yield synthesis of [3]rotaxanes by utilising a macrocycle and an axle which both contained isoquinoline units, examples of which are shown in Figure 9.25a,b. Complexation of these types of system with iron(II) salts leads to formation of octahedrally coordinated complexes containing two axles and a macrocycle which can be capped to give the [3]rotaxane.[67] In this work it did not prove possible to remove the iron centre to give metal-free rotaxanes, but later work developed similar complexes based on cobalt which could be demetallated.[68] These systems all had two identical axles, but recent work has utilised stepwise reaction to synthesise [3]rotaxanes with two different axles.[69]

9.2.5 Rotaxanes as molecular machines

The use of these rotaxanes as nanosized 'molecular machines' has been of great interest recently and has been reviewed by Balzani *et al.*[70] and Kay and Leigh.[71] We have already described some examples of molecular shuttles based on rotaxanes. Others include the pH-sensitive rotaxane (Figure 9.26a,b), where under basic conditions amide–amide hydrogen bonds hold the wheel over the peptide residue but under acidic conditions the amine group is protonated and the ammonium–crown ether binding dominates.[72] Irreversible switches have also been synthesised, as shown schematically in Figure 9.27a, where the presence of a bulky side arm prevents the systems from returning. An early example was a rotaxane with an amide macrocycle (Figure 9.17b) and an axle containing two succinimidyl binding sites (Figure 9.27b) but with a bulky silyl group on the central hydrocarbon chain and different stopper groups on each end.[73] Removal of the silyl group and then reattachment means that the ring can shuttle back and forth between the two systems, so if a pure rotaxane with the ring adjacent to one of the stopper groups is synthesised and then undergoes this process, after reattachment of the silyl stopper a 50:50 mix of the two possible positional isomers is obtained. A more complex system invovles one of the succinimidyl being replaced by an unsaturated group (Figure 9.27c). This can be photochemically switched between the *trans* and *cis* isomers. When the pure rotaxane with the wheel located on the fumaramide group is synthesised and the mid-chain blocking group is removed and then replaced after equilibrium is reached, 85% of the resultant mixture still has the wheel located on the fumaramide group. However, if the reaction mixture is irradiated, a different mixture in which 56% of the compound has the wheel located on the succinimidyl group is obtained.[73] Later work continued this theme and developed a molecular ratchet where instead of utilising a chemical mid-chain blocking process, a rotaxane containing a 24-crown-8 wheel substituted with a benzophenone group and an axle with two ammonium binding sites separated by a stilbene moiety was developed.[74] Irradiation of this system changed the relative populations of the macrocycle along the axle. Other work on these types of system included the incorporation of benzyl ester groups as mid-chain blockers. Benzyl esters are chiral and it was found that if chiral esterification catalysts were used, they

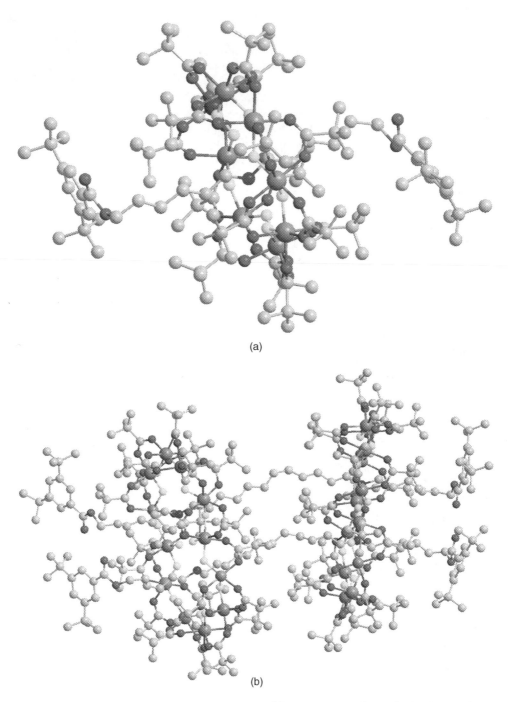

(a)

(b)

Figure 9.24 X-ray crystal structures of (a) a [2]rotaxane and (b) a [4]rotaxane formed using inorganic macrocycles

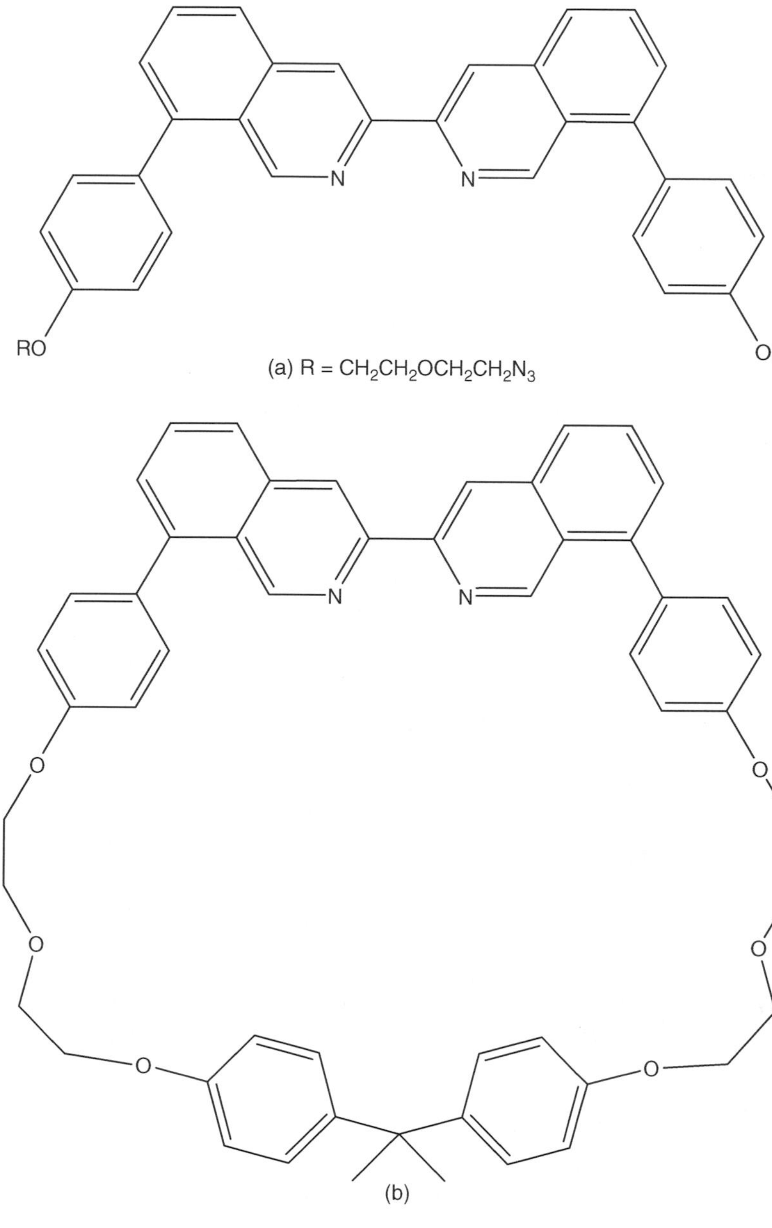

(a) R = CH₂CH₂OCH₂CH₂N₃

(b)

Figure 9.25 *Bipyridyl-containing species: (a) axle and (b) macrocycle*

affected the distribution of the final product away from a 1 : 1 mixture of final enantiomers to a 2 : 1 mixture, controlled by the chirality of the catalyst.[75]

Other controllable shuttles include systems where an amide macrocycle has been threaded onto an axle containing a fullerene group (Figure 9.28a). The macrocycle binds to the amide end of the axle by hydrogen-bonding when in chloroform, but in DMSO this bonding is disrupted and the

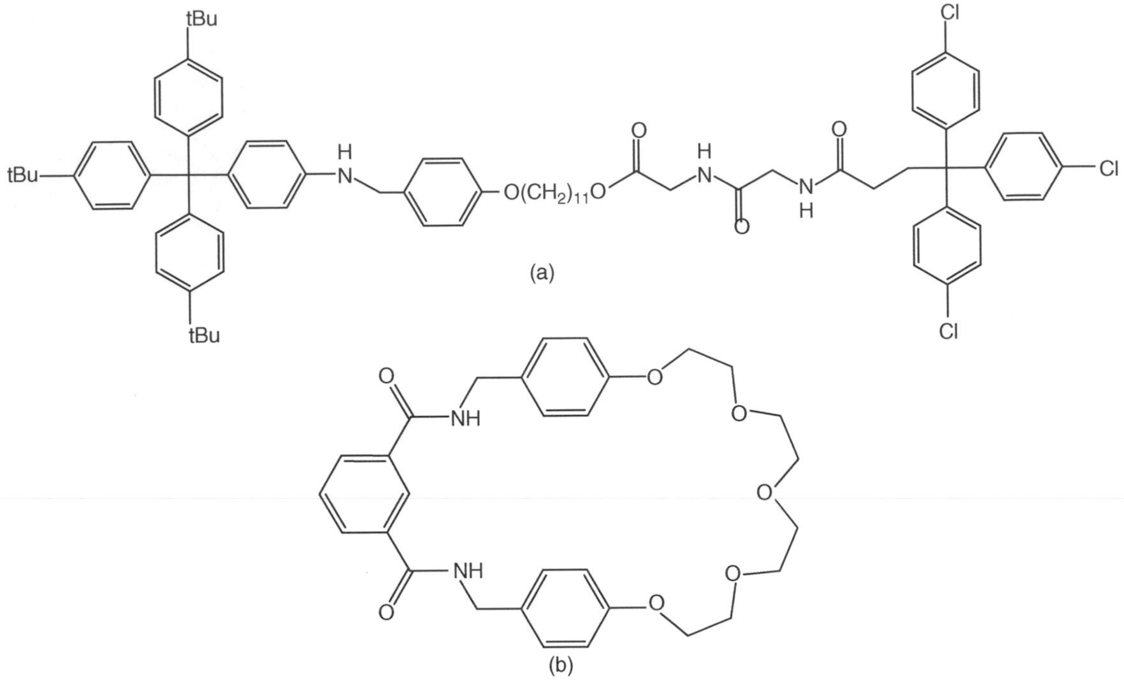

(a)

(b)

Figure 9.26   *pH-responsive molecular shuttle: (a) axle and (b) wheel*

macrocycle shuttles to the fullerene end of the axle.[76] A similar effect can be obtained by electrochemically reducing the fullerene to its trianion. This is unusual since it means the $\pi-\pi$ stacking interactions are stronger than hydrogen-bonding; normally the reverse is true. A more complex system described as a 'molecular elevator' has been designed; it consists of a compound containing three ammonium and three bipyridyl binding sites arranged around a 1,3,5-triphenylbenzene core.[77] This can form a complex with a *tris* crown-ether system which is capped to give the supramolecular system shown in Figure 9.28b. In this system the crown ethers preferentially form complexes with the ammonium units in the triply threaded array; however, under basic conditions these are deprotonated and the crown-ether units slip 'down' the legs to interact with the bipyridyl units.

Anions have also been used to moderate switching; a rotaxane containing an axle with a pyridinium and a triazole binding site along with a crown-ether wheel can be synthesised.[78] When a chloride counterion is used the wheel rests on the pyridinium moiety, but for hexafluorophosphate the wheel shuttles along to the triazole binding site (Figure 9.29a). A rotaxane with an axle containing a tetrathiafulvalene and a triazole binding site along with an α-cyclodextrin wheel has been shown to adopt a conformation in which the cyclodextrin encircles the tetrathiafulvalene unit.[79] However, oxidation of this unit to the radical cation or dication causes the macrocycle to migrate to encircle the triazole unit.

A rotaxane with two potential binding sites has been synthesised with a *bis*-para-phenylene-34-crown-10 wheel and the axle shown in Figure 9.29b,c. The crown ether usually resides on the viologen unit furthest from the ruthenium complex, but irradiation leads to electron transfer from the metal complex to this viologen, converting it to its radical cation.[80] This causes the wheel to shuttle onto the central viologen unit; back transfer of the electron then occurs and the wheel resets itself to its original position. Later work utilised the structure shown in Figure 9.30, which combines a bistable molecular switch with

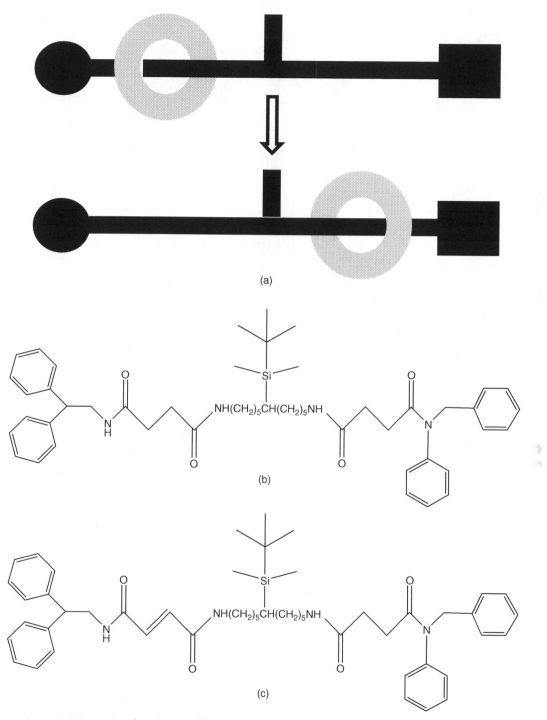

Figure 9.27 *(a) Schematic of an irreversible molecular switch, (b) structure of an axle with central blocking group and (c) fumaroyl-containing axle*

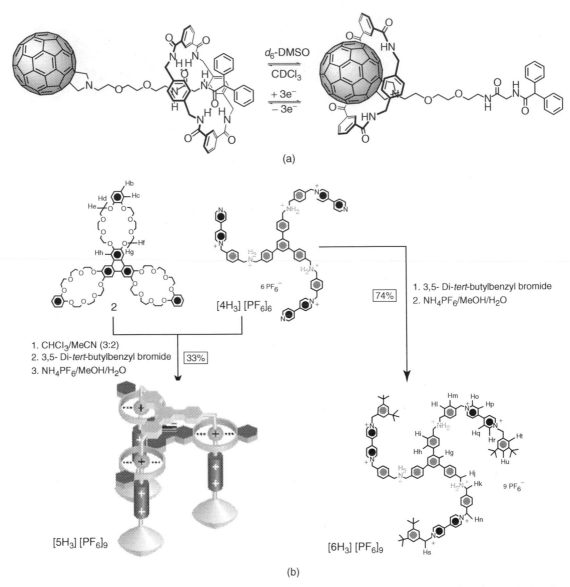

Figure 9.28 *(a) Fullerene-moderated shuttle and (b) schematic synthesis of a 'molecular elevator' (Reprinted with permission from[77]. Copyright 2006 American Chemical Society)*

a light-fueled power generator.[81] Visible light is adsorbed by the porphyrin and leads to formation of a charge-separated state in which the tetrathiafulvalene is oxidised to its radical cation and the electron is accepted by the fullerene unit. Since these are separated by some distance, recombination is relatively slow. The result of this is that the macrocycle shuttles onto the dioxynaphthalene unit. Spectroscopic measurements demonstrate remarkable electronic interactions between the various units, indicating the adoption of folded conformations in solution. As the authors point out, this is reminiscent of natural systems

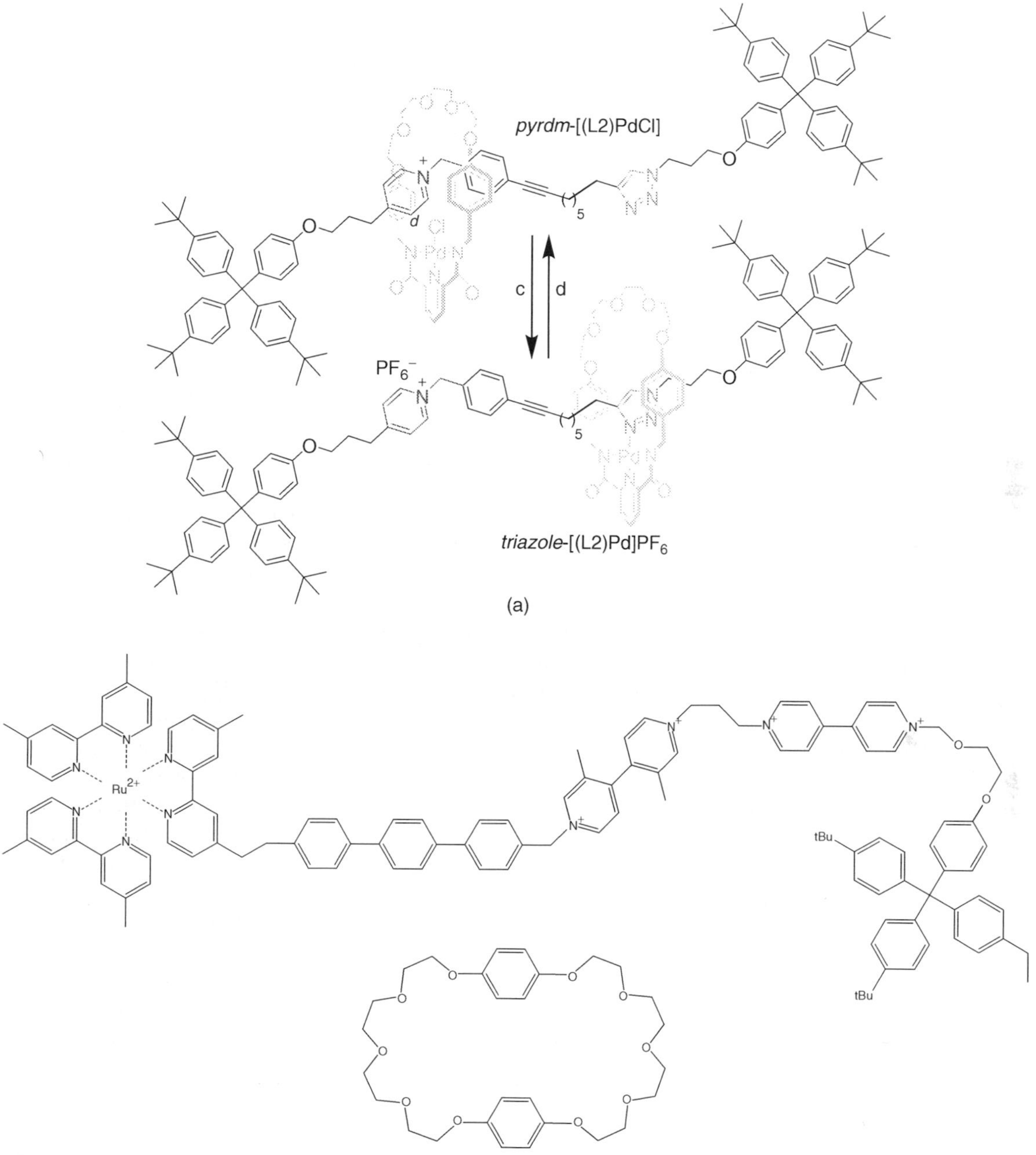
pyrdm-[(L2)PdCl]
c
d
PF6−
triazole-[(L2)Pd]PF6
(a)
Ru2+
tBu
tBu
(b)

Figure 9.29 *(a) An anion-mediated 'molecular shuttle' (Copyright Wiley-VCH Verlag GmbH & Co. KGaA. Reproduced with permission[78]), and (b) axle and (c) wheel of a photoswitchable rotaxane*

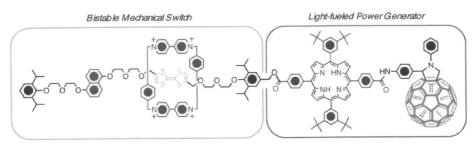

Figure 9.30 *Redox-driven multicomponent molecular shuttle (Reprinted with permission from[81]. Copyright 2007 American Chemical Society)*

in which it is not only the chemical composition that is important in determining the properties of natural compounds but also their secondary and tertiary structures. Metal ions can also be used to mediate switching; for example, the macrocycle shown in Figure 9.25b can be complexed with an axle that contains two potential binding sites, a phenanthroline and a terpyridyl site,[82] with a bipyridyl spacer to aid shuttling of the macrocycle (Figure 9.31a). Binding between the macrocycle and the sites occurs via copper complex formation and the conformation is dependent on the oxidation state of the copper atom. Copper(I) prefers a four-coordinate binding pattern, forming a complex between the phenanthroline unit and the macrocycle, whereas copper(II) binds the macrocycle to the terpyridyl unit in a five-coordinate complex. These two states can be switched electrochemically, as shown in Figure 9.30a.

The possibility of switching these molecules could potentially change many of their properties. Workers have synthesised a [3]rotaxane with two crown-ether wheels substituted with electron-rich pyrene moieties and an axle with a central electron-deficient naphthalene diimide unit.[83] Addition of base to this system causes a change in its conformation, bringing one of the pyrene moieties close to the naphthalene diimide and facilitating formation of a charge-transfer complex, as shown by UV/Vis spectroscopy. Similarly, a [2]rotaxane with a fullerene-stopped axle and a porphyrin-substituted wheel (Figure 9.31b) undergoes a strong photoinduced electron-transfer effect since the conformation of the system allows close interaction between fullerene and porphyrin units.[84] However, when the cationic ammonium unit in the axle is reacted with benzoyl chloride to give a neutral amide, the ring shuttles along the axle and the system adopts a conformation in which they cannot interact (Figure 9.31b).

A four-level fluorescence switch has been developed based on the axle and wheel shown in Figure 9.32a,b. The resultant rotaxane displays strong binding of the crown ether to the axle ammonium as expected and the rotaxane is not fluorescent.[85] Deprotonation of this system causes shuttling of the wheel onto the amide-recognition site and weak fluorescence is then measured (reversible by adding trifluoroacetic acid). Addition of Li^+ to the system leads to complex formation between the alkali metal and the crown-ether and amide units, further increasing fluorescence, with this process being reversible by adding free 12-crown-4. Finally, addition of Zn^{2+} causes a shuttling back of the macrocycle to the amine due to strong coordination of the zinc to the amine group, leading to a high level of fluorescence (this process can be reversed by ethylene diamine tetraacetic acid). This shows the potential for rotaxanes as active components in sensors. Other workers in this field have combined the recognition properties of calixarene macrocycles with the switching abilities of rotaxanes. For example, the axle and wheel structures shown in Figure 9.33a,b could be combined together via a clipping reaction, for which the chloride anion is thought to act as a template, to give a [2]rotaxane (Figure 9.32c).[86] This compound has been shown to respond to the presence of anions in acetone solution, with the addition of chloride, nitrate and especially hydrogen sulfate tetrabutylammonium salts leading to significant enhancement in emission

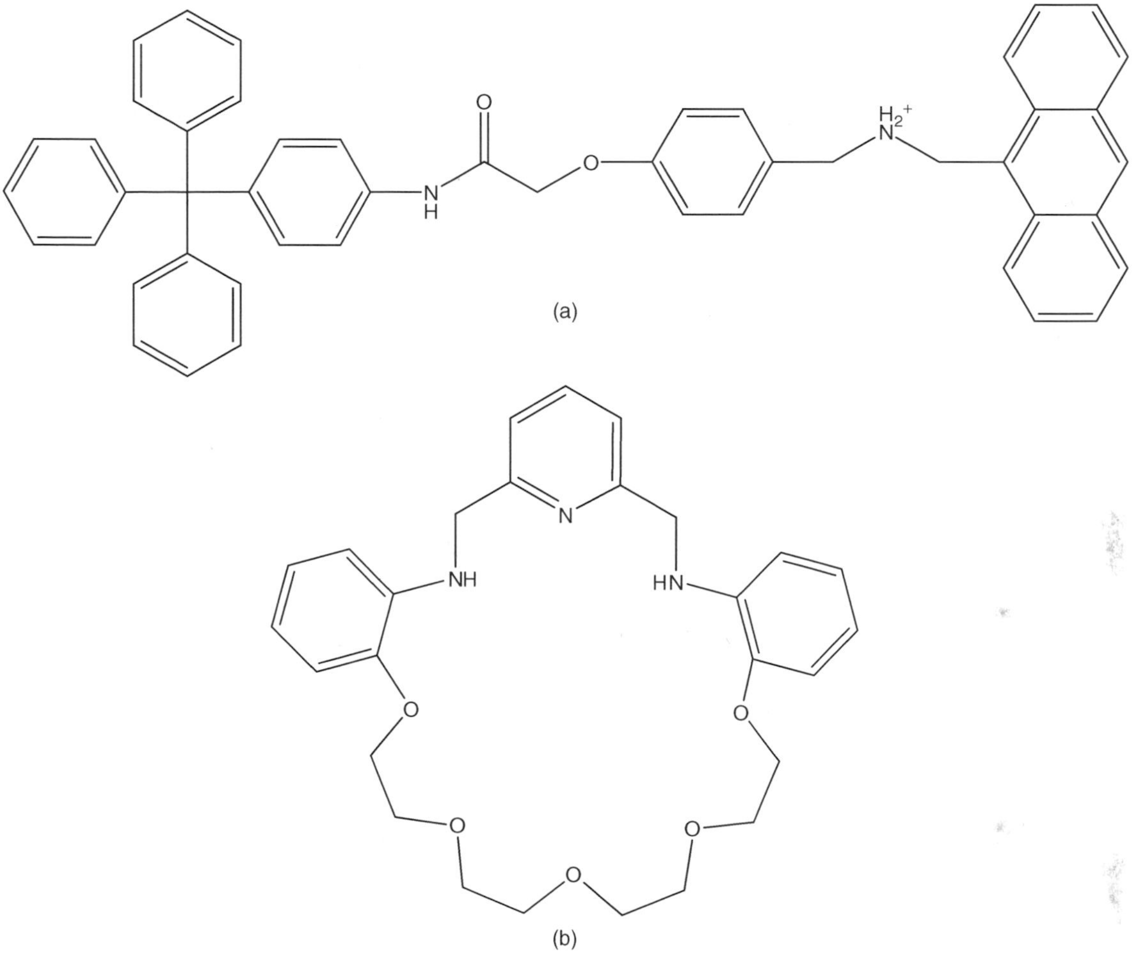

(a)

(b)

Figure 9.31   *(a) Axle and (b) wheel of a rotaxane that acts as a four-level switch*

intensity. This is thought to be due to anion complexation inside the rotaxane structure increasing its rigidity and hindering nonradiative decay processes. Other work by the same group incorporated a calixarene unit into the wheel of [2]rotaxanes,[87] the structure of one of which was proved by X-ray crystallography. The resultant rotaxane bound chloride preferably but also had good affinity for bromide and dihydrogen phosphate, and a lower affinity for acetate. Using a longer, less-rigid axle gave a rotaxane which still bound ions but with lower efficiency and selectivity.

Recent papers have described more complex molecular machines. A molecular crank has been developed where the rotation of two macrocyclic molecules causes movement of a molecular piston (Figure 9.34a). NMR studies of this system show there is movement and that it is a function of the interaction between the axle and wheel of the rotaxane, with strong interactions leading to very slow movement and less strong interactions allowing much more freedom of movement.[88] Other workers[89] have described a rotaxane system in which a long axle containing two ammonium binding sites is threaded through a crown-ether-based cage (Figure 9.34b). The addition and removal of fluoride ions causes conformational changes that

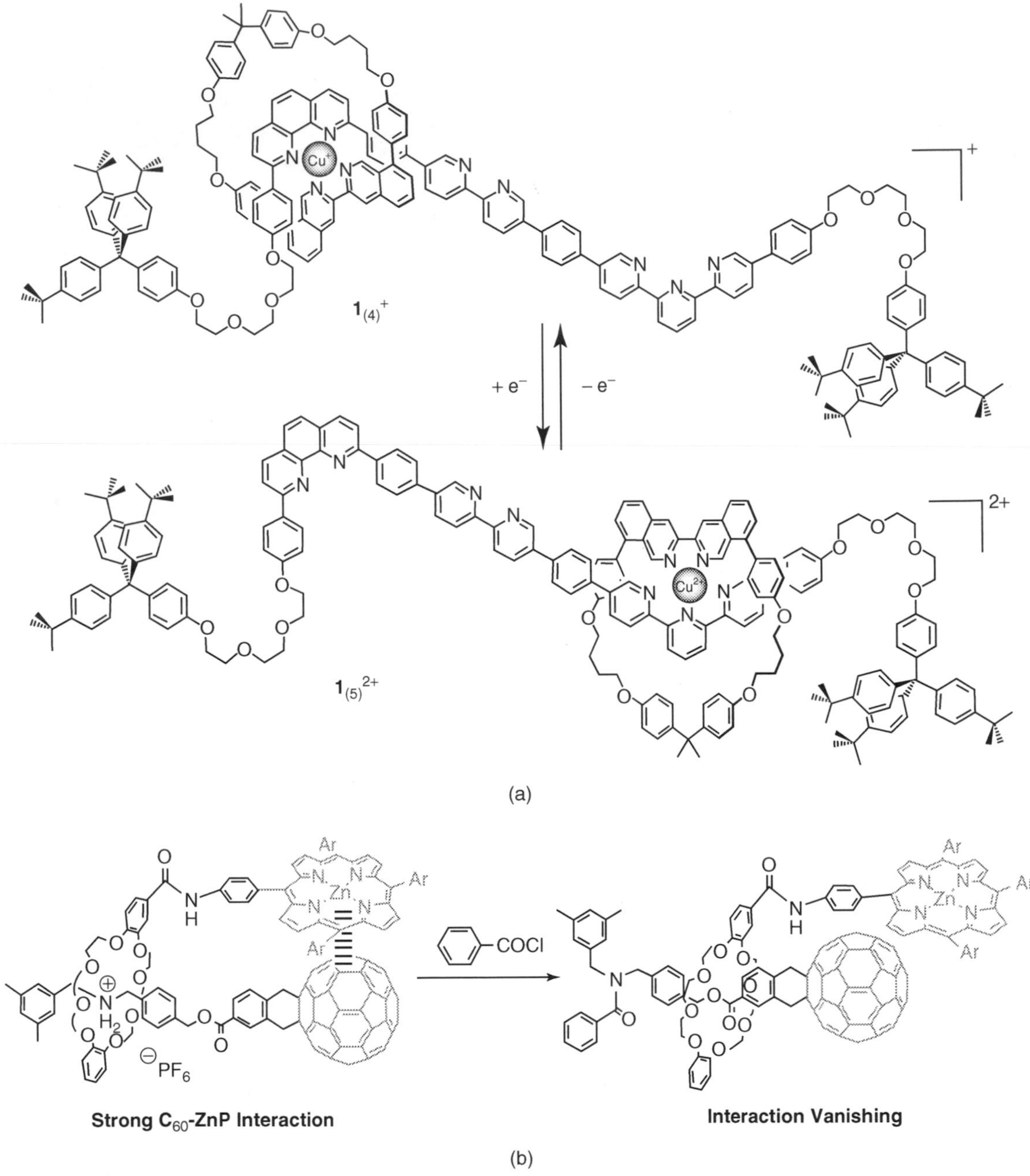
1(4)+
+ e−
− e−
1(5)2+
(a)
COCl
PF6
Strong C60-ZnP Interaction
Interaction Vanishing
(b)

Figure 9.32 *(a) Structure of a copper-mediated 'molecular shuttle' (Copyright Wiley-VCH Verlag GmbH & Co. KGaA. Reproduced with permission[82]) and (b) the axle and wheel of a fullerene/porphyrin-containing rotaxane (Reproduced by permission of the PCCP Owner Societies[84])*

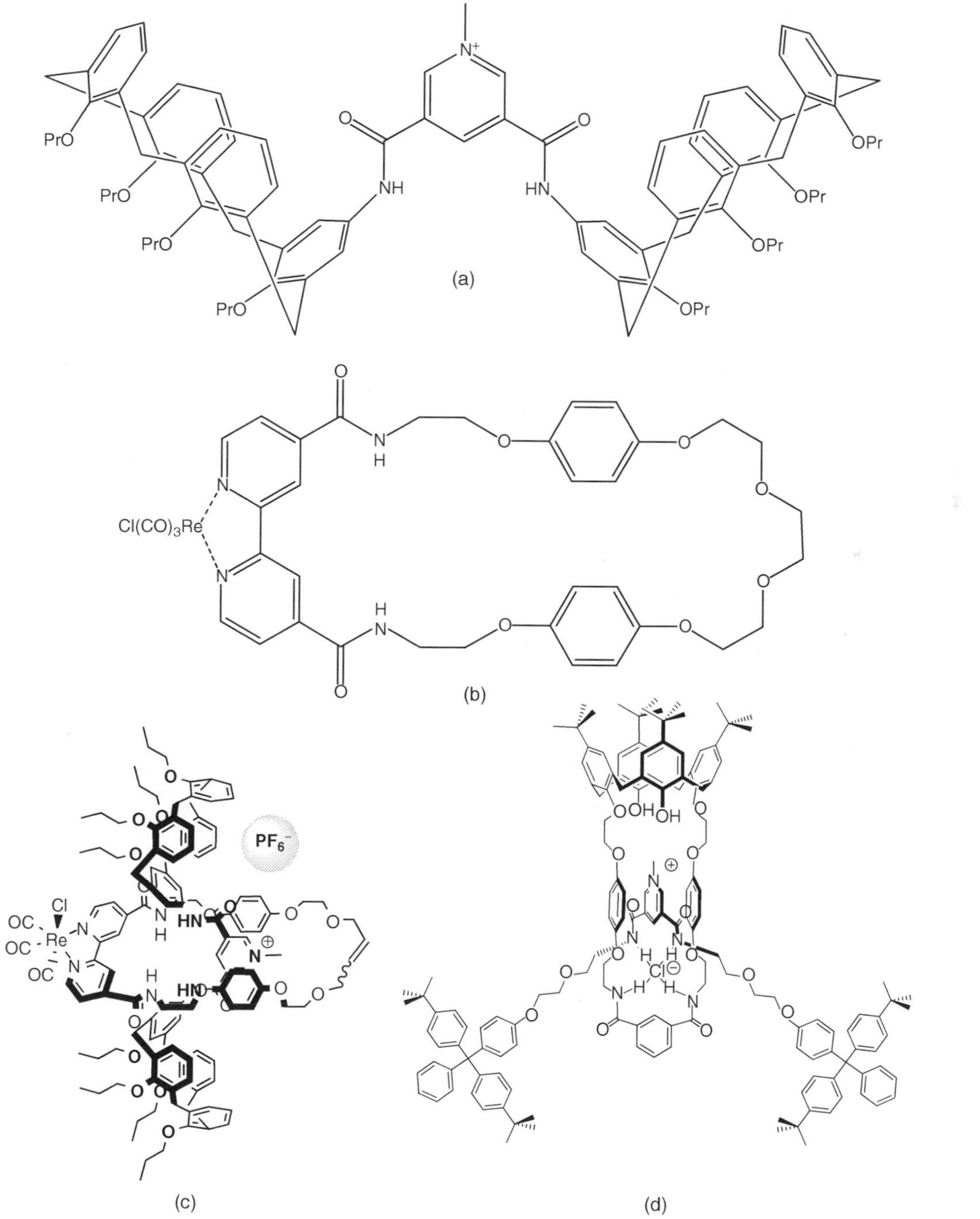

Figure 9.33 *(a) Axle and (b) wheel of a (c) rotaxane that acts as a sensor for hydrogen sulfate ions (Reproduced by permission of the Royal Society of Chemistry[86]) and (d) an anion-sensitive calixarene rotaxane (Copyright Wiley-VCH Verlag GmbH & Co. KGaA. Reproduced with permission[87])*

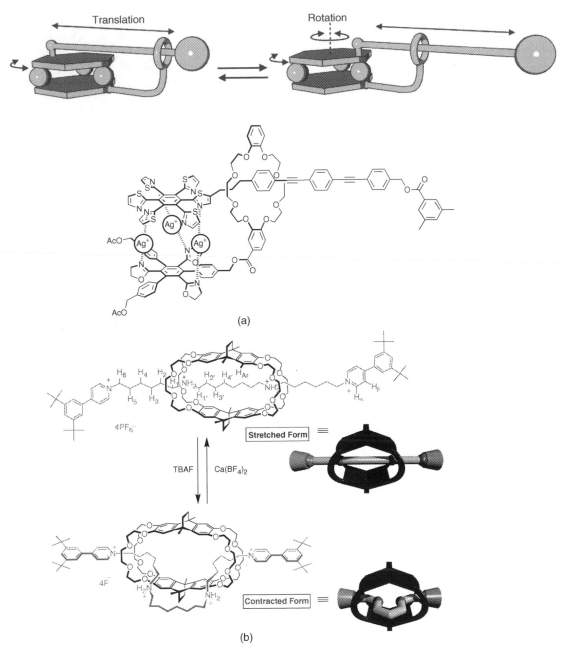

(a)

(b)

Figure 9.34 *(a) Structure and operation of a molecular crank (Reproduced by permission of the Royal Society of Chemistry[88]) and (b) structure and operation of a molecular muscle (Reprinted with permission from[89]. Copyright 2009 American Chemical Society)*

lead the molecule to switch between extended and contracted forms with length changes of about 36%, larger than those found for human muscle (27%). The possibility of using rotaxane molecules as molecular pistons has also been discussed in the literature.[90]

9.2.6 Thin films of rotaxanes

There has been much interest in using rotaxanes in a variety of possible applications, but they will probably need to be immobilised in some kind of device rather than existing in solution or as a powder. This has led to investigations of rotaxanes incorporated into thin films by a variety of methods. We have already discussed how an amine macrocycle can be threaded onto an amide-containing axle and how shuttling of the macrocycle along the axle induced by change of solvent[76] leads to a change in fluorescence properties. A similar axle with an anthracene stopper can be incorporated within a pseudorotaxane and the other end of the axle polymerised along with methyl methacrylate to give a rotaxane-containing polymer film.[91] Cast films of the resultant polymers are not fluorescent when irradiated with a UV lamp, but upon exposure to DMSO vapour a blue fluorescence can be observed under UV irradiation. Masking of the polymer from the vapour allows the production of fluorescent grid patterns. In other work, rotaxanes containing fumaramide moieties in the axles and cyclic amide wheels have simply been cast onto substrates such as pyrolytic graphite.[92] The film morphology depends greatly on the nature of casting solvent and on the structure of the rotaxane.

More controlled methods of deposition of thin films of rotaxanes have already been studied. The rotaxane formed between dibenzo-24-crown-8 and the axle shown in Figure 9.35a can be cast onto water along with a suitable phospholipid and co-deposited as Langmuir–Blodgett multilayers.[93] LB films containing the rotaxane can be deposited on electrodes and cyclic voltammetric measurements can be performed on the films. Exposure to HCl or ammonia vapour causes changes in the films' electrochemical behaviour, which is thought to be due to pH-dependent shuttling of the crown onto and off the bipyridyl unit. Model bipyridyl compounds without a rotaxane structure do not show this behaviour. There has been a detailed study of the structures of some LB films of these rotaxanes. Low-angle X-ray studies on monolayers of these compounds at the air–water interface showed that these rotaxanes were highly hydrated and formed highly tilted or folded monolayers.[94] Increasing the surface pressure both reduced the amount of hydration and increased monolayer thickness. Monolayers of the axles alone displayed much less pronounced hydration and tilting, indicating that hydration and tilting are caused by the presence of the hydrophilic wheel component. It was also shown that injection of an oxidant, ferric perchlorate, into the subphase led to changes in electron density throughout the monolayer compatible with oxidation of the tetrathiafulvalene unit and shuttling of the wheel onto the 1,5-dioxanaphthalene unit.[95] Two isomeric rotaxanes were used in this study, that in Figure 9.36a and an alternative version in which the positions of the tetrathiafulvalene and 1,5-dioxanaphthalene units are reversed; both clearly demonstrated shuttling of the macrocycle in the expected direction. Other work varied the hydrophilic headgroups of these materials, making changes in the length of the oligoethylene chains and utilising both —OMe and —OH headgroups, and studyied the effects of these changes on the monolayer behaviour.[96] All the materials tested gave good-quality deposition to form Langmuir–Blodgett multilayers. Much of this experimental work has been confirmed by extensive molecular dynamics situations,[97] which show that as the area per molecule increases so does the molecular tilt, whilst the thickness of the monolayer decreases.

Self-assembly methods have also proved suitable for immobilising rotaxanes at surfaces; for example, the tripodal rotaxane consisting of the axle shown in Figure 9.35b and a crown ether (Figure 9.7a) forms monolayers on a gold substrate.[98] Electrochemical measurements indicate a surface coverage of 1.1×10^{-10} mol cm^{-2}. The tripodal axle without tetraphenyl stopper groups also adsorbs to form a monolayer on gold and can bind the crown ether from aqueous solution to form a pseudorotaxane. A similar axle, however,

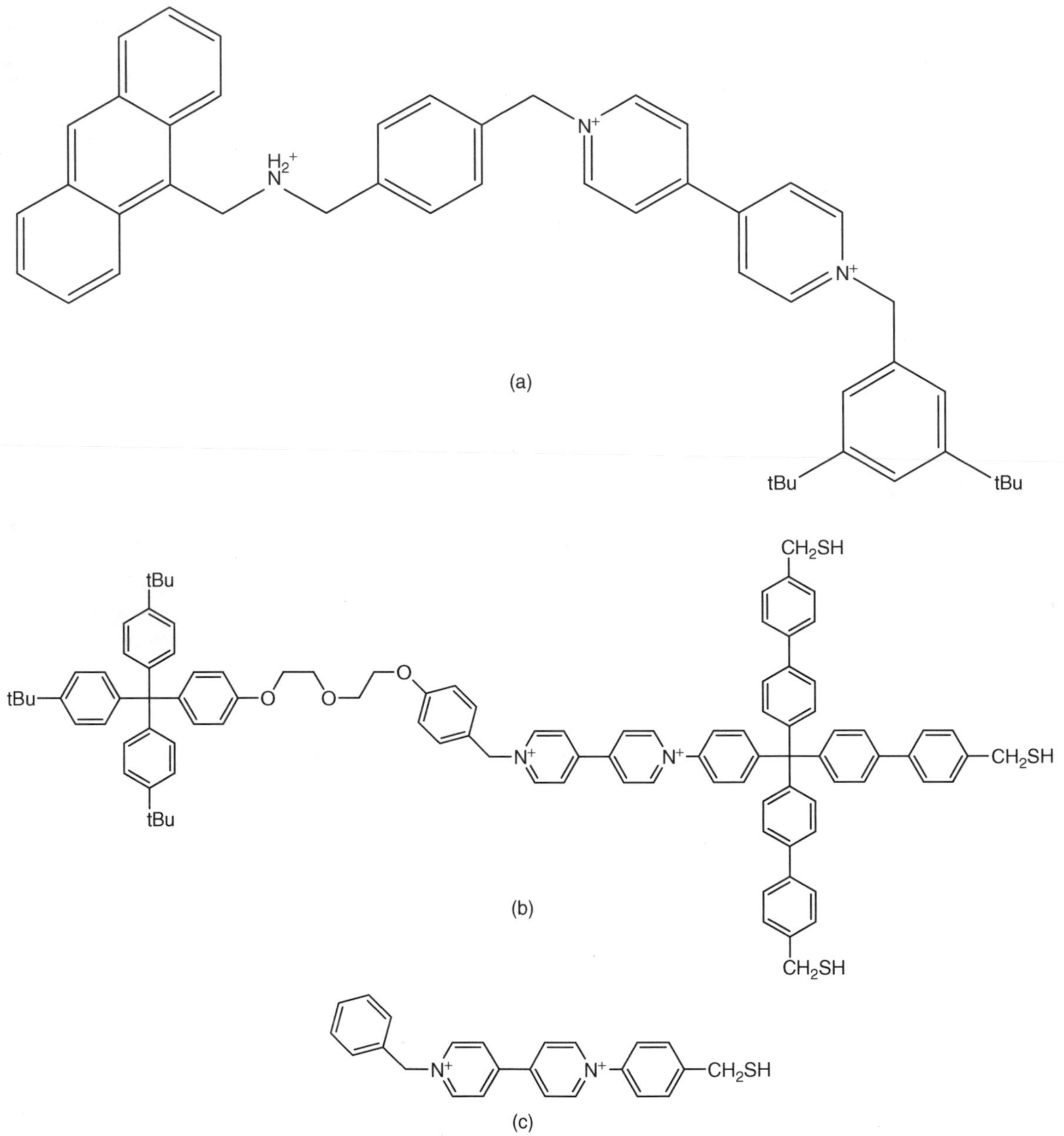

(a)

(b)

(c)

Figure 9.35 *Axles of rotaxanes deposited by (a) the Langmuir–Blodgett technique and (b,c) self-assembly onto gold*

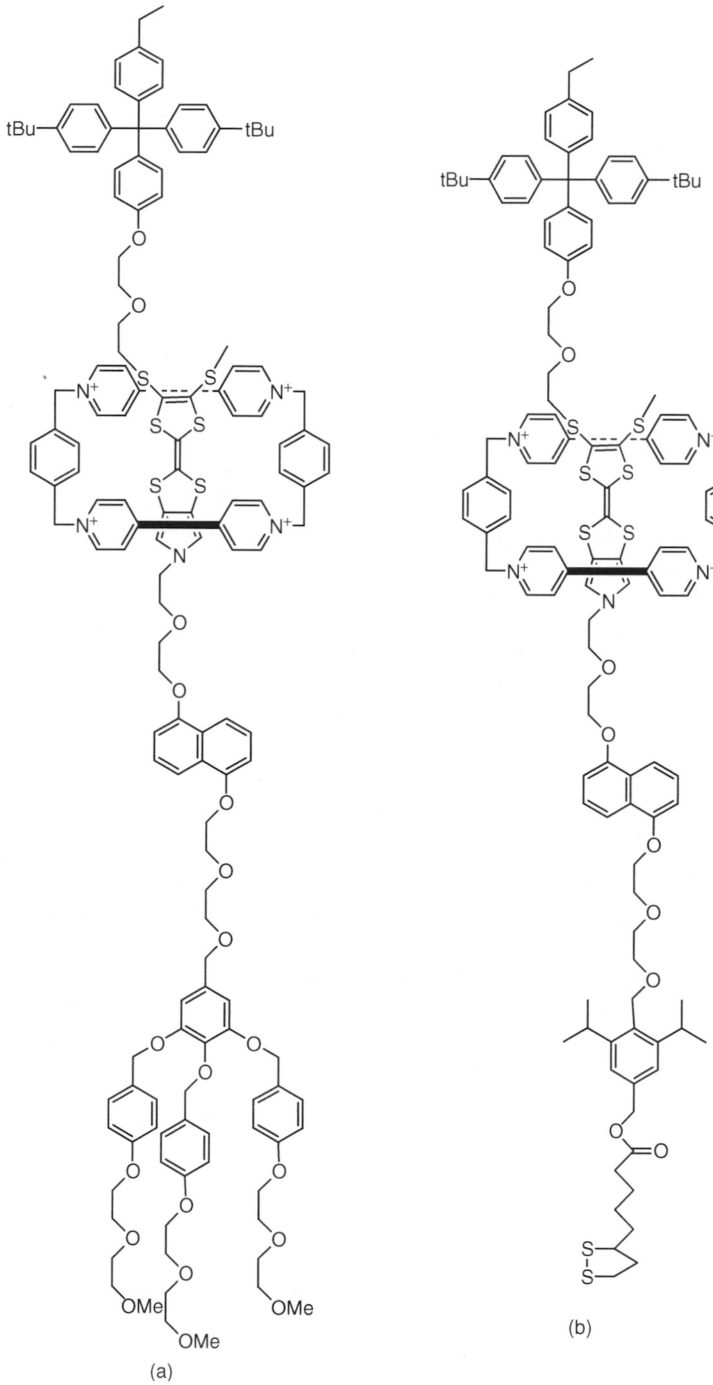

Figure 9.36   Rotaxanes deposited by (a) the Langmuir–Blodgett technique and (b) self-assembly onto gold to give molecular switches (Reprinted with permission from[94,99]. Copyright 2005 and 2006 American Chemical Society)

with a much smaller headgroup (Figure 9.35c), forms a layer with much higher coverage (2.7×10^{-10} mol cm^{-2}) and cannot bind the crown from solution, probably because the axles are too tightly packed to allow encirclement by the macrocycle. Molecular modelling has also been used to simulate these systems and concurs that oxidation of the tetrathiafulvalene occurs along with shuttling.[99]

The use of rotaxanes in molecular electronics devices has been reviewed by Jang and Goddard.[99] Various types of devices including rotaxanes have been proposed. Rotaxanes (shown in Figure 9.36a) can be transferred using the Langmuir–Blodgett technique onto silicon substrates to give smooth monolayers.[100–102] Coating these samples with metals allows electrical contact to be made and the resulting system can be switched electrically. In this system the molecule is initially in the ground state and about 90% of the macrocycles encircle the tetrathiafulvalene unit, but on application of +2.0 V it switches into a metastable excited state with a concurrent increase in conductivity. At this voltage the tetrathiafulvalene is oxidised to the radical cation or dication and the wheel shuttles along the axle to exclusively encircle the 1,5-dioxynaphthalene unit. The rotaxane can be switched from one state to the other and back over at least 35 cycles with no loss of performance. These systems, in conjunction with silicon nanowire electrodes, can be used to fabricate simple 64 bit memory devices[101] with approximately 10^8 molecules per junction. Using these devices, text strings in ASCII code can be written, stored, read and erased. Larger arrays have also been fabricated.[100] Infra-red studies were made on these systems after evaporation of Ti, which showed that in the case of layers deposited from tightly packed monolayers on water, the Ti deposition process has a minimal effect on the IR spectra, indicating that no damage to the films occurs during the Ti deposition process.[103] Monolayers deposited at lower pressure are more loosely packed and demonstrate changes in IR spectra and poor switching behaviour, thereby indicating that in tightly packed films the bulky stoppers 'protect' the monolayer during the Ti deposition process. Similar rotaxanes can be deposited as LB films and written on by applying voltage pulses using a scanning tunnelling microscope.[104] This enables reversible, erasable and rewritable nanorecording with a dot size of 3 nm. Raman spectra confirmed the switching of the rotaxanes. XPS studies of these systems have also confirmed movement of the macrocycle along the axle after chemical and electrochemical switching.[105]

Similar rotaxanes containing sulfur species have also been synthesised and self-assembled as monolayers onto gold surfaces.[106] Cyclic voltametry showed that, as in the Langmuir–Blodgett films, these rotaxanes could be electrochemically switched between different states. In both types of film, relaxation back to the ground state was much slower than in solution. Other workers formed layers of mercaptoundecanoic acid on gold and utilised these to immobilise [2]rotaxanes containing an amide macrocycle and succinimide and naphthalene diimide units in the axle.[107] This material acts as a three-way switch: at equilibrium the wheel is preferentially located on the succinimide unit, at −0.68 V (vs SCE) it switches so that the wheel is in equilibrium between the two sites and at −1.21 V the wheel is located on the naphthalene diimide unit.

As an alternative to adsorbing rotaxanes onto a surface, a different approach is to utilise the surface as part of the rotaxane; that is, as one of the stoppers. One of the earliest examples of this is where the compound shown in Figure 9.37a was synthesised on a gold surface by stepwise chemical reactions. This axle can adsorb from solution a ferrocene-substituted β-cyclodextrin to give a pseudorotaxane in which the macrocycle can slip off the axle.[108] After capping the axle with an anthracene stopper this dissociation is no longer possible and the species can then be thought of as a surface-bound rotaxane. This is an interesting example of optical switching, in that when the azobenzene unit is in the *trans* form the cyclodextrin encircles it and is located close to the gold surface. This allows fast electron transfer between the ferrocene substituent and the gold electrode. Irradiation switches the azobenzene to the *cis* form however and the cyclodextrin disassociates and shuttles along the alkyl chain, increasing the ferrocene–gold distance and lowering the rate of electron transfer. A similar method was used to construct the axle shown in Figure 9.37b, which forms a complex with a *bis*-viologen macrocycle (Figure 9.7e) and can be capped with an adamantyl unit to give the rotaxane.[109] The macrocycle encircles the π-donor

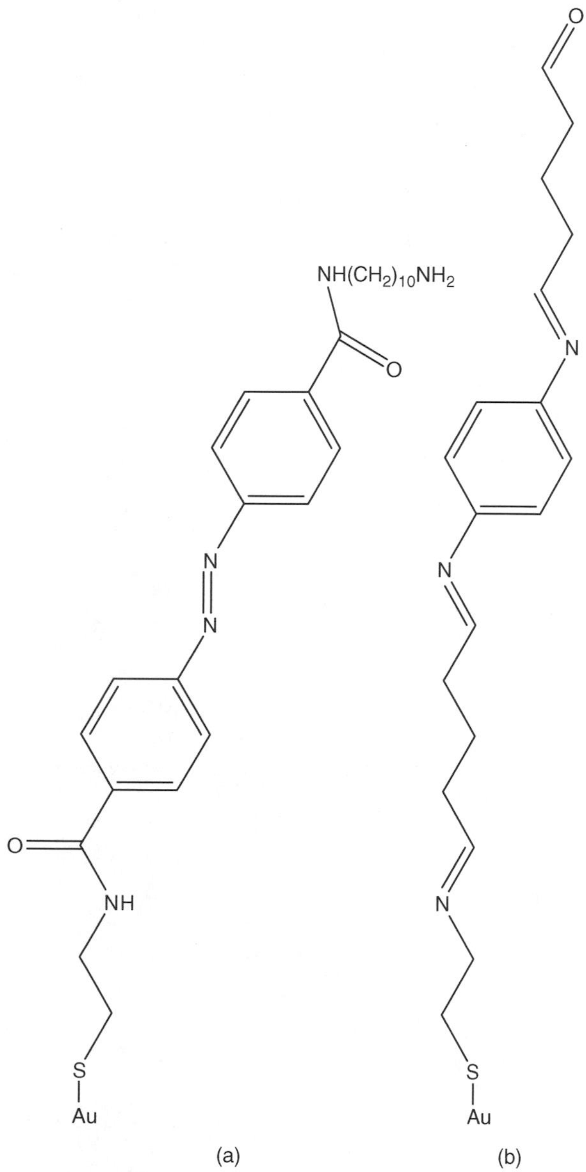

Figure 9.37 *Thiol-substituted axles used to immobilise (a) a cyclodextrin and (b) a cyclic* bis-*viologen on gold*

diiminobenzene unit, but on electrochemical reduction of the macrocycle-respective biradical dication this complex is no longer energetically favourable and the macrocycle shuttles towards the electrode. The same rotaxane was used to bind flavin adenine dinucleotide and immobilise apo-glucose oxidase at the electrode surface.[110] This enabled the detection of glucose at −0.4 V (vs SCE), a much lower potential than is found in many other biosensors, and showed that the rotaxane is facilitating electrical contact between enzyme and electrode. Other workers have also immobilised rotaxanes with ferrocene units in both axle and wheel and demonstrated their ability as anion sensors with a high specificity for chloride.[111]

The switching effect of rotaxanes at a surface has been shown to modify the surface properties of the resultant film. For example, when the rotaxane mentioned above[109] is electrochemically reduced, the hydrophilic *bis*-viologen macrocycle shuttles away from the surface, rendering it more hydrophobic. Other workers have studied light-switchable surfaces, for example by immobilising a cyclodextrin/azobenzene rotaxane at the surface.[112] Switching the azobenzene from *trans* to *cis* caused shuttling of the cyclodextrin away from the surface, increasing the water contact angle from 70° to 120°. Other workers immobilised a fumaramide-containing rotaxane onto a surface and by shining light on one edge of a droplet of diiodomethane were able to change the surface contact angle on that edge, causing the drop to preferentially move to one side.[113] By moving the light spot it was possible to move the droplet in a controlled manner across the surface and even up a 12° incline.

A [3]rotaxane (Figure 9.38) can be immobilised onto a silicon cantilever.[114] When exposed to an oxidant solution, the tetrathiafulvalene units are oxidised, causing migration of the macrocycle and

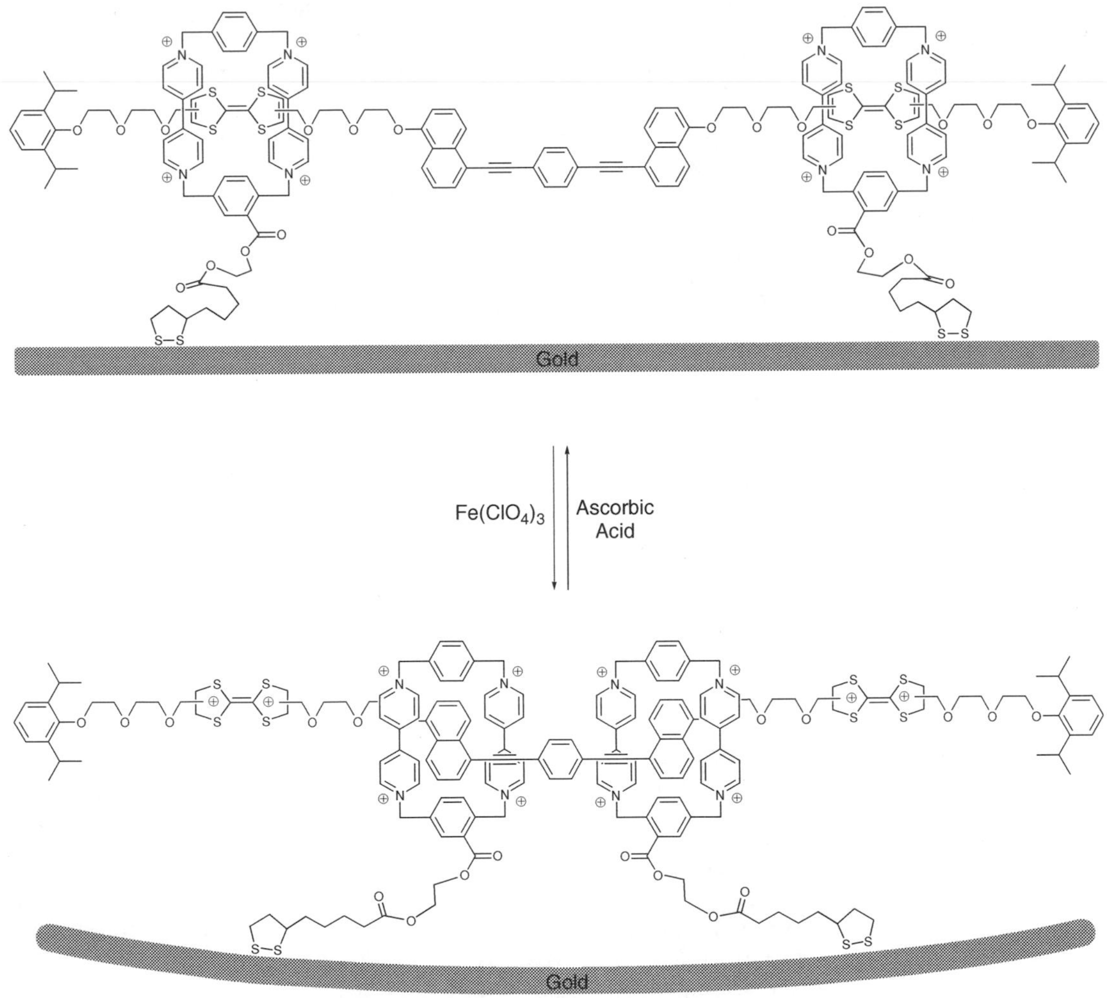

Figure 9.38   *Rotaxanes immobilised on a cantilever act as 'molecular muscles' (Reproduced by permission of the Royal Society of Chemistry*[117]*)*

contraction of the rotaxane. This means each rotaxane acts as a molecular muscle and between them the immobilised molecules cause bending of the cantilever substrate. Reduction with ascorbate reverses the effect. Other workers have immobilised rotaxanes onto porous silica substrates and used the immobilised molecules as nanovalves.[115] Upon switching, these valves open and close and can be used to entrap and release dye molecules.

Rotaxanes can also be dispersed in polymer electrolyte gels and have been shown to be capable of being used in electrochromic devices.[116] In the ground state the rotaxane–polymer composite is green but upon electro-oxidation it switches to a red/purple colour and relaxes back to the ground state after removal of the potential. Many of the applications discussed within this section, and further applications, have recently been reviewed.[117]

9.2.7   Polyrotaxanes

Rotaxanes can be included in a wide range of polymers. Since this book is centred around macrocyclic compounds rather than polymers, we will not give a large number of examples of polyrotaxanes but instead limit ourselves to leading examples of each type. Much of the work on polyrotaxanes up to 1999 has been reviewed by Raymo and Fraser,[118] in the Bibliography and elsewhere in the literature.

Many of the earliest polyrotaxanes were synthesised by threading macrocyclic compounds onto linear polymers, giving 'molecular necklaces' with the basic structure shown in Figure 9.39a. Common examples

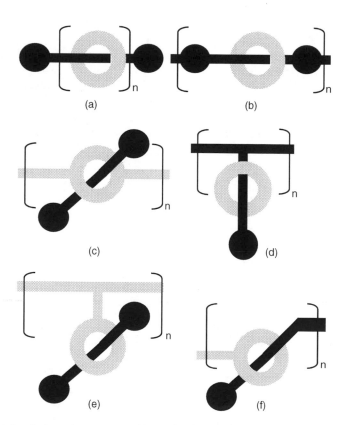

Figure 9.39   *(a,b,c) Main-chain polyrotaxanes, (d,e) side-chain polyrotaxanes and (f) daisy-chain polyrotaxanes*

include the systems obtained with cyclodextrins and cucurbiturils, with polymers such as polyethylene glycol, and many of these systems have already been detailed in the relevant chapters on these macrocycles. Incorporation of macrocycles onto polymers often has major effects on their properties; for example, polyrotaxanes can be synthesised by the condensation of dimethyl sebacate with triethylene glycol in the presence of crown ethers, followed by capping with triarylmethane groups.[119] A polyrotaxane with one crown-ether unit every four polymer repeat units (on average) can be obtained and is a viscous liquid while the parent polymer is a high-melting solid. Many other polymers have been shown to have their glass transition temperatures lowered by rotaxane formation.[1,118]

Other systems include threading the commonly used *bis*-viologen macrocycle (Figure 9.7a) onto a variety of polymers containing either hydroquinone or 1,5-dioxanaphthyl ether units.[120] Alternatively, the macrocycle can be threaded onto a preformed polymer containing both of these electron-rich aromatic groups.[121] In these families of polyrotaxanes the macrocyclic rings are free to interact with each other. There has been interest in the use of macrocycles such as cyclodextrins to encapsulate conductive polymers such as poly(para-phenylene), poly(4,4′-diphenylene vinylene) and polyfluorene.[122] The effects of threading cyclodextrins onto these 'molecular wires' are to increase solution processability and increase their photoluminescence efficiency whilst reducing intermolecular interactions and aggregation. These composite materials have been used in the fabrication of light-emitting diodes. Other workers recently synthesised a permethylated α-cyclodextrin with a polymerisable aromatic side chain, which formed a self-inclusion complex in which the aromatic unit is bound within the cavity.[123] Coupling of this aromatic unit with another aromatic monomer leads to formation of a polyrotaxane containing a conjugated organic polymer axle and cyclodextrin wheels.

Polymers of the type shown in Figure 9.39b are similar to those above except that the macrocyclic rings are held apart by bulky stopper groups and cannot interact with each other. Examples of these types of polyrotaxane include cyclodextrins that have been complexed with stilbene polymers[124] and then photochemically reacted with further stilbene monomers; in the final substance the macrocycles are separated by tetraphenyl cyclobutane moieties (Figure 9.40a).

The two families of polyrotaxanes above both utilise main-chain polymeric axles with the macrocycles attached via 'mechanical' bonds. In another family the macrocycles are part of the main chain, held by covalent bonds and threaded to give the polyrotaxanes. For example, *bis*(5-acetoxymethyl-1, 3-phenylene)-32-crown-10 can be polymerised with sebacoyl chloride to give the poly(crown ether) shown in Figure 9.40b. Self-threading can be a problem in synthesis of these types of material, but this can be avoided by the use of the strong hydrogen-bonding solvent DMSO.[125] Polypseudorotaxanes can then be formed with viologen compounds and give strong orange colours, indicating that charge-transfer interactions are occurring. Melting the polymer above its T_g leads to loss of pseudorotaxane structure. Obviously, capping these systems will give polyrotaxanes.

Another variant on this scheme is to combine the axle and macrocycles shown in Figure 9.41 via a copper(I) template to give a pseudorotaxane and then use an electropolymerisation method to deposit the polypseudorotaxane onto a platinum electrode.[126] Removal of the copper template leads to collapse of the rotaxane structure, but this can be mitigated by the use of lithium salts. Addition of methyl groups *ortho* to the nitrogen atoms in the wheel greatly alters the physical and electrochemical properties of the resultant polypseudorotaxane[127] and gives a material with reversible copper(I) decoordination/recoordination properties. A similar axle containing a bipyridyl rather than a phenanthroline unit can be complexed with the wheel in Figure 9.41b and electrodeposited as an electroactive polymer film.[128] This material's electroactive properties vary with the presence or absence of various metal ions, leading the authors to propose its potential use as a sensor material. Another approach utilises the pyrrole-containing axle shown in Figure 9.41c, which can be used to form a pseudorotaxane.[129] Once again this can be electrodeposited to give a film of electroactive polypyrrole-based polyrotaxane. A substituted hexathiophene compound

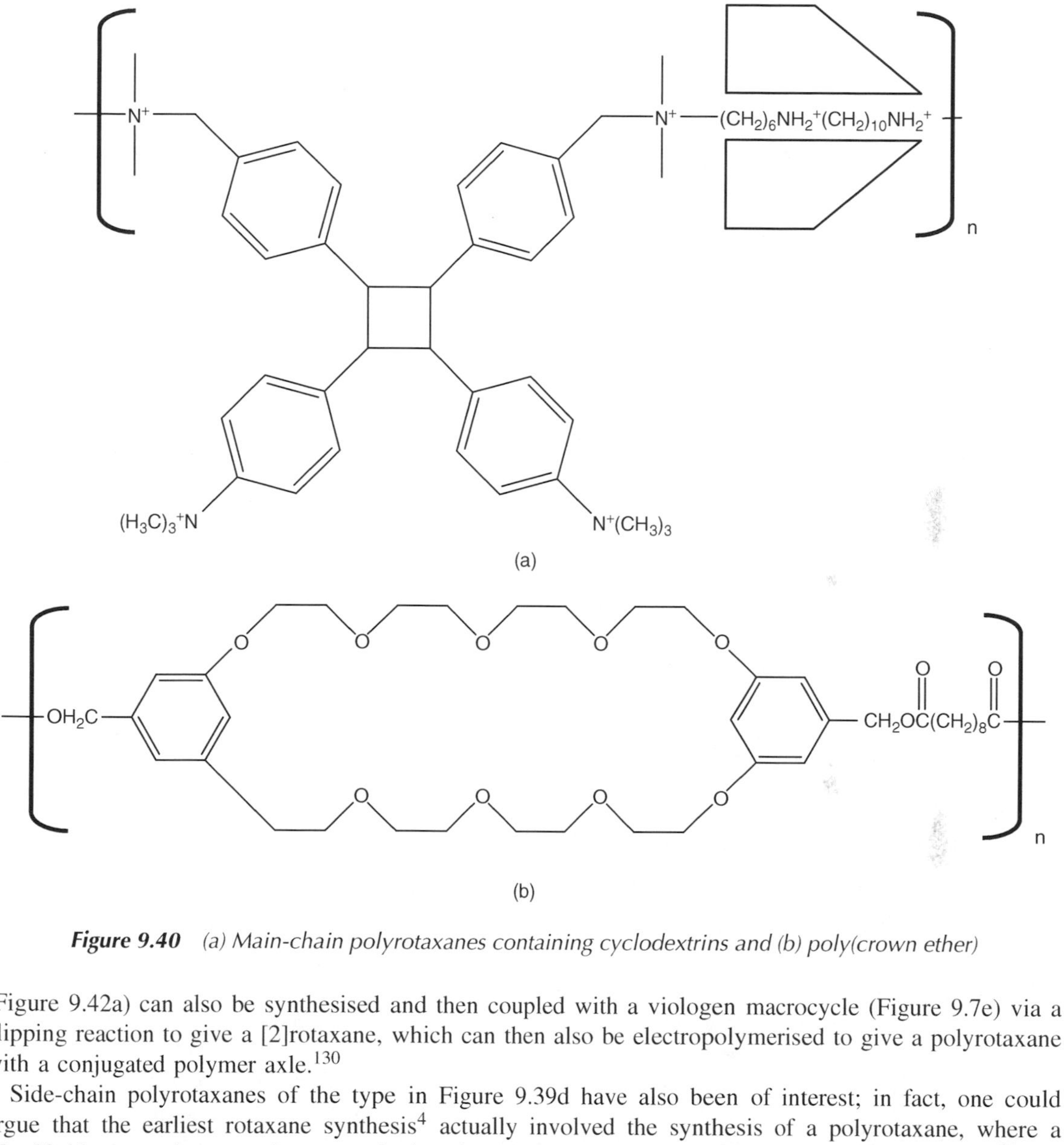

(a)

(b)

Figure 9.40 (a) Main-chain polyrotaxanes containing cyclodextrins and (b) poly(crown ether)

(Figure 9.42a) can also be synthesised and then coupled with a viologen macrocycle (Figure 9.7e) via a clipping reaction to give a [2]rotaxane, which can then also be electropolymerised to give a polyrotaxane with a conjugated polymer axle.[130]

Side-chain polyrotaxanes of the type in Figure 9.39d have also been of interest; in fact, one could argue that the earliest rotaxane synthesis[4] actually involved the synthesis of a polyrotaxane, where a Merrifield-type polymer resin was substituted with rotaxane side chains. Earlier in this chapter[91] we discussed the synthesis of polyrotaxanes via polymerisation of substituted axles, and in Chapters 6 and 8 we discussed polymers containing cyclodextrin and cucurbituril units. Branched polymers have been used to form polyrotaxanes with β-cyclodextrins[131] and the mobility of the cyclodextrin units around the chains is shown to increase upon methylation. A cholic acid moiety containing a long side chain with a terminal acrylamide group forms complexes with β-cyclodextrin and can be polymerised to give the polyrotaxane.[132] Polybenzimidazole can also be functionalised with side chains complexed with cyclodextrins.[133]

Polymers with the macrocycle appended as a side chain and the axle joined by a mechanical bond (Figure 9.39d) have also been synthesised. For example, conjugated polymers based on polythiophene

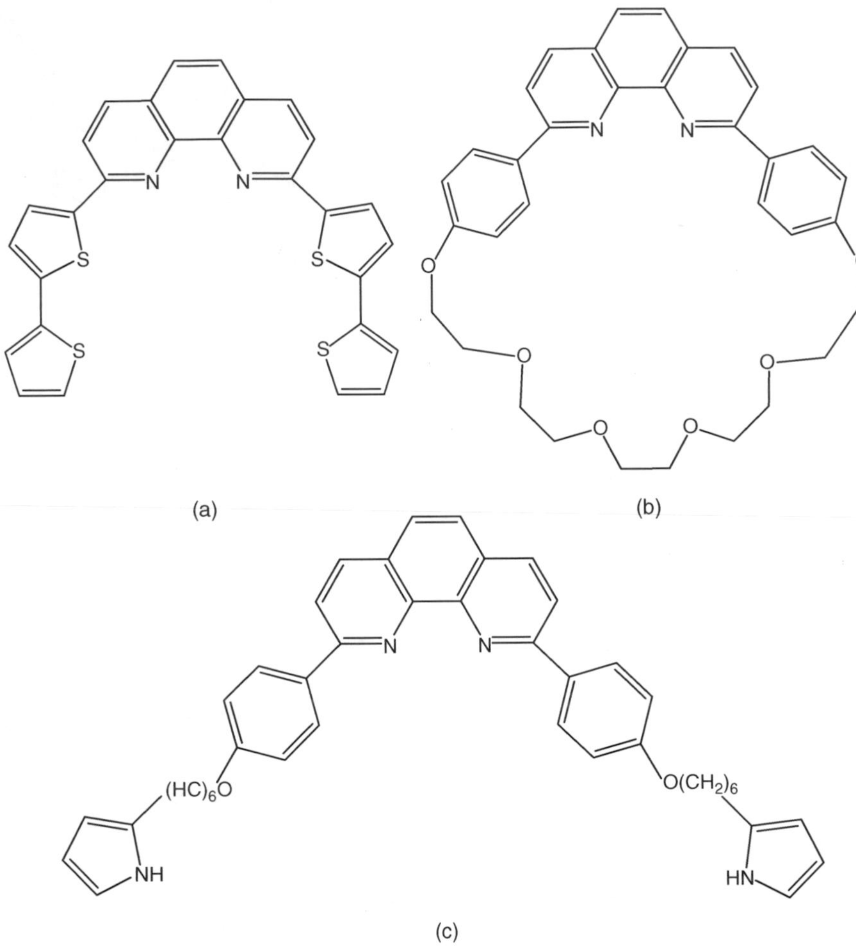

(a)

(b)

(c)

Figure 9.41 *Phenanthroline compounds used in the formation of polyrotaxanes*

or poly(phenyleneethynylenes) have been synthesised with crown ethers bound to the main chain of the polymer (Figure 9.42b). These then form pseudorotaxanes with viologen units.[134]

The final group of polymers are the so-called 'daisy chain' polymers shown schematically in Figure 9.39e. These differ from all the other polymers in that the polymer main chain is not covalent but rather is held together by mechanical bonds. Pseudorotaxane daisy chains have been formed, with for example the monomer shown in Figure 9.42c having been shown to self-associate in solution to give polymers approximately 50 units long.[135] Addition of bulky stopper groups stabilises these assemblies. Daisy-chain polymers containing cyclodextrins and a cyclic daisy-chain trimer have also been reported;[136] these are described in more detail in Section 6.2.2.

9.3   Catenanes

Catenanes were briefly described at the beginning of this chapter and consist a of supramolecular system in which two or more rings are linked together to form a chain-like structure, shown schematically in

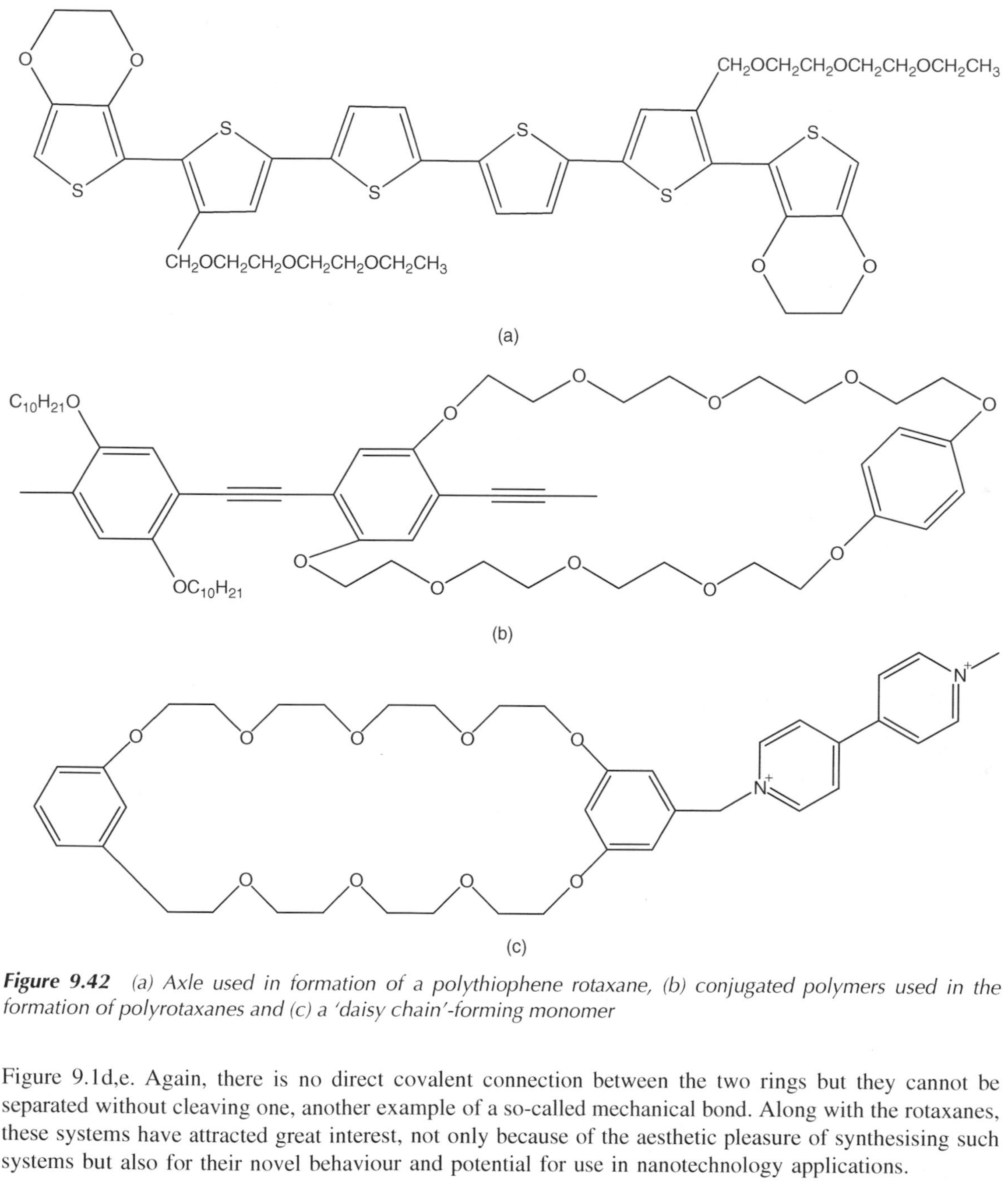

(a)

(b)

(c)

Figure 9.42 *(a) Axle used in formation of a polythiophene rotaxane, (b) conjugated polymers used in the formation of polyrotaxanes and (c) a 'daisy chain'-forming monomer*

Figure 9.1d,e. Again, there is no direct covalent connection between the two rings but they cannot be separated without cleaving one, another example of a so-called mechanical bond. Along with the rotaxanes, these systems have attracted great interest, not only because of the aesthetic pleasure of synthesising such systems but also for their novel behaviour and potential for use in nanotechnology applications.

9.3.1   Synthesis using statistical and directed methods

In 1961 an early paper on potential topological isomers was published, describing such possible ring systems as catenanes and molecular knots.[137] Within this paper is a report that the great German chemist

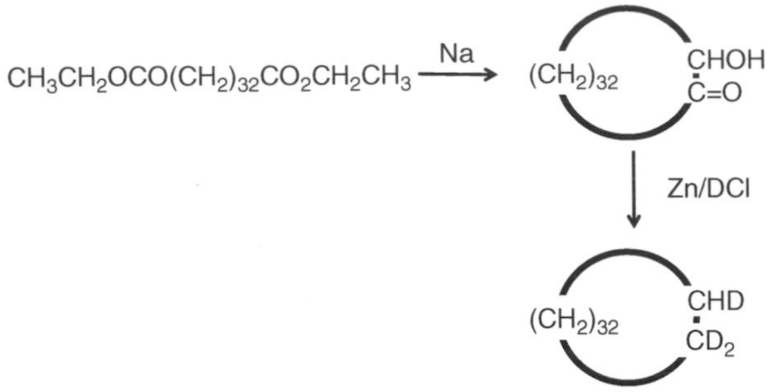

Figure 9.43   *Ring-closure reaction used in the statistical synthesis of catenanes*

and 1915 Nobel Prize winner Richard Willstätter (who elucidated the structure of chlorophyll) discussed the possibility of interlinked-ring systems at seminars in Zurich (he was professor there between 1906 and 1912). The first actual synthesis of a catenane was described in 1960 and involved the cyclisation of a diester of a long-chain carboxylic acid ester.[138] Reaction of the diester with sodium caused the formation of the cyclic acyloin compound shown in Figure 9.43. This compound could be reduced with Zn/DCl to give a deuterated derivative. The cyclisation was then carried out on another sample of the long-chain diester in a 1:1 mixture of toluene and the deuterated macrocycle. Again a cyclic acyloin was obtained, but some product was also isolated by chromatography and was shown by IR spectroscopy to contain both C—D and C═O units. Chromatography proved this was a single product, not just a mixture of the two ring systems. Oxidation of the acyloin ring system back gave the expected linear dicarboxylic acid plus the deuterated macrocycle. However, as in the case of the rotaxanes, these statistical syntheses are highly inefficient, giving poor yields due to only a small proportion of macrocycles being threaded at any one time.

Directed methods such as those published by Schill in his work *Catenanes, Rotaxanes, Knots* have also been applied. Again these are very long multi-step processes and will not be detailed here, though Figure 9.44a shows an example of a typical synthesis of this type.[139] The same group also managed the directed synthesis of the [3]catenane shown in Figure 9.44b, in which the central ring contains two bulky aromatic groups.[140] This compound can exist as two isomers, first as the one shown in the figure, in which the two lateral rings are on the same side of the aromatic groups, and secondly as the corresponding isomer with one lateral ring on either side.

9.3.2   Catenates and catenanes via metal templates

An early breakthrough in catenane chemistry came from Sauvage's group, which developed relatively high-yield synthetic methods for catenanes based on metal-ion templates. We discussed earlier in this chapter how rotaxanes can be made using a combination of phenanthroline units and copper(I) ions. Phenanthroline units form 2:1 complexes with the metal template in which the two aromatic units are orthogonal to each other. Attachment of bridging chains to these aromatic units can lead to formation of a complex with interlinked rings. This is often called a catenate, since the complex is held together by metal coordination. Removal of the metal ion then gives the true catenane, where the physical threading of the chains alone holds the two ring systems together.

In the initial example of this work, the macrocyclic phenanthroline compound (Figure 9.41b) was combined with the phenanthroline-*bis*-phenol (Figure 9.45a) and Cu$^+$ to give the 1:1:1 complex, which was then reacted with a *bis*-iodo-substituted polyether to give a catenate containing two identical units of the type shown in Figure 9.45b, in which the two macrocyclic rings are threaded through each other.[141]

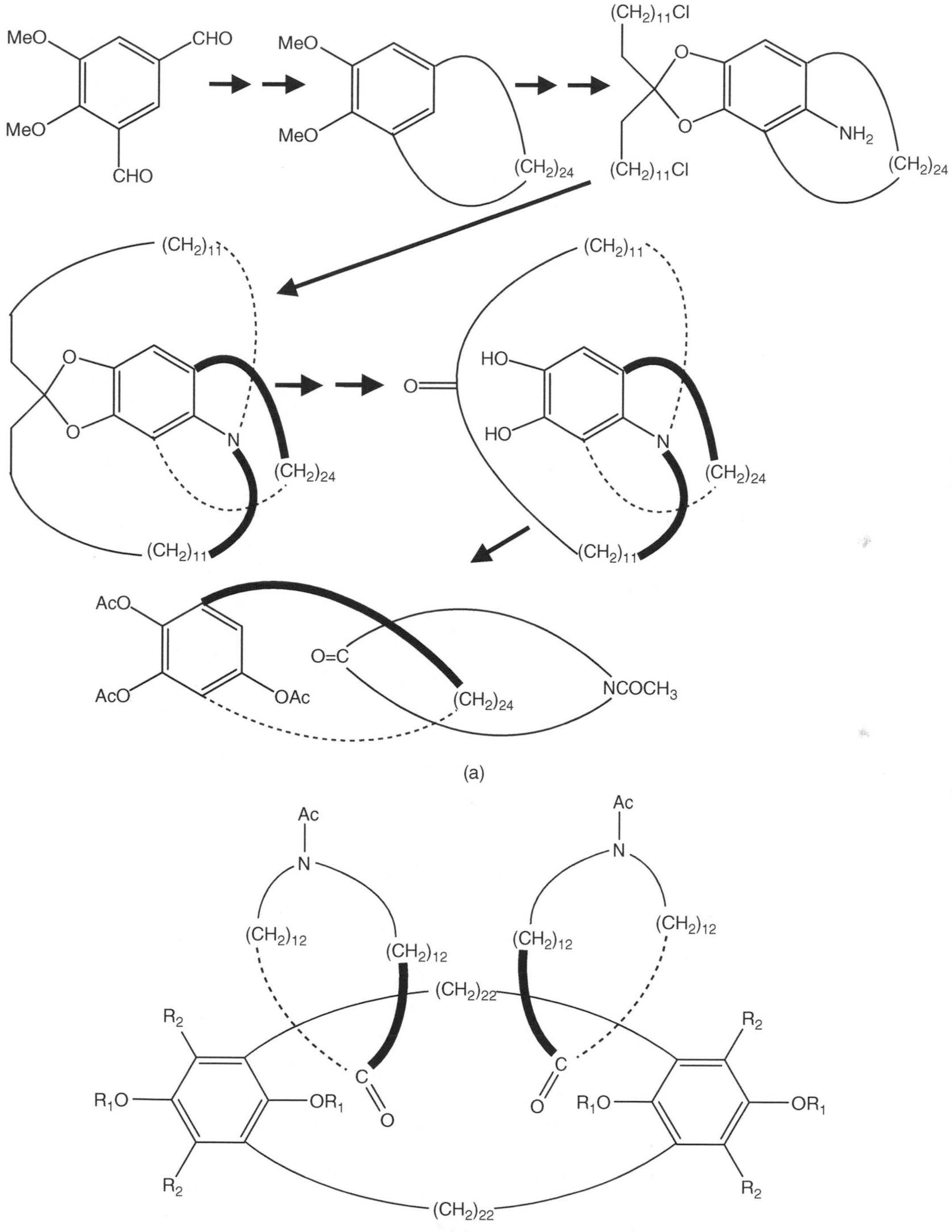

Figure 9.44 (a) *Directed synthesis of a [2]catenane (Copyright Wiley-VCH Verlag GmbH & Co. KGaA. Reproduced with permission[139]) and (b) one isomer of a [3]catenane synthesised by directed synthesis (Copyright Wiley-VCH Verlag GmbH & Co. KGaA. Reproduced with permission[140])*

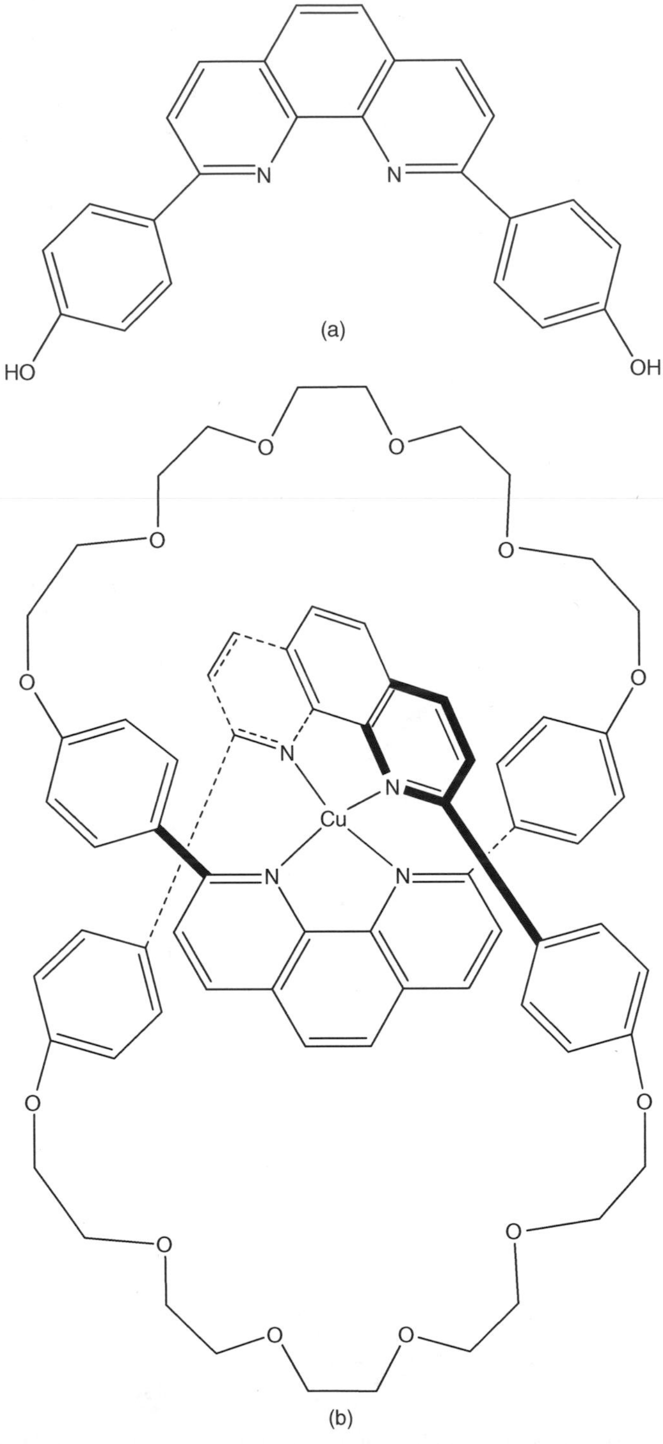

Figure 9.45   *(a) A phenanthroline-bis-phenol and (b) Cu(I) catenate*

Later work successfully developed the simpler process where just the phenolic compound is complexed with Cu^+ to give the 2:1 complex and reacted with two equivalents of 1,14-diiodo-3,6,9,12-tetraoxatetradecane under high-dilution conditions to give the catenate in 27% yield.[142] Treatment with cyanide ions removes the copper template, giving the free catenane. The catenane can be converted back to a catenate by treatment with Cu^+, Ag^+ or Li^+. The complexes have been shown by electrochemical methods to be more stable and difficult to demetallate than the parent macrocycle (Figure 9.41b). The conformation of the catenane has been shown to change greatly depending on presence of a template;[143] when metallated the phenanthroline units are held in close proximity to each other, but on removal of the template, NMR studies show the phenanthroline units to be completely separated. This has been confirmed by X-ray crystallography[144] to also occur in the solid state (Figure 9.46b). Kinetic studies indicate that the formation of complexes of the catenane with many cations occurs as a two-stage process: in the first one of the phenanthroline rings complexes the metal in a process that depends on metal-ion concentration; the second is independent of the cation concentration and is thought to correspond to movement of the two rings relative to each other while the second phenanthroline unit attempts to coordinate to the metal.[145] Interestingly, the catenane is shown to be several orders of magnitude more basic than the single macrocyclic analogue and X-ray studies of the protonated catenane show it to adopt a configuration similar to the copper complex.[146] This arrangement is thought to be stabilised by π-stacking interactions between phenyl and phenanthroline groups. These results demonstrate that chemistry is often not simply a result of the functional groups present in a compound but can also depend on secondary structure and the relationship in which these groups are held to one another. From this we can begin to gain an understanding of how biological systems can have such precise properties, although their structure is immensely more complex than the systems described here.

The catenane has been shown by NMR studies to be a mixture of different conformations which cannot be 'frozen out'; that is, they rapidly interconvert. However, a substituted catenane has been synthesised in which one of the phenanthroline rings bears two phenyl substituents.[147] Figure 9.47a,b shows a schematic of the two isomers available for this ring system in the catenane. Because of the nature of the copper-template process used to synthesise these materials, only the isomer in Figure 9.47a can be synthesised as the catenane (because both phenanthroline groups are complexed to the copper ion), and removal of the cation does not allow interconversion to occur since the additional phenyl substituents are too bulky to pass through the annulus of the companion macrocycle. Reducing the size of the macrocycle by using a smaller 1,11-diiodo-3,6,9-trioxaundecane unit to close it leads to much lower synthetic yield (single macrocycles being the major product, instead of catenanes) and a much more rigid system.[148] Indeed, the catenane formed from two units of a phenanthroline macrocycle with a shorter polyether chain (Figure 9.47c) is chiral due to restricted rotation of the rings and no racemisation is seen up to temperatures as high as 160 °C.

These methods can be modified to produce more complex catenanes.[149] For example, if the 1,14-diiodo-3,6,9,12-tetraoxatetradecane used to bridge the phenanthroline units is replaced in the reaction scheme by shorter linking units such as the dibromides of tri- and tetraethylene glycol, these cannot bridge across a single phenanthroline unit and instead bridge between two separate complexes to give catenates such as that shown in Figure 9.48. A 2-catenate with the same central ring but just one peripheral ring can also be obtained, but yields are low due to formation of the free macrocycles. The catenates can be demetallated to give the free catenanes. NMR and X-ray studies show that there are strong interactions between the phenanthroline aromatic systems of the peripheral rings of these catenanes in both the solid state and solution.[150]

As an alternative synthetic method, the compound in Figure 9.45a was derivatised to give the dipropargyl ether and then coupled together with macrocycles such as that in Figure 9.41b to give a complex which was further reacted to give the catenate shown in Figure 9.49a. These types of catenate could be obtained in up to 68% yields, and the 4-catenate with three complexes coupled together to give a larger central and three peripheral rings could be obtained in up to 22% yield.[151] The 3-catenate could be crystallised and the

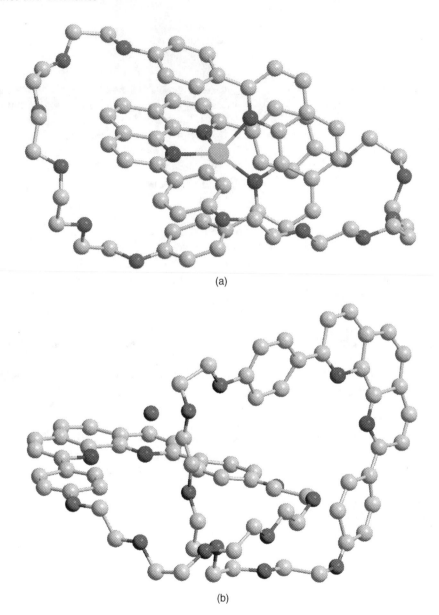

(a)

(b)

Figure 9.46   *Crystal structures of (a) the Cu(I) catenate and (b) free catenane*

structure shows that strong interactions exist between the peripheral rings in the solid state (Figure 9.49b), leading to a compact, curled structure. NMR measurements demonstrate that similar interactions exist in solution. Even larger catenates from this reaction have been detected by mass spectrometry,[152,153] with molecular weights up to 7800. The 3-catenates can be demetallated and then remetallated to give complexes with silver, zinc, cobalt or nickel,[154] and electrochemical studies have demonstrated that the two metal ions interact with each other, confirming the folded state of the catenate. By stepwise demetallation followed by remetallation, it is also possible to produce heteronuclear catenates containing two different metal ions.[155]

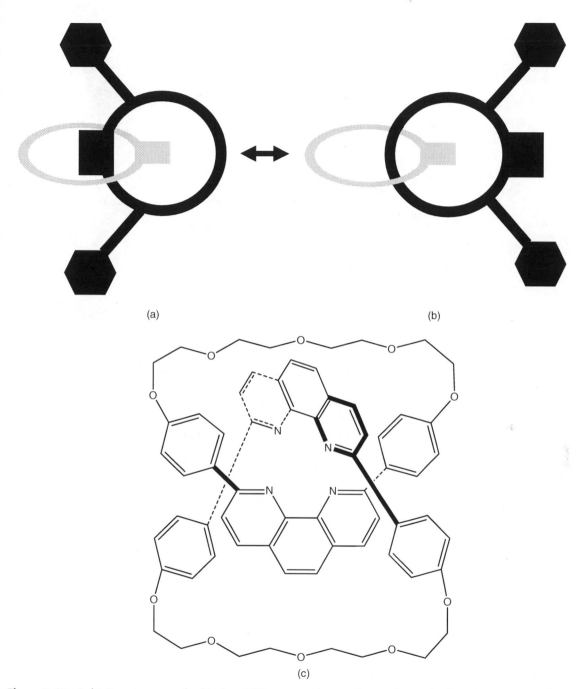

Figure 9.47 *(a,b) Two isomers of a hindered [2]catenane (rectangles = phenanthroline, hexagons = phenyl groups) and (c) a catenane with hindered rotation*

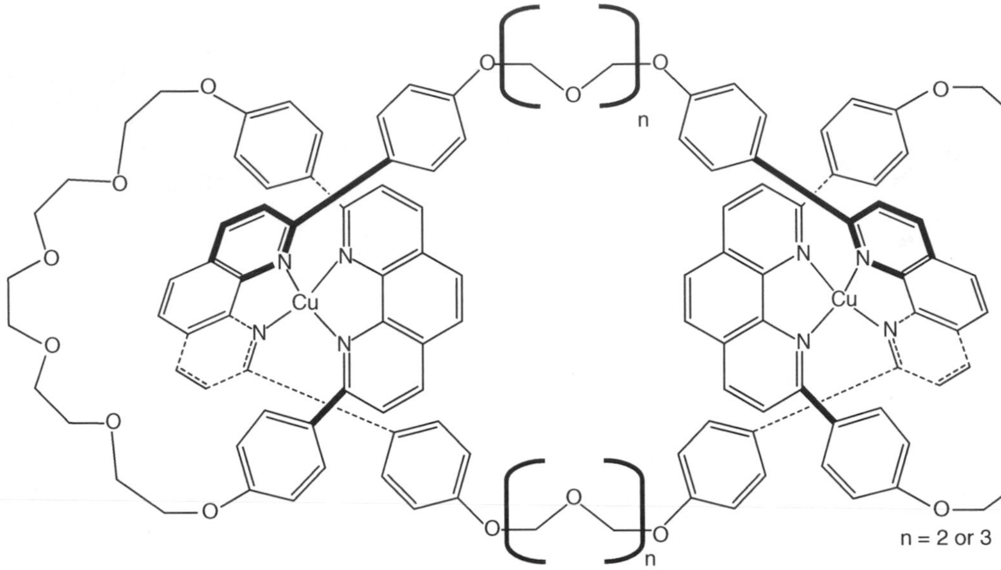

Figure 9.48   *A 3-catenate*

Fluorescence studies have been performed on the catenane obtained from Figure 9.49a, along with its homo- and heterocatenates and the monocatenate with one free and one complexed phenanthroline ring.[156] The photochemistry of these systems has been extensively investigated and shows that interactions exist between the metal ions or between metal-bound and free catenane units.

These methods pioneered by Sauvage's group have been used to assemble a range of organic groups of interest into catenated systems. For example, a porphyrin was synthesised with a bridging chain that contained a phenanthroline unit, a so-called 'basket with a handle' porphyrin.[157] This was complexed with a second phenanthroline unit (Figure 9.45a), which was the reacted with a second porphyrin unit to give the resultant catenane, a double basket with interlinked handles (Figure 9.50). A similar catenate (Figure 9.51a) has also been synthesised containing tetrathiafulvalene units.[158] When a terpyridyl-containing complex is assembled as a catenate with a phenanthroline-containing macrocycle, the resulting structure is as shown in Figure 9.51b. However, when the copper ion is oxidised to Cu(II) there is a slow change of spectra over a few days that is consistent with rearrangement of the catenate so that the Cu(II) ion is complexed with one phenanthroline and one terpyridyl unit, since this ion 'prefers' to be pentacoordinate.[159] This oxidation is reversible and accompanied by a rapid rearrangement to the original complex, giving us a 'switchable' catenane. A different method using a ruthenium template has succeeded in synthesising a *bis*-terpyridyl-containing catenate.[160]

More complex catenates and catenanes have been synthesised by utilisation of components containing multiple binding groups. For example, a linear strand containing three phenanthroline groups can form copper-templated complexes with a similar strand, which upon cyclisation reactions with oligoethylene glycol bridges gives doubly interlocked catenates (Figure 9.52a). A singly interlocked version of this material can also be obtained and in both cases these materials can be converted to the demetallated catenanes.[161] NMR studies show the singly entangled isomer to be more conformationally flexible than the doubly interlocked compound and mass spectral studies show that under high-energy conditions the doubly interlocked compound is less stable than its topological isomer.[162] Another system containing three phenanthroline units has also been synthesised, but the core is bicyclic.[163] Reaction of this compound

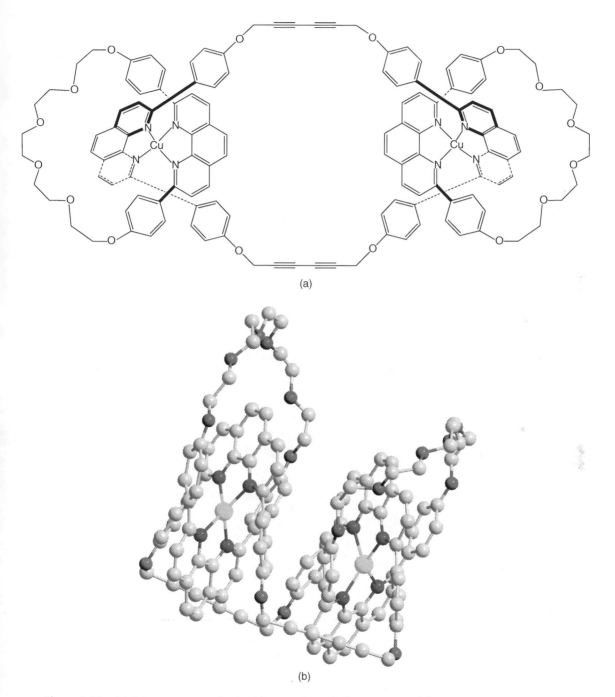

(a)

(b)

Figure 9.49   *(a) A 3-catenate synthesised by coupling of alkyne units and (b) its X-ray crystal structure*

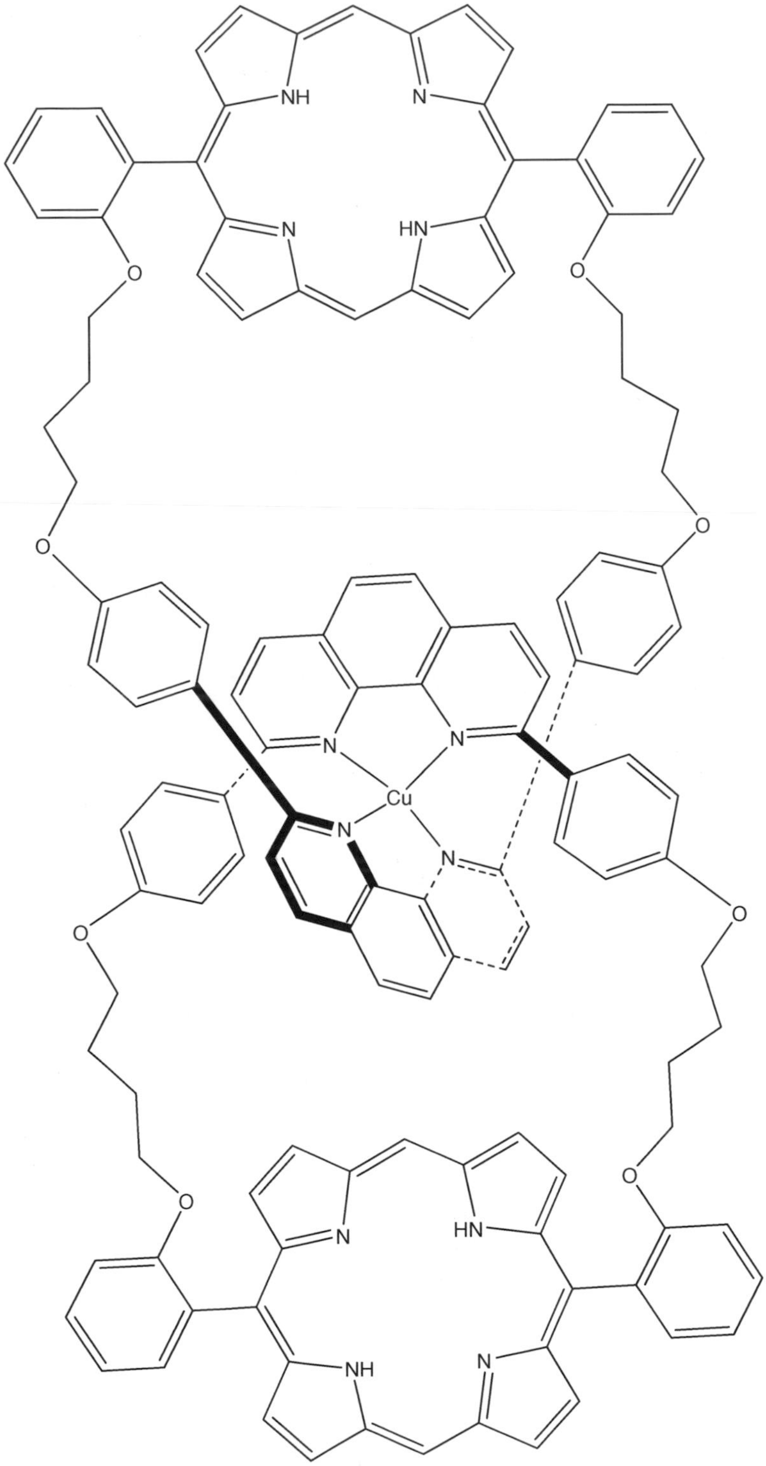

Figure 9.50   *A [2]catenate containing porphyrin units (baskets with handles)*

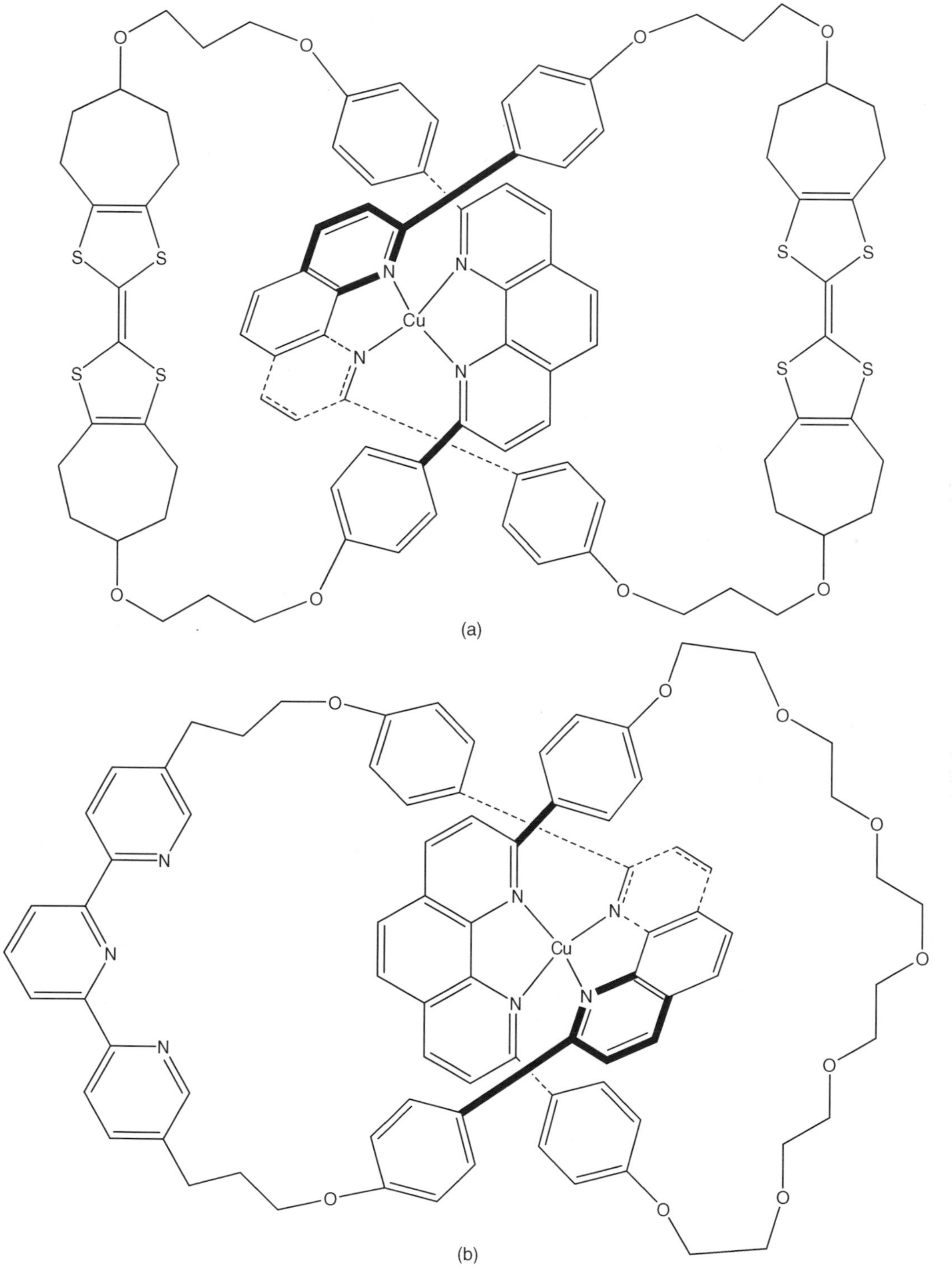

Figure 9.51 *(a) A [2]catenate containing tetrathiafulvalene units and (b) a switchable [2]catenate containing a terpyridine unit*

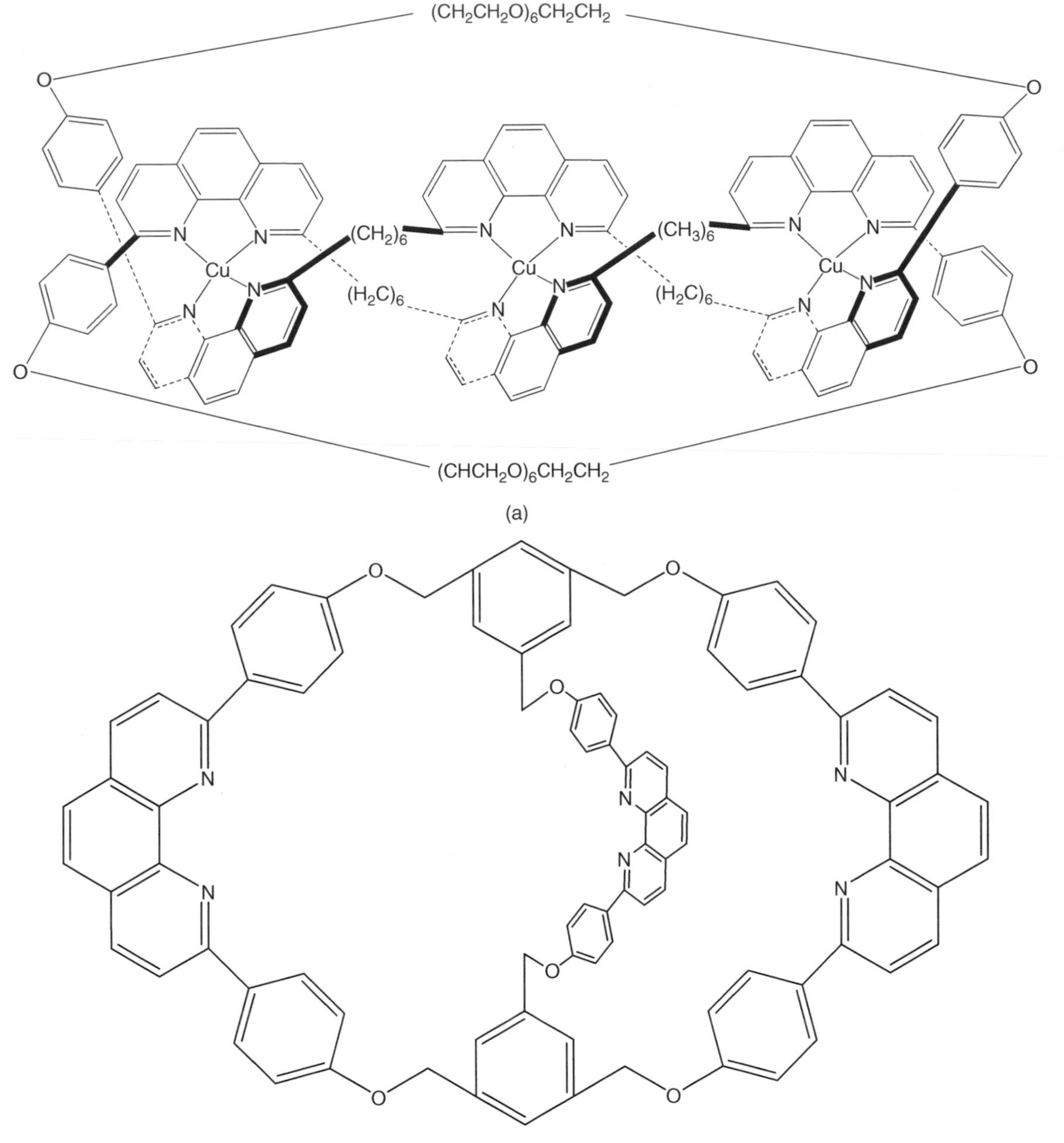

Figure 9.52   *(a) A doubly interlocked catenate and (b)* a tris-phenanthroline macrocycle used as the core in the *synthesis of multiple catenates*

with phenanthroline macrocycles gives the *mono-*, *bis-* and *tris*-catenates, albeit in very low yields. NMR studies demonstrate large differences in rigidity between the three products.

Other systems that have been synthesised include the so-called 'hook and ladder' conformations, where a catenane is synthesised using similar methods to those utilised by Sauvage's group except that the oligoethylene glycol units contain reactive side groups in the centre of the chain.[164] These groups can then be reacted with each other to give a type of structure shown schematically in Figure 9.53a. Other chemistries have been utilised to create metal-containing catenates and catenanes and we will describe several of these here. The first organometallic catenane was synthesised[165] by reacting a bridged diphenyl magnesium cyclophane with a crown ether to give the structure shown in Figure 9.53b. Another system, which as the authors point out is reminiscent of the magician's 'magic rings' trick on a nanoscale, involves the synthesis of the macrocycle in Figure 9.53c by combination of the pyridyl compound with palladium nitrate ethylene diamine complex.[166] At low concentrations the macrocycle exists as a monomeric structure as shown, but at higher concentrations the catenane becomes the dominant species, as proved by mass spectroscopy and NMR. A simple mechanism to allow this to occur is that the pyridine–palladium coordination bonds must break and reform to allow a form of clipping reaction to take place, as shown in Figure 9.54a. An alternative mechanism has been proposed, where two macrocycles come together and undergo interconversion by way of ligand exchange between Pd—N bonds.[167] Normally this reaction would bind the two macrocycles together and then release them, but it is proposed that there is a twisting of the rings (as shown in Figure 9.54b) which leads to catenate formation. Mixed systems have also been synthesised where macrocycles containing hydrogenated or fluorinated aromatic rings are combined and mixed catenanes containing one of each type of ring are formed.[167,168]

The same macrocycle but with platinum ligands (Figure 9.53d) does not show this behaviour at room temperature, indicating stronger pyridyl–platinum binding.[169] However, upon heating to 100 °C, this bond becomes labile and the catenane forms in high yield. Cooling the solution then traps the molecule in the catenated form, behaviour which the authors describe as that of a 'molecular lock'. The catenane can even be crystallised and its structure confirmed by X-ray crystallography.

Mixed systems can also be synthesised, with for example compounds containing both phenanthroline and pyridine units being synthesised and then linked as dimers using copper(I) coordination of the phenanthrolines.[170] These can then be clipped to form mixed metal catenates using palladium to give compounds such as that shown in Figure 9.54c. Both entwining together of two macrocycles and a clipping procedure, which allows the formation of catenanes with two different macrocyclic units, have been utilised and both have been shown to be highly efficient synthetic methods. A similar procedure utilises ruthenium complexes to give catenates containing Cu(I) and Ru(II), and the central copper ion can also be removed.[171] Bimetallic Ru-Cu and trimetallic Ru-Cu-Ru complexes can be obtained and potentially utilised as photoactive multicomponent systems. Other workers have synthesised the 'figure of eight' molecule[172] shown in Figure 9.55a and demonstrated that in solution it can be induced to form the elegant cyclic *tris*-catenanes shown schematically in Figure 9.55b. Combination of this figure-of-eight molecule with other macrocycles containing one palladium or platinum bridging group (Figure 9.55c) gives rise to complex structures such as linear *bis*-[2]catenanes and linear *tris*-[2]catenanes.[173]

Box-like compounds have also been utilised to synthesise catenanes. One of the earliest examples[174] utilised the synthetic scheme shown, where two pyridine-containing compounds are combined along with palladium bridging groups to give the catenane in Figure 9.56a. The catenane displays high stability, not reverting back to individual macrocycles even under high-dilution conditions. Various catenanes of this type can be obtained and the interlocked structure has been proven by NMR and X-ray crystallography. Chiral binaphthyl groups can be included in these structures and it has been found that if *meso* macrocycles containing one of each binaphthyl enantiomer are synthesised, the resulting [2]catenane is chiral.[175] When this catenane is crystallised it is found to assemble via a series of secondary intermolecular Pd–ligand binding

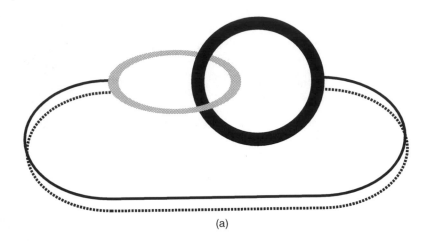

(a)

(b)

(c) M = Md
(d) M = Pt

Figure 9.53   (a) Schematic of a 'hook and ladder' compound, (b) an organometallic catenane, (c) a 'magic rings' macrocycle and (d) a 'molecular lock'

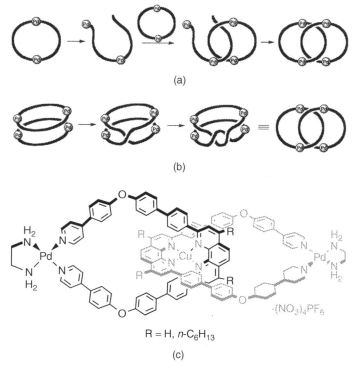

Figure 9.54 *Possible mechanisms for the rapid slippage of two molecular rings: (a) conventional mechanism and (b) ligand-exchange mechanism (Reprinted with permission from[167]. Copyright 1996 American Chemical Society), and (c) a catenate containing both copper and palladium (Reprinted with permission from[170]. Copyright 2003 American Chemical Society)*

interactions to form an infinite one-dimensional polymer. In other work a palladium-linked macrocycle was combined in solution with β-cyclodextrin and shown by NMR spectroscopy to give the catenane shown in Figure 9.56b. This shows that there must be some disassociation and reassociation of the N—Pd bonds.[176] When the larger γ-cyclodextrin is utilised, a pseudorotaxane structure is preferred. High concentrations of the two macrocycles lead to formation of oligomeric products.

A range of octahedrally coordinated metals have been utilised to assemble catenanes. For example, benzylic imine compounds (Figure 9.57a) can be combined with a variety of metal ions such as Mn^{2+}, Co^{2+}, Cu^{2+}, Hg^{2+}, Cd^{2+}, Zn^{2+}, Fe^{2+} and Ni^{2+} to give 2:1 complexes with the metal ion octahedrally coordinated.[177] Cyclisation of the complexes to give the catenates is attained by ring-closing metathesis reactions between the olefinic groups in the complex. The complexes do not undergo simple demetallation reactions; rather reduction of the imines to amines with sodium borohydride allows isolation of free catenanes. The same metal-ion complexation strategy can also be used to assemble rotaxanes.[178] Later work by the same group utilised linear coordination between Au(I) and pyridine units to successfully assemble a gold complex which after ring-closing metathesis gave the catenate shown in Figure 9.57b, which could then be reacted with HCl to give the free catenane.[179] Again, this strategy could also be used in rotaxane synthesis.

Besides acting as a template, copper ions can also catalyse various reactions, such as the reaction of acetylenes and azides to give pyrazole units. This process has been used in catenane synthesis, where the copper ion is complexed within a macrocyclic compound and then combined with the open-chain compound

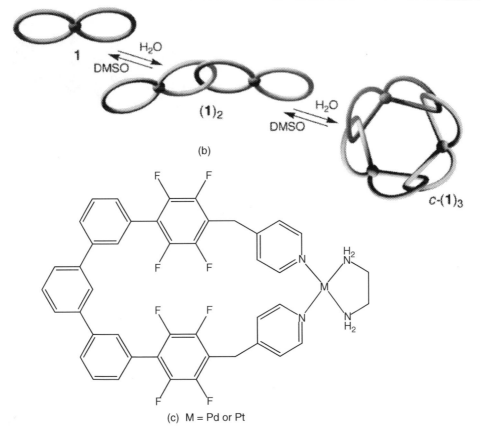

Figure 9.55 *(a) Structure of the figure-of-eight molecule and (b) schematic of its formation of a tris-catenane (Reproduced by permission of the Royal Society of Chemistry[172]), and (c) a macrocycle used in the synthesis of linear bis- and tris-catenanes*

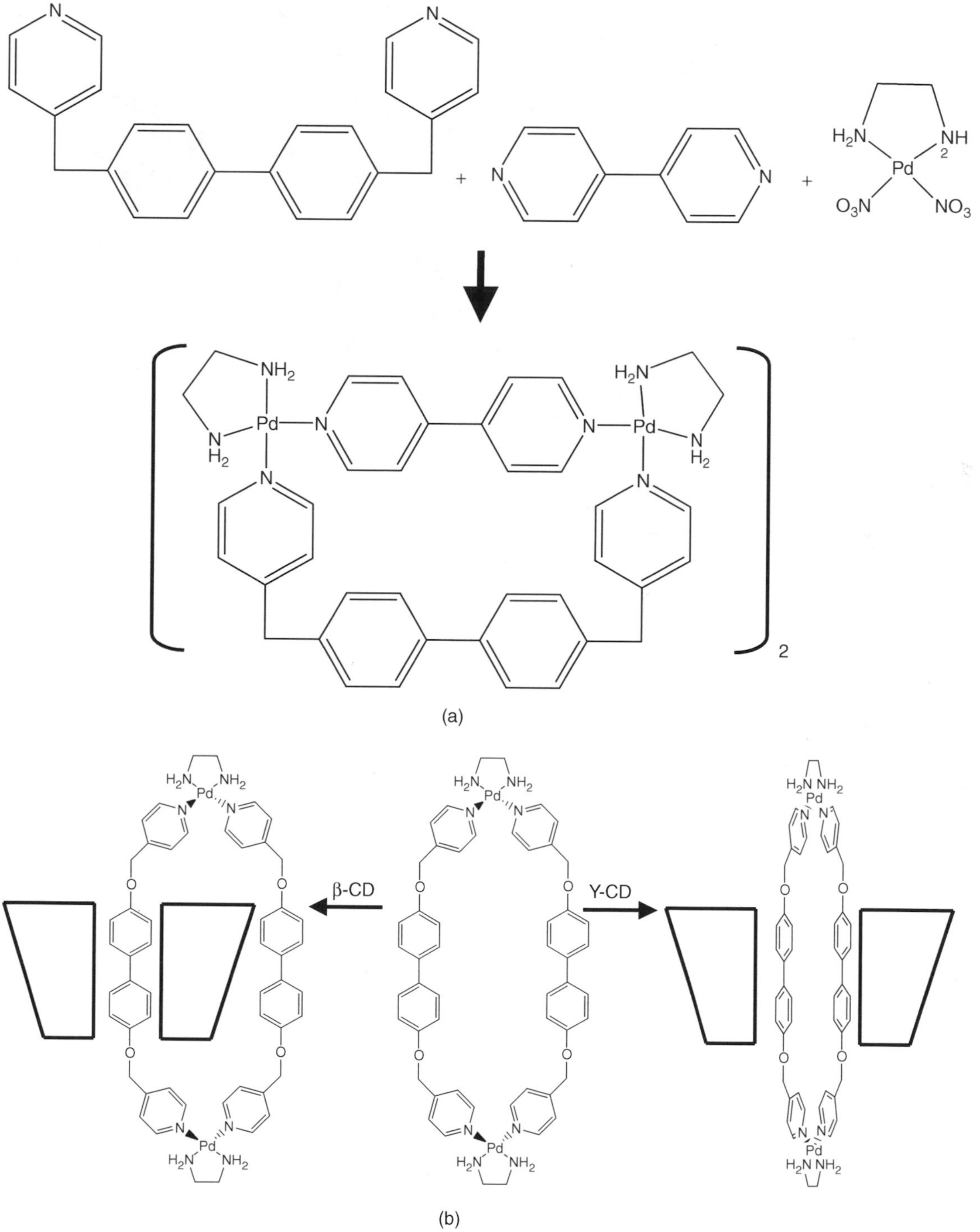

Figure 9.56 (a) Synthesis of the 'box' catenane and (b) formation of rotaxanes and catenanes with cyclodextrins (Reprinted with permission from[176]. Copyright 2004 American Chemical Society)

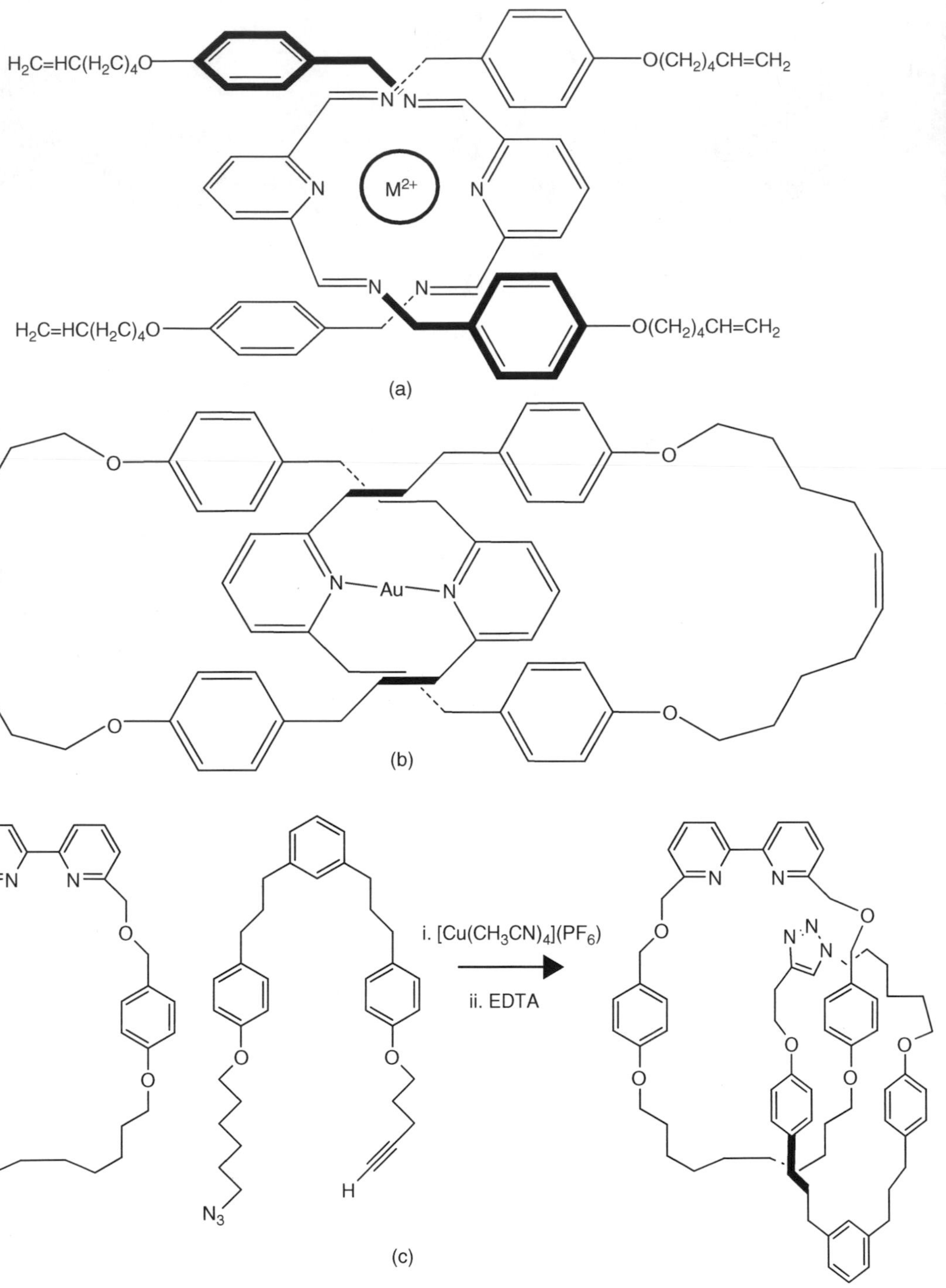

(a)

(b)

(c)

Figure 9.57 (a) Typical precursor for benzylic imine catenates (M = Mn, Fe, Hg, Zn, Co, Ni, Cu, Cd), (b) a gold-based catenate and (c) catenates formed by active metal-template cyclisation (Reprinted with permission from[180]. Copyright 2009 American Chemical Society)

shown in Figure 9.57c. The copper ion catalyses the condensation of the two acetylene groups to give the catenate in 46% yield.[180] Within the same paper another mechanism utilising coupling of substituted acetylenes with the formation of diacetylene compounds was used to form catenates. A double macrocyclisation procedure was also developed that synthesised catenanes from pairs of open-chain molecules.

Other multiple-metal systems that have been synthesised include porphyrin-containing catenates. For example, the catenate shown in Figure 9.58a contains three different metal ions.[181] This species can be synthesised in up to 11.5% yield using a preformed porphyrin macrocycle, which can be complexed with phenanthroline, and then a clipping reaction to incorporate the second porphyrin moiety. NMR studies show that removal of the central copper ion leads to major conformational changes, allowing the porphyrins to come much closer to each other. Later work constructed an even more intricate supramolecular system in which four porphyrin rings were assembled around a catenate core.[182] Further bridging ligands were then

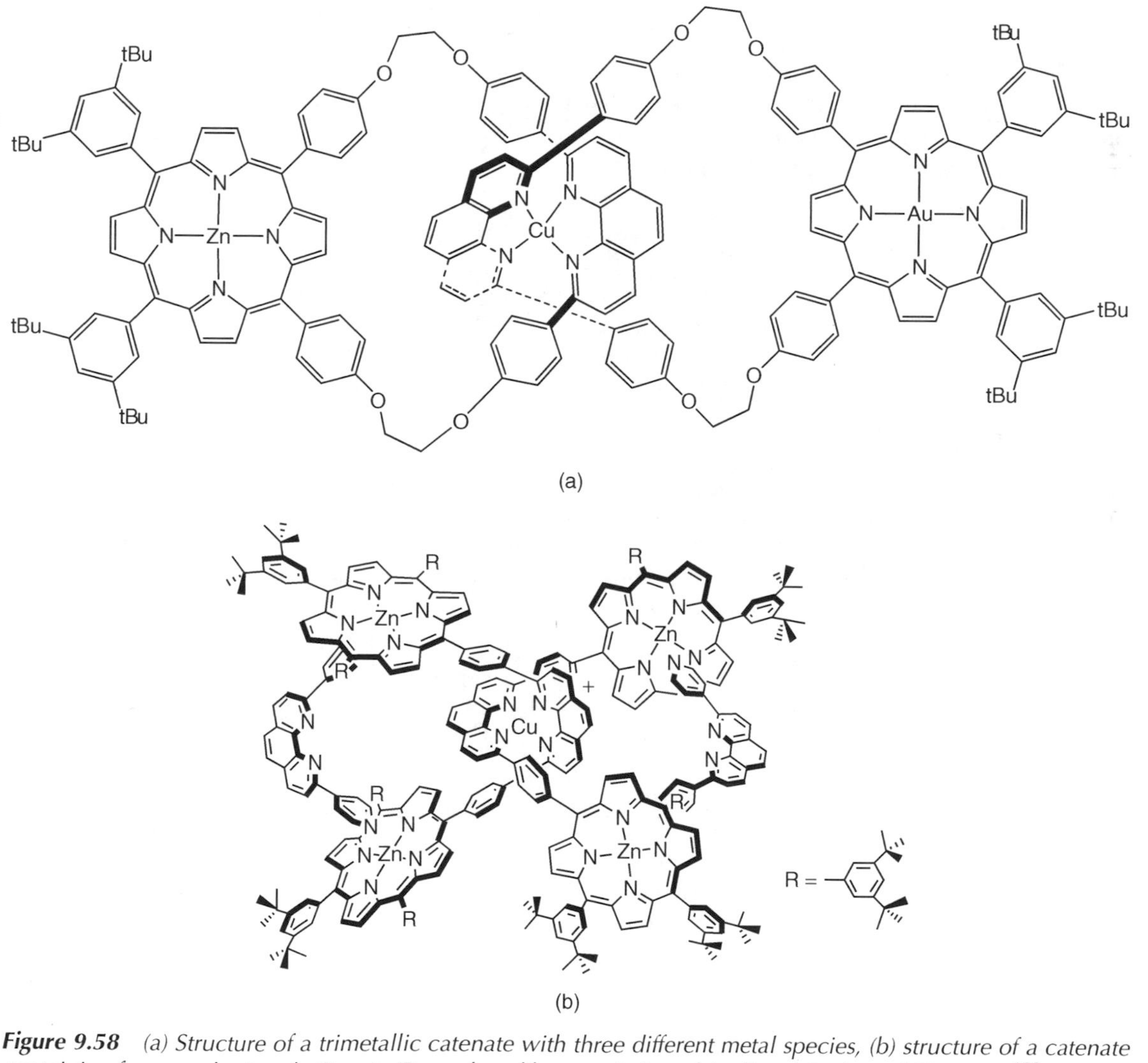

(a)

(b)

Figure 9.58 (a) Structure of a trimetallic catenate with three different metal species, (b) structure of a catenate containing four porphyrin substituents (Reproduced by permission of the Royal Society of Chemistry[182])

introduced between the Zn atoms in the centre of the porphyrin rings to give the complex structure shown in Figure 9.58b.

Other workers have synthesised catenates such as that shown in Figure 9.59a, where porphyrin and fullerene groups have been incorporated onto opposing rings as along with their simpler analogues containing just one of these functional groups.[183] NMR studies show that the fullerene and porphyrin units are as widely separated as possible and photochemical studies show that upon irradiation electron-transfer reactions occur, giving a charge-separated radical pair, with the porphyrin being positively charged and the fullerene negatively charged. Other aromatic species that have been incorporated into these systems include the quaterthiophene-containing catenates shown in Figure 9.59b. These can be demetallated to give the parent catenanes, which like their catenates have the quaterthiophene units in close contact with their phenanthroline units, interacting by through-space donor—acceptor interactions as shown by NMR and quenching experiments.[184] AFM pictures of these systems adsorbed onto pyrolytic graphite clearly show the highly ordered catenane structure (Figure 9.59c). Other workers utilised a cyclotriveratrylene skeleton to assemble three bipyridyl groups and then utilised zinc or cobalt to assemble a triply interlocked catenate structure[185] as shown in Figure 9.60a. This compound can be crystallised and has the X-ray crystal structure shown in Figure 9.60b. Each complex consists of two trigonal bipyramidal cages which interlock through all three windows to form the [2]catenane, which also contains an inner binding core with a central cavity with an estimated volume of 0.2 nm.^{3}

9.3.3   Catenanes based on π—electron interactions

Although both directed synthesis and especially metal-templated synthesis have been successful in the development of catenanes, it is much more intellectually and aesthetically satisfying to develop systems which spontaneously assemble to form intricately linked supramolecular systems rather than forming around a central template. We have already discussed in Section 9.2.3 the assembly of rotaxanes via the use of π—electron interactions, but it is in the field of catenane synthesis that these interactions have really come to the fore.

This field can be thought of as having begun with the synthesis, developed by Fraser Stoddart and other workers, of the catenane shown in Figure 9.61, published in 1989. When the crown ether, *bis*-bipyridyl compound and *p*-xylylene dibromide were reacted together in acetonitrile solution, up to 70% of the catenane could be obtained from this simple one-pot synthesis.[186] Obviously, for such a high yield there must be strong interaction between the cationic macrocycle and the crown ether to preorganise them in such a way that when the dibromo compound clips onto the viologen species, a catenane is formed. The noncyclic *bis*-bipyridyl species shown in Figure 9.61 does not strongly interact with the crown ether, but what is thought to occur is that first one of the bipyridyl units reacts with *p*-xylylene dibromide to give a compound containing one doubly charged viologen unit. The tricationic species does undergo a strong interaction with the crown ether and is therefore encapsulated. NMR does not show the presence of the tricationic species and it is thought that it rapidly ring-closes to give the catenane.[187]

NMR studies show that the catenane in this case is highly ordered in solution. At $-65\,^\circ$C the spectrum shows peaks indicating that rotation is restricted; that is, NMR can distinguish between the viologen that is 'inside' the crown-ether cavity and that which is 'alongside' it. At higher temperatures the cationic macrocycle can rotate around the crown ether and at higher temperatures still both macrocycles can rotate with respect to each other. X-ray crystallographic studies also show the catenane structure (Figure 9.62) and the distances between the aromatic rings indicate strong π—π interactions not only between the two components of the catenane but also between adjacent catenane molecules, which form a stacked structure. Electrochemical measurements also demonstrate that there are interactions between electron-rich and -poor rings in solution. Later work showed that this synthesis could be achieved in 18% yield by combining

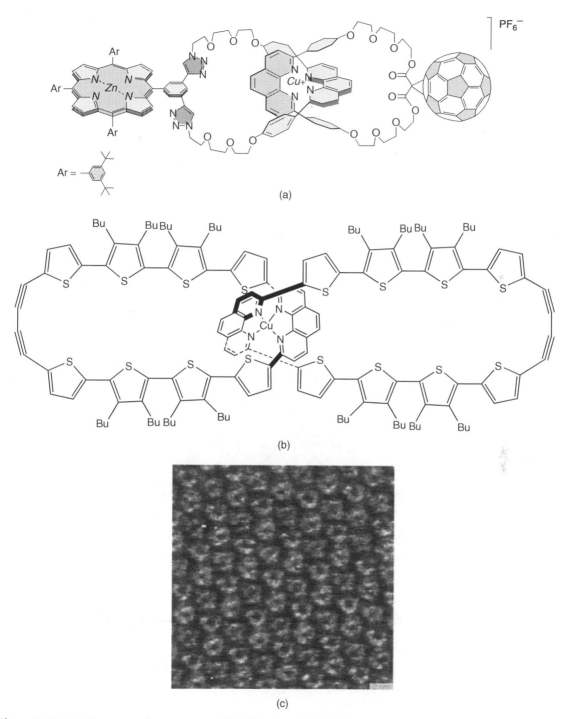

Figure 9.59 (a) Structure of a catenate with fullerene and porphyrin substituents (Reprinted with permission from[183]. Copyright 2010 American Chemical Society), and (b) a thiophene-substituted catenate (Copyright Wiley-VCH Verlag GmbH & Co. KGaA. Reproduced with permission[184]) and (c) STM of a monolayer on pyrolytic graphite

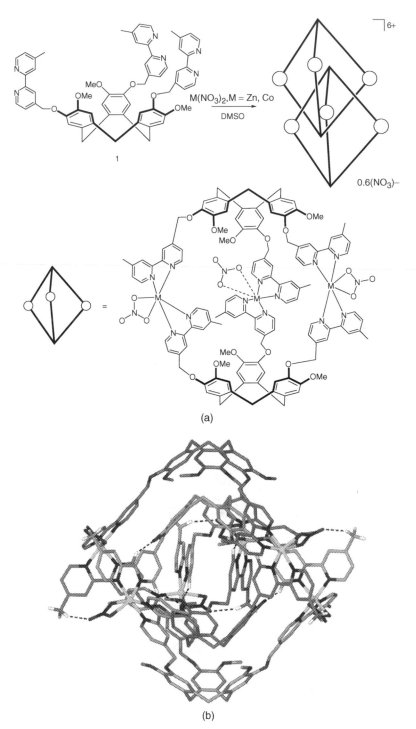

Figure 9.60 *(a) Synthesis of a triply interlocked catenate and (b) its X-ray crystal structure (Reprinted with permission from[185]. Copyright 2008 American Chemical Society)*

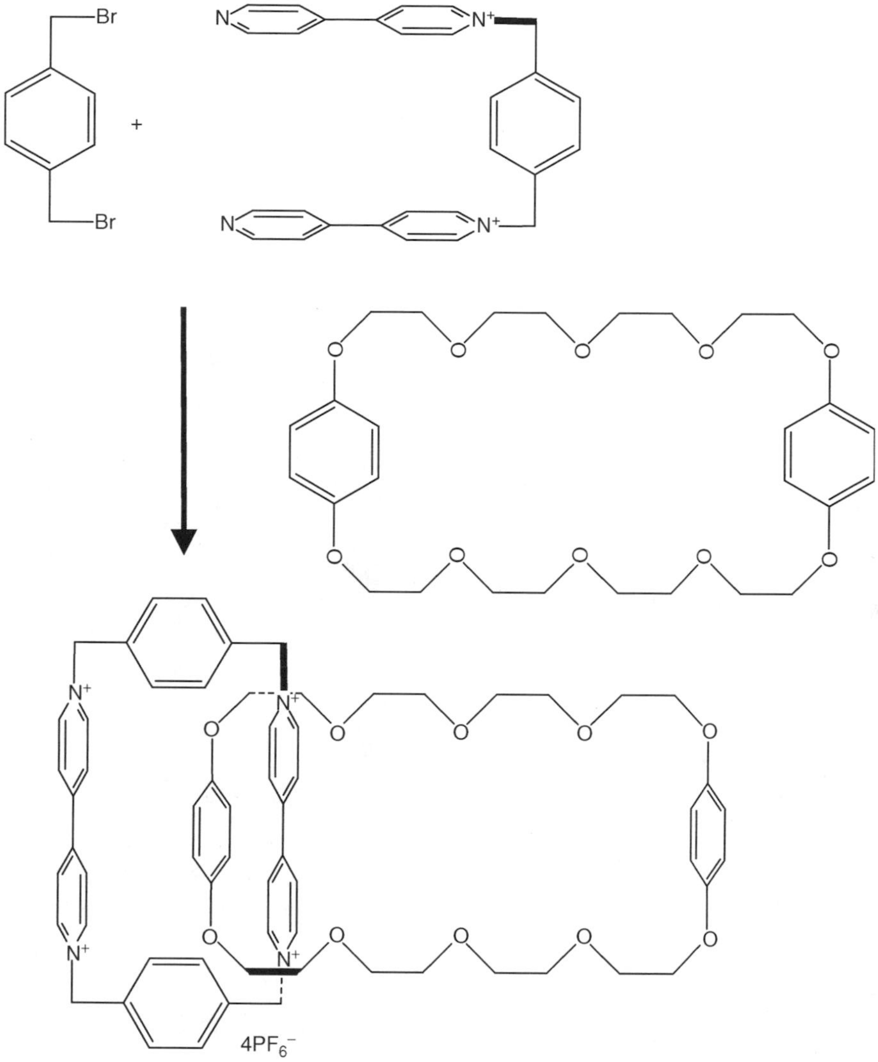

Figure 9.61   *Synthesis of an early catenane by π–electron interactions*

4,4′-bipyridine and *p*-xylylene dibromide in solution in the presence of excess crown ether.[188] Raising the pressure to 12 kBar increased yields to 42%.

Once this synthesis had been achieved, it was inevitable that many phenolic and pyridyl-based systems would be investigated for their abilities to form catenanes. Replacement of the crown ether with a similar resorcinol-based crown ether (Figure 9.63a) and reaction using the same scheme as above allowed construction of the catenane in 17% yield.[189] It is thought that the interactions between the crown and viologen moieties are weaker than for the previous catenane, leading to lower yields. This is borne out by NMR studies, which indicate a much more dynamic system, allowing greater motion of the two macrocycles relative to each other than for the previous catenane. A mixed crown ether containing a hydroquinone and a resorcinol unit (Figure 9.63b) can also be incorporated within the catenane system.[190] NMR studies

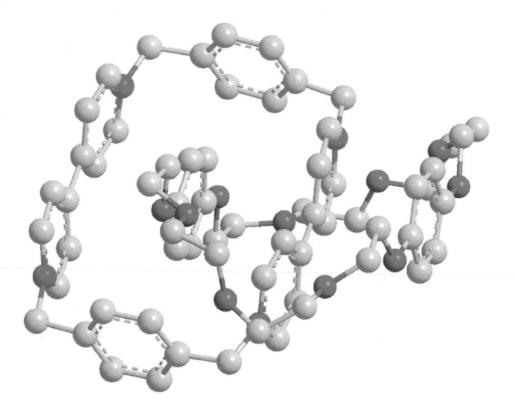

Figure 9.62   *Crystal structure of a catenane*

demonstrate much stronger interactions between the viologens and the hydroquinone unit in that in 98% of the catenane in solution at any one time the hydroquinone unit is encircled by the cationic macrocycle. When one of the hydroquinone units is replaced by a larger naphthalene unit (Figure 9.63c), again there are two possible locations for the cationic ring on the crown ether.[191] The ratio between the two translational isomers in this system is dependent on the polarity of the solvent; in acetone 35% of the catenane has bound the naphthyl unit but in DMSO encapsulation of the naphthyl unit is the preferred option. Since the naphthyl unit is more electron-rich we would expect it to be the preferred occupant of the cationic macrocycle, but solvent-shielding effects also need to be taken into consideration.

A study was also made on increasing the size of the crown-ether spacers.[192] The highest yield (70%) was obtained with the 34-crown-12 compound already described (Figure 9.61); removing one ethyleneoxy unit from the spacer reduced the cavity size and yield fell to 10%. Use of longer spacer chains caused a drop in the yield as well, levelling off at about 40%. Also within this work, larger crown ethers containing more than two hydroquinone units were synthesised, such as *tris(p*-phenylene)-51-crown-15 and *tetrakis(p*-phenylene)-68-crown-20. These could be used to assemble catenanes containing one of the cationic macrocycles, but their larger ring size also allowed binding of two of these bipyridyl macrocycles, in order to form [3]catenanes.

As an obvious counterpoint to using larger crown ethers, a series of larger bipyridyl-based macrocycles can also be synthesised. When a bipyridyl-based compound containing a biphenyl spacer is reacted with 4,4′-*bis*-bromomethyl biphenyl in the presence of crown ethers, up to 25% of [3]catenanes can be isolated. These consist of a bipyridyl macrocycle, as shown in Figure 9.64a, with two crown-ether units of the type

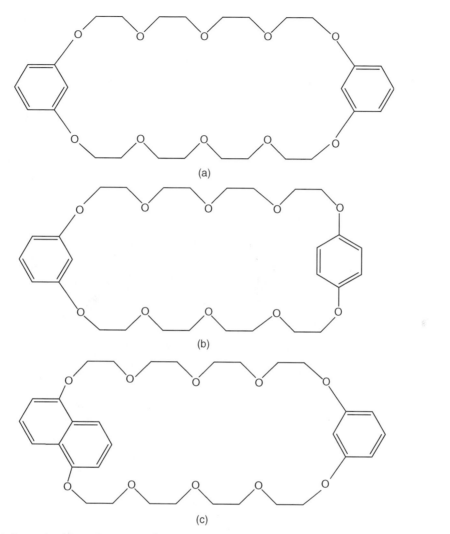

Figure 9.63 *(a) Resorcinol-based crown ether, (b) mixed crown ether and (c) crown ether containing a naphthalene unit*

in Figure 9.60 encircling the bipyridyl groups,[192] along with a low yield of the [4]catenane, consisting of three crown-ether units arranged around the macrocycle shown in Figure 9.64b. If larger crown ethers such as that in Figure 9.65 are used, the resultant [3]catenanes can then be further reacted using the standard clipping procedure (Figure 9.60) to give [4] and [5]catenanes, albeit in low yields. Mass spectrometry has confirmed the formation of the higher catenanes. NMR studies have also been made on the dynamics of these systems and show that when the macrocycles reach sufficient size, there is much more freedom of rotation relative to each other.

Besides the resorcinol and naphthyl systems mentioned earlier, the central hydroquinone units have been replaced with a number of other aromatic systems. In two successive papers, the incorporation of furan or pyridine macrocycles into the crown ether was described. The furan-based system had lowered recognition for the bipyridyl units but still gave the catenane in 40% yield.[193] The pyridine-based crown

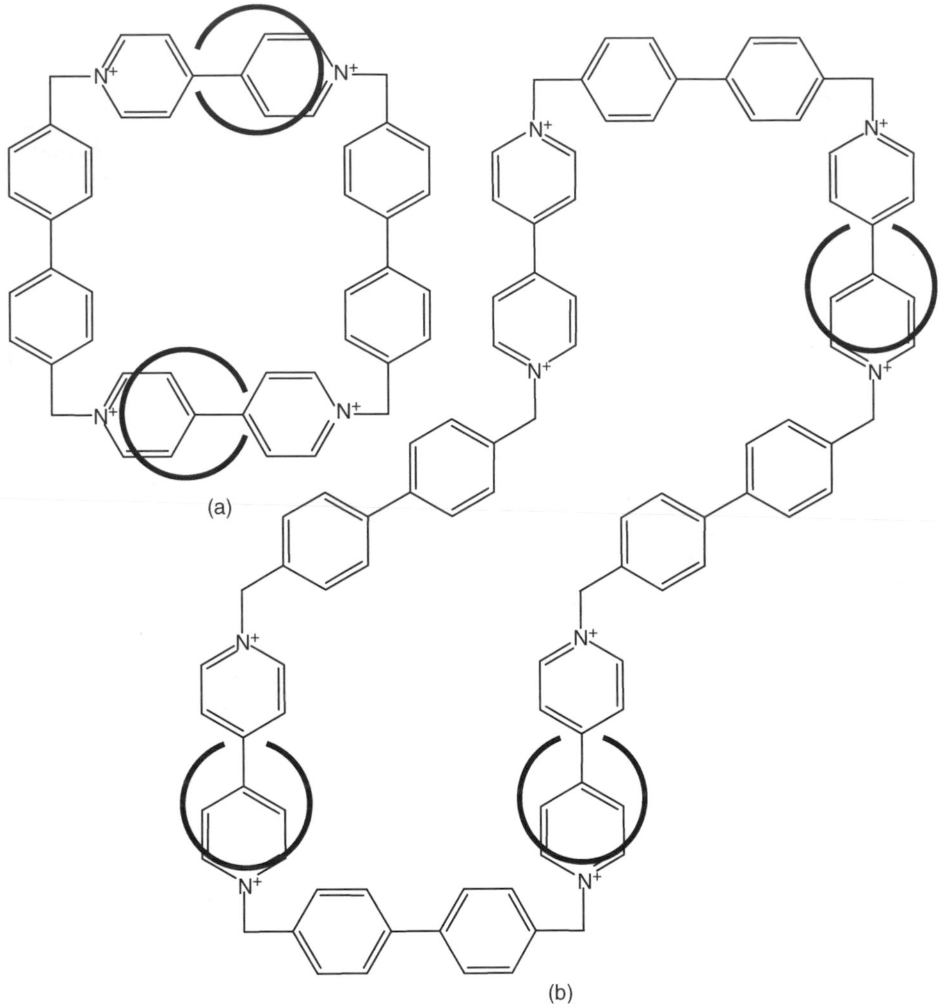

Figure 9.64   *(a,b) Expanded bipyridyl macrocycle-based [3] and [4]catenanes*

ether showed good recognition for the bipyridyl units but gave the catenane in lower (31%) yield.[194] Both systems appeared to be much more mobile than those based on hydroquinone. Further aromatic units could also be added to the crown ethers, as in the structure shown in Figure 9.66a, where two catechol units have been added.[195] When the catenane is created, it can exist in two isomeric forms as shown; NMR studies have shown that the isomer with the cationic macrocycle closer to the catechol units is more stable (64% in this form in acetonitrile solution), probably due to greater electron density for the combined hydroquinone/catechol units. The [3]catenane in Figure 9.65 and the related [2]catenane containing only one cationic macrocycle have been studied by NMR to determine the dynamics of the system.[196] In the [2]catenane the cationic macrocycle is shown to rapidly travel around the crown ether. Similar behaviour has been observed for the [3]catenane in which the macrocycles are always on diametrically opposite hydroquinone rings. The authors of this work liken the behaviour to trains on a track and comment, 'all that remains to be done is to introduce signals to direct the train'.

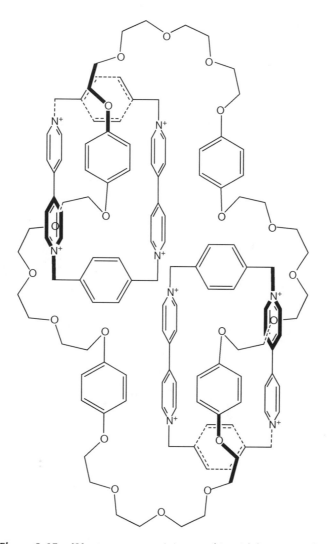

Figure 9.65 *[3]catenane containing two bipyridyl macrocycles*

Porphyrins have also been incorporated into these systems; as an example the porphyrin-based macrocycle shown in Figure 9.66b can be used to template the standard clipping reaction to give the catenane shown.[197] NMR studies show the cationic macrocycle slowly rotates around the hydroquinone unit (accelerating at higher temperatures) and there is significant face-to-face interaction between the bipyridinium rings and the porphyrin. Further work has shown that depending on the catenane structure, rotation rates at room temperature variy between 50 and 2500 rotations per second.[198] These compounds are much more resistant to protonation than their open-chain analogues, requiring strongly acidic conditions.[199] The resultant protonated systems display much less interaction between porphyrin and bipyridyl units due to electrostatic repulsion, which causes a rearrangement of the structure so that the electron-rich hydroquinone bridging groups are nearest the porphyrin. Also, the catenanes can rotate more freely, as demonstrated by a catenane that has a rotation rate of 80 s^{-1} being increased to 1000 s^{-1} upon protonation. A version of this with a 1,5-dihydroxynaphthyl bridging group shows similar behaviour.

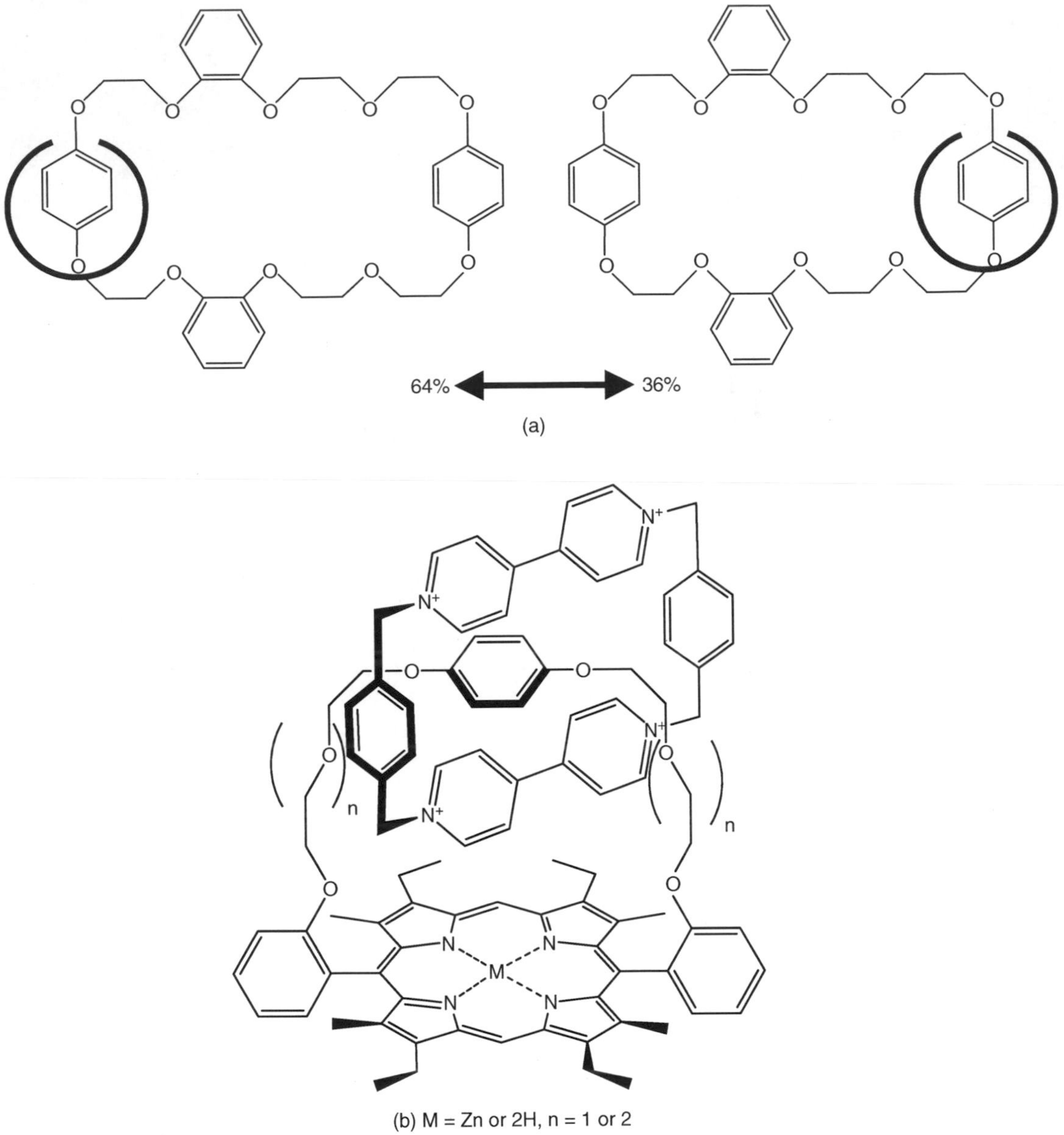

64% ◄────────► 36%

(a)

(b) M = Zn or 2H, n = 1 or 2

Figure 9.66 *(a) Catechol-substituted crown ether and (b) porphyrin-containing catenane (Copyright Wiley-VCH Verlag GmbH & Co. KGaA. Reproduced with permission[197])*

One problem with the formation of multiple catenanes is that many of the early syntheses of catenanes often had poor yields, making multi-step syntheses challenging due to low final yields and presence of many byproducts. The high yields and relatively simple clean-up of many of these π−electron-interaction assembly procedures has allowed the development of multiple ring catenanes. We have already discussed how [4] and [5]catenanes can be synthesised by use of larger macrocycles, but initially only very low yields were obtained.[192] A [5]catenane was synthesised as early as 1994 and its resemblance to the five rings of the

Olympic flag led to the authors suggesting the name 'Olympiadane' for these compounds.[200] By replacing the phenylene units in the crown-ether components with 1,5-dioxanaphthyl units (which led to better interactions with the clipping agents), along with high-pressure reaction conditions, much more efficient syntheses were devised by the group of Fraser Stoddart. First, clipping together a cationic macrocycle containing biphenyl spacers in the presence of crown ethers[192] has been shown to yield a [3]catenane (Figure 9.64a). This process can be applied to crown ethers containing three 1,5-dioxynaphthalene groups to give some [2]catenane but mainly the [3]catenane[201] (Figure 9.67). The [3]catenane can then be further reacted again via a clipping reaction to give the [4] and [5]catenanes shown in Figure 9.68. This procedure can be enhanced by the use of high pressures and good yields of Olympiadane (30%), and even higher catenanes (28% for the [6]catenane and 26% for the [7]catenane) can be obtained, along with a second [5]catenane which is a topological isomer of Olympiadane. Figure 9.68 shows the reaction scheme for all of these compounds. X-ray crystallography studies of Olympiadane and the [7]catenane[201,202] have been obtained and show highly compact structures (Figure 9.69). Extensive NMR investigations have also been made and thse demonstrate that as the size and complexity of the catenanes increase, the dynamics of the system are diminished and translations such as the exchange of cationic macrocycles between the 1,5-dioxynaphthalene units are dramatically slowed.[201]

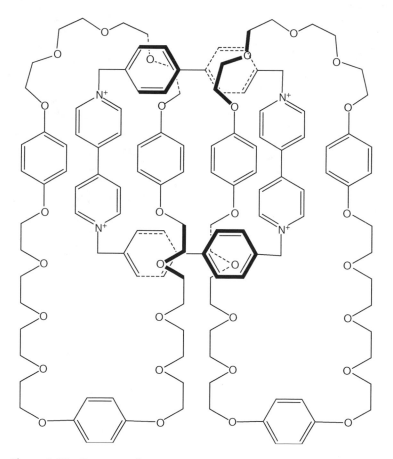

Figure 9.67 *Structure of a [3]catenane containing two crown ether units*

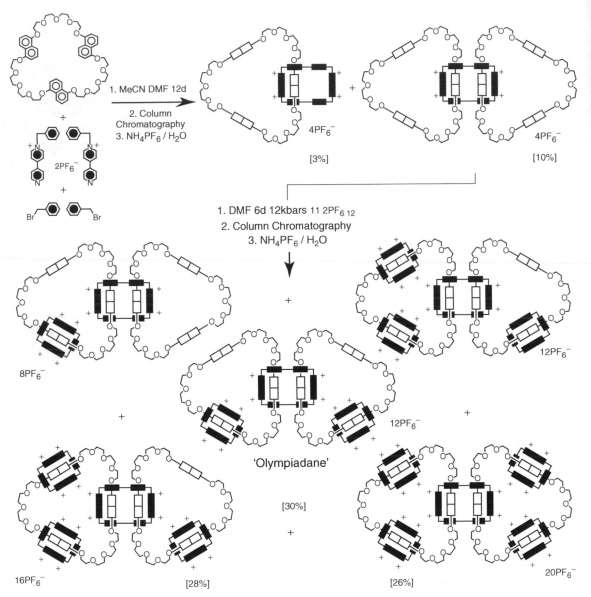

Figure 9.68 *Synthesis of Olympiadane and other catenanes (Reprinted with permission from[201]. Copyright 1998 American Chemical Society)*

Besides varying the constitution of the crown-ether units, it is of course possible to vary the cationic macrocycle structure. There is however one difficulty with this since this is the macrocycle that is actually formed during the clipping reaction. Should the structure of the precursor to the macrocycle be unable to cyclise readily, yields of the resultant catenane may be greatly diminished. The classical synthetic scheme for these reactions involves reacting a dibromo compound with at least two equivalents of a bipyridyl compound and then reacting this precursor with a second dibromo compound to achieve cyclisation[187]

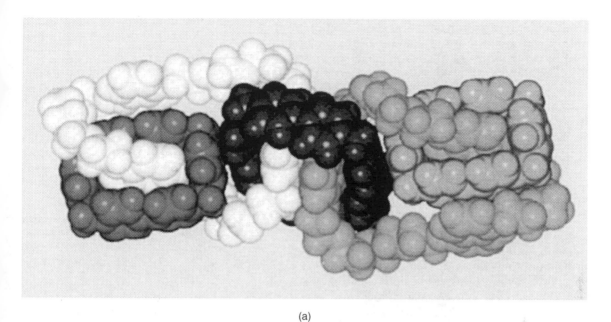

(a)

(b)

Figure 9.69   *Space-filling representations of the X-ray structures of (a) Olympiadane and (b) the [7]catenane (Copyright Wiley-VCH Verlag GmbH & Co. KGaA. Reproduced with permission[202])*

(although single-step procedures have been shown to work [188]). To attempt to determine the factors that affect the efficiency of these synthetic schemes, initial experiments were made to compare two isomers of xylylene dibromide as ring-forming agents.

Two macrocyclic precursors were synthesised by reacting 4,4'-bipyridine with either the *para* or *meta* isomer of xylylene dibromide. The precursors were then mixed with a crown ether (Figure 9.61) and reacted with another equivalent of xylylene dibromide.[203] When both the xylylene units were *para*, a 70% yield of catenane was obtained, whereas when both were *meta*, no catenane was obtained, probably because of the much smaller size of the cavity of the macrocycle formed. However, increasing the pressure to 10 kBar allowed the formation of this catenane in 28% yield. Mixed systems could be synthesised, as shown in Figure 9.70. When the *meta* isomer was used to ring-close the macrocycle, the catenane was only obtained in 18% yield, but the *para* xylylene dibromide was much more efficient, with the yield increasing to 40%. NMR studies of these systems[202] showed that rotation of the cationic macrocycle was easiest in the product with two *para*-xylylene spacers (free energy of activation 15.6 kcal mol^{-1}), increasing to 16.5 kcal mol^{-1} for the mixed system and 17.6 kcal mol^{-1} for the catenane with two *meta*-xylylene spacers.

Similar work replaced synthesised catenanes containing one or two thiophene units in 59% and 36% yields respectively (Figure 9.71a,b). These systems display lower degrees of movement of the hydroquinone rings, and from the X-ray structure (Figure 9.71c) it is clear there is an S—O interaction between the hydroquinone and thiophene units.[204] Photoactive azobenzene groups have also been incorporated into catenanes (Figure 9.72). These systems can be isomerised from the E to the Z form by irradiation.[205] The reisomerisation process back to the Z form is assisted by incorporation in the catenanes; the half-lives for the catenanes in Figure 9.72a,b are 20.5 hours and 12 days, whereas the corresponding azobenzene macrocycles without any crown-ether counterpart have half-lives of 9.3 and 12 days respectively. The effect is most noticeable for the catenane in Figure 9.72a, which is thought to be a function of the more open nature of this catenane.

An intensive and detailed study of a number of catenanes was made with the scheme shown in Figure 9.73a. With these the electron-poor groups can be either bipyridyl or the stilbene-like structures shown, whereas the crown-ether units can contain either electron-rich phenylene or naphthyl moieties.[206] In this case there are nine potential combinations, giving rise to a variety of catenanes. All of these can be synthesised with yields from 19 to 70%. When the catenanes are nonsymmetric in nature, the preference is for the naphthyl residues to be complexed by the tetracationic cyclophane components and a bipyridinium unit inside the crown-10 components. Four of the possible combinations exhibit translational isomerisation in solution, as shown by NMR. One combination has four different aromatic groups, giving four possible isomers. In all cases where crystal structures are obtained the catenanes are found to adopt only one conformation, identical to the preferred combination in solution. Figure 9.73b shows a typical crystal structure, in this case for the catenane containing two dipyridylethylene and two 1,5-dioxanaphthalnene groups. Electrochemical measurements also show novel behaviour, and in this context for example the catenanes containing a bipyridyl and a dipyridylethylene unit in the cationic macrocycle show signs of undergoing electrochemically controlled switching of the isomers. Although the *trans*-bipyridylethylene units can be photochemically isomerised in macrocyclic structures, incorporation into the catenane prevents this process from occurring.

Other workers synthesised the chiral catenane showed in Figure 9.74b, along with a flexible, achiral analogue.[207] NMR studies indicated the system had a degree of mobility and X-ray studies demonstrated there was steric hindrance between the chiral hydrobenzoin units and the bipyridyl systems, as shown in the splaying of the bipyridyl units, which probably explains the low yield (8%) of this reaction. Another method of synthesising chiral systems is to utilise 1,2,4,5-*tetrakis*-bromomethyl benzene as the ring-closing agent instead of xylylene dibromide.[208] This enables the synthesis of *bis*-[2]catenanes (Figure 9.75) with a yield of 13%; these are covalently linked via a common benzene unit and since ring closure is most effective in

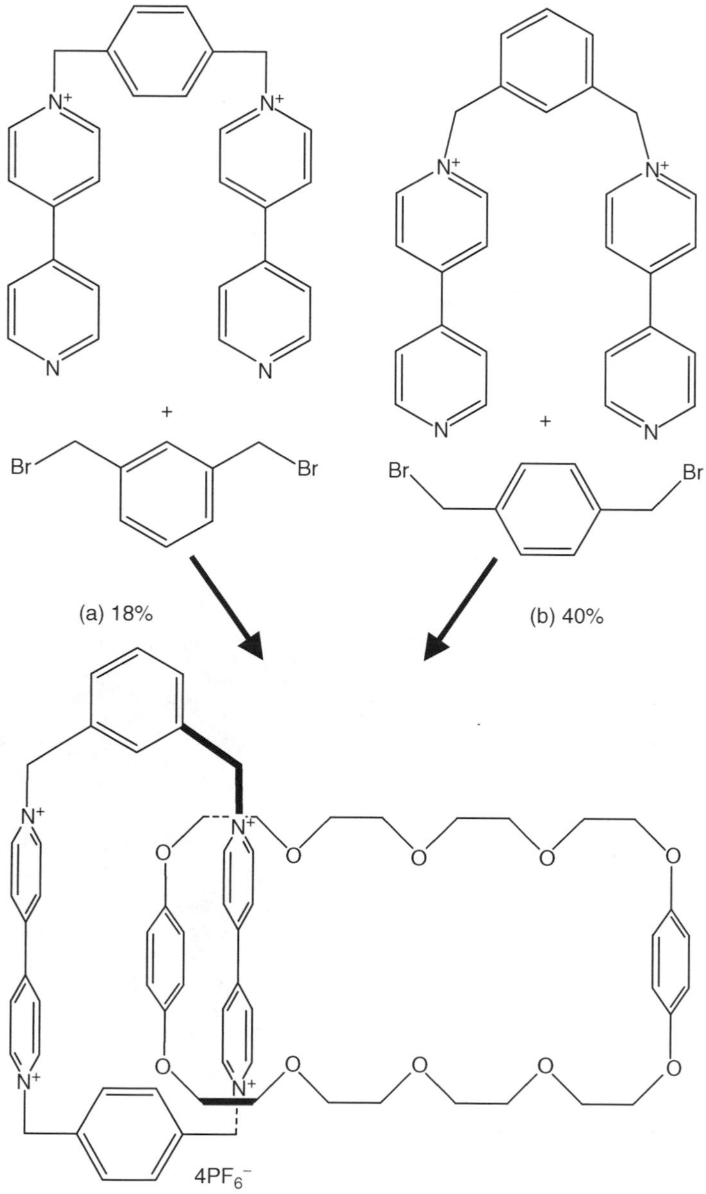

Figure 9.70   *Catenane synthesis by two routes*

the *para* conformation, this linkage introduces a centre of chirality into the system. NMR studies on the resultant catenane indicate that the crown ether groups can freely move around the cationic macrocycles. Interestingly, reacting the 1,2,4,5-*tetrakis*-bromomethyl benzene with 4,4-bipyridine and attempting to cyclise the resultant *tetrakis*-bipyridyl compound with *p*-xylylene dibromide does not afford catenanes. Other workers also used longer linkers to couple together two catenane units, giving structures such as that shown in Figure 9.76c (another version of this synthesis replaces the unsubstituted phenylene ring

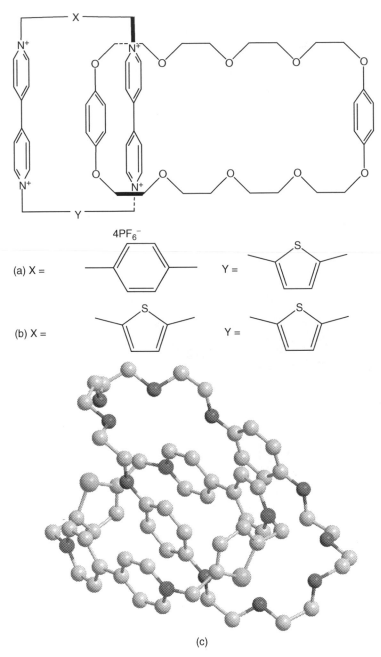

(a) X =   Y =

(b) X =   Y =

(c)

Figure 9.71   *(a,b) Structures of thiophene-containing catenanes and (c) X-ray crystal structure of 9.71b*

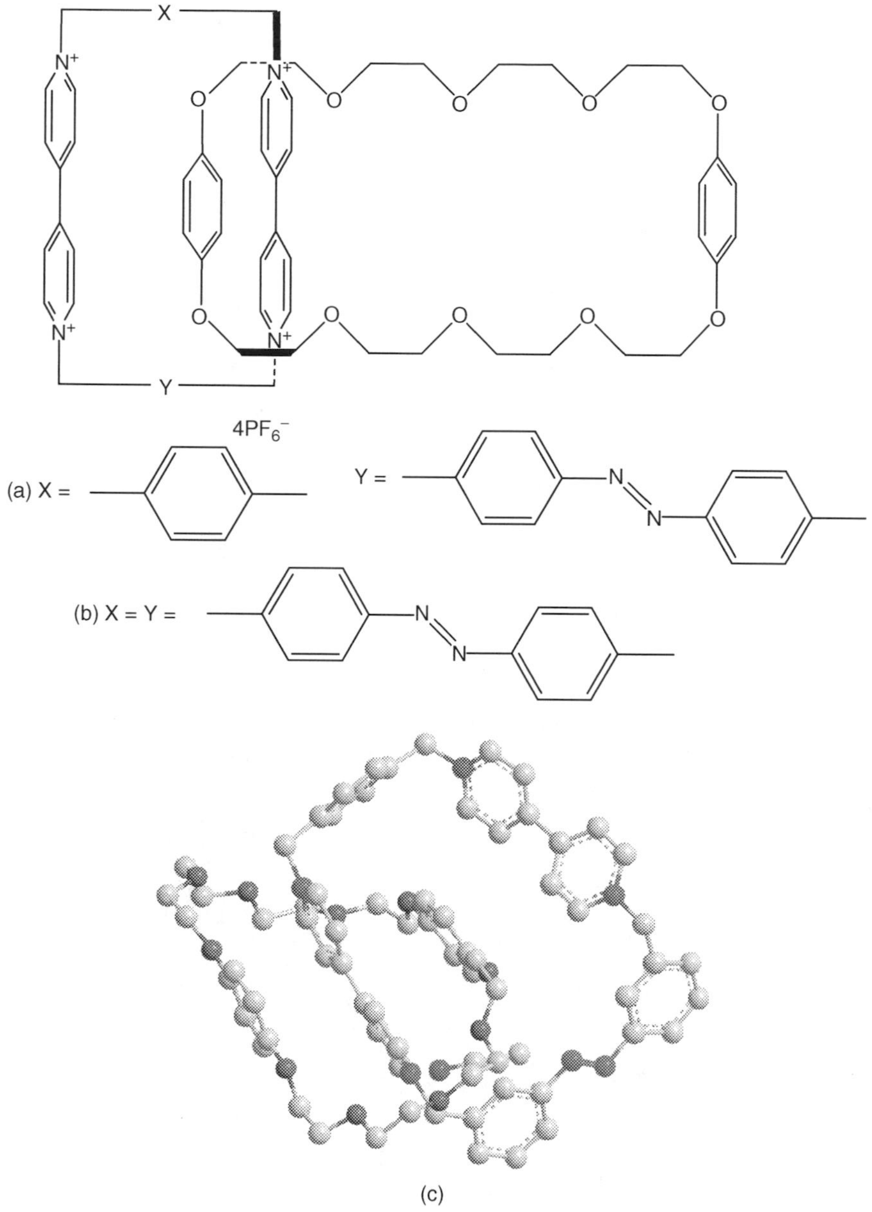

$4PF_6^-$

(a) X =   Y =

(b) X = Y =

(c)

Figure 9.72 *(a,b) Structures of azobenzene-containing catenanes and (c) X-ray structure of 9.72a*

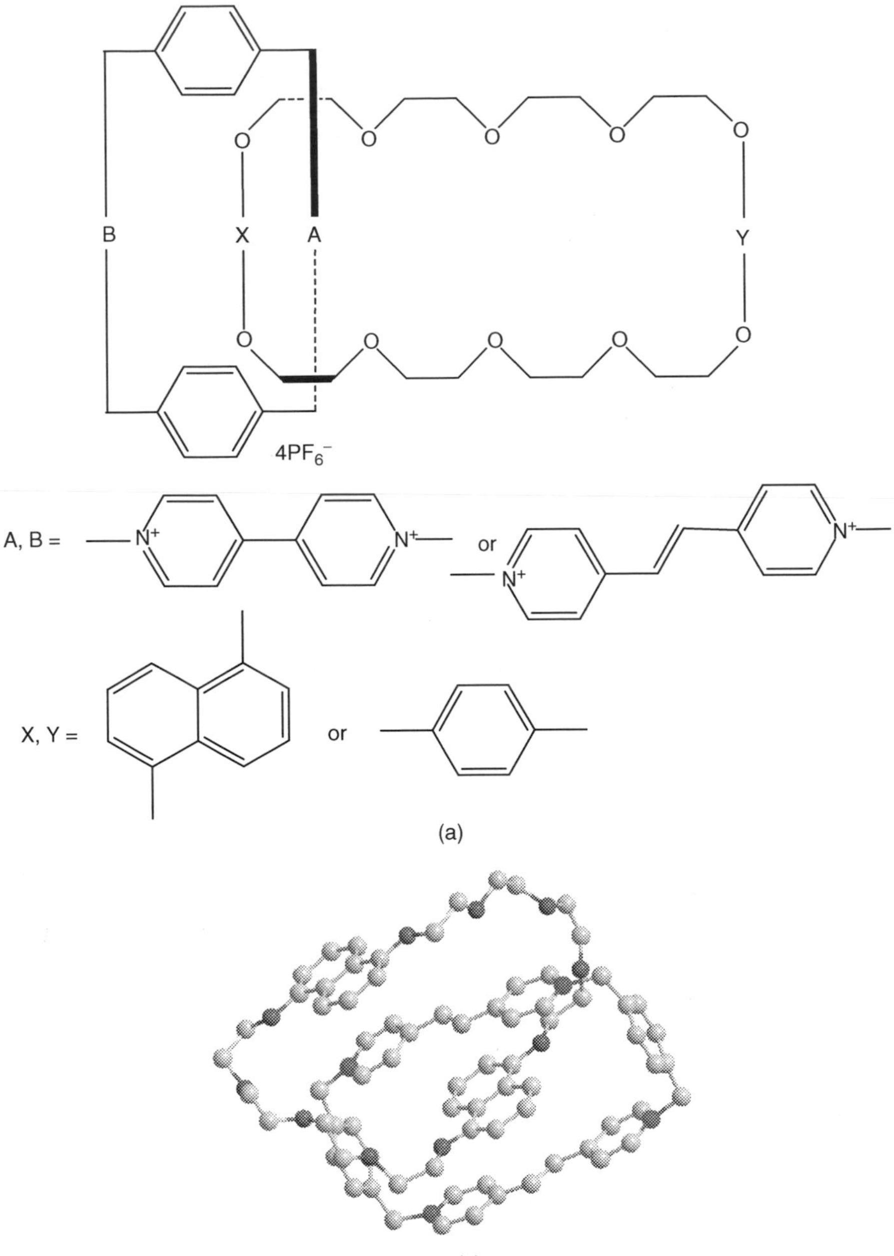

(a)

(b)

Figure 9.73 (a) Structures of catenanes with a variety of aromatic groups and (b) crystal structure of a dipyridylethylene catenane

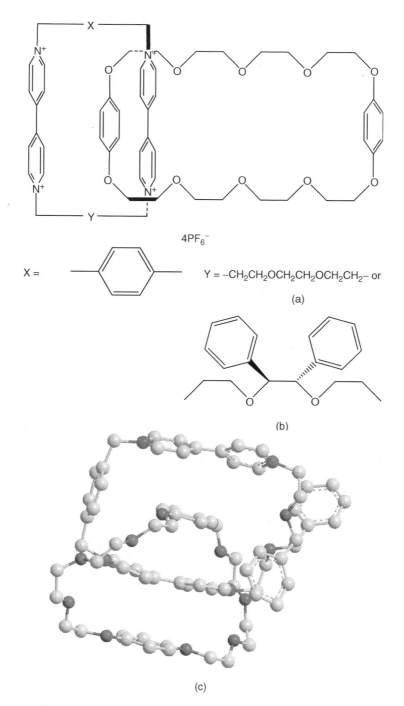

X = [benzene ring structure]   Y = –CH₂CH₂OCH₂CH₂OCH₂CH₂– or

(a)

(b)

(c)

Figure 9.74   *(a,b) Structures of a flexible and a chiral catenane and (c) X-ray structure of 9.74a*

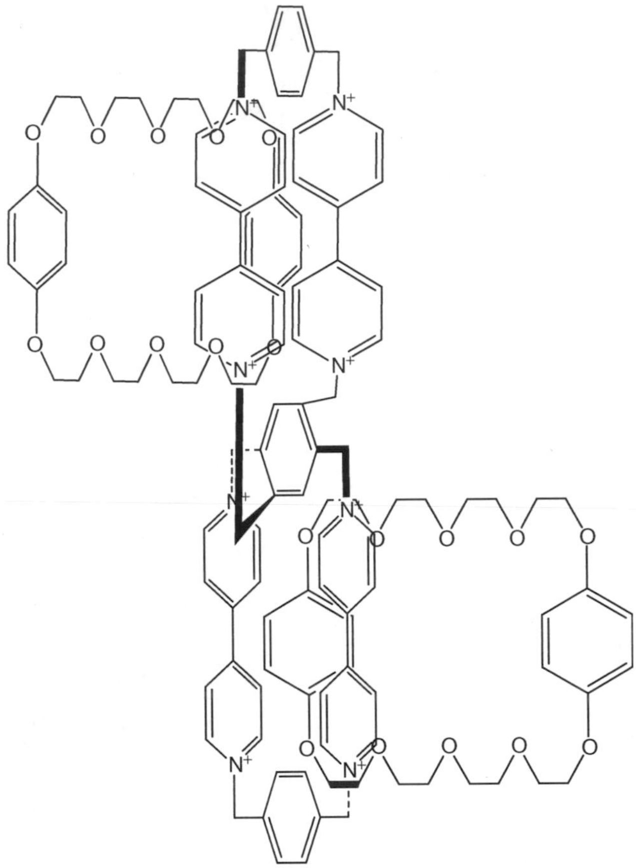

Figure 9.75   Structures of bis-catenanes held together by a common benzene unit

with a 1,5-dioxynaphthyl unit). This provides an example of chemical modification after synthesis of a catenane. A crown ether containing an N-allylphthalimide unit (Figure 9.76a) was synthesised. Reduction of the phthalimide to an amine followed by amidation gave a macrocycle that could be incorporated into a catenane and its structure was proved by X-ray studies (Figure 9.76b). Coupling of two of these moieties gave the structure shown in Figure 9.76c. NMR studies demonstrated the presence of a substantial degree of order characterizing the molecular structure of the catenanes.[209] A second method of synthesising *bis*-[2]catenanes is to utilise the compound shown in Figured 9.77a as the ring-closing linker. Reacting this with a *bis*-bipyridyl compound (Figure 9.61) and a crown ether gives a *bis*-[2]catenane (Figure 9.77b) in 31% yield.[209] These various types of *bis*-catenane are of great interest since if they are suitably substituted, they can undergo condensation reactions to yield polycatenanes.

As an alternative to bipyridyl units in the cationic macrocycles, diazapyrenium units have also been studied.[210] NMR studies indicate that there is a dramatic increase in association constants between diaza-pyrenium moieties and either hydroquinone or 1,5-dioxynaphthalene units compared to those observed with bipyridyl. This is borne out by the study of asymmetric catenanes in which the cationic macrocycle contain a bipyridyl and a diazapyrenium unit (Figure 9.78a). NMR studies show that in solution such a

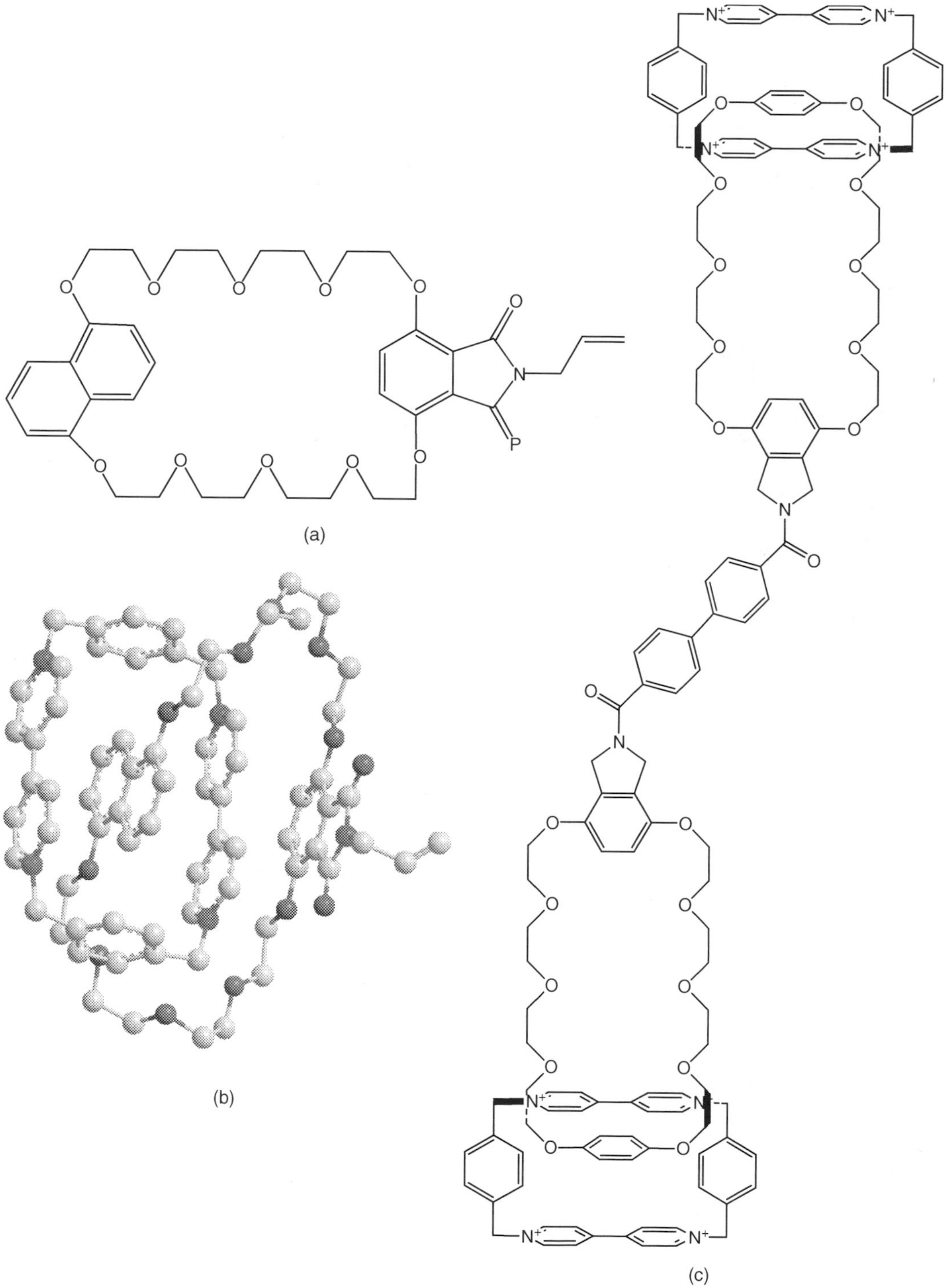

Figure 9.76 (a) Phthalimide-substituted crown ether, (b) its X-ray crystal structure and (c) a bis-[2]catenane synthesised from 9.76a

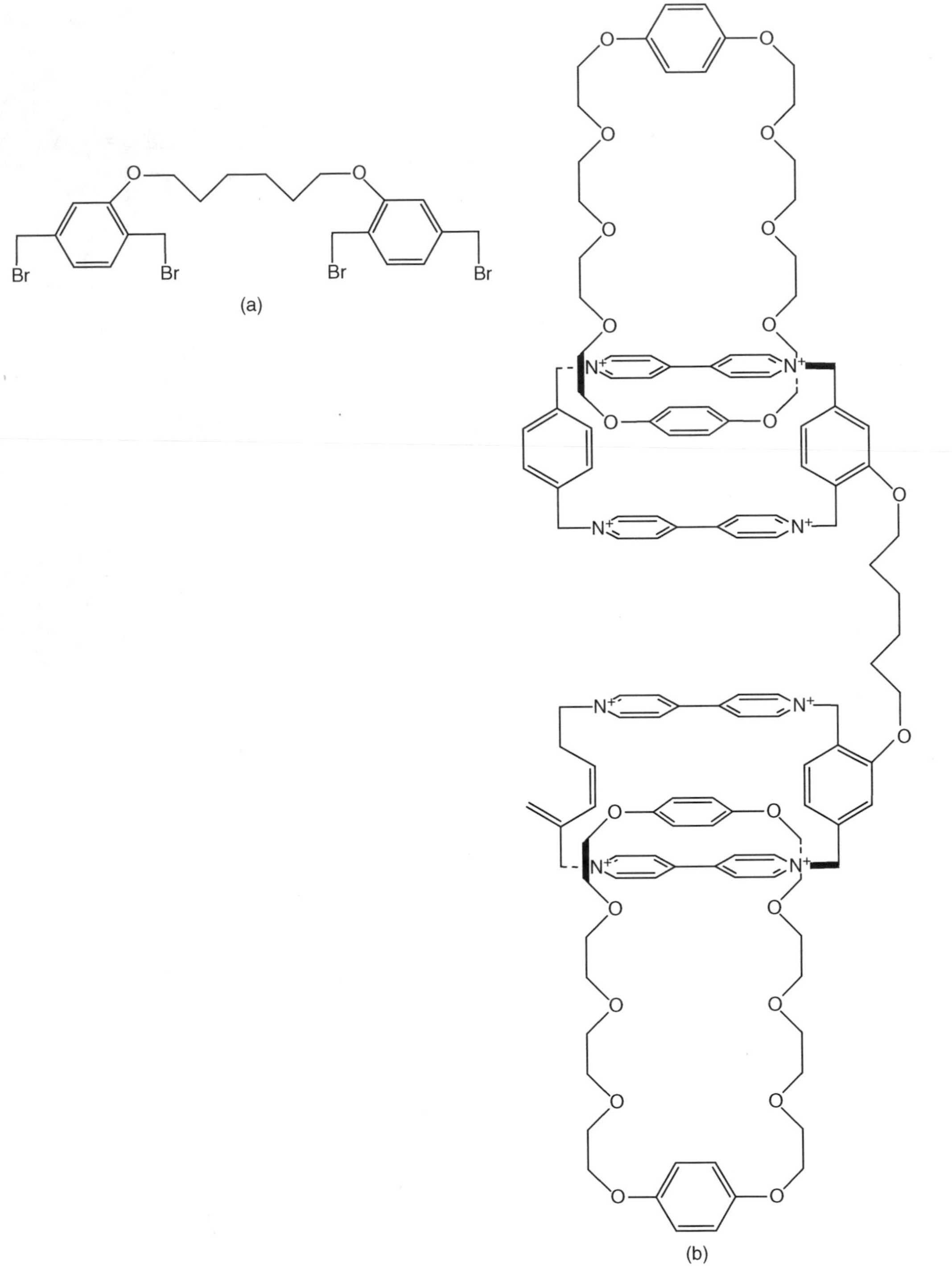

(a)

(b)

Figure 9.77   (a) A bis-xylylene dibromide compound and (b) the bis-[2]catenane formed using this linker

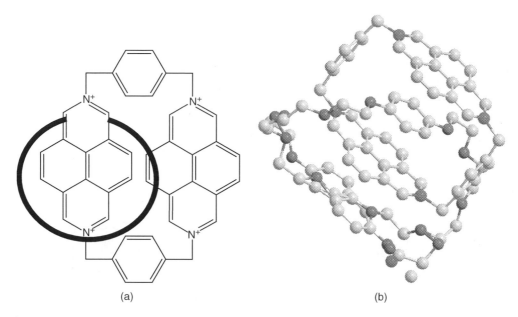

Figure 9.78   (a) Diazapyrenium macrocycle/crown-ether catenane and (b) crystal structure of 9.78a

catenane exists as a mixture of isomers, but the diazapyrenium is preferably encapsulated by the crown ether; this has been confirmed to occur exclusively in the solid state by a crystal structure (Figure 9.78b). NMR studies also demonstrate that the barriers to rotation are higher in diazapyrenium-containing systems, probably due to the stronger intercomponent interactions.

A variety of other molecular units have been incorporated into the catenane system; for example, the commonly used bipyridyl macrocycle (Figure 9.61) can be incorporated into a catenane along with the macrocyclic crown ether in Figure 9.79a. The cationic unit encapsulates the hydroquinone unit of the crown ether rather than the pyrrole/tetrathiafulvalene moiety.[211] A series of pseudorotaxanes have also been made using this compound and have been shown to dethread when the tetrathiafulvalene unit is oxidised. In the case of the catenanes, oxidation produces strong spectral changes but of course cannot destroy the catenane structure. Other workers utilised a strapped tetrathiafulvalene derivative (Figure 9.79b) and formed a catenane with the bipyridyl macrocycle (Figure 9.61). In this compound in the solid state the tetrathiafulvalene was shown to preferentially exist in the *cis* conformation with types of charge-transfer interaction.[212] The X-ray structure (Figure 9.79c) shows the hydroquinone unit is encapsulated by the bipyridyl macrocycle and that there are charge-transfer interactions, but the tetrathiafulvalene unit also appears to participate in charge-transfer interactions with the pyridinium moeities, leading to a strong green colour for the complex.

Other workers have successfully incorporated porphyrin and naphthalene diimide units into catenanes such as that shown in Figure 9.80a. These are synthesised by taking a porphyrin macrocycle and clipping together two units of *bis*-N-propargyl naphthalene diimide to give structures like that shown in the figure.[213] NMR studies show that for the longer-strapped catenane the naphthalene diimide macrocycle rotates around the dialkoxy axis at a rate of 420–450 revolutions per second; in the shorter-strapped catenanes no rotation is observed. UV/Vis spectra indicate that there are charge-transfer interactions and electronic communication between the two components of the catenane. Other workers synthesised catenanes between the classical cationic macrocycle (Figure 9.61), a triazole-containing crown ether (Figure 9.80b), and

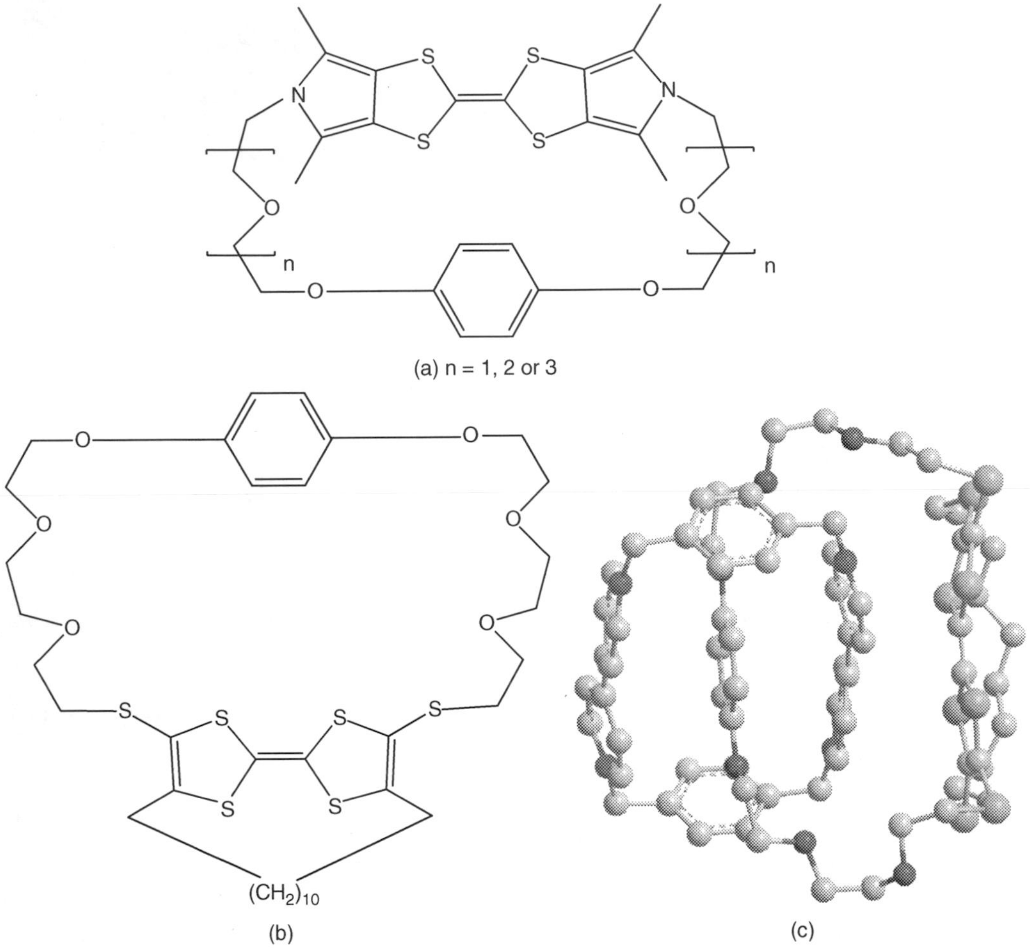

Figure 9.79   *(a) A pyrrole/TTF macrocycle, (b) A TTF macrocycle and (c) X-ray structure of the catenane formed with 9.79b*

successfully crystallised the product.[243] X-ray crystallography revealed the catenane structure, where the naphthalene unit is encapsulated by the cationic macrocycle. What is interesting is that this catenane is chiral and crystallisation spontaneously resolves the enantiomers. The two enantiomers are present overall in equal amounts but each individual crystal consists of a single enantiomer. Earlier work has also shown that the *bis*-N-propargyl naphthalene diimide can be assembled with a crown ether containing two 1,5-dioxanaphthalene units and is preorganised so that oxidative coupling of the acetylenic groups leads to catenane formation.[215] Later work also utilised two *bis*-acetylenic monomers which preorganised in solution and upon oxidative coupling formed catenanes.[216]

Other members of this family of supramolecular systems include a catenane in which one of the macrocycles contains fluorene and azobenzene groups (Figure 9.80c). This can be used as the template for the clipping reaction (Figure 9.61) to give a small (8%) yield of catenane.[217] X-ray crystallography shows that the cationic macrocycle encapsulates one benzene ring of the azobenzene unit.

A number of other methods which utilise various noncovalent interactions to preorganise various moieties before reaction to give catenanes have been developed. Metathesis reactions of various olefins have

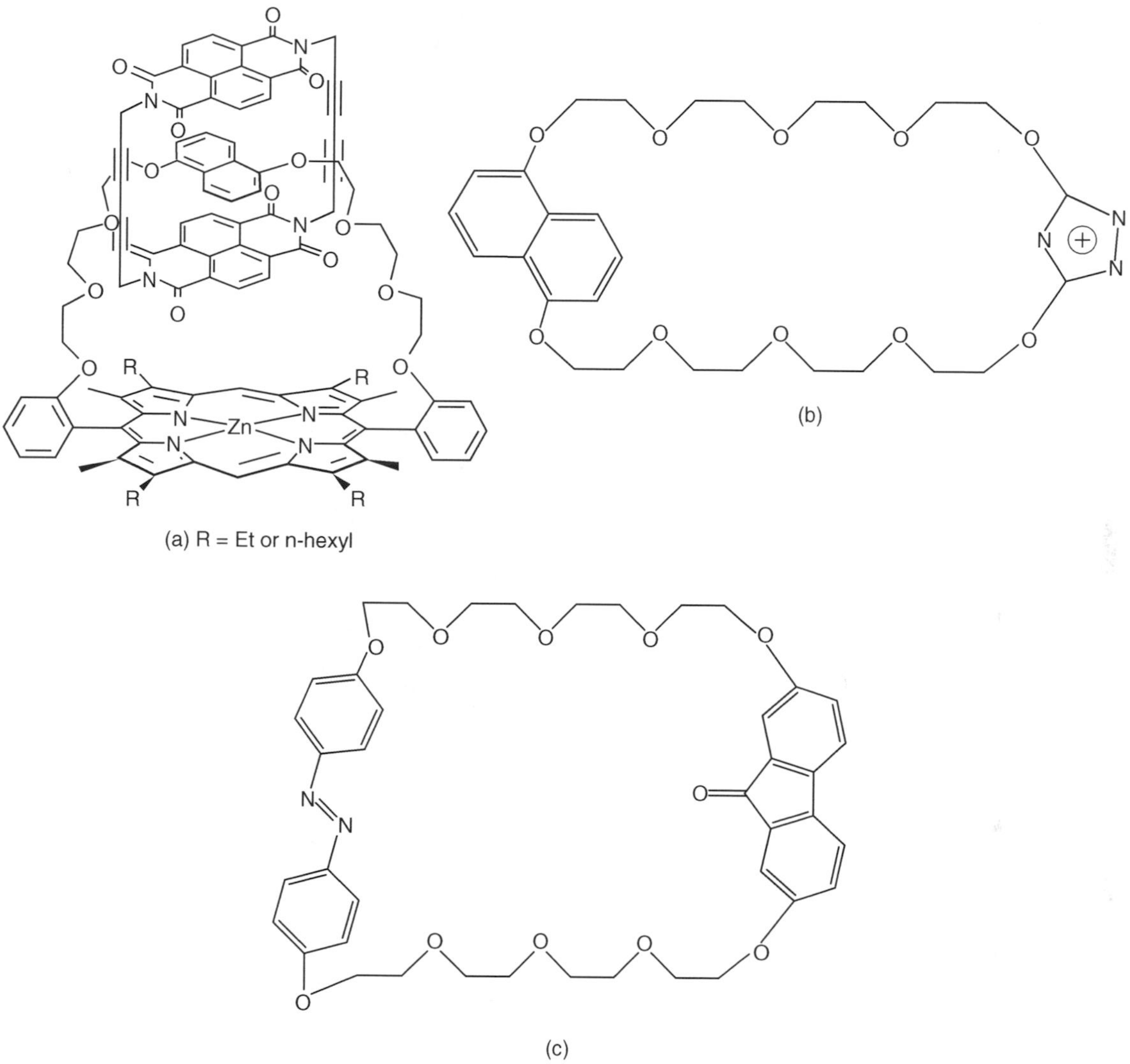

Figure 9.80 *(a) A porphyrin–naphthalene diimide-containing catenane, (b) a triazole-containing macrocycle and (c) a fluoreno/azobenzo macrocycle (Reproduced by permission of the Royal Society of Chemistry[213])*

been utilised – one elegant synthesis involves an unsaturated crown ether being combined with a macrocycle containing an ammonium centre (Figure 9.81a). When catalysed by ruthenium, the crown ether reversibly ring-opens and – closes, and whilst in acyclic form can thread through the ammonium macrocycle and then recyclise to form the catenane.[218] Alternatively, acyclic precursors can be used in this synthetic method. Another method of using reversible ring-open reactions is to take the classic bipyridyl macrocycle (Figure 9.61) and subject it to attack by a catalytic amount of tetrabutyl ammonium iodide.[219] This reagent reversibly reacts with the macrocycle to give the acyclic derivative shown in Figure 9.82a and then ring-closes again. If this is done in the presence of a suitable crown ether, such as that shown in Figure 9.61, catenanes can be obtained in high yield without any byproducts. Reaction of dialdehydes

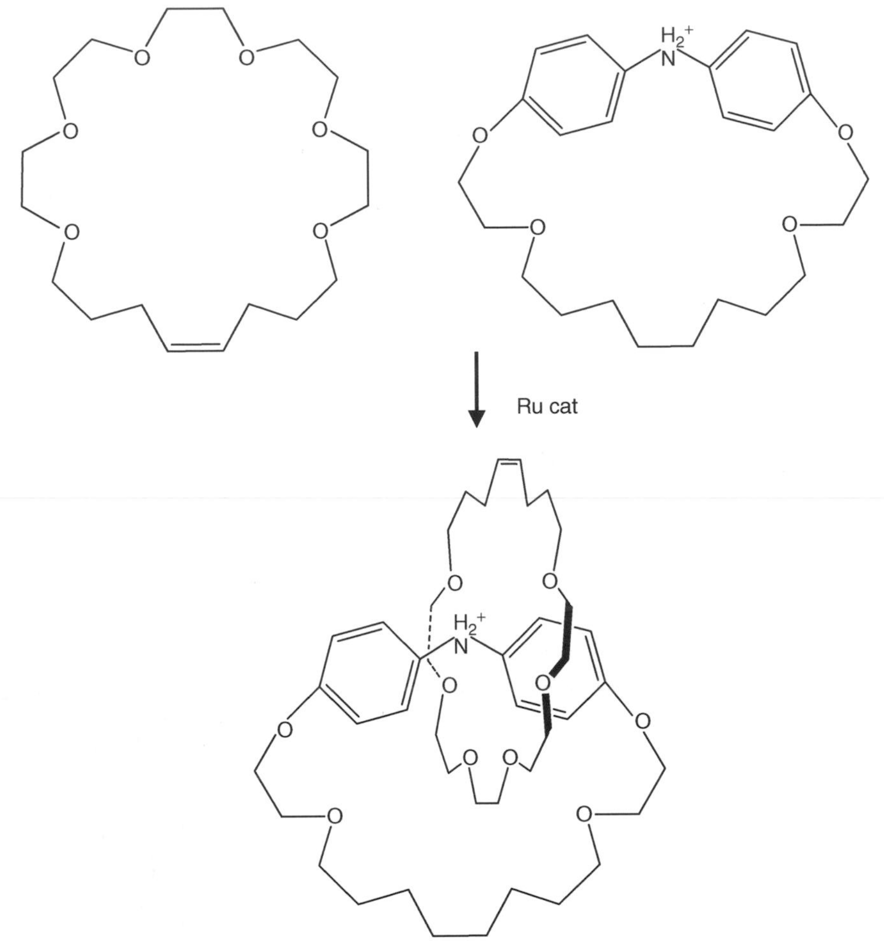

Figure 9.81   Synthesis of catenanes by olefin metathesis

with a diamino ethylene glycol in the presence of the bipyridyl macrocycle gives catenanes. What is of interest is that if a phenylene and a naphthalene dialdehyde are used in a 1 : 1 ratio, the resultant asymmetric catenane (Figure 9.82b) is formed in 90% yield.[220] This indicates a high degree of preorganisation before chemical reactions take place. Other workers combined thiol-substituted naphthalene diimide and 1,5-dioxanaphthalene units and oxidised the thiols to form disulfides to give a mixture of products from which catenanes could be isolated.[221]

9.3.4   Catenanes based on amide hydrogen-bonding interactions

The first catenane of this type was reported in 1992 and once again provides an example of serendipity in these types of system. Workers at Sheffield were attempting to make the macrocycle shown in Figure 9.83 by condensation of isophthaloyl chloride and a diamino compound.[222] When attempting a two-stage synthesis, they obtained three products: the dimer shown in Figure 9.83, a cyclic tetramer and the catenane. This compound consisted of two of the dimers interlocked with each other and could be obtained in 34%

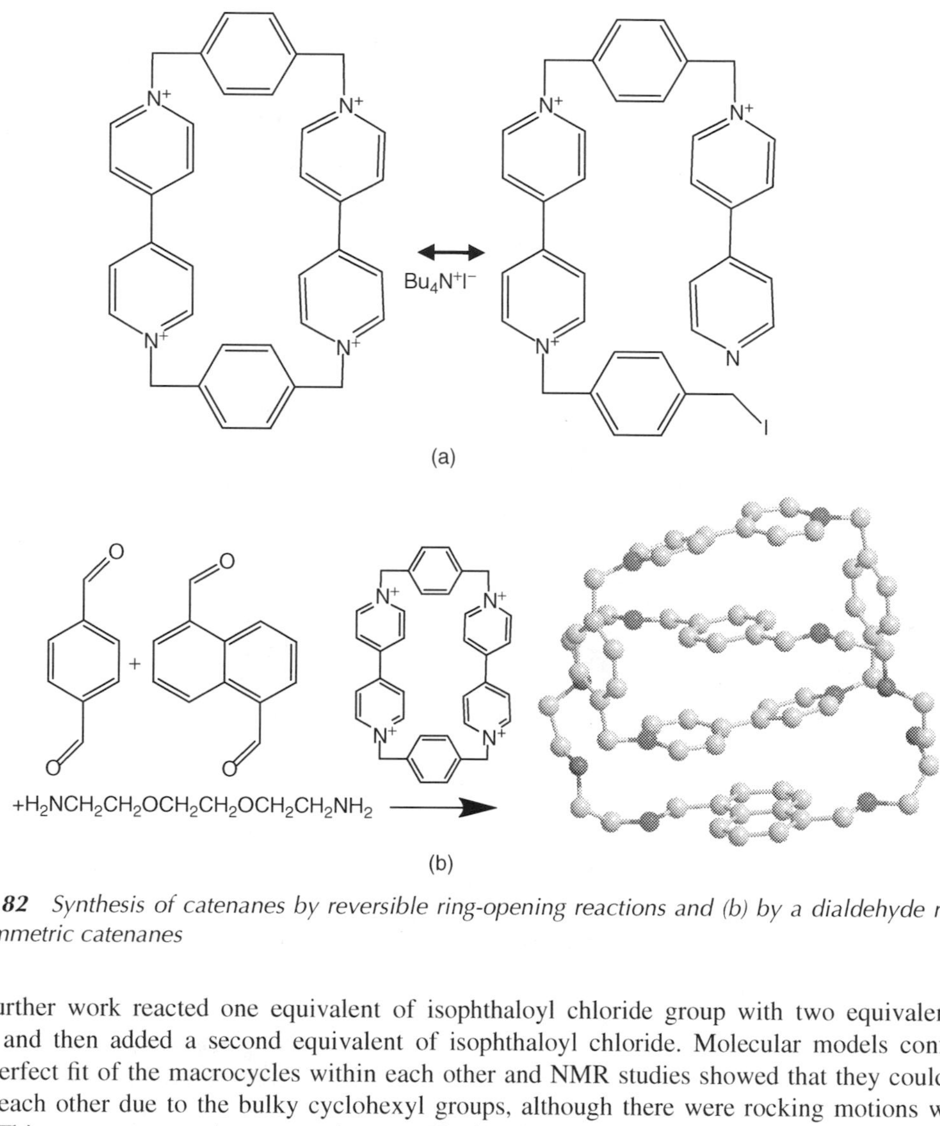

Figure 9.82 Synthesis of catenanes by reversible ring-opening reactions and (b) by a dialdehyde reaction to give asymmetric catenanes

yield. Further work reacted one equivalent of isophthaloyl chloride group with two equivalents of the diamine and then added a second equivalent of isophthaloyl chloride. Molecular models confirmed an almost perfect fit of the macrocycles within each other and NMR studies showed that they could not pass through each other due to the bulky cyclohexyl groups, although there were rocking motions within this system. This catenane was shown to exist as a 50:50 mixture of enantiomeric forms as shown by NMR studies with chiral shift reagents. The synthesis of the catenane in such high yield again indicates a high level of preorganisation, probably due to hydrogen-bonding between amide groups. Finally, after a series of failed attempts, crystals were obtained of sufficient quality to allow a determination of the catenane structure and proof of chirality by X-ray crystallography.[223] Replacement of the second isophthaloyl group with a 2,6-pyridinyl group led to no formation of the dimer catenane; instead a [2]catenane formed in which one unit was the cyclic dimer and one the cyclic tetramer, along with traces of the dimer–dimer–tetramer [3]catenane.[224] If the initial isophthaloyl unit was replaced with a 2,6-pyridinyl group, no catenane formation could be observed. This was thought to be due to the templating effect of amide hydrogen-bonding being disrupted by the presence of the pyridine nitrogen.

Other workers also utilised this synthetic strategy; for example, a similar catenane in which a methoxy group is in the 5-position of the isophthaloyl unit was synthesised in 8.4% yield and was shown to have

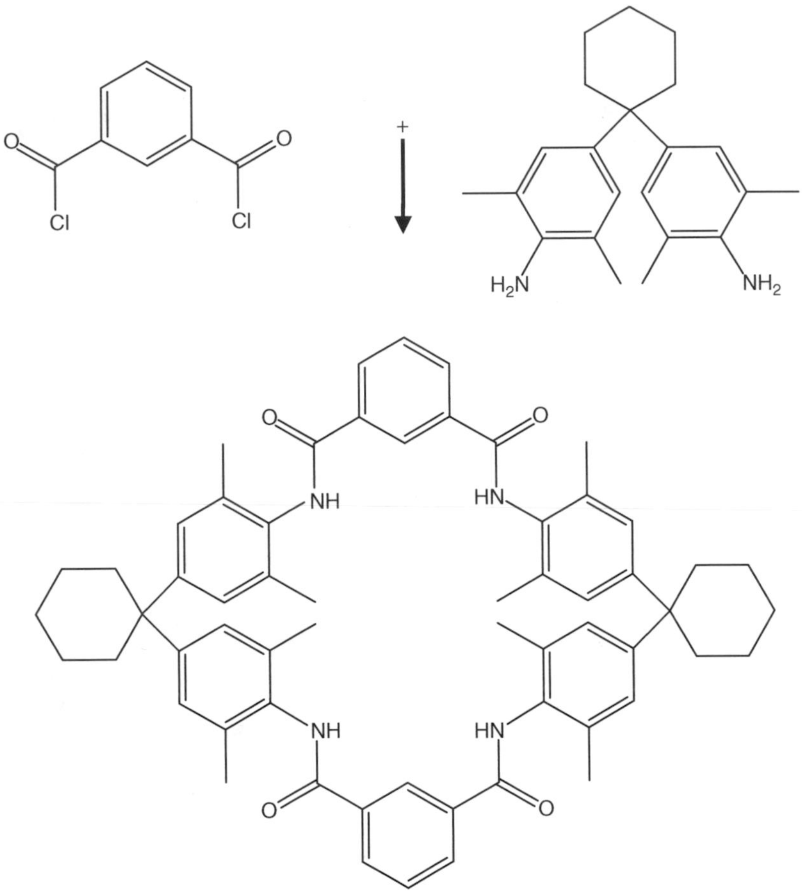

Figure 9.83   *Synthesis of macrocycles by reaction of acid chlorides and amines*

a fixed conformation with inner and outer methoxy groups.[225] Further work achieved the synthesis of asymmetric lactam catenanes in which the two isophthaloyl groups bear different substituents.[226] NMR studies showed different isomers could be obtained for these molecules, with the methoxy groups being either 'inside' or 'outside' the macrocycle.[227] Using these more complex synthetic strategies allowed a much more detailed determination of the mechanism of reaction. This group also developed a series of catenanes incorporating furan groups (Figure 9.84a) and managed to isolate several isomers.[228] Since these systems cannot freely rotate there is a definite 'inside' aromatic group and 'outside' aromatic group. Three possible isomers exist, with the furan being in–in, in–out or out–out, and the in–out and out–out were isolated. A crystal structure was obtained for the out–out isomer.

An even more elegant synthesis, which starts with common, commercially available materials, was first described in 1995. Isophthaloyl chloride and *p*-xylylene diamine were reacted together to give up to 20% of the catenane (Figure 9.84b). This is a surprising result since a mixture of oligomeric and polymeric products was the expected product; to have a reaction that requires not only the interlinking of the ring systems but the formation of eight covalent bonds to progress in such high yield, demonstrates the amount of preorganisation that occurs in these systems.[229] Unlike the earlier amide-based catenanes, the two rings

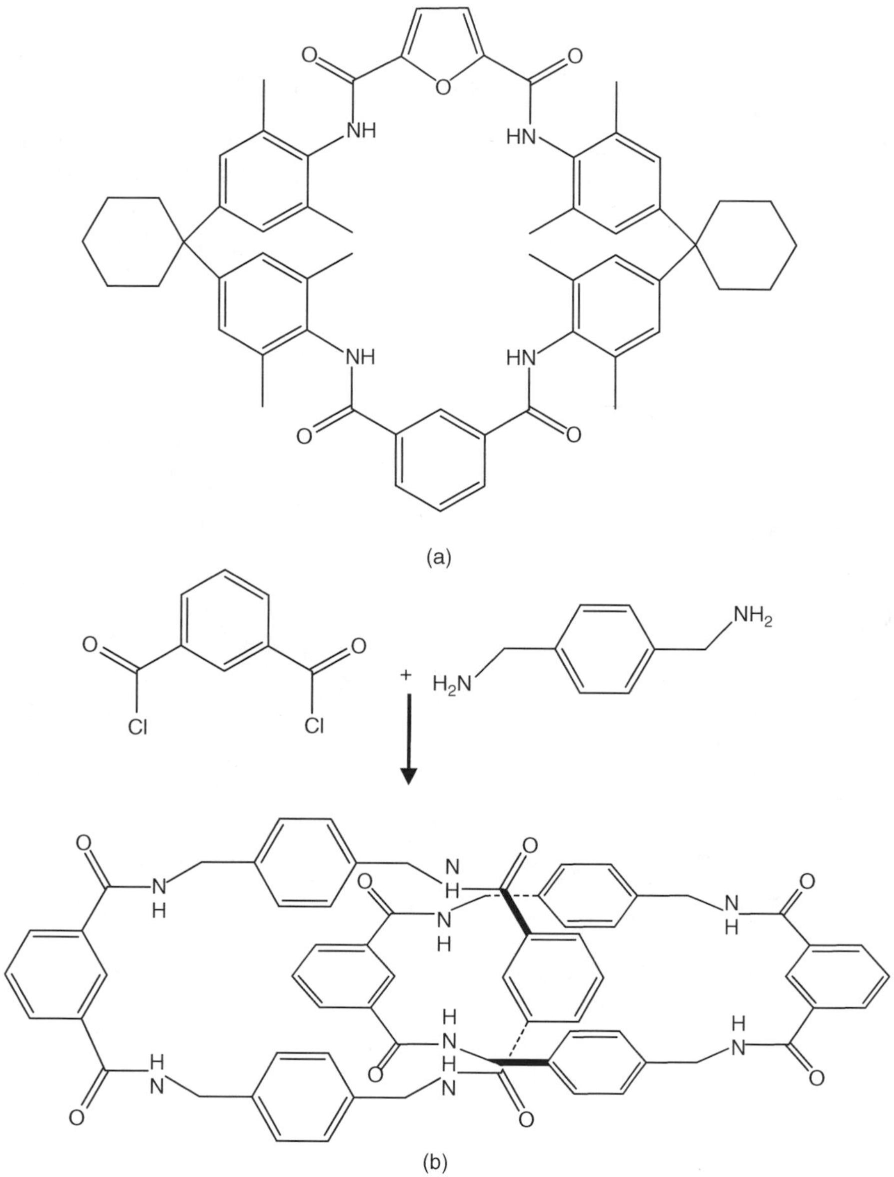

(a)

(b)

Figure 9.84 (a) *Structure of a furan-containing macrocycle and (b) facile synthesis of catenanes from acid dichlorides and diamines*

in this system rotate quite rapidly around each other in solution since there are no bulky side groups. X-ray crystal structures confirmed the catenane structure and demonstrated a layered structure of the catenanes, with extensive hydrogen-bonding between layers. In another paper published simultaneously,[230] the authors described the synthesis of a series of catenanes in yields from 15 to 27% containing a variety of functional groups, such as 5-substituted isophthaloyl groups, naphthalene units, pyridines and so on. Sulfonamide groups have also been introduced into these systems, such as in the macrocycles shown in Figure 9.83, but with one —CONH— linking group replaced by —SO$_2$ NH— a 10% yield of the catenane can be synthesised.[231] Use of a single —SO$_2$NCH$_3$— instead of —SO$_2$ NH— increased the yield of catenane to 19%. These catenanes were topologically chiral. Much of this group's early work in the field is reviewed by Jäger and Vögtle[232] and by Vögtle *et al.*[233]

Other workers synthesised the system shown in Figure 9.85, where a bipyrrole unit is utilised to synthesise a catenane, albeit in only 2% yield.[234] This can be improved slightly, to 4%, by use of a step-wise process. This compound acts as an anion receptor with high affinities to several species, especially dihydrogen phosphate and chloride. Interestingly, the catenane displays much higher binding coefficients than the open-chain precursor, perhaps indicating that the catenane is preorganised for binding.

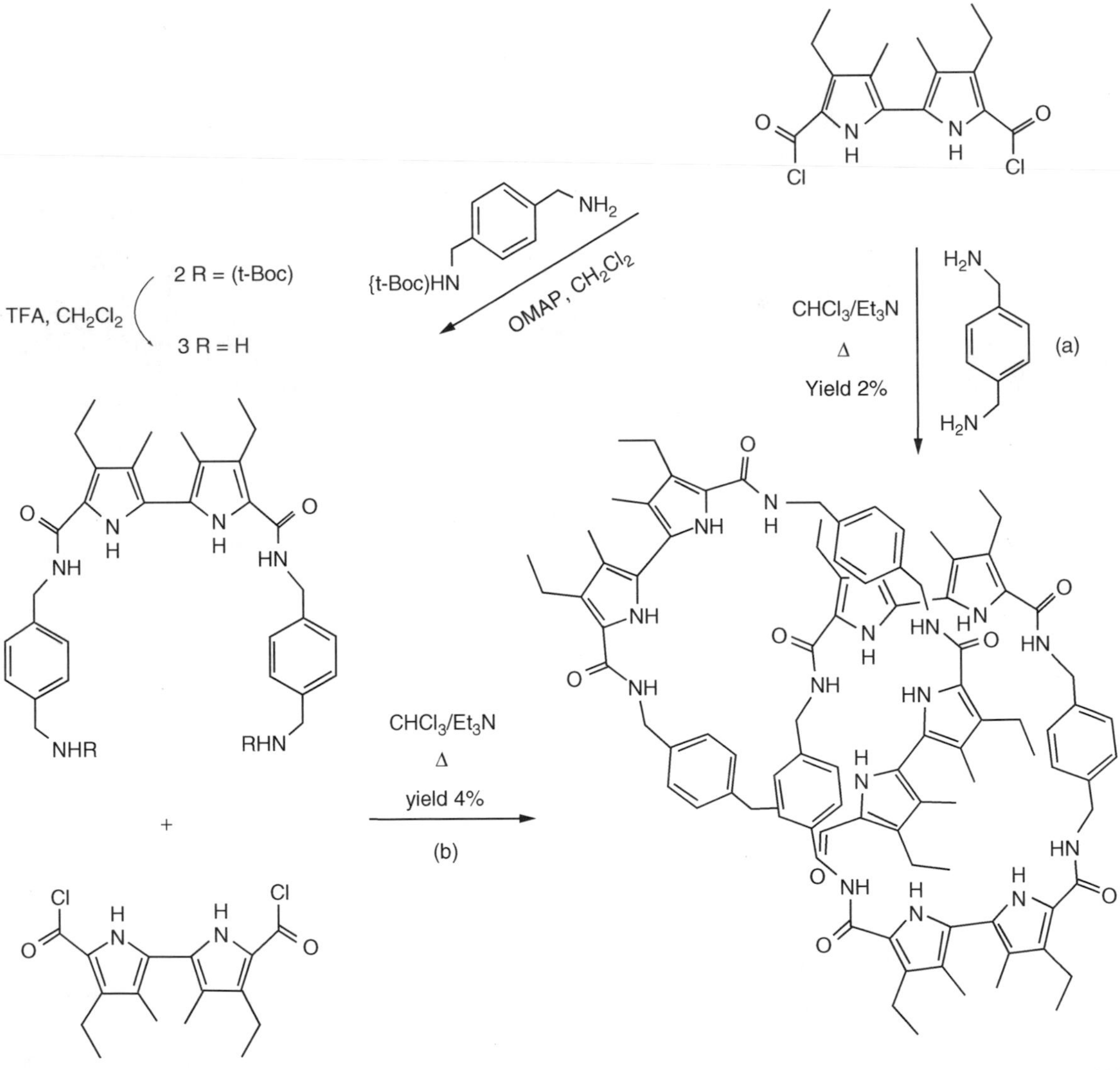

Figure 9.85 *Synthesis and structure of a bipyrrole-containing catenane (Reprinted with permission from*[234]. *Copyright 1998 American Chemical Society)*

Most early work utilised arene-linking groups, but it proved possible to incorporate long alkyl chains into these systems. When 1,12-diaminododecane was reacted with 3-chlorosulfonyl benzoyl chloride in the presence of a macrocyclic compound (Figure 9.83), the catenane shown in Figure 9.86 could be obtained in 21% yield.[235] This is probably due to binding of an amide intermediate within the macrocycle cavity to form a pseudorotaxane, followed by reaction to give ring closure. This indicates that preorganisation afforded by fixed and angular building units is unnecessary for catenane formation. Because of the large size of the aliphatic macrocycle, the two units of the catenane can rotate freely even down to $-80\,^{\circ}$C, but by substituting one of the sulfonamide nitrogens with a bipyridine group it is possible to prevent this circumrotation. By utilising a larger aromatic macrocycle it is possible to synthesise [3]catenanes containing two aliphatic and one aromatic macrocycle, albeit in low (3–4%) yields.[236] It is also possible to synthesise a variety of [2]catenanes with different lengths of aliphatic chains, from 8 to 14 carbons, in yields from 8 to 24%. Other workers synthesised catenanes containing aliphatic chains (Figure 9.87a) and showed that their conformations in solution depended on the solvent.[237] In nonpolar solvents there is strong interaction between the amide units, which are in close proximity to maximise hydrogen-bonding, a conformation which is also found in the crystal form. However, polar solvents disrupt this intramolecular bonding and the rings rotate relative to each other in order to allow the amides to interact with the solvent and bury the lipophilic chains in the centre of the catenane.

The presence of sulfonamide groups in the macrocycle shown in Figure 9.86 allows selective substitution of these groups; for example, the —SONH— can be converted to —SONCH$_3$—. This has allowed the synthesis of several catenane derivatives, for example by bridging between the two —SONH— groups of Figure 9.86 by reaction with an I(CH$_2$CH$_2$O)$_2$CH$_2$CH$_2$I chain to give an interlocked compound[238] with a yield of 11%. This molecule, due to its shape, has been nicknamed a 'pretzelane'. Metal-binding groups have also been incorporated into catenanes, such as the system described in Figure 9.88, which upon ring closure can give the tetralactam structure shown (10% yield), the larger octalactam macrocycle (31%), or the catenanes formed from two tetralactams (1%) or one tetra and one octalactam (7%).[239] Mass spectral studies have shown that these macrocycles and catenanes can form complexes with copper(I). A flavin unit can also be incorporated into a catenane (Figure 9.89).[240] A flavin macrocycle was complexed with a tetralactam macrocycle via a clipping synthetic scheme. X-ray crystallography showed the presence of hydrogen bonds between amide groups of the two macrocycles. NMR studies showed that in chloroform there are again these hydrogen-bond interactions, but they are disrupted in DMSO and the tetralactam encapsulates the alkyl chain of the other macrocycle, not the succinamide site. Electrochemical studies showed that the electrochemical transitions in the catenane occurred at significantly lowered potential compared to the free macrocycle, indicating stabilisation of the reduced state by hydrogen-bonding in the catenane.

9.3.5   Catenanes containing other macrocyclic units

A variety of other macrocyclic species have been included into catenanes. One of the earliest examples is the use of cyclodextrins in catenane synthesis. We have already described how many linear molecules can thread through a cyclodextrin cavity, and earlier in this chapter we showed how this procedure, along with capping of the guest, can be used to form rotaxanes. Should it be possible to join together the ends of these linear guests rather than just capping them, a catenane will be formed. The first successful study on these systems utilised a linear oligoethylene glycol with a central bitolyl ring system, which was incorporated into the cavity of heptakis(2,6-di-O-methyl)-β-cyclodextrin.[241] Reaction of this with terephthaloyl chloride gave two catenated products, as shown in Figure 9.90a, with the cyclodextrin including the aromatic bitolyl moiety in the cavity. A later, more extensive paper[242] described the synthesis of a series of [2] and [3]catenanes, which are shown in Figure 9.90b. The two [3]catenanes (Figures 9.90c,d) are actually topological isomers of each other and can be separated chromatographically and distinguished by NMR spectroscopy.

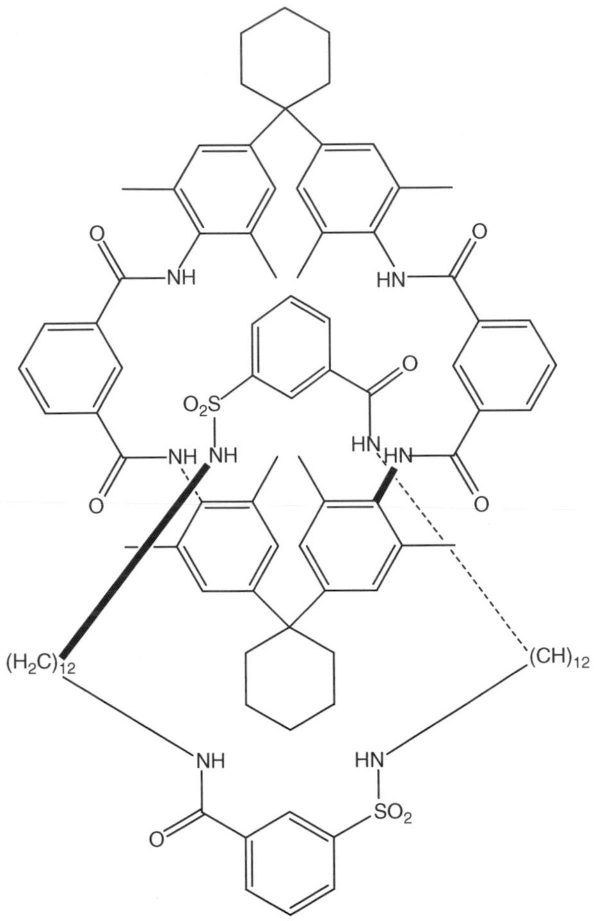

Figure 9.86   *Structure of an aliphatic chain-containing catenane*

NMR studies all indicated that the bitolyl units were located within the cyclodextrin cavity, and this was confirmed by a crystal structure of Figure 9.90a. Other workers have also combined cyclodextrins with other macrocyclic rings to give catenanes, as mentioned earlier in this chapter.[176]

Calixarenes have also been successfully incorporated into some catenanes. We have discussed how multiple-looped calixarenes can serve as a base for a fourfold rotaxane.[24] Using the same method, calixarenes substituted with four alkenyl units can be used to form a hydrogen-bonded dimer with a second calixarene.[243] Ring-closing reactions between the alkenyl groups lead to formation of two more loops, which interlink with those of the first calixarene, leading to formation of a *bis*-[2]catenane. The schematic of this reaction is shown in Figure 9.91a. The resultant catenane is chiral and only the heterodimer is formed, with no homodimers being detected. Later work synthesised a range of these *bis*-[2]catenanes and resolved their enantiomers using chiral HPLC.[244] Usually (but not in every case) the calixarenes adopt the cone conformation, as shown by NMR, and the structure is that of a molecular capsule. However, in DMSO the hydrogen-bonding between the urea moieties is disrupted and the capsule is destroyed, although the two calixarenes are of course still held together by the loops. One of the *bis*-[2]catenanes was successfully crystallised and its structure elucidated (Figure 9.91b). When a heterodimer of a tetralooped

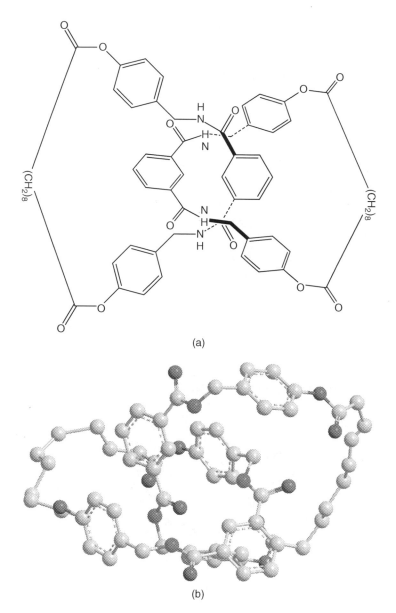

Figure 9.87   *(a) Structure of an aliphatic chain-containing catenane and (b) its X-ray crystal structure*

calixarene with a second calixarene substituted with eight alkenyl groups is subjected to the same reaction, four more loops are formed, which pass through and interlock with the two adjacent loops, thereby forming an [8]catenane.[245] The structure of this multiple catenane has been demonstrated by NMR and X-ray crystallography (Figure 9.92a). A series of these looped catenanes have been synthesised now, including *bis*-[3]catenanes, and these compounds, their NMR, binding properties and crystal structures have been extensively reviewed.[246] A variety of calixarenes and loop sizes were utilised to make a range of catenane materials. Guest incorporation into the bowl of these calixarene catenanes was studied and it was found that

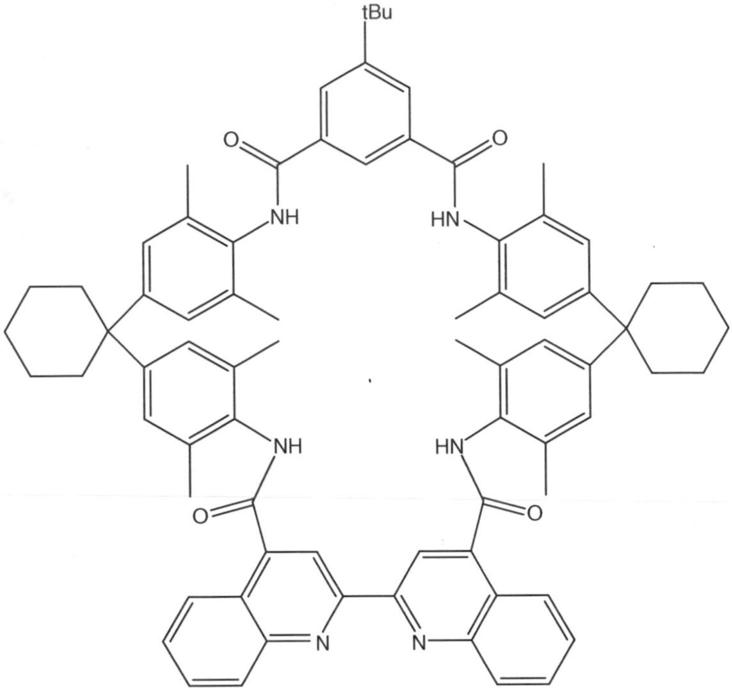

Figure 9.88   *Structure of a tetralactam that can form catenanes*

for the *bis*-[3]catenanes, long loops meant that guests exchanged freely but smaller loops slowed guest mobility; for example, replacement of chloroform by benzene took almost two days for one catenane, while the exchange of C_6H_6 and C_6D_6 took seven to eight days. Ammonium compounds could also be incorporated. The kinetics were slowed even further for the more sterically hindered [8]catenanes.

Other workers synthesised a chiral calixarene and substituted it with two bipyridyl units. A classical clipping reaction with *p*-xylylene bromide and a crown ether led to formation of a catenane.[247] The resultant catenanes were more flexible and mobile than classical bipyridyl/crown-ether catenanes. Other workers reversed this synthesis by taking a calixarene which had been bridged by a number of crown ethers, like that shown in Figure 9.92b, and then either complexing these systems with bipyridine to form pseudorotaxanes or performing a clipping reaction to give catenanes.[248]

Anions have been shown to template the formation of calixarene catenanes. For example, the macrocycle shown in Figure 9.93a can be synthesised and has been shown to bind anions.[249] This can be combined with an acyclic amide derivative substituted with alkenyl-substituted chains and then ring-closed. The catenane is only formed in the presence of chloride ions, with a yield of 29% (or bromide, with a yield of 8%); no catenanes are formed if iodide or hexafluorophosphate ions are used instead. If the templating chloride is removed, the rigidity of the catenane means that it retains a high binding affinity for chloride. As an alternative approach the same group synthesised an upper-rim macrocyclic derivative, which could also be utilised in an anion-templated clipping reaction with an acyclic isophthaloyl derivative.[250] When chloride ion was used as the template, yields of up to 60% of the catenane could be obtained. The templating ion could be displaced by the nonbinding hexafluorophosphate anion to give a catenane with a binding affinity of 2050 M^{-1} for chloride, 840 M^{-1} for bromide and lower affinities for dihydrogen phosphate, acetate and fluoride.

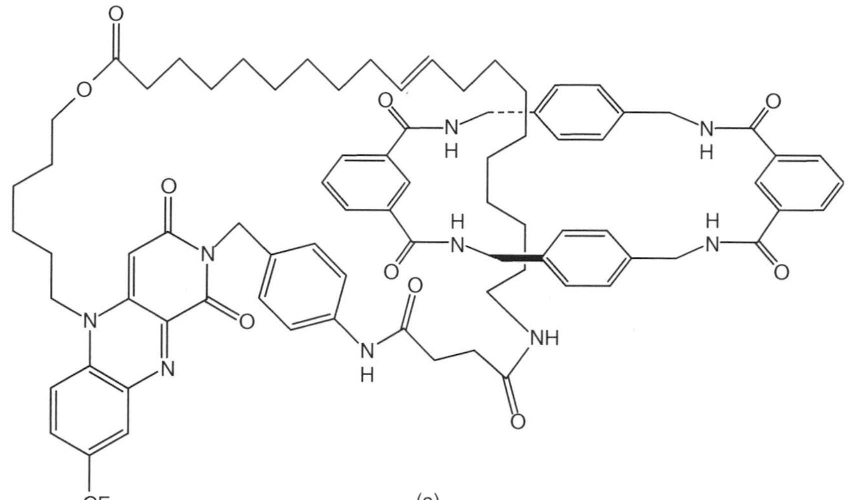

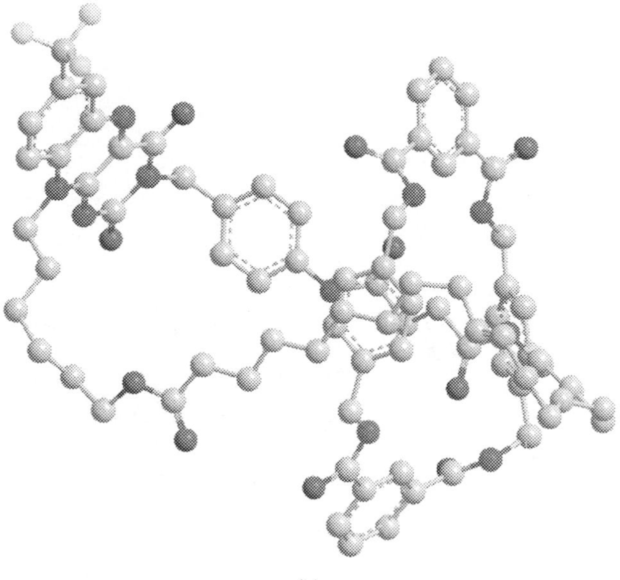

Figure 9.89 *(a) Structure of a flavin-containing catenane and (b) its X-ray crystal structure*

In fact, noncalixarene-containing analogues of these systems have been successfully demonstrated to undergo chloride-templated cyclisation to give catenanes. For example, the acyclic compound shown in Figure 9.94a can by cyclised by a metathesis reaction between the double bonds to give a catenated system in 34% yield when templated by chloride ion.[251] Other forces such as $\pi-\pi$ stacking and hydrogen-bonding must have an effect, however, since a yield of 16% is obtained even when a hexafluorphosphate counter-ion is used. The crystal structure of the chloride complex of the catenane (Figure 9.94c) clearly shows the chloride ion residing in a central cavity, stabilised by hydrogen-bonding. NMR studies show that both

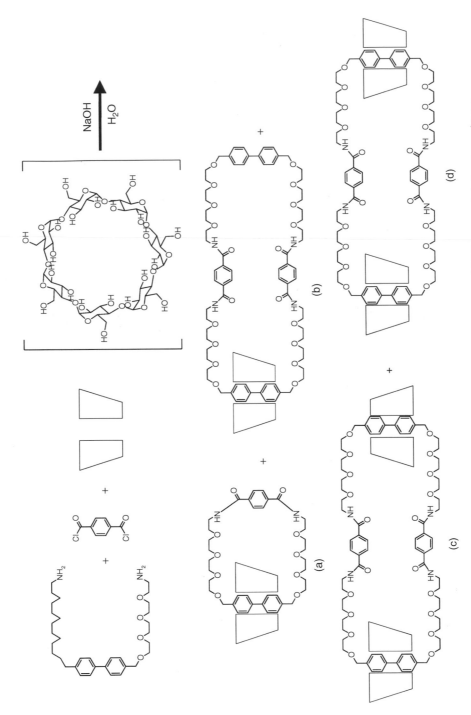

Figure 9.90 (a–d) Synthesis of cyclodextrin-containing catenanes and (e) X-ray crystal structure of 9.90a

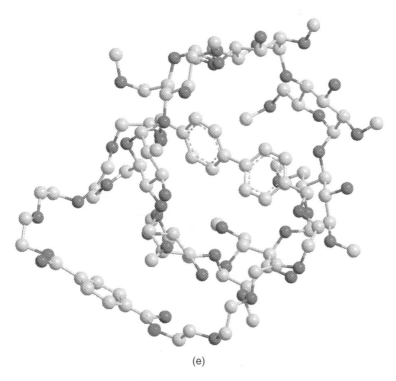

(e)

Figure 9.90   *(continued)*

amide hydrogens and the hydrogen para to the pyridinium nitrogens bind to the chloride ion. Again the templating ion can be displaced by hexafluorphosphate anion to give a catenane with a binding affinity of 9240 M^{-1} for chloride, 790 M^{-1} for bromide and 420 M^{-1} for acetate ions. A second guest ion can be bound, but much more weakly. A slightly different precursor (Figure 9.95a) was synthesised and its ring closure reaction shown to be templated by sulfate ions with an exceptional yield of 80% of catenane.[252] It is proposed that besides favourable electrostatic interactions, the orthogonal arrangement of the four amide hydrogen-bond-donating groups in the catenane complements the sulfate anion's tetrahedral shape. No catenated product was obtained with chloride, bromide or hexafluorophosphate counter-ions. The resultant catenane could again have its sulfate ion displaced by hexafluorophosphate, with the resultant material showing good affinity for sulfate (2200 M^{-1}). This was higher towards chloride (780 M^{-1}) and other anions and twice that of sulfate for the simple macrocycle formed by the metathesis reaction. Much of the work on anion-binding and anion-templated reactions has recently been reviewed.[253]

Fullerenes have also been included into catenanes, one example being the molecule shown in Figure 9.95b. This can be utilised to form rotaxanes or catenanes with crown ether-containing naphthalene units, which form due to association between the electron-rich naphthalene units and the electron-poor naphthalene diimide moiety.[254] This is followed by a capping reaction to form rotaxanes or a reaction between the fullerene and the 1,3-diketone end group to form a catenane.

9.3.6   Switchable catenanes

We have already discussed within this chapter how rotaxanes can be made switchable; that is, they can exist in two or more isomeric forms and can be converted from one to another by a suitable stimulus, such

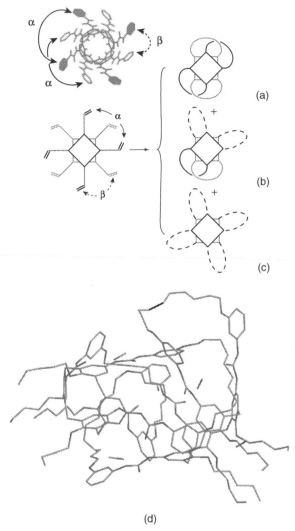

(a)

(b)

(c)

(d)

Figure 9.91 *Schematic of the formation of (a)* bis-[2]catenanes *and (b,c) other potential products not isolated (Copyright Wiley-VCH Verlag GmbH & Co. KGaA. Reproduced with permission*[243]*), and (d) X-ray structure of a chiral* bis-[2]catenane

as a pH change or electrochemical or photochemical transitions. The same procedure can often be applied to catenanes and they can be synthesised with the rings bound together in a particular arrangement, which then changes upon a chemical or other modification. Examples already discussed include the Sauvage-type compounds, which have quite different structures depending on the presence or absence of the templating metal. A range of other switchable catenanes will now be described. One early example of a catenane which can exist in more than one state is shown in Figure 9.96. A crown ether/ammonium-type rotaxane can be reacted in a 2 + 2 addition with a suitably protected dialdehyde to give the catenane in 12% yield.[255] Deprotection of the catenane affords a resultant mixture of catenanes in which each unit has four potential ammonium binding sites but only two crown-ether wheels. The wheels can reside on either adjacent ammonium groups (*proximate*) or ammonium groups on opposite sides of the macrocycle (*distant*). NMR

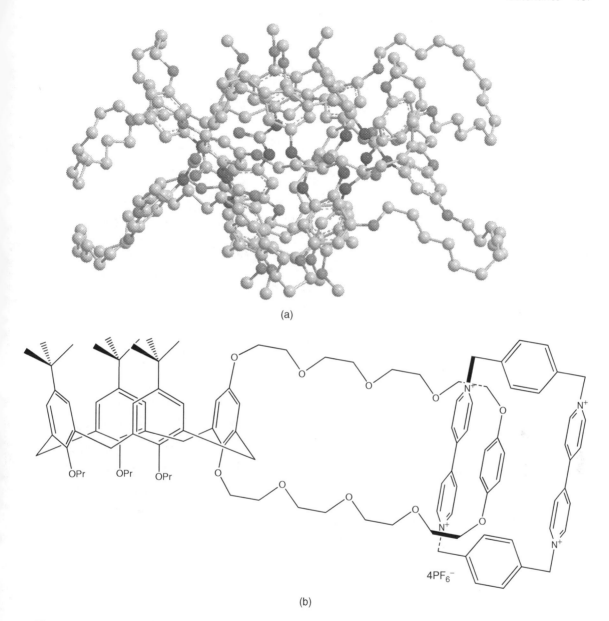

(a)

(b)

Figure 9.92   *(a) Crystal structure of an [8]catenane and (b) structure of a calixarene/crown-ether catenane*

studies show both isomers to be present. A [3]rotaxane can also be obtained using these synthetic methods and again has been shown to exist as a mixture of isomers.

Other workers developed a catenane capable of circumrotation upon addition of base.[256] When synthesised, the catenane in Figure 9.97a exists in the conformation shown, but upon the addition of base, the phenol group deprotonates to give phenolate. The catenane can then undergo a topological rotation to give the structure in Figure 9.97b, where the phenolate anion is stabilised by hydrogen-bonding. Addition of acid reverses this process. What is also of great interest is that this family of catenanes has been shown

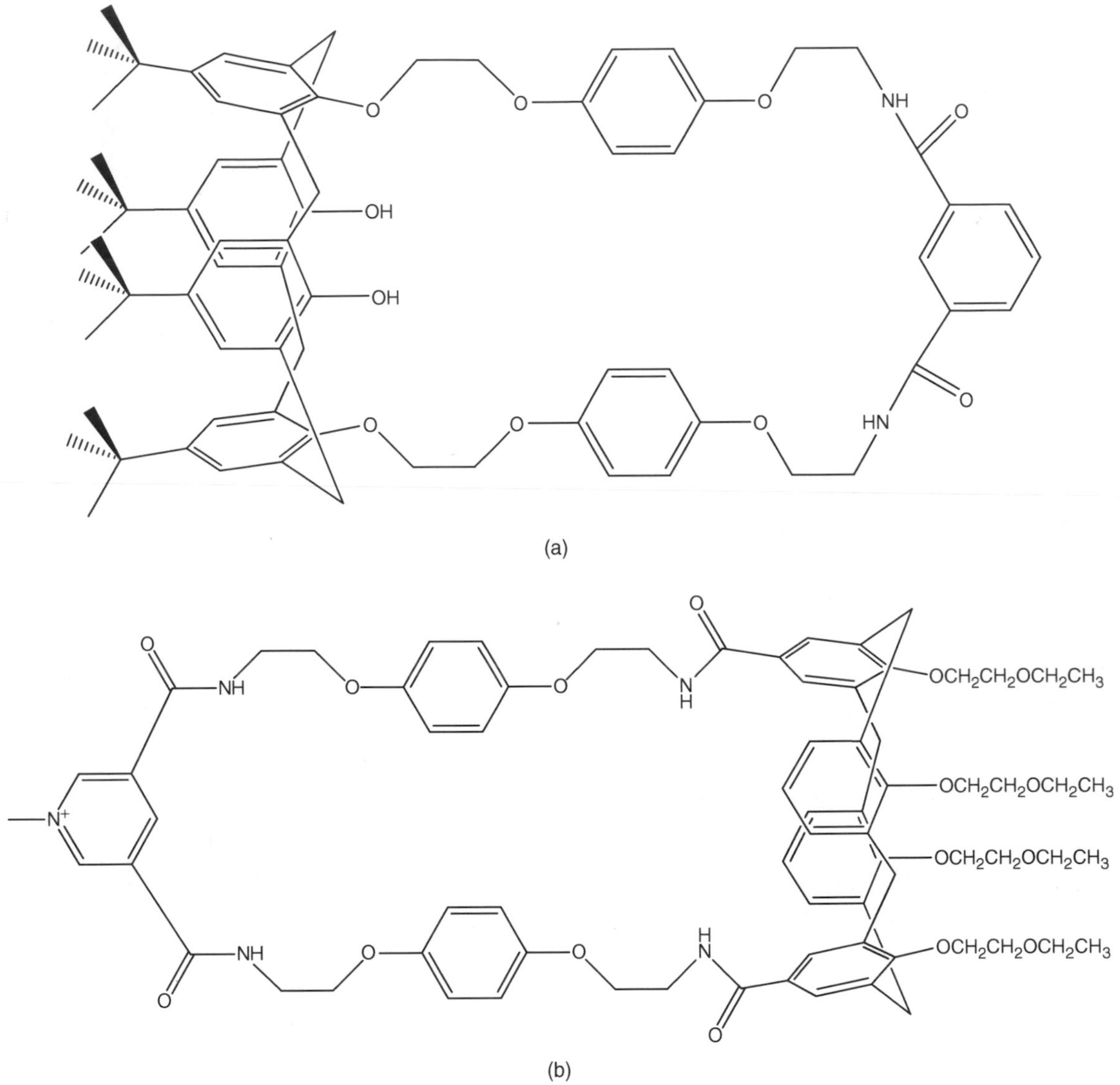

(a)

(b)

Figure 9.93 *Structures of (a) a lower-rim calixarene macrocycle and (b) an upper-rim calixarene macrocycle suitable for catenane synthesis*

to form strong complexes with chloride ion, and this molecular system is no exception. Upon addition of chloride, NMR experiments confirm formation of a complex; further addition of base does not lead to any spectroscopic change, demonstrating that the presence of chloride in the binding pocket of the catenane inhibits molecular rotation. Other workers synthesised the catenane structure shown in Figure 9.97b, in which a crown ether is mechanically interlocked with an electron-poor naphthalene diimide unit.[257] This structure also contains a viologen moiety and between these two binding sites there are bulky tetraaryl-methane units, described by the authors as 'speed bumps'. At room temperature this conformation is stable since the crown ether cannot get past the energy barrier of the 'speed bumps'. However, at 70 °C this

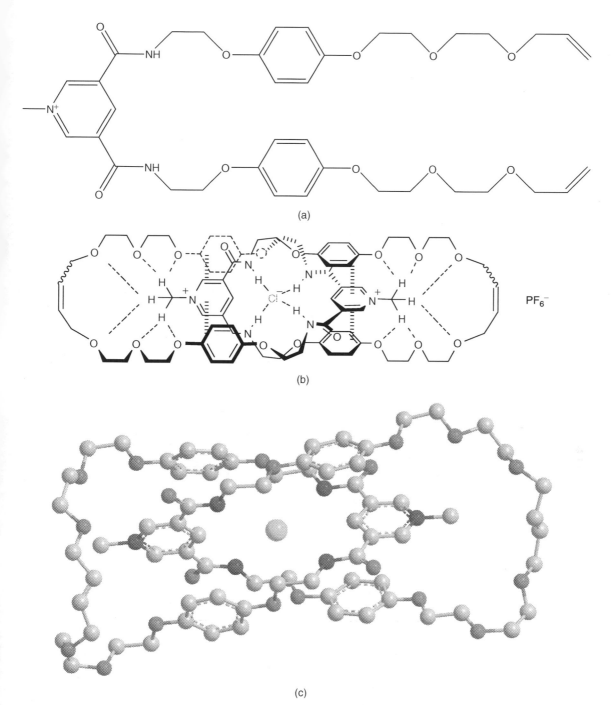

(a)

(b)

PF$_6^-$

(c)

Figure 9.94 (a) Structure of a macrocycle precursor, (b) the catenane formed by metathesis and (c) its crystal structure (Reproduced by permission of the Royal Society of Chemistry[251])

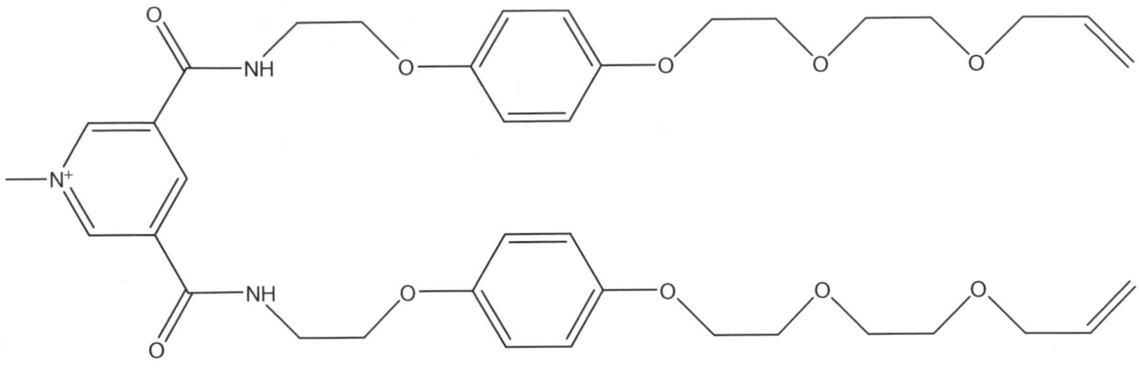

(a)

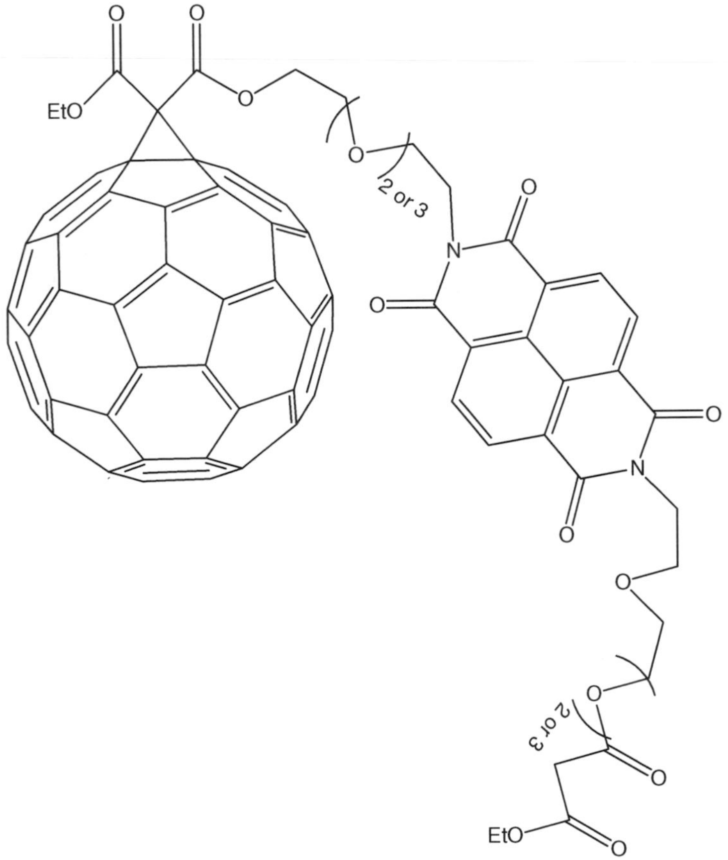

(b)

Figure 9.95   *(a) Structure of a macrocycle precursor templated by sulfate ions and (b) a fullerene axle suitable for further reaction to give catenanes or rotaxanes*

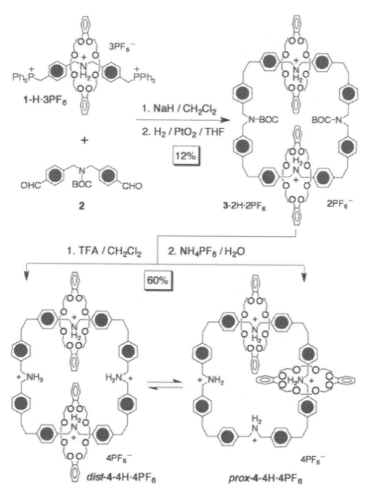

Figure 9.96   *Synthesis of catenanes that can exist in either* distant *or* proximate *forms (Reprinted with permission from[255]. Copyright 2002 American Chemical Society)*

barrier can be circumvented and the equilibrium state at this temperature has a roughly equal amount of the conformer, where the crown encircles the bipyridyl unit. Electrochemical switching of the conformers is not possible, because of the high-energy barrier.

Electrochemical switching of catenanes offers one of the most useful ways of controlling the behaviour of these molecules. For example, a simple catenane of the donor–acceptor type containing a tetrathiafulvalene unit (Figure 9.98) can be synthesised.[258] Compounds of this type adopt a conformation in the ground state where the tetrathiafulvalene unit is encircled by the cationic macrocycle. Chemical or electrochemical oxidation of the tetrathiafulvalene unit gives cationic species that no longer interact strongly with the cationic macrocycle, which instead moves around the ring to occupy a phenylene or naphthalene site, concurrent with a change in adsorption spectra and colour. Other work utilised similar catenanes but containing diazapyrenium rather than viologen units. However, electrochemical switching of these catenanes can also be obtained, as discussed for the similar rotaxanes already described.[116] Switching can also occur within a viscous polymer matrix, allowing the development of electrochromic devices.

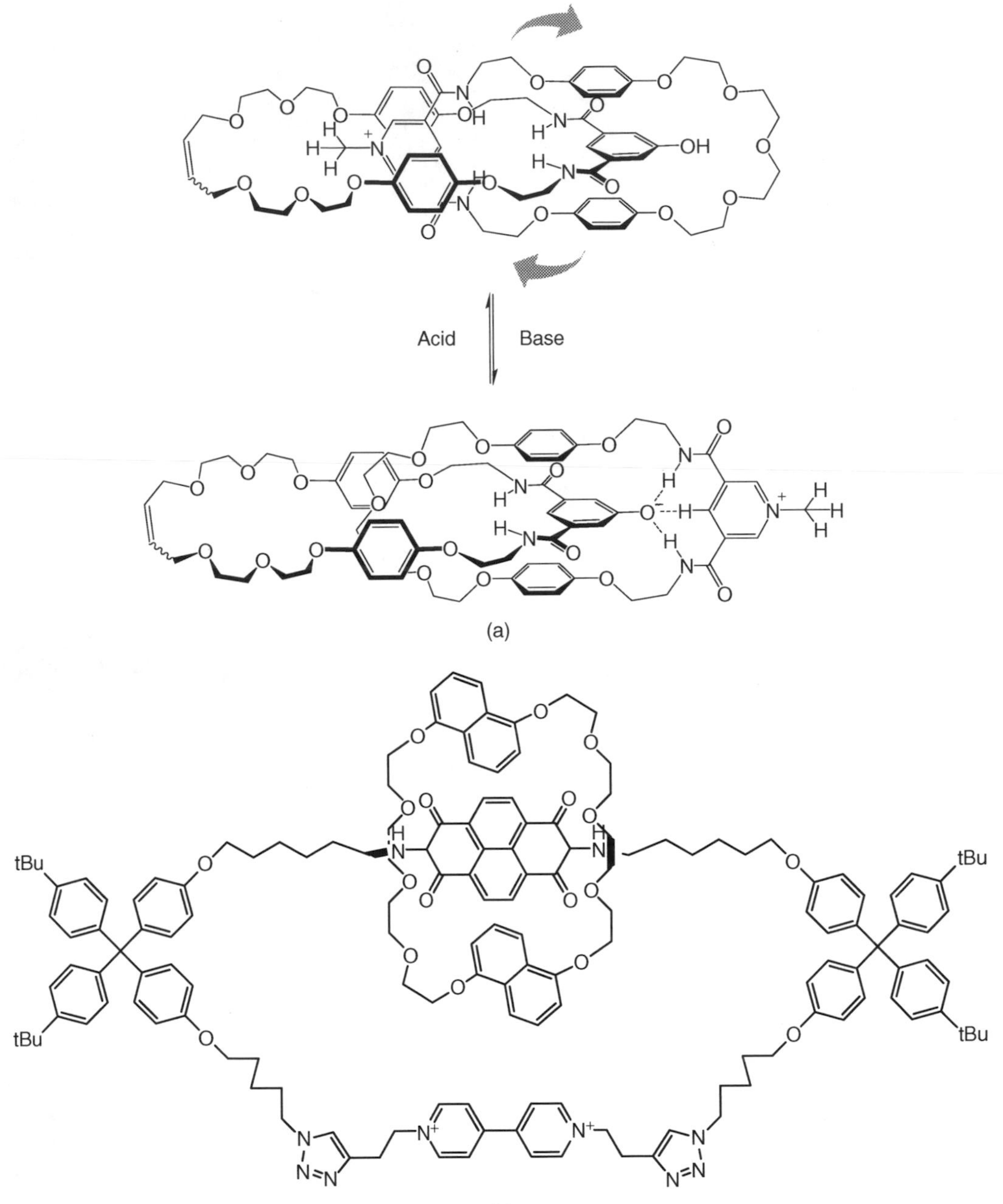

Acid | Base

(a)

(b)

Figure 9.97 Structures of (a) a catenane capable of anion-induced circumrotation (Reproduced by permission of the Royal Society of Chemistry[256]) and (b) a bistable catenane (Reprinted with permission from[257]. Copyright 2008 American Chemical Society)

These catenanes are bistable; further work has given tristable materials. A material that can exist in any of the three primary colours, red, green and blue, would potentially be of great use, and an 'electrochromic paper' device has been proposed which would be based on electrochromism of electrochemically controllable materials (Figure 9.99a). Potentially, three-station [2]catenanes capable of exhibiting red, green, blue colours could be utilised within the pixel layer of such a device. The colours of these materials could be manipulated by applying different voltages. We have already discussed how catenanes and rotaxanes based on naphthalene or phenylene ethers complexed with viologen-type derivatives undergo charge-transfer reactions, which give the material a red-orange colour, and tetrathiafulvalene–viologen interactions lead to a green colour. A catenane based on the classical viologen macrocycle complexed with fluorobenzidine units has been synthesised and the charge-transfer interactions lead to a blue colouration.[259] A material has been proposed (Figure 9.99b) based on a macrocyclic polyether containing three binding stations, catenated with a tetracationic cyclophane.[260] Charge-transfer adsorption bands that are dependent on which station the tetracationic cyclophane occupies are the basis of the colour changes. At a neutral state the cyclophane resides on the tetrathiafulvalene unit, leading to a green colour. Oxidation of the tetrathiafulvalene and benzidine systems leads to translocation of the macrocycle onto the dioxynaphthalene unit and a change to red. Careful reduction can reduce the benzidine first, leading to a translocation of the macrocycle onto this unit and the formation of a blue colour. Unfortunately, this state is not completely stable so the catenane is only quasi-tristable. These materials can be incorporated into polymethylmethacrylate gels, which slow down the transitions and allow some control over colour retention time. This use of catenanes in electrochromic materials has been extensively reviewed.[261]

9.3.7   Catenanes on surfaces

We have already postulated that for many applications of rotaxanes they must be immobilised on surfaces, and the same argument holds for catenanes, as reviewed by Coronado *et al.*[117] Some of the earliest work used a fairly simple switchable catenane (Figure 9.98a), which could be spread along with dimyristoyl phosphatidic acid as a Langmuir monolayer and then transferred onto a silicon substrate.[102] A Ti/Al top electrode could be evaporated on top of this and then the layer switched between states in the same manner as demonstrated earlier for a rotaxane.[101,102] As an alternative method, a copper-templated catenate of the Sauvage type was synthesised in which one of the oligoether chains contained a disulfide (—S—S—) bridge.[262] This material adsorbed onto a gold surface via breaking of the disulfide bridge and formation of strong Au-S bonds to give a homogeneous layer with a slightly grainy texture, as visualised by STM and AFM. IR spectroscopy indicated that the molecules were orientated perpendicularly to the gold surface. Clear evidence of electrochemical switching could be observed, but in the case of the catenane containing a phenanthroline and a terpyridyl unit, no evidence of molecular motion was seen, unlike the compounds in solution, where the macrocycles rotated around each other. However, these types of catenane (when copper-free) could be deposited onto a silver surface via vacuum sublimation.[263] Spectroscopic studies indicated that no decomposition of the macrocycles occurred during deposition. Addition of copper(I) ions to the film led to complexation of the metal at the surface, indicating copper catenates were forming and that therefore the macrocyclic rings of the catenanes must be able to slide past each other. STM studies showed that the catenanes packed in a dimer-chain structure on the surface (Figure 9.100), probably via π–π stacking interactions, and that this was disrupted to give isolated species on copper complexation.

Rotaxanes and catenanes of the type shown in Figure 9.98 but containing pyrrole substituents could be synthesised and then deposited onto a platinum electrode using electrochemical methods,[264] as was demonstrated earlier for pseudorotaxanes.[129] Electrochemical studies showed deposition was occurring and proved that the cationic macrocycles were immobilised on the platinum surface, but no redox wave for polypyrrole could be seen, indicating that at best short oligomers were formed, possibly due to steric

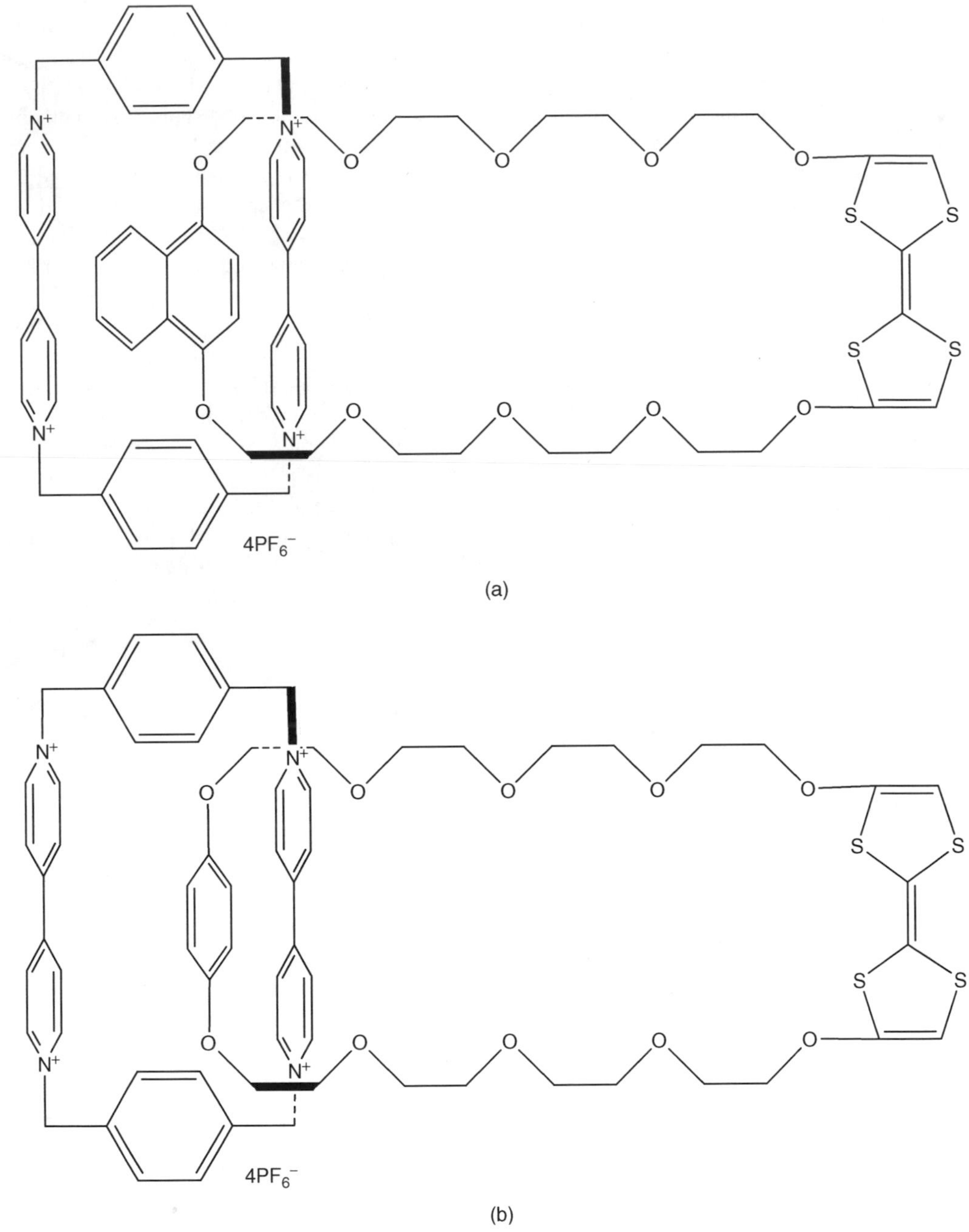

(a)

(b)

Figure 9.98   *Structures of electrochemically switchable catenanes*

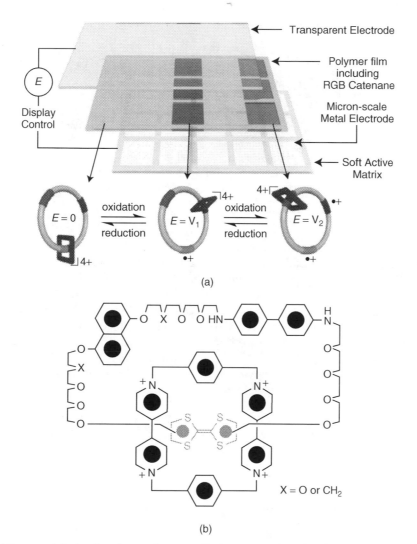

Figure 9.99 (a) Proposed design for electronic paper displays (Reproduced with permission from the Institute of Physics) and (b) an electrochemically switchable catenane (Copyright Wiley-VCH Verlag GmbH & Co. KGaA. Reproduced with permission[260])

hindrance inhibiting the polymerisation. Other work also studied incorporation of catenanes in conducting polymers.[265] Both the catenane shown in Figure 9.101a and the parent macrocycle without the tetracationic cyclophane could be electrodeposited onto electrodes and gave cyclic voltammagrams consistent with the formation of a conductive polymer.

The systems above have all immobilised catenanes onto planar surfaces, but another approach is to immobilise these molecules onto nanoparticles. When salts of such metals as gold, platinum and palladium are reduced in the presence of thiolate ligands, nanoparticles of these metals are formed, coated with the thiolate species. Other ligands can also stabilise these nanoparticles, such as tetraalkyl ammonium salts, which coordinate weakly to the metal surface. The group of Stoddart synthesised nanoparticles

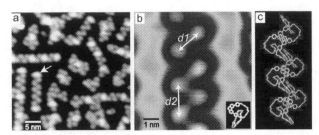

Figure 9.100 *(a) STM topography of dimer-chain structures of a catenane on a Ag(111) surface; the arrow points to a molecule sitting on top of a chain, (b) high-resolution STM topography of a single structure and (c) a tentative model of the dimer-chain structure (Reprinted with permission from[265]. Copyright 2007 American Chemical Society)*

of these three metals and then exposed them to solutions of sulfur-substituted catenanes.[266] The strong binding of sulfur for these noble-metal surfaces displaced the alkylammonium capping groups. Best results were obtained when a mixture of disulfide-substituted catenane (Figure 9.101b) and an 'inert' ligand such as a polyethylene glycol with a disulfide substituent were used. The resultant metal nanoparticles were approximately 2–5 nm in size, displayed enhanced solubility in polar solvents and were coated with a mixture of catenated and inert disulfides. The tetrathiafulvalene units of the catenanes could be switched between the neutral and cationic oxidised state by use of perchlorate and ascorbate, resulting in a change in the zeta potential of the nanoparticles,[266] which supported the premise that the catenanes change conformation with the polyether moving from the tetrathiafulvalene to the dioxanaphthalene unit. Nanoparticles substituted with linear polyethers containing dioxanaphthalene groups were also shown to bind the tetracationic macrocycle (Figure 9.61) from solution. Threading/dethreading of these rotaxanes could be controlled by the application of certain potentials. It was found that the redox potential for switching could be regulated by the composition of the self-assembled layer. An extensive review has recently been published on these switchable rotaxane and catenane moieties as well as their immobilisation in thin films.[267]

9.3.8   Polycatenanes and catenated polymers

As might be expected, chemists have attempted to incorporate these catenane species into a variety of polymer types, some of which are shown schematically in Figure 9.102. Many of these species are unusual in that the polymer chains are held together not only by covalent but by mechanical bonds. Examples of polycatenanes must exist in crosslinked polymer networks since the high density of polymer chains and multiple crosslinks will almost guarantee the presence of interlocked rings. However, due to the intractability and lack of ring-size control and relative conformations, these systems will not be considered here. Neither will the numerous inorganic, organometallic and organic crystals which inherently contain what can be thought of as interdigitated rings be discussed, as we feel they are outside of the scope of this work. Instead we will concentrate on linear polymers which are held together by a combination of covalent and mechanical bonds.

The most tempting polymers must be the polycatenanes, shown schematically in Figure 9.102a, which are a series of interlinked rings held together only by mechanical bonds. However, these types of material have so far proved elusive. We have already described the sequential build-up of molecules such as Olympiadane, which can be thought of as an oligomer of this series, but this process would surely prove too long and costly should we wish to construct molecules that were true polymers. Imagine trying

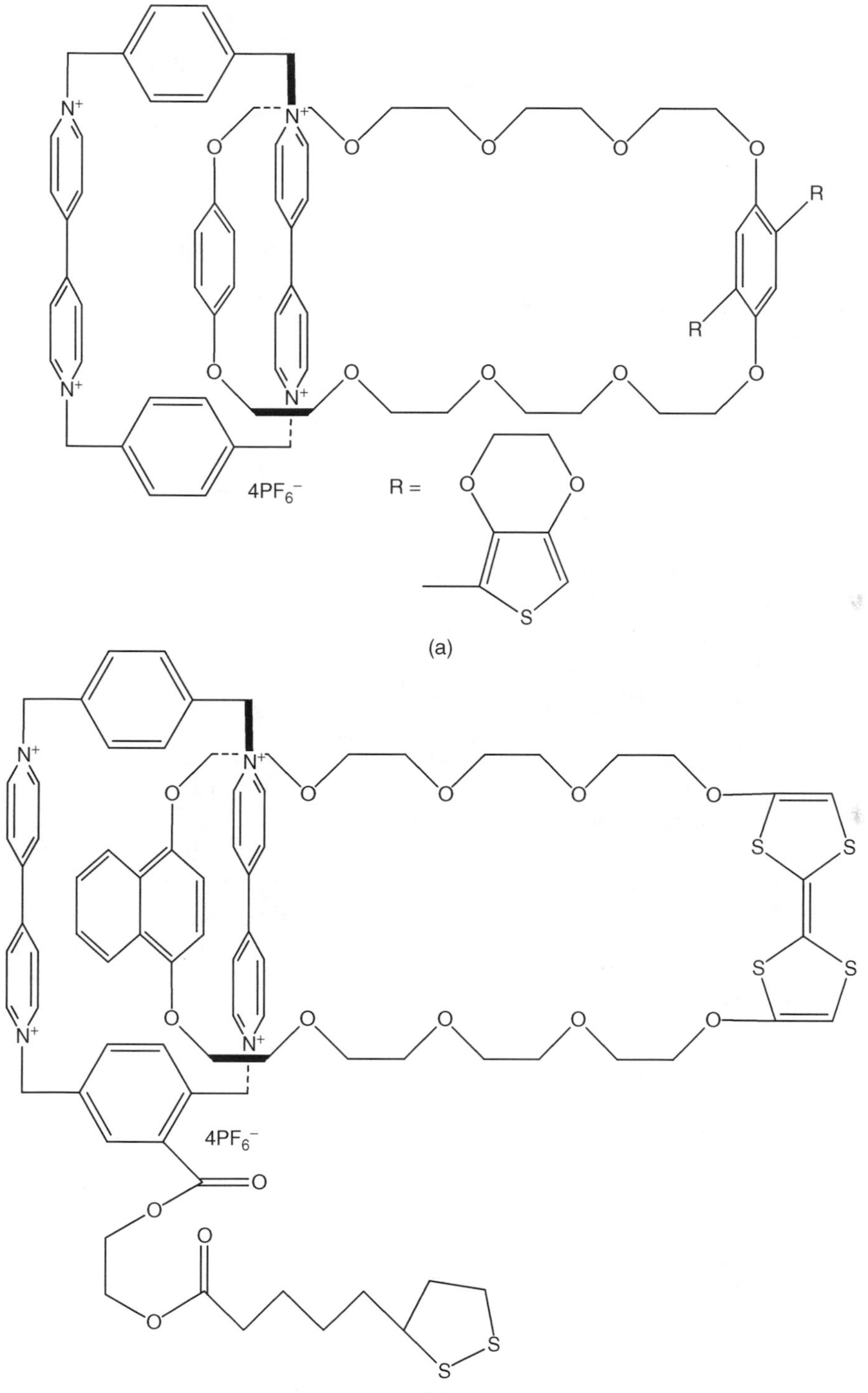

(a)

(b)

Figure 9.101 (a) A catenane that can be incorporated into a conducting polymer and (b) a switchable catenane that can bind to noble-metal nanoparticles

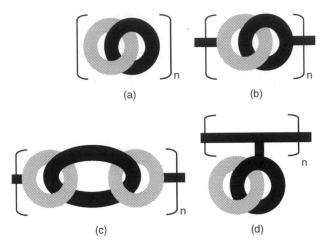

Figure 9.102   (a) Polycatenane, (b) poly[2]catenane, (c) poly[3]catanane and (d) side-chain catenane polymer

to synthesise 'Centurydane', with a hundred rings! Constructing polymers of this type requires multiple ring-closure or clipping reactions to occur simultaneously, but most such reactions do not progress with high enough yields to allow the build-up of more than a few rings. Also, many of these reactions require high-dilution conditions, again hindering polymer formation. Increasing concentrations would probably not solve this problem since crosslinking reactions could then occur, rendering only intractable networks as products.

A more promising approach is to synthesise catenanes in which each ring bears a reactive group. In this way the resultant [2]catenane bears two reactive groups, which can then be used in the formation of condensation polymers using classical polymer chemistry, giving as a product a poly[2]catenane, held together by alternating covalent and mechanical bonds (Figure 9.102b). Early work took catenanes of the classic amide type synthesised by Hunter and Vogtle and utilised palladium-catalysed coupling reactions with *bis*-hexyloxy benzene or acetylenic co-monomers to give oligo[2]catenanes, but only degrees of polymerisation as high as eight were attained.[268] Other workers synthesised amide catenanes in which one of the isophthaloyl units on each macrocycle contained a benzyl ether substituent in the 5-position.[269] This catenane proved quite insoluble, so the amide functions were methylated to enhance solubility and reduce molecular motion (in fact, steric hindrance was such that it proved possible to methylate only seven of the eight amide functions). Acid hydrolysis removed the benzyl groups to give free hydroxyls, which then underwent a condensation polymerisation with a highly soluble terephthalic acid derivative to give the copolyester shown in Figure 9.103a. Gel-permeation chromatography indicated molecular weights of about 60 000 and MALDI mass spectral studies clearly demonstrated polymers of up to 52 000 Daltons. Thermal studies showed the polymer was highly stable up to 380 °C and displayed a glass transition temperature of 265 °C. Later work synthesised substituted catenanes of this type in both the in–out and out–out conformations and utilised the same polymerisation method to synthesise polymers.[270] What was interesting was that in the case of the in–out conformer, where the functional groups are at an angle of 60° to each other, a fraction was obtained that was a cyclic trimer and which did not occur in the out–out catenane copolymer. The in–out catenane was also shown to give a polymer with a much more compact structure in solution.

A number of poly[2]catenanes based on the Sauvage-type catenanes have also been synthesised. Again, early work synthesised catenanes or catenates with two free hydroxyl groups (one on each macrocycle) and

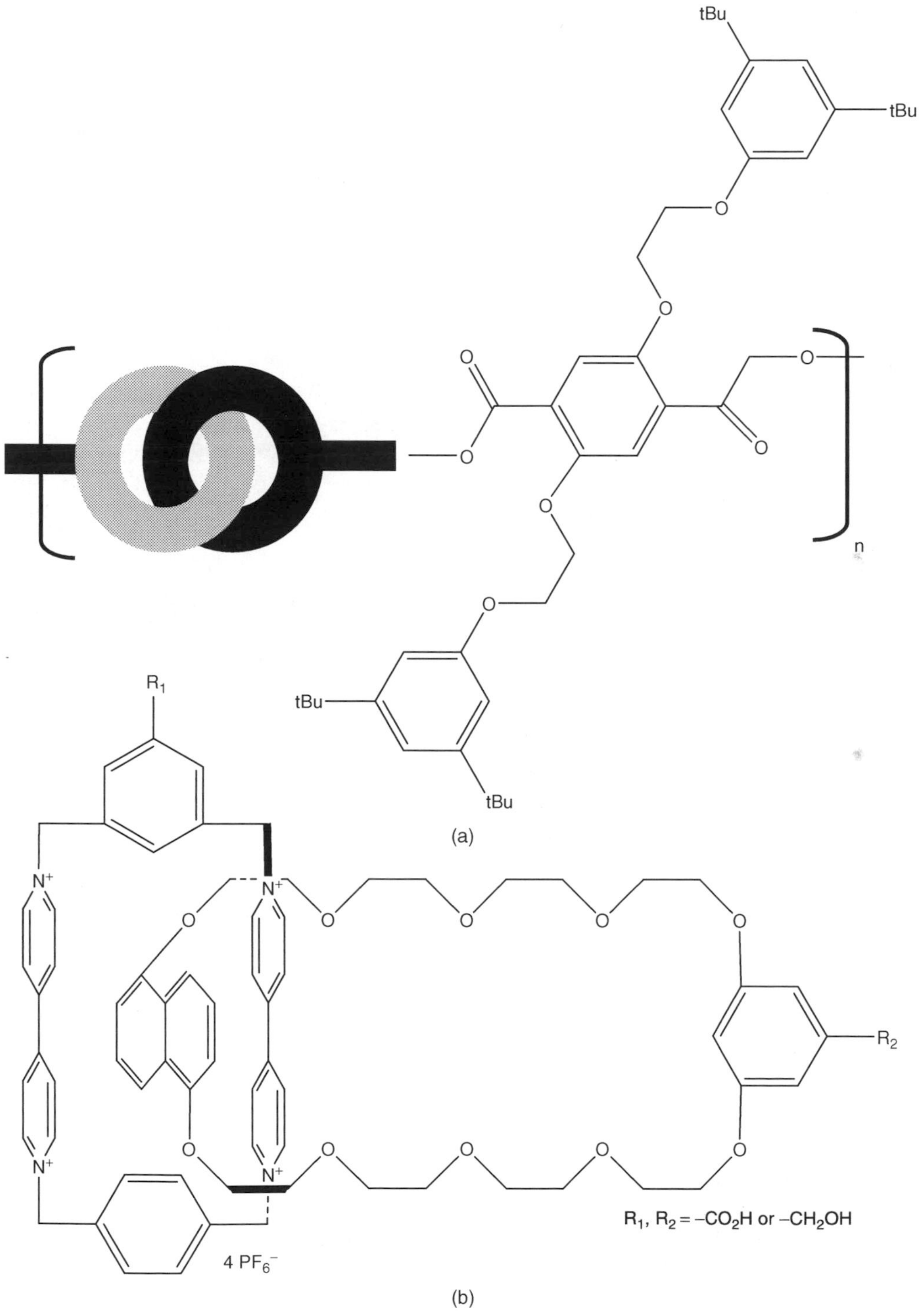

Figure 9.103 *(a) Poly[2]catenane polyester and (b) catenanes bearing reactive groups*

copolymerised them with the same terephthalic acid derivative to give polyesters.[271] Molecular weights were found to be quite high (600 000 for the polycatenate and 55 000 for the catenane) and the polymers displayed good thermal stability. Later work suggested however that these molecular weights for the catenanes were overestimations and only oligomers of up to eight or nine units were formed.[272] Interestingly it appears the catenate forms a high-molecular-weight linear polymer, whereas the catenane without the central copper ion forms cyclic oligomers. However, a linear polycatenane can be obtained by removing the metal ion from the polycatenate.

Other works also synthesised Sauvage-type catenates and catenanes, but these contained an amino group in the bridging oligoether chains.[273] These amino groups could then be reacted with a diacid chloride (adipoyl chloride) to give polyamides in the case of the copper-complexed catenates, but the free catenane did not polymerise and instead the adipoyl chloride reacted with both amino groups of the same catenane to give a pretzelane. Again the polycatenane could be obtained by removal of the copper from the catenate, with a molecular weight of 81 000.

Another method of forming polycatenanes involved the synthesis and full characterisation of the catenanes shown in Figure 9.103b, where the substituents could be acid or alcohol groups.[274] Attempts were made to homopolymerise these materials to form polyesters, but these were unsuccessful, possibly due to stereoelectronic effects. However, the catenane-containing —CH_2OH groups on both macrocycles could be copolymerised with *bis*(4-isocyanatophenyl) methane to give a polyurethane which by GPC had an average degree of polymerisation of 17.

An easier way to incorporate catenanes into polymeric systems is to synthesise a catenane containing a reactive group which can be modified to render it polymerisable and then to polymerise that group. In this case the polymer chain is wholly covalent in nature, as is its bonding to one of the rings. The only mechanical bond is that between the two rings. A schematic of this type of material is shown in Figure 9.102d. We have already discussed some examples of this, where catenanes were substituted with pyrrole or thiophene groups and electropolymerised onto electrode surfaces.[263,264] Chemical methods have also been utilised; for example, a bistable catenane was synthesised in which one ring was a crown ether containing a naphthalene and a tetrathiafulvalene ring (Figure 9.98a), clipped with a tetracationic macrocycle of the type widely discussed within this work, which in this case bore a side chain with a terminal propargyl unit.[275] This was then reacted with a methacrylate polymer bearing azide-substituted side chains, the resultant 'click' chemistry forming pyrazole links between polymer and catenane. What is interesting about this polymer is that it self-assembles into a variety of nanostructures, including hollow spheres of about 200 nm diameter and 20 nm shell thickness. Electrochemical and spectroscopic studies indicate that switching of the catenane units is still possible in the polymer and within these aggregates. Other workers used a similar approach, utilising a crown ether with a propargyl substituent, clipped with a cationic macrocycle.[276] These catenanes and similar rotaxanes could then be reacted with a azide-functionalised polystyrene to give polycatenanes and polyrotaxanes. We have already described how iodide-catalysed ring-opening and -closing of tetracationic macrocycles can be used to construct catenanes.[219] This method can also be used in the synthesis of side-chain polycatenanes, where a polymer containing numerous side chains terminated with tetracationic macrocycles can be reacted with a typical crown ether containing two dioxanaphthalene units under iodide-catalysis conditions to give the polycatenane.[277] What is amzing is that the yield of catenated side chains is essentially quantitative, and this is thought to be due to favourable π−π interactions along the polymer chain as the catenanes form.

Besides polycatenanes there has also been some work on catenated polymers; that is, on using cyclic polymers as the units in a catenane. It is possible that polymer catenanes form spontaneously during many polymerisations, but the yields must be extremely low and would be extremely difficult to separate from the main body of polymers. However, it has proved possible to deliberately synthesise catenated polymers. For example, 2-vinylpyridine can be polymerised using an anionic protocol to give a linear polymer with a narrow molecular-weight distribution and two anionic end groups. Reacting this under high-dilution conditions with *p*-xylylene dibromide gives a cyclic polymer, and if performed in the presence of a previously synthesised cyclic polystyrene can give the catenane.[278] The resultant catenated polymer, with an overall molecular weight of about 10 000, can be isolated from the simple individual macrocycles. Multiple catenation cannot be detected. Other workers have also used the anionic polymerisation method to synthesise a catenane with a cyclic polystyrene and a cyclic polyisoprene unit.[279] The polymer has an overall molecular weight of 37 000 and is shown by transmission electron microscopy to have a nanoscale phase-separated structure in the bulk. Normally polystyrene and polyisoprene are not miscible, but of course they cannot phase-separate on a large scale since the rings are mechanically linked together. These syntheses often give less than 1% of catenated polymer, but more recently other workers have synthesised catenated polyethers.[280] A cyclic polytetrahydrofuran can be synthesised with an isophthaloyl benzylic amide unit located in the ring. A second linear polymer containing the same central unit has also been synthesised and then cyclised in the presence of the first ring. This leads to up to 7% yield of the catenated polymer due to hydrogen-bonding and electrostatic interactions preorganising the polymers to some extent. MALDI has proven the successful synthesis of the catenane.

Macromolecular polymers usually have a range of molecular weights, but other large ring catenanes have been synthesised without any spread of molecular weight. The large macrocycle (Figure 9.104a) and a linear precursor (Figure 9.104b) can be combined so that the phenoxide and carbonyl chloride react together to link the two moieties via a carbonate group.[281] The trimethylsilyl groups on the linear precursor are then reacted to cyclise it and the carbonate group is hydrolysed away to give a mixture of the two macrocycles and their resultant catenane. The use of the carbonate linkage holds the two portions together during the cyclisation reaction, thereby increasing the amount of threading and the yield of the catenane (Figure 9.105) to 63%. NMR studies show that these two rings rotate freely with respect to each other. Later work extended this synthesis to give catenanes with rings containing 63 or 147 atoms.[282] A small review of these compounds has recently been published.[282] This work also details the reduction of the alkyne links to simple hydrocarbon chains by reaction with hydrogen/Pd and the incorporation of these large catenanes into polymeric systems, albeit with low degrees of polymerisation (degrees of polymerisation of about 10 units). These catenanes and the non-entwined macrocycles display thermotropic liquid-crystalline behaviour.

Catenanes have also been incorporated into more complex three-dimensional systems. For example, crown ethers substituted with benzene carboxylic acid units have been synthesised and utilised to build up three-dimensional metal-organic frameworks with Zn and Cu.[284] Not only the crown ethers but their catenanes and rotaxanes are also suitable for incorporation into these systems. The catenane shown in Figure 9.105a can be reacted with copper(II) nitrate to give a metal-organic framework. During this reaction Cu(II) is reduced to Cu(I) and a two-dimensional layered structure is formed.[285] X-ray crystallography studies show the regular arrangement of the catenanes within this layered framework (Figure 9.105b,c) and suggest that there is enhanced $\pi-\pi$ stacking within this extended system.

9.3.9 Natural catenanes

Although this chapter has demonstrated the inventiveness of many chemists in designing synthetic schemes for some of these intricate structures, as usual we find that Nature has beaten them to it. A review on the catenanes and other structures that are found in the natural world is once again beyond the scope of this work but we will show a few examples in an attempt to whet the reader's appetite.

In 1967 the first example of catenated closed circular mitochondrial DNA molecules was identified in extracts of HeLa cells.[286] Another paper published immediately afterwards studied human leukemic leucocytes and showed evidence[287] for the presence of [3]catenanes. Electron micrographs of DNA extracted from sea urchin eggs has also showed the presence of catenanes.[288] Other work on mutant bacterial

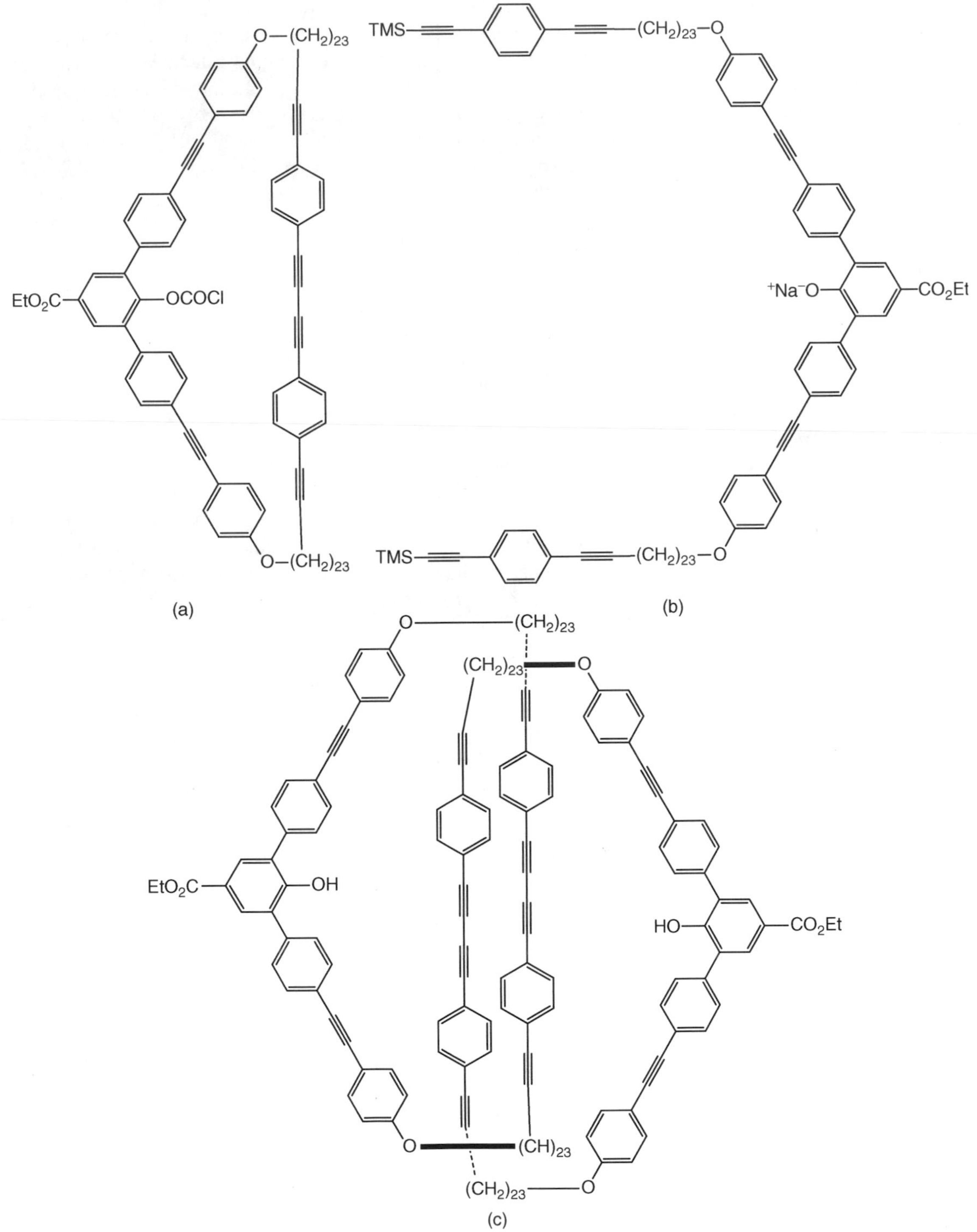

Figure 9.104   *(a) Cyclic and (b) acyclic precursors to (c) an 87-atom ring [2]catenane*

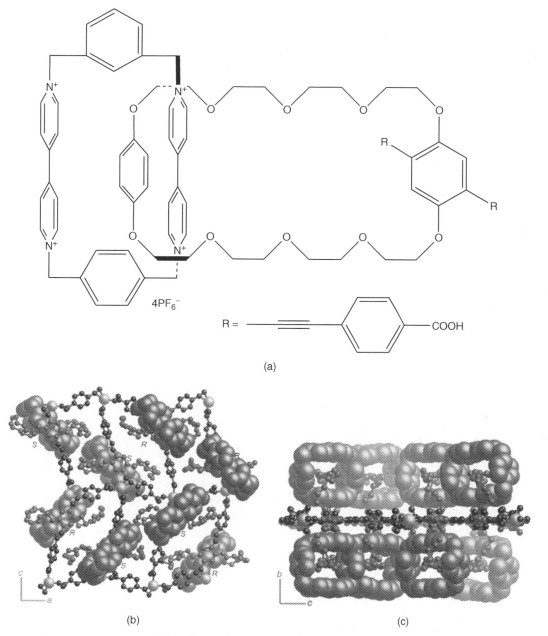

R = ———≡———⟨benzene⟩———COOH

(a)

(b)   (c)

Figure 9.105   *(a) Catenane, and (b) plan and (c) side-on view of the crystal structure of the metal-organic framework formed (Reproduced by permission of the Royal Society of Chemistry[285])*

topoisomerase DNA showed the presence of catenanes and trefoil knot structures.[289] Proteins have also been shown to display catenated structures.[290] A survey of the Brookhaven protein database showed that quinoprotein methylamine dehydrogenase, cytochrome and human chorionic gonadotropin all contained catenanes, as well as the presence of knotted and catenated closed loops in ascorbate oxidase and human lactoferrin.[291] Besides these natural systems, it has also proved possible to synthesise polymer catenanes. For example Yan and Dawson utilised a polypeptide that forms an interlocked dimer;[292] the end groups of these two peptides could then be ligated together to give a protein catenane.

9.4 Conclusions

The synthesis of structures like those described throughout this chapter has advanced immensely over the last few decades. Initially rotaxanes and catenanes were only available in tiny quantities using statistical methods of synthesis. The work of Schill and others advanced the field, but directed synthesis of catenanes and rotaxanes required multiple steps, was time-consuming and expensive, and often produced poor overall yields. The big advances came when synthetic schemes that made use of preorganisation of the components before chemical reaction were developed.

Several major groups of catenenes and rotaxanes make use of different physical and chemical effects. Metal-ion templating was one of the first effects to be employed, as shown by the work of Jean-Pierre Sauvage and others, where copper ions were used to hold phenanthroline residues in an arrangement that facilitated the synthesis of catenanes and rotaxanes. Alternative methods such as the interactions between electron-rich and electron-poor aromatic systems were pioneered by Fraser Stoddart. Besides these, we also have the amide macrocycles, where hydrogen-bonding is one of the driving forces towards the formation of complex supramolecular systems such as those obtained by Hunter, Vogtle and Leigh, along with those systems whose formation is templated by anions.

One of the main effects of these techniques is that it has become possible to obtain designed molecular superstructures in good yields, often with volumes of many grams, at reasonable expense and within sensible time frames. This has enabled the construction of molecular devices from these systems. Molecular shuttles, switches, elevators, muscles, sensors and valves have all been demonstrated. The potential applications for these catenanes and rotaxanes, such as data-storage devices and molecular computers, along with a host of other possible nanotechnological applications, are becoming more apparent and we may soon see these systems being used in real-world applications.

Bibliography

Steed JW, Atwood JL. Supramolecular Chemistry. John Wiley and Sons; 2000, 2009.

Atkinson IM, Lindoy LF, Stoddart JF. Self-assembly in Supramolecular Systems. Monographs in Supramolecular Chemistry. Royal Society of Chemistry; 2000.

Sauvage J-P, Dietrich-Buchecker C. Molecular Catenanes, Rotaxanes and Knots: A Journey Through the World of Molecular Topology. Wiley VCH; 1999.

Schill G. Catenanes, Rotaxanes, Knots. Academic Press Inc; 1971.

Coronado E, Gaviña P, Tatay S. Catenanes and threaded systems: from solution to surfaces. *Chem Soc Rev.* 2009; **38**: 1674–1689.

Chemical Society Reviews. 2009 themed issue dedicated to Professor Jean-Pierre Sauvage. Volume 38, Issue 6. Includes the papers below:

Balzani V, Credi A, Venturi M. Light powered molecular machines. *Chem Soc Rev.* 2009; **38**: 1542–1550.

Crowley JD, Goldup SM, Lee A-L, Leigh DA, McBurney RT. Active metal template synthesis of rotaxanes, catenanes and molecular shuttles. *Chem Soc Rev.* 2009; **38**: 1530–1541.

Coronado E, Gaviña P, Tatay S. Catenanes and threaded systems: from solution to surfaces. *Chem Soc Rev.* 2009; **38**: 1674–1689.

Stoddart JF. The chemistry of the mechanical bond. *Chem Soc Rev.* 2009; **38**: 1802–1819.

References

1. Amabilino DB, Stoddart JF. Interlocked and intertwined structures and superstructures. *Chem Rev.* 1995; **95**: 2725–2828.
2. Raymo FM, Stoddart JF. Interlocked macromolecules. *Chem Rev.* 1999; **99**: 1643–1663.
3. Stoddart JF. The master of chemical topology. *Chem Soc Rev.* 2009; **38**: 1521–1529.
4. Harrison IT, Harrison S. The synthesis of a stable complex of a macrocycle and a threaded chain. *J Am Chem Soc.* 1967; **89**: 5723–5724.
5. Harrison IT. The effect of ring size on threading reactions of macrocycles. *J Chem Soc Chem Commun.* 1972; 231–232.
6. Harrison IT. Preparation of rotaxanes by the statistical method. *J Chem Soc. Perkin Trans.* 1974; I: 301–304.
7. Agam G, Zilkha A. Synthesis of a catenane by a statistical double-stage method. *J Am Chem Soc.* 1976; **98**: 5214–5216.
8. Schill G, Beckmann W, Schweickert N, Fritz H. Studies on the statistical synthesis of rotaxanes. *Chem Ber-Recueil.* 1986; **119**: 2647–2655.
9. Schill G, Henschel A. Rotaxane compounds.2. A diansa compound of 5-amino-6-methoxy-4.7-dimethyl-benzodioxole as a model of catenanes and rotaxanes. *Liebig Ann Chem.* 1970; **731**: 113–199.
10. Ogino H. Relatively high-yield syntheses of rotaxanes: syntheses and properties of compounds consisting of cyclodextrins threaded by α,ω-diaminoalkanes coordinated to cobalt(III) complexes. *J Am Chem Soc.* 1981; **103**: 1303–1304.
11. Yamanari K, Shimura Y. Stereoselective formation of rotaxanes composed of polymethylenebridged dinuclear cobalt(III) complexes and α- or β-cyclodextrin. *Bull Chem Soc Jpn.* 1983; **56**: 2283–2289.
12. Rao TVS, Lawrence DS. Self-assembly of a threaded molecular loop. *J Am Chem Soc.* 1990; **112**: 3614–3615.
13. Manka JS, Lawrence DS. Template-driven self-assembly of a porphyrin-containing supramolecular complex. *J Am Chem Soc.* 1990; **112**: 2440–2442.
14. Kim K. Mechanically interlocked molecules incorporating cucurbituril and their supramolecular assemblies. *Chem Soc Rev.* 2002; **31**: 96–107.
15. Jun SI, Lee JW, Sakamoto S, Yamaguchi K, Kim K. Rotaxane-based molecular switch with fluorescence signalling. *Tet Lett.* 2000; **41**: 471–475.
16. Lee JW, Kim K, Kim K. A kinetically controlled molecular switch based on bistable [2]rotaxane. *J Chem Soc Chem Commun.* 2001; 1042–1043.
17. Tuncel D, Özsar Ö, Tiftik HB, Salih B. Molecular switch based on a cucurbit[6]uril containing bistable [3]rotaxane. *J Chem Soc Chem Commun.* 2007; 1369–1371.
18. Tuncel D, Katterle M. pH-triggered dethreading-rethreading and switching of cucurbit[6]uril on bistable [3]pseudorotaxanes and [3]rotaxanes. *Chem Eur J.* 2008; **14**: 4110–4116.
19. Tuncel D, Cindir N, Koldemir U. [5]Rotaxane and [5]pseudorotaxane based on cucurbit[6]uril and anchored to a meso-tetraphenyl porphyrin. *J Inclus Phenom Macro Chem.* 2007; **55**: 373–380.
20. Liu S, Wu X, Huang Z, Yao J, Liang F, Wu C. Construction of pseudorotaxanes and rotaxanes based on cucurbit[n]uril. *J Inclus Phenom Macro Chem.* 2004; **50**: 203–207.
21. Lee JW, Ko YH, Park S-H, Yamaguchi K, Kim K. Novel pseudorotaxane-terminated dendrimers: supramolecular modification of dendrimer periphery. *Angew Chem Int Ed.* 2001; **40**: 746–749.
22. Lee JW, Han SC, Kim JH, Ko YH, Kim K. Formation of rotaxane dendrimers by supramolecular click chemistry. *Bull Korean Chem Soc.* 2007; **28**: 1837.

23. Arduini A, Bussolati R, Credi A, Faimani G, Garaud S, Pochini A, Secchi A, Semeraro M, Silvi S, Venturi M. Towards controlling the threading direction of a calix[6]arene wheel by using nonsymmetric axles. *Chem Eur J.* 2009; **15**: 3230–3242.

24. Gaeta C, Vysotsky MO, Bogdan A, Böhmer V. Fourfold [2]rotaxanes based on calix[4]arenes. *J Am Chem Soc.* 2005; **127**: 13136–13137.

25. Fischer C, Nieger M, Mogck O, Böhmer V, Ungaro R, Vögtle F. Calixarenes as stoppers in rotaxanes. *Eur J Org Chem.* 1998; 155–161.

26. Anelli PL, Ashton PR, Ballardini R, Balzani V, Delgado M, Gandolfi MT, Goodnow TT, Kaifer AE, Philp D, Pietraszkiewicz M, Prodi L, Reddington MV, Slawin AMZ, Spencer N, Stoddart JF, Vicent SC, Williams DJ. Molecular meccano. 1. [2]rotaxanes and a [2]catenane made to order. *J Am Chem Soc.* 1992; **114**: 193–218.

27. Anelli PL, Spencer N, Stoddart JF. A molecular shuttle. *J Am Chem Soc.* 1991; **113**: 5131–5133.

28. Sun XQ, Amabilino DB, Ashtonr PR, Parsons IW, Stoddart JF, Tolley MS. Towards the self-assembly of polyrotaxanes. *Macromol Symp.* 1994; **77**: 191–207.

29. Ashton PR, Bissell RA, Spencer N, Stoddart JF, Tolley MS. Towards controllable molecular shuttles. 3. *Synlett.* 1992; **11**: 923–926.

30. Bissell RA, Córdova E, Kaifer AE, Stoddart JF. A chemically and electrochemically switchable molecular shuttle. *Nature.* 1994; **369**: 133–137.

31. Ashton PR, Philp D, Spencer N, Stoddart JF. A new design strategy for the self-assembly of molecular shuttles. *J Chem Soc Chem Commun.* 1992; 1124–1128.

32. Ashton PR, Belohradsky M, Philp D, Spencer N, Stoddart JF. The self assembly of [2]- and [3]-rotaxanes by slippage. *J Chem Soc Chem Commun.* 1993; 1274–1277.

33. Amabilino DB, Ashton PR, Bglohradsky M, Raymo FM, Stoddart JF. The controlled self-assembly of a [3]rotaxane incorporating three constitutionally different components. *J Chem Soc Chem Commun.* 1995; 747–750.

34. Amabilino DB, Ashton PR, Bglohradsky M, Raymo FM, Stoddart JF. The self-assembly of branched [n]rotaxanes: the first step towards dendritic rotaxanes. *J Chem Soc Chem Commun.* 1995; 751–753.

35. Kolchinski AG, Busch DH, Alcock NW. Gaining control over molecular threading: benefits of second coordination sites and aqueous–organic interfaces in rotaxane synthesis. *J Chem Soc Chem Commun.* 1995; 1289–1291.

36. Kolchinski AG, Roesner RA, Busch DH, Alcock NW. Molecular riveting: high yield preparation of a [3]-rotaxane. *J Chem Soc Chem Commun.* 1998; 1437–1438.

37. Ashton PR, Raymo FM, Chrystal EJT, Glink PT, Menzer S, Philip D, Spencer N, Stoddart JF, Tasker PA, Williams DJ. Dialkylammonium ion/crown ether complexes: the forerunners of a new family of interlocked molecules. *Angew Chem Int Ed.* 1995; **34**: 1865–1869.

38. Ashton PR, Chrystal EJT, Glink PT, Menzer S, Sciavo C, Stoddart JF, Tasker PA, Williams DJ. Doubly encircled and double-stranded pseudorotaxanes. *Angew Chem Int Ed.* 1995; **34**: 1869–1871.

39. Fyfe MCT, Glink PT, Menzer S, Stoddart JF, White AJP, Williams DJ. Anion-assisted self-assembly. *Angew Chem Int Ed.* 1997; **36**: 2068–2070.

40. Ashton PR, Glink PT, Stoddart JF, Menzer S, White AJP, Williams DJ. Pseudorotaxanes formed between secondary dialkylammonium salts and crown ethers. *Chem Eur J.* 1996; **2**: 729–736.

41. Martinez-Diaz MV, Spencer N, Stoddart JF. The self-assembly of a switchable [2]rotaxane. *Angew Chem Int Ed.* 1997; **36**: 1904–1907.

42. Ashton PR, Baxter I, Fyfe MCT, Spencer N, Stoddart JF, White AJP, Williams DJ. Multiply stranded and multiply encircled pseudorotaxanes. *J Am Chem Soc.* 1998; **120**: 2297–2307.

43. Handel M, Plevoets M, Gestermann S, Vögtle F. Synthesis of rotaxanes by brief melting of wheel and axle components. *Angew Chem Int Ed.* 1997; **36**: 1199–1201.

44. Lane AS, Leigh DA, Murphy A. Peptide-based molecular shuttles. *J Am Chem Soc.* 1997; **119**: 11092–11093.

45. Johnston AG, Leigh DA, Murphy A, Smart JP, Deegan MD. The synthesis and solubilization of amide macrocycles via rotaxane formation. *J Am Chem Soc.* 1996; **118**: 10662–10663.

46. Vögtle F, Händel M, Meier S, Ottens-Hildebrandt S, Ott F, Schmidt T. Template synthesis of the first amide-based rotaxanes. *Liebigs Ann.* 1995; 739–743.

47. Vögtle F, Jäger R, Händel M, Ottens-Hildebrandt S, Schmidt W. Amide-based rotaxanes with terephthal, furan, thiophene and sulfonamide subunits. *Synthesis.* 1996; 353–366.

48. Dünnwald T, Parham AH, Vögtle F. Non-ionic template synthesis of amide-linked rotaxanes: axles with benzophenone and cinnamic acid units. *Synthesis*. 1998; 339–348.

49. Vögtle F, Dunnwald T, Handel M, Jager R, Meier S, Harder G. A [3]rotaxane of the amide type. *Chem Eur J*. 1996; **2**: 640–643.

50. Diinnwald T, Jager R, Vögtle F. Synthesis of rotaxane assemblies. *Chem Eur J*. 1997; **3**: 2043–2051.

51. Sugiyama J, Tomita I. Novel approach to stabilize unstable molecular wires by simultaneous rotaxane formation: synthesis of inclusion complexes of oligocarbynes with cyclic host molecules. *Eur J Org Chem*. 2007; 4651–4653.

52. Suzaki Y, Osakada K. Ferrocene-containing [2]- and [3]rotaxanes. Preparation via an end-capping cross-metathesis reaction and electrochemical properties. *J Chem Soc. Dalton Trans*. 2007; 2376–2383.

53. Suzaki Y, Chihara E, Takagi A, Osakada K. Rotaxanes of a macrocyclic ferrocenophane with dialkylammonium axle components. *J Chem Soc. Dalton Trans*. 2009; 9881–9891.

54. Hsu C-C, Chen N-C, Lai C-C, Liu Y-H, Peng S-M, Chiu S-H. Solvent-free synthesis of the smallest rotaxane prepared to date. *Angew Chem Int Ed*. 2008; **47**: 7475–7478.

55. Hsueh S-Y, Cheng K-W, Lai C-C, Chiu S-H. Efficient solvent-free syntheses of [2]- and [4]rotaxanes. *Angew Chem Int Ed*. 2008; **47**: 4436–4439.

56. Nakazono K, Kuwata S, Takata T. Crown ether–tert-ammonium salt complex fixed as rotaxane and its derivation to nonionic rotaxane. *Tet Lett*. 2008; **49**: 2397–2401.

57. Suzuki S, Nakazono K, Takata T. Selective transformation of a crown ether/sec-ammonium salt-type rotaxane to N-alkylated rotaxanes. *Org Lett*. 2010; **12**: 712–715.

58. Megiatto JD, Spencer R, Schuster DI. Efficient one-pot synthesis of rotaxanes bearing electron donors and [60]fullerene. *Org Lett*. 2009; **11**: 4152–4155.

59. Sandanayaka ASD, Sasabe H, Araki Y, Kihara N, Furusho Y, Takata T, Ito O. Axle length effect on photoinduced electron transfer in triad rotaxane with porphyrin, [60]fullerene, and triphenylamine. *J Phys Chem A*. 2010; **114**: 5242–5250.

60. Wang J-Y, Han J-M, Yan J, Ma Y, Pei J. A mechanically interlocked [3]rotaxane as a light-harvesting antenna: synthesis, characterization, and intramolecular energy transfer. *Chem Eur J*. 2009; **15**: 3585–3594.

61. Mateo-Alonso A, Prato M. Synthesis of fullerene-stopped rotaxanes bearing ferrocene groups on the macrocycle. *Eur J Org Chem*. 2010; 1324–1332.

62. Yin J, Dasgupta S, Wu J. Synthesis of [n]rotaxanes by template-directed clipping: the role of the dialkylammonium recognition sites. *Org Lett*. 2010; **12**: 1712–1715.

63. Suzaki Y, Taira T, Osakada K, Horie M. Rotaxanes and pseudorotaxanes with Fe-, Pd- and Pt-containing axles: molecular motion in the solid state and aggregation in solution. *J Chem Soc. Dalton Trans*. 2008; 4823–4833.

64. Jeong K-S, Choi JS, Chang S-Y, Chang H-Y. Self-assembly of rotaxane-like compounds with macrocycles containing reversible coordinate bonds. *Angew Chem Int Ed*. 2000; **39**: 1692–1695.

65. Lee C-F, Leigh DA, Pritchard RG, Schultz D, Teat SJ, Timco GA, Winpenny REP. Hybrid organic–inorganic rotaxanes and molecular shuttles. *Nature*. 2009; **458**: 314–318.

66. Klotz EJF, Claridge TDW, Anderson HL. Homo- and hetero-[3]rotaxanes with two ð-systems clasped in a single macrocycle. *J Am Chem Soc*. 2006; **128**: 15374–15375.

67. Prikhodko AI, Durola F, Sauvage J-P. Iron(II)-templated synthesis of [3]rotaxanes by passing two threads through the same ring. *J Am Chem Soc*. 2008; **130**: 448–449.

68. Prikhodko AI, Sauvage J-P. Passing two strings through the same ring using an octahedral metal center as template: a new synthesis of [3]rotaxanes. *J Am Chem Soc*. 2009; **131**: 6794–6807.

69. Goldup SM, Leigh DA, McGonigal PR, Ronaldson VE, Slawin AMZ. Two axles threaded using a single template site: active metal template macrobicyclic [3]rotaxanes. *J Am Chem Soc*. 2010; **132**: 315–320.

70. Balzani V, Credi A, Silvi S, Venturi M. Artificial nanomachines based on interlocked molecular species: recent advances. *Chem Soc Rev*. 2006; **35**: 1135–1149.

71. Kay ER, Leigh DA. Beyond switches: rotaxane- and catenane-based synthetic molecular motors. *Pure Appl Chem*. 2008; **80**: 17–29.

72. Leigh DA, Thomson AR. Switchable dual binding mode molecular shuttle. *Org Lett*. 2006; **8**: 5377–5378.

73. Chatterjee MN, Kay ER, Leigh DA. Beyond switches: ratcheting a particle energetically uphill with a compartmentalized molecular machine. *J Am Chem Soc.* 2006; **128**: 4058–4073.

74. Serreli V, Lee C-F, Kay ER, Leigh DA. A molecular information ratchet. *Nature.* 2007; **445**: 523–527.

75. Alvarez-Perez M, Goldup SM, Leigh DA, Slawin AMZ. A chemically-driven molecular information ratchet. *J Am Chem Soc.* 2008; **130**: 1836–1838.

76. Mateo-Alonso A, Fioravanti G, Marcaccio M, Paolucci F, Rahman GMA, Ehli C, Guldi DM, Prato M. An electrochemically driven molecular shuttle controlled and monitored by C_{60}. *J Chem Soc Chem Commun.* 2007; 1945–1947.

77. Badjic JD, Ronconi CM, Stoddart JF, Balzani V, Silvi S, Credi A. Operating molecular elevators. *J Am Chem Soc.* 2006; **128**: 1489–1499.

78. Barrell MJ, Leigh DA, Lusby PJ, Slawin AMZ. An ion-pair template for rotaxane formation and its exploitation in an orthogonal interaction anion-switchable molecular shuttle. *Angew Chem Int Ed.* 2008; **47**: 8036–8039.

79. Zhao Y-L, Dichtel WR, Trabolsi A, Saha S, Aprahamian I, Stoddart JF. A redox-switchable R-cyclodextrin-based [2]rotaxane. *J Am Chem Soc.* 2008; **130**: 11294–11296.

80. Raiteri P, Bussi G, Cucinotta CS, Credi A, Stoddart JF, Parrinello M. Unravelling the shuttling mechanism in a photoswitchable multicomponent bistable rotaxane. *Angew Chem Int Ed.* 2008; **47**: 3536–3539.

81. Saha S, Flood AH, Stoddart JF, Impellizzeri S, Silvi S, Venturi M, Credi A. A redox-driven multicomponent molecular shuttle. *J Am Chem Soc.* 2007; **129**: 12159–12171.

82. Collin J-P, Durola F, Lux J, Sauvage J-P. A rapidly shuttling copper-complexed [2]rotaxane with three different chelating groups in its axis. *Angew Chem Int Ed.* 2009; **48**: 8532–8535.

83. Jiang Q, Zhang H-Y, Han M, Ding Z-J, Liu Y. Charge-transfer behavior in bistable [3]rotaxane. *Org Lett.* 2010; **12**: 1728–1731.

84. Sasabe H, Sandanayaka ASD, Kihara N, Furusho Y, Takata T, Arakid Y, Ito O. Axle charge effects on photoinduced electron transfer processes in rotaxanes containing porphyrin and [60]fullerene. *Phys Chem Chem Phys.* 2009; **11**: 10908–10915.

85. Zhou W, Li J, He X, Li C, Lv J, Li Y, Wang S, Liu H, Zhu D. A molecular shuttle for driving a multilevel fluorescence switch. *Chem Eur J.* 2008; **14**: 754–763.

86. Curiel D, Beer PD. Anion directed synthesis of a hydrogensulfate selective luminescent rotaxane. *J Chem Soc Chem Commun.* 2005; 1909–1911.

87. McConnell AJ, Serpell CJ, Thompson AL, Allan DR, Beer PD. Calix[4]arene-based rotaxane host systems for anion recognition. *Chem Eur J.* 2010; **16**: 1256–1264.

88. Okuno E, Hiraoka S, Shionoya M. A synthetic approach to a molecular crank mechanism: toward intramolecular motion transformation between rotation and translation. *J Chem Soc. Dalton Trans.* 2010; **39**: 4107–4116.

89. Chuang C-J, Li W-S, Lai C-C, Liu Y-H, Peng S-M, Chao I, Chiu S-H. A molecular cage-based [2]rotaxane that behaves as a molecular muscle. *Org Lett.* 2009; **11**: 385–388.

90. Sevick EM, Williams DRM. Piston-rotaxanes as molecular shock absorbers. *Langmuir.* 2010; **26**: 5864–5868.

91. Leigh DA, Morales MAF, Perez EM, Wong JKY, Saiz CG, Slawin AMZ, Carmichael AJ, Haddleton DM, Brouwer AM, Jan Buma W, Wurpel GWH, Leon S, Zerbetto F. Motion: rotaxane-based switches and logic gates that function in solution and polymer films. *Angew Chem Int Ed.* 2005; **44**: 30623067.

92. Farrell AA, Kay ER, Bottari G, Leigh DA, Jarvis SP. The effect of solvent upon molecularly thin rotaxane film formation. *Appl Surf Sci.* 2007; **253**: 6090–6095.

93. Clemente-León M, Credi A, Martínez-Díaz M-V, Mingotaud C, Stoddart JF. Towards organization of molecular machines at interfaces: Langmuir films and Langmuir–Blodgett multilayers of an acid–base switchable rotaxane. *Adv Mater.* 2006; **18**: 1291–1296.

94. Nørgaard K, Jeppesen JO, Laursen BW, Simonsen JB, Weygand MJ, Kjaer K, Stoddart JF, Bjørnholm T. Evidence of strong hydration and significant tilt of amphiphilic [2]rotaxane molecules in langmuir films studied by synchrotron X-ray reflectivity. *J Phys Chem B.* 2005; **109**: 1063–1066.

95. Nørgaard K, Laursen BW, Nygaard S, Kjaer K, Tseng H-R, Flood AH, Stoddart JF, Bjørnholm T. Structural evidence of mechanical shuttling in condensed monolayers of bistable rotaxane molecules. *Angew Chem Int Ed.* 2005; **44**: 7035–7039.

96. Lee IC, Frank CW, Yamamoto T, Tseng H-R, Flood AH, Stoddart JF, Jeppesen JO. Langmuir and Langmuir–Blodgett films of amphiphilic bistable rotaxanes. *Langmuir.* 2004; **20**: 5809–5828.

97. Jang SS, Jang YH, Kim Y-H, Goddard WA, Choi JW, Heath JR, Laursen BW, Flood AH, Stoddart JF, Nørgaard K, Bjørnholm T. Molecular dynamics simulation of amphiphilic bistable [2]rotaxane Langmuir monolayers at the air/water interface. *J Am Chem Soc.* 2005; **127**: 14804–14816.

98. Nikitin K, Lestini E, Lazzari M, Altobello S, Fitzmaurice D. A tripodal [2]rotaxane on the surface of gold. *Langmuir.* 2007; **23**: 12147–12153.

99. Jang YH, Goddard WA. Mechanism of oxidative shuttling for [2]rotaxane in a Stoddart–Heath molecular switch: density functional theory study with continuum-solvation model. *J Phys Chem B.* 2006; **110**: 7660–7665.

100. Dichtel WR, Heath JR, Stoddart J. Designing bistable [2]rotaxanes for molecular electronic devices. *Phil Trans R Soc A.* 2007; **365**: 1607–1625.

101. Luo Y, Collier CP, Jeppesen JO, Nielsen KA, DeIonno E, Ho G, Perkins J, Tseng H-R, Yamamoto T, Stoddart JF, Heath JR. Two-dimensional molecular electronics circuits. *Chem Phys Chem.* 2002; **3**: 519–525.

102. Collier CP, Mattersteig G, Wong EW, Luo Y, Beverly K, Sampaio J, Raymo FM, Stoddart JF, Heath JR. A [2]catenane-based solid state electronically reconfigurable switch. *Science.* 2000; **289**: 1172–1175.

103. DeIonno E, Tseng H-R, Harvey DD, Stoddart JF, Heath JR. Infrared spectroscopic characterization of [2]rotaxane molecular switch tunnel junction devices. *J Phys Chem B.* 2006; **110**: 7609–7612.

104. Feng M, Gao L, Deng Z, Ji W, Guo X, Du S, Shi D, Zhang D, Zhu D, Gao H. Reversible, erasable, and rewritable nanorecording on an H2 rotaxane thin film. *J Am Chem Soc.* 2007; **129**: 2204–2205.

105. Huang TJ, Tseng H-R, Sha L, Lu W, Brough B, Flood AH, Yu B-D, Celestre PC, Chang JP, Stoddart JF, Ho C-M. Mechanical shuttling of linear motor-molecules in condensed phases on solid substrates. *Nano Lett.* 2004; **4**: 2065–2071.

106. Tseng H-R, Wu D, Fang NX, Zhang X, Stoddart JF. The metastability of an electrochemically controlled nanoscale machine on gold surfaces. *Chem Phys Chem.* 2004; **5**: 111–116.

107. Fioravanti G, Haraszkiewicz N, Kay ER, Mendoza SM, Bruno C, Marcaccio M, Wiering PG, Paolucci F, Rudolf P, Brouwer AM, Leigh DA. Three state redox-active molecular shuttle that switches in solution and on a surface. *J Am Chem Soc.* 2008; **130**: 2593–2601.

108. Willner I, Pardo-Yissar V, Katz E, Ranjit KT. A photoactivated 'molecular train' for optoelectronic applications: light-stimulated translocation of a β-cyclodextrin receptor within a stoppered azobenzene-alkyl chain supramolecular monolayer assembly on a Au-electrode. *J Electroanal Chem.* 2001; **497**: 172–177.

109. Katz E, Lioubashevsky O, Willner I. Electromechanics of a redox-active rotaxane in a monolayer assembly on an electrode. *J Am Chem Soc.* 2004; **126**: 15520–15532.

110. Katz E, Sheeney-Haj-Ichia L, Willner I. Electrical contacting of glucose oxidase in a redox-active rotaxane configuration. *Angew Chem Int Ed.* 2004; **43**: 3292–3300.

111. Bayly SR, Gray TM, Chmielewski MJ, Davis JJ, Beer PD. Anion templated surface assembly of a redox-active sensory rotaxane. *Chem Commun.* 2007; 2234–2236.

112. Wan P, Jiang Y, Wang Y, Wang Z, Zhang X. Tuning surface wettability through photocontrolled reversible molecular shuttle. *J Chem Soc Chem Commun.* 2008; 5710–5712.

113. Berna J, Leigh DA, Lubomska M, Mendoza SM, PeRez EM, Rudolf P, Teobaldi G, Zerbetto F. Macroscopic transport by synthetic molecular machines. *Nat Mater.* 2005; **4**: 704–710.

114. Huang TJ, Brough B, Ho C-M, Liu Y, Flood AH, Bonvallet PA, Tseng H-R, Stoddart JF, Baller M, Magonov S. A nanomechanical device based on linear molecular motors. *Appl Phys Lett.* 2004; **85**: 5391–5393.

115. Nguyen TD, Tseng H-R, Celestre PC, Flood AH, Liu Y, Stoddart JF, Zink JI. A reversible molecular valve. *Proc Natl Acad Sci USA.* 2005; **102**: 10029–10034.

116. Steuerman DW, Tseng H-R, Peters AJ, Flood AH, Jeppesen JO, Nielsen KA, Stoddart JF, Heath JR. Molecular-mechanical switch-based solid-state electrochromic devices. *Angew Chem Int Ed.* 2004; **43**: 6486–6491.

117. Coronado E, Gaviña P, Tatay S. Catenanes and threaded systems: from solution to surfaces. *Chem Soc Rev.* 2009; **38**: 1674–1689.

118. Raymo FM, Fraser JF. Interlocked macromolecules. *Chem Rev.* 1999; **99**: 1643–1663.

119. Wu C, Bheda MC, Lim C, Shen YX, Sze J, Gibson H. Synthesis of polyester rotaxanes via the statistical threading method. *Polym Commun.* 1991; **32**: 204–207.

120. Hodge P, Monvisade P, Owen GJ, Heatley F, Pang Y. 1H NMR spectroscopic studies of the structures of a series of pseudopolyrotaxanes formed by threading. *New J Chem.* 2000; **24**: 703–709.

121. Mason PE, Parsons LW, Tolley MS. Dynamic behaviour of a pseudo[n]polyrotaxane containing a bipyridyl-based cyclophane: spectroscopic observations. *Polymer.* 1998; **39**: 3981–3991.

122. Cacialli F, Wilson JS, Michels JJ, Daniel C, Silva C, Friend RH, Severin N, Samorì P, Rabe JP, O'Connell MJ, Taylor PN, Anderson HL. Cyclodextrin-threaded conjugated polyrotaxanes as insulated molecular wires with reduced interstrand interactions. *Nature Mater.* 2002; **1**: 160–164.

123. Terao J, Tsuda S, Tanaka Y, Okoshi K, Fujihara T, Tsuji YI, Kambe N. Synthesis of organic-soluble conjugated polyrotaxanes by polymerization of linked rotaxanes. *J Am Chem Soc.* 2009; **131**: 16004–16005.

124. Herrmann W, Schneider M, Wenz G. Photochemical synthesis of polyrotaxanes from stilbene polymers and cyclodextrins. *Angew Chem Int Ed.* 1997; **36**: 2511–2514.

125. Gong C, Balanda PB, Gibson HW. Supramolecular chemistry with macromolecules: new self-assembly based main chain polypseudorotaxanes and their properties. *Macromolecules.* 1998; **31**: 5278–5289.

126. Vidal PL, Billon M, Divisia-Blohorn B, Bidan G, Kern JM, Sauvage JP. Conjugated polyrotaxanes containing coordinating units: reversible copper(I) metallation–demetallation using lithium as intermediate scaffolding. *J Chem Soc Chem Commun.* 1998; 629–630.

127. Sauvage J-P, Kern J-M, Bidan G, Divisia-Blohorn B, Vidal P-L. Conjugated polyrotaxanes: improvement of the polymer properties by using sterically hindered coordinating units. *New J Chem.* 2002; **26**: 1287–1290.

128. Zhu SS, Carroll PJ, Swager TM. Conducting polymetallorotaxanes: a supramolecular approach to transition metal ion sensors. *J Am Chem Soc.* 1996; **118**: 8713–8714.

129. Kern J-M, Sauvage J-P, Bidan G, Billon M, Divisia-Blohorn B. Electroactive films with a polyrotaxane organic backbone. *Adv Mater.* 1996; **8**: 580–582.

130. Ikeda T, Higuchi M, Kurth DG. From thiophene [2]rotaxane to polythiophene polyrotaxane. *J Am Chem Soc.* 2009; **131**: 9158–9159.

131. Born M, Ritter H. Pseudo-polymer analogous reactions: methylation of alcohol groups of non-covalently anchored 2,6-dimethyl-β-cyclodextrin components located in branched side chains of a poly(tandem-rotaxane). *Adv Mater.* 1996; **8**: 149–151.

132. Noll O, Ritter H. Synthesis of new side-chain polyrotaxanes via free radical polymerization of a water-soluble semi-rotaxane monomer consisting of 2,6-dimethyl-b-cyclodextrin and 3-O-(11-acryloylaminoun-decanoyl)cholic acid. *Macromol Chem Phys.* 1998; **199**: 791–794.

133. Yamaguchi I, Osakada K, Yamamoto T. Introduction of a long alkyl side chain to poly(benzimidazole)s: N-alkylation of the imidazole ring and synthesis of novel side chain polyrotaxanes. *Macromolecules.* 1997; **30**: 4288–4294.

134. Zhou Q, Swager TM. Fluorescent chemosensors based on energy migration in conjugated polymers: the molecular wire approach to increased sensitivity. *J Am Chem Soc.* 1995; **117**: 12593–12602.

135. Yamaguchi N, Nagvekar DS, Gibson HW. Self-organization of a heteroditopic molecule to linear polymolecular arrays in solution. *Angew Chem Int Ed.* 1998; **37**: 2361–2364.

136. Hoshino T, Miyauchi M, Kawaguchi Y, Yamaguchi H, Harada A. Daisy chain necklace: tri[2]rotaxane containing cyclodextrins. *J Am Chem Soc.* 2000; **122**: 9876–9877.

137. Frisch HL, Wasserman E. Chemical topology. *J Am Chem Soc.* 1961; **83**: 3789–3795.

138. Wasserman E. The preparation of interlocking rings: a catenane. *J Am Chem Soc.* 1960; **82**: 4433–4434.

139. Logemann E, Rissler K, Schill G, Fritz H. Synthesis and spectra of 2 novel catenanes. *Chem Ber-Rec.* 1981; **114**: 2245–2260.

140. Schill G, Rissler K, Fritz H, Vetter W. Synthesis, isolation, and identification of translationally isomeric [3]catenanes. *Angew Chem Int Ed.* 1985; **20**: 187–189.

141. Dietrich-Buchecker CO, Sauvage JP, Kintzinger JP. Une nouvelle famille de molecules: les metallo-catenanes. *Tet Lett.* 1983; **24**: 5095–5098.

142. Dietrich-Buchecker CO, Sauvage JP, Kern JM. Templated synthesis of interlocked macrocyclic ligands: the catenands. *J Am Chem Soc.* 1984; **106**: 3043–3045.

143. Dietrich-Buchecker C, Sauvage J-P. Templated synthesis of interlocked macrocyclic ligands, the catenands. preparation and characterization of the prototypical bis-30 membered ring system. *Tetrahedron.* 1990; **46**: 503–512.

144. Cesario M, Dietrich-Buchecker CO, Guilhem J, Pascard C, Sauvage JP. Molecular structure of a catenand and its copper(I) catenate: complete rearrangement of the interlocked macrocyclic ligands by complexation. *J Chem Soc Chem Commun.* 1985; 244–247.

145. Albrecht-Gary AM, Dietrich-Buchecker C, Saad Z, Sauvage J-P. Topological kinetic effects: complexation of interlocked macrocyclic ligands by cationic species. *J Am Chem Soc.* 1988; **110**: 1467–1472.

146. Cesario M, Dietrich CO, Edel A, Guilhem J, Kintzinger J-P, Pascard CE, Sauvage J-P. Topological enhancement of basicity: molecular structure and solution study of a monoprotonated catenand. *J Am Chem Soc.* 1986; **108**: 6250–6254.

147. Dietrich-Buchecker CO, Sauvage J-P, Weiss J. Interlocked macrocyclic ligands: a catenand whose rotation of one ring into the other is precluded by bulky substituents. *Tet Lett.* 1986; **27**: 2257–2260.

148. Dietrich-Bucheckera CO, Edel A, Kintzinger JP, Sauvage J-P. Synthese et etude d'un catenate de cuivre chiral comportant deux anneaux coordinant a 27 atomes. *Tetrahedron.* 1987; **43**: 333–344.

149. Sauvage J-P, Weiss J. Synthesis of biscopper(I) [3]-catenates: multiring interlocked coordinating systems. *J Am Chem Soc.* 1985; **107**: 6108–6110.

150. Guilhem J, Pascard C, Sauvage J-P, Weiss J. Solution study and molecular structure of a [3]-catenand. Intramolecular interaction between the two peripheral rings. *J Am Chem Soc.* 1988; **110**: 8711–8713.

151. Dietrich-Buchecker CO, Guilhem J, Khemiss AK, Kintzinger J-P, Pascard C, Sauvage J-P. Molecular structure of a [3]-catenate: curling up of the interlocked system by interaction between the two copper complex subunits. *Angew Chem Int Ed.* 1987; **26**: 661–663.

152. Bitsch F, Dietrich-Buchecker CO, Khemiss AK, Sauvage J-P, Van Dorsselaer A. Multiring interlocked systems: structure elucidation by electrospray mass spectrometry. *J Am Chem Soc.* 1991; **113**: 4023–4025.

153. Bitsch F, Hegy G, Dietrich-Buchecker C, Leize E, Sauvage J-P, Vandorsselaer A. Fast-atom-bombardment and electrospray-ionization mass-spectrometry of coordination-compounds: application to the analysis of multiring catenates. *New J Chem.* 1994; **18**: 801–807.

154. Dietrich-Buchecker CO, Hemmert C, Khemiss AK, Sauvage J-P. Synthesis of dicopper [3]-catenates and [3]-catenands by acetylenic oxidative coupling: preparation and study of corresponding homodimetallic [3]-catenates [silver(1+), zinc(2+), cobalt(2+), and nickel(2+)]. *J Am Chem Soc.* 1990; **112**: 8002–8008.

155. Dietrich-Buchecker C, Hemmert C, Sauvage J-P. Disymmetrical dimetallic [3]-catenates: long-range electrostatic interaction between the metal centers through stacked aromatic ligands. *New J Chem.* 1990; **14**: 603–605.

156. Armaroli N, Balzani V, Barigelletti F, De Cola L, Flamigni L, Sauvage J-P, Hemmert C. Supramolecular photochemistry and photophysics: a [3]-catenand and its mononuclear and homo- and heterodinuclear [3]-catenates. *J Am Chem Soc.* 1994; **116**: 5211–5217.

157. Montenteau M, Le Bras F, Loock B. Synthesis of interlocked basket handle porphyrins. *Tet Lett.* 1994; **35**: 3289–3292.

158. Jørgensen T, Becher J, Chambron J-C, Sauvage J-P. A copper(I) [2]-catenate incorporating a tetrathiafulvalene unit. *Tet Lett.* 1994; **35**: 4339–4342.

159. Livoreil A, Dietrich-Buchecker CO, Sauvage J-P. Electrochemically triggered swinging of a [2]-catenate. *J Am Chem Soc.* 1994; **116**: 9399–9400.

160. Sauvage J-P, Ward M. A bis(terpyridine)ruthenium(II) catenate. *Inorg Chem.* 1991; **30**: 3869–3874.

161. Nierengarten J-F, Dietrich-Buchecker CO, Sauvage J-P. Synthesis of a doubly interlocked [2]-catenane. *J Am Chem Soc.* 1994; **116**: 375–376.

162. Dietrich-Buchecker C, Leize E, Nierengarten J-F, Sauvage J-P, Van Dorsselaer A. Singly and doubly interlocked [2]-catenanes: influence of the degree of entanglement on chemical stability as estimated by fast atom bombardment (FAB) and electrospray ionization (ESI) mass spectrometries (MS). *J Chem Soc Chem Commun.* 1994; 2257–2258.

163. Dietrich-Buchecker C, Frommberger B, Lüer I, Sauvage J-P, Vögtle F. Multiring catenanes with a macrobicyclic core. *Angew Chem Int Ed.* 1993; **32**: 1434–1437.

164. Walba DM, Zheng QY, Schilling K. Topological stereochemistry. 8. Experimental studies on the hook and ladder approach to molecular knots: synthesis of a topologically chiral cyclized hook and ladder. *J Am Chem Soc.* 1992; **114**: 6259–6260.

165. Gruter GJM, de Kanter FJJ, Markies PR, Nomoto T, Akkerman OS, Bickelhaupt F. Formation of the first organometallic catenane. *J Am Chem Soc.* 1993; **115**: 12179–12180.

166. Fujita M, Ibukuro F, Hagihara H, Ogura K. Quantitative self-assembly of a [2]catenane from two preformed molecular rings. *Nature.* 1994; **367**: 720–723.

167. Fujita M, Ibukuro F, Seki H, Kamo O, Imanari M, Ogura K. Catenane formation from two molecular rings through very rapid slippage: a Möbius strip mechanism. *J Am Chem Soc.* 1996; **118**: 899–900.

168. Fujita M, Nagao S, Iida M, Ogata K, Ogura K. Palladium(II)-directed assembly of macrocyclic dinuclear complexes composed of (en)Pd2+ and bis(4-pyridyl)-substituted bidentate ligands: remarkable ability for molecular recognition of electron-rich aromatic guests. *J Am Chem Soc.* 1993; **115**: 1574–1576.

169. Fujita M, Ibukuro F, Yamaguchi K, Ogura K. A molecular lock. *J Am Chem Soc.* 1995; **117**: 4175–4176.

170. Dietrich-Buchecker C, Colasson B, Fujita M, Hori A, Geum N, Sakamoto S, Yamaguchi K, Sauvage J-P. Quantitative formation of [2]catenanes using copper(I) and palladium(II) as templating and assembling centers: the entwining route and the threading approach. *J Am Chem Soc.* 2003; **125**: 5717–5725.

171. Colasson BX, Sauvage J-P. Irreversible but noncovalent Ru(II)–pyridine bond: its use for the formation of [2]-catenanes. *Inorg Chem.* 2004; **43**: 1895–1901.

172. Hori A, Yamashita K-I, Kusukawa T, Akasaka A, Biradha K, Fujita M. A circular tris[2]catenane from molecular figure-of-eight. *J Chem Soc Chem Commun.* 2004; 1798–1799.

173. Yamashita K-I, Horia A, Fujita M. Formation of bis- and tris[2]catenanes via the cross-catenation of Pd(II)- and Pt(II)-linked coordination rings. *Tetrahedron.* 2007; **63**: 8435–8439.

174. Fujita M, Aoyagi M, Ibukuro F, Ogura K, Yamaguchi K. Made-to-order assembling of [2]catenanes from palladium(II)-linked rectangular molecular boxes. *J Am Chem Soc.* 1998; **120**: 611–612.

175. Burchell TJ, Eisler DJ, Puddephatt RJ. A chiral [2]catenane self-assembled from meso-macrocycles of palladium(II). *J Chem Soc. Dalton Trans.* 2005; 268–272.

176. Lim CW, Sakamoto S, Yamaguchi K, Hong J-I. Versatile formation of [2]catenane and [2]pseudorotaxane structures: threading and noncovalent stoppering by a self-assembled macrocycle. *Org Lett.* 2004; **6**: 1079–1082.

177. Leigh DA, Lusby PJ, Teat SJ, Wilson AJ, Wong JKY. Benzylic imine catenates: readily accessible octahedral analogues of the sauvage catenates. *Angew Chem Int Ed.* 2001; **40**: 1538–1543.

178. Hogg L, Leigh DA, Lusby PJ, Morelli A, Parsons S, Wong JKY. A simple general ligand system for assembling octahedral metal–rotaxane complexes. *Angew Chem Int Ed.* 2004; **43**: 1218–1221.

179. Goldup SM, Leigh DA, Lusby PJ, McBurney RT, Slawin AMZ. Gold(I)-template catenane and rotaxane synthesis. *Angew Chem Int Ed.* 2009; **47**: 6999–7003.

180. Goldup SM, Leigh DA, Long T, McGonigal PR, Symes MD, Wu J. Active metal template synthesis of [2]catenanes. *J Am Chem Soc.* 2009; **131**: 15924–15929.

181. Linke M, Fujita N, Chambron J-C, Heitz V, Sauvage J-P. A [2]-catenane whose rings incorporate two differently metallated porphyrins. *New J Chem.* 2001; **25**: 790–796.

182. Beyler M, Heitz V, Sauvage J-P. Quantitative formation of a tetraporphyrin [2]catenane via copper and zinc coordination. *J Chem Soc Chem Commun.* 2008; 5396–5398.

183. Megiatto JD, Schuster DI, Abwandner S, de Miguel G, Guldi DM. [2]catenanes decorated with porphyrin and [60]fullerene groups: design, convergent synthesis, and photoinduced processes. *J Am Chem Soc.* 2010; **132**: 3847–3861.

184. Bäuerle P, Ammann M, Wilde M, Götz G, Mena-Osteritz E, Rang A, Schalley CA. Oligothiophene-based catenanes: synthesis and electronic properties of a novel conjugated topological structure. *Angew Chem.* 2007; **119**: 367–372.

185. Westcott A, Fisher J, Harding LP, Rizkallah P, Hardie MJ. Self-assembly of a 3-D triply interlocked chiral [2]catenane. *J Am Chem Soc.* 2008; **130**: 2950–2951.

186. Ashton PR, Goodnow TT, Kaifer AE, Reddington MV, Slawin AMZ, Spencer N, Stoddart JF, Vicent C, Williams DJ. A [2] catenane made to order. *Angew Chem Int Ed.* 1989; **28**: 1396–1399.

187. Brown CL, Philp D, Spencer N, Stoddart JF. The mechanisms of making molecules to order. *Israel J Chem.* 1992; **32**: 61–67.

188. Brown CL, Philp D, Stoddart JF. The self-assembly of a [2]catenane. *Synlett.* 1991; **7**: 459–461.

189. Amabilino DB, Ashton PR, Stoddart JF. Macrocyclic polyethers incorporating resorcinol residues as templates for cyclobis(paraquat-p-phenylene) in the self-assembly of [2]catenanes. *Supramol Chem.* 1995; **5**: 5–8.

190. Stoddart JF, Williams DJ, Amabilino DB, Anelli P-L, Ashton PR, Brown GR, Cordova E, Godinez LA, Hayes W. Molecular meccano. 3. Constitutional and translational isomerism in [2]catenanes and [n]pseudorotaxanes. *J Am Chem Soc.* 1995; **117**: 11142–11170.

191. Ashton PR, Blower M, Philp D, Spencer N, Stoddart JF, Tolley MS, Ballardini R, Ciano M, Blazani V, Gandolfi MT, Prodi L, McLean CH. The control of translational isomerism in catenated structures. *New J Chem.* 1993; **17**: 689–695.

192. Amabilino DB, Ashton PR, Brown CL, Cordova E, Godinez LA, Goodnow TT, Kaifer AE, Newton SP, Pietraszkiewicz M. Molecular meccano. 2. Self-assembly of [n]catenanes. *J Am Chem Soc.* 1995; **117**: 1271–1293.

193. Ashton PR, Blower MA, Iqbal S, McLean CH, Stoddart JF, Tolley MS, Williams DJ. The design and self-assembly of a furan-containing [2]catenane. *Synlett.* 1994; **12**: 1059–1062.

194. Ashton PR, Blower MA, Iqbal S, McLean CH, Stoddart JF, Tolley MS. The design and self-assembly of a pyridine-containing [2]catenane. *Synlett.* 1994; **12**: 1063–1066.

195. Amabilino DB, Stoddart JF. New approach to controlling catenated structures. *Red Trau Chim Pays-Bas.* 1993; **112**: 429–430.

196. Ashton PR, Brown CL, Chrystal EJT, Parry KP, Pietraszkiewicz M, Spencer N, Stoddart JF. Molecular trains: the self-assembly and dynamic properties of two new catenanes. *Angew Chem Int Ed.* 1991; **30**: 1042–1045.

197. Gunter MJ, Johnston MR. Towards molecular scale mechano-electronic devices: porphyrin catenanes. *J Chem Soc Chem Commun.* 1992; 1163–1165.

198. Gunter MJ, Hockless DCR, Johnston MR, Skelton BW, White AH. Self-assembling porphyrin [2]-catenanes. *J Am Chem Soc.* 1994; **116**: 4810–4823.

199. Gunter MJ, Johnston MR. Porphyrin [2]catenanes: dynamic control through protonation. *J Chem Soc Chem Commun.* 1994; 829–830.

200. Amabilino DB, Ashton PR, Reder AS, Spencer N, Stoddart JF. Olympiadane. *Angew Chem Int Ed.* 1994; **33**: 1286–1290.

201. Amabilino DB, Ashton PR, Balzani V, Boyd SE, Credi A, Lee JY, Menzer S, Stoddart JF, Venturi M, Williams DJ. Oligocatenanes made to order. *J Am Chem Soc.* 1998; **120**: 4295–4307.

202. Amabilino DB, Ashton PR, Boyd SE, Lee JY, Menzer S, Stoddart JF, Williams DJ. The five-stage self-assembly of a branched heptacatenane. *Angew Chem Int Ed.* 1997; **36**: 2070–2072.

203. Amabilino DB, Ashton PR, Tolley MS, Stoddart JF, Williams DJ. Isomeric self-assembling [2]catenanes. *Angew Chem Int Ed.* 1993; **32**: 1297–1301.

204. Ashton PR, Preece JA, Stoddart JF, Tolley MS, White AJP, Williams DJ. The self-assembly and dynamic properties of thiophene-containing [2]catenanes. *Synthesis.* 1994; **12**: 1344–1352.

205. Bauer M, Muller WM, Muller U, Rissanen K. Azobenzene-based photoswitchable catenanes. *Justus Liebigs Ann.* 1995; 649–656.

206. Ashton PR, Ballardini R, Balzani V, Credi A, Gandolfi MT, Menzer S, Prrez-Garcia L, Prodi L, Stoddart JF, Venturi M, White AJP, Williams DJ. Molecular meccano. 4. The self-assembly of [2]catenanes incorporating photoactive and electroactive n-extended systems. *J Am Chem Soc.* 1995; **117**: 11171–11197.

207. Ashton PR, Iriepa I, Reddington MV, Spencer N, Slawin AMZ, Stoddart JF, Williams DJ. An optically-active [2]catenane made to order. *Tet Lett.* 1994; **35**: 4835–4838.

208. Ashton PR, Reder AS, Spencer N, Stoddart JF. Self-assembly of a chiral bis[2]catenane. *J Am Chem Soc.* 1993; **11**: 5286–5287.

209. Ashton PR, Huff J, Menzer S, Parsons IW, Preece JA, Stoddart JF, Tolley MS, White AJP, Williams DJ. Bis[2]catenanes and a bis[2]rotaxane-model compounds for polymers with mechanically interlocked components. *Chem Eur J.* 1996; **2**: 31–44.

210. Ashton PR, Boyd SE, Brindle A, Langford SJ, Menzer S, Perez-Garcia L, Preece JA, Raymo FM, Spencer N, Stoddart JF, White AJP, Williams DJ. Diazapyrenium-containing catenanes and rotaxanes. *New J Chem.* 1999; **23**: 587–602.

211. Ballardini R, Balzani V, Di Fabio A, Gandolfi MT, Becher J, Lau J, Brondsted Nielsen M, Stoddart JF. Macrocycles, pseudorotaxanes and catenanes containing a pyrrolo-tetrathiafulvalene unit: absorption spectra, luminescence properties and redox behaviour. *New J Chem.* 2001; **25**: 293–298.

212. Nygaard S, Hansen SW, Huffman JC, Jensen F, Flood AH, Jeppesen JO. Two classes of alongside charge-transfer interactions defined in one [2]catenane. *J Am Chem Soc.* 2007; **129**: 7354–7363.

213. Gunte MJ, Farquhar SM. Neutral π-associated porphyrin [2]catenanes. *Org Biomol Chem.* 2003; **1**: 3450–3457.

214. Alcalde E, Pérez-García L, Ramos S, Stoddart JF, White AJP, Williams DJ. Spontaneous resolution in a family of [2]catenanes containing proton-ionisable 1H-1,2,4-triazole subunits. *Mendeleev Commun.* 2004; **14**: 233–235.

215. Hamilton DG, Sanders JKM, Davies JE, Clegg W, Teat SJ. Neutral [2]catenanes from oxidative coupling of -stacked components. *J Chem Soc Chem Commun.* 1997; 897–898.

216. Hamilton DG, Feeder N, Prodi L, Teat SJ, Clegg W, Sanders JKM. Tandem hetero-catenation: templating and self-assembly in the mutual closure of two different interlocking rings. *J Am Chem Soc.* 1998; **120**: 1096–1097.

217. Lukyanenko NG, Lyapunov AY, Kirichenko TI, Zubatyuk RI, Shishkin OV. Self-assembly of a [2]catenane incorporating a fluorenonophane-containing azobenzene moiety. *Mendeleev Commun.* 2006; **16**: 143–145.

218. Guidry EN, Cantrill SJ, Stoddart JF, Grubbs RH. Magic ring catenation by olefin metathesis. *Org Lett.* 2005; **7**: 2129–2132.

219. Miljanic OS, Stoddart JF. Dynamic donor–acceptor [2]catenanes. *Proc Nat Acad Sci.* 2007; **104**: 12966–12970.

220. Koshkakaryan G, Cao D, Klivansky LM, Teat SJ, Tran JL, Liu Y. Dual selectivity expressed in [2 + 2 + 1] dynamic clipping of unsymmetrical [2]catenanes. *Org Lett.* 2010; **12**: 1528–1531.

221. Au-Yeung HY, Pantos GD, Sanders JKM. Dynamic combinatorial synthesis of a catenane based on donor–acceptor interactions in water. *Proc Nat Acad Sci.* 2009; **106**: 10466–10470.

222. Hunter CA. Synthesis and structure elucidation of a new [21]-catenane. *J Am Chem Soc.* 1992; **114**: 5303–5311.

223. Adams H, Carver FJ, Hunter CA. [2]catenane or not [2]catenane? *J Chem Soc Chem Commun.* 1995; 809–810.

224. Carver FJ, Hunter CA, Shannon RJ. High dilution directed macrocyclisation reactions. *J Chem Soc Chem Commun.* 1994; 1277–1280.

225. Vögtle F, Meier S, Hoss R. One-step synthesis of a fourfold functionalized catenane. *Angew Chem Int Ed.* 1992; **31**: 1619–1622.

226. Ottens-Hildebrandt S, Meier S, Schmidt W, Vögtle F. Isomeric lactam catenanes and the mechanism of their formation. *Angew Chem Int Ed.* 1994; **33**: 1767–1770.

227. Vögtle F, Jager R, Hiindel M, Wens-Hildebrandt S. Catenanes and rotaxanes of the amide type. *Pure Appl Chem.* 1996; **68**: 225–232.

228. Ottens-Hildebrandt S, Nieger M, Rissanen K, Rouvinen J, Meier S, Harder G, Vögtle F. Amide-based furano-catenanes: regioselective template synthesis and crystal structure. *J Chem Soc Chem Commun.* 1995; 777–778.

229. Johnston AG, Leigh DA, Pritchard RJ, Deegan MD. Facile synthesis and solid-state structure of a benzylic amide [2]catenane. *Angew Chem Int Ed.* 1995; **34**: 1209–1212.

230. Johnston AG, Leigh DA, Nezhat L, Smart JP, Deegan MD. Structurally diverse and dynamically versatile benzylic amide [2]catenanes assembled directly from commercially available precursors. *Angew Chem Int Ed.* 1995; **34**: 1212–1216.

231. Ottens-Hildebrandt S, Schmidt T, Harren J, Vögtle F. Sulfonamide-based catenanes: regioselective template synthesis. *Liebigs Ann.* 1995; 1855–1860.

232. Jäger R, Vögtle F. A new synthetic strategy towards molecules with mechanical bonds: nonionic template synthesis of amide-linked catenanes and rotaxanes. *Angew Chem Int Ed.* 1997; **36**: 930–944.

233. Vögtle F, Dünnwald T, Schmidt T. Catenanes and rotaxanes of the amide type. *Acc Chem Res.* 1996; **29**: 451–460.

234. Andrievsky A, Ahuis F, Sessler JL, Vögtle F, Gudat D, Moini M. Bipyrrole-based [2]catenane: a new type of anion receptor. *J Am Chem Soc.* 1998; **120**: 9712–9713.

235. Baumann S, Jäger R, Ahuis F, Kray B, Vögtle F. Flexible, long-chain alkanediamines as building blocks for catenanes: steric hindrance of circumrotation by derivatization. *Liebigs Ann.* 1997; 761–766.

236. Safarowsky O, Vogel E, Vögtle F. Amide-based [3]catenanes and [2]catenanes with aliphatic chains. *Eur J Org Chem.* 2000; 499–505.

237. Leigh DA, Moody K, Smart JP, Watson KJ, Slawin AMZ. Catenane chameleons: environment-sensitive translational isomerism in amphiphilic benzylic amide [2]catenanes. *Angew Chem Int Ed.* 1996; **35**: 306–310.

238. Jäger R, Schmidt T, Karbach D, Vögtle F. The first pretzel-shaped molecules: via catenane precursors. *Synlett.* 1996; 723–725.

239. Li X-Y, Illigen J, Nieger M, Michel S, Schalley CA. Tetra- and octalactam macrocycles and catenanes with exocyclic metal coordination sites: versatile building blocks for supramolecular chemistry. *Chem Eur J.* 2003; **9**: 1332–1347.

240. Caldwell ST, Cooke G, Fitzpatrick B, Long D-L, Rabania G, Rotello VM. A flavin-based [2]catenane. *J Chem Soc Chem Commun.* 2008; 5912–5914.

241. Armspach D, Ashton PR, Moore CP, Spencer N, Stoddart JF, Wear TJ, Williams DJ. The self-assembly of catenated cyclodextrins. *Angew Chem Int Ed.* 1993; **32**: 854–858.

242. Armspach D, Ashton PR, Ballardini R, Balzani V, Godi A, Moore CP, Prodi L, Spence N, Stoddart JF, Tolley MS, Wear TJ, Williams DJ. Catenated cyclodextrins. *Chem Eur J.* 1995; **1**: 33–55.

243. Bogdan A, Vysotsky MO, Ikai T, Okamoto Y, Bohmer V. Rational synthesis of multicyclic bis[2]catenanes. *Chem Eur J.* 2004; **10**: 3324–3330.

244. Wang L, Vysotsky MO, Bogdan A, Bolte M, Bohmer V. Multiple catenanes derived from calix[4]arenes. *Science.* 2004; **304**: 1312–1314.

245. Molokanova O, Bogdan A, Vysotsky MO, Bolte M, Ikai T, Okamoto Y, Bohmer V. Calix[4]arene-based bis[2]catenanes: synthesis and chiral resolution. *Chem Eur J.* 2007; **13**: 6157–6170.

246. Molokanova O, Podoprygorina G, Bolte M, Bohmer V. Multiple catenanes based on tetraloop derivatives of calix[4]arenes. *Tetrahedron.* 2009; **65**: 7220–7233.

247. Okada Y, Miao Z, Akibaa M, Nishimura J. Synthesis and characterization of chiral catenanes based on rigid calix[4]arene. *Tet Lett.* 2006; **47**: 2699–2702.

248. Lu L-G, Li G-K, Peng X-X, Chen C-F, Huang Z-T. Synthesis and self-assembly of novel calix[4]arenocrowns: formation of calix[4]areno[2]catenanes. *Tet Lett.* 2006; **47**: 6021–6025.

249. Lankshear MD, Evans NH, Bayly SR, Beer PD. Anion-templated calix[4]arene-based pseudorotaxanes and catenanes. *Chem Eur J.* 2007; **13**: 3861–3870.

250. Phipps DE, Beer PD. A [2]catenane containing an upper-rim functionalized calix[4]arene for anion recognition. *Tet Lett.* 2009; **50**: 3454–3457.

251. Ng K-Y, Cowley AR, Beer PD. Anion templated double cyclization assembly of a chloride selective [2]catenane. *J Chem Soc Chem Commun.* 2006; 3676–3678.

252. Huang B, Santos SM, Felix V, Beer PD. Sulfate anion-templated assembly of a [2]catenane. *J Chem Soc Chem Commun.* 2008; 4610–4612.

253. Chmielewski MJ, Davis JJ, Beer PD. Interlocked host rotaxane and catenane structures for sensing charged guest species via optical and electrochemical methodologies. *Org Biomol Chem.* 2009; **7**: 415–424.

254. Nakamura Y, Minami S, Iizuka K, Nishimura J. Preparation of neutral [60]fullerene-based [2]catenanes and [2]rotaxanes bearing an electron-deficient aromatic diimide moiety. *Angew Chem Int Ed.* 2003; **42**: 3158–3162.

255. Chiu S-H, Elizarov AM, Glink PT, Stoddart JF. Translational isomerism in a [3]catenane and a [3]rotaxane. *Org Lett.* 2002; **4**: 3561–3564.

256. Ng K-Y, Felix, V. Santos SM, Rees NH, Beer PD. Anion induced and inhibited circumrotation of a [2]catenane. *Chem Commun.* 2008; 1281–1283.

257. Coskun A, Saha S, Aprahamian I, Stoddart JF. A reverse donor–acceptor bistable [2]catenane. *Org Lett.* 2008; **10**: 3187–3190.

258. Balzani V, Credi A, Mattersteig G, Matthews OA, Raymo FM, Stoddart JF, Venturi M, White AJP, Williams DJ. Switching of pseudorotaxanes and catenanes incorporating a tetrathiafulvalene unit by redox and chemical inputs. *J Org Chem.* 2000; **65**: 1924–1936.

259. Ikeda I, Aprahamian I, Stoddart JF. Blue-colored donor–acceptor [2]rotaxane. *Org Lett.* 2007; **9**: 1481–1484.

260. Ikeda T, Saha S, Aprahamian I, Leung KC-F, Williams A, Deng W-Q, Flood AH, Goddard WA, Stoddart JF. Toward electrochemically controllable tristable three-station [2]catenanes. *Chem Asian J.* 2007; **2**: 76–93.

261. Ikeda T, Stoddart JF. Electrochromic materials using mechanically interlocked molecules. *Sci Technol Adv Mater.* 2008; **9**: 014104.

262. Raehm L, Kern J-M, Sauvage J-P, Hamann C, Palacin S, Bourgoin J-P. Disulfide- and thiol-incorporating copper catenanes: synthesis, deposition onto gold, and surface studies. *Chem Eur J.* 2002; **8**: 2153–2162.

263. Cooke G, Daniels LM, Cazier F, Garety JF, Hewage SG, Parkin A, Rabani G, Rotello VM, Wilson CC, Woisel P. The synthesis of a pyrrole-functionalized cyclobis(paraquatp-phenylene) derivative and its corresponding [2]rotaxane and [2]catenane and their subsequent deposition onto an electrode surface. *Tetrahedron.* 2007; **63**: 11114–11121.

264. Simone DL, Swager TM. A conducting poly(cyclophane) and its poly([2]-catenane). *J Am Chem Soc.* 2000; **122**: 9300–9301.

265. Payer D, Rauschenbach S, Malinowski N, Konuma M, Virojanadara C, Starke U, Dietrich-Buchecker C, Collin J-P, Sauvage J-P, Lin N, Kern K. Toward mechanical switching of surface-adsorbed [2]catenane by in situ copper complexation. *J Am Chem Soc.* 2007; **129**: 15662–15667.

266. Klajn R, Fang L, Coskun A, Olson MA, Wesson PJ, Stoddart JF, Grzybowski BA. Metal nanoparticles functionalized with molecular and supramolecular switches. *J Am Chem Soc.* 2009; **131**: 4233–4235.

267. Davis JJ, Orlowski GA, Rahman H, Beer PD. Mechanically interlocked and switchable molecules at surfaces. *J Chem Soc Chem Commun.* 2010; **46**: 54–63.

268. Geerts Y, Muscat D, Mullen K. Synthesis of oligo[2]catenanes. *Macromol Chem Phys.* 1995; **196**: 3425–3435.

269. Muscat D, Witte A, Kohlel W, Mullen K, Geerts Y. Synthesis of a novel poly[2]-catenane containing rigid catenanes. *Macromol Rapid Commun.* 1997; **18**: 233–241.

270. Muscat D, Kohler W, Rader HJ, Martin K, Mullins S, Muller B, Mullen K, Geerts Y. Synthesis and characterization of poly[2]-catenanes containing rigid catenane segments. *Macromolecules.* 1999; **32**: 1737–1745.

271. Weidmann J-L, Kern J-M, Sauvage J-P, Geerts Y, Muscat D, Millen K. Poly[2]-catenanes containing alternating topological and covalent bonds. *J Chem Soc Chem Commun.* 1996; 1243–1244.

272. Weidmann J-L, Kern J-M, Sauvage J-P, Muscat D, Mullins S, Köhler W, Rosenauer C, Räder HJ, Martin K, Geerts Y. Poly[2]catenanes and cyclic oligo[2]catenanes containing alternating topological and covalent bonds: synthesis and characterization. *Chem Eur J.* 1999; **5**: 1841–1851.

273. Shimada S, Ishikawa K, Tamaoki N. Synthesis and switchable condensation reaction of bifunctional catenane. *Acta Chem Scand.* 1998; **52**: 374–376.

274. Menzer S, White AJP, Williams DJ, Belohradsky M, Hamers C, Raymo FM, Shipway AN, Stoddart JF. Self-assembly of functionalized [2]catenanes bearing a reactive functional group on either one or both macrocyclic components: from monomeric [2]catenanes to polycatenanes. *Macromolecules.* 1998; **31**: 295–307.

275. Olson MA, Braunschweig AB, Fang L, Ikeda T, Klajn R, Trabolsi A, Wesson PJ, Benitez D, Mirkin CA, Grzybowski BA, Stoddart JF. A bistable poly[2]catenane forms nanosuperstructures. *Angew Chem Int Ed.* 2009; **48**: 1792–1797.

276. Bria M, Bigot J, Cooke G, Lyskawa J, Rabani G, Rotello VM, Woisel P. Synthesis of a polypseudorotaxane, polyrotaxane, and polycatenane using 'click' chemistry. *Tetrahedron.* 2009; **65**: 400–407.

277. Olson MA, Coskun A, Fang L, Basuray AN, Stoddart JF. Polycatenation under thermodynamic control. *Angew Chem Int Ed.* 2010; **49**: 3151–3156.

278. Gan Y, Dong D, Hogen-Esch TE. Synthesis and characterization of a catenated polystyrene-poly (2-vinylpyridine) block copolymer. *Macromolecules.* 2002; **35**: 6799–6803.

279. Ohta Y, Kushida Y, Kawaguchi D, Matsushita Y, Takano A. Preparation, characterization, and nanophase-separated structure of catenated polystyrene-polyisoprene. *Macromolecules.* 2008; **41**: 3957–3961.

280. Ishikawa K, Yamamoto T, Asakawa M, Tezuka Y. Effective synthesis of polymer catenanes by cooperative electrostatic/hydrogen-bonding self-assembly and covalent fixation. *Macromolecules.* 2010; **43**: 168–176.

281. Ünsal Ö, Godt A. Synthesis of a [2]catenane with functionalities and 87-membered rings. *Chem Eur J.* 1999; **5**: 1728–1733.

282. Duda S, Godt A. The effect of ring size on catenane synthesis. *Eur J Org Chem.* 2003; 3412–3420.

283. Godt A. Non-rusty [2]catenanes with huge rings and their polymers. *Eur J Org Chem.* 2004; 1639–1654.

284. Zhao Y-L, Liu L, Zhang W, Sue C-H, Li Q, Miljanic OS, Yaghi OM, Stoddart JF. Rigid-strut-containing crown ethers and [2]catenanes for incorporation into metal–organic frameworks. *Chem Eur J.* 2009; **15**: 13356–13380.

285. Li Q, Zhang W, Miljani OS, Knobler CB, Stoddart JF, Yaghi OM. A metal–organic framework replete with ordered donor–acceptor catenanes. *J Chem Soc Chem Commun.* 2010; **46**: 380–382.

286. Hudson B, Vinograd J. Catenated circular DNA molecules in HeLa cell mitochondria. *Nature.* 1967; **216**: 647–652.

287. Clayton DA, Vinograd J. Circular dimer and catenate forms of mitochondrial DNA in human leukaemic leucocytes. *Nature.* 1967; **216**: 652–657.

288. Pikó L, Blair DG, Tyler A, Vinograd J. Cytoplasmic DNA in the unfertilized sea urchin egg: physical properties of circular mitochondrial DNA and the occurrence of catenated forms. *Proc Nat Acad Sci.* 1968; **59**: 838–845.

289. Adams DE, Shekhtman EM, Zechiedrich EL, Schmid MB, Cozzarelli NR. The role of topoisomerase IV in partitioning bacterial replicons and the structure of catenated intermediates in DNA replication. *Cell.* 1992; **71**: 277–288.

290. Liang C, Mislow K. Topological chirality of proteins. *J Am Chem Soc.* 1994; **116**: 3588–3592.

291. Liang C, Mislow K. Topological features of protein structures: knots and links. *J Am Chem Soc.* 1995; **117**: 4201–4213.

292. Yan LZ, Dawson PE. Design and synthesis of a protein catenane. *Angew Chem Int Ed.* 2001; **40**: 3625–3627.

10

Other Supramolecular Systems, Molecular Motors, Machines and Nanotechnological Applications

10.1 Introduction

Within this chapter we will discuss a few other molecular systems that we feel are of interest but that have not been the subjects of enough research to warrant chapters of their own. Following this we will expand on the potential functionalities of the molecular systems discussed within this book and report on the strides that are being made in the fields of molecular machines and motors, along with other nanotechnological applications. Finally, we will conclude with an overview of this work and briefly discuss the visions of the future that macrocyclic chemistry has the potential to make possible.

10.2 Other Molecular Systems

10.2.1 Aromatic and cyclacene compounds

Within this work we have detailed a number of aromatic compounds consisting of linked benzene rings. Aromaticity 'starts' with simple molecules such as benzene and can extend through both flat aromatic structures such as the larger aromatics, including for example naphthalene, pyrene, ovalene and so on, as far as the essentially 'infinite' aromatic systems of graphene and graphite. Aromaticity is not limited to flat systems either, as shown by molecules such as fullerene and carbon nanotubes. We include below two examples of large aromatic systems that we feel exemplify the potential structures available.

One of the largest quasi-rigid macrocyclic aromatic systems that we have found which is still a pure compound rather than an 'infinite' lattice is a cyclotetraicosaphenylene derivative,[1] the structure of which is shown in Figure 10.1a. This material could be crystallised and X-ray studies (Figure 10.1b) showed it to have a 'chair' conformation, with the centre of the ring being filled with hexyl chains and disordered solvent molecules. The molecules stack together so that the holes in the centre of the macrocycle lead to the formation of tubes through the crystal. This material could conceivably act as an organic zeolite-type

Macrocycles: Construction, Chemistry and Nanotechnology Applications, First Edition.   Frank Davis and Séamus Higson.
© 2011 John Wiley & Sons, Ltd. Published 2011 by John Wiley & Sons, Ltd.

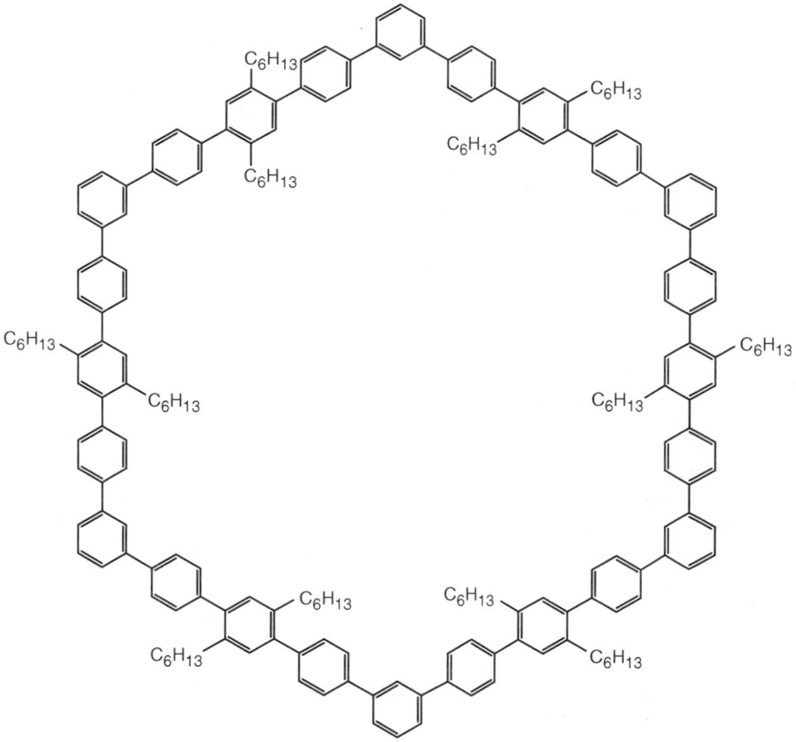

(a)

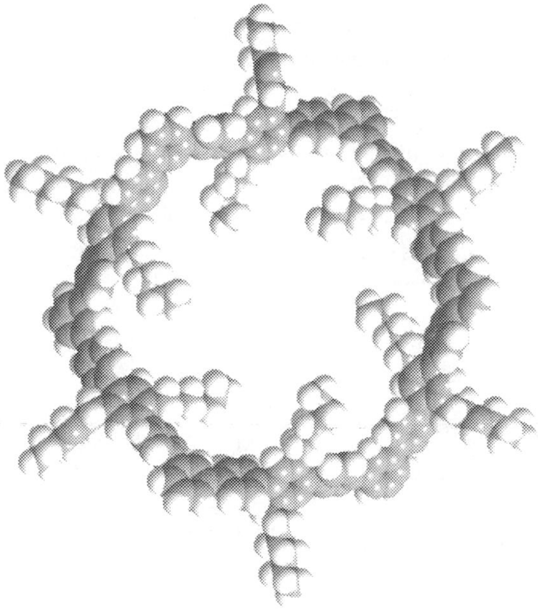

(b)

Figure 10.1   (a) Structure and (b) crystal structure of a cyclotetraicosaphenylene

compound, or perhaps metal ions could be complexed within the tubes and then reduced to the metal, giving rise to the formation of quantum wires.

Other workers have synthesised a series of compounds that they describe as an analogue of 'cubic graphite'; the structure of one of their compounds is shown in Figure 10.2a. This compound could be crystallised and was shown to exist in a form in which each individual hexaphenylbenzene unit exists in a six-bladed 'propeller' formation.[2] These two structures show how organic chemistry is starting to meet the nanoworld.

Carbon nanotubes are exciting much interest at the time of writing. The structure of a nanotube is shown in Figure 10.3a; it can be thought of as a stack of repeating belts made up of benzene rings. There have been a series of attempts to synthesise these carbon belts, known as cyclacenes. One of the earliest involved utilising a number of Diels–Alder reactions to synthesise the compound shown in Figure 10.3b. This compound was first obtained by Stoddart and coworkers and can be thought of as a precursor to [12]cyclacene (Figure 10.3c). However, unfortunately as yet there has been no successful conversion of the precursor into the cyclphane.[3,4] This could well be due to the strain involved in these systems and theoretical studies have indicated that such linearly annulated molecules display small energy gaps between triplet and singlet states, which leads to the expectation that they would be unstable species.[5] It has also been proposed that the 'aromaticity' of a fused or annulated system depends on the number of aromatic sextets that can be drawn within it.[5] Benzene has one aromatic sextet per ring, naphthalene has one sextet in two, tetracene one in four and so on, and an [n]cyclacene has none and so could be thought of rather as a conjugated olefin. So far no cyclacenes consisting solely of benzene rings have been synthesised, but a related [10]cyclophenacene system within a cage molecule (Figure 10.3d) has been synthesised by selective reduction of a fullerene.[6] These compounds form stable yellow crystals and X-ray and other studies have shown them to be aromatic.

Although cyclacenes containing just six-membered rings have not been obtained as yet, other ring systems have been successfully cyclised. A fully conjugated molecular belt consisting of alternating six- and eight-membered rings has been successfully synthesised[7] and its structure shown by X-ray crystallography (Figure 10.4a). The authors use the terminology [6,8]$_3$cyclacene for this compound. Further work on this synthetic scheme and the synthesis of precursors for the [6,8]$_4$cyclacene has also been published.[8] The same group has also managed to synthesise similar systems with alternating four- and eight-membered rings, albeit with stabilisation of the butadiene moieties by complexation with organometallic systems.[9] For example, irradiation of a solution of 5,6,11,12-tetradehydro-dibenzo[a,e]cyclooctatetraene in the presence of a cobalt reagent led to the synthesis of a [4,8]$_3$cyclacene stabilised by cyclopentadienyl cobalt (Figure 10.4c). Similar rhodium species could be obtained. A crystal structure of the cobalt species is shown in Figure 10.4d. Some of the species isolated during this work displayed near-planarity of the cyclooctatetraene rings.

10.2.2   Other cavitands

We have described a wide range of cavitands within this work and wish to add just a few more that we think are worthy of consideration. A cyclic compound containing four triazole units (Figure 10.5a) has been developed utilising 'click' condensation chemistry between alkyne and azide groups.[10] The resultant triazolophane could be synthesised in an overall yield of 27% and displayed a high affinity for chloride ions. The association constant was approximately 130 000 M^{-1} and NMR studies indicated that the ion resides in the central cavity and shows hydrogen bonding between the eight internal hydrogens and the ion. Substitution of the phenyl rings allowed increases of binding coefficient up to at best 11 000 000 M^{-1} for chloride, 7 500 000 M^{-1} for bromide and 280 000 M^{-1} for fluoride.[11] Iodide bound much less effectively (19 000 M^{-1} at best); NMR results indicated that hydrogen bonding by the triazoles was the major factor

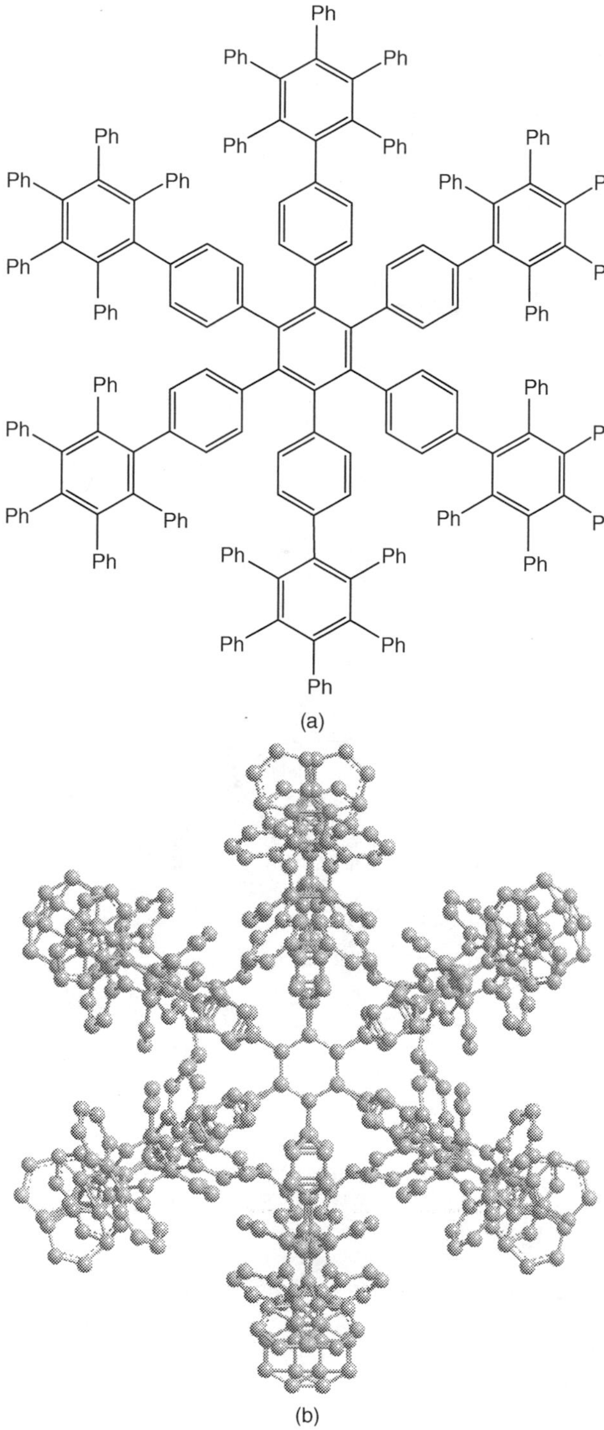

Figure 10.2 *(a) Structure and (b) crystal structure of a polyphenylene dendrimer*

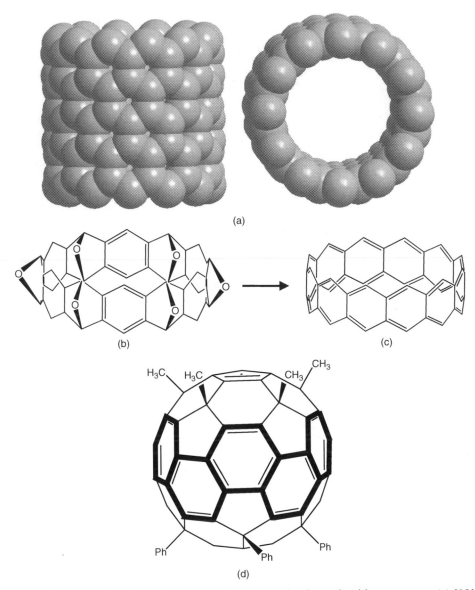

Figure 10.3 *(a) Side-on and end-on views of a carbon nanotube (b) Diels-Alder precursor, (c) [12]cyclacene, (d) [10]cyclophenacene derivative from C_{60} (phenacene system shown in bold) (Reprinted with permission from[6]. Copyright 2003 American Chemical Society)*

in determining binding. Replacement of two of the phenylene units with pyridine rings gave macrocycles which bound iodide by far the most strongly (binding coefficient of 8.6×10^{10} M^{-2}) but exclusively as a 2:1 sandwich complex.[12] Recent work has described the incorporation of these materials into PVC membranes for use within potentiometric sensors.[13]

Two isomeric Schiff-base macrocycles have been synthesised and shown by molecular modeling to have large internal cavities (1.00 × 1.05 nm for Figure 10.5b and 1.23 × 0.93 nm for Figure 10.5c), much wider than those of most of the commonly used macrocyclic hosts.[14] These were shown by NMR studies to adopt

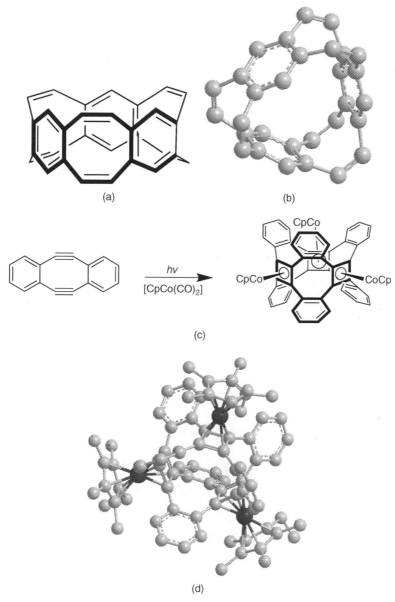

Figure 10.4 *(a) [6,8]3cyclacene (b) crystal structure (c) synthesis of [4,8]3cyclacene (Copyright Wiley-VCH Verlag GmbH & Co. KGaA. Reproduced with permission[9]) (d) crystal structure*

a cone-like conformation and to strongly bind alkyl pyridinium salts. A similar compound with pyrogallol units replacing the resorcinol units was found to be a good host for the classic catenane-forming macrocycle, cyclobis(paraquat-p-phenylene). Another interesting species is the so-called 'ouroborand', named after the legendary snake Ourobouros, which eats its own tail.[15] This is based on a resorcinarene macrocycle substituted with a side chain that binds within the cavity of the same molecule (Figure 10.6), thereby rendering the cavity inaccessible for binding of other guests. However, on binding of metal ions such as

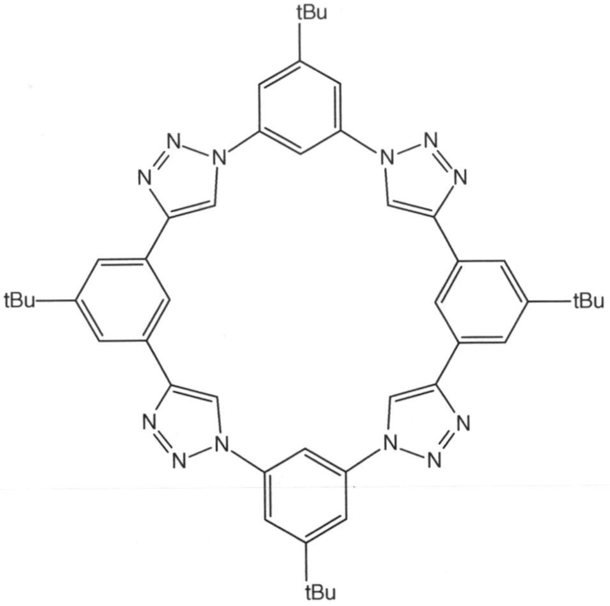

(a)

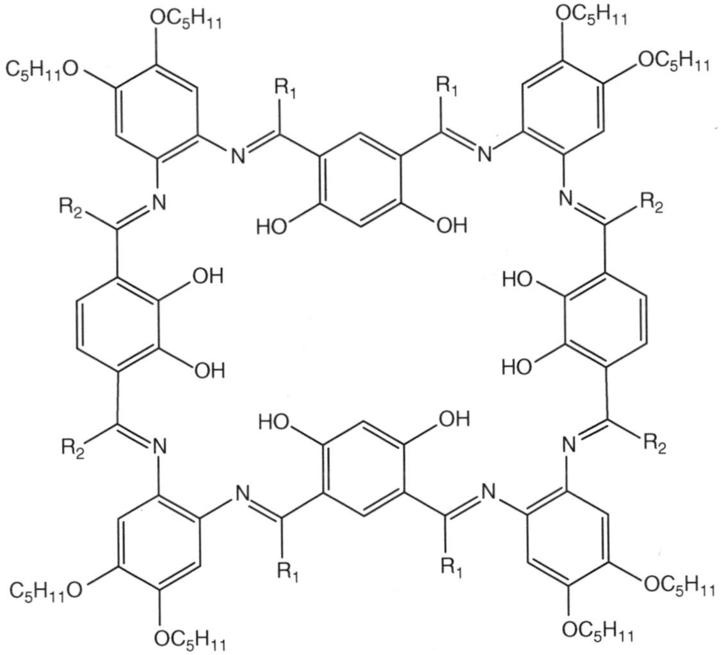

(b) $R_1 = H, R_2 = Ph$

(c) $R_1 = Ph, R_2 = H$

Figure 10.5   *(a) Structure of a triazolophane (b,c) cyclic Schiff bases*

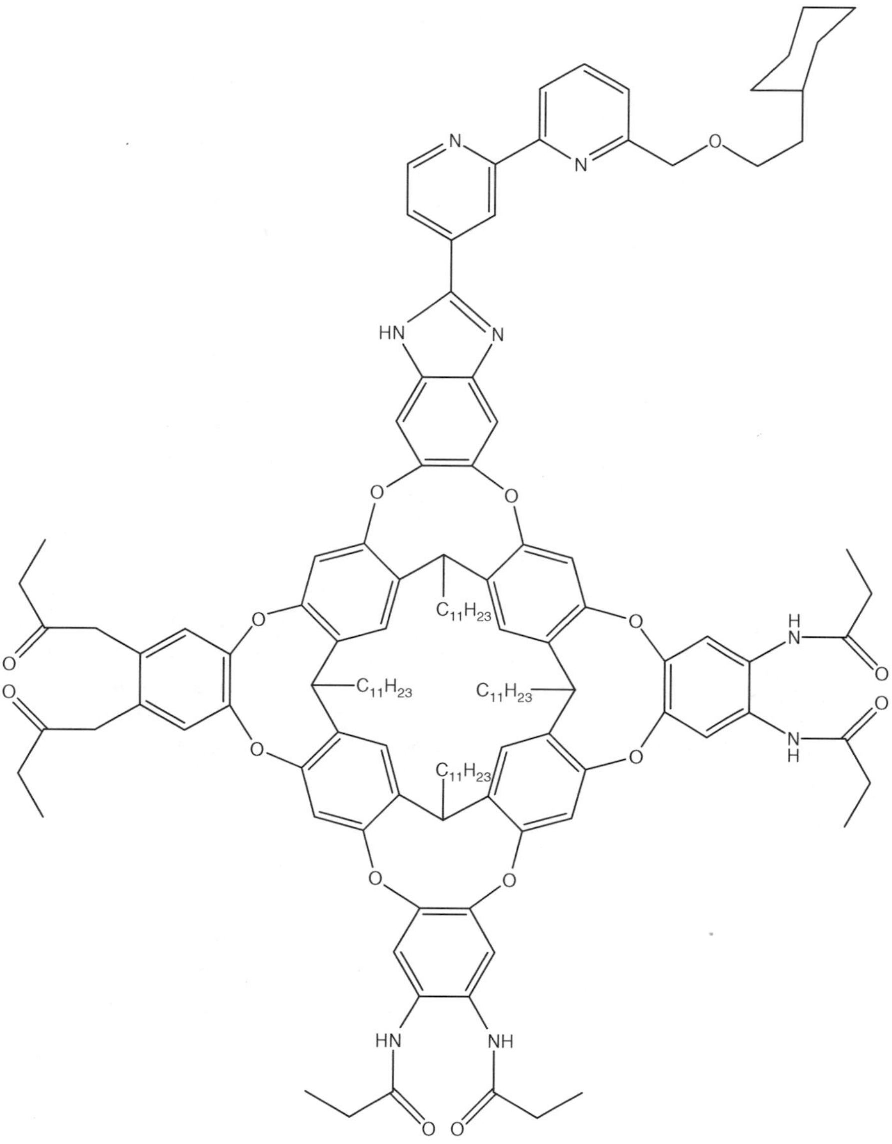

Figure 10.6 *An "Ouroborand"*

zinc to the bipyridyl unit, this unit switches from the *anti* to the *syn* form, thereby forcing the tail away from the cavity and rendering the molecule capable of binding guests such as adamantanes. Removal of the zinc leads to the 'tail' displacing the guest from the cavity.

10.2.3 Pretzelanes, bonnanes and [1]rotaxanes

We have already mentioned pretzelanes, named because of their shape, in Chapter 9; they are catenanes in which the two interlinked rings are then also covalently joined together by a linked group. Several varieties of these systems have been studied, one of the earliest being the amide-type catenanes, in which one amide on each ring is replaced by a sulfonamide group (Figure 9.86). The sulfonamide is much more acidic than an amide and can therefore be deprotonated by a strong base and substituted with alkyl halides.[16] If dibromoalkanes are used they will bridge between the two catenane rings; the yield has been shown to decrease with shorter bridge lengths, the minimum bridge that gives reasonable yields being one of six methylene groups. Rotaxanes with sulfonamide-bearing axles and rings have also been synthesised and then converted into a so-called [1]rotaxane, in which there is a bridge between axle and ring.[16] In this case bridges as short as three methylene groups can be incorporated. Both pretzelanes and [1]rotaxanes are chiral and capable of being separated by chromatography into their enantiomers. The same group synthesised an amide catenane which bore a hydroxyl group on one isophthaloyl group of each ring.[17] These hydroxyls could be reacted with xylylene dibromide to bridge the two rings, meaning again that they are linked by both a mechanical and a covalent bond. The same group has recently successfully synthesised amide-based 3- and 4-rotaxanes, again containing sulfonamide groups, and successfully linked both the rings of a 3-rotaxane.[18] Bridging groups could be either decamethylene, *bis*-(4-bromomethylphenyl)methane or an oligoethylene glycol chain, and good yields (65–90%) were obtained. These systems were conformationally chiral and could be separated by chromatographic methods. The 4-rotaxane exists in two chiral and two *meso* forms. The name 'bonnanes' has been proposed for these linked rotaxanes.

Pretzelanes can also be synthesised based on $\pi-\pi$ interactions.[19] One interesting system is shown in Figure 10.7a, where a substituted crown ether is 'clipped' to give the pretzelane shown. Two enantiomers exist for pretzelanes of this type, but in the second compound shown (Figure 10.7b) there is also an additional chiral centre located between the two ester groups.[19] Since we have a chiral centre in an inherently chiral pretzelane, two diastereoisomers must be formed. NMR studies have shown that this is the case; two diastereoisomers were detected in a 9:1 ratio in acetonitrile and a 6:1 ratio in DMSO. In both cases the M-conformer was preferred (Figure 10.7c). Molecular modelling studies were carried out to ascertain the absolute configurations of the diastereoisomers. The same reaction scheme was used to assemble a pretzelane containing a tetrathiafulvalene group.[20] This system could be switched electrochemically between two states where the cationic macrocycle binds either to the tetrathiafulvalene or the 1,5-dioxanaphthalene units. The behaviour of the pretzelane and a similar catenane were compared and interestingly on switching the catenane, the cationic macrocycle can move in either direction around the ring. However, in the case of the pretzelane, movement can only occur in a preferred direction; that is, back and forth rather than the random circling of the bipyridyl ring around the crown ether that occurs on switching of the catenane.

We have discussed at length the behaviour of catenanes and rotaxanes and it was inevitable that researchers would attempt to combine both systems. One example of this is a 'rotacatenane' synthesised by Stoddert's group which combines both of these systems.[21] We have already shown how a biphenyl containing cationic macrocycle can be utilised to synthesised 3-catenates such as those shown in Figure 9.67. In this case however the initial step was a clipping reaction to give the 2-catenane containing one cationic macrocycle and one cyclic crown ether, followed by interaction with a polyether derivative of 1,5-dioxanaphthalene which threaded the cationic macrocycle. Capping of this system with silyl groups gave the rotacatenane shown in Figure 10.7d.

10.2.4   Trefoil knots

Another series of interlocked molecular systems that have provoked great interest and research effort involves the chains of a macrocycle being knotted together, sometimes known as 'knotanes'. One of the most widely researched has been the synthesis of a trefoil knot, shown schematically in Figure 10.8a. Different techniques have been used to synthesise these systems. The Schill group attempted to use the directed synthesis method, which had previously been used to synthesise rotaxanes and catenanes, to

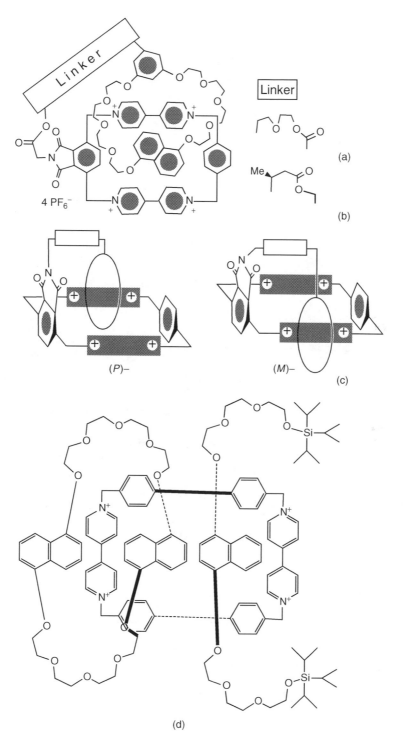

Figure 10.7   *(a) Chiral pretzelane based on π–π interactions (b) with chiral linker (c) P- and M- conformers (Reproduced by permission of the Royal Society of Chemistry[19]) (d) a rotacatenane*

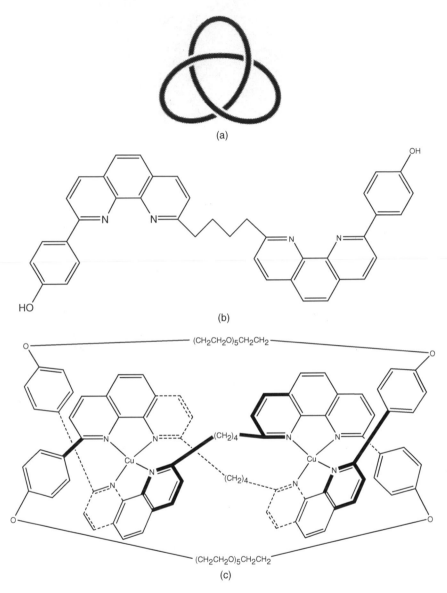

Figure 10.8   *(a) A trefoil knot (b) a bis-phenanthroline ligand (c) a copper templated molecular trefoil knot*

construct molecular knots. Although a strategy was devised, the necessity for a multistep synthesis combined with the low yields of some of the reactions meant that not enough of the final product could be synthesised and purified for a successful proof of structure.[22]

The template-based approach of the Sauvage group proved much more successful and in 1989 they reported the synthesis of a trefoil knot.[23] They initially synthesised a *bis*-phenanthroline ligand (Figure 10.8b) and then complexed it with copper ions to form a helical structure. Linking together the terminal hydroxyl groups with oligoethylene glycol chains gave the knotane shown in Figure 10.8c.

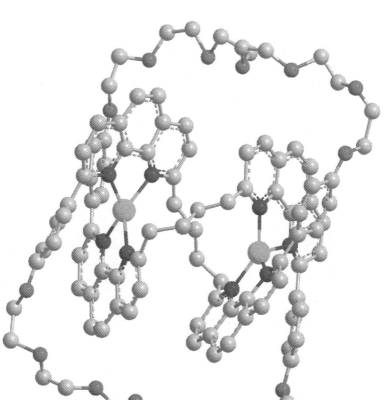

Figure 10.9   *(a) Crystal structure of a copper templated molecular trefoil knot*

Removal of the copper ions to give the free knotane was possible and NMR spectroscopy with chiral chemical-shift reagents confirmed the chirality of the knotane. The knotane was also successfully crystallised and Figure 10.9 shows the crystal structure of the *bis*-copper(I) complex.[24] Although the compound is chiral and synthesised as a racemate, each individual crystal is enantiomerically pure, as is found for most of this series of compounds. A series of knotanes based on this synthesis with different linkers and oligoethylene chain lengths was synthesised and yields of up to 24% were obtained.[25] Later work utilised a phenylene linker between the two phenanthroline units and substituted the hydroxyl groups with oligoethylene linkers terminated with allyl groups.[26] This allowed ring closure by a metathesis reaction and gave knotane yields of up to 74%. More complex systems such as the double knotane compound shown in Figure 10.10 can exist in three forms, since each individual knot is chiral: two enantiomers plus a *meso* form.[27]

Studies were carried out on the demetallation reactions of these knots.[28] It is of interest that the spacer unit between the two phenanthroline units has a major effect on the disassociation kinetics of these systems. In the case of a more flexible oligomethylene spacer, the first copper ion is released very slowly and the second much faster, whereas for a more rigid *m*-phenylene spacer the reverse is true. NMR studies indicate that this is due to structural rearrangements, which either facilitate or hinder the access of

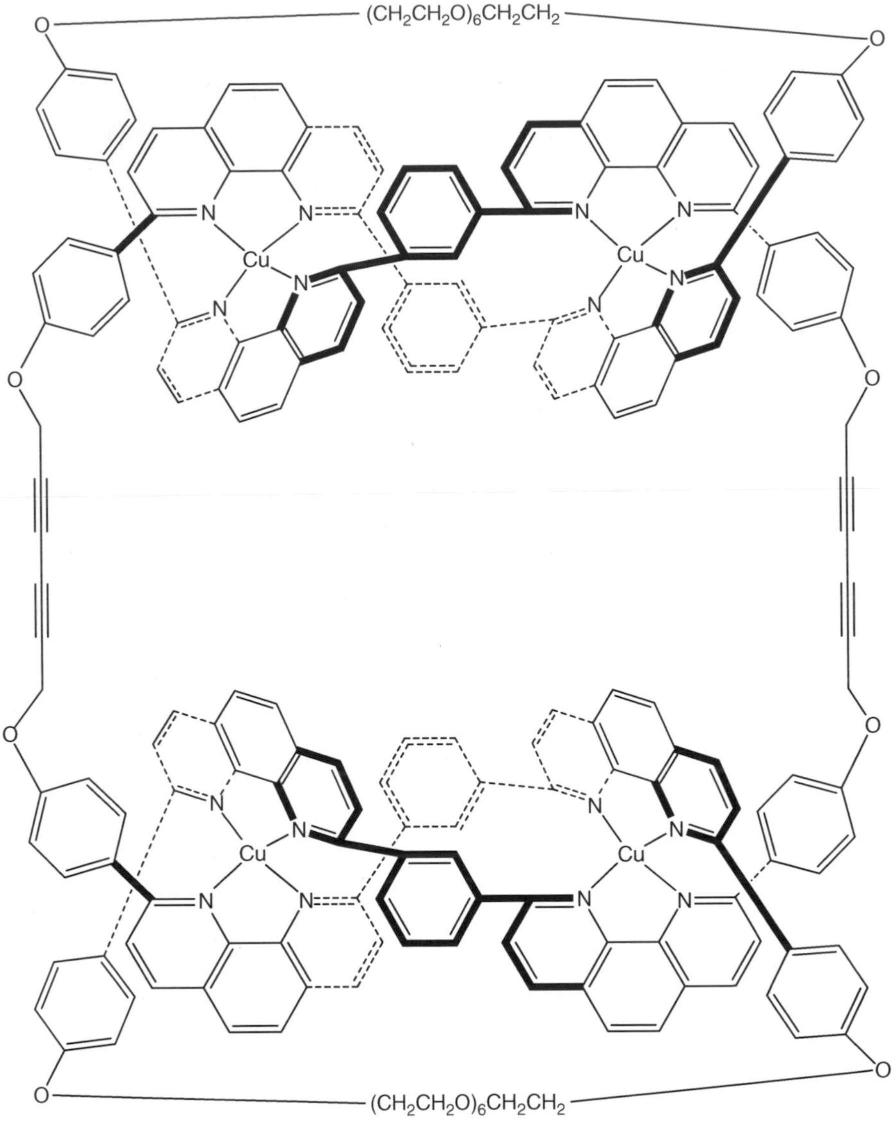

Figure 10.10 *Structure of a copper templated double molecular trefoil knot*

the demetallating agent (cyanide) to the metal ions. NMR studies have shown that the free knotanes are much more conformationally mobile in solution than their copper complexes. Once demetallated, these complexes can also be remetallated. The ability to remove the first copper ion much faster than the second also allows the preparation of heteronuclear knotane complexes with two different metal ions.[29]

Other metal ions have been utilised to template knot formation, for example in the work of Adams *et a.*,[30] where a bipyridine-containing oligomer (Figure 10.11a) was complexed with zinc ions. X-ray crystallography confirmed the formation of a knotted structure templated around a central zinc ion (Figure 10.11b). This is not a true knot since removal of the template ion will cause the knot to unravel, but later work substituted

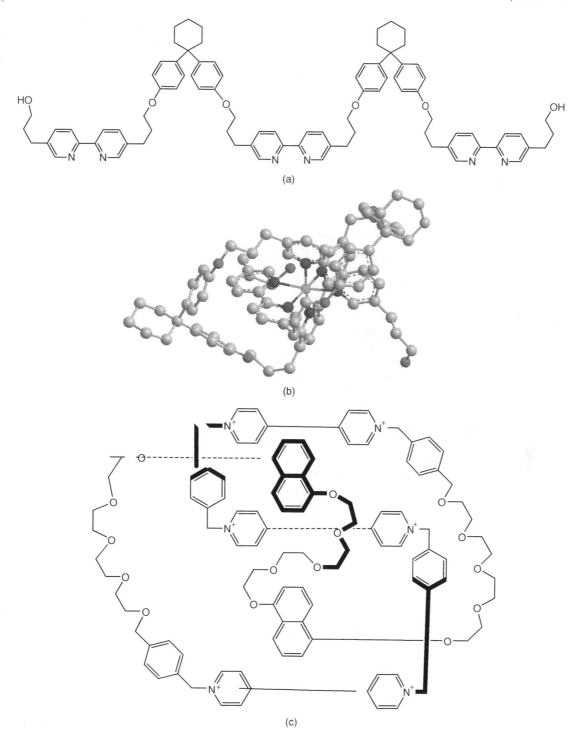

Figure 10.11 *(a) Structure of a bipyridinyl compound that forms a molecular trefoil knot (b) crystal structure (c) π–π based knot*

the end groups with alkenes, which were then subjected to a metathesis reaction to close the knot.[31] The zinc ion could then be removed to leave the free knotane. Knots have also been synthesised by Stoddart's group utilising the $\pi-\pi$ interactions between electron-rich and -poor aromatic systems which they applied so successfully to catenane and rotaxane synthesis.[32] A complex was synthesised between two acyclic compounds containing either electron-poor bipyridyl groups or electron-rich 1,5-dioxanaphthalene groups, which adopted a helical structure. Ring-closure reactions gave the knotted compound (Figure 10.11c), albeit in low yield.

We have already discussed the extensive range of threaded and interlinked structures formed using cyclic amides, and these hydrogen-bonded systems have also proved successful in the synthesis of molecular knots. The synthesis and behaviour of these materials have been reviewed by Lukin and Vogtle,[33] and we will present a few examples of these systems. A large series of amide molecular knots with the general structure shown in Figure 10.12 have been synthesised by condensation of a 2,6-pyridine dicarboxylic acid dichloride with a diamino compound. The earliest example of this reaction (Figure 10.12) gave up to 20% yield of the 3 + 3 condensation product, along with 1 + 1, 2 + 2 and 4 + 4 condensation products.[34] However, when the 3 + 3 product was isolated and crystallised, X-ray studies showed it adopted a knotted conformation, as shown in Figure 10.13a (the structure shown is for derivative 7, Figure 10.12). Again, this is proposed to be caused by formation of hydrogen bonds during the condensation reaction, leading to folding of the growing oligoamide and acting as a template for knot formation. Later work varied the composition of the knot by varying both the diamino compound and the dicarbonyl chloride to give a number of different chiral knots in varying yields.[35] Separation of the chiral knots into their enantiomers was achieved using chiral chromatography.

The possibility of incorporating functional groups in these knotanes is also important for assembling them into larger systems. A number of synthetic schemes were utilised to render these knots amenable to further reactions. It was found from X-ray crystal studies that the pyridinyl units are located at the 'outside' of the knot, whereas the isophthaloyl groups are buried within the system, this being reflected by the fact that substituting the isophthaloyl groups can greatly diminish knotane yields, often to zero. Larger substituents can be incorporated into the 4-positions of the pyridinyl units and knots containing benzyl ethers and larger dendrimers have been successfully synthesised by direct synthesis using substituted pyridinyl units.[36] However, this technique often gives low yields and does not work for large dendrimers, and therefore an indirect method in which a knot containing three pyridinyl units bearing benzyl ether substituents in the 4-positions was synthesised, with the ethers being removed by catalytic hydrogenation. This scheme was partially successful, giving a mixture of knots that contained one, two or three hydroxyl groups. These were then substituted with dendrimers and separated by chromatography. The purified racemates could be separated into their enantiomers by chiral chromatography.[36]

A better synthetic scheme involved the use of 4-allyloxy protecting groups, which could be converted to the hydroxyl compound by the use of tributyl tin hydride and a palladium catalyst.[37] The *tris*-allyloxy knot can be isolated in 8% yield from a one-pot reaction and then reduced to give either the *mono*- or *tris*-hydroxy knot by controlling the amount of reducing agent. The *bis*-hydroxy unit can also be obtained as a mixture, with the *mono*-hydroxy compound and its derivatives separated. The hydroxyl groups could be phosphorylated with diethylchlorophosphate or sulfonated with *p*-toluene sulfonyl chloride to give knotanes with one, two or three types of substituent. NMR studies showed that in DMSO the knots existed in a relatively rigid, nonsymmetrical conformation, whereas in other solvents a more flexible behaviour was observed, which was slow on the NMR timescale, giving rise to broad signals. Heating the DMSO solutions to 80 °C led to signal coalescence, indicating higher flexibility at increased temperatures. Using mixed solvents could help 'freeze' the molecular conformations, with for example just 10% of DMSO in chloroform inducing formation of the rigid structure. It is thought that this could be due to inclusion of DMSO molecules within the 'loops' of the knotane, stabilising the rigid structure by formation of hydrogen

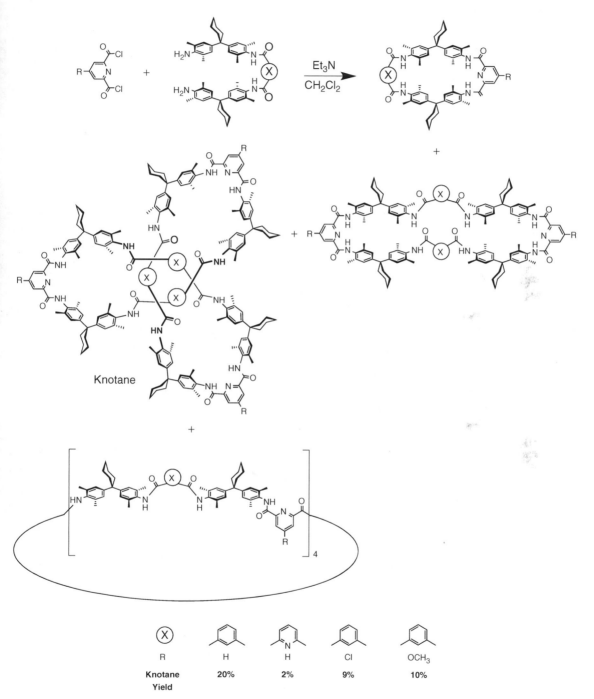

Figure 10.12   *Synthetic scheme and yields of some amide molecular knots (Copyright Wiley-VCH Verlag GmbH & Co. KGaA. Reproduced with permission[33])*

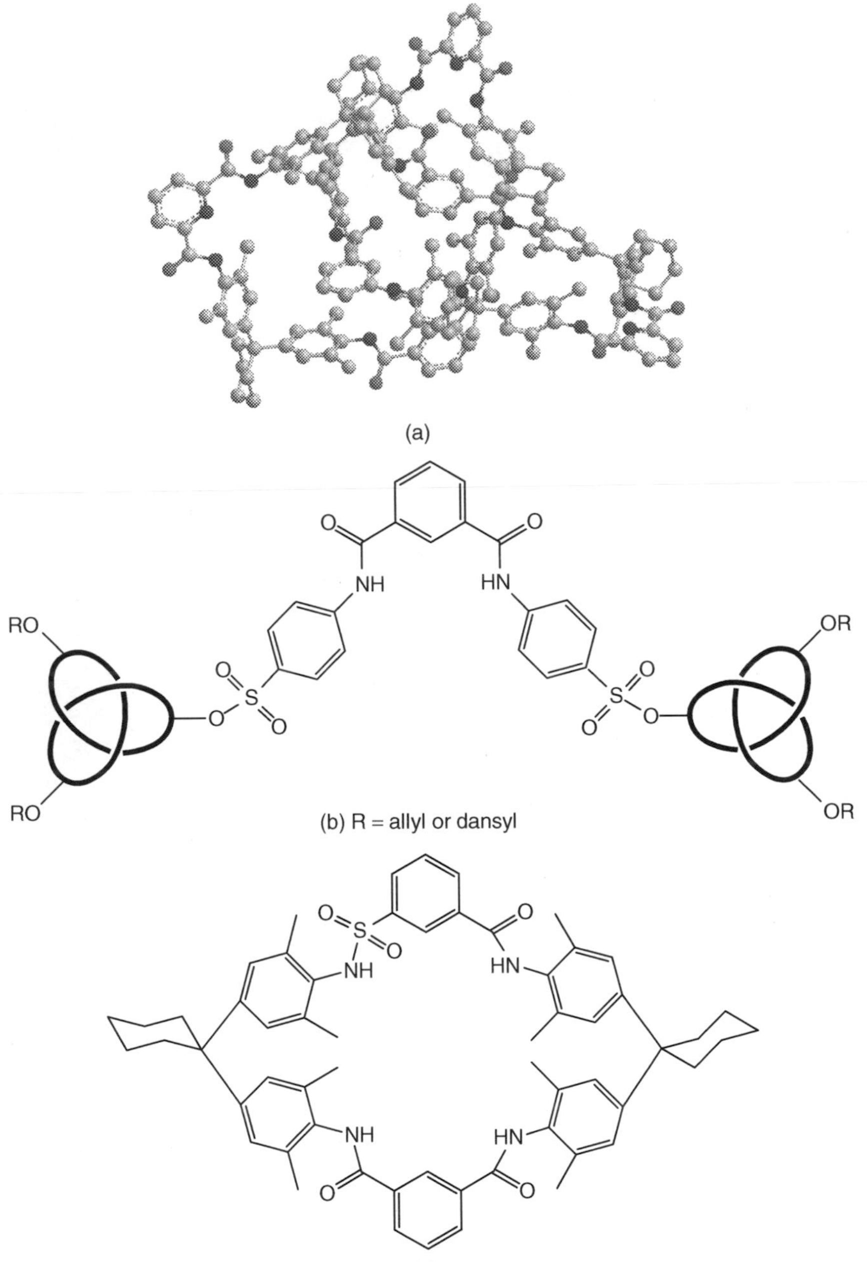

(a)

(b) R = allyl or dansyl

(c)

Figure 10.13 *(a) Crystal structure of an amide knot (b) A bis-knotane (c) macrocyclic wheel of knotaxane*

bonds with the amides. This has been borne out by molecular modelling, which shows the perfect fit of two DMSO molecules inside the knot.[33]

The *mono-* and *tris*-hydroxy knots could also be reacted with (1S)-(+)camphor-10-sulfonyl chloride, which since both the knot and the camphor group are chiral, gave rise to the formation of diastereoisomers.[38] These could then be separated using standard silica gel chromatographic techniques without requiring chiral chromatography. When a *mono*-hydroxy knot was reacted with biphenyl-4,4′-disulfonyl chloride, a 'molecular dumbbell' was the result, containing two knots linked by a biphenyl-4,4′-disulfonyl unit.[39] This existed in one *meso* and two enantiomeric forms, all of which could be separated by chiral chromatography. A longer axle could also be synthesised and used to form a *bis*-knotane, as shown in Figure 10.13b. If this reaction was carried out in the presence of the macrocycle however (Figure 10.13c), both the *bis*-knotane and the 'knotaxane' could be obtained in 55% and 19% yield, respectively (for the allyl-substituted knotane). The knotaxane consisted of the *bis*-knotane as the axle threaded through the cavity of the macrocycle.[40] The shorter isophthaloyl chloride linker was also utilised, but in this case no knotaxane was formed, probably due to steric effects. The knotaxane could be resolved somewhat into its enantiomers by chiral HPLC, but it was found to be quite insoluble in organic solvents and so a *bis*-dansyl derivative was used instead, which could be resolved much more easily. One topic of interest is whether, if enantiomerically pure knotaxanes were obtained, the revolution of the ring around the axle would have a preferred direction.

It was found that larger assemblies of knotanes could be obtained and that, for example, the dumbbell-shaped compound[39] mentioned above could be deprotected to remove just one allyl group. Two equivalents of this compound could then be reacted with biphenyl-4,4′-disulfonyl chloride to give a linear tetramer[41] (Figure 10.14a). Alternatively, a *tris*-allyloxy knot could have just one of its allyl groups removed, which could be reacted with excess biphenyl-4,4′-disulfonyl chloride. This substituted knot could then be reacted with a completely deprotected knot with three hydroxy groups to give the star-like tetramer shown in Figure 10.14b.[41] Cyclic oligomers could be synthesised by taking the *tris*-hydroxy knot, protecting one of the groups with a tosyl group and then reacting with biphenyl-4,4′-disulfonyl chloride to give the 4 + 4 addition product (Figure 10.14c) along with the 2 + 2 and 3 + 3 cyclic addition products in approximately equal amounts.[41] MALDI and HPLC proved the composition of these macrocycles and showed no higher condensation products were formed. Unfortunately, the complexity of these systems meant that in most cases complete enantiomeric resolution could not be attained.

In an attempt to further understand the knotting procedure, a series of oligoamides corresponding to potential intermediates in the formation of a trefoil knot was synthesised.[42] Studies indicated that as an oligoamide becomes longer it adopts a twisted conformation in solution. There was an attempt to confirm this by reacting the amide with bulky stoppers that would give an 'open knot' but would hinder dethreading; however, it is thought that the stoppers used were not bulky enough and an equilibrium mixture of knotted and linear polyamide resulted, which could not be separated. In other work the same group synthesised knotanes in which the 4-positions of the pyridinyl groups were substituted with long alkyl ether or oligoethylene glycol chains (Figure 10.15). These compounds displayed enhanced solubilities and could easily be separated into their enantiomers.[43] The presence of long alkyl side chains allowed the knotanes of this type to be deposited as Langmuir–Blodgett films. Good isotherms were obtained for butyloxy, hexyloxy, decyloxy and *p*-bromobenzyloxy substituents (Figure 10.15a–d). These were shown to have a thickness of 1.6 nm, which is approximately what would be expected for monolayers. They could be transferred onto mica substrates and were shown by AFM to form homogenous films. Octenyloxy-substituted knots (Figure 10.15g) were also synthesised and were shown to undergo metathesis reactions with styrene, or with each other to give triple-bridged double knots. The *p*-bromobenzyloxy-substituted knot underwent Suzuki coupling to give *tris*-pyrene-substituted knots (Figure 10.15h).

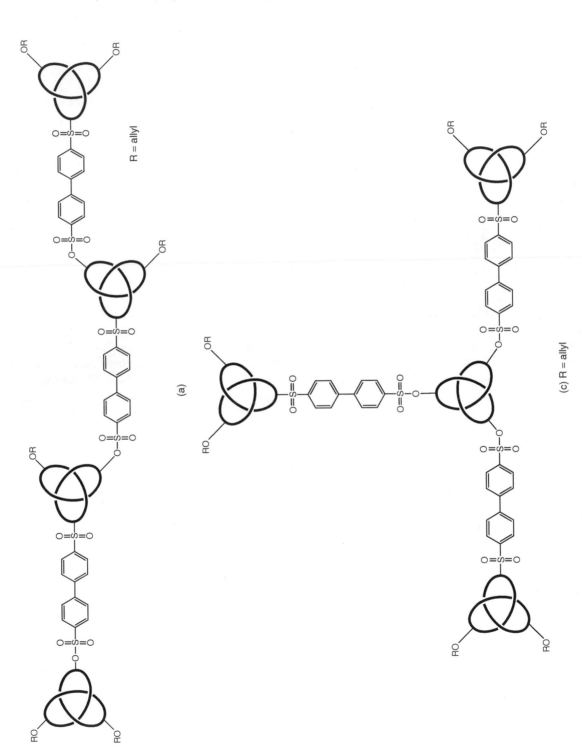

Figure 10.14   *Structures of three tetra-knotanes*

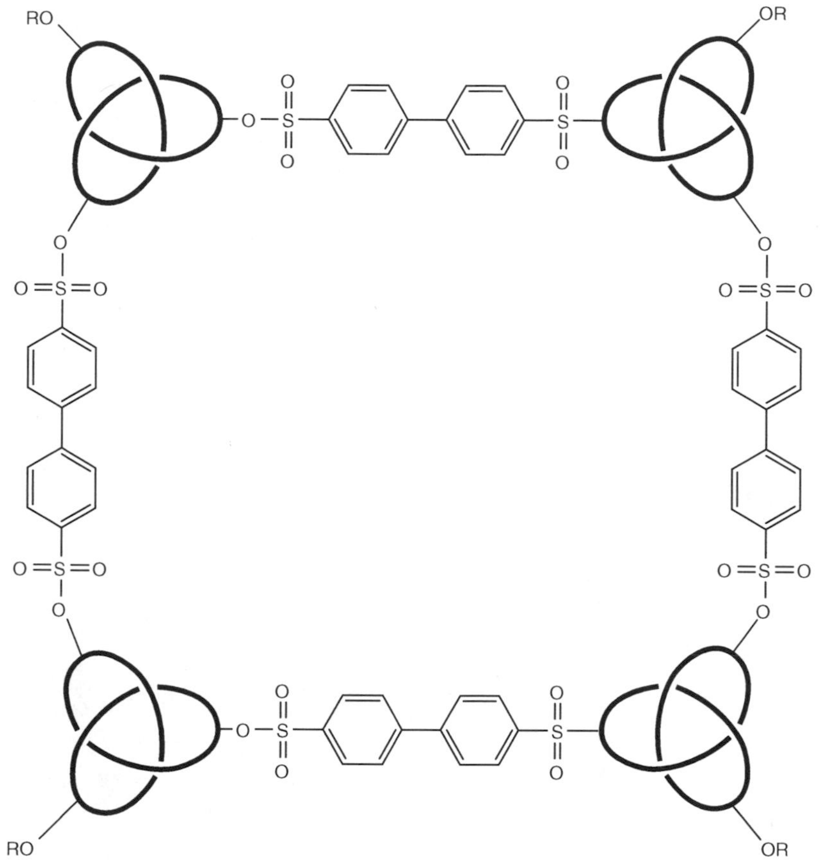

(b) R = tosyl

Figure 10.14 *(continued)*

Amide knots with mixtures of allyloxy, dansyl and pyrene sulfonyl groups synthesised from the readily available *tris*-allyloxyknotane have also had their fluorescence behaviour studied.[44] Fluorescence quantum yields and lifetimes were partially quenched by the knotane. In the compound containing one pyrene sulfonyl and two dansyl groups (Figure 10.15i), the pyrene fluorescence was quenched and the dansyl fluorescence sensitised, demonstrating energy transfer from pyrene to dansyl units. Protonation of the dansyl group led to an enhancement of the pyrene fluorescence.

Other workers have synthesised cyclooligoamides consisting of valine and aminodeoxycholanic acid (Figure 10.16a) and have shown that they adopt knotted structures.[45] When the hexapeptide was made and cyclised, two isomers were found, one being the simple hexapeptide macrocycle and one a trefoil knot containing twelve peptide units. The knot structure was proved by X-ray crystallography (Figure 10.16b). There is a possibility of diastereoisomers forming, since both peptides used are chiral, but only one diastereoisomer was isolated, indicating that the chirality of the amino acids determines the chirality of the knot. The knots of this type are bowl-shaped and have a nonpolar outer and a polar inner surface.

One of the most direct visualisations of knot formation comes from a recent paper in which workers took ABC triblock copolymers where the central block is a chloroethyl vinyl ether polymer and the end

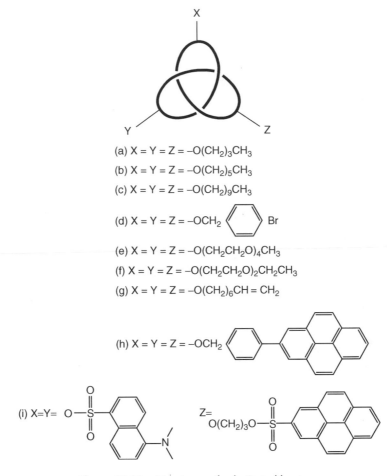

(a) X = Y = Z = –O(CH$_2$)$_3$CH$_3$

(b) X = Y = Z = –O(CH$_2$)$_5$CH$_3$

(c) X = Y = Z = –O(CH$_2$)$_9$CH$_3$

(d) X = Y = Z = –OCH$_2$ Br

(e) X = Y = Z = –O(CH$_2$CH$_2$O)$_4$CH$_3$

(f) X = Y = Z = –O(CH$_2$CH$_2$O)$_2$CH$_2$CH$_3$

(g) X = Y = Z = –O(CH$_2$)$_6$CH = CH$_2$

(h) X = Y = Z = –OCH$_2$

(i) X=Y= O–S

Z= O(CH$_2$)$_3$O–S

Figure 10.15   *Structures of substituted knotanes*

A and C blocks are polystyrene.[46] These polymers could be cyclised to form macrocyclic materials and when cast onto pyrolytic graphite could then be imaged by AFM. The results are some stunning pictures of interlinked rings. Trefoil knots (Figure 10.17a) can be visualised along with simple macrocycles, catenanes and figure-of-eight polymers.

We have already mentioned how natural DNA can adopt catenated and knotted structures, as can several peptides, which can adopt a knotted conformation via the formation of cystine bonds.[47] Other workers have also taken specific synthetic single-stranded DNA and ligated it together to form knotted structures using T4 DNA ligase.[48] Linear, circular and knotted DNA could all be obtained and separated. This same group undertook much research in the field of DNA topology and much of it is reviewed by Seeman.[49,50] Shapes that have been synthesised include cubic multi-catenated structures, Borromean rings and truncated octahedrons.

The field of molecular knots has led to some of the most elegant syntheses of macrocyclic compounds to date. Much of the work on knots up to 2005 is reviewed by Vogtle and Lukin,[51] and for more detail, especially of the history and mathematics of knots, the reader is referred to the Bibliography.

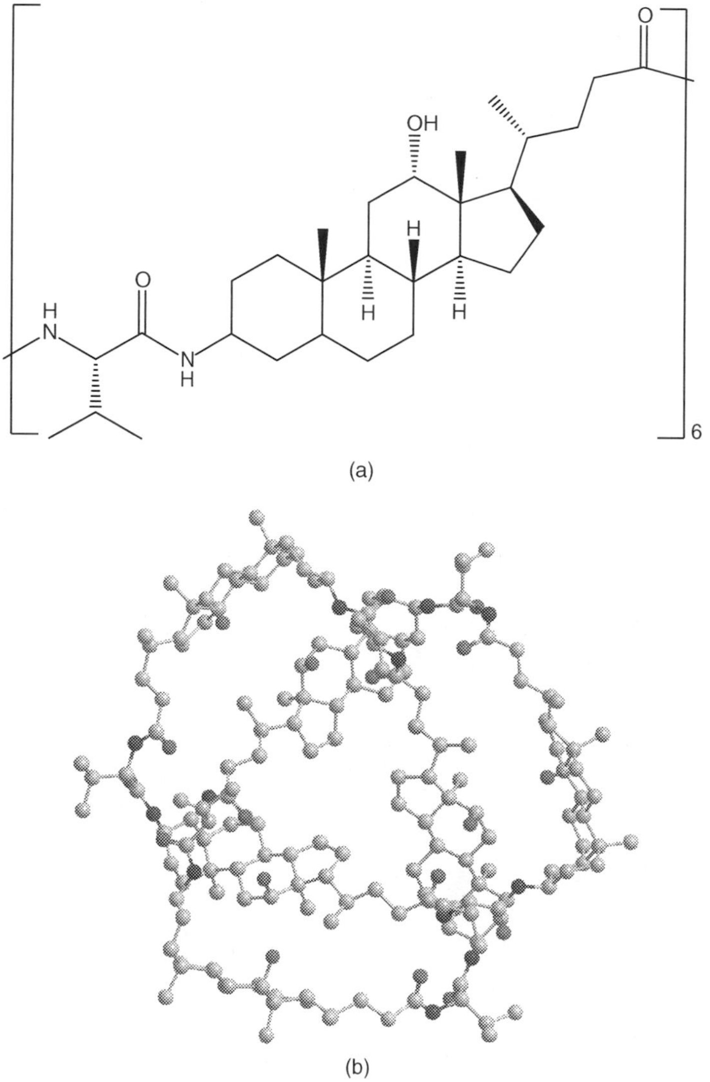

(a)

(b)

Figure 10.16 *(a) Schematic and (b) crystal structure of a steroid amino acid knotane*

10.2.5 Solomon links and Borromean rings

We have already described in Chapter 9 the synthesis and structure of a doubly interlinked catenane.[52] These types of structure can also be thought of as Solomon links (also known as Solomon's knots), doubly interlinked systems shown schematically in Figure 10.18. One of the earliest systems of this type was synthesised by the reaction of gold acetylides with a diphosphane ligand, $Ph_2P(CH_2)_4PPh_2$, to give a variety of compounds including macrocyclic rings, catenanes and a doubly braided catenane or Solomon link – depending on the structure of the gold acetylide.[53] A schematic of the double-braid catenane is shown in Figure 10.18b. This compound could be crystallised and its structure proved by X-ray studies. The crystal structure demonstrates the high amount of binding within this system, both Au-Au and $\pi-\pi$ interactions

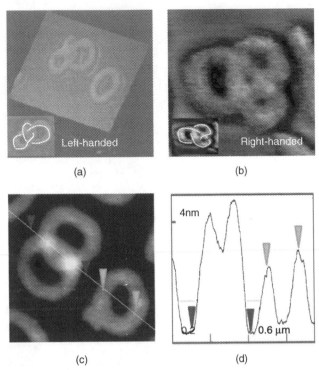

Figure 10.17 *AFM images of (a, b) enantiomers of polymeric trefoil knots, (c) a catenane structure (d) depth profiling AFM showing increased thickness of catenane compared to a simple ring (Copyright Wiley-VCH Verlag GmbH & Co. KGaA. Reproduced with permission[46])*

being present. NMR studies indicate the rings rock with respect to each other at room temperature, although actual rotation of one ring through the other is probably hindered by the bulky diphenyl phosphine groups.

A Solomon link has also been synthesised using the attractions between electron-rich and -poor aromatic rings, as described at length in Chapter 9 for catenanes and rotaxanes. The synthetic scheme used in Figure 10.19 was developed for the combination of bipyridyl or diazapyrene moieties with a crown ether and a palladium or platinum complex.[54] An intense red colouration demonstrated the presence of electron-transfer interactions, and mass spectra proved the formation of a $2:2:1$ aza compound–metal complex–crown ether species. Proof of the Solomon-link structure with the metallocycle doubly encircled by the crown ether could be obtained from NMR studies and X-ray crystal structures (shown for the compound 5c in Figure 10.19b). In the bipyridine-based system there was fast rotation (on the NMR timescale) of the metallocycle at room temperature, but lower temperatures (188 K) gave a frozen structure. The diazapyrene-based system was found to be much less conformationally mobile.

An even more complex system is that of the Borromean rings. These are shown schematically in Figure 10.20a. An elegant templating procedure for the formation of these molecules was first reported in 2004, when the combination of zinc ions with pyridine-containing species gave the Borromean ring.[55] This compound was synthesised by the reaction of 2,6-diformylpyridine with a diamine containing a 2, 2′-bipyridine group, catalysed by trifluoroacetic acid. In the absence of any template, a complex mixture of macrocyclic and polymeric structures would be the likely outcome. Extensive molecular modelling was undertaken and these structures were chosen to be optimal. Zinc ions were chosen as the template since they bind to pyridyl and bipyridyl species- and are relatively kinetically labile. Therefore zinc acetate was

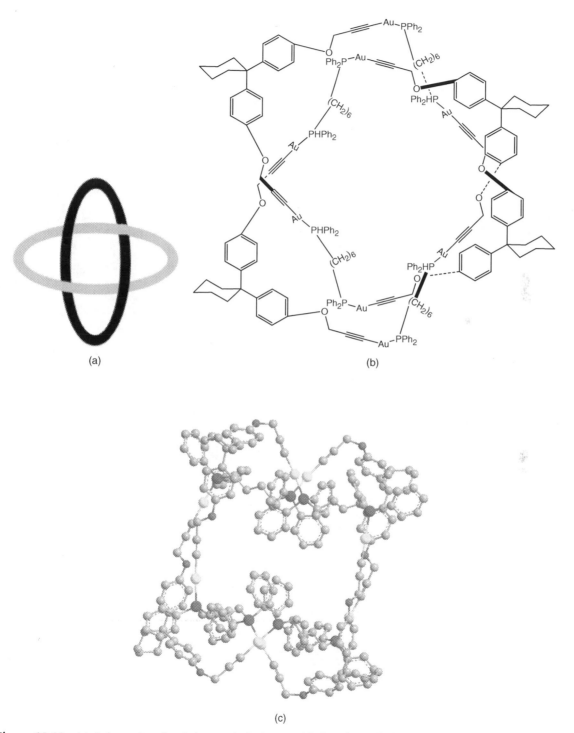

Figure 10.18   (a) Schematic of a Solomon link (b) a gold-phosphine double braided catenane and (c) its crystal structure

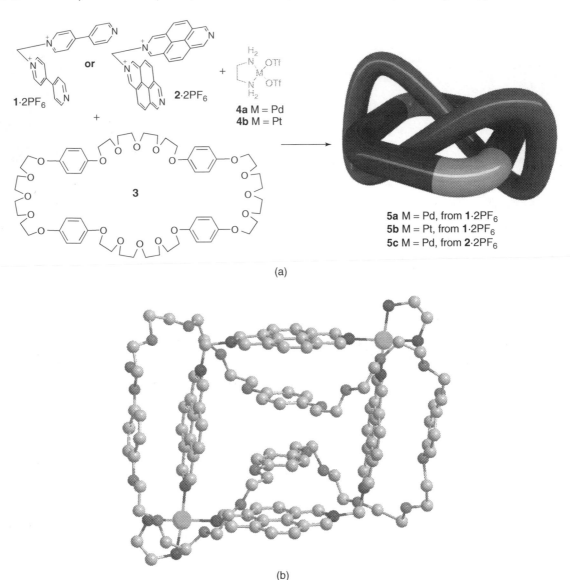

Figure 10.19 *(a) Synthetic scheme of a Solomon link (Reprinted with permission from[54]. Copyright 2009 American Chemical Society) (b) crystal structure of the compound 5c*

utilised as the template for the reaction, resulting in one zinc atom in each of a total of six pentacoordinate complexation sites (Figure 10.20b). This preparation can only work if there is an extremely high degree of preorganisation, since 18 precursor molecules have to come together to form the tri-ring Borromean system. Molecular modelling indicates that the building blocks self-assemble through 12 aromatic $\pi-\pi$ interactions and 30 zinc–nitrogen dative bonds. These multiple interactions render the Borromeate as the thermodynamically most stable reaction product out of potentially many others. All of the condensation reactions are reversible and the system is in equilibrium- and it follows therefore that the borromeate should be the predominant reaction product.

(a)

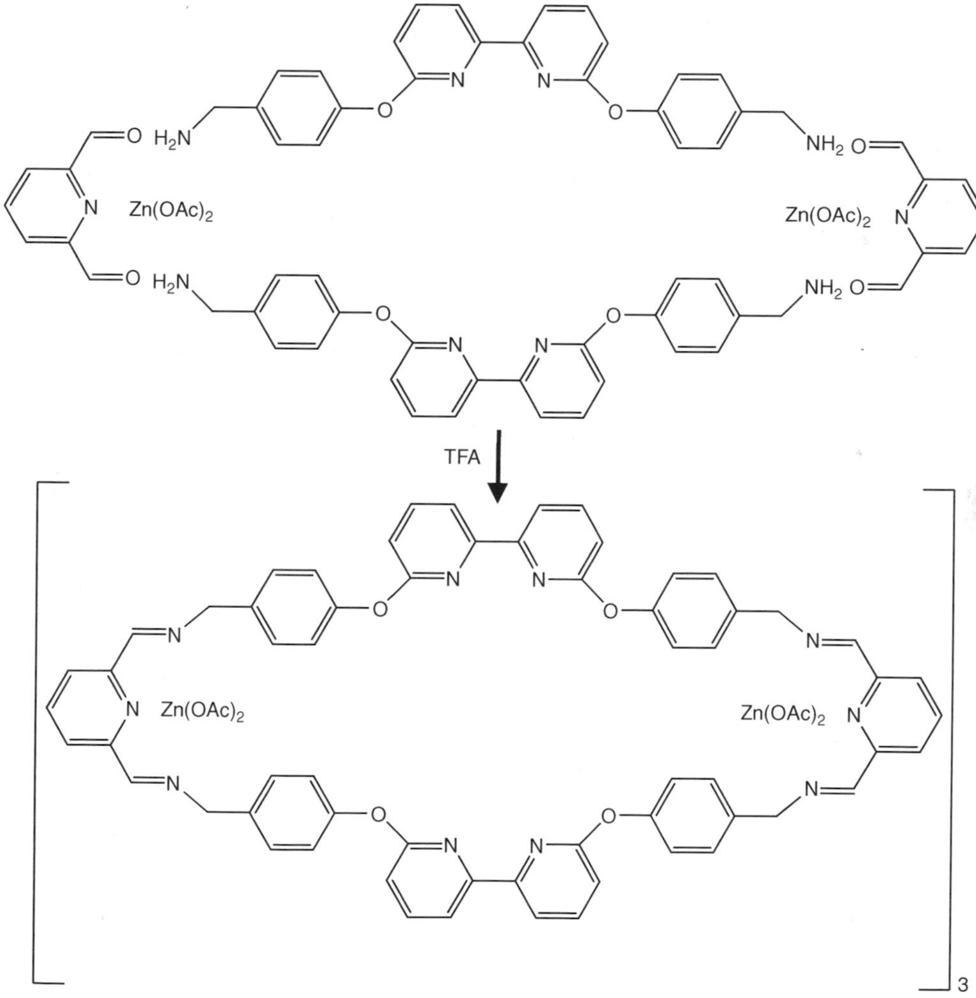

(b)

Figure 10.20 (a) Schematic of a Borromean ring system (b) zinc templated condensation reaction to give a Borromean ring

When the reagents are combined, up to 90% of a single product can be obtained.[55] The mass spectrum is consistent with a trimeric structure and the NMR indicates the system is highly symmetric (in a Borromean ring system all rings are equivalent) and fluxional. The X-ray crystal structure is shown in Figure 10.21 and clearly shows the Borromean structure. The system also contains an inner chamber with a diameter of

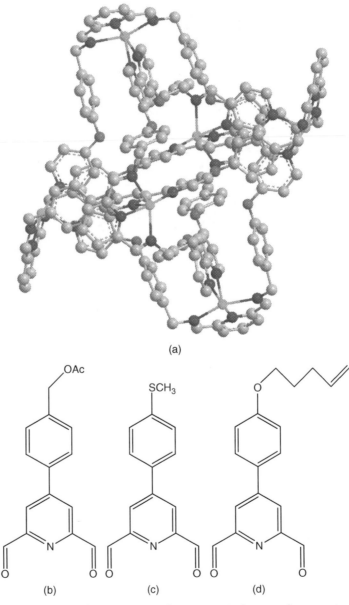

(a)

(b)   (c)   (d)

Figure 10.21   *(a) Crystal structure of a Borromean ring system (solvent and acetate ions removed for clarity) (b-d) substituted 2,6-diformyl pyridines*

2.5 nm, which is essentially lined with the oxygen atoms of the rings, which form a cuboctahedral array. As the authors point out, this system is somewhat like Cram's spherands in appearance and there is evidence that a seventh Zn atom is complexed within this cavity. This reaction is so efficient the authors have modified it as a lab instruction experiment suitable for undergraduate students, requiring seven four-hour blocks of time and synthesising Borromean rings with 90% on a scale of grams.[56] The name 'borromeates' has been suggested for these systems.

The Borromean ring described above is still held together in part by zinc ions and also contains potentially hydrolysable imine units. These imines could be reduced to amines by sodium borohydride in ethanol to give the metal-free borromeate compound.[57] Under more drastic conditions some of the ring compounds cleave, thereby releasing the other macrocycles, since if you cleave one ring, the whole Borromean ring structure falls apart. It has also proved possible to synthesise substituted borromeates by utilising 2, 6-diformylpyridine building blocks containing either 4-acetoxymethylphenyl or 4-methylthiophenyl groups in the 4-position (Figure 10.21b,c) in yields of 97% and 89% respectively.[58] Likewise, the substituted pyridine compound (Figure 10.21d) can be utilised to synthesise a borromeate in 85% yield.[59] The olefin functions can then be reacted with styrene under metathesis conditions to give borromeate with between one and five aromatic substituents, but none of the fully hexasubstituted compound can be obtained. However, this can be synthesised in good yields by metathesising the pyridinyl compound first and then assembling into the borromeate.

Since borromeates are again inherently chiral, synthesis from chiral building blocks can impose a preferred chirality on the final ring system. Chiral bipyridines (Figure 10.22a) can be used in the synthesis of borromeates.[60] When condensed with diformylpyridines, borromeates are obtained in good yield. The chirality of the bipyridine unit has been shown to affect the overall borromeate chirality. From X-ray crystal structures it appears that chirality is imposed upon the zinc ligands. Variation of the templating metal is also possible[61] and although Zn(II) gives the best results, other divalent ions that have been used to template borromeates include Ni, Cu, Mn, Co and Cd in yields from 36 to 85%. One unexpected result occurred however when the borromeate synthesis was carried out with a 50:50 mixture of Zn(II) and Cu(II) ions. A green crystalline compound was obtained by recrystallisation from methanol/diethyl ether and its crystal structure was determined (Figure 10.22b). Instead of the expected borromeate trimer, a two-ring system was in fact produced with the structure of a Solomon link.[61] NMR studies indicate that a mixture of Solomon link and Borromean ring structures exists in solution and the crystallisation of the Solomon link is kinetically controlled.

Other workers have attempted a step-by-step synthesis of Borromean rings. This method has the advantage that it proves possible to incorporate different macrocycles in a Borromean-type structure. For this to be feasible, one intermediate has to be a stable ring inside a ring complex, which can then be threaded with the third ring. Stable complexes of this type have been obtained, for example by using terpyridine containing rings and complexing them with ruthenium to give ring-in-ring complexes.[62] Other workers have utilised the classic interactions between 1,5-dioxanaphthalene crown ethers and paraquat-based macrocycles to give a stable ring-in-ring complex.[63] However, as yet we are unaware of any group taking these synthetic schemes to completion by incorporation of a third ring.

We have attempted to present here examples of some of the most interesting and intricate linked-ring systems to be constructed to date. Much greater detail on the construction of these novel architectures is available from the Bibliography and from Stoddart's review.[64] The use of the reversible imine-forming reactions utilised in the Borromean-ring synthesis and in reactions to give other interlocked molecules has also been reviewed.[65] One big advantage of this method is that the imine-forming reaction is reversible and the products are in equilibrium, so any 'errors' in the formation of these linked systems can be reversed and corrected. This is probably why such high yields are possible for such complex reactions.

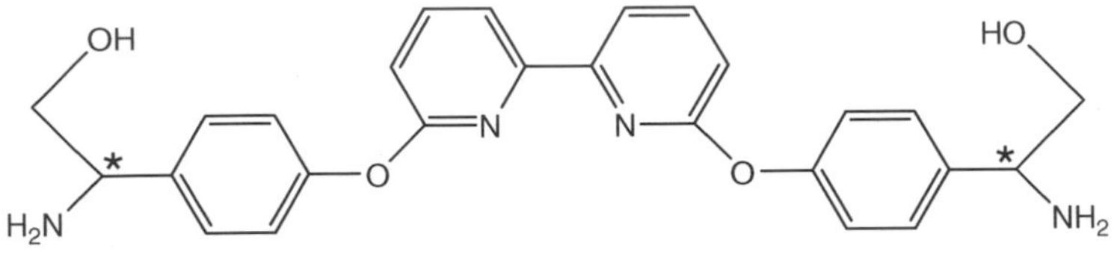

(a)

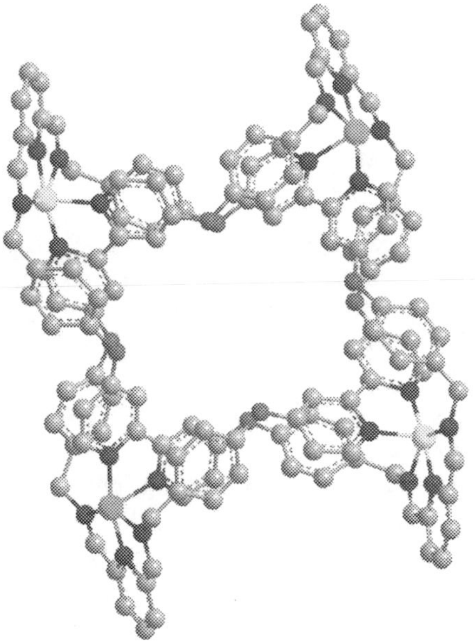

(b)

Figure 10.22 *(a) Chiral bipyridine unit (b) crystal structure of a Solomon knot from a mixed Cu(II)/Zn(II) catalysed reaction (solvent and acetate ions removed for clarity)*

10.2.6   Other systems

A so-called 'molecular bundle' can be synthesised by complexing a *tris*-ammonium compound with three equivalents of crown ethers containing one or two aldehyde units to give a 4-pseudorotaxane.[66] Reaction of these pseudorotaxanes with melamine leads to the formation of singly and doubly capped interlocked species, as shown in Figure 10.23a. NMR and mass spectrometry have been used to prove the structure of these moieties but as yet there has been no success in growing X-ray-quality crystals. A similar singly capped system has been synthesised by utilising crown ethers substituted with alkenyl chains and metathesis reactions.[67]

A very intimately linked molecular architecture has been devised in which one macrocyclic material is almost encapsulated in another.[68,69] A 3-rotaxane can be synthesised by threading two crown ether wheels containing aldehyde substituents – as shown above – onto an axle (Figure 10.23b). These crowns can then be joined together by reaction with *p*-diaminobenzene to form imine groups. The resultant structure then has the axle encapsulated inside a 'suit' formed of the two linked crown ether units, leading the name 'suitane' to be proposed for these systems. When combined with the crown ethers, an almost quantitative yield of the 3-rotaxane was formed. Addition of two equivalents of the diamine led initially to a dynamic mixture

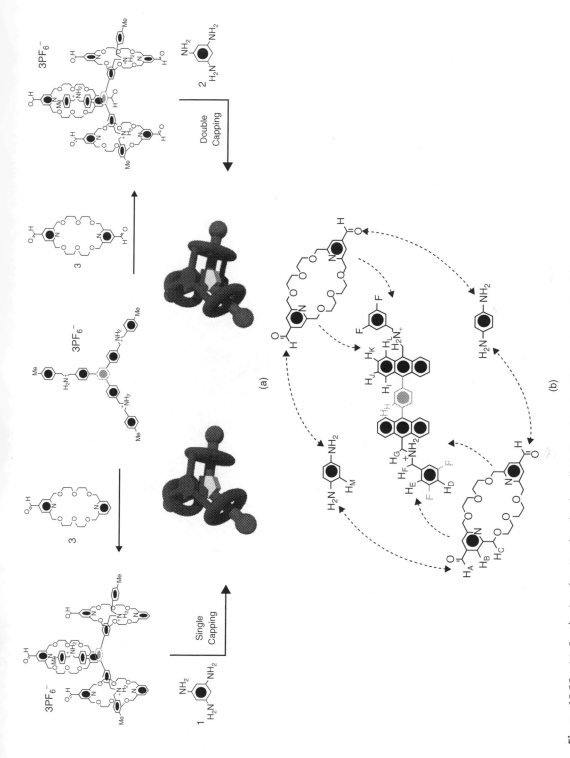

Figure 10.23 (a) Synthesis of a "molecular bundle" (Reprinted with permission from[66]. Copyright 2006 American Chemical Society) (b) synthesis of a "suitane" (Copyright Wiley-VCH Verlag GmbH & Co. KGaA. Reproduced with permission[68])

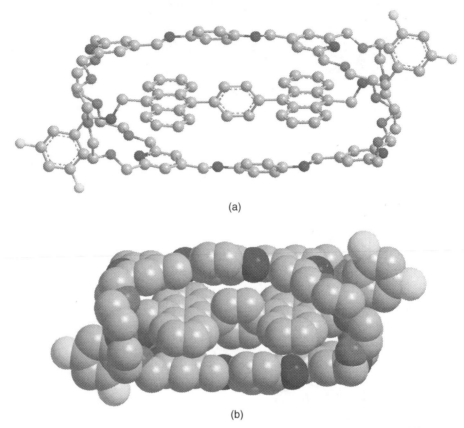

(a)

(b)

Figure 10.24   *(a) X-ray crystal structures of the ''suitane'' ball and stick and (b) space filling representation of the ''suitane''*

of products, which after 29 days had reached equilibrium. Crystals could be grown and the structure of the suitane elucidated (Figure 10.24a) and shown to be identical to that obtained from molecular modelling studies.[67] The space-filling representation in Figure 10.24b demonstrates the intimate contact between 'body' and 'suit'. Heating the sample in acetonitrile to 70 °C for 30 days caused no changes in the NMR spectrum, indicating that a truly interlocked structure had been formed.

Besides these covalently bound assemblies, there have also been a wide range of studies on self-assembly of capsules in solution. We have already mentioned this for such systems as calixarenes and other macrocycles in earlier chapters, and cucurbiturils have had a chapter devoted to them, but there are also some linear versions of these materials that self-assemble in solution to form molecular capsules. One of the earliest reports came from the group of Julius Rebek, who synthesised a glycouril derivative (Figure 10.25a), which dimerises in solution.[70] This dimer is held together by eight hydrogen bonds and has a structure in solution that has been described as being like a tennis ball. The molecule has a central cavity and has been shown to complex species such as methane. Using an analogue of this system in which there are two differently substituted glycoluril functions on the monomer gives rise to a chiral system.[71]

A larger glycouril derivative (Figure 10.25b) also dimerises to give a molecular capsule capable of incorporation of larger guests such as ferrocene or adamantane in solution.[72] A smaller system also self-assembles in solution (Figure 10.25c) to form a capsule, and in this case four units are required.[73] Again

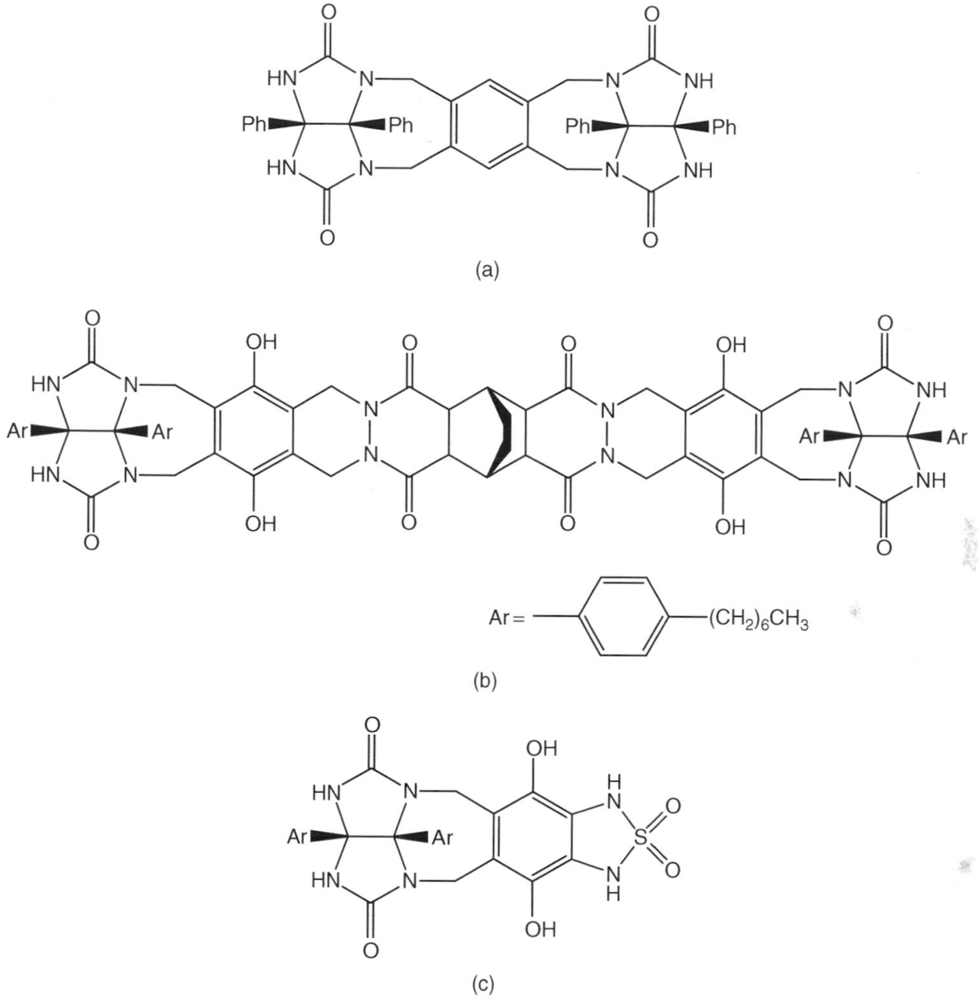

(a)

Ar = [benzene ring]—(CH$_2$)$_6$CH$_3$

(b)

(c)

Figure 10.25   *(a) Structure of the "tennis ball" monomer (b) the "softball" monomer and (c) the tetrameric "football" monomer*

adamantane is included as a guest within this capsule. Generally the capsules only form in nonhydrogen-bonding solvents (being monomeric in solvents such as DMSO) and in many cases they often only dissolve if there is a hydrophobic guest present. These capsules can also be substituted to render them chiral and have been shown to strongly bind tetralkyl ammonium salts.[74] A review on these systems and other molecular capsules and their guest binding has been published.[75]

10.3   Molecular Devices, Motors and Machines

The last 50 years have seen an unprecedented change in the size and complexity of the machines we use. For example, early computers were huge devices, containing thousands of valves and occupying entire rooms. We have seen shrinkage of this down to desktop, laptop and palmtop computers, to say nothing of the revolution in communications technology, where the mobile phone has become ubiquitous within society. In 1965 Gordon E. Moore wrote his paper in which he predicted that the number of transistors on a single circuit would double every year,[76] although he later revised this to every two years. Moore's Law has proven reasonably accurate, with the number of transistors on a chip continuing to increase exponentially – at the time of writing, for example, the Intel Itanium processor contains two billion transistors.

Obviously an increase in transistor number can only occur if the transistors become smaller, or else processors would of course become larger and unwieldy. At present commercial transistors are fabricated by a top-down approach from silicon, usually using lithographic techniques. However, the photolithographic process does have its limitations in resolution (becoming much more economically and technically limited for features smaller than 100 nm), and as silicon structures become smaller and closer together there is also the possibility of electron 'leakage'. Due to the limitations of these methods it appears that Moore's Law will eventually cease to be valid.

One way to circumvent this problem would be to use discrete molecular or supramolecular systems, which could operate by electronic or physical rearrangements. This is one of the driving forces for the construction of such devices as the molecular switches described in previous chapters. These could potentially be utilised to construct devices via bottom-up approaches, in which simple components spontaneously assemble to form a complex device. We have seen this occur on molecular and supramolecular levels, as examplified by the spontaneous assembly of rotaxane and catenanes in solution by templating, hydrogen bonding and $\pi-\pi$ interactions. It is possible that these types of interaction could also be used to cause these components to self-assemble into larger devices.

There has also been a huge interest in nanotechnology. The ability to develop miniaturised devices on the same scale as biological moieties such as enzymes, antibodies and cells could revolutionise medicine. Other potential uses for these supramolecular systems include energy harvesting from sources such as solar energy. A detailed description of this field warrants a whole book, if not a series of books, so we will concentrate on the areas in which macrocycles have shown promise in molecular machines and devices; some of these have already been covered in earlier chapters.

10.3.1   Molecular and supramolecular switches

Many macrocyclic materials display structural changes upon complexation of a specific guest, thereby demonstrating molecular recognition. This can also modify the host properties. Molecular recognition has been widely discussed in previous chapters, for example in the recognition of crown ethers towards metal ions, cucurbituril towards diamines and so on (Chapters 3 and 8). Other systems we thought to be of interest include the *bis*-crown ether shown in Figure 10.26a, where complexation of a zinc atom changes the conformation of the host such that the two crown-ether rings are moved into an arrangement in which they can cooperatively bind a diammonium compound,[77] and the so-called 'molecular syringe', in which a calix[4]arene binds a silver ion as shown in Figure 10.26b; upon protonation, the silver–crown complex is destabilised and the metal ion is bound instead by the terminal ethers.[78] Deprotonation reverses this process and the silver ion passes back through the calixarene macrocycle and returns to being bound by the bridging crown ether.

Structural changes within a single unit can also be observed, as shown within the compound in Figure 10.27a, which has a structure similar to a pseudorotaxane, where the bipyridyl unit is complexed within the cavity of the crown ether[79] to give what the authors describe as a 'scorpion-like' structure. This means this system does not form complexes with guests such as the diazastilbene compound (Figure 10.27b). Deprotonation of the bipyridyl group with tributylamine to the monocation leads to dethreading of this internal pseudorotaxane and in the presence of the diazastilbene leads to the formation

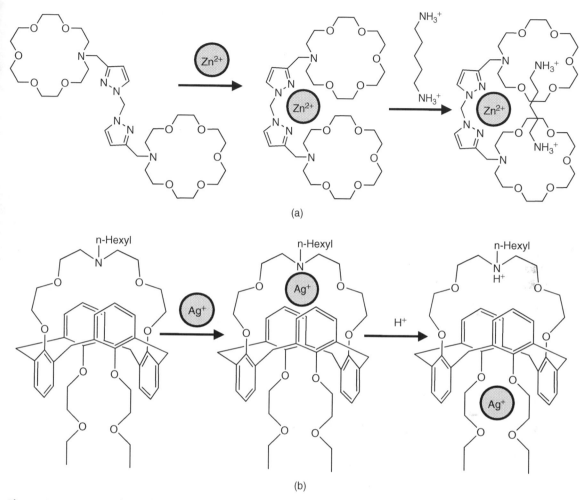

(a)

(b)

Figure 10.26 (a) Binding of zinc ion generates a favourable conformation to bind the diammonium compound (b) a calixarene that can act as a molecular syringe

of a 1 : 1 pseudorotaxane. Solvent and electrochemical effects can also control the threading–dethreading equilibrium. Photochemical effects can also be utilised; the complex between the bipyridyl crown ether and a ruthenium(*bis*-phenanthroline) unit (Figure 10.27c) for example can be decomposed photochemically but recombined quantitatively upon heating.[80]

Other internal rearrangements can also be switched photochemically, such as that shown for the azobenzene-substituted crown ether (Figure 10.28a). In solution this system cannot form an intramolecular complex, but in the *trans* form photochemical isomerisation to the *cis* form allows 'tail-biting' to occur, where the ammonium group is complexed within the crown ether.[81] More complex behaviour for these systems also exists, such as the anthracene-substituted crown ether (Figure 10.28b). The compound is essentially nonfluorescent except in the presence of protons and sodium ions, which protonate the amine and form a complex with the crown ether respectively.[82] Without protons, the amine donates electrons to the anthracene and quenches fluorescence; without sodium the crown ether behaves in the same way. If both protons and sodium are present, the compound becomes highly fluorescent and can be thought of as

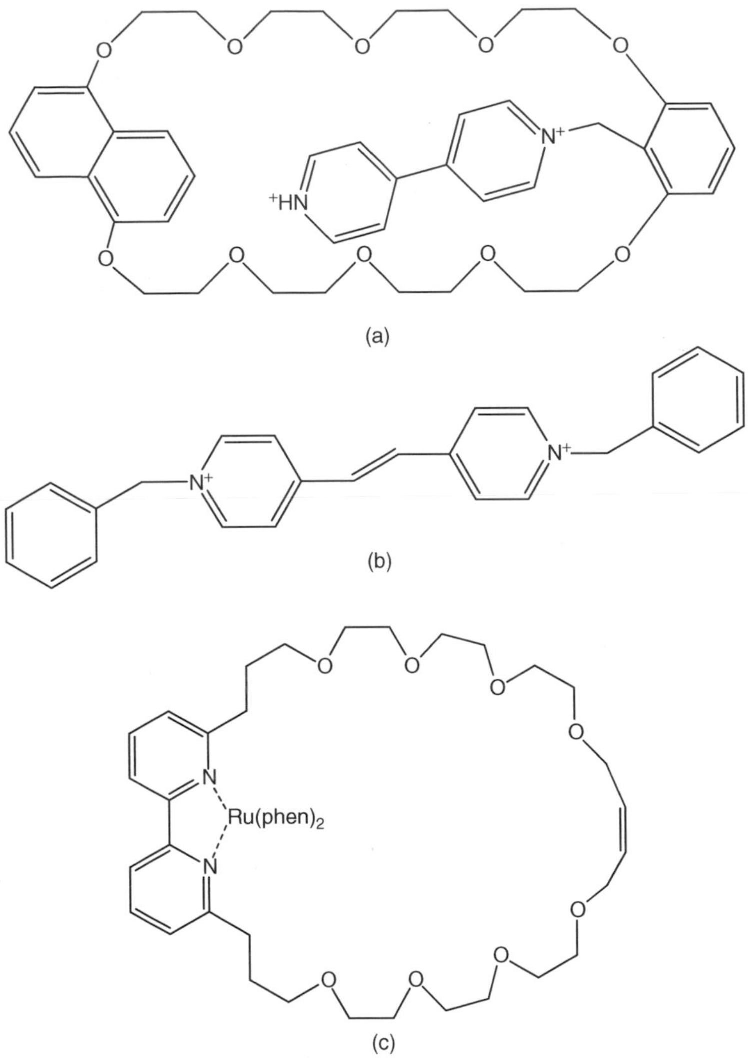

Figure 10.27 (a) A crown ether derivative that forms an "internal" pseudorotaxane (b) a diazastilbene compound

acting as a molecular AND gate. Another system that behaves in this way is that shown in Figure 10.28c, which acts as an AND gate and is also affected by pH. In acidic solution this system acts as a fluorescent sensor for caesium and in alkaline media it acts as a fluorescent sensor for potassium ions.[83]

Other examples of simple logic gates based on macrocyclic systems have been developed. Some are very simple, such as the compound shown in Figure 10.29a, which can act both as a YES and a NOT gate for the presence of acid.[84] When irradiated, the neutral compound demonstrates a broad emission centred at 438 nm, which is attributed to charge-transfer interactions between naphthyl and amine units; however, upon addition of acid this disappears and is replaced by intense fluorescence centred at 342 nm. Therefore, when monitored at 342 nm this system gives a positive (YES) response to acid, and at 438 nm it gives a negative (NOT) response. An XOR logic gate could be synthesised by combining the two materials shown in Figure 10.29b,c to give a pseudorotaxane.[85] If the fluorescence output of the diazapyrene compound

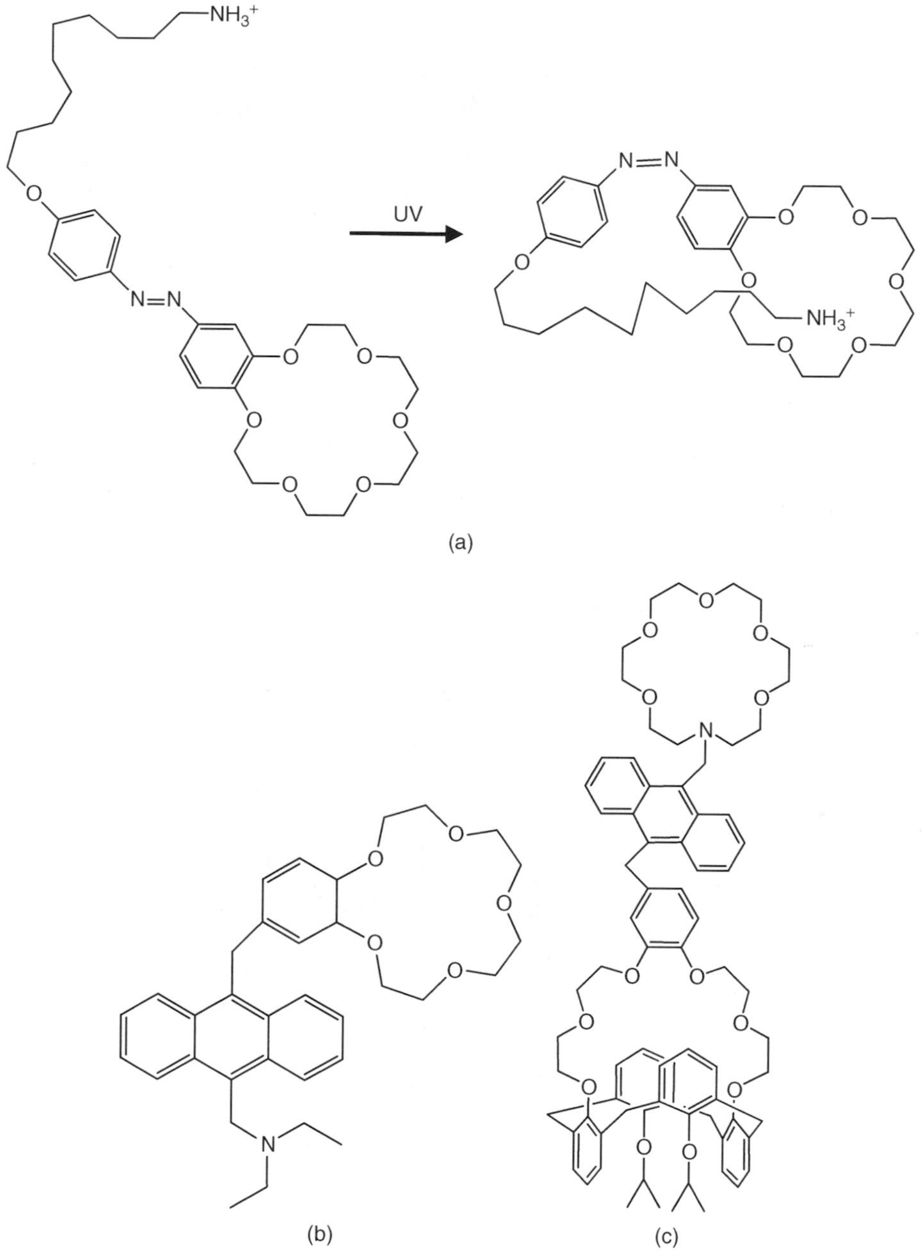

(a)

(b)

(c)

Figure 10.28   *(a) A azobenzene-crown ether derivative that forms an "internal" complex upon photochemical stimulation (b) a anthracene-crown ether (c) calixarene based sensor*

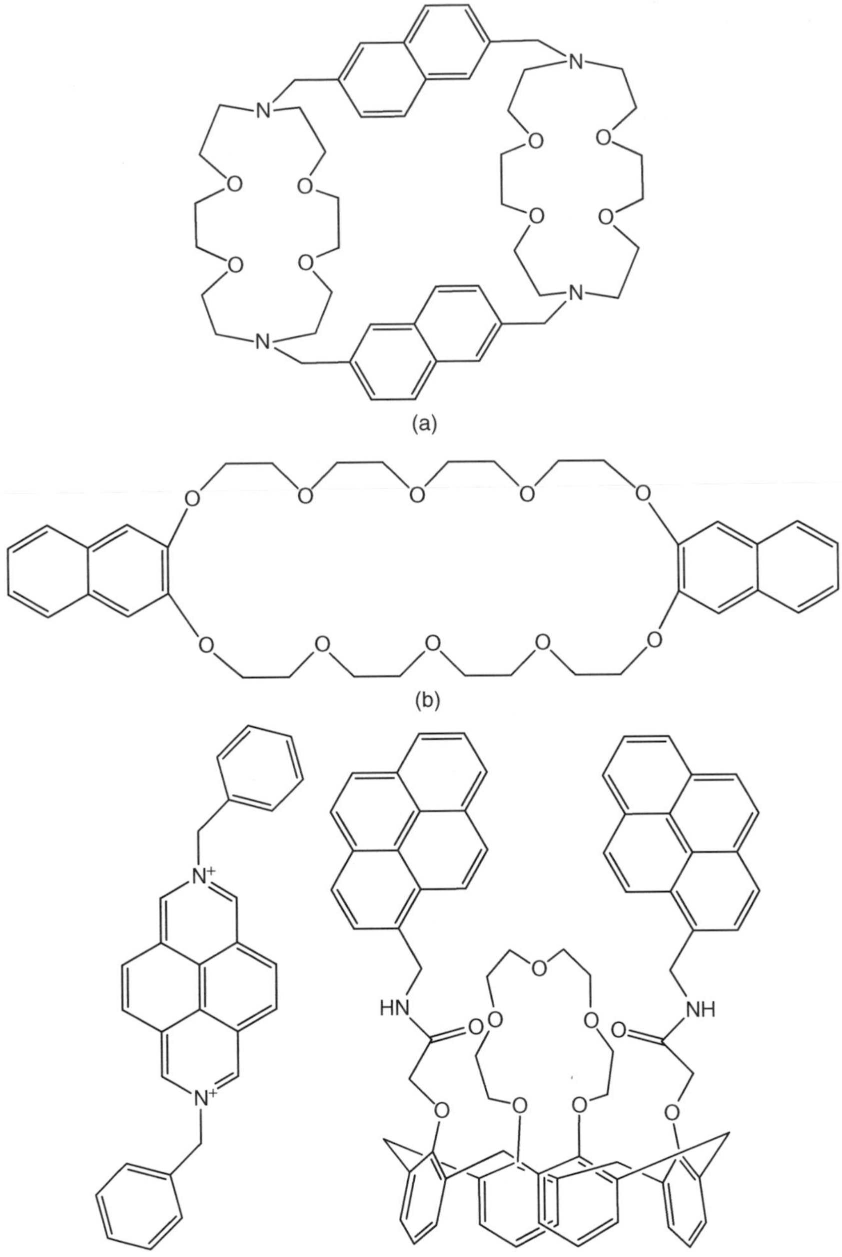

Figure 10.29 *(a) A bis-azacrown ether derivative that acts as a YES or NOT gate (b, c) crown ether/diazapyrene complex (d) calixarene based species that acts as a logic gate*

is monitored at 343 nm, the system gives a positive response in the presence of either a strong acid (trifluoroacetic acid) or tributylamine, both of which cause dethreading. In the absence of these systems the two components form a pseudorotaxane and fluorescence is quenched; in the presence of both materials in 1:1 ratio, again there is no fluorescence since the acid and base simply neutralise each other. Other workers have also appended pyrenyl units to a calixarene-based scaffold to give structures like that shown in Figure 10.29d and have demonstrated that they can act as NOR, XNOR and INHIBIT gates in the presence of such species as acid, base and lead ions.[86]

Many of the systems described within the literature are of interest because they have electron-transfer processes that can be influenced by various chemical or electrochemical inputs. Examples include the ammonium–crown ether interactions described in Chapter 9 for various rotaxanes and catenanes, which are influenced by acid–base inputs. For example, two substituted crown-ether systems which form a pseudorotaxane are shown schematically in Figure 10.30a. The ruthenium-substituted crown ether forms a pseudorotaxane with the protonated ammonium group of the second crown ether under acidic conditions and this complex disassociates on addition of base.[87] A separate bipyridyl moiety can also interact with the second crown-ether unit, with this interaction being capable of being controlled electrochemically. The ruthenium complex and the bipyridyl can then be thought of as being linked together by a molecular 'extension cable'. On irradiation the ruthenium complex acts as an electron donor and the bipyridyl unit as the acceptor. Photochemically activated electron transfer (as demonstrated by the quenching of the ruthenium emission) only occurs when all three components are assembled together. An earlier variation on this assembly in which the ammonium and bipyridyl units are located on the 'extension cable' and bind to two different crown ethers (Figure 10.30b) was also constructed and was shown to display similar behaviour.[88]

10.3.2   Molecular motion

One area that has been reviewed intensely is the conversion of molecular chemical or electronic changes to mechanical motion. In Chapter 9 we discussed at length the motions of molecular species, such as the shuttling of rings along axles in rotaxanes and the relative rotation of rings in catenanes. Species which acted as artificial muscles, molecular elevators and pistons were also discussed. Within this section we will discuss other systems of these and other types that have caught our attention in their relation to macrocyclic components. Other types of molecular systems which have been incorporated into devices such as molecular rotors have been reviewed extensively[89] and have formed a special edition within *Topics in Current Chemistry*.[90]

One simple example of a mobile macrocyclic system is a molecular oscillator like that in Figure 10.31a. In this system the two porphyrin rings are shown by NMR studies to oscillate relative to each other in a rotational manner around the cerium centre.[91] In similar systems without the crown-ether linkers the two porphyrins can freely rotate with respect to each other. Other systems with rotational ability include the paddlewheel-like compounds shown in Figure 10.31b, where triptycene moieties can rotate around the place of the crown ether rings.[92] NMR studies demonstrated that rotation of the paddlewheels was dependent on the temperature and the size of the crown-ether macrocycle.[93] A molecular turnstile was also synthesised and was shown to undergo rapid rotation at ambient temperature when R = —CH_2OCH_3 (Figure 10.31c). This rotation could be frozen out at low temperature ($-54\,°C$) and when a bulky aromatic substituent was used, no rotation could be observed, even at elevated temperatures.[94] Other workers used similar systems to synthesise molecular gyroscopes[95,96] containing aromatic spindles – demonstrating that these underwent rotation.

Other applications include the possibility of utilising these types of supramolecular system to selectively capture guests; that is, to act as molecular tweezers. A simple system involves an azobenzene substituted with two crown-ether moieties (Figure 3.21a) which upon irradiation converts to the *cis* form and binds

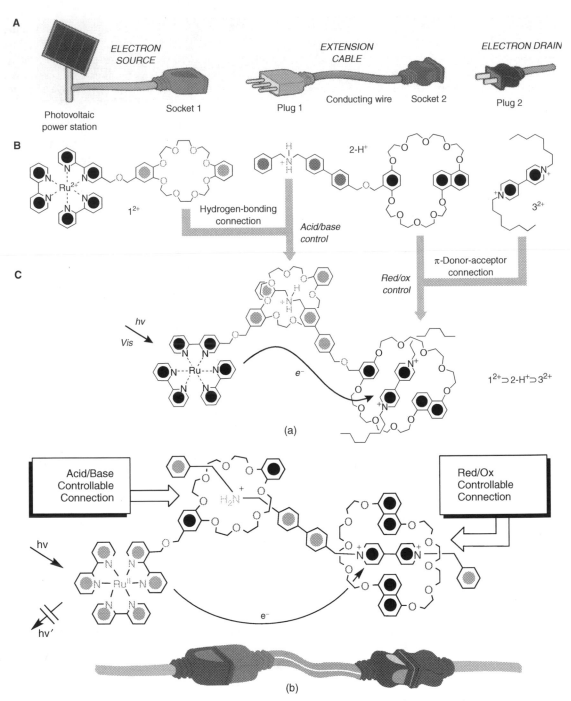

Figure 10.30 (a) The three components of the self-assembling system that mimics a molecular "cable extension" (Reprinted with permission from[87]. Copyright 2006 National Academy of Sciences, U.S.A) (b) a second version of this system (Reprinted with permission from[84]. Copyright 1994 American Chemical Society)

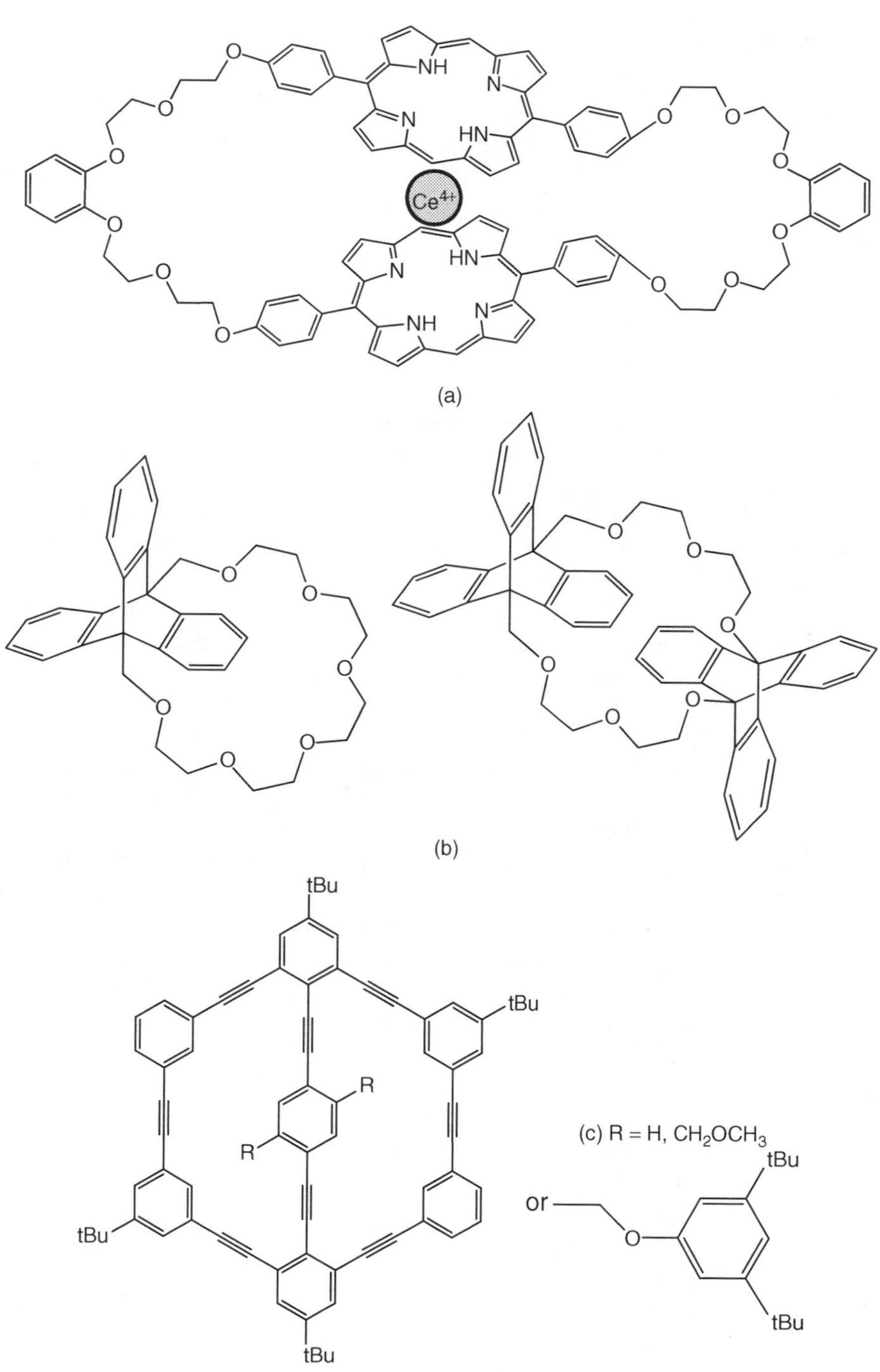

(a)

(b)

(c) R = H, CH$_2$OCH$_3$

Figure 10.31   (a) A molecular oscillator (b) molecular paddlewheels (c) a molecular turnstile

alkali metals in a cooperative manner (Chapter 3, references 101 and 102). Another photoresponsive system utilised a dithienylethene unit substituted with crown ethers such as that shown in Figure 10.31a. This material bound large metal ions in a cooperative manner but on irradiation at >480 nm underwent a ring-closure reaction, with a concurrent conformational change that prevented cooperative binding and then released the metal.[97] Cyclodextrin systems have also been used; for example, an azobenzene-bridged β-cyclodextrin was not capable of binding 4,4-dipyridyl in solution, but isomerisation of the azobenzene to the *cis* form expanded the cavity and allowed binding.[98] Upon reisomerisation back to the *trans* form, the guest was expelled.

Another photoresponsive system is that shown in Figure 10.31b. When irradiated, the azobenzene unit of this system undergoes isomerisation to the *cis* form.[99] This then causes the ferrocene rings to rotate relative to each other and the whole molecule changes conformation, forcing the porphyrin units further apart and twisting the *bis*-azanaphthyl guest. This supramolecular system has been likened to a 'molecular pedal'.

Redox-responsive systems have also been described, such as a calix[4]arene with urea groups at the upper rim and substituted with ferrocene groups at the lower rim.[100] In chloroform this system formed a stable dimer, but upon treatment with an oxidising agent the ferrocene units became cationic ferricinium moieties and the dimer disassociated to the monomer. Disassociation could also be achieved by the addition of hydrogen bond-disrupting solvents such as DMSO. We have already described in detail in Chapter 9 many electrochemically switchable systems, such as rotaxanes and catenanes containing the tetrathiafulvalene (TTF) unit. This system can also be used to form a three-way switch in solution by mixing TTF, a *bis*-1,5-dioxanaphthyl crown ether (Figure 9.10b) and a tetracationic macrocycle (Figure 9.7e). In solution TTF in its neutral form forms a complex with the tetracationic macrocycle.[101] Reduction of the TTF to its monocation caused dissociation of the complex and the formation of free TTF$^+$. A second reduction step gave TTF^{2+}, which then formed a stable complex with the crown ether. Formation of these complexes could be detected by X-ray crystallography in the solid state, NMR in solution and mass spectroscopy in the gas phase. A simpler ditopic system was formed by combining a 1,5-dioxanaphthalene compound, the tetracationic macrocycle and TTF.[102] The TTF–cationic macrocycle complex which formed preferentially could be dissociated by addition of ascorbic acid, and the macrocycle then formed a complex with the naphthalene derivative. Similar redox-controlled threading and dethreading processes could also be observed for example between cyclodextrins and substituted viologens.[103]

Redox-controlled binding of β-cyclodextrins to dendrimers containing ferrocene[104] or cobalticenium[105] has also been demonstrated. Oxidation of the ferrocene residues to ferricinium disrupted this binding and caused disassociation of the supercomplex. With the cobalticenium dendrimers no initial binding was seen but reduction of the organometallic moieties led to their complexation by the β-cyclodextrin.[105] Redox switching of guests between two different macrocyclic host systems has also been observed in solution, such as in the case of cobalticenium, which forms a strong complex with calix[6]arene hexasulfonate in water in the presence of β-cyclodextrin.[106] However, when the cobalticenium is reduced to the neutral cobaltocene, this complex disassociates and 1:1 complexes between cobaltocene and β-cyclodextrin are formed.

A photochemically based association/disassociation reaction has also been studied. A diazapyrenium macrocycle (similar to that shown in Figure 9.78) forms a pseudorotaxane with the azobenzene compound shown in Figure 10.32c. Irradiation (at 365 nm) of this system leads to isomerisation of the azobenzene to its *cis* isomer.[107] This then dethreads, as shown by fluorescence measurements – a process that can be reversed by irradiation at 436 nm.

Metal-ion binding has also been used to chemically drive motion. For example, the structure shown in Figure 10.33a was synthesised[108] and then complexed with copper(II) ions to give a dimer containing two phenanthroline and two terpyridyl groups. This could then be stoppered with bulky tetraphenylmethane derivatives to give the structure shown in Figure 10.33b. In this system the copper complexed with the phenanthroline units to give a system approximately 8.5 nm in length. However, removal of the copper ions

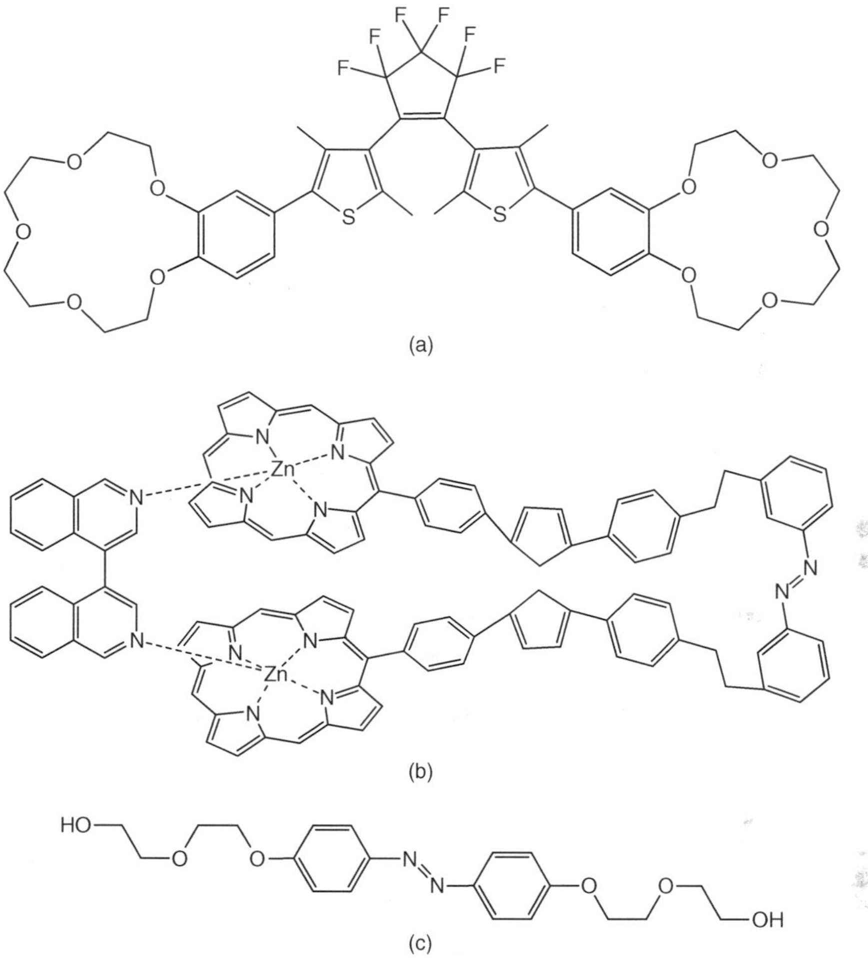

Figure 10.32 *(a) Dithienylethene based bis-crown ether (b) azobenzene containing macrocycle that acts as a molecular "pedal" (c) a photoswitchable azobenzene guest*

with cyanide and then remetallation with zinc led to a different structure since zinc formed a phenanthroline/terpyridyl complex (Figure 10.33b), causing the 'molecular muscle' to contract to 6.5 nm in length. The same group have used other metal-based systems such as ruthenium based compounds. These systems have been studied because they have been shown to undergo various photochemical rearrangements.[109,110] For example, the catenane shown in Figure 10.34a can be synthesised and upon irradiation undergoes a rearrangement reaction in which the bipyridyl unit decomplexes from the ruthenium moiety and the rings rotate with respect to each other[110] – a process that can be reversed thermally. Similar pseudorotaxane systems have also been synthesised and shown to undergo light-stimulated dethreading.[109]

A porphyrin-based macrocycle containing a manganese unit (Figure 10.34b) has been synthesised and shown to thread onto a number of polymers to form a pseudorotaxane.[111] When threaded onto polybutadiene, the manganese unit catalyses the oxidation of the double bonds to epoxide units. As the catalyst oxidises each double bond it moves on, although the initial work did not show whether the movement is sequential or random. This can be thought of as being a synthetic mimic of such natural processes as the

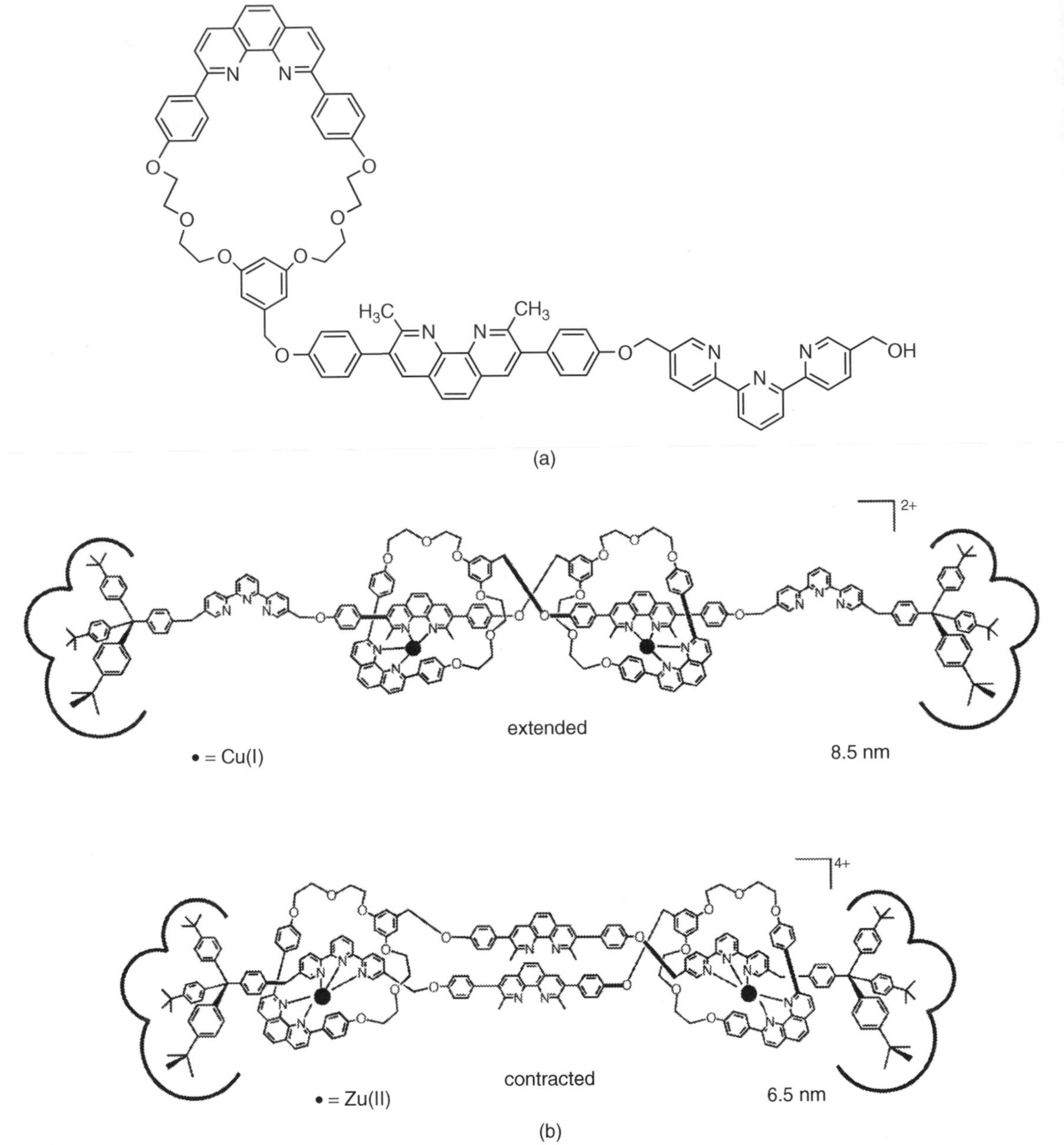

(a)

(b)

Figure 10.33 *(a) Component of molecular muscle (b) interaction of muscle when complexed with copper or zinc (Reproduced by permission of the Royal Society of Chemistry[108])*

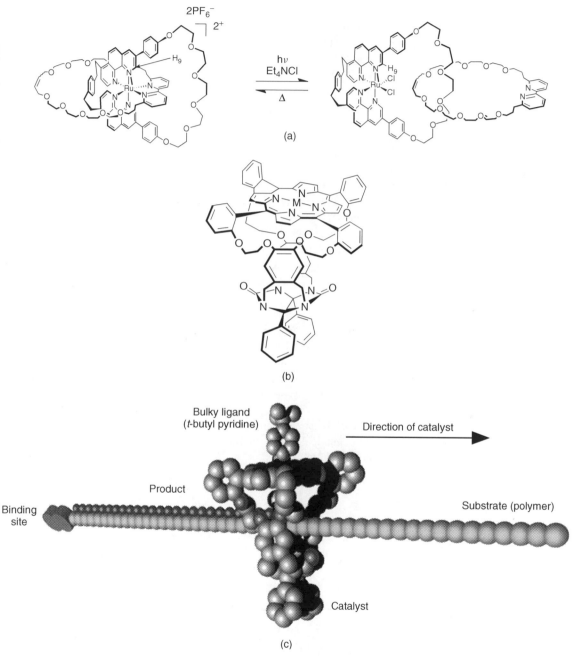

Figure 10.34 *(a) A light switchable catenane (Copyright Wiley-VCH Verlag GmbH & Co. KGaA. Reproduced with permission[110]), (b) porphyrin structure (c) movement of catalyst along polymer (Reprinted with permission from[112]. Copyright 2007 American Chemical Society)*

action of DNA polymerase. Later work using a variety of polymers showed that the speed of the threading process is related to the porphyrin structure.[112]

A variety of different movements of macrocycles along rotaxane axles were discussed in Chapter 9. Several others are also of interest, such as the anthracene-based systems shown in Figure 10.35. These form rotaxanes with the tetracationic macrocycles previously described (Figure 9.7e). In the case of the 2, 6-dioxyanthracene-containing axle (Figure 10.35a), the cationic macrocycle is clipped around the anthracene unit.[113] Oxidation of the 2-rotaxane led to the anthracene being converted to a radical cation, with concurrent migration of the macrocycle onto the hydroquinone units. When a 9,10-dioxyanthracene-containing axle is used (Figure 10.35b), both the 2- and 3-rotaxanes can be isolated and in both cases the wheels reside solely on the hydroquinone units, probably due to the increased steric hindrance that occurs with the 9,10-dioxyanthracene moiety. A 3-rotaxane with an axle containing both stilbene and azobenzene with two α-cyclodextrin wheels (Figure 10.35c-e) has also been synthesised and was shown to undergo light-initiated molecular motion.[114] This system actually displayed three stable states, as synthesised it existed with one cyclodextrin (CD1) encompassing an azobenzene unit and one (CD2) encompassing the stilbene unit (Figure 10.35c). Irradiation at 280 nm caused isomerisation of the stilbene unit and migration of CD2 onto the biphenyl moiety of the axle; this could be reversed by irradiation at 313 nm. Similarly, irradiation at 380 nm caused isomerisation of the azobenzene unit and migration of CD1 onto the biphenyl moiety of the axle; this could be reversed by irradiation at 450 nm or by heating. A slightly simpler system containing the same axle and only one cyclodextrin unit could also be synthesised and was found to be able to act as a 'molecular abacus', being able to add together two photochemical stimulation events.[115] The cyclodextrin freely shuttles between the azobenzene and stilbene binding sites when they are both in the *trans* conformation. Switching of either binding site to the *cis* form causes the cyclodextrin to bind exclusively to the other site. Isomerisation of both of these to the *cis* form causes migration of the cyclodextrin to the biphenyl binding site, meaning that in effect this system mimics a half-adder with distinct AND and XOR logic gates. A half-adder can carry out elementary addition by using the XOR gate to generate the sum digit and the AND gate to generate the carry digit.[115]

There has also been a great deal of study on movement in catenanes caused by various switching processes. These processes can be electrochemical, photochemical or chemical in nature. For example, we have already discussed (Chapter 9, reference 258) how a 2-catenane made up of the classic tetracationic macrocycle (Figure 9.98) and a polyether containing a TTF unit and either a phenoxy or naphthyloxy unit has a conformation in which the TTF unit is encapsulated by the cationic macrocycle. Electrochemical or chemical oxidation of the TTF leads to migration of the cationic macrocycle onto the phenoxy or naphthyloxy binding site. Similarly, the catenane shown in Figure 10.36a can be switched from its normal state, where the diazapyrenium unit is encapsulated inside the crown ether, to a conformation where the bipyridyl unit is encapsulated by addition of hexylamine.[116] This amine forms a charge-transfer adduct with the diazapyrenium unit, leading to migration of the crown ether. This process can be reversed by addition of trifluoroacetic acid to protonate the amine and disrupt the adduct. Catenanes can also be synthesised containing expanded cationic macrocycles such as that shown in Figure 10.36b, along with phenoxy- or naphthyloxy-containing crown ethers.[117] Both 2-catenanes and 3-catenanes containing two crown ether units can be synthesised. The crown ethers preferentially reside on the bipyridyl units but can be switched onto ammonium binding sites by electrochemical reduction of the bipyridyl units. One system also acts as an AND gate, being switchable both electrochemically and by variation of pH.

In most of these systems, when one ring moves relative to the other it can migrate in either direction around the ring. It has however proven possible to synthesise systems in which this movement is in a preferred direction. One example is shown in Figure 10.37, where an amide-type catenane is synthesised which contains two blocking groups to prevent rotation of the smaller ring around the larger.[118] The compound is synthesised with the amide ring bound at the fumaramide site, as shown in Figure 10.37.

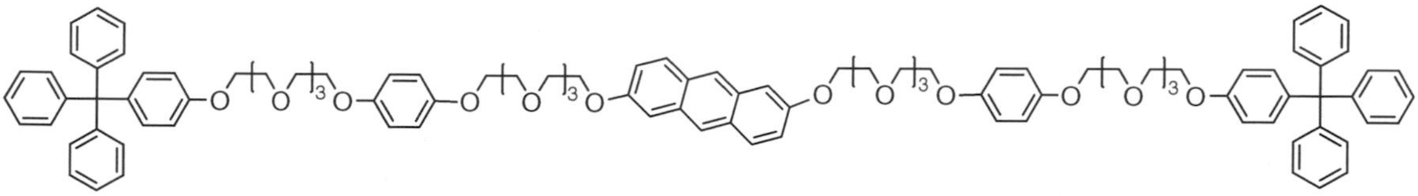

Figure 10.35 *(a) 2,6-dioxyanthracene containing axle (b) 9,10-dioxyanthracene containing axle (c-e) cyclodextrin based 3-rotaxanes with photoswitchable units*

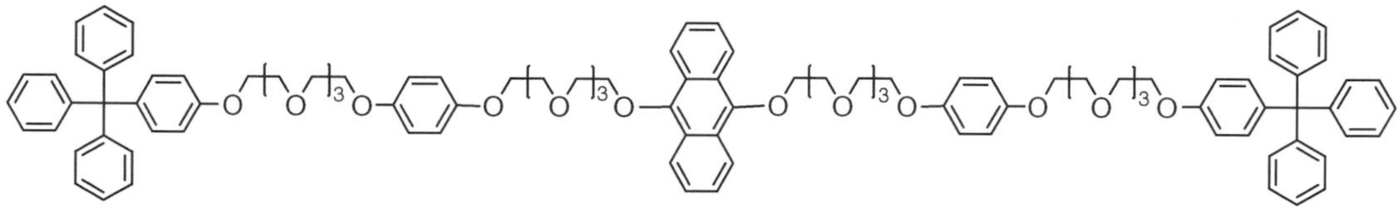

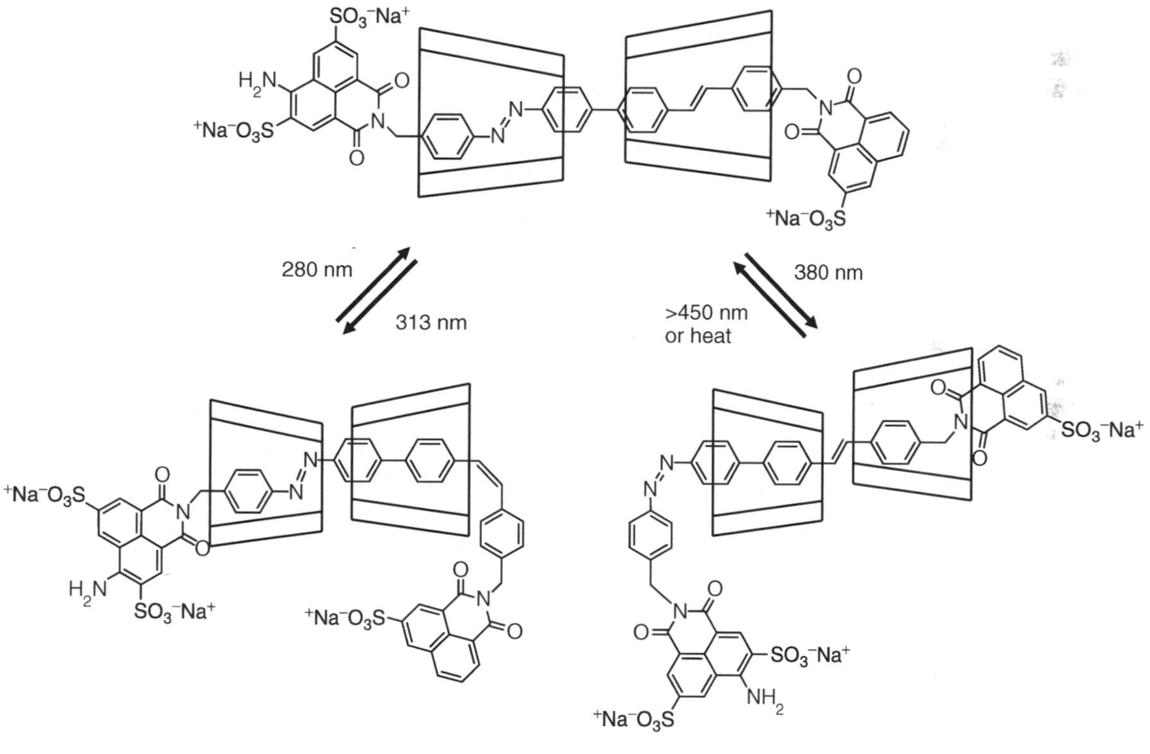

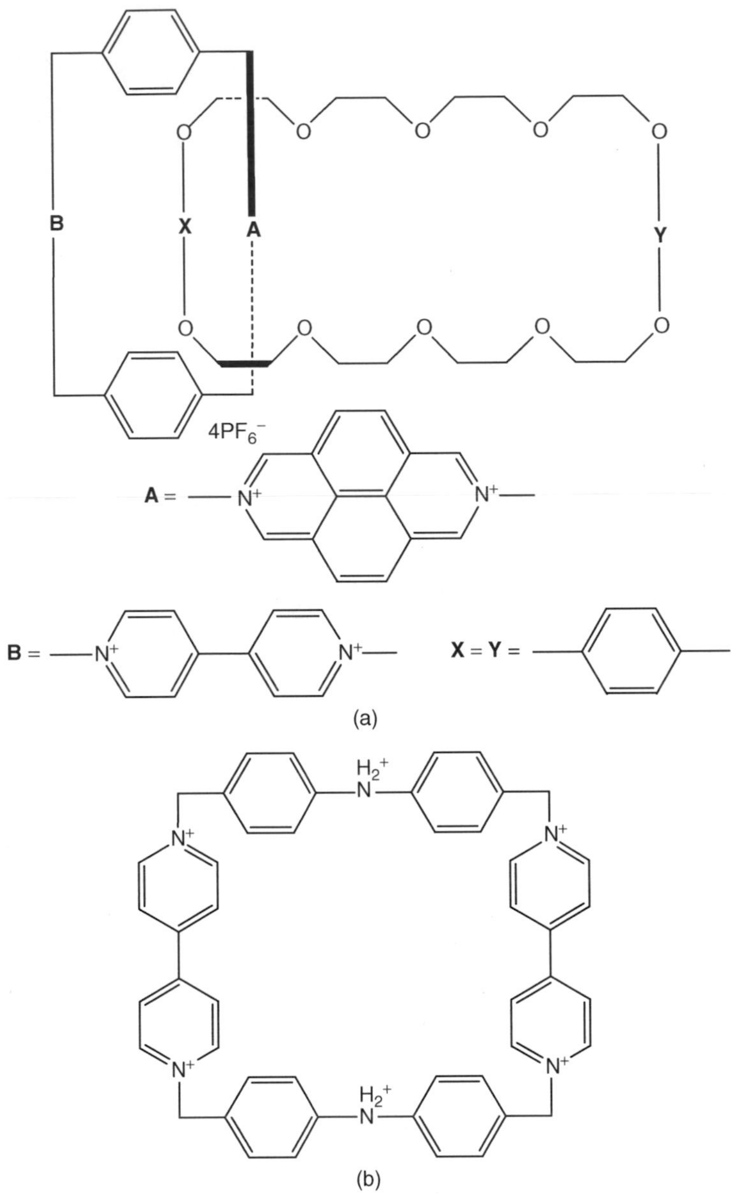

Figure 10.36 *(a) Structures of chemically switchable catenanes (b) macrocycle containing ammonium and bipyridyl binding sites*

The silyloxy-protecting group is removed and the fumaramide unit is photoisomerised to a maleamide. The amide ring then migrates to the succinimide binding site because of unfavourable steric effects. The silyloxy unit is then replaced. Addition of piperidine causes the maleamide to reisomerise back to the fumaramide. Removal of the trityl-protecting group allows free migration of the amide back to its original site and replacement of the trityl group gives the original compound, with the amide ring having moved in one direction around the larger ring (clockwise in Figure 10.37).

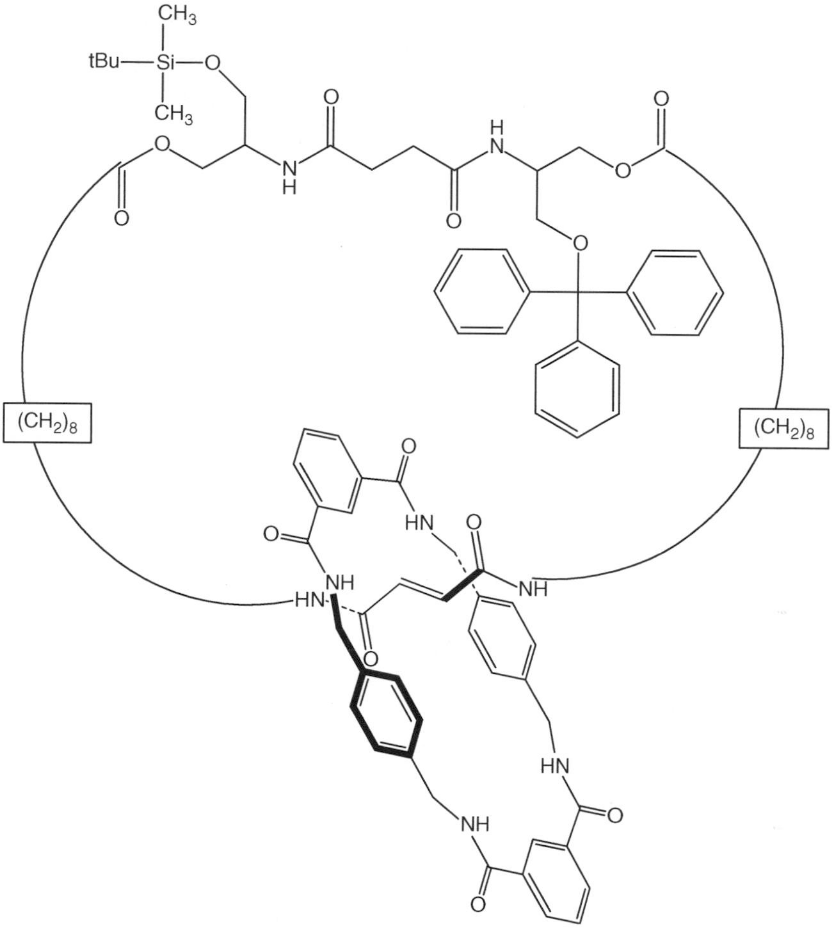

Figure 10.37   *Structure of a "unidirectional" catenane*

Although the system above does give unidirectional motion, it requires several deprotection and reprotection steps, which would make it unusable in any device. A more elegant method would involve different photochemical stimuli 'pushing' one ring around the other. Such a system has been developed, in which the amide macrocycle shown in Figure 10.37 is used to form a catenane with the macrocycle shown in Figure 10.38. In this system the amide macrocycle resides on the nonmethylated fumaramide site.[119] However, this group is sensitised by the benzophenone unit and can be isomerised by irradiation at 350 nm. This causes the amide macrocycle to move 'clockwise' onto the methylated fumaramide site. Irradiation at 254 nm causes this site to isomerise and the macrocycle then moves onto the succinimide ester site. Finally, heating the macrocycle causes reisomerisation and the amide relocates back to its original position. A 3-catenane bearing two amide groups can also be synthesised and shows similar unidirectional motion.[119]

10.3.3   Light harvesting

Other macrocyclic systems have been used to harvest energy from light. A light-harvesting antenna is a system in which multiple chromophores capture light and transfer the energy to a single acceptor system.

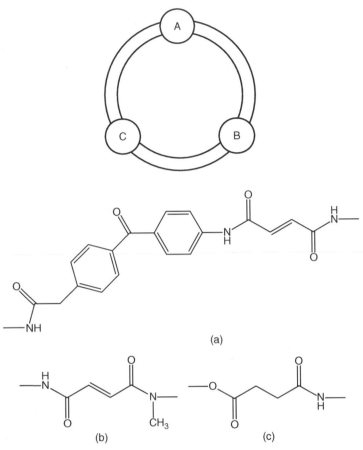

Figure 10.38   *Structure of a "unidirectional" catenane without protecting groups containing (a) benzophenone sensitised fumaramide (b) methylated fumaramide (c) succinamide ester binding sites*

For example, a water-soluble β-cyclodextrin containing seven naphthyloxy substituents has been shown to form a 1:1 complex with a merocyanine dye in solution.[120] Energy transfer from the naphthyloxy units to the dye has been shown to occur with 100% efficiency. Similarly, a polyrotaxane can be synthesised from polyethylene glycol- and naphthyloxy-substituted cyclodextrins and then stoppered with anthracene units.[121] Upon irradiation, energy transfer from naphthyl to anthracyl units is observed, with transfer efficiencies as high as 79%.

10.3.4   Ion channels

In natural systems, ion channels are pore-forming proteins that reside within cell membranes and facilitate and regulate ion diffusion through these membranes. Attempts have been made to synthesise artificial versions of these systems, many of which have been based on helical peptides. Since crown ethers bind sodium and potassium ions strongly, they have been studied as potential ion-channel materials. A number of papers have studied crown ether/peptide systems, an example of which is shown in Figure 10.39a. This material is a helical 21-mer peptide with a structure in which the six crown-ether rings stack above each other, forming a 'column' of crown ethers.[122] This material transports ions such as caesium across vesicle membranes with an effectiveness approximately the same as that of gramicidin A. A much simpler series of compounds is the urea-substituted crown ethers (Figure 10.39b), which hydrogen-bond together to form stacks containing columnar channels of crown ethers.[123] These materials transport ions across bilayer lipid membranes and form a mixture of crown-ether pores which can also associate together to form structures with a larger central pore. A series of resorcinarene compounds with attached crown ethers (Figure 10.39c) have also been synthesised and were shown to transport ions, especially K^+, across both liquid and lipid bilayer membranes.[124] Their effectiveness was of a level comparable to that of natural systems (e.g. gramicidin) and they displayed outstanding K^+/Na^+ flux selectivity. It is thought the resorcinarenes aggregate to form pores or channels within the membrane.

10.3.5   Other uses of interlinked systems

We briefly discussed in Chapter 9 how some workers immobilised rotaxanes onto porous silica substrates and used the immobilised molecules as nanovalves.[125] Upon switching, these valves opened and closed, and they could be used to entrap and release dye molecules (Figure 10.40a). Later work optimised these nanovalves, which could be opened using ascorbic acid, and showed that the most effective valves had the active component of the rotaxane anchored deep within the pores and utilised short covalent linking groups between rotaxane and nanoparticle.[126] The construction and use of these systems has recently been reviewed[127] Other nanoparticles that have been modified using these types of system include silver nanoparticles, which were surface-functionalised by a thiol-substituted crown ether.[128] Addition of *bis*-dibenzylammonium dication led to formation of pseudorotaxanes, which initiated aggregation of the silver nanocrystal dispersion, opening up the potential for controlled assembly of nanoparticles into complex constructions.

A bistable rotaxane has been substituted with mesogenic groups and shown to form a stable smectic A liquid-crystal phase over a wide temperature range.[129] A smectic A phase is a liquid-crystal phase in which the molecules are aligned in the same direction and form a layered structure ('smectic' is derived from the Greek word for soap and many soaps do adopt this type of structure). These types of material could potentially display electrochemically switchable liquid-crystal behaviour, with the mechanical switching affecting the properties of the liquid-crystal phase as a whole.

Very recently, workers have combined porphyrins substituted with eight β-cyclodextrin units and porphyrins substituted with eight adamantane units in water and shown that the multiple β-cyclodextrin/adamantane interactions actually lead to the spontaneous formation of porphyrin nanowires, 100 nm in width and with lengths of a few microns.[130] Use of more dilute solutions gives nanowires just 3.5 nm in diameter and hundreds of nanometres long.

10.4   Conclusions

In writing this work, we have attempted to demonstrate how simple materials can often be assembled into more complex structures. As bricks, beams, slates and so on all go together to make a house, so simple chemical structures can be fitted together to make more complex ones. Sometimes this fitting together is a long, complex process, as can be seen for example in the total-synthesis schemes recorded in the literature for many natural compounds. We started this work with simple chemistry and described the rings and chains that go on to make up so many compounds within the text. The assembly of these materials to make more complex regular structures is one of the most fascinating fields in chemical research. The assembly of small units to form macrocycles makes up a large section of this work, before we go on to

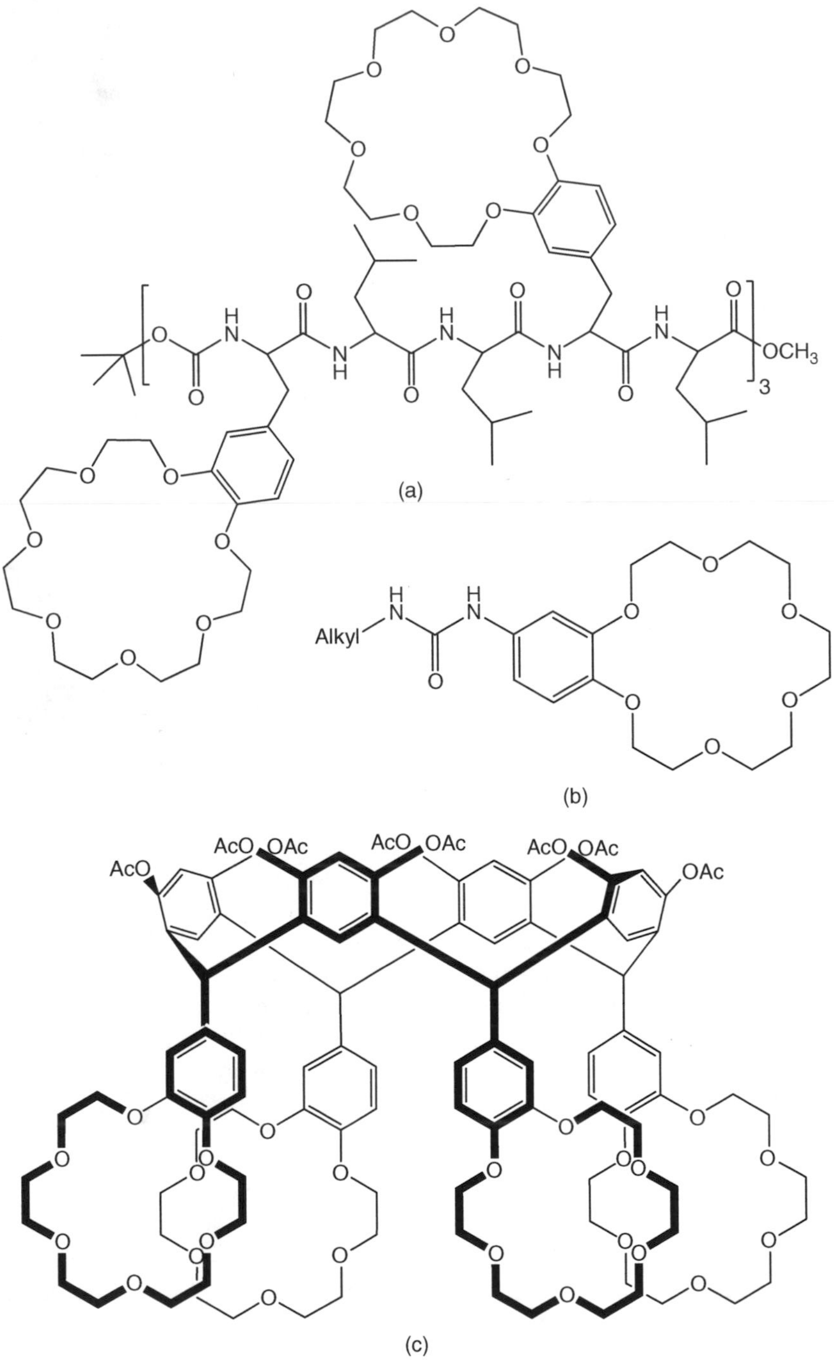

Figure 10.39 *(a) Peptide based poly-crown ethers (b) urea crown ethers (c) resorcinarene crown ethers used in ion channels*

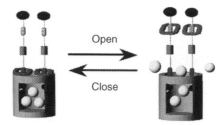

Figure 10.40 *Schematic of rotaxane "nanovalves" (Reprinted with permission from[126]. Copyright 2007 American Chemical Society)*

even more complex systems such as the mechanically linked molecules described in Chapters 9 and 10, in which macrocycles are threaded or interlinked together. The structure of these compounds is often so complex that first impressions are that they would be impossible to synthesise in any reasonable yield. However, one theme that we have returned to many times is how various interactions can be utilised to preorganise these systems and facilitate their synthesis in good yields.

We first described this effect, often known as the template effect, in the synthesis of crown ethers, where the presence of alkali metal ions preorganises the reagents in such a way that ring closure occurs to give the cyclic oligomers. Throughout this book different templating effects are seen to occur. This is especially significant in the later chapters, where a range of interactions such as metal templating, hydrogen bonding and $\pi - \pi$ interactions are used to construct supramolecular systems of increasing beauty and complexity. It is these effects combined with serendipity that has actually led to the discovery of many of these systems. Crown ethers and calixarenes were both isolated from reactions intended to create other products and it was only the patience and dedication of workers that allowed the structural details of these systems to be obtained. Similarly, cucurbiturils were synthesised many years before their structures were actually elucidated and cyclodextrins were obtained from natural sources as mixtures in small amounts and only a great deal of hard work allowed their separation and structural determination. As an example of simple components forming a highly complex system, a possible trefoil-knot structure has been proposed for the macrocucle compound t-butylcalix[20]arene, a simple condensation product of trioxan and t-butylphenol, although as yet no crystallographic evidence for this has been obtained.[131]

The macrocyclic compounds described within this work are becoming less of an academic curiosity and more prominent in practical and commercial applications. For example, crown ethers are widely used as phase-transfer catalysts and as selective ionophores, whereas cyclens are used within MRI contrast agents. Calixarenes have been used within the oil industry and their ion-binding properties have led to their use in ion-selective electrodes and potentially as encapsulents for radionuclides within the nuclear industry. Cucurbiturils are being investigated for applications such as waste remediation and gas purification. One of the most used families of macrocycles though is the cyclodextrins. The ability to make these compounds in large amounts at reasonable prices, combined with their strong binding abilities, has led to widespread applications in the food industry, as air fresheners, as encapsulents for perfumes and in a range of potential therapeutic roles in the field of drug delivery. Although the most complex architectures we have described, the physically interlinked molecules such as catenanes, rotaxanes and knots, have not yet been exploited commercially, there is great potential for their use in future applications such as computing and nanotechnology. What makes these systems of such interest is that they are intermediate in size between our macroscopic world and the quantum world of the atom. Further suggested applications include controlled drug-release devices, memory storage and molecular computers. Whether these applications will come about, only time will tell.

Whilst writing this book we have studied the work of a large number of notable scientists. Many of these are sadly no longer with us and it is a matter of conjecture what would be the reaction of the workers who made the initial steps in the discovery of the various systems if they could see how far the field has come. We can wonder what scientists like Baeyer, Baekeland, Villiers and Schardinger, Ewins and Robinson, Behrend and Willstätter, and more recently Pedersen and Cram would make of the current state of macrocyclic chemistry. We feel that not only would they see how far we had come but also how much, much further we have to go before we can construct supramolecular structures anywhere close to the size, specificity and intricacy found in the natural world.

Bibliography

Steed JW, Atwood JL. Supramolecular Chemistry. John Wiley and Sons; 2000, 2009.

Atkinson IM, Lindoy LF, Stoddart JF. Self-assembly in Supramolecular Systems. Monographs in Supramolecular Chemistry. Royal Society of Chemistry; 2000.

Sauvage J-P, Dietrich-Buechecker C. Molecular Catenanes, Rotaxanes and Knots: A Journey Through the World of Molecular Topology. Wiley VCH; 1999.

Schill G. Catenanes, Rotaxanes, Knots. Academic Press Inc; 1971.

Balzani V, Credi A, Venturi M. Molecular Devices and Machines. Wiley VCH; 2008.

Kelly TR. Molecular Machines. Topics in Current Chemistry. Vol **262**. Springer; 2005.

References

1. Müller P, Uson I, Hensel V, Schlüter AD, Sheldrick GM. Crystal structure of a cyclotetraicosaphenylene. *Helv Chim Acta.* 2001; **84**: 778–785.
2. Shen X, Ho DM, Pascal RA. Synthesis of polyphenylene dendrimers related to cubic graphite. *J Am Chem Soc.* 2004; **126**: 5798–5805.
3. Kohnke FH, Slawin AMZ, Stoddart JF, Williams DJ. Molecular belts and collars in the making: a hexaepoxy-octacosahydro[12]cyclacene derivative. *Angew Chem Int Ed.* 1987; **26**: 892–894.
4. Ashton PR, Girreser U, Giuffrida D, Kohnke FH, Mathias JP, Raymo FM, Slawin AMZ, Stoddart JF, Williams DJ. Molecular belts. 2. Substrate-directed syntheses of belt-type and cage-type structures? *J Am Chem Soc.* 1993; **115**: 5422–5429.
5. Gleiter R, Esser B, Kornmayer SC. Cyclacenes: hoop-shaped systems composed of conjugated rings. *Acc Chem Res.* 2009; **42**: 1108–1116.
6. Nakamura E, Tahara K, Matsuo Y, Sawamura M. Synthesis, structure, and aromaticity of a hoop-shaped cyclic benzenoid [10]cyclophenacene. *J Am Chem Soc.* 2003; **125**: 2834–2835.
7. Esser B, Rominger F, Gleiter R. Synthesis of [6.8]₃cyclacene: conjugated belt and model for an unusual type of carbon nanotube. *J Am Chem Soc.* 2008; **130**: 6716–6717.
8. Esser B, Bandyopadhyay A, Rominger F, Gleiter R. From metacyclophanes to cyclacenes: synthesis and properties of [6.8](3)cyclacene. *Chem Eur J.* 2009; **15**: 3368–3379.
9. Kornmayer SC, Hellbach B, Rominger F, Gleiter R. Synthesis, properties and formation of (RCp)Co- and (RCp)Rh-stabilized [4.8](3)cyclacene derivatives. *Chem Eur J.* 2009; **15**: 3380–3389.
10. Li Y, Flood AH. Pure C-H hydrogen bonding to chloride ions: a preorganized and rigid macrocyclic receptor. *Angew Chem Int Ed.* 2008; **47**: 2649–2652.
11. Li Y, Flood AH. Strong, size-selective, and electronically tunable C—H· · ·halide binding with steric control over aggregation from synthetically modular, shape-persistent [34]triazolophanes. *J Am Chem Soc.* 2008; **130**: 12111–12122.
12. Li Y, Pink M, Karty JA, Flood AH. Dipole-promoted and size-dependent cooperativity between pyridyl-containing triazolophanes and halides leads to persistent sandwich complexes with iodide. *J Am Chem Soc.* 2008; **130**: 17293–17295.

13. Zahran EM, Hua Y, Li Y, Flood AH, Bachas LG. Triazolophanes: a new class of halide-selective ionophores for potentiometric sensors. *Anal Chem.* 2010; **82**: 368–375.

14. Jiang J, MacLachlan MJ. Cationic guest inclusion in widemouthed Schiff base macrocycles. *J Chem Soc Chem Commun.* 2009; 5695–5697.

15. Durola F, Rebek J. The ouroborand: a cavitand with a coordination-driven switching device. *Angew Chem Int Ed.* 2010; **49**: 3189–3191.

16. Reuter C, Mohry A, Sobanski A, Vögtle F. [1]rotaxanes and pretzelanes: synthesis, chirality and absolute configuration. *Chem Eur J.* 2000; **6**: 1674–1682.

17. Li QY, Vogel E, Parham AH, Nieger M, Bolte M, Fröhlich R, Saarenketo P, Rissanen K, Vögtle F. Synthesis and X-ray structure of amide-based macrocycles, catenanes and pretzelane. *Eur J Org Chem.* 2001; 4041–4049.

18. Kishan MR, Parham A, Schelhase F, Yoneva A, Silva G, Chen X, Okamoto Y, Vögtle F. Bridging rotaxanes wheels–cyclochiral bonnanes. *Angew Chem Int Ed.* 2006; **45**: 7296–7299.

19. Liu Y, Vignon SA, Zhang X, Houk KN, Stoddart JF. Conformational diastereoisomerism in a chiral pretzelane. *J Chem Soc Chem Commun.* 2005; 3927–3929.

20. Zhao Y-L, Trabolsia A, Stoddart JF. A bistable pretzelane. *J Chem Soc Chem Commun.* 2009; 4844–4846.

21. Amabilino DB, Ashton PR, Bravo JA, Raymo FM, Stoddart JF, White AJP, Williams DJ. Molecular meccano. 52. Template-directed synthesis of a rotacatenane. *Eur J Org Chem.* 1999; 1295–1302.

22. Boeckmann J, Schill G. Knotenstrukturen in der Chemie. *Tetrahedron.* 1974; **30**: 1945–1957.

23. Dietrich-Buchecker CO, Sauvage J-P. A synthetic molecular trefoil knot. *Angew Chem Int Ed.* 1989; **28**: 189–192.

24. Dietrich-Buchecker CO, Guilhem J, Pascard C, Sauvage J-P. Structure of a synthetic trefoil knot coordinated to two copper(I) centers. *Angew Chem Int Ed.* 1990; **29**: 1154–1156.

25. Dietrich-Buchecker C, Nierengarten J-F, Sauvage J-P, Armaroli N, Balzani V, De Cola L. Dicopper(I) trefoil knots and related unknotted molecular systems: influence of ring size and structural factors on their synthesis and electrochemical and excited-state properties. *J Am Chem Soc.* 1993; **115**: 11237–11244.

26. Dietrich-Buchecker C, Rapenne G, Sauvage J-P. Efficient synthesis of a molecular knot by copper(i)-induced formation of the precursor followed by ruthenium(ii)-catalysed ring closing metathesis. *J Chem Soc Chem Commun.* 1997; 2053–2054.

27. Carina RF, Dietrich-Buchecker C, Sauvage J-P. Molecular composite knots. *J Am Chem Soc.* 1996; **118**: 9110–9116.

28. Meyer M, Albrecht-Gary A-M, Dietrich-Buchecker CO, Sauvage J-P. Dicopper(I) trefoil knots: topological and structural effects on the demetalation rates and mechanism. *J Am Chem Soc.* 1997; **119**: 4599–4607.

29. Dietrich-Buchecker CO, Sauvage J-P, Armaroli N, Ceroni P, Balzani V. Knotted heterodinuclear complexes. *Angew Chem Int Ed.* 1996; **35**: 1119–1121.

30. Adams H, Ashworth E, Breault GA, Guo J, Hunter CA, Mayers C. Knot tied around an octahedral metal centre. *Nature.* 2001; **411**: 763.

31. Guo J, Mayers PC, Breault GA, Hunter CA. Synthesis of a molecular trefoil knot by folding and closing on an octahedral coordination template. *Nature Chem.* 2010; **2**: 218–222.

32. Ashton PR, Matthews OA, Menzer S, Raymo FM, Spencer N, Stoddart JF, Williams DJ. Molecular meccano. 27. A template-directed synthesis of a molecular trefoil knot. *Liebigs Ann.* 1997; 2485–2494.

33. Lukin O, Vögtle F. Knotting and threading of molecules: chemistry and chirality of molecular knots and their assemblies. *Angew Chem Int Ed.* 2005; **44**: 1456–1477.

34. Safarowsky O, Nieger M, Fröhlich R, Vögtle F. A molecular knot with twelve amide groups: one-step synthesis, crystal structure, chirality. *Angew Chem Int Ed.* 2000; **39**: 1616–1618.

35. Recker J, Vögtle F. Amide-based molecular knots. *J Inclus Phenom Macrocy Chem.* 2001; **41**: 3–5.

36. Recker J, Muller WM, Muller U, Kubota T, Okamoto Y, Nieger M, Vögtle F. Dendronized molecular knots: selective synthesis of various generations, enantiomer separation, circular dichroism. *Chem Eur J.* 2002; **8**: 4434–4442.

37. Lukin O, Muller WM, Muller U, Kaufmann A, Schmidt C, Leszczynski J, Vögtle F. Covalent chemistry and conformational dynamics of topologically chiral amide-based molecular knots. *Chem Eur J.* 2003; **9**: 3507–3517.

38. Lukin O, Yoneva A, Vögtle F. Diastereoisomeric molecular knots by combination of central and topological chiralities. *Eur J Org Chem.* 2004; 1236–1238.

39. Lukin O, Recker J, Bohmer A, Muller WM, Kubota T, Okamoto Y, Nieger M, Frohlich R, Vögtle F. A topologically chiral molecular dumbbell. *Angew Chem Int Ed.* 2003; **42**: 442–445.

40. Lukin O, Kubota T, Okamoto Y, Schelhase F, Yoneva A, Muller WRM, Muller U, Vögtle F. Knotaxanes: rotaxanes with knots as stoppers. *Angew Chem Int Ed.* 2003; **42**: 4542–4545.

41. Lukin O, Kubota T, Okamoto Y, Kaufmann A, Vögtle F. Topologically chiral covalent assemblies of molecular knots with linear, branched, and cyclic architectures. *Chem Eur J.* 2004; **10**: 2804–2810.

42. Bruggemann J, Bitter S, Muller S, Muller WM, Muller U, Maier NM, Lindner W, Vögtle F. Spontaneous knotting: from oligoamide threads to trefoil knots. *Angew Chem Int Ed.* 2007; **46**: 254–259.

43. Böhmer A, Brüggemann J, Kaufmann A, Yoneva A, Müller S, Müller WM, Müller U, Vergeer FW, Chi L, De Cola L, Fuchs H, Chen X, Kubota T, Okamoto Y, Vögtle F. Long chain-substituted and triply functionalized molecular knots: synthesis, topological chirality and monolayer formation. *Eur J Org Chem.* 2007; 45–52.

44. Passaniti P, Ceroni P, Balzani V, Lukin O, Yoneva A, Vögtle F. Amide-based molecular knots as platforms for fluorescent switches. *Chem Eur J.* 2006; **12**: 5685–5690.

45. Feigel M, Ladberg R, Engels S, Herbst-Irmer R, Frohlich R. A trefoil knot made of amino acids and steroids. *Angew Chem Int Ed.* 2006; **45**: 5698–5702.

46. Schappacher M, Deffieux A. Imaging of catenated, figure-of-eight, and trefoil knot polymer rings. *Angew Chem Int Ed.* 2009; **48**: 5930–5933.

47. Epand RM, Vogel HJ. Diversity of antimicrobial peptides and their mechanisms of action. *Biochim Biophys Acta.* 1999; **1462**: 11–28.

48. Mueller JE, Du SM, Seeman NC. Design and synthesis of a knot from single-stranded DNA. *J Am Chem Soc.* 1991; **113**: 6306–6308.

49. Seeman NC. Nucleic acid nanostructures and topology. *Angew Chem Int Ed.* 1998; **37**: 3220–3238.

50. Seeman NC. DNA components for molecular architecture. *Acc Chem Res.* 1997; **30**: 357–363.

51. Vögtle F, Lukin O. Molecular knots. *Angew Chem Int Ed.* 2005; **44**: 1456–1477.

52. Nierengarten J-F, Dietrich-Buchecker CO, Sauvage J-P. Synthesis of a doubly interlocked [2]-catenane. *J Am Chem Soc.* 1994; **116**: 375–376.

53. McArdle CP, Vittal JJ, Puddephatt RJ. Molecular topology: easy self-assembly of an organometallic doubly braided [2]catenane. *Angew Chem Int Ed.* 2000; **39**: 3819–3822.

54. Peinador C, Blanco V, Quintela JM. A new doubly interlocked [2]catenane. *J Am Chem Soc.* 2009; **131**: 920–921.

55. Chichak KS, Cantrill SJ, Pease AR, Chiu S-H, Cave GWV, Atwood JL, Stoddart JF. Molecular Borromean rings. *Science.* 2004; **304**: 1308–1312.

56. Pentecost CD, Tangchaivang N, Cantrill SJ, Chichak KS, Peters AJ, Fraser Stoddart J. Making molecular Borromean rings: a gram-scale synthetic procedure for the undergraduate organic lab. *J Chem Educ.* 2007; **84**: 855.

57. Peters AJ, Chichak KS, Cantrill SJ, Stoddart JF. Nanoscale Borromean links for real. *J Chem Soc Chem Commun.* 2005; 3394–3396.

58. Chichak KS, Peters AJ, Cantrill SJ, Stoddart JF. Nanoscale Borromeates. *J Org Chem.* 2005; 7956–7962.

59. Yates CR, Benítez D, Khan SI, Stoddart JF. Hexafunctionalized Borromeates using olefin cross metathesis. *Org Lett.* 2007; **9**: 2433–2436.

60. Pentecost CD, Peters AJ, Chichak KS, Cave GWV, Cantrill SJ, Stoddart JF. Chiral Borromeates. *Angew Chem Int Ed.* 2006; **45**: 4099–4104..

61. Pentecost CD, Chichak KS, Peters AJ, Cave GWV, Cantrill SJ, Stoddart JF. A molecular Solomon link. *Angew Chem Int Ed.* 2007; **46**: 218–222.

62. Loren JC, Yoshizawa M, Haldimann RF, Linden A, Siegel JS. Synthetic approaches to a molecular borromean link: two-ring threading with polypyridine templates. *Angew Chem Int Ed.* 2003; **42**: 5702–5705.

63. Forgan RS, Spruell JM, Olsen JC, Stern CL, Stoddart JF. Towards the stepwise assembly of molecular borromean rings: a donor–acceptor ring-in-ring complex. *J Mex Chem Soc.* 2009; **53**: 134–138.

64. Stoddart JF. The chemistry of the mechanical bond. *Chem Soc Rev.* 2009; **38**: 1802–1819.

65. Meyer CD, Joiner CS, Stoddart JF. Template-directed synthesis employing reversible imine bond formation. *Chem Soc Rev.* 2007; **36**: 1705–1723.

66. Northrop BH, Aricó F, Tangchiavang N, Badjić JD, Stoddart JF. Template-directed synthesis of mechanically interlocked molecular bundles using dynamic covalent chemistry. *Org Lett.* 2006; **8**: 3899–3902.

67. Hou H, Leung KC-F, Lanari D, Nelson A, Stoddart JF, Grubbs RH. Template-directed one-step synthesis of cyclic trimers by ADMET. *J Am Chem Soc.* 2006; **128**: 15358–15359.

68. Williams AR, Northrop BH, Chang T, Stoddart JF, White AJP, Williams DJ. Suitanes. *Angew Chem Int Ed.* 2006; **45**: 6665–6669.

69. Northrop BH, Spruell JA, Stoddart JF. Efficient routes to novel molecular architectures: template-directed synthesis of mechanically interlocked suitanes. *Chim Oggi.* 2007; **25**: 4–7.

70. Branda N, Wyler R, Rebek J. Encapsulation of methane and other small molecules in a self-assembling super-structure. *Science.* 1994; **263**: 1267–1268.

71. Szabo T, Hilmersson G, Rebek J. Dynamics of assembly and guest exchange in the tennis ball. *J Am Chem Soc.* 1998; **120**: 6193–6194.

72. Kang J, Rebek J. Entropically driven binding in a self-assembling molecular capsule. *Nature.* 1996; **382**: 239–241.

73. Martín T, Obst U, Rebek J. Molecular assembly and encapsulation directed by hydrogen-bonding preferences and the filling of space. *Science.* 1998; **281**: 1842–1845.

74. Hof F, Nuckolls C, Rebek J. Diversity and selection in self-assembled tetrameric capsules. *J Am Chem Soc.* 2000; **122**: 4251–4252.

75. Palmer LC, Rebek J. The ins and outs of molecular encapsulation. *Org Biomol Chem.* 2004; **2**: 3051–3059.

76. Moore GE. Cramming more components onto integrated circuits. *Electronics Magazine.* 1965; **38**: 4.

77. Brunet E, Juanes O, José de la Mata M, Rodríguez-Ubis JC. A simple polyheterotopic molecular receptor derived from bispyrazolylmethane showing ambivalent allosteric cooperation of zinc(II). *Eur J Org Chem.* 2000; 1913–1922.

78. Ikeda A, Tsudera T, Shinkai S. Molecular design of a 'molecular syringe' mimic for metal cations using a 1,3-alternate calix[4]arene cavity. *J Org Chem.* 1997; **62**: 3568–3574.

79. Balzani V, Ceroni P, Credi A, Gómez-López M, Hamers C, Stoddart JF, Wolf R. Controlled dethread-ing/rethreading of a scorpion-like pseudorotaxane and a related macrobicyclic self-complexing system. *New J Chem.* 2001; **25**: 25–31.

80. Collin JP, Laemmel A-C, Sauvage J-P. Photochemical expulsion of a Ru(phen)2 unit from a macrocyclic receptor and its thermal recoordination. *New J Chem.* 2001; **25**: 22–24.

81. Shinkai S, Ishihara M, Ueda K, Manabe O. Photoresponsive crown ethers. Part 14. Photoregulated crown–metal complexation by competitive intramolecular tail(ammonium)-biting. *J Chem Soc. Perkin Trans.* 1985; **2**: 511–518.

82. Prasanna de Silva A, Nimal Gunaratne HQ, McCoy CP. Molecular photoionic AND logic gates with bright fluorescence and 'off–on' digital action. *J Am Chem Soc.* 1997; **119**: 7891–7892.

83. Ji H-F, Dabestani R, Brown GM. A supramolecular fluorescent probe, activated by protons to detect cesium and potassium ions, mimics the function of a logic gate. *J Am Chem Soc.* 2000; **122**: 9306–9307.

84. Ballardini R, Balzani V, Credi A, Gandolfi MT, Kotzyba-Hibert F, Lehn J-M, Prodi L. Supramolecular pho-tochemistry and photophysics: a cylindrical macrotricyclic receptor and its adducts with protons, ammonium ions, and a Pt(II) complex. *J Am Chem Soc.* 1994; **166**: 5741–5746.

85. Credi A, Balzani V, Langford SJ, Stoddart JF. Logic operations at the molecular level: an XOR gate based on a molecular machine. *J Am Chem Soc.* 1997; **119**: 2679–2681.

86. Lee SH, Kim JY, Kim SK, Leed JH, Kim JS. Pyrene-appended calix[4]crowned logic gates involving normal and reverse PET: NOR, XNOR and INHIBIT. *Tetrahedron.* 2004; **60**: 5171–5176.

87. Ferrer B, Rogez G, Credi A, Ballardini R, Gandolfi MT, Balzani V, Liu Y, Tseng H-R, Stoddart JF. Photoinduced electron flow in a self-assembling supramolecular extension cable. *Proc Nat Acad Sci.* 2006; **103**: 18411–18416.

88. Ballardini R, Balzani V, Clemente-León M, Credi A, Gandolfi MT, Ishow E, Perkins J, Stoddart JF, Tseng H-R, Wenger S. Photoinduced electron transfer in a triad that can be assembled/disassembled by two different external inputs: toward molecular-level electrical extension cables. *J Am Chem Soc*. 2002; **124**: 12786–12795.

89. Kottas GS, Clarke LI, Horinek D, Michl J. Artificial molecular rotors. *Chem Rev*. 2005; **105**: 1281–1376

90. Kelly TR. Molecular Machines. Topics in Current Chemistry. Vol **262**. Springer; 2005.

91. Tashiro K, Fujiwara T, Konishi K. Rotational oscillation of two interlocked porphyrins in cerium bis(5, 15-diarylporphyrinate) double-deckers. *J Chem Soc Chem Commun*. 1998; 1121–1122.

92. Gakh AA, Sachleben RA, Bryan JC, Moyer BA. A facile synthesis and X-ray structure determination of the first triptycenocrown ethers. *Tet Lett*. 1995; **36**: 8163–8166.

93. Gakh AA, Sachleben RA, Bryan JC. Molecular gearing systems. *Chem Tech*. 1997; **26**: 27–33.

94. Bedard TC, Moore JS. Design and synthesis of molecular turnstiles. *J Am Chem Soc*. 1995; **117**: 10662–10671.

95. Dominguez Z, Dang H, Strouse MJ, Garcia-Garibay MA. Molecular 'compasses' and 'gyroscopes'. I. Expedient synthesis and solid state dynamics of an open rotor with a bis(triarylmethyl) frame. *J Am Chem Soc*. 2002; **124**: 2398–2399.

96. Godinez CE, Zepeda G, Garcia-Garibay MA. Molecular 'compasses' and 'gyroscopes'. II. Synthesis and characterization of molecular rotors with axially substituted bis[2-(9-triptycyl)ethynyl]arenes. *J Am Chem Soc*. 2002; **124**: 4701–4707.

97. Takeshita M, Irie M. Photoresponsive tweezers for alkali metal ions: photochromic diarylethenes having two crown ether moieties. *J Org Chem*. 1998; **63**: 6643–6649.

98. Ueno A, Yoshimura H, Saka R, Osa T. Photocontrol of binding ability of capped cyclodextrin. *J Am Chem Soc*. 1979; **101**: 2779–2780.

99. Muraoka T, Kinbara K, Aida T. Mechanical twisting of a guest by a photoresponsive host. *Nature*. 2006; **440**: 512–515.

100. Moon K, Kaifer AE. Dimeric molecular capsules under redox control. *J Am Chem Soc*. 2004; **126**: 15016–15017.

101. Ashton PR, Balzani V, Becher J, Credi A, Fyfe MCT, Mattersteig G, Menzer S, Nielsen MB, Raymo FM, Stoddart JF, Venturi M, Williams DJ. A three-pole supramolecular switch. *J Am Chem Soc*. 1999; **121**: 3951–3957.

102. Credi A, Montalti M, Balzani V, Langford SJ, Raymo FM, Stoddart JF. Simple molecular-level machines. Interchange between different threads in pseudorotaxanes. *New J Chem*. 1998; **22**: 1061–1065.

103. Mirzoian A, Kaifer AE. Electrochemically controlled self-complexation of cyclodextrin–viologen conjugates. *J Chem Soc Chem Commun*. 1999; 1603–1604.

104. Castro R, Cuadrado I, Alonso B, Casado CM, Morán M, Kaifer AE. Multisite inclusion complexation of redox active dendrimer guests. *J Am Chem Soc*. 1997; **119**: 5760–5761.

105. González B, Casado CM, Alonso B, Cuadrado I, Morán M, Wang Y, Kaifer AE. Synthesis, electrochemistry and cyclodextrin binding of novel cobaltocenium-functionalized dendrimers. *J Chem Soc Chem Commun*. 1998; 2569–2570.

106. Wang Y, Kaifer AE. Redox control of host–guest recognition: a case of host selection determined by the oxidation state of the guest. *J Chem Soc Chem Commun*. 1998; 1457–1458.

107. Balzani V, Credi A, Marchioni F, Stoddart JF. Artificial molecular-level machines: dethreading–rethreading of a pseudorotaxane powered exclusively by light energy. *J Chem Soc Chem Commun*. 2001; 1860–1861.

108. Consuelo Jimenez-Molero M, Dietrich-Buchecker C, Sauvage J-P. Towards artificial muscles at the nanometric level. *J Chem Soc Chem Commun*. 2003; 1613–1616.

109. Bonnet S, Collin J-P, Koizumi M, Mobian P, Sauvage J-P. Transition-metal-complexed molecular machine prototypes. *Adv Mater*. 2006; **18**: 1239–1250.

110. Mobian P, Kern J-M, Sauvage J-P. Light-driven machine prototypes based on dissociative excited states: photoinduced decoordination and thermal recoordination of a ring in a ruthenium(II)-containing [2]catenane. *Angew Chem Int Ed*. 2004; **43**: 2392–2395.

111. Thordarson P, Bijsterveld EJA, Rowan AE, Nolte RJM. Epoxidation of polybutadiene by a topologically linked catalyst. *Nature*. 2003; **424**: 915–918.

112. Ramos PH, Coumans RGE, Deutman ABC, Smits JMM, de Gelder R, Elemans JAAW, Nolte RJM, Rowan AE. Processive rotaxane systems. studies on the mechanism and control of the threading process. *J Am Chem Soc*. 2007; **129**: 5699–5702.

113. Ballardini R, Balzani V, Dehaen W, Dell'Erba AE, Raymo FM, Stoddart JF, Venturi M. Molecular meccano. 56. Anthracene-containing [2]rotaxanes: synthesis, spectroscopic, and electrochemical properties. *Eur J Org Chem*. 2000; 591–602.

114. Qu D-H, Wang Q-C, Ma X, Tian H. A [3]rotaxane with three stable states that respondsto multiple-inputs and displays dual fluorescence addresses. *Chem Eur J*. 2005; **11**: 5929–5937.

115. Qu D-H, Wang Q-C, Tian H. A half adder based on a photochemically driven [2]rotaxane. *Angew Chem Int Ed*. 2005; **44**: 5296–5299.

116. Credi A, Langford SJ, Raymo FM, Stoddart JF, Venturi M. Constructing molecular machinery: a chemically-switchable [2]catenane. *J Am Chem Soc*. 2000; **122**: 3542–3543.

117. Ashton PR, Baldoni V, Balzani V, Credi A, Hoffmann HDA, Martínez-Díaz M-V, Raymo FM, Stoddart JF, Venturi M. Dual-mode 'co-conformational' switching in catenanes incorporating bipyridinium and dialkylam-monium recognition sites. *Chem Eur J*. 2001; **7**: 3482–3493.

118. Hernandez JV, Kay ER, Leigh DA. A reversible synthetic rotary molecular motor. *Science*. 2004; **306**: 1532–1537.

119. Leigh DA, Wong JKY, Dehez F, Zerbetto F. Unidirectional rotation in a mechanically interlocked molecular rotor. *Nature*. 2003; **424**: 174–179.

120. Jullien L, Canceill J, Valeur B, Bardez E, Lefèvre J-P, Lehn J-M, Marchi-Artzner V, Pansu R. Multichromophoric cyclodextrins. 4. Light conversion by antenna effect. *J Am Chem Soc*. 1996; **118**: 5432–5442.

121. Tamura M, Gao D, Ueno A. A polyrotaxane series containing cyclodextrin and naphthalene-modified-cyclodextrin as a light-harvesting antenna system. *Chem Eur J*. 2001; **7**: 1390–1397.

122. Voyer N, Robitaille M. Novel functional artificial ion channel. *J Am Chem Soc*. 1995; **117**: 6599–6600.

123. Cazacu A, Tong C, van der Lee A, Fyles TM, Barboiu M. Columnar self-assembled ureido crown ethers: an example of ion-channel organization in lipid bilayers. *J Am Chem Soc*. 2006; **128**: 9541–9548.

124. Wright AJ, Matthews SE, Fischer WB, Beer PD. Novel resorcin[4]arenes as potassium-selective ion-channel and transporter mimics. *Chem Eur J*. 2001; **7**: 3474–3481.

125. Nguyen TD, Tseng H-R, Celestre PC, Flood AH, Liu Y, Stoddart JF, Zink JI. A reversible molecular valve. *Proc Natl Acad Sci USA*. 2005; **102**: 10029–10034.

126. Nguyen TD, Liu Y, Saha S, Leung KC-F, Stoddart JF, Zink JI. Design and optimization of molecular nanovalves based on redox-switchable bistable rotaxanes. *J Am Chem Soc*. 2007; **129**: 626–634.

127. Saha S, Leung KC-F, Nguyen TD, Stoddart JF, Zink JI. Nanovalves. *Adv Funct Mat*. 2007; **17**: 685–693.

128. Ryan D, Rao SN, Rensmo H, Fitzmaurice D, Preece JA, Wenger S, Stoddart J, Zaccheroni N. Heterosupramolec-ular chemistry: recognition initiated and inhibited silver nanocrystal aggregation by pseudorotaxane assembly. *J Am Chem Soc*. 2000; **122**: 6252–6257.

129. Aprahamian I, Yasuda T, Ikeda T, Saha S, Dichtel WR, Isoda K, Kato T, Stoddart JF. A liquid-crystalline bistable [2]rotaxane. *Angew Chem Int Ed*. 2007; **46**: 4675–4679.

130. Fathalla M, Neuberger A, Li S-C, Schmehl R, Diebold U, Jayawickramarajah J. Straightforward self-assembly of porphyrin nanowires in water: harnessing adamantane/β-cyclodextrin interactions. *J Am Chem Soc*. 2010; **132**: 9966–9967.

131. Stewart DR, Gutsche CD. Isolation, characterization, and conformational characteristics of p-tert-butylcalix [9-20]arenes. *J Am Chem Soc*. 1999; **121**: 4136–4146.

Index

Note: Figures, Schemes, structures and Tables are indicated by *italic page numbers*, Boxes by **emboldened numbers**, and footnotes by suffix 'n'.
Abbreviations: CTV = cyclotriveratrylene; CB = cucurbituril

Macrocycles: Construction, Chemistry and Nanotechnology Applications, First Edition.   Frank Davis and Séamus Higson.
© 2011 John Wiley & Sons, Ltd. Published 2011 by John Wiley & Sons, Ltd.